INTERNATIONAL SERIES OF MONOGRAPHS IN PURE AND APPLIED BIOLOGY

Division: **ZOOLOGY**

GENERAL EDITOR: G. A. KERKUT

VOLUME 50

EMBRYOLOGY AND PHYLOGENY IN ANNELIDS AND ARTHROPODS

OTHER TITLES IN THE ZOOLOGY DIVISION

General Editor: G. A. KERKUT

Vol. 1. RAVEN—*An Outline of Developmental Physiology*
Vol. 2. RAVEN—*Morphogenesis: The Analysis of Molluscan Development*
Vol. 3. SAVORY—*Instinctive Living*
Vol. 4. KERKUT—*Implications of Evolution*
Vol. 5. TARTAR—*The Biology of Stentor*
Vol. 6. JENKIN—*Animal Hormones—A Comparative Survey. Part 1. Kinetic and Metabolic Hormones*
Vol. 7. CORLISS—*The Ciliated Protozoa*
Vol. 8. GEORGE—*The Brain as a Computer*
Vol. 9. ARTHUR—*Ticks and Disease*
Vol. 10. RAVEN—*Oogenesis*
Vol. 11. MANN—*Leeches (Hirudinea)*
Vol. 12. SLEIGH—*The Biology of Cilia and Flagella*
Vol. 13. PITELKA—*Electron-microscopic Structure of Protozoa*
Vol. 14. FINGERMAN—*The Control of Chromatophores*
Vol. 15. LAVERACK—*The Physiology of Earthworms*
Vol. 16. HADZI—*The Evolution of the Metazoa*
Vol. 17. CLEMENTS—*The Physiology of Mosquitoes*
Vol. 18. RAYMONT—*Plankton and Productivity in the Oceans*
Vol. 19. POTTS and PARRY—*Osmotic and Ionic Regulation in Animals*
Vol. 20. GLASGOW—*The Distribution and Abundance of Tsetse*
Vol. 21. PANTELOURIS—*The Common Liver Fluke*
Vol. 22. VANDEL—*Biospeleology—The Biology of Cavernicolous Animals*
Vol. 23. MUNDAY—*Studies in Comparative Biochemistry*
Vol. 24. ROBINSON—*Genetics of the Norway Rat*
Vol. 25. NEEDHAM—*The Uniqueness of Biological Materials*
Vol. 26. BACCI—*Sex Determination*
Vol. 27. JØRGENSEN—*Biology of Suspension Feeding*
Vol. 28. GABE—*Neurosecretion*
Vol. 29. APTER—*Cybernetics and Development*
Vol. 30. SHAROV—*Basic Arthropodan Stock*
Vol. 31. BENNETT—*The Aetiology of Compressed Air Intoxication and Inert Gas Narcosis*
Vol. 32. PANTELOURIS—*Introduction to Animal Physiology and Physiological Genetics*
Vol. 33. HAHN and KOLDOVSKY—*Utilization of Nutrients during Postnatal Development*
Vol. 34. TROSHIN—*The Cell and Environmental Temperature*
Vol. 35. DUNCAN—*The Molecular Properties and Evolution of Excitable Cells*
Vol. 36. JOHNSTON and ROOTS—*Nerve Membranes. A Study of the Biological and Chemical Aspects of Neuron–Glia Relationships*
Vol. 37. THREADGOLD—*The Ultrastructure of the Animal Cell*
Vol. 38. GRIFFITHS—*Echidnas*
Vol. 39. RYBAK—*Principles of Zoophysiology—Vol. 1*
Vol. 40. PURCHON—*The Biology of the Mollusca*
Vol. 41. MAUPIN—*Blood Platelets in Man and Animals—Vols. 1–2*
Vol. 42. BRØNDSTED—*Planarian Regeneration*
Vol. 43. PHILLIS—*The Pharmacology of Synapses*
Vol. 44. ENGELMANN—*The Physiology of Insect Reproduction*
Vol. 45. ROBINSON—*Genetics for Cat Breeders*
Vol. 46. ROBINSON—*Lepidoptera Genetics*
Vol. 47. JENKIN—*Control of Growth and Metamorphosis*
Vol. 48. CULLEY—*The Pilchard: Biology and Exploitation*
Vol. 49. BINYON—*Physiology of Echinoderms*

EMBRYOLOGY AND PHYLOGENY IN ANNELIDS AND ARTHROPODS

D. T. ANDERSON

Professor of Biology, University of Sydney

PERGAMON PRESS

OXFORD • NEW YORK • TORONTO
SYDNEY • BRAUNSCHWEIG

Pergamon Press Ltd., Headington Hill Hall, Oxford
Pergamon Press Inc., Maxwell House, Fairview Park, Elmsford, New York 10523
Pergamon of Canada Ltd., 207 Queen's Quay West, Toronto 1
Pergamon Press (Aust.) Pty. Ltd., 19a Boundary Street, Rushcutters Bay, N.S.W. 2011, Australia
Vieweg & Sohn GmbH, Burgplatz 1, Braunschweig

First edition 1973

Library of Congress Cataloging in Publication Data

Anderson, Donald Thomas.
Embryology and phylogeny in annelids and arthropods.

(International series of monographs in pure and applied biology. Division: zoology, v. 50)
1. Embryology—Arthropoda. 2. Embryology—Annelida. 3. Phylogeny. I. Title.
QL958.A5 1973 595 73-1019
ISBN 0-08-017069-2

Printed in Hungary

To

DR. SIDNIE M. MANTON, F.R.S.

CONTENTS

PREFACE AND ACKNOWLEDGEMENTS

THIS work is an attempt to bring unity and order to the descriptive embryology of the segmented invertebrates and to utilize the morphological facts of embryonic development in these animals in the furtherance of speculations on their phylogenetic relationships. To this end, I have eschewed reference to the causal analysis of embryonic development in annelids and arthropods. The experimental embryology of several of the groups of animals included here has been fully discussed in recent reviews (see Reverberi (1971), Annelida; Counce (1972), Pterygota; Costlow (1968) and Green (1971), Crustacea; the references are included in the reference list appended to Chapter 1). That of the other groups is either poorly known (Arachnida) or unknown (Onychophora, Myriapoda and Apterygota) at the present time. Since the groups that have been studied are notoriously variable in their causal embryological processes, the embryological arguments for phylogenetic relationships are little affected by the results of these experimental studies. At the same time, the interchange of ideas between descriptive and experimental embryology is essential to both disciplines. My debt to the concepts established as a result of experimentation is obvious throughout this book, since I have attempted, among other things, to establish a uniform terminology for annelid and arthropod embryology which takes cognizance of underlying causal processes. I hope that, in turn, experimental embryologists will find value in this unification as a basis for posing further significant questions.

Although the main purpose of the work is phylogenetic, it is anticipated that the book will be able to be used in a number of ways. The essence of the phylogenetic appraisal can be extracted by reading the first and last chapters only, with some assistance from the illustrations. A summary of the basic mode of embryonic development of each of the major groups of animals discussed is given at the end of the chapter describing the group and can be used, together with the illustrations, as a brief outline of the essential facts. Finally, each chapter provides a full account of developmental variation among the species to which it pertains and embodies all of the details of developmental morphology on which the phylogenetic conclusions are based.

In writing this book, I have received assistance and encouragement from many sources. Above all, my gratitude goes to my wife, Joanne, without whose patience, understanding and practical help in innumerable ways, the task would have never come to fruition. She has borne single-handed the burden of typing the manuscript and has carried out much of the preparatory work required for the illustrations. I am also indebted to Mrs. Patricia Spalding,

M.SC., who has redrawn all of the final illustrations and whose zoological knowledge and technical skill have been invaluable. My work has been supported for a number of years by grants from the University of Sydney, the Nuffield Foundation and the Australian Research Grants Committee, whose generous provision of assistance has allowed me to carry out the research studies from which the ideas embodied here have emerged.

To the many colleagues who have sent me reprints of their work, and whose ideas and illustrations I have drawn upon so freely, I extend my thanks. I have no doubt that they will, collectively and individually, disagree with some of my conclusions. I shall welcome both their advice on any point in which I have erred and the new studies which will bring us all closer to a proper appraisal of the subject.

The preparation of the book has been carried out mainly at the University of Sydney. Professor L. C. Birch, Challis Professor of Biology and Head of the School of Biological Sciences of that University, has been unfailing in his support of the project. Much assistance has also been given by the Fisher Library of the University of Sydney, the library of the Australian Museum, Sydney, and the library of the Linnean Society of New South Wales. An important part of the work was also completed during a year of study leave in 1970 as Visiting Professor of Zoology at the University of London, King's College, at the kind invitation of Professor D. R. Arthur. The facilities of the Zoology Department of King's College and the many opportunities for discussion with Professor Arthur provided the knowledge necessary to write the chapter on chelicerates (Chapter 9). During the same year, I was also greatly helped by Dr. J. P. Harding, then Keeper of Zoology in the British Museum (Natural History), who arranged access for me to the superb library of that institution. The preparation of Chapter 7 on the Pterygota was greatly facilitated by the advice and trenchant criticism accorded to me by Associate Professor S. J. Counce of the Zoology Department of Duke University, North Carolina, U.S.A. Chapter 3 on clitellate annelids owes much to the encouragement I have received from Professor R. O. Brinkhurst of the University of Toronto. To all these persons and institutions I extend my sincere thanks.

Finally, I wish to acknowledge the continued advice and encouragement I have received over many years from Dr. S. M. Manton, F.R.S. (Mrs. J. P. Harding), to whom this book is dedicated. My greatest satisfaction is that the conclusions presented herein support her views on arthropod phylogeny. The inspiration of the work is hers. The errors and omissions, which her discriminating eye will immediately detect, are my own.

Elanora D. T. ANDERSON

CHAPTER 1

INTRODUCTION—EMBRYOLOGY AND PHYLOGENY

ANNELID and arthropod embryology has always been beset more by diversity than by unity. The development of the embryo is extremely specialized in most arthropods and the specializations, while often superficially alike in members of the different arthropod classes, reveal fundamental and seemingly irreconcilable differences when attempts are made to compare the development of one class and another. In the century that has passed since the first histological descriptions were made of annelid and arthropod embryos, the frustrations engendered by early attempts to find a common ground for the development of these animals have fostered a fragmentation of the subject between groups of workers, each devoted to a single major group of animals. An analysis of the resulting literature reveals that each class of annelids and arthropods has had its own retinue of embryologists, almost none of whom has ventured into work on embryos of another class. The only prominent exceptions to this rule have been the French embryologist C. Dawydoff, whose *Traité d'Embryologie Invertebré* (1928) has been a source book in comparative embryology for many years and whose more recent contributions in Grassé's *Traité de Zoologie* have been a valuable supplement to this classic work, and the English zoologist S. M. Manton, who has made fundamental contributions to both crustacean embryology (1928, 1934) and onychophoran embryology (1949).

In consequence of the restricted attention accorded by most annelid and arthropod embryologists to their circumscribed group of animals, different concepts and different terminologies have evolved in each line of study and communication between them has become ever more difficult. Due to the method of description, it is effectively impossible, for example, to discern any relationship between the embryonic development of polychaetes and that of pterygote insects, or between that of Crustacea and that of Onychophora, in the papers on these different groups. Consequently, comparative embryology has until recently had little impact on concepts of the interrelationships between the major arthropod classes and between the arthropods and the annelids. Yet the advances recently attained in the understanding of the comparative morphology of the annelids and arthropods (e.g. Tiegs and Manton, 1958; Manton, 1964, 1970, 1972; Clark, 1964) pose a crucial question for comparative embryology. If, as the morphological evidence suggests, the ancestors of the arthropods were not annelids, and if, furthermore, the arthropod grade of morphological organization has evolved several times, yielding the Uniramia (= the onychophoran–

myriapod–hexapod assemblage, Manton, 1972), the Crustacea and the Chelicerata independently, why is it that the great wealth of embryological evidence accumulated in a century of intensive investigation cannot be employed effectively to test these ideas? On general principles, comparative development must form an integral part of the evidence of phylogenetic relationships between animal groups (see, for example, Anderson, 1967) and the conclusions from comparative development must accord with those from comparative morphology. If the arthropods are a phylum, their embryology should reveal an underlying unity, as it does in all other phyla. If, on the other hand, the arthropods are polyphyletic, their embryology should underscore the morphological dissimilarities which separate the arthropods into at least two, or possibly three, phyla. Finally, if any or all of the arthropod classes are related to the annelids, fundamental similarities must surely exist between annelid embryonic development and certain kinds of arthropod embryonic development.

The purpose of the present book is to re-examine the facts of morphological development of the embryos of annelids and arthropods in relation to the alternatives set out above. In order to do this, it is necessary to find a method of descriptive analysis which will avoid the difficulties arising from varied terminologies and will at the same time emphasize the functional steps by which new organization arises during embryonic development. The functional interrelationships between parts of an embryo are equally as important as those between parts of an adult animal and can vary only in ways which maintain functional unity. Related animals will therefore show fundamentally similar functional patterns of development, in which each developing part retains its place and role within the whole, while unrelated animals will have different functional configurations of developing parts. It has been found during the last few years (Anderson, 1966 a, b, c, 1969, 1971, 1972 a, b; see also Cather, 1971) that the most useful functional interpretation that can be applied to annelid and arthropod embryos is the one long used by vertebrate embryologists, namely, the formation and fates of presumptive areas of the blastula. The blastula, or its equivalent stage of embryonic development, has a greater stability of functional configuration than any stage that precedes or follows it, no doubt because this is the stage at which the fundamental framework of bodily organization is established. Furthermore, the configuration can be epitomized in a fate map which effectively summarizes all that is important about the embryonic development of the animal in question from a comparative point of view. In what follows, therefore, the embryonic development of each major class of annelids and arthropods is summarized in terms of the formation and fates of the presumptive areas of the blastula or blastoderm and the results are then compared and analysed from the aspect of functionally feasible phylogeny.

The background for the present interpretation lies, of course, in the history of invertebrate embryology's first hundred years. Perhaps the most important step taken in nineteenth-century annelid and arthropod embryology was the substantiation of the fact that annelid embryonic development has its basis in spiral cleavage and the formation of a trochophore, while arthropod embryonic development does not. The arthropods soon became renowned for their frequent occurrence of superficial cleavage and a blastoderm, and the few attempts that were made to interpret the total cleavage of certain arthropods as a modified spiral cleavage (e.g. in some crustaceans) met with no success. Controversy was rife and many hundreds of pages were devoted to attempts to make inadequate facts and misleading concepts satisfy the requirements of a simple, preconceived phylogeny. At the same time, the classic papers of that period still abound with evidence of the technical virtuosity and precision of observation of their authors. E. B. Wilson on polychaetes, Whitman and Vedjovsky on oligochaetes and leeches, Woltereck on archiannelids, Sedgwick and Kennel on Onychophora, Heymons on myriapods and insects, Claus on Crustacea and Kingsley on Chelicerata are papers whose factual contents are fresh and significant even today. In fact, as we shall see, some of the crucial information pertaining to a unified interpretation of annelid and arthropod embryology is to be found in these works.

Between the turn of the century and the onset of the First World War, descriptive annelid and arthropod embryology suffered a gradual decline. By this time, other aspects of zoology had begun to clamour for attention and the few workers who retained an interest in describing annelid and arthropod development did little more than add erratically to the then existing knowledge. An exception must be made, however, for the works of Delsman on polychaetes, Schleip on clitellates, and Fuchs and Kühn on Crustacea, whose studies were in the best classical tradition. During the nineteen-twenties, new histological procedures began to become available and interest in annelid and arthropod embryology was restored. Although by now the divergence of workers between different groups of animals had become marked, this period saw the first modern accounts of development of the clitellate annelids by Penners and Schmidt, of the Crustacea by Cannon and Manton and of the pterygote insects by Seidel, and Leuzinger and Wiesmann. Many investigators followed their lead during the nineteen-thirties, when the insects, especially, were popular subjects of embryological description; but the next fundamental advance, producing a massive increase in knowledge and understanding, came only with the work of Tiegs (1940, 1947) on myriapods and Manton (1949) on the Onychophora. Much has followed during the last twenty years, highlighted by Anderson (1959) and Åkesson (1961, 1962, 1963, 1967 a, b, 1968) on polychaetes, Weygoldt (1958, 1960, 1961), Scholl (1963), Stromberg (1965, 1967), Anderson (1969) and Dohle (1970) on Crustacea, Striebel (1960), Ando (1962), Anderson (1962, 1963, 1964, 1966c) and Ullman (1964, 1967) on Pterygota, and Holm (1941, 1947, 1952, 1954), Yoshikura (1954, 1955, 1958, 1961) and Weygoldt (1964, 1965, 1968) on Chelicerata. The clitellates and onychophorans have once again been neglected, though some further progress in the interpretation of myriapod and apterygote embryos has been achieved by Dohle (1964), Larink (1969) and others. As I have already partially shown in previous reviews (Anderson 1966 a, b, 1969, 1971, 1972 a, b) the work of the authors mentioned above dominate the theme of a unified interpretation of annelid and arthropod embryology at the present stage of knowledge, though the contributions of many other workers have all added significantly to the total sum of understanding. A critical reassessment can now be made in the light of modern interpretations of arthropod phylogeny (Tiegs and Manton, 1958; Manton, 1964, 1970, 1972) and modern understanding of development as a functional process, testing the proposition that the arthropods are a polyphyletic group of animals in which the Onychophora, Myriapoda and Hexapoda form an assemblage, the Uniramia, probably related to annelids, while the Crustacea and Chelicerata are unrelated to this assemblage, to each other or to the annelids, except in so far as all are components of a larger spiral–cleavage assemblage embracing many phyla.

References

ÅKESSON, B. (1961) On the histological differentiation of the larvae of *Pisione remota* (Pisionidae, Polychaeta). *Acta Zool., Stockholm* **42,** 177–226.

ÅKESSON, B. (1962) The embryology of *Tomopteris helgolandica. Acta Zool., Stockholm* **43,** 135–99.

ÅKESSON, B. (1963) The comparative morphology and embryology of the head in scale worms (Aphroditidae, Polychaeta). *Ark. Zool.,* **16,** 125–63.

ÅKESSON, B. (1967a) The embryology of the polychaete *Eunice kobiensis. Acta Zool., Stockholm* **48,** 141–92.

ÅKESSON, B. (1967b) On the nervous system of the *Lopadorhynchus* larva (Polychaeta). *Ark. Zool.* **20,** 55–78.

ÅKESSON, B. (1968) The ontogeny of the glycerid prostomium. *Acta Zool., Stockholm* **49,** 203–18.

ANDERSON, D. T. (1959) The embryology of the polychaete *Scoloplos armiger. Q. Jl. microsc. Sci.* **100,** 89–166.

ANDERSON, D. T. (1962) The embryology of *Dacus tryoni* (Frogg.) (Diptera, Trypetidae (= Tephritidae)), the Queensland fruit fly. *J. Embryol. exp. Morph.* **10,** 248–92.

ANDERSON, D. T. (1963) The embryology of *Dacus tryoni.* 2. Development of imaginal discs in the embryo. *J. Embryol. exp. Morph.* **11,** 339–51.

ANDERSON, D. T. (1964) The embryology of *Dacus tryoni.* 3. Origins of imaginal rudiments other than the principal discs. *J. Embryol. exp. Morph.* **12,** 65–75.

ANDERSON, D. T. (1966a) The comparative embryology of the Polychaeta. *Acta Zool., Stockholm* **47,** 1–42.

ANDERSON, D. T. (1966b) The comparative early embryology of the Oligochaeta, Hirudinea and Onychophora. *Proc. Linn. Soc. N.S.W.* **91,** 10–43.

ANDERSON, D. T. (1966c) The comparative embryology of the Diptera. *Ann. Rev. Entomol.* **11,** 23–46.

ANDERSON, D. T. (1967) Larval development and segment formation in the branchiopod crustaceans *Limnadia stanleyana* King (Conchostraca) and *Artemia salina* (L.) (Anostraca). *Aust. J. Zool.* **15,** 47–91.

ANDERSON, D. T. (1969) On the embryology of the cirripede crustaceans *Tetraclita rosea* (Krauss), *Tetraclita purpurascens* (Wood), *Chthamalus antennatus* Darwin and *Chamaesipho columna* (Spengler), and some considerations of crustacean phylogenetic relationships. *Phil. Trans. R. Soc.* B, **256,** 183–235.

ANDERSON, D. T. (1971) The embryology of aquatic oligochaetes. In BRINKHURST, R. O. and JAMIESON, B. G. (eds.), *Aquatic Oligochaeta of the World,* Oliver & Boyd, Edinburgh.

ANDERSON, D. T. (1972a) The development of hemimetabolous insects. In COUNCE, S. J. (ed.), *Developmental Systems—Insects,* Academic Press, New York.

ANDERSON, D. T. (1972b) The development of holometabolous insects. In COUNCE, S. J. (ed.), *Developmental Systems—Insects,* Academic Press, New York.

ANDO, H. (1962) The comparative embryology of Odonata with special reference to the relic dragonfly *Epiophlebia superstes* Selys. *Jap. Soc. Prom. Sci.,* Tokyo.

CATHER, J. N. (1971) Cellular interactions in the regulation of development in annelids and molluscs. *Adv. Morphog.* **9,** 67–125.

CLARK, R. B. (1964) *Dynamics in Metazoan Evolution,* Clarendon Press, Oxford.

COSTLOW, J. D., Jr. (1968) Metamorphosis in crustaceans. In ETKIN, W. and GILBERT, L. I. (eds.), *Metamorphosis,* North Holland Publishing Co., Amsterdam.

COUNCE, S. J. (1972) The analysis of development. In COUNCE, S. J. (ed.), *Developmental Systems—Insects,* Academic Press, New York.

DAWYDOFF, C. (1928) *Traité d'embryologie comparée des Invertebrés,* Masson, Paris.

DOHLE, W. (1964) Die Embryonalentwicklung von *Glomeris marginata* (Villers) im Vergleich zur Entwicklung anderer Diplopoden. *Zool. Jb. Anat. Ont.* **81,** 241–310.

DOHLE, W. (1970) Über Eiablage und Entwicklung von *Scutigera coleoptrata* (Chilopoda). *Bull. Mus. Nat. d'Hist. Nat.,* ser. 2, **41,** 53–57.

GREEN, J. (1971) Crustaceans. In REVERBERI, G. (ed.), *Experimental Embryology of Marine and Freshwater Invertebrates,* North Holland Publishing Co., Amsterdam.

HOLM, A. (1941) Studien über die Entwicklung und Entwicklungsbiologie der Spinnen, *Zool. Bidrag., Uppsala* **19,** 1–214.

HOLM, A. (1947) On the development of *Opilio parietinus* Deg. *Zool. Bidrag., Uppsala* **25,** 409–22.

HOLM, A. (1952) Experimentelle Untersuchungen über die Entwicklung and Entwicklungsphysiologie der Spinnen. *Zool. Bidrag., Uppsala* **29,** 293–424.

HOLM, A. (1954) Notes on the development of an orthognath spider, *Ischnothele karschi* Bos. and Lenz. *Zool. Bidrag., Uppsala* **30,** 199–222.

LARINK, O. (1969) Zur Entwicklungsgeschichte von *Petrobius brevistylis* (Thysanura, Insecta). *Helgolander wiss. Meeresunters.* **19,** 111–55.
MANTON, S. M. (1928) On the embryology of a mysid crustacean, *Hemimysis lamornae. Phil. Trans. R. Soc.* B, **216,** 363–463.
MANTON, S. M. (1934) On the embryology of the crustacean, *Nebalia bipes. Phil. Trans. R. Soc.* B, **233,** 163–238.
MANTON, S. M. (1949) Studies on the Onychophora VII. The early embryonic stages of *Peripatopsis* and some general considerations concerning the morphology and phylogeny of the Arthropoda. *Phil. Trans. R. Soc.* B, **233,** 483–580.
MANTON, S. M. (1964) Mandibular mechanisms and the evolution of the arthropods. *Phil. Trans. R. Soc.* B, **247,** 1–183.
MANTON, S. M. (1970) Arthropods: Introduction. *Chemical Zoology* **5,** 1–34.
MANTON, S. M. (1972) The evolution of arthropodal locomotory mechanisms, Part 10. *J. Linn. Soc. Zool.* **51,** 203–400.
REVERBERI, G. (1971) Annelids. In REVERBERI, G. (ed.), *Experimental Embryology of Marine and Freshwater Invertebrates*, North Holland Publishing Co., Amsterdam.
SCHOLL, G. (1963) Embryologische Untersuchungen an Tanaidaceen. *Zool. Jb. Anat. Ont.* **80,** 500–54.
STRIEBEL, H. (1960) Zur Embryonalentwicklung der Termiten. *Acta trop.* **13,** 193–260.
STROMBERG, J.-O. (1965) On the embryology of the isopod *Idotea. Ark. Zool.* **17,** 421–73.
STROMBERG, J.-O. (1967) Segmentation and organogenesis in *Limnoria lignorum. Ark. Zool.* **20,** 91–139.
TIEGS, O. W. (1940) The embryology and affinities of the Symphyla, based on a study of *Hanseniella agilis. Q. Jl. microsc. Sci.* **82,** 1–225.
TIEGS, O. W. (1947) The development and affinities of the Pauropoda, based on a study of *Pauropus sylvaticus. Q. Jl. microsc. Sci.* **88,** 165–267 and 275–336.
TIEGS, O. W. and MANTON, S. M. (1958) The evolution of the Arthropoda. *Biol. Rev.* **33,** 255–337.
ULLMAN, S. L. (1964) The origin and structure of the mesoderm and the formation of the coelomic sacs in *Tenebrio molitor* L. (Insecta, Coleoptera). *Phil. Trans. R. Soc.* B, **248,** 245–77.
ULLMAN, S. L. (1967) The development of the nervous system and other ectodermal derivatives in *Tenebrio molitor* L. (Insecta, Coleoptera). *Phil. Trans. R. Soc.* B, **252,** 1–25.
WEYGOLDT, P. (1958) Die Embryonalentwicklung des Amphipoden *Gammarus pulex pulex* L. *Zool. Jb. Anat. Ont.* **77,** 51–110.
WEYGOLDT, P. (1960a) Beiträge zur Kenntnis der Malakostrakenentwicklung: Die Keimblatterbildung bei *Asellus aquaticus* (L.). *Z. wiss. Zool.* **163,** 340–54.
WEYGOLDT, P. (1960b) Embryologische Untersuchungen an Ostrakoden: Die Entwicklung von *Cyprideis littoralis* (G. S. Brady) (Ostracoda, Podocopa, Cytheridae). *Zool. Jb. Anat. Ont.* **78,** 369–426.
WEYGOLDT, P. (1961) Beitrag zur Kenntnis des Ontogenie der Dekapoden: Embryologische Untersuchungen an *Palaemonetes varians* (Leach). *Zool. Jb. Anat. Ont.* **79,** 223–70.
WEYGOLDT, P. (1964a) Vergleichend-embryologische Untersuchungen an Pseudoscorpionen (Chelonethi). *Z. Morph. Ökol. Tiere* **54,** 1–106.
WEYGOLDT, P. (1964b) Vergleichende-embryologische Untersuchungen an Pseudoscorpionen II. Das zweite Embryonalstadium von *Lasiochernes pilosus* Ellingsen und *Cheiridium museorum* Leach. *Zool. Beitr.* N.F. **10,** 353–68.
WEYGOLDT, P. (1965) Vergleichende-embryologische Untersuchungen an Pseudoscorpionen III. Die Entwicklung von *Neobisium muscorum* Leach (Neobisiinea, Neobisiidae). Mit dem Versuch einer Deutung der Evolution des Embryonalen Pumporgans. *Z. Morph. Ökol. Tiere* **55,** 321–82.
WEYGOLDT, P. (1968) Vergleichende-embryologische Untersuchungen an Pseudoscorpionen IV. Die Entwicklung von *Chthonius tetrachelatus* Preyssl., *Chthonius ischnocheles* Herman (Chthoniinea, Chthoniidae) und *Verrucaditha spinosa* Banks (Chthoniinea, Tridenchthoniidae). *Z. Morph. Tiere* **63,** 111–54.
YOSHIKURA, M. (1954) Embryological studies of the liphistid spider *Heptathela kumurai. Kumamoto J. Sci.* B, **3,** 41–50.
YOSHIKURA, M. (1955) Embryological studies on the liphistid spider, *Heptathela kumurai.* II, *Kumamoto J. Sci.* B, **2,** 1–86.
YOSHIKURA, M. (1958) On the development of a purse-web spider, *Atypus karschi* Donitz, *Kumamoto J. Sci. B(2) Biol.* **3,** 73–86.
YOSHIKURA, M. (1961) The development of the whip-scorpion *Typopeltis stimpsonii* Wood. *Acta arachnol.* **17,** 19–24.

CHAPTER 2

POLYCHAETES

Any consideration of the comparative embryology of the annelids and arthropods must begin with the polychaetes. In spite of well-founded doubts about the interpretation of modern polychaetes as representative of a generalized adult morphology in annelids, it is abundantly clear that their developmental morphology is rooted in a primitive pathway. Total spiral cleavage, a hollow blastula, a trochophore, serial addition of segments from a growth zone, and a greater or lesser degree of metamorphosis (Fig. 1), are a constant theme in almost all species, and none has escaped from this into new and more specialized pathways of development.

Much of the existing knowledge of the early phases of morphological change in polychaete eggs, through cleavage and gastrulation, was set out in the classical "cell lineage" studies of the eighteen-nineties and the early years of this century. Models of meticulous and painstaking observation and description, these investigations, by their authors' own admission, left many facts unelucidated and species unstudied, and the subsequent climate of embryological research has been inimical to further studies of this type. The later phases of polychaete development have been best described, in contrast, in papers of recent origin, since in general the organogeny of polychaetes is a series of small-scale events which only the better optical qualities of modern light microscopes have rendered visible. In particular, the Swedish embryologist, Bertil Åkesson, has contributed greatly to the recent progress of descriptive polychaete embryology (Åkesson, 1961, 1962, 1963, 1967 a, b, 1968; Åkesson and Melander, 1967). The pages which follow will make detailed reference to his work, carried out mainly at the University of Lund, but I have not deemed it necessary to refer more than cursorily to the names of earlier contributors to the elucidation of the present story, since these are listed in detail in a previous review (Anderson, 1966).

Cleavage

Spiral cleavage, as all students of zoology discover at some stage in their early career, is not only difficult to see, but even more difficult to grasp and, most difficult of all, to describe. As early as the eighteen-seventies, various attempts were made to describe spiral cleavage in polychaetes, but the first accurate description was given by E. B. Wilson (1892), working at the Woods Hole Biological Station on the nereid species *Neanthes succinea* (= *Nereis limbata*). Wilson perfected a numerical notation which enabled the origin and division of

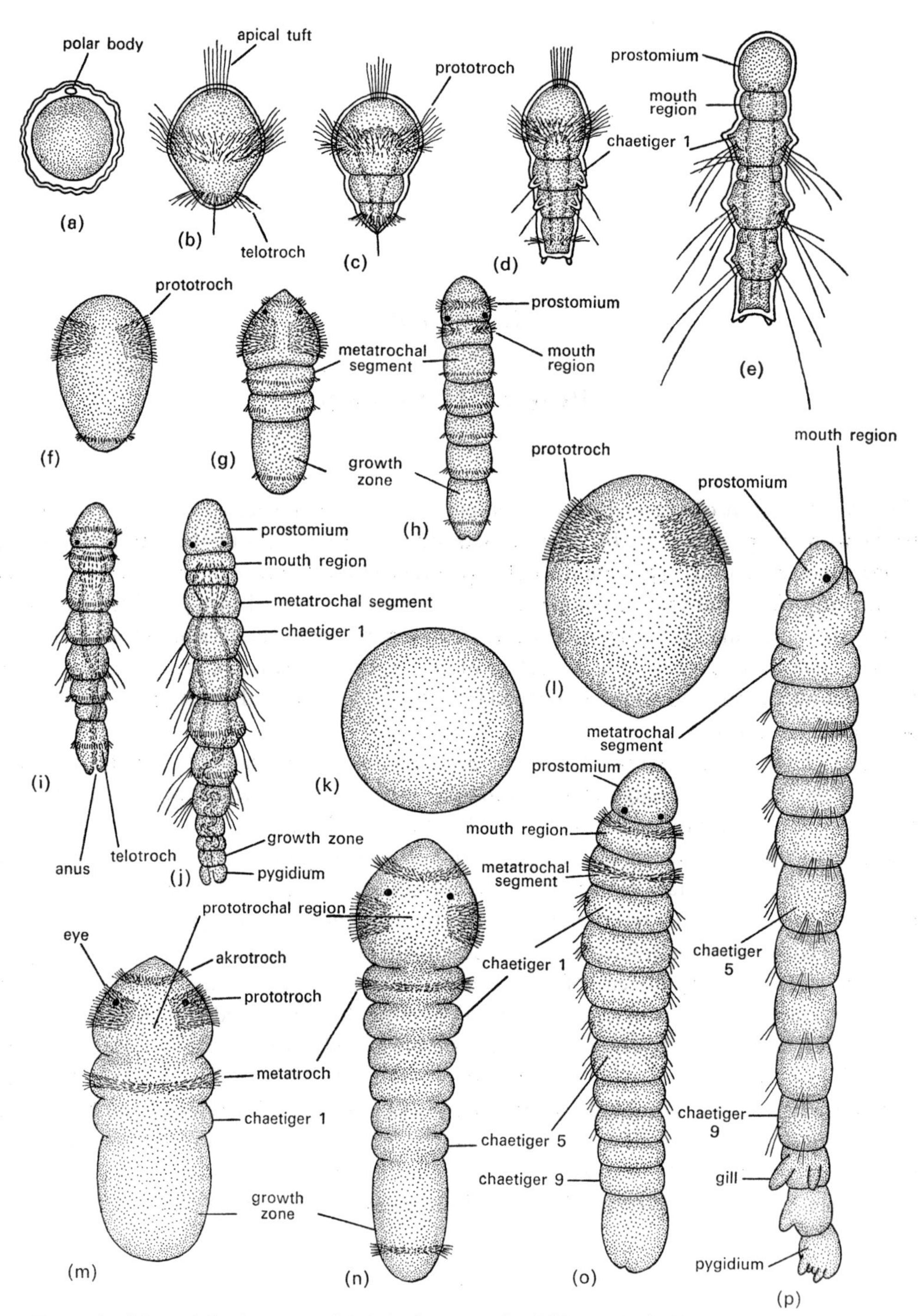

FIG. 1. (a–e) Larval development of *Ophelia bicornis*, after Wilson (1948). (a) Egg, 95 μ in diameter; (b) planktonic lecithotrophic trochophore; (c) segment delineation before metamorphosis; (d) early metamorphosing planktonic larva; (e) fully metamorphosed, first benthic stage. (f–j) Larval development of *Haploscoloplos fragilis*, after Anderson (1961). (f) Lecithotrophic trochophore, 200 μ long from gelatinous egg mass; (g) early segment delineation; (h) planktonic lecithotrophic stage; (i) early benthic stage undergoing metamorphosis; (j) first benthic feeding stage, after completion of metamorphosis and resumption of segment formation. (k–p) Development of *Scoloplos armiger*, after Anderson (1959). (k) Egg, 250 μ in diameter; (l) lecithotrophic trochophore; (m) early segment delineation; (n) further segment delineation; (o) early metamorphosis; (p) first benthic feeding stage after completion of metamorphosis. Escape from the gelatinous cocoon takes place between stages (o) and (p) and there is no planktonic phase.

individual cleavage blastomeres to be expressed easily, so that the entire sequence of cleavage divisions could be described and illustrated. Once this notation was available, other workers applied the same method to the description of cleavage in other species. Mead (1897), working on *Amphitrite ornata* and *Clymenella torquata*, Child (1900), working on *Arenicola cristata*, and Treadwell (1901), investigating *Podarke pugettensis*, were the main contributors to polychaete studies in the American school of classical cell-lineage workers. They were followed in later years by Woltereck (1902, 1904) on *Polygordius*, Soulier (1902, 1916 a, b, 1918, 1920) in France and Shearer (1911) in England on serpulids and, most significantly, by the Dutch embryologist H. C. Delsman (1916) on the ariciid polychaete *Scoloplos armiger*. In a sense, Delsman, who also worked with similar care on molluscan and cirripede cleavage (Delsman, 1914, 1917), was out of step with the fashionable biology of his time, but his culminating study in the classical phase of polychaete embryology remains the most important descriptive paper yet published on polychaete cleavage. Later workers have touched on the description of cleavage in several species not treated by Delsman and his predecessors (Okada, 1941, *Arenicola*; Bookhout and Horn, 1949, *Axiothella*; Newell, 1951, *Clymenella*; Bookhout, 1957, *Dasybranchus*; Allen, 1959, 1964, *Diopatra* and *Autolytus*; George, 1966, *Scolecolepis*; Åkesson and Melander, 1967, *Tomopteris*; Guérrier, 1970, *Sabellaria*); but none has emphasized the full sequence of cleavage and cell lineage in these species, the authors usually being more concerned with the later stages of development.

Differing opinions as to the precise application of Wilson's method of description, and differences in the cleavage sequences among the eggs of different species, led to a number of wordy and entertaining polemics as the description of new species was pursued. In the immediate outcome, none of these was satisfactorily resolved, because the mode of enumeration employed by Wilson and his successors, while descriptively useful, contains an intrinsic flaw. By its very nature, the system emphasizes the segregation of the blastomeres into quartets and the lineage of individual blastomeres stemming from these quartets. Unfortunately, the segregation of developmentally functional components in the blastula wall of different species does not obey these rules. Similarly designated blastomeres have different fates in different species, depending among other things on the extent to which the cells of the blastula develop as temporary larval organs in the ensuing trochophore. The segregation of quartets of cells bears no strict relationship to their subsequent development, and one has to think rather, as I will discuss below, of the segregation of the cleavage blastomeres among the various presumptive rudiments of the major components of the developing animal, each rudiment making up a local area of the wall of the blastula. It is essential at the same time, however, to understand spiral cleavage in polychaetes as a phenomenon of spatial subdivision, a process through which an egg becomes a hollow blastula after a minimum number of cell divisions, and for this purpose the numerical description of Wilson and his successors still provides an essential key.

When a polychaete egg begins to divide, commencing as a spherical egg with a notional, though not usually visible, anteroposterior axis, the first two cleavage planes extend along this axis and approximately bisect the angles between the sagittal and frontal planes (Fig. 2). The four resulting cleavage blastomeres are consequently left-lateral, ventral, right-lateral and dorsal in relation to the major axes of the developing embryo. In *Neanthes succinea*, with an egg 120 μ in diameter, the first two cleavages are unequal and, of the resulting

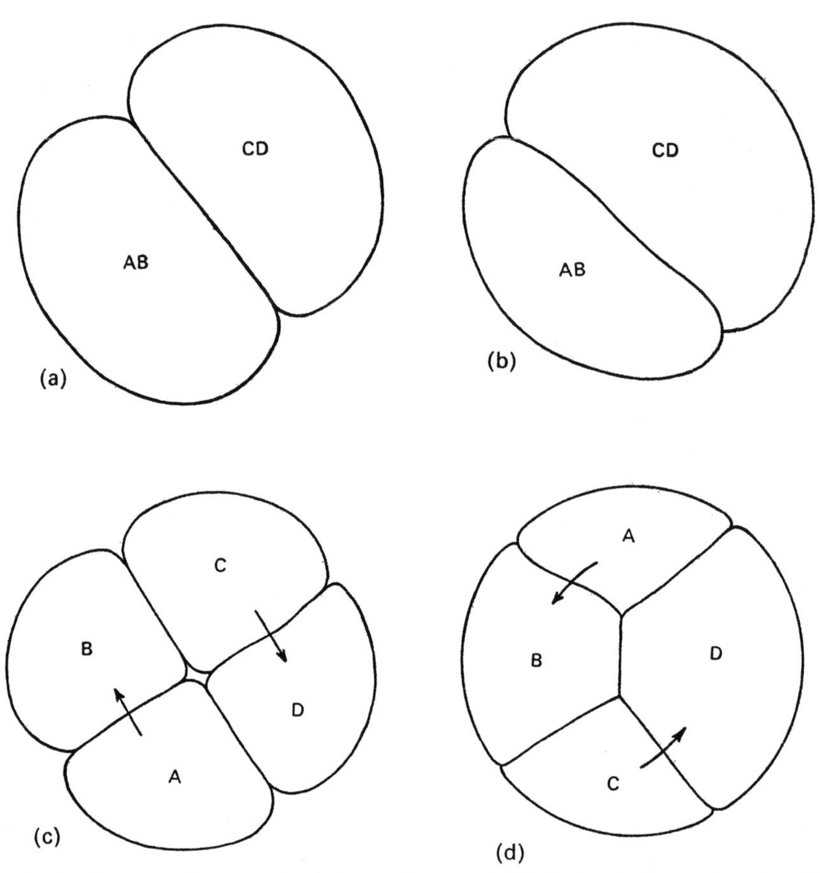

FIG. 2. The 2-cell and 4-cell stages of polychaete cleavage. (a) *Podarke*, 2-cell, anterior view; (b) *Amphitrite*, 2-cell, anterior view; (c) *Podarke*, 4-cell, anterior view; (d) *Arenicola*, 4-cell, posterior view. [(a) and (c) after Treadwell, 1901; (b) after Mead, 1897; (d) after Child, 1900; modified from Anderson, 1966].

blastomeres, the dorsal is the largest and its sister cell, the right-lateral, is the second largest. Wilson, who recognized this, designated the dorsal cell D, the right-lateral cell C, the ventral cell B, and its sister left-lateral cell A, thus enumerating four quadrants of the embryo. He also established that A and C make contact along a sagittal line at the anterior end of the egg, while B and D make contact along a frontal line at the posterior end of the egg. The same inequality of the first two cleavage divisions has subsequently been found in all polychaete eggs with a diameter of more than 70 μ, enabling the four quadrants to be easily identified. It occurs, for instance, in *Amphitrite ornata*, with a 100-μ egg, *Arenicola cristata*, with a 120-μ egg, *Clymenella torquata*, with a 150-μ egg, *Scoloplos armiger*, with a 250-μ egg, and many other species.

In those polychaetes which retain a small, 50–70-μ egg, such as *Polygordius* and some polynoids, hesionids and serpulids, the first two cleavages are equal and the four quadrants are more difficult to distinguish. In spite of this, we can reasonably infer from the work of Soulier, Shearer and especially Treadwell and Woltereck, that the next four cleavage divisions in small polychaete eggs proceed in the same manner as that more easily and constantly discerned in eggs of larger diameter, namely, the manner of typical spiral cleavage. The third, fourth, fifth and sixth cleavage divisions occur more or less synchronously in all the blastomeres present at those stages, so that the 4-cell stage is succeeded by distinct 8-cell, 16-cell, 32-cell and 64-cell stages. Between 64 cells and the onset of gastrulation, many of the blastomeres divide one or more times, so that the wall of the blastula is composed of more than 64 cells, but it is clear from both descriptive and experimental evidence that the 64-cell stage is a crucial one, the stage at which the segregation of the major embryonic presumptive rudiment between distinct groups of blastomeres is completed. If we are to correctly interpret the peculiarities of cleavage in animals in which the basic spiral sequence has become secondarily modified, it is essential to know the details of polychaete cleavage up to the 64-cell stage.

In each of the third, fourth, fifth and sixth cleavage divisions, the plane of division of the cells is perpendicular to those of the first two cleavages, that is, transverse in relation to the anteroposterior axis. At the same time, the mitotic spindles of the dividing cells are inclined obliquely to the anteroposterior axis, either all in a clockwise direction when viewed from the anterior pole, or all in an anticlockwise direction when viewed from the same pole. The third cleavage division has a clockwise rotation. The fourth is perpendicular to the third, with an anticlockwise rotation. The fifth is then clockwise, the sixth anticlockwise. As a result of these perpendicular alternations in the orientation of successive mitotic spindles, the 64 cells resulting from the divisions interlock with one another to form a compact, hollow sphere.

At the third cleavage division (Fig. 3), the four blastomeres A, B, C and D divide with a clockwise twist into four anterior and four posterior cells. The anterior four lie in the furrows between the posterior four, each cell being displaced to the left with respect to the sister cell behind it. Sagittal contact is retained between the A and C quadrant cells at the anterior pole and frontal contact persists between the B and D quadrant cells at the posterior pole. In the species first studied by Wilson, *Neanthes succinea*, the third division is highly unequal, the anterior quartet of cells being small and yolkless while the posterior four cells are large and yolky. It no doubt, therefore, seemed reasonable to distinguish the anterior cells as a first quartet of micromeres cut off from four stem cells, to designate the first quartet as 1a, 1b, 1c and 1d, and to define the stem cells (previous A, B, C and D) as 1A, 1B, 1C and 1D. Herein, however, lies one of the difficulties of grasping the spiral cleavage notation. In the next or fourth cleavage division, both quartets of cells divide in a similar way, but the daughter cells of the anterior quartet retain the prefix 1 and are obviously daughter cells of the first quartet, while the products of division of the posterior quartet are now given a new prefix, 2, and are said to comprise a second quartet 2a–2d and group of four persistent posterior stem cells 2A–2D. This manner of description unfortunately creates entirely the wrong kind of emphasis. It suggests that the anterior products of the third division 1a–1d are different from and subordinate to the posterior products 1A–1D. It further implies that the next division of 1A–1D produces a second subordinate group of cells 2a–2d, equivalent to 1a–1d, and that the posterior cells continue as stem cells but undergo a change which necessitates giving them a new notation, 2A–2D. The difficulty is compounded when the same emphases are pursued during the fifth and sixth cleavage divisions. In fact, as we shall see, the successive quartets and the stem cells designated by this notation are not meaningful

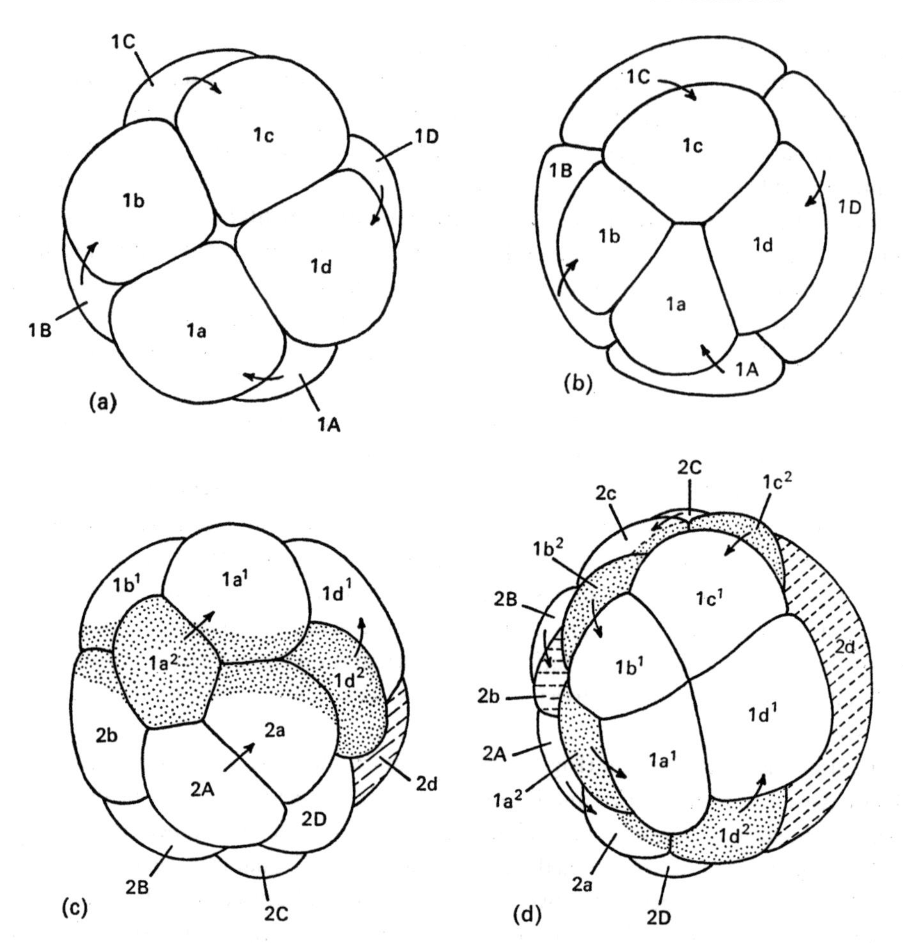

FIG. 3. The 8-cell and 16-cell stages of polychaete cleavage. (a) *Podarke*, 8-cell, anterior view; (b) *Neanthes*, 8-cell, anterior view; (c) *Podarke*, 16-cell, left lateral view; (d) *Scoloplos*, 16-cell, anterior view. [(a) and (c) after Treadwell, 1901; (b) after E. B. Wilson, 1892; (d) after Delsman, 1916; modified from Anderson, 1966].

entities in terms of the segregation of presumptive embryonic rudiments between the cells of the blastula. Furthermore, in the equal cleavage of small polychaete eggs, the posterior, so-called stem cells differ neither in size nor yolk content from the cells anterior to them, so that to talk of the "budding off of successive quartets" in these eggs is manifest nonsense.

What, then, is the point of continuing the usage of this system of spiral cleavage nomenclature? One good reason is that the system has been applied in all the detailed descriptions of spiral cleavage available in the literature, so that its use must continue in some degree if this valuable work is to remain accessible. A second, perhaps more important reason is that the same system provides a workable notation for the expression of two constant aspects of spiral cleavage in polychaetes up to the 64-cell stage, namely the cell lineage of every cell and the position of every cell relative to the major axes. If we wish to compare cleavage in other groups with that of polychaetes, these geometrical constants form an important base line, and I shall therefore outline them once again.

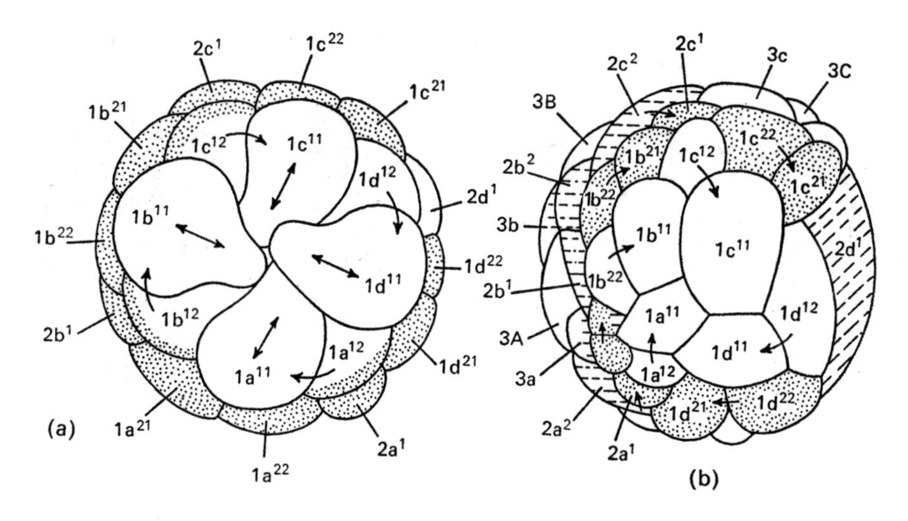

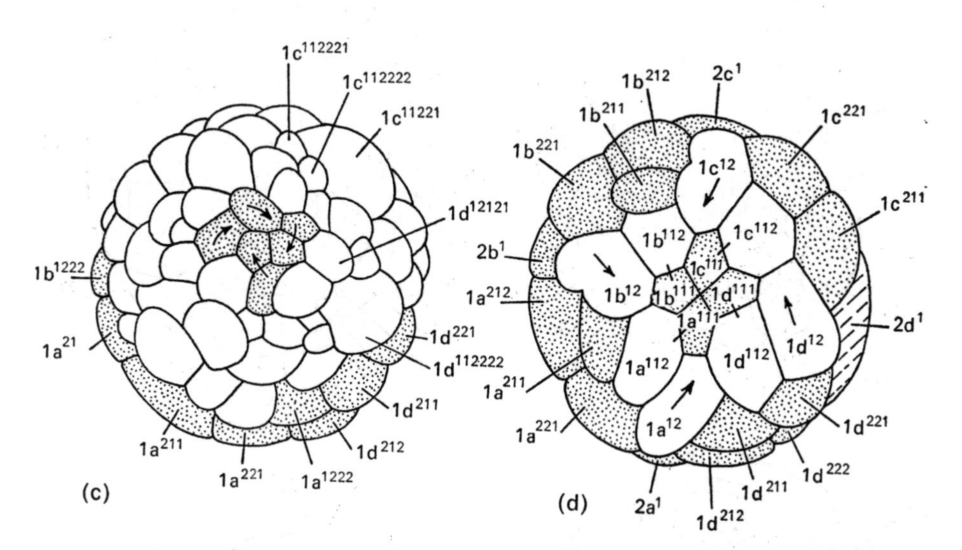

FIG. 4. 32-cell and 64-cell stages of polychaete cleavage. (a) *Podarke*, 32-cell, anterior view; (b) *Scoloplos*, 32-cell, anterior view; (c) *Podarke*, 64-cell, anterior view; (d) *Amphitrite*, 64-cell, anterior view [(a) and (c) after Treadwell, 1901; (b) after Delsman, 1916; (d) after Mead, 1897; modified from Anderson, 1966.]

In the fourth cleavage division, 8–16 cells (Fig. 3), when the first quartet cells 1a–1d divide into anterior and posterior daughter cells, the anterior cells are designated $1a^1$–$1d^1$. By this stage, the sagittal contact between the A-quadrant cells and C-quadrant cells at the anterior pole is no longer obvious. The anticlockwise twist of the division displaces the anterior cells to the right relative to their sister cells and brings them back into the line of their respective quadrants, left-lateral, ventral, right-lateral and dorsal respectively. Their sister cells, behind them and still displaced to the left in each quadrant, are designated $1a^2$–$1d^2$. In a similar manner at the opposite pole, the stem cells 2A–2D retain their original

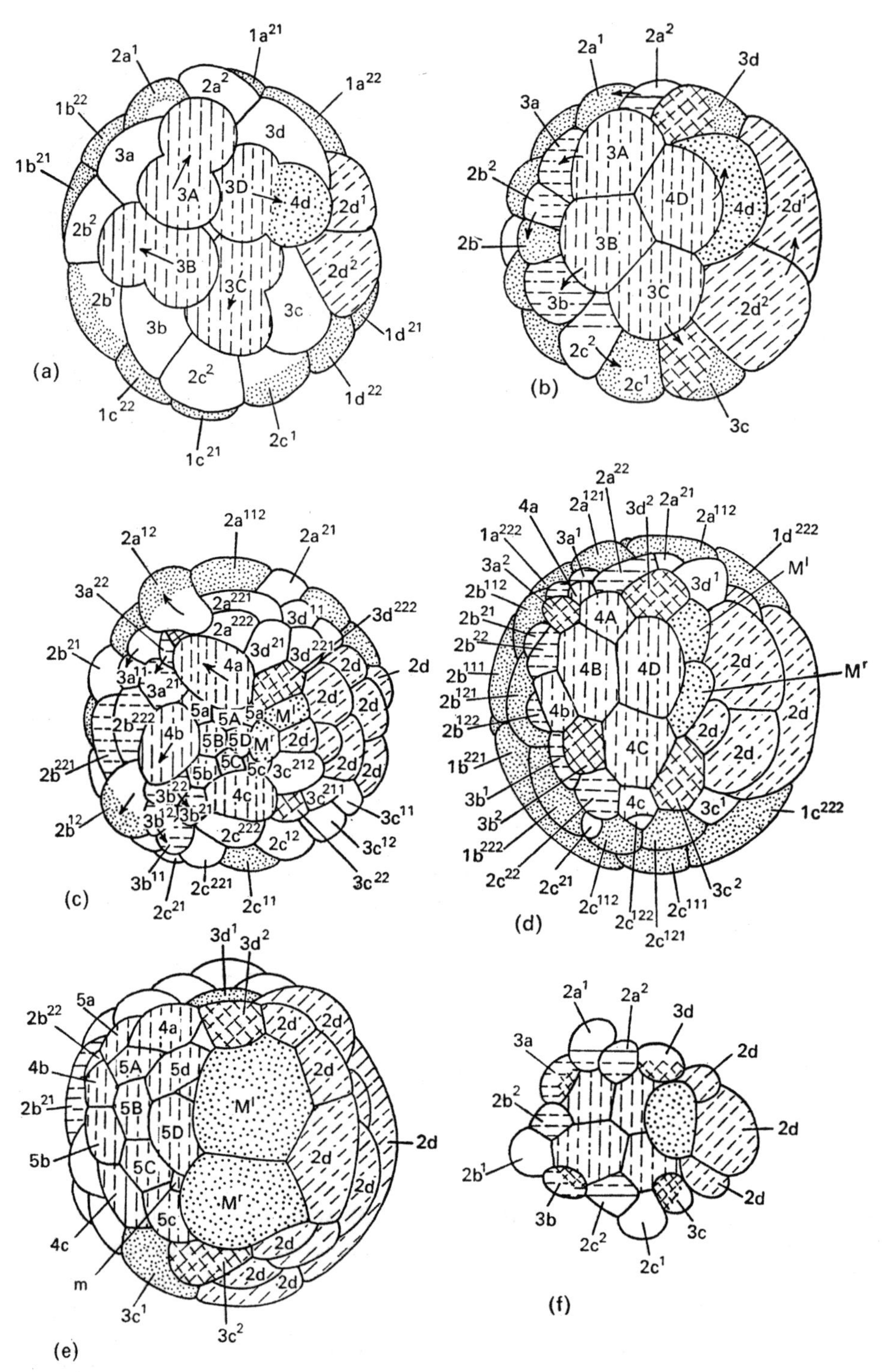

FIG. 5. 32-cell and 64-cell stages of polychaete cleavage (continued). (a) *Podarke*, 32-cell, posterior view; (b) *Arenicola*, 32-cell, posterior view; (c) *Podarke*, 64-cell, posterior view; (d) *Amphitrite*, 64-cell, posterior view; (e) *Scoloplos*, 64-cell, posterior view; (f) *Neanthes*, posterior end after the fifth cleavage division. [(a) and (c) after Treadwell, 1901; (b) after Child, 1900; (d) after Mead, 1897; (e) after Delsman, 1916; (f) after E. B. Wilson, 1892; modified from Anderson, 1966.]

quadrant alignments and frontal contact between the B and D quadrants, but their sister cells, the second quartet cells 2a–2d, lie in front of them and displaced to the right in each quadrant. 2a–2d thus interlock closely with $1a^2$–$1d^2$ around the equator of the egg and the spherical outline of the embryo is preserved.

When the cells of a quartet divide, then, their anterior daughter cells are given the index^1 and their posterior daughter cells the index^2. When the posterior stem cells divide, their anterior daughter cells form a new quartet while the stem cells themselves have their prefix designation augmented by 1. This system can now be applied to the events of the fifth cleavage division, proceeding with a clockwise twist (Figs. 4 and 5). The anterior pole cells $1a^1$–$1d^1$ and their sister cells behind them $1a^2$–$1d^2$ divide to form four tiers of smaller cells. At the anterior pole lie $1a^{11}$–$1d^{11}$, with sister cells $1a^{12}$–$1d^{12}$ behind them and displaced to the right. Interlocked with $1a^{12}$–$1d^{12}$ are $1a^{21}$–$1d^{21}$, followed by their sister cells $1a^{22}$–$1d^{22}$, also displaced to the right. As a result of this division, $1a^{22}$–$1d^{22}$ interlock closely with the anterior daughter cells of the second quartet, $2a^1$–$2d^1$. The posterior daughters of the second quartet, $2a^2$–$2d^2$, are again displaced to the right relative to their anterior sister cells, interlocking with the cells of the third quartet, 3a–3d, budded off from the posterior stem cells, now 3A–3D. The frontal contact between 3B and 3D persists at the posterior pole.

Finally, in the sixth cleavage division (Figs. 4 and 5), each cell now divides with an anticlockwise twist which places its posterior daughter slightly to the left of its anterior daughter. As a result, the successive tiers of daughter cells continue to interlock closely with one another. The anterior pole is occupied by $1a^{111}$–$1d^{111}$, still dorsally, ventrally and laterally placed, forming a group known as the apical rosette. Behind them, making up the entire anterior hemisphere of the embryo in small eggs but a lesser proportion in larger eggs, lie the remaining descendants of the first quartet, $1a^{112}$–$1d^{112}$, $1a^{121}$–$1d^{121}$, $1a^{122}$–$1d^{122}$, $1a^{211}$–$1d^{211}$, $1a^{212}$–$1d^{212}$, $1a^{221}$–$1d^{221}$ and $1a^{222}$–$1d^{222}$. Behind these again, as a more or less equatorial girdle, lie the products of the second quartet, $2a^{11}$–$2d^{11}$, $2a^{12}$–$2r^{12}$, $2a^{21}$–$2d^{21}$ and $2a^{22}$–$2d^{22}$, followed by the third quartet cells $3a^1$–$3d^1$ and $3a^2$–$3d^2$. Finally, as a result of the further division of the residual stem cells, a fourth quartet, 4a–4d, is cut off, leaving the stem cells as 4A–4D. The frontal contact between 4B and 4D persists at the posterior pole even at this stage.

Presumptive Areas of the Blastula

From a comparative point of view, the most important outcome of the investigation of cleavage in polychaetes using the Wilsonian notation is the demonstration that all species retain the same spatial framework of cell division up to the 64-cell stage. Each cell has the same lineage and placement in every species. Therefore, when an egg is cut up into blastomeres through a sequence of divisions different from that of polychaetes, it becomes an easy matter to discern whether this sequence is a modification of the basic spiral sequence or an entirely unrelated sequence, a question of particular importance in the interpretation of certain arthropod cleavage patterns. In order to comprehend cleavage in polychaetes as a process of segregation of the presumptive rudiments of the embryo into different cells, however, we have to approach its outcome at the 64-cell stage from another angle of attack. At this stage, as can be discerned from the subsequent fate of the cells, the major rudiments become fully separated for the first time and the blastula wall is sharply zoned as a composite

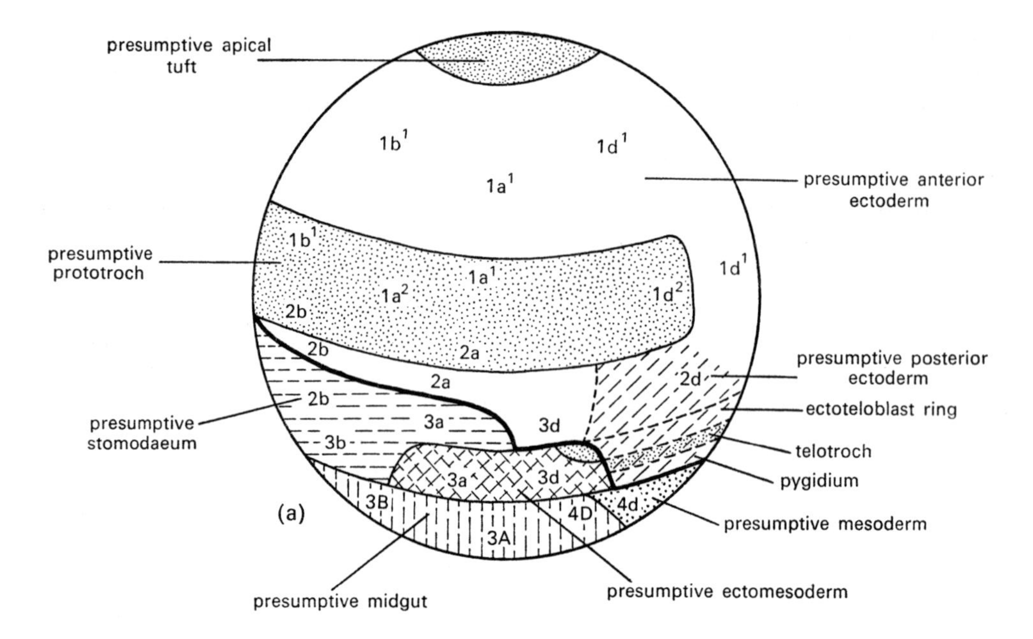

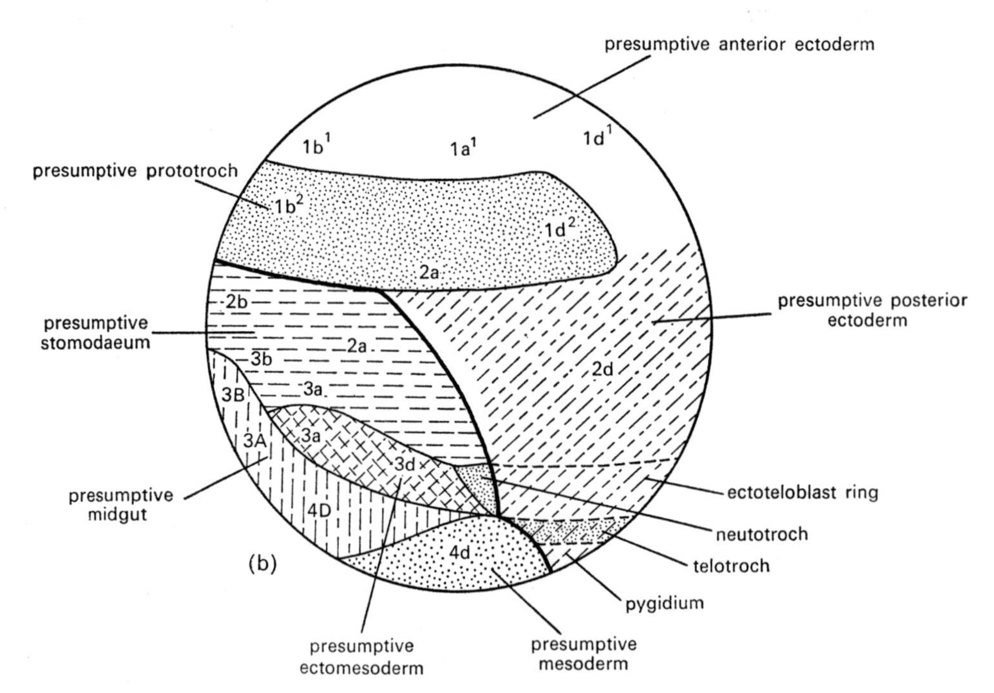

FIG. 6. Fate maps of the polychaete blastula, drawn in left lateral view. (a) *Podarke*, based on data of Treadwell (1901); (b) *Scoloplos*, based on data of Delsman (1916) and Anderson (1959). [After Anderson, 1966.]

of presumptive areas. A comparison of the distribution and composition of the presumptive areas of the blastula of different species (Anderson, 1966; Cather, 1971) immediately reveals two fundamental facts. One is that the arrangement of the presumptive areas in the wall of the blastula is constant in all species, so that we can speak of a presumptive area pattern or fate map for polychaetes. The second is that the same presumptive area of the blastula wall need not include the same cells in all species.

The presumptive areas themselves are as follows (Fig. 6):

1. An equatorial or sub-equatorial area of presumptive prototroch.
2. An area of presumptive anterior ectoderm in front of the prototroch.
3. An area of presumptive posterior ectoderm dorsally and laterally behind the prototroch.
4. An area of presumptive stomodaeum ventrally behind the prototroch.
5. Paired posterolateral areas of presumptive ectomesoderm.
6. A posterodorsal area of presumptive mesoderm.
7. A posterior to posteroventral area of presumptive midgut.

The variations in the cellular composition of these areas, in so far as they are known from detailed cell lineage studies, are twofold. In different species, cells of similar notation may be components of different areas. Secondly, cells of the same area may vary in different species in their relative size, yolk content and subsequent detailed fate, depending on the extent to which development retains functional expression of the larval features of the trochophore, or is modified in association with yolk. We can examine each area in turn with these points in mind.

The presumptive prototroch

Demonstrably for several polychaetes (*Podarke*, *Protula*, *Polygordius*, *Amphitrite*, *Arenicola*, *Neanthes*, *Clymenella* and *Scoloplos*) and presumably in all, the major part of the presumptive prototroch comprises the girdle of 16 cells formed by further division of the posterior daughters of the first quartet: $1a^{211}$–$1d^{211}$, $1a^{212}$–$1d^{212}$, $1a^{221}$–$1d^{221}$ and $1a^{222}$–$1d^{222}$. The remainder of the area is formed by a variety of adjacent blastomeres. In the moderately yolky embryos of *Amphitrite*, *Arenicola* and *Clymenella*, these are the nine cells $2a^{111}$–$2c^{111}$, $2a^{112}$–$2c^{112}$ and $2a^{121}$–$2c^{121}$ lying behind the primary prototroch cells ventrally and laterally. In the large, yolky trochophore of *Scoloplos*, in contrast, only the two cells $2a^{1}$ and $2c^{1}$ are included as additional prototroch cells. On the other hand, in *Podarke*, whose development proceeds from a small, microlecithal egg towards full functional expression of the trochophore larva, not only the same nine cells as in *Amphitrite* but also three cells, $1a^{1222}$, $1b^{1222}$ and $1c^{1222}$ in front of the primary prototroch cells, form part of the presumptive prototroch. No doubt other variations occur in other species.

The presumptive anterior ectoderm

Anterior to the presumptive prototroch is a cap of cells formed by most or all of the cells stemming from further divisions of the anterior daughters of the first quartet, $1a^{1}$–$1d^{1}$. At the 64-cell stage, as pointed out above, the most anterior of these cells form an apical rosette, $1a^{111}$–$1d^{111}$. In several species (*Podarke*, *Lepidonotus*, *Amphitrite*, *Neanthes* and *Clymenella*) the apical rosette cells are known to give rise to the apical tuft of the trochophore, but the apical tuft varies greatly in extent and cell number in other species, its origin is in most cases obscure and it is often entirely absent, as in *Scoloplos* and *Tomopteris* (Anderson, 1959; Åkesson, 1962). In general the cells $1a^{111}$–$1d^{111}$, $1a^{112}$–$1d^{112}$, $1a^{121}$–$1d^{121}$ and $1a^{122}$–$1d^{122}$ give rise to numerous small cells which develop as the prostomium, cerebral ganglion and surface epithelium of the trochophore in front of the prototroch, all of which can be generalized as anterior ectoderm. The presumptive anterior ectoderm, then, unlike the presumptive prototroch, has an almost constant cellular composition.

The presumptive posterior ectoderm

All of the superficial structures behind the prototroch, together with the ventral nerve cord, take origin from a group of cells lying dorsally and laterally behind the presumptive prototroch as presumptive posterior ectoderm. In embryos such as that of *Scoloplos armiger*, in which the D cell and its daughter, the 2d cell, are disproportionately large, the major part of the presumptive posterior ectoderm is formed by division products of 2d, that is, $2d^{11}$, $2d^{12}$, $2d^{21}$ and $2d^{22}$. The area, however, always incorporates cells which lie at the margin of the 2d group, and in small embryos in which 2d is not disproportionately large, these marginal cells make a significant contribution to the area. Division products of $2a^2$, $2b^2$ and $2c^2$ usually form the lateral margins of the presumptive posterior ectoderm, while minor contributions may also be made by divisions of $3a^1$–$3d^1$ and $3a^2$–$3d^2$. Anteriorly, $1d^{222}$, immediately anterior to the 2d group in the mid-dorsal line, is also part of the presumptive posterior ectoderm in *Podarke*, *Amphitrite*, *Arenicola* and perhaps other genera. The mode of segregation of the area is thus quite variable. Nevertheless, irrespective of this variability, the presumptive posterior ectoderm displays a constant pattern of sub-areas as follows:

1. Near the posterior margin of the area, a transverse band of presumptive ectoteloblasts of the trunk segments. When the D cell is large, these teloblasts are formed exclusively as division products of 2d and are also large, setting up the capacity to form the ectoderm of several trunk segments before feeding begins. When the D cell is relatively small, the ectoteloblasts are formed by descendants of 3c and 3d as well as 2d.
2. Immediately behind the presumptive ectoteloblasts, a transverse band of presumptive telotroch cells.
3. At the lateral ends of the presumptive telotroch, presumptive neurotroch.
4. Behind the presumptive telotroch, presumptive pygidial ectoderm.

The remainder of the presumptive posterior ectoderm, in front of the ectoteloblast band, gives rise to surface epithelium of the trochophore behind the prototroch.

The presumptive stomodaeum

The presumptive stomodaeum is formed by a number of ventrally placed descendants of the second and third quartet cells, in various ways in different species. In *Scoloplos*, its components are 2b midventrally, with $2a^2$ and $2c^2$ on either side of it and a number of small, adjacent cells cut off from 3a and 3b. *Polygordius*, *Amphitrite*, *Neanthes* and *Clymenella* develop their presumptive stomodaeum in an approximately similar way. In *Podarke*, *Arenicola* and *Capitella*, 3a and 3b join 2b in making major contributions to the stomodaeal area, while $2a^2$ and $2c^2$ cut off only small cells as stomodaeal components.

The presumptive ectomesoderm

The evidence for a paired posterolateral area of presumptive ectomesoderm in polychaetes is still unsatisfactory in many ways. *Scoloplos* certainly develops such an area, formed by the cells 3a, 3b, $3c^2$ and $3d^2$, which otherwise make only a small contribution to the presumptive stomodaeum and to the margin of the presumptive posterior ectoderm. Similarly, the cells $3a^{222}$, $3c^{222}$ and $3d^{222}$ are presumptive ectomesoderm cells in *Podarke*, and cells of corresponding position and fate have been hinted at for several other species (e.g. descendants of 2a, 2c, 3a and 3b in *Polygordius*). Whether the fate of these cells is to form the larval musculature of the trochophore, as in *Podarke* and *Polygordius*, or later post-larval musculature, as in *Scoloplos*, in which larval musculature of the trochophore is eliminated in association with direct development, the point can still be made that different blastomeres comprise the same ectomesodermal areas in different species.

The presumptive mesoderm

The source of all the definitive mesodermal cells in polychaetes, in contrast, is a presumptive area of more exact and constant composition. The presumptive mesoderm comprises a pair of posterodorsal blastomeres, M^l and M^r, always formed by equal bilateral division of the cell 4d of the fourth quartet. These cells subsequently become internal and give rise teloblastically to paired ventrolateral mesodermal bands. 4d, like 2d, is relatively large in yolky eggs (e.g. *Arenicola, Scoloplos*), reflecting its subsequent substantial contribution to the formation of trunk segments before feeding begins.

In *Arenicola* and *Amphitrite*, M^l and M^r are exclusively mesodermal. In other species, as will be mentioned below, they have been found to cut off one or two small cells against the presumptive midgut before assuming their mesodermal teloblastic role.

The presumptive midgut

Like the presumptive mesoderm, the presumpitve midgut in polychaetes always has a constant cell composition being formed by the stem cells 3A, 3B, 3C and 4D. The number of divisions subsequently performed by these cells in the blastula wall depends on their relative size and yolk content. In nereids, where the cells are large and dense with yolk, no further divisions occur. In *Aricia*, *Amphitrite*, *Clymenella*, *Capitella* and *Scoloplos*, in which 3A, 3B, 3C and 4D are quite large but not disproportionately packed with yolk, the first three undergo two divisions and the last one division, yielding 11 cells. In *Podarke*, *Polygordius* and *Arenicola*, with posterior blastomeres little different in size from anterior blastomeres, an extra division ensues in the fourth quartet cells 4a–4c, so that the final number of presumptive midgut cells is 14.

As mentioned above, the presumptive midgut is augmented in some species by small cells cut off from M^l and M^r as presumptive posterior midgut cells. Described for *Podarke*, *Spio*, *Aricia* and *Scoloplos*, the nereids *Neanthes* and *Platynereis*, and *Tomopteris*, this may be a widespread event in polychaetes. The role of the posterior presumptive midgut cells is of particular interest in the nereids and *Tomopteris* (Åkesson, 1962), whose presumptive midgut cells are large and yolky, creating problems in their subsequent transition into a midgut epithelium. There is evidence that the posterior midgut cells lying behind the yolky cells are the source of a major part of the midgut epithelium in these embryos.

By recognizing a common pattern of presumptive areas in the blastula wall at the 64-cell stage of polychaetes and by examining the blastomere composition of these areas, we can attain a useful generalization. Each area is segregated during cleavage in such a way that it has the same location, relative to all other areas, in every species. The cells of the same area, however, can vary in size, number and yolk content in different species. Furthermore, without alteration in the sequence of spiral cleavage divisions, cells of the same lineage and notation can become components of different, adjacent areas in different species. Even within the polychaetes, therefore, we can recognize that the boundaries of the presumptive areas established in the blastula wall are set independently of the sequence of cleavage planes that subdivides the egg. Provided that the lines of cleavage finally coincide with the presumptive area boundaries, as they do at the 64-cell stage in polychaetes, it matters little how previous divisions proceed. It follows that an homologous pattern of presumptive areas can be postulated even when the basic sequence of spiral cleavage divisions has become modified or even eliminated, provided that the relative locations and fates of the areas in the blastula wall remain constant. It also follows that spiral cleavage, in the sense of a particular sequence of spindle orientations and cleavage planes, is indicative of phylogenetic affinity only when taken in conjunction with comparative presumptive area patterns. If presumptive area boundaries can vary independently of spiral cleavage planes, then more than one basic presumptive area pattern could have evolved in association with spiral cleavage. As we shall see later, there are good reasons for supposing this to be the case among the modern arthropods.

Gastrulation

Presumptive areas in the wall of the blastula have not only a location but a fate, the expression of which begins with the onset of gastrulation. This phase of development, in fact, is best defined as the process of migration of the presumptive areas of the blastula into their organ-forming positions. We can then view it operationally as a series of interrelated movements rather than as a single process with fixed characteristics.

In polychaetes, the presumptive rudiments of the internal organs lie posterior to the prototroch, so that gastrulation movements are most in evidence in this region. The manner in which the movements occur is closely bound to the number, relative size and yolk content of the presumptive midgut cells (Figs. 7–9). Most of the detailed information on polychaete gastrulation was provided by the same workers who described cell lineages during cleavage, namely E. B. Wilson (1892), Mead (1897), Child (1900), Treadwell (1901), Woltereck (1904), Shearer (1911) and Delsman (1916), though some further interesting observations on this process have recently been made by Åkesson (1962, 1967a), working on *Tomopteris* and

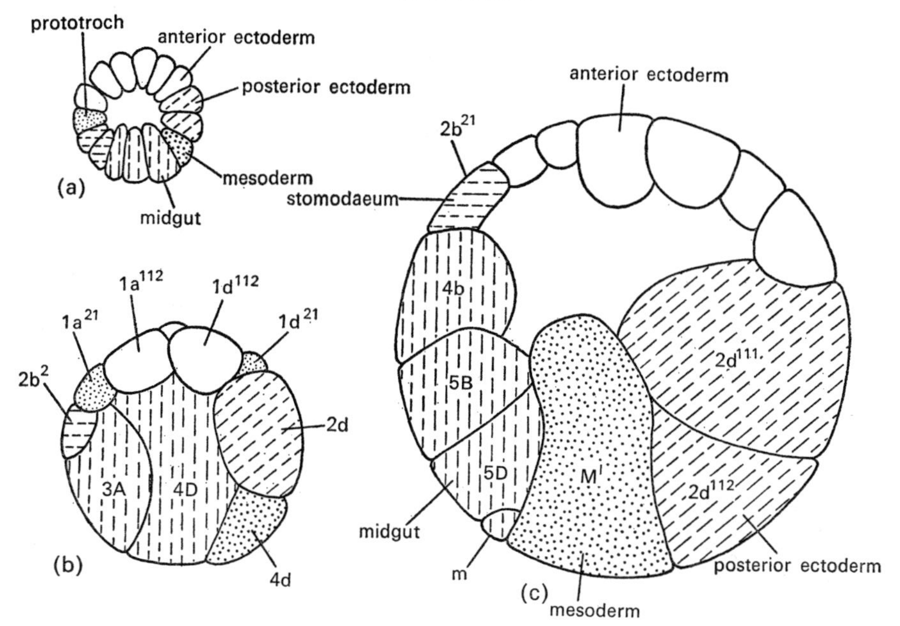

FIG. 7. Polychaete blastulae in slightly parasagittal section. (a) The serpulid *Eupomatus*, after Shearer (1911); (b) the nereid *Neanthes* after E. B. Wilson (1892); (c) the ariciid *Scoloplos*, after Delsman (1916). [Modified from Anderson, 1966.]

Eunice. The events of gastrulation in *Tomopteris*, as we shall see, are in line with expectations for a yolky polychaete embryo, but the peculiarities of *Eunice* are unique and will be deferred until the general features of polychaete gastrulation have been discussed.

In species whose small eggs contain little yolk, such as *Podarke*, *Polygordius* and the serpulid *Eupomatus*, the blastula has a relatively large blastocoel and the presumptive midgut cells are small, numerous and posterior in location. The first sign of gastrulation here is a movement of the presumptive midgut cells into the interior as a typical invagination. Division of the midgut cells continues as invagination proceeds. The invagination cavity opens to the surface posteroventrally, the margin of the opening being occupied by presumptive stomodaeal cells anteroventrally, presumptive ectomesoderm laterally and presumptive mesoderm posterodorsally.

When the egg is larger but the yolk is more or less evenly distributed among the blastomeres during cleavage, as in *Amphitrite*, *Arenicola*, *Clymenella* and *Scoloplos*, the several presumptive midgut cells are relatively large and lie more ventrally than posteriorly. A conspicuous blastocoel is retained in embryos of this type. Usually, the midgut cells move into the blastocoel by a mass amoeboid migration during which the cells become temporarily long and narrow. The exposed area of midgut at the surface, bordered by stomodaeum anteriorly, ectomesoderm laterally and mesoderm posteriorly, gradually diminishes and some tendency to invagination is observed, but no invagination cavity is formed.

In species in which the presumptive midgut cells are large, densely yolky and few in number, the cells of the remaining presumptive areas are cut off during cleavage as a superficial layer which covers the yolky cells anteriorly, laterally and dorsally. The yolky, presumptive midgut cells are still exposed at the surface of the blastula ventrally to posteriorly,

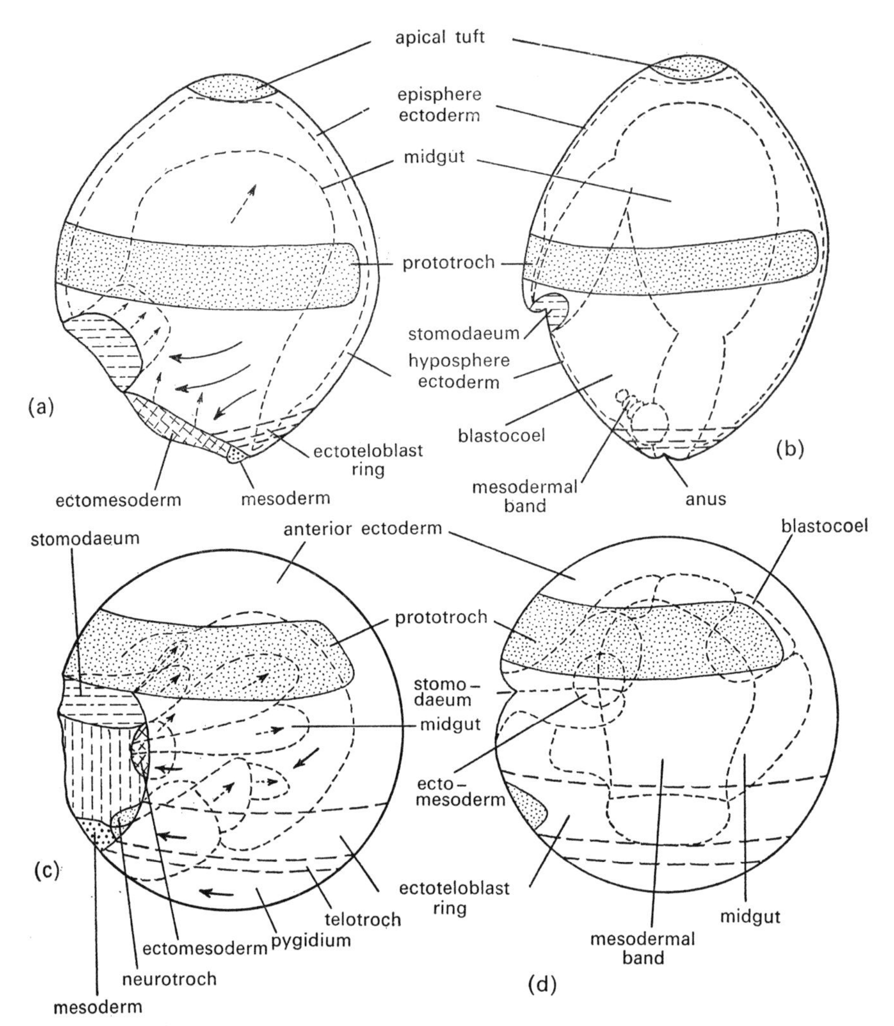

FIG. 8. Diagrams summarizing modes of gastrulation in polychaetes. (a) and (b) Stages in the gastrulation of *Podarke*, based on data of Treadwell (1901); (c) and (d) stages in the gastrulation of *Scoloplos*, based on data of Delsman (1916) and Anderson (1959) [Modified from Anderson, 1966.]

but due to the gross inequality of their previous division, cleavage has already placed them partially internally, occluding what otherwise would have been the blastocoel. This condition is characteristic of nereids and has recently been described for *Tomopteris* by Åkesson (1962). The complete enclosure of the midgut cells follows as a result of further spread of the presumptive ectoderm, without any gastrulation movement of the midgut cells themselves. When the presumptive midgut cells are excessively large and yolky in polychaetes, therefore, they become internal partially as a result of modification in cleavage and partially due to overgrowth by gastrulating ectoderm, but lose any intrinsic gastrulation movement. As we shall see in later chapters, this modification has arisen repeatedly in annelids and arthropods in association with increased yolk.

As the presumptive midgut cells become internal, the presumptive mesoderm cells M^1 and

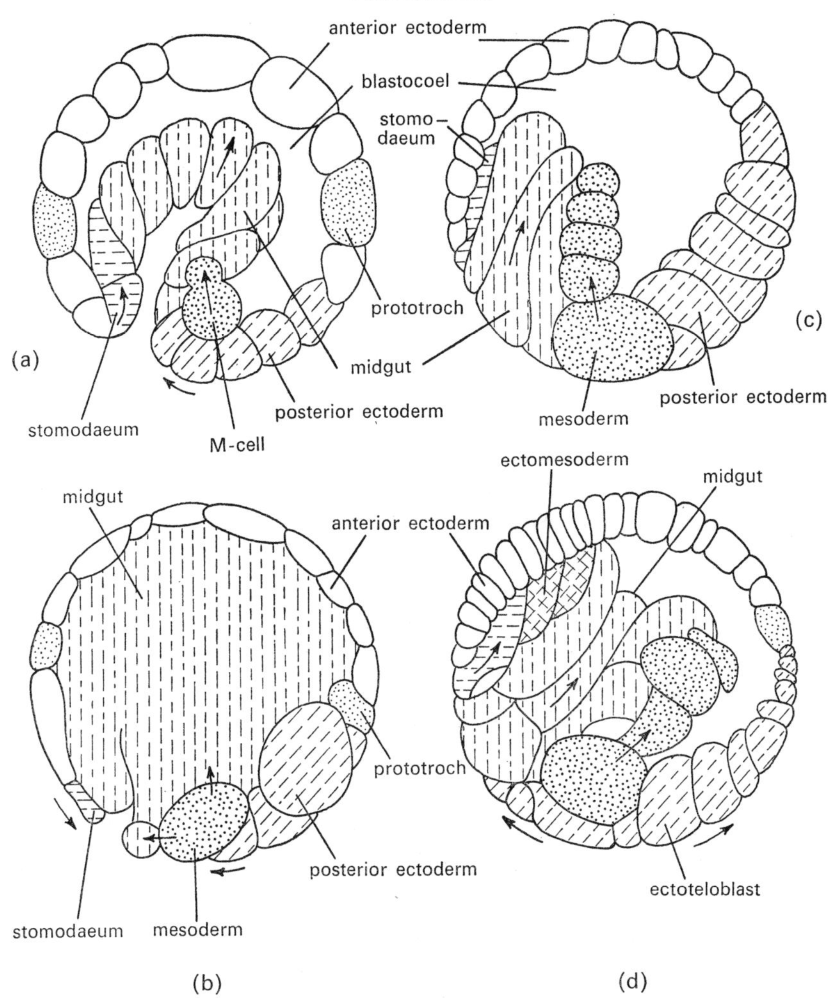

FIG. 9. Polychaete gastrulae in slightly parasagittal section. (a) *Eupomatus*, after Shearer (1911); (b) *Neanthes*, after E. B. Wilson (1892); (c) and (d) *Scoloplos*, after Delsman (1916). [Modified from Anderson, 1966.]

M^r also move into the interior, to lie on either side of the anteroposterior axis, behind the presumptive midgut. The degree to which this shift is due to amoeboid migration of the cells themselves, and the possible importance of pushing and overgrowth movements by other cells in their translocation, are not yet established for any species. Amoeboid immigration seems likely to be significant in species which retain a blastocoel, such as *Podarke* and *Scoloplos*, but may be minimal or absent when the blastocoel is filled by the midgut cells, as in *Neanthes*.

In small embryos, like those of *Podarke* and *Eupomatus*, in which the M-cells are no larger than other cells, the M-cells complete their gastrulation movement into the interior before they begin to act as teloblasts. In larger embryos, such as *Arenicola*, *Neanthes*, *Clymenella* and *Scoloplos*, the relatively large M-cells begin to bud off their corresponding mesodermal bands while still at the surface of the embryo and continue budding as they become internal.

Åkesson (1962) has recently discovered an interesting further modification of this trend in *Tomopteris*. As explained above, *Tomopteris* has a large, yolky presumptive midgut which fills the blastocoel before gastrulation begins, and its remaining presumptive areas are composed of small cells containing little or no yolk. The presumptive mesoderm, instead of remaining a pair of M-cells which become internal, divides at the surface into paired postero-dorsal groups of small cells which then proliferate the mesodermal bands directly into the interior, without becoming internal themselves. Each superficial site of proliferation develops a small, temporary invagination. As we shall again see below, the modification of presumptive mesoderm as small cells proliferating at the surface and performing little or no gastrulation movement into the interior is another frequent feature of yolk-adapted arthropod embryos, although as far as is known it has not evolved in any annelid other than the specialized *Tomopteris*.

Movement of presumptive ectomesoderm cells into the interior, accompanying entry of the presumptive midgut and M-cells, takes place by amoeboid immigration. The cells enter the space between the presumptive midgut and the outer epithelial wall of the embryo. Their immigration was studied closely by Delsman (1916) in *Scoloplos armiger*, where the ecto-mesoderm cells finally cluster around the stomodaeum. In other species, the movements of the ectomesoderm cells once internal have not been satisfactorily described.

The presumptive stomodaeal cells become internal in a constant manner in all species except *Eunice*. First sinking beneath the surface by amoeboid immigration in front of the presumptive midgut in the ventral midline, the presumptive stomodaeal cells either form the anterior ventral wall of the midgut invagination or lie as an arc around the anterior face of the solid midgut rudiment. In the specialized embryo of *Tomopteris*, Åkesson (1962) has described a paired stomodaeal rudiment, separated in the ventral midline, but it seems likely (see below, p. 34) that this condition is secondary. Once below the surface, the ends of the arc of stomodaeal cells curve towards the ventral midline and the stomodaeal arc closes to form a tube in front of the midgut. The tube retains external connection with the surface of the embryo and its superficial entrance becomes the mouth. In small embryos developing as feeding trochophores, as in *Podarke*, *Polygordius* and *Eupomatus*, the tube has an open lumen, but in larger, yolky embryos (e.g. *Neanthes*, *Scoloplos*), the lumen is occluded by the large cells making up the stomodaeal wall. Only *Tomopteris* once again provides an exception among yolky embryos, having an open stomodaeal lumen from the start.

In polychaetes generally, the gastrulation movements of the midgut, mesoderm, ecto-mesoderm and stomodaeum proceed simultaneously. Accompanying them, the presumptive posterior ectoderm spreads at the surface, replacing the other rudiments as they move into the interior. In the main, the increase in surface area of the presumptive ectoderm is attained through cell division, in sharp contrast to the other presumptive rudiments, whose migratory activities are accompanied by a temporary reduction in mitotic activity. In small embryos with invaginate midgut entry, the main direction of spread of the posterior presumptive ectoderm is lateral to posteroventral, pushing the lateral lips of the invagination opening together in the posteroventral midline. Once contact is established between these lips, the hollow midgut rudiment becomes tubular and separates from the outer epithelial wall. In *Podarke*, the posterior end of the invagination aperture remains open, the outer epithelial and midgut walls retain continuity around the opening, and the opening itself becomes the anus. In *Polygordius* and *Eupomatus*, the same region becomes temporarily closed, but soon reopens as the anus.

Larger polychaete embryos retain a similar spread of the presumptive posterior ectoderm towards the ventral midline, but simply as a spread across the surface of the immigrating midgut and mesoderm rudiments. As might be expected from the spherical shape of the embryo, coverage is completed first at the posterior end, then progressively in an anterior direction along the ventral midline towards the mouth. The anus in these species opens late, after all gastrulation movements have been completed.

As part of the gastrulation spread of the posterior presumptive ectoderm, the presumptive ectoteloblasts and presumptive telotroch extend as transverse bands around the embryo,

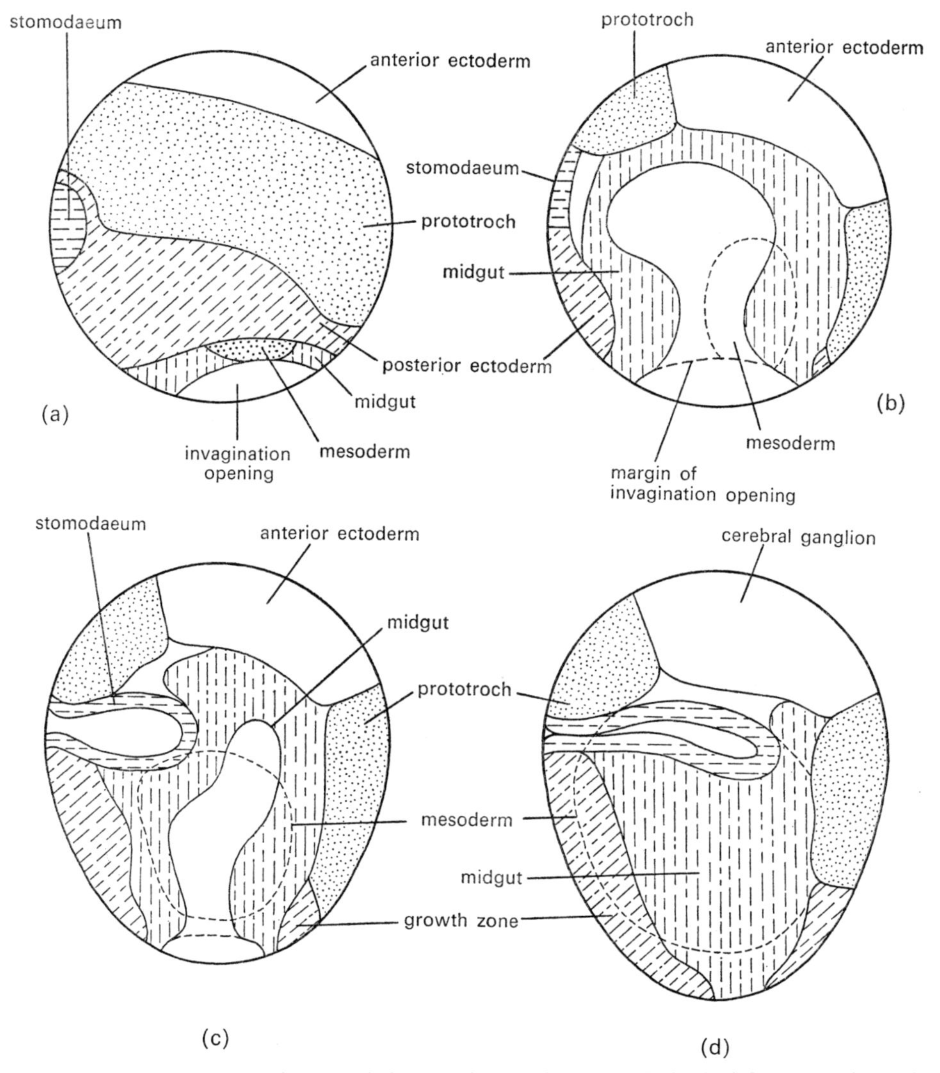

FIG. 10. The gastrulation of *Eunice kobiensis*, drawn diagrammatically in left lateral view, after Åkesson (1967a). (a) Early gastrula, showing presumptive areas (compare Fig. 6); (b) early gastrula in sagittal section; (c) and (d) sagittal sections showing completion of gastrulation.

just in front of the posterior end. Behind them lies the pygidial ectoderm, around the present or future anus. In *Scoloplos* and *Arenicola*, and probably in other species, the paired presumptive neurotroch cells are carried down to the ventral midline between the ends of the expanding ectoteloblast ring.

As mentioned above, Åkesson (1967a) has recently described a unique mode of gastrulation in the large, yolky embryo of *Eunice kobiensis*. The early phases of the process unfortunately proved rather recalcitrant of study, but it is obvious from the details of the later phases that the process is most unusual (Fig. 10). In spite of an egg diameter of 370 μ and an unequal cleavage, the blastula retains a blastocoel. The presumptive area pattern of the embryo differs little from that of other polychaetes and yet, uniquely for a yolky embryo, the presumptive midgut invaginates deeply at the posterior end of the embryo, creating a large temporary archenteron whose posterior opening is wholly separate from the presumptive stomodaeum. Spread of the presumptive posterior ectoderm in a posterior direction closes the opening of the archenteron at approximately the site at which the anus will later form. Meanwhile, the midgut cells come together to fill the inerior of the embryo and occlude the archenteron, while the presumptive stomodaeum invaginates and penetrates deeply into the interior by an independent gastrulation movement.

Although it is not difficult to interpret the mode of gastrulation of *Eunice* as a specialization of the basic fate map and gastrulation patterns of small-egged species, the interest of this species lies in the fact that it cannot be generalized in the same manner as the other yolky, polychaete embryos previously described. *Eunice* represents an alternative mode of specialization of the polychaete blastula and gastrulation in relation to yolk. It shows that even within the polychaetes, the formation of the mouth may be divorced from "closure of the blastopore". If further shows that yolky polychaete eggs do not display a single pattern of modification in their early development, raising the possibility that yet other patterns remain to be discovered.

Completion of the Trochophore

Once gastrulation is complete, or, in the case of the anterior ectoderm, while gastrulation is proceeding, the various rudiments begin to show changes preliminary to their further progress into organogeny. These are the first steps in the organogenetic development of the trochophore. In contrast to cleavage and gastrulation in polychaetes, which were described in detail by only a few classical workers and have been virtually ignored descriptively for the last fifty years, the organogenetic development of the polychaete trochophore has been a continuing source of interest. Even before the cell lineage of polychaete spiral cleavage was properly understood, the development and structure of the trochophore was examined histologically by a number of European workers. Salensky (1882, 1883) made the first significant contribution in a comparative study of the trochophores of several species (*Protula*, *Spirorbis*, *Polymnia*, *Perinereis* and *Aricia*), and was followed by Meyer (1888, *Protula*), Wistinghausen (1891, *Platynereis*), Häcker (1894, *Polynöe*), Schively (1897, *Spirorbis*) and Eisig (1898, *Capitella*). The workers on cell lineage in polychaetes also provided additional information on the formation of the trochophore, beginning with E. B. Wilson (1892, *Neanthes*), Mead (1897, *Amphitrite* and *Clymenella*), Child (1900, *Arenicola*) and Treadwell (1901, *Podarke*), with especially important contributions from Woltereck

(1902, 1904) and Shearer (1911) on the planktotrophic trochophores of *Polygordius* and *Eupomatus* and from Delsman (1916) on the yolky trochophore of *Scoloplos*. In the years between the two World Wars, the Russian embryologist P. P. Iwanoff (1928), as part of his since refuted proof of the existence of a fundamental distinction between primary and secondary segments in annelids and arthropods, made further studies on *Eupomatus*, while Segrove (1941), working in England, also described in detail the trochophore of another serpulid, *Pomatoceros*. More significantly, in a study of outstanding technical brilliance, the English worker D. P. Wilson, whose past and continuing investigations at the Plymouth Marine Laboratory are a unique contribution to the analysis of polychaete larval development (see the Zoological Record for almost any year since 1928), described in detail the development of the delicate, highly specialized planktotrophic trochophore of *Owenia fusiformis* (D. P. Wilson, 1932). Other aspects of this work will be mentioned in later paragraphs. Wilson also (1936) extended his studies to the trochophore of the sabellid *Branchiomma*. More recently, further details of trochophore development have been elucidated by Dales (1952, *Phragmatopoma*), Korn (1958, *Harmothöe*), Anderson (1959, *Scoloplos*) and especially Åkesson (1961, *Pisione*; 1962, *Tomopteris*; 1963, *Harmothoe*; 1967a, *Eunice*; 1967b, *Lopadorhynchus*; 1967c, *Ophryotrocha*; 1968, *Glycera*). We can now consider the major conclusions arising from these studies, concerning the further development of the presumptive rudiments in the polychaete embryo between the completion of gastrulation (Figs. 9 and 10) and the completion of the trochophore (Figs. 11 and 12).

The simplest development is expressed in the prototroch, the presumptive cells of which differentiate, without further cell divisions, into large ciliated cells, irrrespective of whether

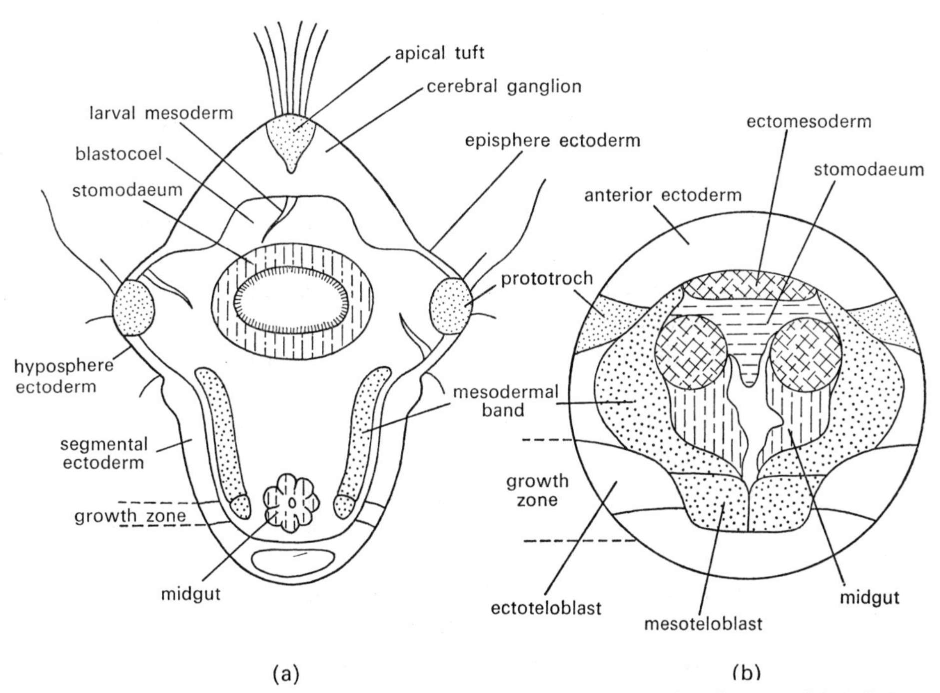

FIG. 11. Diagrammatic frontal sections of polychaete trochophores. (a) The serpulid *Galeolaria*; (b) the ariciid *Scoloplos*. [Modified from Anderson, 1966.]

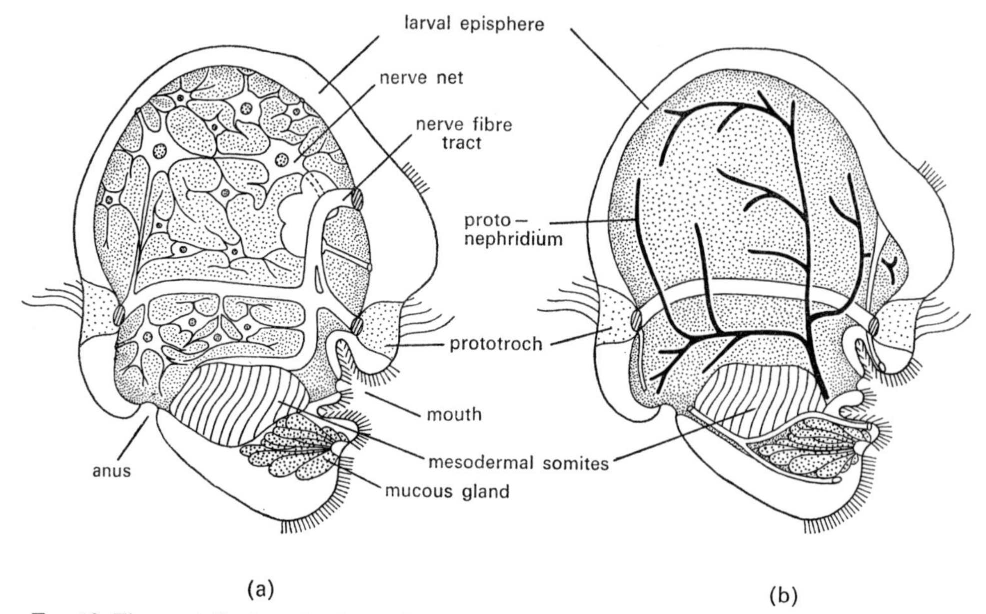

FIG. 12. The specialized trochophore of the phyllodocid *Lopadorhynchus*, drawn diagrammatically in right lateral view, after Åkesson (1967b). (a) The larval nervous system; (b) the larval protonephridia.

the embryo is planktonic, demersal or in a cocoon at this stage, and of whether it is developing towards an actively feeding or a lecithotrophic trochophore. The variations in the structure and function of the prototroch in different species, e.g. in the length of the cilia and the shape of the ciliated cells, result from different pathways of differentiation of the prototroch cells themselves.

The anterior embryonic ectoderm, in contrast, shows more complex changes, differing in embryos of different types. In small embryos developing as planktotrophic trochophores with an apical tuft, the apical tuft cells differentiate in the same manner as the prototroch cells. Adjacent to them, either as a single or, in errant polychaetes (Aphroditidae, Pisionidae, Phyllodocidae, Nephthyidae, etc.), a paired structure, the cerebral ganglion is formed by local proliferation of cells. The remainder of the anterior ectoderm develops as flattened larval epithelium, gland cells and elements of the larval nervous system. In errant species, localized groups of cells within this epithelium also form the rudiments of tentacles and palps.

In larger, yolky embryos, little or no larval epithelium is developed from the cells of the anterior ectoderm. Almost all of the cells take part in formation of the adult prostomial and cerebral ganglion rudiments, manifested at this stage as widespread cell proliferation without differentiation. In eunicids, nereids and tomopterids, secondarily yolky errant forms, a paired origin of the cerebral ganglion is retained.

The posterior embryonic ectoderm also differentiates mainly as flattened larval epithelium in small trochophores such as those of hesionids, oweniids and serpulids, save for the ring of ectoteloblasts, which remain large and undifferentiated, and the telotroch cells behind them, which differentiate in the same manner as the prototroch and apical tuft cells. When the embryo is yolky, the differentiation of the telotroch cells persists but the general epithelium as well as the ectoteloblasts remains undifferentiated (e.g. *Amphitrite*, *Arenicola*, *Clymenella*, *Capitella*, *Scoloplos*, *Tomopteris*, *Eunice*).

The epithelial stomodaeal wall, if it surrounds a lumen in a potentially planktotrophic embryo, differentiates rapidly as a ciliated larval epithelium. If, on the other hand, the embryo is yolky and the stomodaeal lumen is virtual, the stomodaeal epithelium remains undifferentiated and shows cellular proliferation which increases the stomodaeal length (Fig. 10).

The midgut develops in a similar way. In the completion of a planktotrophic trochophore, the midgut rudiment differentiates rapidly into the functional epithelium of a swollen stomach and an intestine opening at the anus. When the midgut cells are yolky but relatively numerous, as in *Scoloplos*, they form a loose agglomeration in the centre of the embryo and then, with further cell divisions, become arranged as an epithelium around a central lumen. The epithelium remains functionally undifferentiated, continuity is not yet established between the stomodaeal and midgut cavities, and the anus is not yet formed. In cases of extreme yolkiness, e.g. *Eunice*, *Platynereis*, *Capitella*, the midgut lumen may be absent at the trochophore stage (Fig. 10).

As part of the development of planktotrophic trochophores, cells which are probably descendants of the presumptive ectomesoderm give rise to a variety of larval muscle strands crossing the blastocoel from the surface epithelium to the gut, and also form muscles on the wall of the stomodaeum. In yolky trochophores, the ectomesoderm persists as large, undifferentiated cells around and in front of the stomodaeum.

The M-cells of all developing trochophores give rise teloblastically to a pair of ventrolateral mesodermal bands. In planktotrophic trochophores, the mesodermal bands are small and inconspicuous. In yolky trochophores with precociously active M-cells, the mesodermal bands fill the spaces laterally between the surface epithelium and the gut as thick bands of undifferentiated cells which stretch forward on either side of the stomodaeum.

A pair of simple protonephridia is also characteristic of planktotrophic trochophores (Fig. 12b). Each is composed of a few cells pierced by an intracellular lumen closed by a flame cell, and lies at the side of the intestine, opening to the exterior near the anus. Each protonephridium develops from a single cell. There is evidence that the protonephridial cells originate independently of the M-cells (Woltereck, 1904), but their exact origin is still obscure. In yolky trochophores, as far as is known, protonephridia are absent.

Whether small or large, planktotrophic or lecithotrophic, all polychaete trochophores contain the rudiment of their future trunk segments as a growth zone, a ring of ectoteloblasts enclosing paired M-cells, just in front of the telotroch. Accordingly, the trochophore can be divided into three regions, each of which makes a distinct contribution to the subsequent adult organization:

1. The prototrochal region, in front of the growth zone, containing the major part of the gut.
2. The growth zone itself.
3. The pygidium.

We can now concern ourselves with the continuing development of these parts and the emergence of the adult organization.

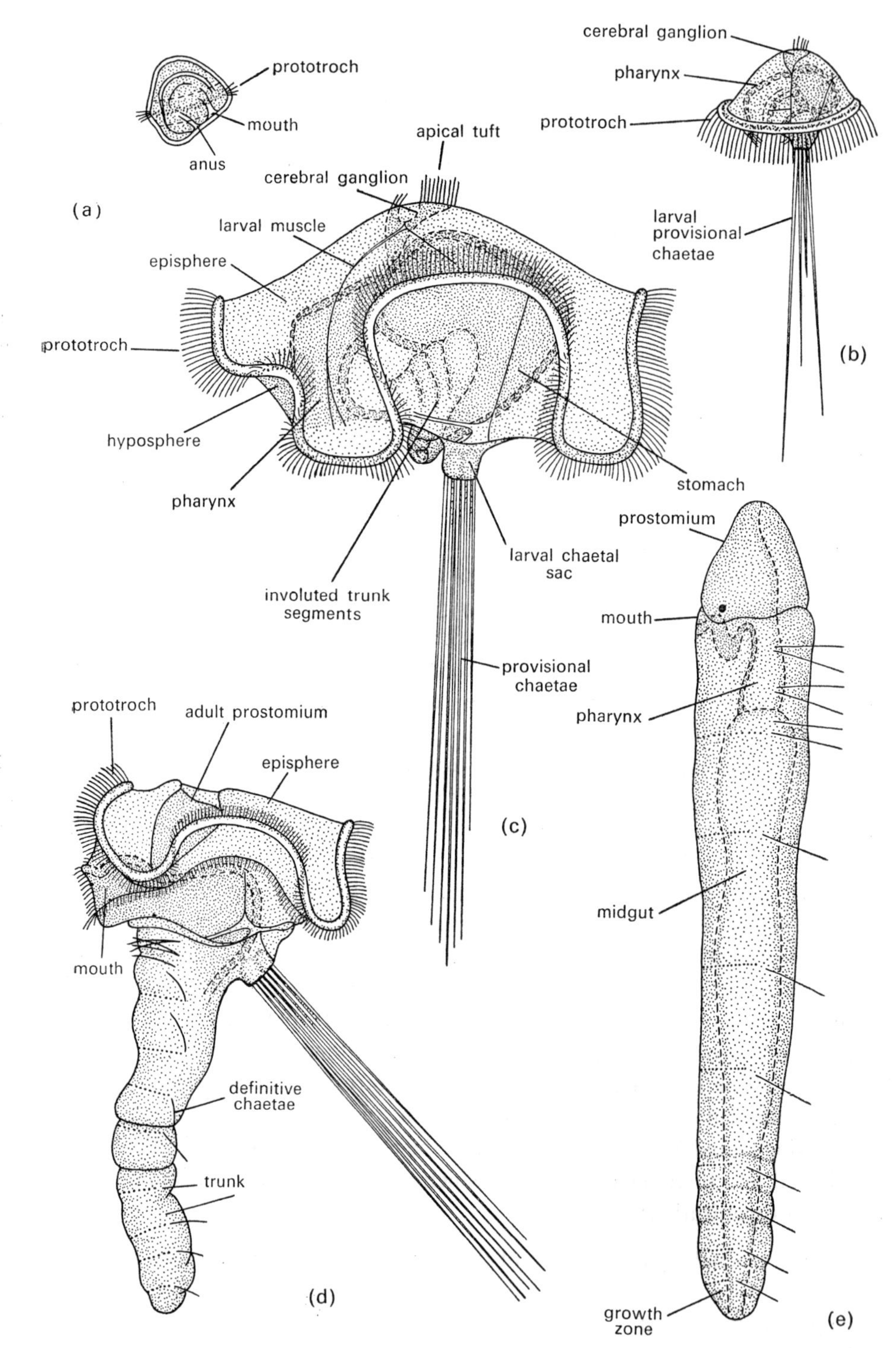

FIG. 13. The larval development and metamorphosis of *Owenia fusiformis*, after D. P. Wilson (1932). (a) Planktotrophic trochophore, 110 μ long; (b) young mitraria larva; (c) fully developed mitraria, about to metamorphose; (d) mitraria 15 seconds after the onset of metamorphosis; (e) first benthic stage, 15–20 minutes after the onset of metamorphosis.

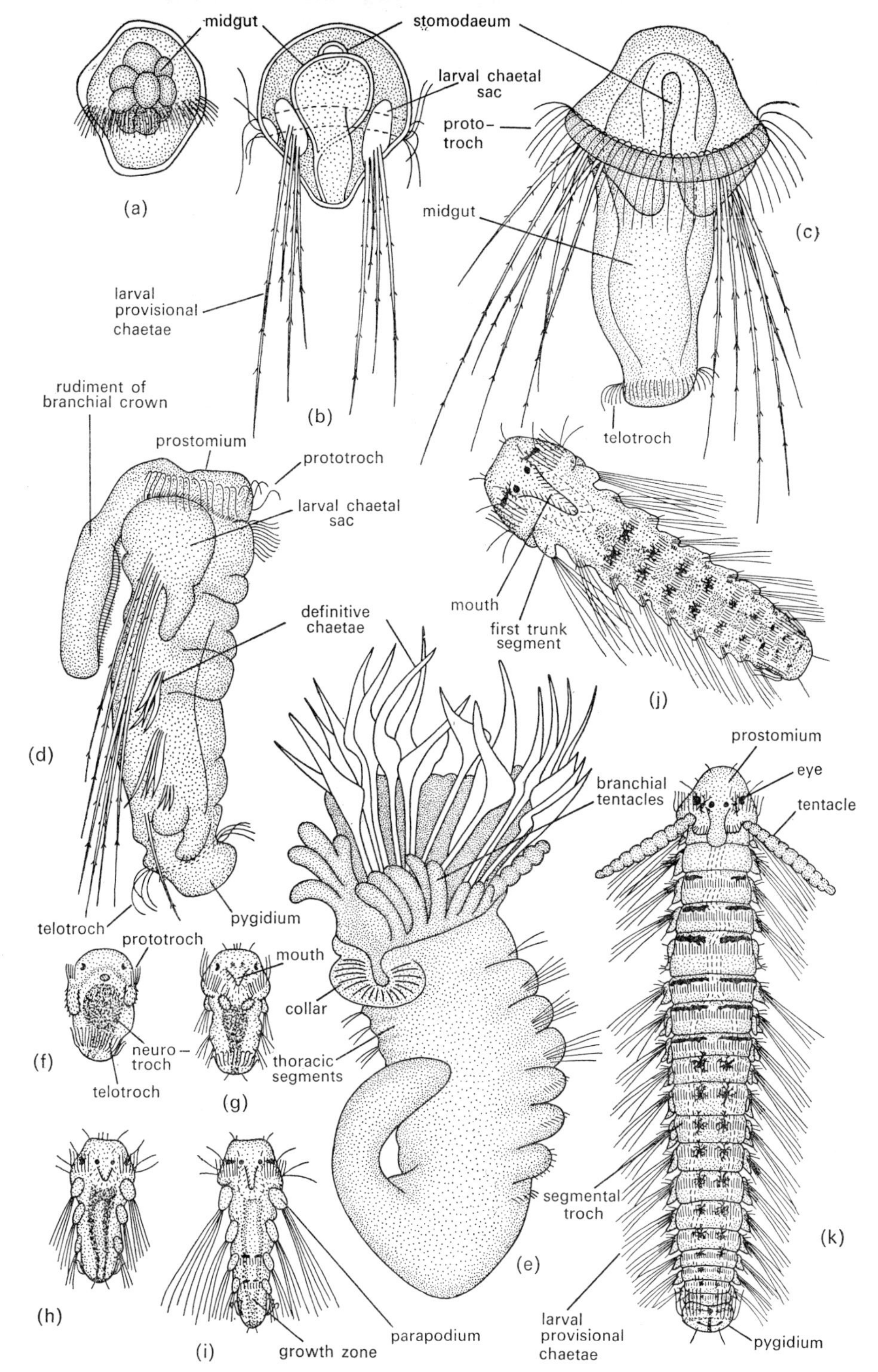

FIG. 14. (a–e) The larval development and metamorphosis of the sabellariid *Phragmatopoma californica*, after Dales (1952). (a) Planktonic, lecithotrophic trochophore developed from a 75-μ egg; (b) young chaetigerous larva, first planktotrophic stage; (c) planktotrophic larva with elongating trunk; (d) stage in metamorphosis; (e) metamorphosed, settled stage in which benthic feeding is about to begin. (f–k) The larval development of the spionid *Polydora ciliata*, after D. P. Wilson (1928). (f) Planktonic, lecithotrophic trochophore, 170 μ long; (g) delineation of first three trunk segments; (h) further development of first three trunk segments; (i) young planktotrophic larva in which further segment formation has begun; (j) planktotrophic larva with nine trunk segments; (k) fully developed planktotrophic larva just before metamorphosis.

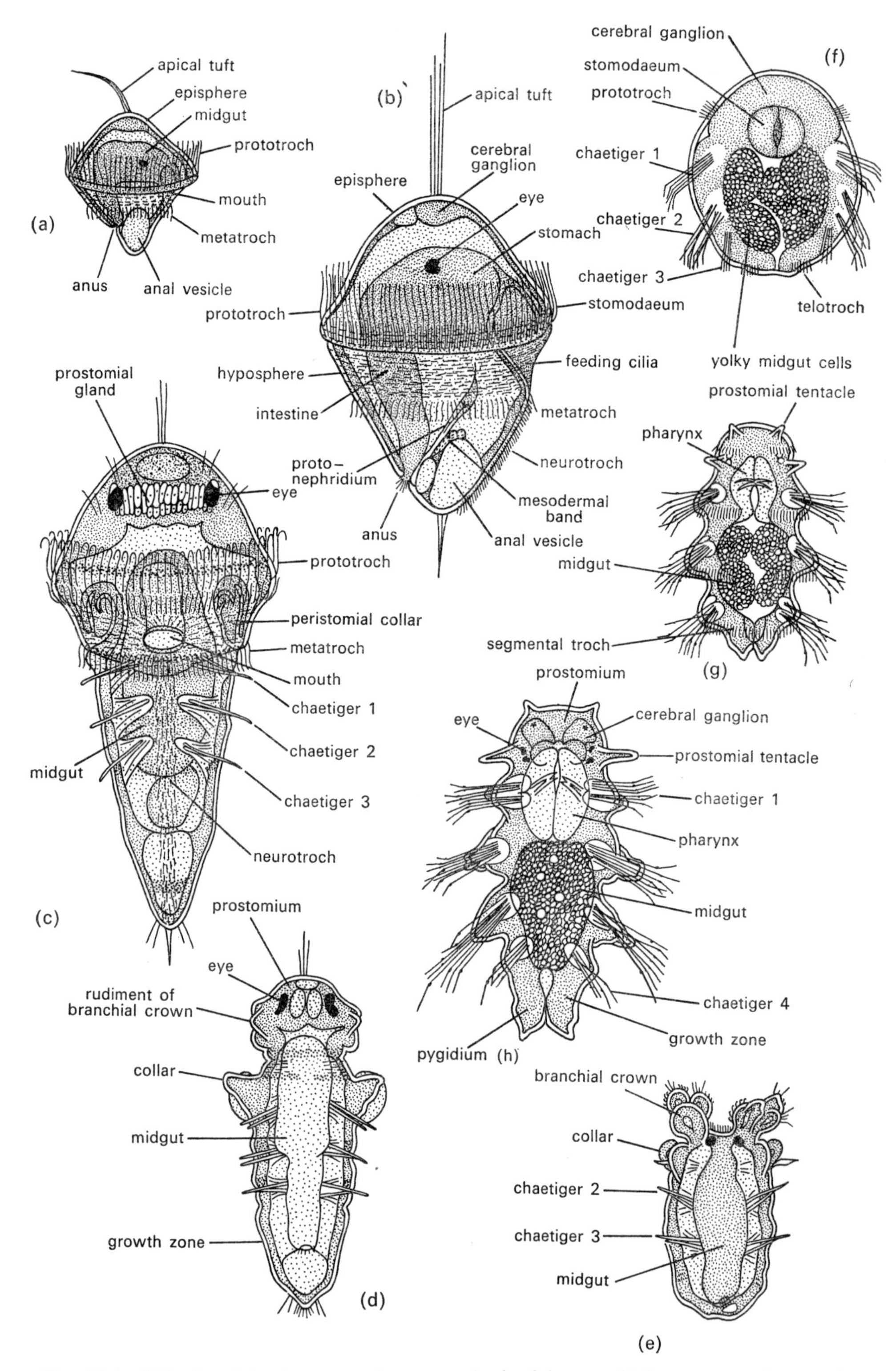

FIG. 15. (a–e) The larval development and metamorphosis of the serpulid *Pomatoceros triqueter*, after Segrove (1941). (a) Young planktotrophic trochophore, 80 μ long; (b) advanced trochophore; (c) fully developed, three-segment larva, about to metamorphose; (d) early settled stage, undergoing metamorphosis; (e) metamorphosis complete. (f–h) the larval development of the nereid *Nereis diversicolor*, after Dales (1951). (f) Newly hatched, demersal, lecithotrophic stage with three larval segments; (g) fully developed three-segment larval; (h) metamorphosis complete, segment formation resumed.

Larval Organs and Metamorphosis

Before examining the further development of the definitive adult structure, a brief reference can be made to the fate of the larval organs of polychaetes. Planktotrophic trochophores (Figs. 12, 13, 14 and 15) have the majority of their cells differentiated as functional larval cells. At the surface, these comprise the cells of the apical tuft, prototroch, neurotroch and telotroch, the anterior and posterior surface epithelia and the cerebral ganglion. Internally lie the larval nervous system, larval muscles, protonephridia and a functional stomodaeum and midgut. The only undifferentiated, directly adult rudiments are the prostomial rudiments other than the brain, the ectoteloblast ring and M-cells of the growth zone and the short mesodermal bands.

The most important studies of metamorphosis in planktotrophic polychaete larvae have been those of Woltereck (1904, 1905) on *Polygordius*, D. P. Wilson (1932, 1936) on *Owenia* and *Branchiomma*, Segrove (1941) on *Pomatoceros*, Dales (1952) on *Phragmatopoma* and Korn (1958) on *Harmothöe*.

In general, the results of this work show that the cerebral ganglion, stomodaeum and midgut are the only larval organs to persist through metamorphosis, undergoing varying degrees of redifferentiation. The remaining larval organs are histolysed and resorbed after a period of growth and larval function, so that the prototrochal region of the trochophore makes little direct contribution to the subsequent adult structure. This is especially obvious in the cataclysmic metamorphosis of *Owenia* (Fig. 13).

A similar histolysis and resorption of the prototroch, neurotroch and telotroch take place during the further development of yolky trochophores (Anderson, 1959; Åkesson, 1967a). Larval nervous system, musculature and protonephridia are, of course, absent, while the cerebral ganglion, anterior and posterior surface epithelia, stomodaeum and midgut, like the growth zone and mesodermal bands, proceed directly into adult development. Here, then, the prototrochal region of the trochophore has become secondarily diverted for the most part from a larval to an adult fate, as shown in Fig. 1.

The development of the various additional larval trochs and provisional chaetae which arise on the anterior trunk segments of many polychaete larvae before metamorphosis (Figs. 1, 13 and 14) has not been studied in detail in more than one or two species. Provisional chaetae appear to be secreted by the chaetal sacs in the same manner as the definitive chaetae (D. P. Wilson, 1932; Dales, 1952) and are simply shed at metamorphosis. Additional larval trochs, in *Scoloplos* at least (Anderson, 1959), are differentiated from rings of cells within the segmental trunk ectoderm as minor secondary specializations of the latter and are histolysed and resorbed in the same manner as the prototroch. Clearly, the formation and subsequent loss of these additional larval organs make little structural impact on the development of the trunk segments in polychaetes.

Further Development of the Stomodaeum and Midgut

Bearing in mind that the stomodaeum and midgut rudiments of polychaetes may or may not pass through a phase of temporary larval function before attaining their definitive adult form, we can now examine their further development to that form. Stomodaeal and midgut development attracted some interest among the early polychete embryologists (Salensky, 1883, 1908; Kleinenberg, 1886; Häcker, 1894, 1896; Eisig, 1898; Meyer, 1901; Woltereck, 1902, 1904, 1905; Lillie, 1905; Schaxel, 1912; Schneider, 1913) and were also studied by D. P. Wilson (1932) in *Owenia* and Segrove (1941) in *Pomatoceros*, but have only been comprehensively described in the last twenty years, by Bookhout and Horn (1949, *Dasybranchus*), Horn and Bookhout (1950, *Haploscoloplos*) and Allen (1964, *Autolytus*) in America, Newell (1951, *Clymenella*) and Anderson (1959, *Scoloplos*) at the University of London, Korn (1958, *Harmothöe*) at Kiel and Åkesson (1961, 1962, 1963, 1967 a, b, 1968, *Pisione*, *Tomopteris*, *Harmothöe*, *Eunice*, *Lopadorhynchus*, *Glycera*) at Lund.

When the stomodaeum and midgut are differentiated as a functional larval pharynx and larval stomach and intestine, the two latter components usually transform directly, at metamorphosis, into the adult midgut epithelium. Only in the case of *Polygordius*, in which the larval stomach is unusually capacious, has a disintegration of the larval stomach at metamorphosis been observed (Woltereck, 1905). In some species (e.g. *Owenia*, *Pomatoceros*) the larval pharynx also transforms directly into the adult pharyngeal epithelium. The larval pharynx of aphroditids, pisionids, phyllodocids, nephthyids, spionids and *Polygordius*, however, produces a pair of evaginations (Fig. 19) which act as imaginal discs for the replacement of the larval pharyngeal epithelium at metamorphosis. It seems likely that the paired origin reported for the stomodaeum of yolky nereid and tomopterid embryo is a secondary modification of this condition, in which only the adult paired rudiments are retained.

In contrast, in the further development of the gut of yolky, polychaete embryos, the stomodaeum develops directly into the adult pharyngeal epithelium (Anderson, 1959; Åkesson, 1967a). A direct development of the adult midgut also takes place in species, such as *Scoloplos*, in which a relatively large number of midgut cells lies as a yolky epithelium around a midgut cavity in the trochophore. The yolk in the cells is gradually resorbed as the midgut tube lengthens by generalized cell divisions, and the posterior end of the tube becomes confluent with an anal opening developed on the pygidium (Fig. 16). When the midgut cells are large and dense with yolk, however, as in *Eunice* and *Platynereis*, a distinction can be made between anterior and posterior midgut components. In *Eunice*, as described by Åkesson (1967a), most of the yolk-filled cells of the midgut rudiment of the trochophore continue their role as yolk-digesting cells with little or no further cell division. As the yolk is digested, these cells then give rise to the anterior part of the midgut. Meanwhile, the cells at the posterior end of the midgut mass of the trochophore undergo rapid proliferation to form a cylindrical posterior midgut, extending along the growing trunk. The posterior midgut later opens to the anus through a very short proctodaeum. The functional specialization of the midgut cells in part for yolk digestion and in part for proliferative activity is displayed in a more extreme form in *Platynereis*. According to Schneider (1913), the anterior midgut of *Platynereis* exhibits a typical vitellophage specialization. The four, large midgut cells lose their cell boundaries, while their nuclei divide repeatedly within the unified yolk mass. Each nucleus, surrounded by a small halo of cytoplasm, acts as a vitellophage. The stomodaeum makes the usual contact with the yolk mass anteriorly, but at the posterior end a posterior midgut tube is proliferated behind the yolk mass along the growing trunk. The posterior midgut tube originates from the small midgut cells budded off by the M-cells

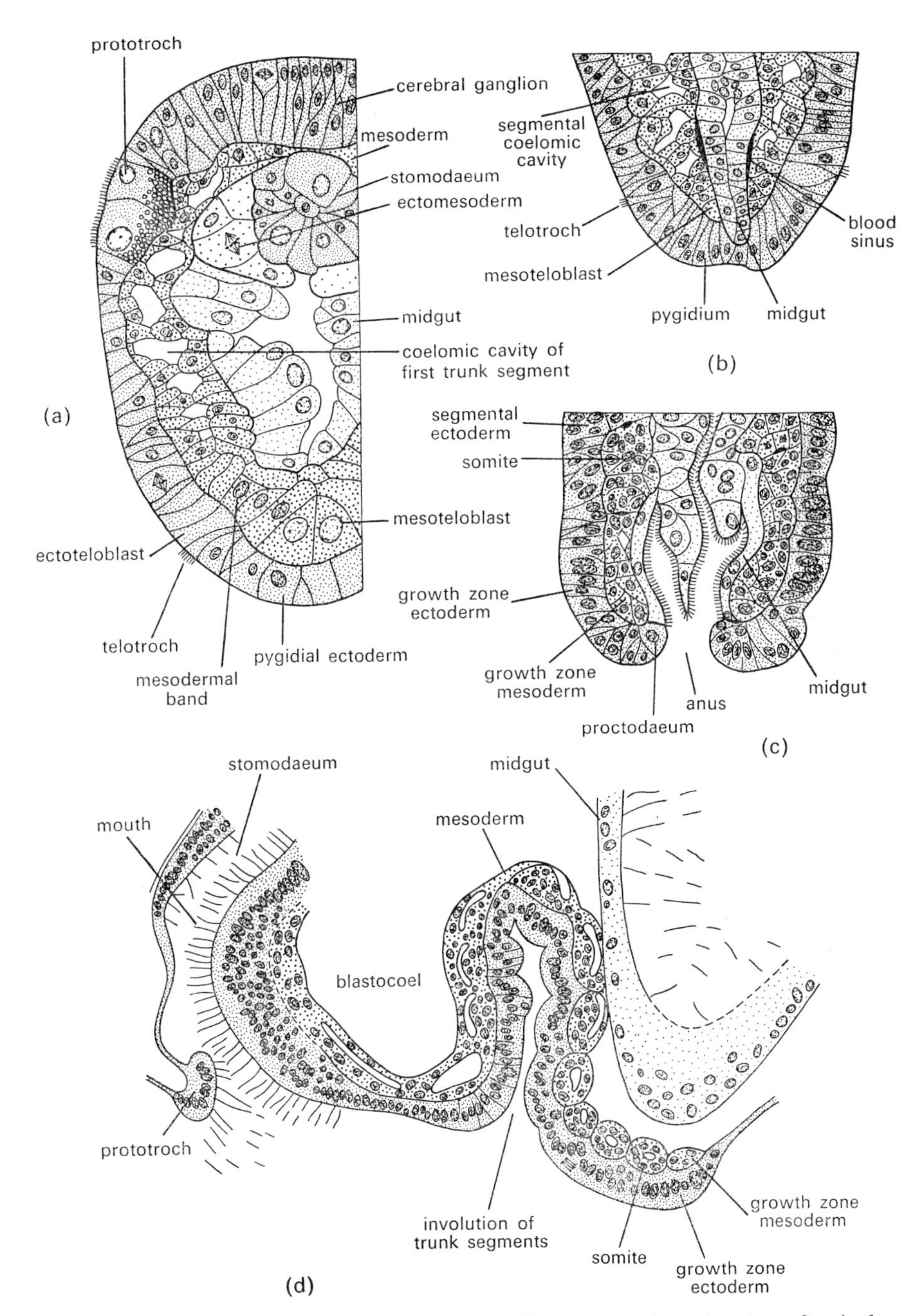

FIG. 16. (a–c) Trunk segment formation in *Scoloplos*, in diagrammatic frontal section, after Anderson (1959). (a) Early post-trochophore; (b) segment-forming region, shortly before metamorphosis; (c) segment-forming region after metamorphosis. (d) Trunk segment formation in the mitraria larva of *Owenia*, in diagrammatic parasagittal section, after D. P. Wilson (1932).

during cleavage. The yolk mass is gradually resorbed and the adult gut epithelium is formed by union of the stomodaeum with the posterior midgut.

The specialization of the anterior midgut cells for vitellophage function and the formation of most or all of the definitive midgut epithelium from a group of small, posterior midgut cells lying behind the yolk mass is, as we shall see, a frequent convergent answer in annelids and arthropods to the problem of containing and digesting yolk and forming a midgut at the same time. It is not, of course, the only answer. Even within the polychaetes, at least one alternative vitellophage specialization has been described, though the source is old (Eisig, 1898) and renewed investigation would be desirable. According to Eisig, the embryo of *Capitella* has four large, yolky midgut cells which fuse together in the same manner as those of *Platynereis*. Their nuclei also divide several times, but then cluster around a small central lumen formed within the yolk. As the yolk is resorbed, the clustered vitellophages give rise to an epithelial tube which extends in a posterior direction to meet the pygidium and eventually gives rise to the entire midgut. Analogous specializations in which the midgut tube develops inside the yolk mass are well known in millipedes (Chapter 5) and some cirripede crustaceans (Chapter 8).

The further development of the stomodaeum and midgut of polychaetes can thus be summarized as follows. Basically, the stomodaeal and midgut epithelia become functional, first in a larval pattern, then after transformation during metamorphosis, in an adult pattern. In yolky-egged species, the first phase is eliminated but in some cases the midgut shows vitellophage specialization. In some planktotrophic larvae, in contrast, the stomodaeum develops imaginal discs, a specialization in the opposite mode.

Further Development of the Mesoderm

As has already been mentioned, any functionally differentiated larval mesoderm in the polychaete trochophore is histolysed during metamorphosis, so that our discussion of the further development of the mesoderm need concern only the mesodermal bands. The few ectomesoderm cells of yolky embryos make little contribution to later structure, though they have been shown by Anderson (1959) to be important in the development of the proboscis of *Scoloplos* and by Åkesson (1961, 1962, 1963, 1967a, b, 1968) to contribute to the prostomial mesoderm of several errant species.

As the growth zone of the trochophore becomes active and begins to proliferate a series of incipient trunk segments, the mesodermal bands increase in length. Within each incipient trunk segment, the newly proliferated mesoderm usually becomes individuated as a pair of mesodermal somites (Fig. 16). Three or more trunk segments are formed before metamorphosis takes place. A pause in segment formation accompanies metamorphosis, after which the proliferation of further trunk segments continues, usually at a different rate.

In several polychaete families, such as the Serpulidae, Nereidae, Eunicidae and Tomopteridae, the three anterior segments developed before metamorphosis are specialized as larval segments (Fig. 15). They arise in rapid succession, sometimes almost simultaneously, and exhibit ectodermal segmentation, including the precocious formation of chaetal sacs and chaetae, while their mesoderm is still unsegmented. The segments added after metamorphosis in these species, in contrast, develop in slower, serial succession and lack the extreme larval specializations of the anterior segments. Species which exhibit this pattern of growth are said to exhibit heteronomy of segment formation.

Iwanoff (1928) put forward the view that heteronomy is a fundamental feature of the segment formation of annelids and arthropods, basing much of his argument on a study of the serpulid *Eupomatus*. The fundamental nature of heteronomy in polychaete development was thought to be supported by the fact that heteronomous development persists in the segments of yolky serpulid and nereid embryos, in which the larval specializations of the first three segments are retained even though they are no longer functionally significant. At that time, segment formation in other polychaetes had not been adequately studied.

Manton (1949) convincingly dismissed Iwanoff's views on heteronomy in arthropod development by demonstrating that heteronomy occurs only in the Crustacea, where it is manifestly a secondary phenomenon. The naupliar larval segments of Crustacea remain a distinctive feature in the development of yolky crustacean embryos (see Chapter 8), but no trace of heteronomy exists in the Onychophora, myriapods, hexapods and chelicerates. The segments develop in these arthropods in strict anteroposterior succession (see Chapters 4, 5, 6, 7 and 9).

Anderson (1959) showed that heteronomy of segment formation is similarly absent in the development of the yolky embryo of the polychaete *Scoloplos armiger* (Fig. 1), as it is in the elaborate planktotrophic larvae of *Owenia* (Fig. 13) and *Polygordius* (Woltereck, 1905; D. P. Wilson, 1932). In these species, the mesoderm of all trunk segments exhibits somite formation before the overlying trunk ectoderm becomes segmentally delineated and there is no vestige of larval specialization in the first three segments. Anderson argued that heteronomy is not a fundamental feature of polychetes since, if it were, it should be recognizable vestigially in the development of all species, as it is in the Crustacea. It seemed more reasonable to suggest that heteronomy has arisen as a secondary larval specialization only in certain families of polychaetes and is retained vestigially with yolk in these families alone.

Åkesson (1967a), however, has recently provided evidence which indicates that the situation may be more complex. In the Eunicidae, the larvae of *Eunice kobiensis* and one species of *Diopatra* are rich in yolk but retain a free-swimming stage and show specialization of the first three trunk segments as larval segments. *Marphysa* and another species of *Diopatra*, on the other hand, which have a yolky, non-pelagic development, do not retain heteronomy of segment formation. The eunicids thus indicate that heteronomy of segment formation in polychaetes can be secondarily lost when it becomes functionally irrelevant. Åkesson argues, on the basis of this evidence, that heteronomy is a fundamental feature of polychaetes and that all cases in which heteronomy is absent are secondary.

There is no doubt that Åkesson's results necessitate a modification of the view put forward by Anderson (1959), but I am now of the opinion that both arguments have attempted to generalize too broadly on a rather variable phenomenon. Perhaps the true picture is one of compromise, allowing for the following possibilities:

1. That the basic mode of development of the trunk segments in polychaetes is for the trochophore to develop the first three trunk segments before metamorphosis, the remainder after metamorphosis, without any larval specialization of the first three segments.

2. That in certain species, metamorphosis has simply been delayed, in association with either elaboration of the trochophore larval organs (e.g. *Owenia*) or increased yolk in the egg (e.g. *Scoloplos*), so that more than three segments develop before metamorphosis occurs.
3. That in other species, the first three trunk segments have become specialized as larval segments, initially in association with a planktonic larval life, and that these specializations persist in some families when development is non-pelagic (e.g. serpulids, nereids) but not in others.

Larval studies also indicate other modes of specialization, such as those of sabellariids and spionids (Fig. 14), which would repay further embryological study, but the basic question of whether polychaete development is fundamentally heteronomous or not devolves upon an investigation of segment formation in species like *Ophelia bicornis* (Fig. 1), which appear to be basic in the sense indicated above. Unfortunately, at the present time, no species of this type has received adequate embryological attention. Our present information on segment development, as we shall see below, has been drawn entirely from species which are specialized in one way or another.

Mesodermal proliferation in the growth zone and the subdivision of the mesodermal bands into somites has been studied for several species during the period before metamorphosis (Salensky, 1882; Meyer, 1888; E. B. Wilson, 1892; Mead, 1897; Eisig, 1898; Child, 1900; Woltereck, 1902, 1904, 1905; Lillie, 1905; Iwanoff, 1928; D. P. Wilson, 1932; Segrove, 1941; Anderson, 1959; Åkesson, 1967a), but in only a few scattered species after this event (Lillie, 1905, *Arenicola*; Iwanoff, 1928, *Eupomatus*; Anderson, 1959, *Scoloplos*; Åkesson, 1962, 1967a, *Tomopteris* and *Eunice*). The somites proliferated before metamorphosis are all products of the teloblastic activity of the M-cells. This fact has been established in a sufficient number of polychaetes (*Polygordius*, *Eupomatus*, *Owenia*, *Amphitrite*, *Arenicola*, *Perinereis*, *Neanthes*, *Capitella* and *Scoloplos*) to sustain it as a basic feature of polychaetes, whether metamorphosis occurs early or late and whether or not the first three trunk segments exhibit larval specialization. At metamorphosis, the M-cells of *Scoloplos*, *Arenicola*, *Eunice* and *Tomopteris* give rise to a circum-enteric ring of small mesoderm cells around the posterior end of the midgut (Fig. 16). These cells then continue to proliferate somites after metamorphosis is completed. Iwanoff (1928), as part of his evidence of heteronomous segment development in *Eupomatus*, claimed that the M-cell mesoderm develops wholly as the mesoderm of the three specialized larval segments formed before metamorphosis and that the remaining somites are proliferated from a separate source in the ectoderm of the growth zone (Fig. 17). This proposition, although not established convincingly by Iwanoff, has never been subsequently tested for either serpulids or nereids. Studies of serpulid and nereid development after metamorphosis present considerable technical difficulties. Nereids have an inordinately long pause in segment formation at this stage and do not seem to survive long enough in culture to produce more segments (e.g. Dales, 1952). Serpulids settle at metamorphosis and begin to secrete a tube, but their settlement requirements are highly specific (e.g. Wisely, 1958; Andrews and Anderson, 1962; Vuillemin, 1967) and the young animals are notoriously difficult to sustain in culture through this difficult phase. Since all of the well-studied polychaete species, including the heteronomous *Eunice* and *Tomopteris*,

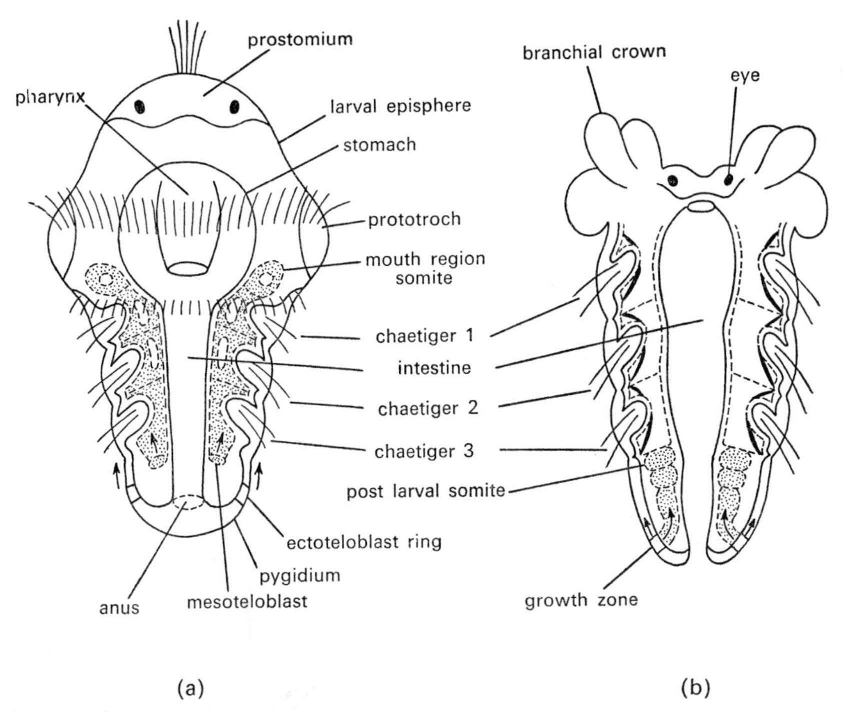

FIG. 17. Formation of primary and secondary segments in serpulids, as interpreted by Iwanoff (1928). (a) Formation of primary trunk segments; (b) formation of secondary trunk segments. [Modified from Anderson, 1966.]

develop all of their somites by proliferation from the M-cells or their descendants, it seems likely that Iwanoff's contrary view for *Eupomatus* was in error, but further study of species with planktotrophic larvae and heteronomous development would be of interest.

In many polychaetes, the mesodermal bands segregate into paired somites in strict antero-posterior succession. Exceptions to this are known only for the larval segments of heteronomous species. The first three trunk segments of serpulids, nereids and tomopterids exhibit simultaneous somite formation. The mesoderm of the larval segments of *Eunice* foregoes somite formation in favour of direct development as segmental musculature, etc. The omission of somite formation as a stage in development is a mesodermal specialization also found in certain types of arthropod development (e.g. in some pterygotes, Chapter 7). The mode of formation of the somites of polychaetes depends on the relative size of the blastocoelic cavity of the embryo (Fig. 16). When this cavity is large, as in the planktotrophic larvae of *Eupomatus*, *Pomatoceros*, *Owenia* and *Polygordius*, the mesodermal bands divide into a paired series of solid segmental blocks, which then hollow out as paired somites. Conversely, when the egg is yolky and the blastocoel is soon obliterated, as in *Arenicola*, *Scoloplos*, *Tomopteris*, *Perinereis* and *Spirorbis*, the paired cavities of the somites appear first in the undivided mesodermal bands and the mesoderm cells become arranged as somite walls around the cavities.

Setting aside for the moment the question of what happens to those parts of the mesodermal bands that invade the prototrochal region in front of the trunk segments, in what

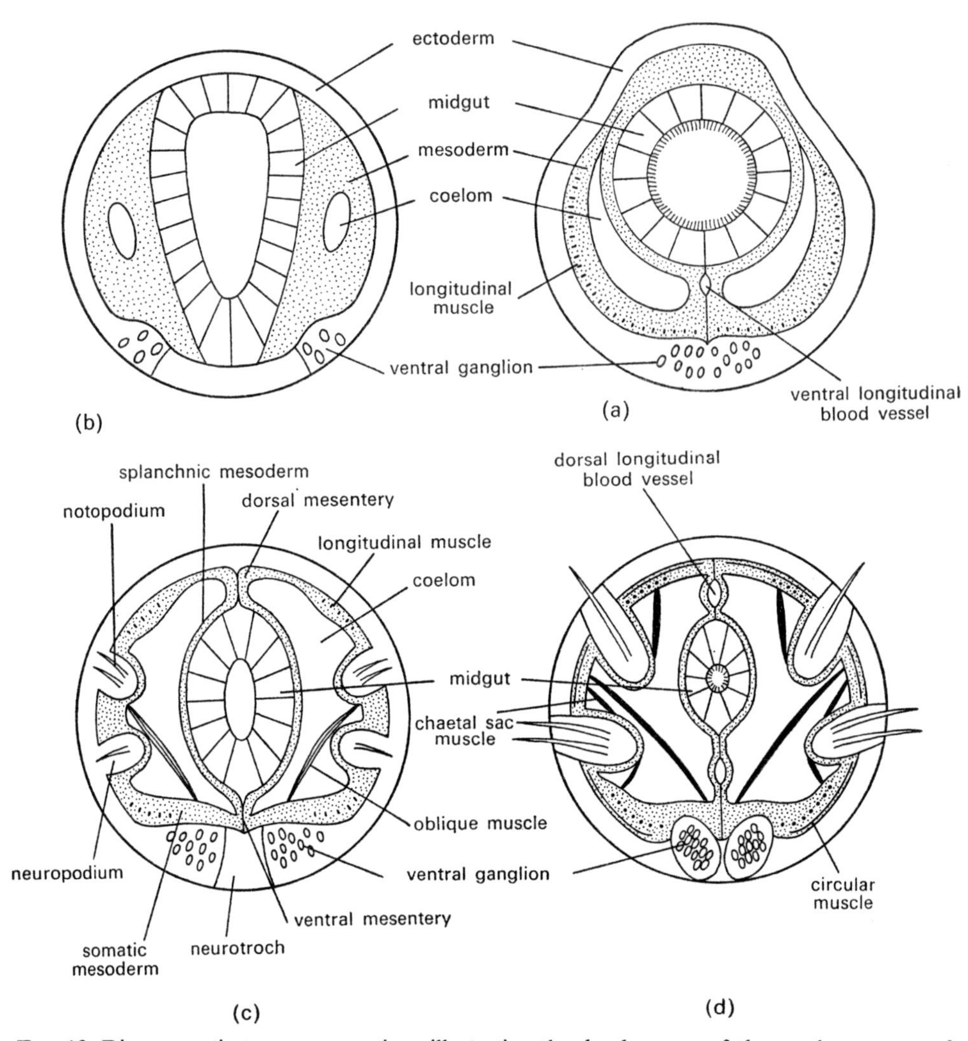

FIG. 18. Diagrammatic transverse sections illustrating the development of the trunk segments of polychaetes. (a) *Owenia*, early histodifferentiation; (b–d) *Scoloplos*, stages in differentiation. [Modified from Anderson, 1966.]

manner do the paired somites contribute to the structure of the trunk segments? The early students of this problem (Salensky, 1882, 1883; Meyer, 1901; Lillie, 1905; Sokolov, 1911; Schaxel, 1912) had little success in resolving the fate of the somite walls of polychaetes, but the combined studies of D. P. Wilson (1932, 1936), Segrove (1941), Anderson (1959) and Åkesson (1962, 1963, 1967a) now permit a reasonably accurate account to be given of this process. The somite walls (Fig. 18) give rise to the somatic musculature, intersegmental septa, dorsal and ventral mesenteries, peritoneum, visceral musculature and the walls of the major blood vessels. They also probably give rise to the coelomoducts, although we have only the work of Meyer (1901) and Lillie (1905) to support this possibility. The gonads are usually associated with the peritoneum, but their germ cells may originate independently of the mesodermal bands in the early embryo (see below).

The somatic musculature is formed by cells of the outer, somatic walls of the hollow somites. Four longitudinal rows of somatic myoblasts, dorsolaterally and ventrolaterally placed, differentiate as longitudinal trunk muscles. Laterally placed myoblasts on the outer surface of the somatic wall extend in a transverse direction between the developing longitudinal muscles and the surface epithelium and give rise to the circular muscles of the trunk. Other cells of the somatic walls form oblique, transverse, dorsoventral, septal and mesenterial muscles. As the chaetal sacs grow in from the surface epithelium, somatic myoblasts adjacent to them differentiate as chaetal sac muscles. The peritonium, lining the coelom and forming the double epithelial wall of the septa and mesenteries, arises directly from the cells surrounding the coelomic cavity of the somite. The coelom, of course, is enlarged and forms the body cavity, while the blood system is established anew in the site of the former blastocoel.

Often the major blood space is a circum-enteric sinus formed through separation of the inner, splanchnic walls of the somites from the gut epithelium (e.g. *Protula*, *Arenicola*, *Owenia*). When definite longitudinal vessels are formed, they result either from partial separation of the splanchnic mesoderm from the gut epithelium (e.g. dorsal vessel of *Scoloplos*, ventral vessel of *Arenicola*) or by a more direct separation of the apposed epithelia of a mesentery (e.g. ventral vessel of *Scoloplos*, dorsal vessel of *Arenicola*). The circum-enteric sinus is perhaps the primitive condition.

One outstanding problem in polychaete embryology is the origin and development of the musculature of the gut. Anderson (1959) was unable to discern muscle fibres in the splanchnic mesoderm of *Scoloplos* by normal histological means even when peristalsis of the gut had become obvious. D. P. Wilson (1932) described the formation of circular muscle fibres in the splanchnic mesoderm of *Owenia*, but found that they served merely to move the blood in the circum-enteric sinus. The splanchnic walls of the somites seem the most likely source of the gut musculature, but for polychaetes this supposition has yet to be proved. The stomodaeal musculature presents another problem. Åkesson (1962, 1963) has shown that the stomodaeal muscles of *Tomopteris* and *Harmothöe* develop at an early stage, probably from cells not derived from the mesodermal bands, but the origin of these cells is still unknown. They may be ectomesodermal. On the other hand, the pharyngeal muscles of *Eunice* and *Glycera* develop from the anterior ends of the mesodermal bands (Åkesson, 1967a, 1968).

Very little is known of the origin and early development of the germ cells of polychaetes. Due to their later association with the somatic peritoneum, the primordial germ cells are usually taken to originate within this epithelium, but no direct proof of this assertion has been made. One problem in investigating germ cell origin in polychaetes is that the anterior segments, the ones usually studied in the embryo, are agametic in most species; but even in *Tomopteris*, which has gonads in all segments, Åkesson (1962) could not trace the source of the germ cells. We are left only with the tantalizing description by Malaquin (1925, 1934) of a pair of cells in the trochophore of the serpulid *Salmacina*, budded off by the M-cells, and said to proliferate the gonocytes. According to Malaquin, these cells multiply in the pygidium and eventually become distributed through the somites. Whether this is right or wrong, specialized or universal among polychaetes, remains to be seen. As discussed in later chapters, an association between primordial germ cells and presumptive mesoderm is commonly found in other annelids and arthropods.

Finally, a brief consideration can be made of the anterior ends of the mesodermal bands, lying in front of the first trunk segment as mesoderm of the prototrochal region. The extent to which the mesodermal bands penetrate this part of the body depends on the degree to which the prototrochal region is functionally differentiated as part of a planktotrophic trochophore and on the stage in development at which the M-cells begin to proliferate the mesodermal bands. Few workers have offered pertinent information. In the serpulids and

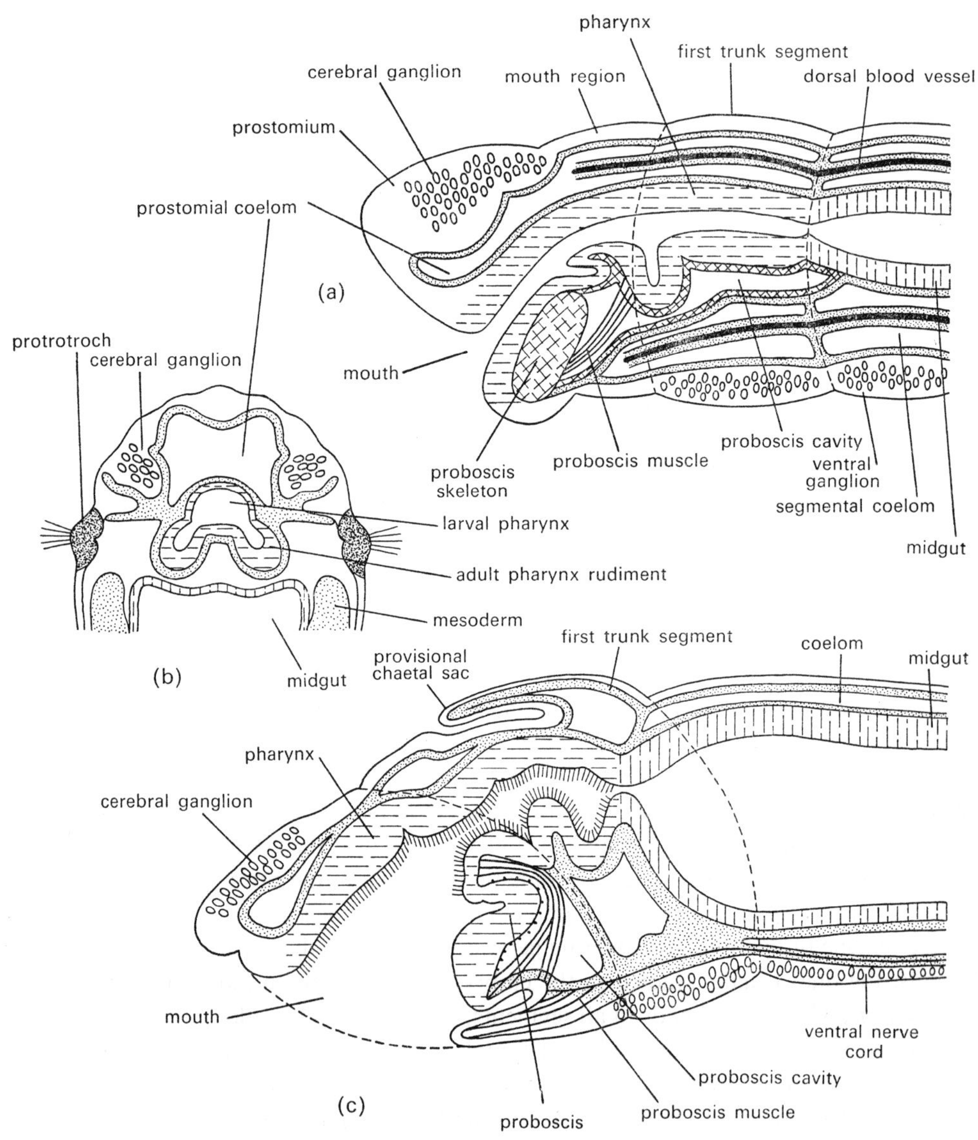

FIG. 19. Development of the head in polychaetes. (a) Diagrammatic sagittal section through the anterior end of a newly metamorphosed embryo of *Scoloplos*; (b) diagrammatic frontal section of the anterior end of the metatrochophore of *Harmothöe*, after Åkesson (1963); (c) diagrammatic sagittal section of the anterior end of a newly metamorphosed larva of *Owenia*, after D. P. Wilson (1932). [Modified from Anderson, 1966.]

Polygordius, according to Meyer (1901) and Woltereck (1902, 1905), the anterior ends of the mesodermal bands penetrate only a short distance into the prototrochal region and are cut off as a single pair of small somites (Fig. 20). In *Owenia*, with a much expanded larval blastocoel, the mesodermal bands do not penetrate the prototrochal region (D. P. Wilson, 1932). In *Scoloplos* and *Capitella*, in contrast, an early onset of teloblastic activity by the large M-cells results in a precocious penetration of the mesodermal bands forwards through the embryonic prototrochal region, pushing past the stomodaeum to meet in the anterior midline beneath the anterior ectoderm (Eisig, 1898; Anderson, 1959). An anterior median and three (*Scoloplos*) or four (*Capitella*) pairs of lateral coelomic cavities develop in the mesoderm in front of the first trunk segment and all contribute to the coelomic cavity of the head. Finally, in several species of errant polychaete, the anterior ends of the mesodermal bands also push forwards past the stomodaeum but do not develop coelomic cavities (Åkesson, 1961, 1962, 1963, 1967 a, b, 1968). The head coelom in these species is a secondary development. Further discussion of the composition of the polychaete head will be deferred until the later development of the ectoderm of the prototrochal regions has been described.

Further Development of the Ectoderm

In considering the further development of the ectoderm and its contribution to the adult structure in polychaetes, we have to take into account the ectoderm of the prototrochal region, the ectoteloblasts of the growth zone and the pygidium. It should also be borne in mind that the anterior presumptive ectoderm forms only that part of the prototrochal region in front of the prototroch, while the posterior presumptive ectoderm contributes the remainder of the ectoderm of the prototrochal region and also the ectoteloblasts and pygidium.

The formation and further development of the ectoderm of the trunk segments is synchronized with that of the somites. A number of workers (E. B. Wilson, 1892; Mead, 1897; Eisig, 1898; Child, 1900; D. P. Wilson, 1932; Anderson, 1959) have shown that the ectodermal sheath lying outside each pair of trunk somites formed before metamorphosis is a product of forward proliferation by the ectoteloblast ring. Lillie (1905) and Anderson (1959) have further shown that in *Arenicola* and *Scoloplos* the same mode of formation of trunk segment ectoderm continues after metamorphosis. The ectodermal sheath is not marked into segmental subdivisions when first laid down. Usually, demarcation occurs in anteroposterior succession, after the corresponding pair of somites has been formed. Secondarily, in serpulids, nereids, eunicids and tomopterids, the anterior trunk ectoderm shows precocious delineation of the boundaries of the larval segments while the underlying mesoderm is still unsegmented (Salensky, 1882; Wistinghausen, 1891; E. B. Wilson, 1892; Meyer, 1901; Iwanoff, 1928; Segrove, 1941; Åkesson, 1962, 1967a). Whether the somites of the more posterior segments of serpulids and nereids are delineated before their overlying ectoderm is still not known. In eunicids and tomopterids, precocious delineation of the segmental ectoderm continues after metamorphosis (Åkesson, 1962).

The subsequent development of the segmental ectoderm is more easily traced than that of the mesoderm, since its products, the surface epithelium, chaetal sacs and ventral ganglia of the segments are sharply distinct (Fig. 18). It is not surprising, therefore, that most of the workers already mentioned in this chapter have had something to say about ectoderm development (see Anderson, 1966). There is general agreement that the chaetal sacs develop as simple ingrowths of the ectoderm, usually as two pairs, dorsolateral notopodial and ventrolateral neuropodial, in each segment. In the larval segments of serpulids, nereids, eunicids and tomopterids the chaetal sacs develop precociously and simultaneously, but an anteroposterior succession is more usual. The ventral ganglia follows the same pattern, originating by repeated divisions of cells lying on either side of the ventral mid-line. Successive ganglia become distinct as a result of early cessation of these divisions intersegmentally. Neuropile begins to form while the ganglia are still superficial, appearing first as paired ganglionic zones which then join across the segment and between successive segments. As the segmental ganglia become more compact, they concentrate towards the ventral midline and migrate into the interior, leaving a superficial layer of ectoderm cells as ventral surface epithelium. The remainder of the surface epithelium is formed directly from the lateral to dorsal ectoderm.

A brief mention can be made, at this stage, of the development of the nephridia of polychaetes. Like the coelomoducts with which they are so often associated (Goodrich, 1946), and the gonads, polychaete nephridia have been sadly neglected by embryologists. Only Meyer (1888) and Lillie (1905), in fact, have attempted to trace their origin. The results suggest that each pair of nephridia develops from a pair of large nephridioblasts differentiated in the ectoderm of each intersegment. Each nephridioblast buds off a chain of cells within which an intracellular lumen develops, later becoming extracellular as the number of cells increases. The outer end of the tube gains an aperture at the body surface while the inner end penetrates the associated septal wall and usually becomes associated with a coelomoduct funnel, although it may open directly into the coelom or terminate in an array of solenocytes. Only fresh investigation can determine the validity of this interpretation.

Behind the segmental growth zone, the pygidial ectoderm develops in a simple way as a thickened mass of cells surrounding the anus. Filamentous outgrowths are sometimes proliferated, forming anal cirri, but there is little or no intucking of cells of this region to form a proctodaeum in any polychaete (Fig. 16).

Anterior to the first trunk segment, the contribution of the ectoderm of the prototrochal region to the prostomium and peristomium of the polychaete head varies according to the extent of larval differentiation of the prototrochal region in the trochophore. Reliable information on the development of the adult head ectoderm in polychaetes has been gained only in recent studies (D. P. Wilson, 1932, 1936; Segrove, 1941; Korn, 1958, 1960; Anderson, 1959; Åkesson, 1961, 1962, 1963, 1967 a, b, 1968; Allen, 1964). It is now well established that the prostomium is the adult product of the presumptive anterior ectoderm, lying in front of the prototroch. In planktotrophic trochophores such as those of *Pomatoceros*, *Branchiomma*, *Owenia* and *Polygordius*, only the apical part of the presumptive anterior ectoderm, comprising the rudiment of cerebral ganglion and its overlying epithelium, gives rise to the prostomium and prostomial appendages. Errant polychaetes such as *Harmothöe*, *Lopadorhynchus*, *Glycera* and *Eunice* show a more complex development of several prostomial rudiments (Fig. 19), comprising paired ventrolateral rudiments of the cerebral ganglia and palps and a number of dorsal groups of cells developing as tentacles, but all originate within the presumptive anterior ectoderm, the remainder of which forms temporary larval epithelium. In other polychaetes such as *Scoloplos* and *Autolytus*, the presumptive anterior ectoderm develops entirely as adult tissue and all except its posterior margin contributes to the prostomium.

Behind the prototroch, a greater or lesser proportion of the ectoderm of the prototrochal region persists through metamorphosis as the ectoderm of what can be called the mouth region (Fig. 20). In planktotrophic trochophores, e.g. *Owenia*, *Polygordius*, the mouth region ectoderm comprises only a narrow band in front of the first trunk segment formed

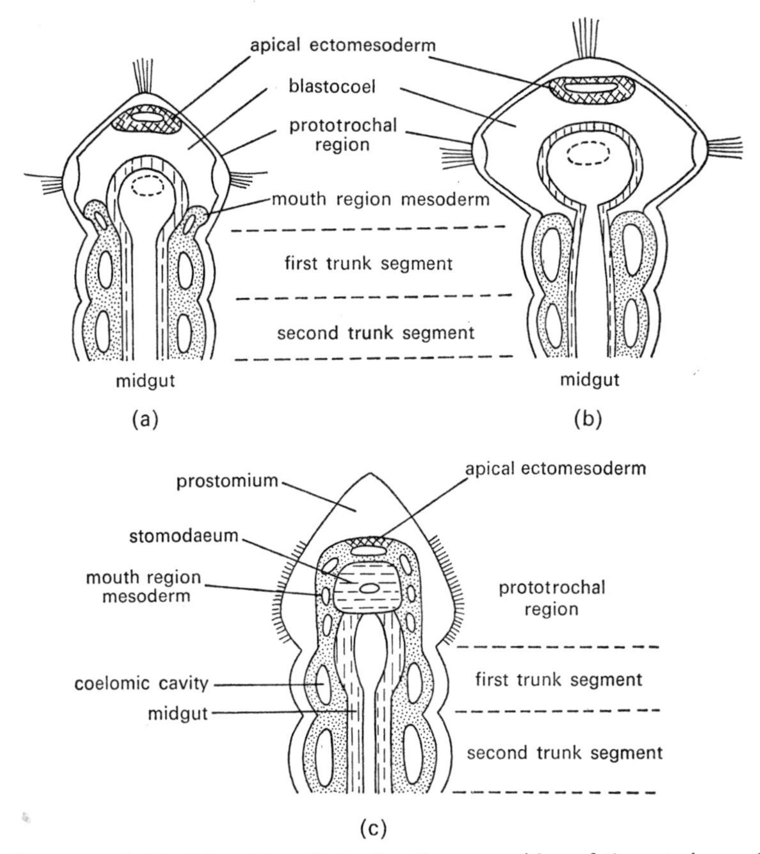

Fig. 20. Diagrammatic frontal sections illustrating the composition of the anterior end in polychaetes. (a) Serpulid; (b) *Owenia*; (c) *Scoloplos*. [After Anderson, 1959, modified from Anderson, 1966.]

by the growth zone, and most of the presumptive posterior ectoderm of the prototrochal region differentiates as larval epithelium, lost at metamorphosis. In *Scoloplos* and other species with a relatively large, yolky embryo, all of the ectoderm between the prototroch and the first trunk segment develops directly as mouth region ectoderm. In addition, the posterior border of the presumptive anterior ectoderm of these species, in front of the prototroch, also has this fate.

The mouth region ectoderm develops only into surface epithelium, differing fundamentally from the ectoderm of the trunk segments in that it lacks chaetal sacs and ventral ganglia. As far as is known at present, the ectoderm of the mouth region gives rise to the entire surface epithelium of the peristomium in serpulids and eunicids, but to only the anterior part of the surface epithelium of the peristomium in most polychaetes. Usually, the polychaete peristomium also includes part or all of the first, and sometimes the second, trunk segments. These units become cephalized as development proceeds.

Composition of the Polychaete Head

The combined evidence of development of the ectoderm and mesoderm at the anterior end of the embryo in polychaetes permits a tentative interpretation of the basic composition of the polychaete head (Anderson, 1959, 1966; Åkesson, 1967a, 1968). The prostomium, as we have seen, is both preoral and presegmental in origin and development. The peristomium, lying mainly in a postoral position, may be formed directly by the mouth region or may include the mouth region together with one or more cephalized segments. In the latter case, the peristomium is obviously a composite structure, not serially homologous with the trunk segments behind it, but the question still remains, whether the mouth region itself is a cephalized segment or something different.

The evidence is conflicting (Anderson, 1959, 1966; Åkesson, 1962, 1963, 1967a; Pilgrim, 1966). Since the mouth region ectoderm not only lacks all developmental expression of the ectodermal criteria of a segment, but also develops in front of the ectoteloblast ring from which the segmental ectoderm is proliferated, it is difficult to avoid the conclusion that the mouth region ectoderm is topographically and phylogenetically presegmental. Yet, in the basic condition, the most anterior pair of somites formed by the mesodermal bands comes to lie beneath the mouth region ectoderm, while secondarily, in large, yolky embryos such as those of *Scoloplos*, more than one pair of somites penetrates into this region. In errant polychaetes with yolky eggs, such as *Eunice*, the anterior ends of the mesodermal bands in the prototrochal region do not exhibit somite formation (Åkesson, 1967a), but it seems likely that this is a secondary simplification. The same animals also lack somite formation in the anterior trunk segments.

One solution to the dilemma of segmented mesoderm beneath pre-segmental ectoderm is to adopt the view that the mesoderm of polychaetes normally expresses somite formation as the first phase of its development, whether it lies beneath pre-segmental ectoderm or incipient segmental ectoderm. There is then no difficulty in maintaining the view that the mouth region is part of a topographically pre-segmental anterior end. The basic structural composition of polychaetes, on this view, is pre-segmental prostomium and mouth region (=basic peristomium), trunk segments and post-segmental pygidium. No doubt this interpretation of the composition of the polychaete head will be modified as further detailed studies are carried out on other species, but we are unlikely ever to see again a revival of the idea that the polychaete peristomium is serially homologous with the trunk segments.

One implication arising from these thoughts has a bearing on the interpretation of the composition of the head in arthropods. If mesoderm can form somites beneath topographically and phylogenetically presegmental ectoderm, the presence of a pair of somites alone is not a sufficient criterion of a segment.

The Basic Pattern of Development in Polychaetes

The visualization of a basic pattern of embryonic development for a major group of animals on the basis of comparisons of the various developmental sequences known for the group is an entirely speculative procedure, but one which is essential if embryological evidence is to be brought to bear on phylogenetic relationships between one major group and another. In spite of the arbitrary nature of the higher grades of taxonomy, it is possible to recognize distinct major groups whose members obviously share a basic morphological organization throughout development, the expression of which in existing species constitutes variations on a theme. For such groups, of which the polychaetes are one, a speculative appraisal of the developmental theme basic to the group is not too difficult to attain, once the variations are generalized in the appropriate terms. We are not concerned here with the establishment of an archetypal polychaete embryo, but with an appreciation of the common features of development in all polychaetes, a basic pattern of emergent organization, which may then be weighed against similar basic patterns in other groups of animals in order to judge the feasibility of ancestral relationships between them.

For polychaetes, the theme of development is sharply defined. It begins with a sequence of spiral cleavage to the 64-cell stage, based on a quadrant orientation, D dorsal, B ventral, A left-lateral and C right-lateral, and on a fixed sequence of spindle orientations and cleavage planes. The resulting blastula wall has a common distribution of presumptive areas relative to major axes. Each area then proceeds to a common fate in all species, displaying functional variations on a theme. The presumptive anterior ectoderm, anterior to the presumptive prototroch, gives rise wholly or in part to the prostomial epithelium and cerebral ganglion and to the temporary larval epithelium of the episphere where this is retained, and in some species contributes to the anterior marginal epithelium of the mouth region. The presumptive posterior ectoderm, dorsally to laterally behind the prototroch, gives rise to the temporary larval epithelium of the hyposphere where this is retained, to the epithelium of the mouth region, to a ring of ectoteloblasts which bud off the ectoderm of the trunk segments in anteroposterior succession, with a pause in budding during metamorphosis, and to the pygidium behind the ectoteloblast ring. The anus forms as an aperture in the centre of the pygidial epithelium. The presumptive stomodaeum, a transverse arc of cells ventrally behind the prototroch, invaginates and tubulates directly as the stomodaeal (pharyngeal) epithelium, opening at the mouth. The presumptive midgut, posteriorly placed, invaginates or immigrates into the blastocoel and gives rise directly to the lining epithelium of the midgut. Temporarily, these cells act as the main modules of yolk storage and release. The presumptive mesoderm, posterodorsally placed, becomes internal to the ectoteloblast ring as paired M-cells and proliferate paired mesodermal bands. The bands segment into an anteroposterior succession of paired somites which become hollow and encircle the gut to meet in the dorsal and ventral midlines. At least one pair of somites comes to lie beneath the ectoderm of the mouth region (secondarily, none, or more than one). The ectoderm proliferated by the ectoteloblast ring becomes serially demarcated as segmental ectoderm outside each pair of somites that it covers, and develops as paired ventral ganglia ventrally, paired chaetal sacs dorsolaterally and ventrolaterally, and surface epithelium elsewhere. The nephridia originate from a single pair of ectodermal nephridioblast cells, differentiated ventrolaterally in each intersegment. The somatic walls of the somites give rise to dorsolateral and ventrolateral longitudinal muscles, to external circular muscles spreading from a paired lateral focus, to chaetal sac and other trunk musculature, and to peritoneum. Gonads are formed in the somatic peritoneum of the segments, but the primordial germ cells may be segregated from the M-cells by an early division and subsequently proliferate and spread to the segmental peritoneum. The somatic peritoneum also forms the intersegmental septa and the dorsal and ventral mesenteries. Dorsal and ventral longitudinal blood vessels arise by separation of apposed peritoneal epithelia of the mesenteries. The splanchnic walls of the somites develop as splanchnic peritoneum and perhaps as musculature of the midgut. The presumptive ectomesoderm, a paired posterolateral area below the presumptive mesoderm and behind the presumptive posterior ectoderm, develops as temporary larval musculature or stomodaeal (proboscis) and prostomial musculature or both. In the completion of functional organization, the prostomial cerebral ganglion becomes united with the ventral ganglia of the first trunk segment by commissures that traverse the mouth region, encircling the stomodaeum, but the mouth region develops no ventral ganglia or chaetal sacs. The paired somites in the mouth region develop as peristomial musculature. The peristomium may develop from the mouth region alone, but is frequently completed by fusion of the first (sometimes also the second) trunk segment with the mouth region. No components of these fused segments migrate forwards into a pre-oral position.

An increase in the volume of yolk in the egg of polychaetes is accompanied by certain basic modifications of embryonic development. The cells of the presumptive posterior ectoderm and presumptive mesoderm, hence the ectoteloblast ring and the M-cells, are large. Differentiation of presumptive anterior and posterior ectoderm cells and of ectomesoderm cells as temporary larval cells is reduced or eliminated. The presumptive midgut cells are large and, associatedly, show emphasis on the temporary function of yolk storage and release and a delay in differentiation as functional midgut epithelial cells. Ventral spread of the posterior ectoderm during gastrulation is also emphasized, as a means of enclosing the large mesodermal and midgut rudiments. In general, these modifications result in a more direct development of the adult organization, minimal metamorphosis and the assumption of an active feedling life only when several trunk segments are formed and functional.

This then, is the essential sequence of morphological construction of a polychaete from the egg. We are now in a position to consider, firstly, how far the clitellate annelids have departed from this mode and, secondly, whether any trace of the annelid sequence of development is discernible in any living arthropod.

References

Åkesson, B. (1961) On the histological differentiation of the larvae of *Pisione remota* (Pisionidae, Polychaeta). *Acta Zool., Stockholm* **42,** 177–226.

Åkesson, B. (1962) The embryology of *Tomopteris helgolandica*. *Acta Zool., Stockholm* **43,** 135–99.

Åkesson, B. (1963) The comparative morphology and embryology of the head in scaleworms (Aphroditidae, Polychaete). *Ark. Zool.* **16,** 125–63.

Åkesson, B. (1967a) The embryology of the polychaete *Eunice kobiensis*. *Acta Zool., Stockholm* **48,** 141–92.

Åkesson, B. (1967b) On the nervous system of the *Lopadorhynchus* larva (Polychaeta). *Ark. Zool.* **20,** 55–78.

ÅKESSON, B. (1967c) On the biology and larval morphology of *Ophryotrocha puerilis* Claparéde and Metschnikoff (Polychaeta). *Ophelia* **4,** 111–19.

ÅKESSON, B. (1968) The ontogeny of the glycerid prostomium. *Acta Zool., Stockholm* **49,** 203–18.

ÅKESSON, B. and MELANDER, Y. (1967) A preliminary report on the early development of the polychaete *Tomopteris helgolandica. Ark. Zool.* **20,** 141–6.

ALLEN, M. J. (1959) Embryological development of *Diopatra cuprea. Biol. Bull., Woods Hole* **116,** 339–61.

ALLEN, M. J. (1964) Embryological development of the syllid *Autolytus fasciatus* (Bosc.). *Biol. Bull., Woods Hole* **127,** 187–205.

ANDERSON, D. T. (1959) The embryology of the polychaete *Scoloplos armiger. Q. Jl. microsc. Sci.* **100,** 89–166.

ANDERSON, D. T. (1961) The development of the polychaete *Haploscoloplos fragilis. Q. Jl. microsc. Sci.* **102,** 257–72.

ANDERSON, D. T. (1966) The comparative embryology of the Polychaeta. *Acta Zool., Stockholm* **47,** 1–41.

ANDREWS, J. C. and ANDERSON, D. T. (1962) The development of the polychaete *Galeolaria coespitosa* Lamarek (Fam. Serpulidae). *Proc. Linn. Soc. N.S.W.* **87,** 185–8.

BOOKHOUT, C. G. (1957) The development of *Dasybranchus caducus* (Grube) from the egg to the preadult. *J. Morph.* **100,** 141–86.

BOOKHOUT, C. G. and HORN, E. C. (1949) The development of *Axiothella mucosa* Andrews. *J. Morph.* **84,** 145–83.

CATHER, J. N. (1971) Cellular interactions in the regulation of development in annelids and molluscs. *Adv. Morphog.* **9,** 67–125.

CHILD, C. M. (1900) The early development of *Arenicola* and *Sternaspis. Arch. Entw. Mech. Org.* **9,** 587–723.

DALES, R. P. (1950) The reproduction and development of *Nereis diversicolor. J. mar. biol. Ass. U.K.* **29,** 321–60.

DALES, R. P. (1952) Development and structure of the anterior region of the body in the Sabellariidae, with special reference to *Phragmatopoma. Q. Jl. microsc. Sci.* **96,** 435–52.

DELSMAN, H. C. (1914) Entwicklungsgeschichte von *Littorina obtusa. Tijdschr. ned. dierk. Vereen.* (2) **13,** 170.

DELSMAN, H. C. (1916) Eifurchung und Keimblatterbildung bei *Scoloplos armiger* O. F. Müller. *Tijdschr. ned. dierk. Vereen.* (2) **14,** 383–498.

DELSMAN, H. C. (1917) Die Embryonalentwicklung von *Balanus balanoides* Linn. *Tijdschr. ned. dierk. Vereen.* (2) **15,** 419–520.

EISIG, H. D. (1898) Zur Entwicklungsgeschichte der Capitelliden. *Mitt. zool. Sta. Neap.* **13,** 1–292.

GEORGE, J. D. (1966) Reproduction and early development of the spionid polychaete, *Scolecolepides viridis* (Verrill). *Biol. Bull., Woods Hole* **130,** 76–93.

GOODRICH, E. S. (1946) The study of nephridia and genital ducts since 1895. *Q. Jl. microsc. Sci.* **86,** 113–301.

GUÉRRIER, P. (1970) Les caractères de la segmentation et la determination de la polarité dorsoventrale dans le développement de quelques Spiralia. II. *Sabellaria alveolata* (Annelide polychète). *J. Embryol. exp. Morph.* **23,** 639–55.

HÄCKER, V. (1894) Die spätere Entwicklung der Polynöe-Larve. *Zool. Jb. Anat. Ont.* **8,** 245–86.

HÄcker, V. (1896) Pelagische Polychaeten larven. *Z. wiss. Zool.* **62,** 74–168.

HORN, E. C. and BOOKHOUT, C. G. (1950) The early development of *Haploscoloplos bustoris. J. Elishah Mitchell Sci. Soc.* **66,** 1–11.

IWANOFF, P. P. (1928) Die Entwicklung der Larvalsegmente bei den Anneliden. *Z. Morph. Ökol. Tiere* **10,** 62–161.

KLEINENBERG, N. (1886) Die Entstehung des Annelids aus der Larve von *Lopadorhynchus. Z. wiss. Zool.* **44,** 1–227.

KORN, H. (1958) Vergleichende-embryologische Untersuchungen an *Harmothöe* Kinberg, 1857 (Polychaeta, Annelida). *Z. wiss. Zool.* **161,** 346–443.

KORN, H. (1960) Das larvale Nervensystem von *Pectinaria* Lamarck und *Nepthys* Cuvier (Annelida, Polychaeta). *Zool. Jb. Anat. Ont.* **78,** 427–56.

LILLIE, R. S. (1905) The structure and development of the nephridia of *Arenicola cristata* Stimpson. *Mitt. zool. Sta. Neap.* **17,** 341–405.

MALAQUIN, S. (1925) La ségregation, au cours de l'ontogénèse, de deux cellules sexuelles primordiales, souches de la lignée germinale, chez *Salmacina dysteri* (Huxley). *C.R. Acad. Sci., Paris* **180,** 324–7.

MALAQUIN, A. (1934) Nouvelles observations sur la lignée germinale de l'annelide *Salmacina dysteri* Huxley. *C.R. Acad. Sci., Paris* **198,** 1804–6.

MANTON, S. M. (1949) Studies on the Onychophora VII. The early embryonic stages of *Peripatopsis* and some general considerations concerning the morphology and phylogeny of the Arthropoda. *Phil. Trans. R. Soc.* B **233,** 483–580.

MEAD, A. D. (1897) The early development of marine annelids. *J. Morph.* **17,** 229–326.

MEYER, E. (1888) Studien über den Körperbau de Anneliden. IV. Der Körperform der Serpulaceen und Hermellen. *Mitt. zool. Sta. Neap.* **8,** 462–662.

MEYER, E. (1901) Studien über den Körperbau der Anneliden. V. Das Mesoderm der Ringelwurmer. *Mitt. zool. Sta. Neap.* **14,** 247–585.

NEWELL, G. E. (1951) The life history of *Clymenella torquata* (Leidy) (Polychaeta). *Proc. zool. Soc. Lond.* **121,** 561–86.

OKADA, K. (1941) The gametogenesis, the breeding habits and the early development of *Arenicola cristata* Stimpson, a tubicolous polychaete. *Sci. Rep. Tohoku Imp. Univ.*, ser. 4 *(Biol.)*, **16,** 99–146.

PILGRIM, M. (1966) The morphology of the head, thorax, proboscis apparatus and pygidium of the maldanid polychaetes *Clymenella torquata* and *Euclymene öerstedi. J. Zool.* **148,** 453–75.

SALENSKY, W. (1882) Étude sur le développement des annelides. I, II. *Arch. Biol., Paris* 6, 345–78 and 561–604.

SALENSKY, W. (1883) Étude sur le développement des annelides. III. *Arch. Biol., Paris* **4,** 188–220.

SALENSKY, W. (1908) Ueber den Bau und die Entwicklung der Schlundtaschen der Spioniden. *St. Petersb. Bull. Ac. Sci.*, ser. 6, **1908,** 686–708.

SCHAXEL, J. (1912) Versuch einer cytologischen Analysis der Entwicklungsvorgange. Teil 1. Die Geschlechtscellenbildung und die normal Entwicklung von *Aricia foetida* Clap. *Zool. Jb. Anat. Ont.* **34,** 381–472.

SCHIVELY, M. A. (1897) Structure and development of *Spirorbis borealis. Proc. Nat. Acad. Sci. Philadelphia,* **1897,** 153–60.

SCHNEIDER, J. (1913) Zur postembryonalen Entwicklung der nereidogenen Form von *Nereis Dumerilii* unter besonderer Berücksichtigung der Darmtractus. *Mitt. zool. Sta. Neap.* **20,** 529–646.

SEGROVE, F. (1941) The development of the serpulid *Pomatoceros triqueter* L. *Q. Jl. microsc. Sci.* **82,** 467–540.

SHEARER, C. (1911) The development of the trochophore of *Hydroides (Eupomatus) uncinatus. Q. Jl. microsc. Sci.* **56,** 543–90.

SOKOLOV, I. (1911) Über eine neue *Ctenodrilus* art und ihre Vermehrung. *Z. wiss. Zool.* **97,** 547–603.

SOULIER, A. (1902) Les premiers stades embryologiques de la serpule. *Trav. Inst. Zool. Univ. Montpellier* **9,** 1–78.

SOULIER, A. (1916a) La membrane vitelline des serpulides. *Arch. Zool. exp. gen.* **56,** *notes et revue* **1,** 16–20.

SOULIER, A. (1916b) La cinquième stade de segmentation (trente-deux cellules) chez *Protula Meilhaci. Arch. Zool. exp. gen.* **56,** *notes et revue* **4,** 100–4.

SOULIER, A. (1918) Le croix et la rosette chez *Protula Meilhaci. Arch. Zool. exp. gen.* **57,** *notes et revue* **1,** 14–20.

SOULIER, A. (1920) La couronne équitoriale ciliée de la trochosphère chez *Protula Meilhaci. Arch. Zool. exp. gen.* **59,** 1–4.

TREADWELL, A. L. (1901) The cytogeny of *Podarke obscura. J. Morph.* **17,** 399–486.

VUILLEMIN, S. (1967) Développement post-larvaire d'*Hydroides uncinata* (Phillipi), annelide polychète. *Bull. Soc. zool. Fr.* **92,** 647–65.

WILSON, D. P. (1928) The larvae of *Polydora ciliata* Johnston and *Polydora hoplura* Claparéde. *J. mar. biol. Ass. U.K.* **15,** 567–603.

WILSON, D. P. (1932) On the mitraria larva of *Owenia fusiformis. Phil. Trans. R. Soc.* B **221,** 231–334.

WILSON, D. P. (1936) The development of the sabellid *Branchiomma vesiculosum. Q. Jl. microsc. Sci.* **78,** 543–603.

WILSON, D. P. (1948) The larval development of *Ophelia bicornis* Savigny. *J. mar. biol. Ass. U.K.* **27,** 540–53.

WILSON, E. B. (1892) The cell lineage of *Nereis. J. Morph.* **6,** 361–480.

WISELY, B. (1958) The development and settling of a serpulid worm, *Hydroides norvegica* Gunnerus (Polychaeta). *Aust. J. mar. f.w. Res.* **9,** 351–61.

WISTINGHAUSEN, C. von (1891) Untersuchungen über die Entwicklung von *Nereis dumerilii. Mitt. zool. Sta. Neap.* **10,** 41–74.

WOLTERECK, R. (1902) Trochophora-Studien I.: Über die Histologie der Larve und die Entstehung des Annelides bei den *Polygordius*—Arten der Nordsee. *Zoologica, Leipzig* **13,** Hft. 34.

WOLTERECK, R. (1904) Beiträge zur praktischen Analyse der *Polygordius* Entwicklung nach dem "Nordsee"- und dem "Mittelmeer"-Typus. I. Die für beide Typen gleichverlaufende Entwicklungsabschnitt: Vorm Ei bis zum jungsten Trochophora-Stadium. *Arch. EntwMech. Org.* **18,** 377–403.

WOLTERECK, R. (1905) Zur Kopffrage der Anneliden. *Verh. dtsch. zool. Ges.* **15,** 154–86.

CHAPTER 3

OLIGOCHAETES AND LEECHES

THE polychaetes (Chapter 2) sustain a basic pattern of embryonic development with little modification, even when the egg is relatively large and yolky. Since the most obvious feature of arthropod development is a more extreme modification of cleavage, gastrulation and organogeny in association with yolk, we need to examine the question, whether any tendency in the same direction is manifested among the other major annelid groups, the oligochaetes and leeches?

In spite of marked differences in their modes of life, the oligochaetes and leeches have long been recognized as related clitellate annelids whose eggs develop in a protective cocoon secreted by the clitellum and hatch from the cocoon as juveniles with numerous segments. In the tubificid and lumbriculid oligochaetes and the glossiphoniid leeches, the egg is relatively large, ranging in size in different species from about 300 μ to 1 mm in diameter, and the yolk of the egg is the major nutrient source for the developing embryo. A quantity of albumen fills the cocoon and bathes the eggs and embryos, but appears to be of little consequence as a food source for development. All the remaining families of clitellates, however, with the partial exception of the Branchiobdellidae, lay smaller eggs with little yolk and exploit the ambient albumen as a major source of food during embryonic development. As will be seen below, the mode of development of these species gives evidence of secondary loss of yolk and reduction in the size of the egg, accompanied by the evolution of a variety of temporary feeding organs in the embryo. The large egg, then, is basic in clitellates and provides an appropriate example of annelid development modified in association with an increase of yolk beyond the polychaete level. We shall see that a total, obviously spiral cleavage persists, leading to a spherical blastula, but that the trochophore is absent as a stage in development, as is any trace of the larval organs or metamorphosis seen in polychaetes. Gastrulation is prolonged and is accompanied and followed by the formation of numerous segments from the posterior growth zone, leading to direct development of the adult organization. Only after elucidating the theme of clitellate development as expressed in these yolky eggs can we make a comprehensive comparison between the embryos of annelids and arthropods.

The value of clitellate embryos from the point of view of comparative embryology has only recently been appreciated (Anderson, 1966, 1971; Cather, 1971). Although a glossiphoniid leech was the first annelid to be described in what was to become the classical school of American cell lineage studies, by Whitman (1878, 1887), the peculiarities of cleavage of the large yolky egg proved extremely difficult to interpret. The classical workers on clitellate development were constantly plagued either by the problem of too much yolk or by the converse problem of too little yolk, both of which obscure the basis of cleavage and gastrulation in these animals. The later stages of development, relatively large and simple in structure, proved more amenable to classical techniques, but even these presented problems in the absence of a proper understanding of the preceding stages.

Clitellate cleavage finally begun to yield to descriptive attack only when the cleavage and cell lineage of polychaetes has been well elucidated and had almost succumbed as a fashionable discipline. The German embryologist Schliep (1913, 1914) was the first to describe the cell lineage of a yolky clitellate egg in a meaningful way. From this beginning, clitellate embryology began to flower, but more or less in isolation. Furthermore, although excellent progress was made between 1920 and 1940, mainly due to the efforts of the German embryologist Penners and the Russian G. A. Schmidt, there has been little augmentation since. Only Schoumkine (1953), who described the development of *Hirudo*, Inase (1967), who gave a new account of cleavage in *Tubifex*, and Devries (1968), who recently described cleavage in *Eisenia*, have provided new facts in the last thirty years. It is a testimony to the workers of the period between the two World Wars, therefore, that a telling appraisal can yet be made of the theme and variations of clitellate embryonic development. Those aspects that have recently been reviewed in detail by Anderson (1966, 1971) are only broadly outlined here, but are rounded out by the inclusion of three previously omitted topics, the peculiarities of cleavage and cell lineage in piscicolid leeches, the gastrulation of the secondarily yolkless embryos of earthworms, gnathobdellid and pharyngobdellid leeches and piscicolid leeches and the further development of earthworms and leeches after gastrulation.

Cleavage

In the early years of descriptive embryology, before spiral cleavage was fully understood, a number of attempts were made by European and American workers to trace cleavage in clitellates. Whitman (1878, 1887), Vedjovsky (1886, 1888–92) and Bergh (1890) gave accounts of cleavage of the large yolky eggs of *Glossiphonia* and the oligochaete *Rhynchelmis*, while Hatschek (1878), Vedjovsky (1888–92) and E. B. Wilson (1887, 1889) briefly described the cleavage of the small eggs of earthworms, but to little avail. Subsequently, after the successful cell-lineage studies of polychaete cleavage of the eighteen-nineties, Sukatschoff (1903) essayed a new interpretation of cleavage in the pharyngobdellid leech, *Erpobdella*, but was again troubled by the peculiarities of the secondarily small egg. By the time that Schliep (1914) succeeded in following the entire cleavage sequence of the yolky egg of *Glossiphonia* and identifying the blastomeres in accordance with Wilson's system of notation for spiral cleavage, interest in the cell lineage of polychaetes had almost ceased, so that in the twenty-five years following Schliep's work, as clitellate cleavage became better understood, its relationship to polychaete cleavage was never seriously discussed. We shall return to this matter later in the present chapter.

Schliep's account of *Glossiphonia* was soon followed by that of Dimpker (1917), who interpreted the cleavage of *Erpobdella* as a modification of the glossiphoniid sequence, and Schmidt (1917; see also Schmidt, 1925a) who confirmed Schliep's interpretation for another glossiphoniid leech, *Theromyzon*. During the same period, Tannreuther (1915), working in America, had achieved an initial success in describing the cleavage of the moderately yolky egg of the branchiobdellid *Bdellodrilus*. Other important work soon followed. Penners (1922, 1924) and the Russian worker Swetloff (1923b) found that the yolky eggs of the oligochaetes *Tubifex* and *Rhynchelmis* cleave in a manner closely allied to that of the glossiphoniid leeches. Swetloff (1923a, 1926, 1928) also studied the peculiarities of cleavage in the earthworms and naidid oligochaetes, while Schmidt (1921, 1924, 1925b, 1930, 1939, 1941, 1944) pursued a series of detailed studies on the cleavage modifications of the small eggs of piscicolid leeches. Since that time, only Dawydoff (1942), who investigated naidid cleavage in Indo-China, Schoumkine (1953), Inase (1967) and Devries (1968) have added further to the description of clitellate cleavage.

When described according to the Wilsonian notation, clitellate cleavage exhibits a wide range of modifications, each difficult to reconcile with the other and with the basic spiral cleavage sequence retained in polychaetes. In all cases, the number of cell divisions leading to formation of the blastula is less than in polychaetes and the blastula has fewer than 64 cells. In order to elicudate and compare the segregation of presumptive areas in different species (Anderson, 1966, 1971; Cather, 1971), it is necessary to review the important features of cleavage in the species that have been investigated.

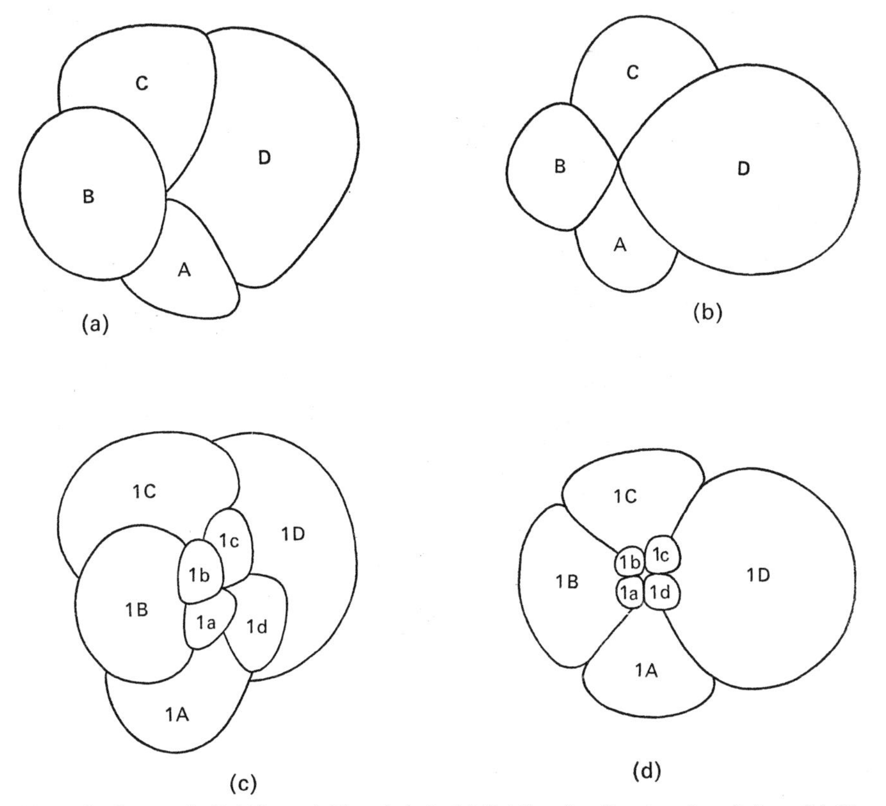

FIG. 21. Early cleavage in *Tubifex* and *Rhynchelmis*. (a) *Tubifex*, 4-cell, anterodorsal view; (b) *Rhynchelmis*, 4-cell, dorsal view; (c) *Tubifex*, 8-cell, anterodorsal view; (d) *Rhynchelmis*, 8-cell, dorsal view. [(a) and (c) after Penners, 1922; (b) and (d) after Swetloff, 1923b; modified from Anderson, 1966.]

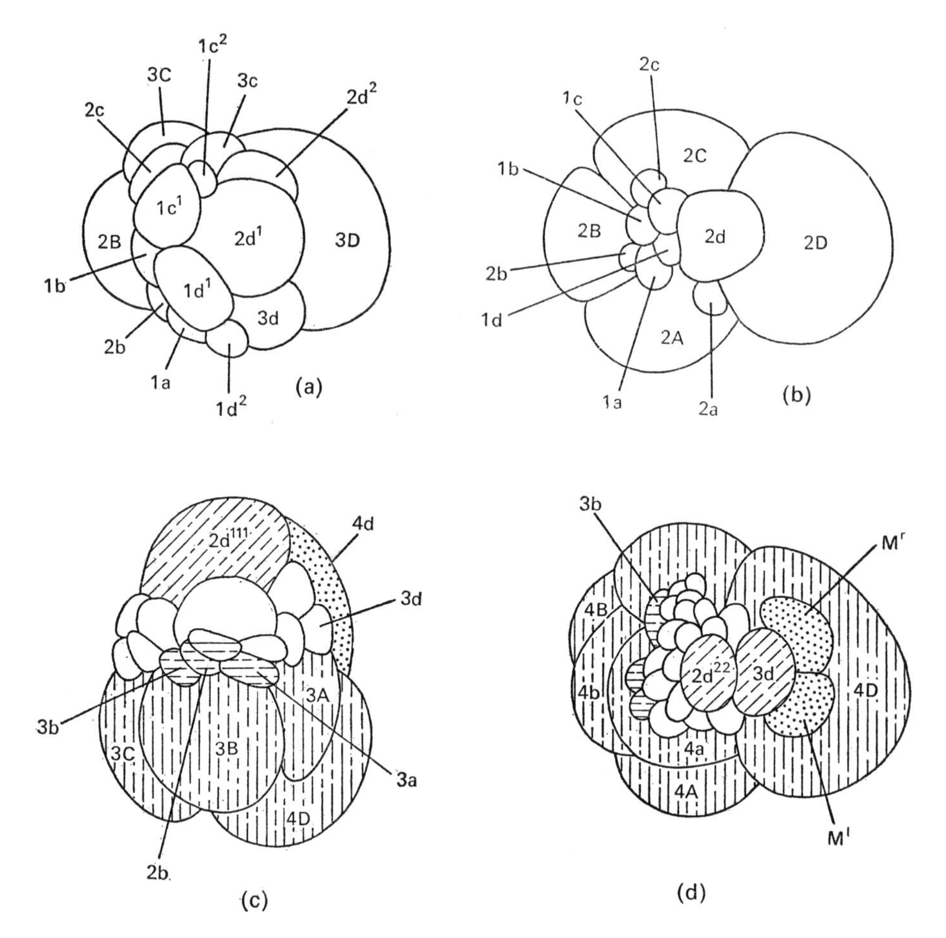

FIG. 22. Later cleavage in *Tubifex* and *Rhynchelmis*. (a) *Tubifex*, 17-cell, dorsal view; (b) *Rhynchelmis*, 12-cell, dorsal view; (c) *Tubifex*, 22-cell, anterolateral view; (d) *Rhynchelmis*, 30-cell, dorsal view. [(a) and (c) after Penners, 1922; (b) and (d) after Swetloff, 1923b; modified from Anderson, 1966.]

Cleavage is least modified in the large yolky eggs of tubificid and lumbriculid oligochaetes and glossiphoniid leeches. These families retain the basic spiral sequence of spindle orientations and cleavage planes through the first six division cycles, even though the outcome is not 64 cells. The first two divisions occur at right angles along the animal vegetal axis, while the next four are perpendicular to this axis and, except for a precocious onset of bilateral divisions in the D-quadrant, successively clockwise, anticlockwise, clockwise, then anticlockwise in the usual way. Associated with the relatively large size of the egg, e.g. *Tubifex*, 300–500 μ, *Rhynchelmis*, 1000 μ; *Glossiphonia*, 500–1000 μ, *Theromyzon*, 600 μ, the first cleavage division is unequal, AB being smaller than CD, and the second division is also unequal in the CD cell, C being smaller than D (Figs. 21 and 23). As in polychaetes, but to a greater extent, the large D cell heralds the formation of a growth zone capable of producing many segments, e.g. 34 in *Tubifex*, before the yolk of the egg is exhausted.

Relative to the anteroposterior and dorsoventral axes of the embryo, the orientation of the four quadrants of the *Tubifex* egg at the 4-cell stage remains as in polychaetes, D dorsal,

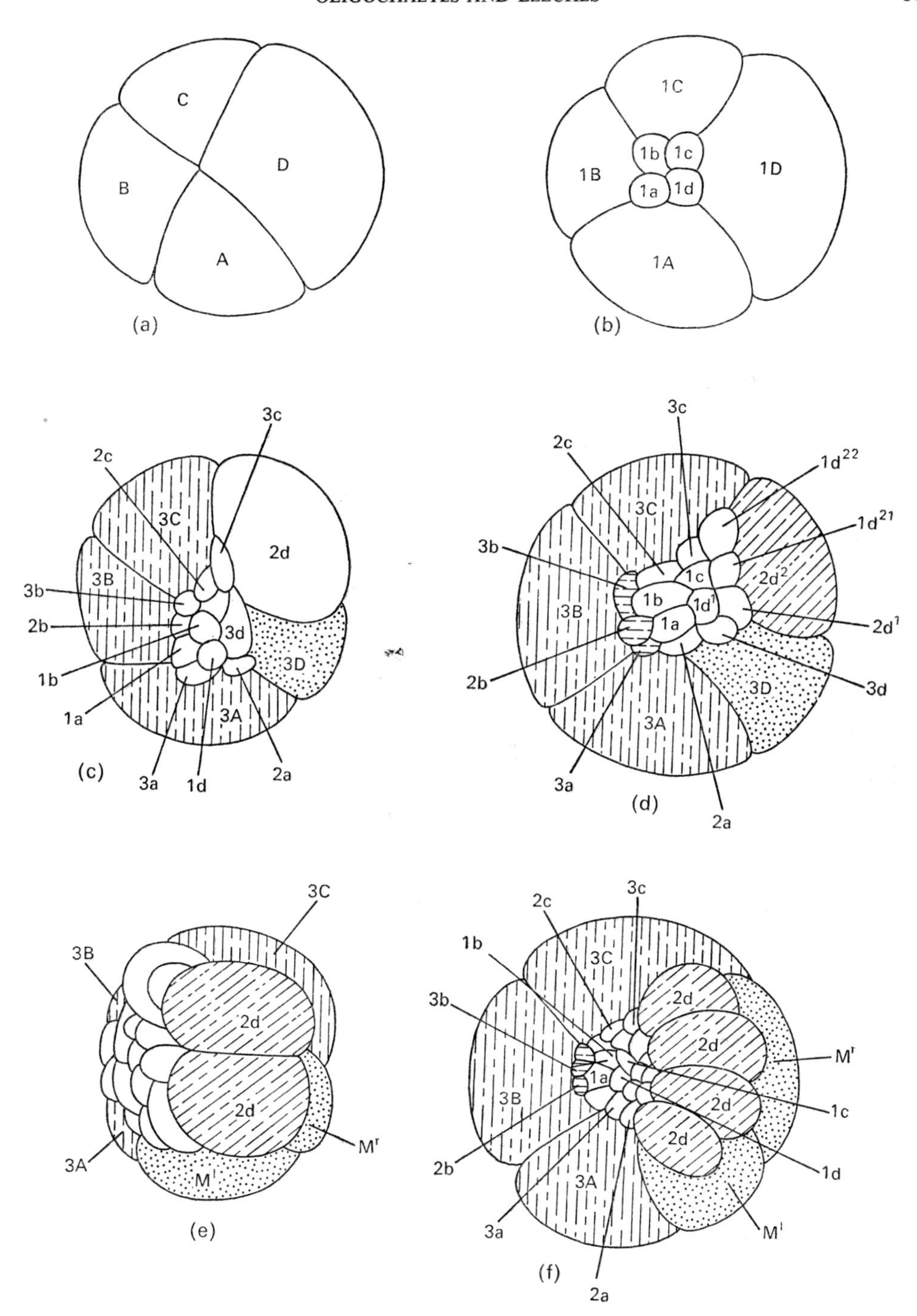

FIG. 23. Cleavage in *Glossiphonia* and *Theromyzon*. (a) *Glossiphonia*, 4-cell, dorsal view; (b) *Glossiphonia*, 8-cell, dorsal view; (c) *Glossiphonia*, 16-cell, dorsal view; (d) *Theromyzon*, 19-cell, dorsal view; (e) *Glossiphonia*, stage with paired M-cells, posterodorsal view; (f) *Theromyzon*, stage during ectoteloblast formation, dorsal view. [(a), (b), (c) and (e) after Schleip, 1914; (d) and (f) after Schmidt, 1917, 1925a, 1944; modified from Anderson (1966).]

A left-lateral, B ventral and C right-lateral. In other clitellates, a new orientation is manifested. The first two cleavage planes cut perpendicular to the anteroposterior axis in such a way that B is anterior and D posterior at the 4-cell stage. The quadrants of *Rhynchelmis*, *Glossiphonia* and *Theromyzon* thus attain, during the first two cleavage divisions, an orientation from which further development can proceed with precise bilateral symmetry (see below). In *Tubifex*, the same shift of quadrant orientations occurs gradually as a result of later cleavage divisions.

In the third and subsequent cleavage divisions of yolky clitellate eggs, the specialization of cleavage in association with yolk begins to become more obvious (Figs. 21, 22 and 23). The clockwise third division is highly unequal, 1a–1d being cut off as yolk-free micromeres much smaller than the stem cells 1A–1D. The same inequality persists in the formation of micromeres 2a, 2b and 2c at the anticlockwise fourth division, but not in the formation of 2d. The latter is cut off as a much larger cell in a dorsal position, with the first and the remaining second quartet cells as an arc of micromeres around its anterior face.

The clockwise fifth division proceeds in slightly different ways in different species. 3a, 3b and 3c are added as small cells to the micromere arc. 3d is formed in the same way in *Tubifex*, *Glossiphonia* and *Theromyzon*, but the 3d cell of *Rhynchelmis* is much larger than the other cells of the third quartet and makes a pair with the 2d cell in front of it. 2d, meanwhile, in all species, divides unequally at the fifth cleavage, adding a further small cell to the micromere arc in front of it. This process is repeated once or twice in subsequent divisions, but a large cell of 2d origin always persists on the dorsal surface, slightly displaced posteriorly and to the right in leeches, and accompanied by a partner cell, 3d, in *Rhynchelmis*.

After the formation of the third quartet, the four stem cells 3A–3D are still large and yolky. In *Rhynchelmis* and leeches these cells retain their initial quadrant orientation, 3D posterior, 3B anterior, 3A left-lateral and 3C right-lateral, with the smaller products of cleavage sitting on them as a dorsal cap. The leeches, as we shall see below, have completed their segregation of presumptive areas by this stage. *Rhynchelmis* retains one additional division typical of spiral cleavage, the division of 3D into a posterodorsal 4d cell above a larger, posteroventral, 4D cell. *Tubifex* also retains this division as a crucial step in the segregation of presumptive areas but, at the same time, undergoes the quadrant reorientation that is exhibited at the first and second cleavages in *Rhynchelmis* and leeches. The 4d cell of *Tubifex* is larger than its associated stem cell 4D, so that the division pushes 4D into a ventral position. This, in turn, displaces 3A, 3B and 3C forwards. 3B becomes anterior and the micromere arc in front of 2d is lifted into a dorsal position. In its gradual attainment of a new quadrant orientation during cleavage, *Tubifex* illustrates a functional transition between the basic spiral cleavage orientation, D dorsal, B ventral, and those species in which B is anterior and D posterior at the 4-cell stage.

In yolky, clitellate eggs, therefore, the basic spiral cleavage sequence of spindle orientations and cleavage planes is retained, but is accompanied by a quadrant shift, a reduction in the number of cell divisions leading to the blastula and an exaggerated inequality of all divisions except those yielding 4d in *Tubifex* and *Rhynchelmis*, 3d in *Rhynchelmis* and 2d in all species. The resulting blastula has a characteristic form. Four large, yolky cells lie ventrally, one anterior, 3B, one posterior, 4D or 3D, and one on each side, 3A and 3C.

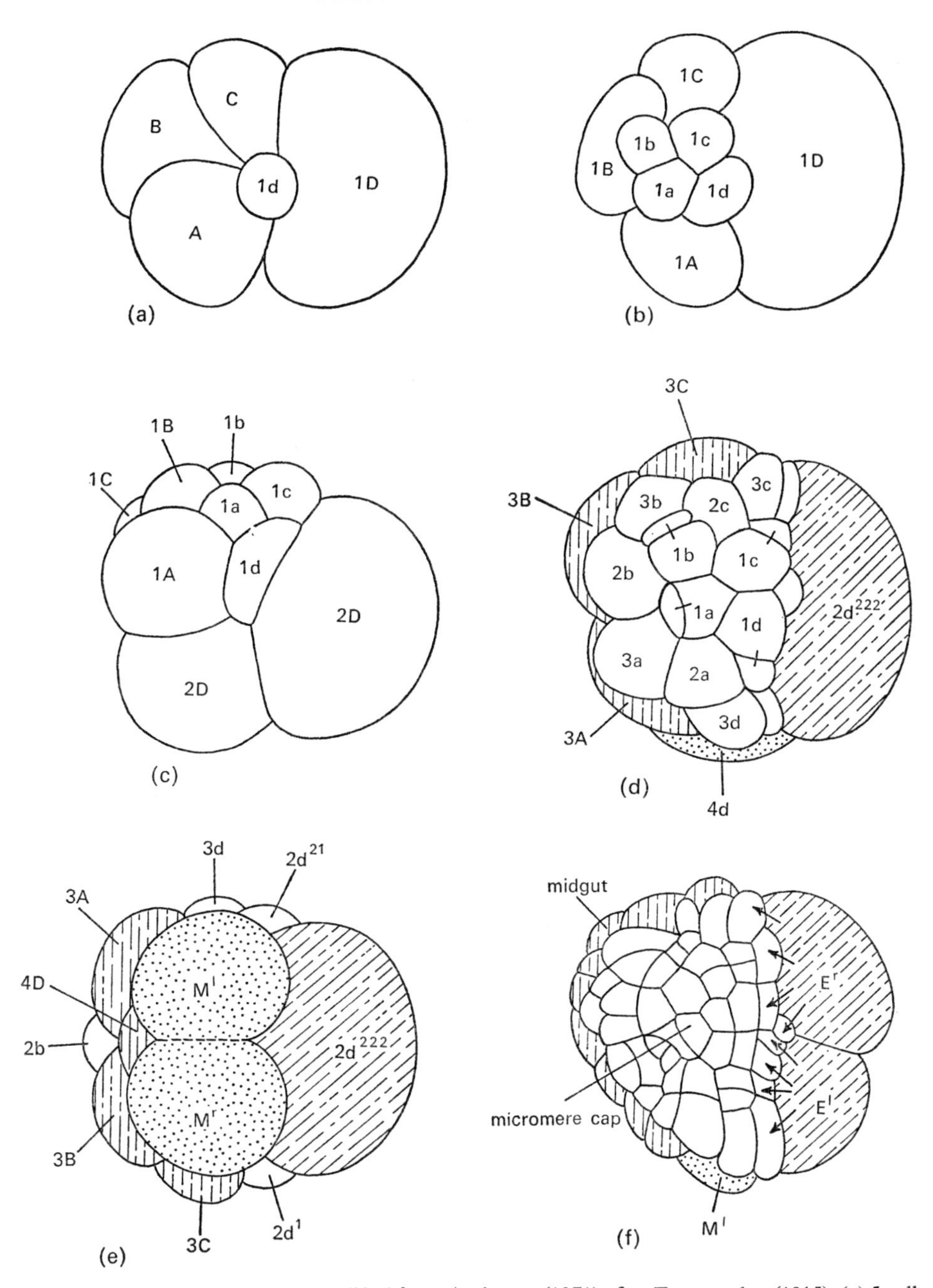

FIG. 24. Cleavage in *Bdellodrilus*, modified from Anderson (1971) after Tannreuther (1915). (a) 5-cell, dorsal view; (b) 8-cell, dorsal view; (c) 9-cell, left lateral view; (d) stage following formation of 4d, dorsal view; (e) stage of division of 4d into paired M-cells, ventral view; (f) stage of early ectoteloblast formation, dorsal view.

Above them, in a dorsal to posterodorsal position, lie a relatively large 2d cell, accompanied by a similar 3d cell in *Rhynchelmis* and, in oligochaetes, a large 4d cell. The micromeres of the first three quartets make up a dorsal arc of small cells around the anterior face of the 2d cell. Obviously, this arrangement constitutes a particular modification of spiral cleavage in association with increased yolk, different from any of the modifications seen in yolky polychaete eggs. What the modification means in terms of functional adaptation to direct, lecithotrophic development in annelids will become apparent when we discuss clitellate fate maps in the next section.

We need now to examine the ways in which the basic sequence of clitellate cleavage discussed above has become further modified in association with a secondary reduction of yolk and precocious onset of feeding on the ambient albumen of the cocoon. A useful intermediate is provided by *Bdellodrilus*, whose egg is about 260 μ in diameter.

Cleavage in *Bdellodrilus* (Fig. 24) follows precisely the same sequence of divisions as that of *Tubifex*, but is modified in such a way that 2d and 4d are the largest cells of the blastula,

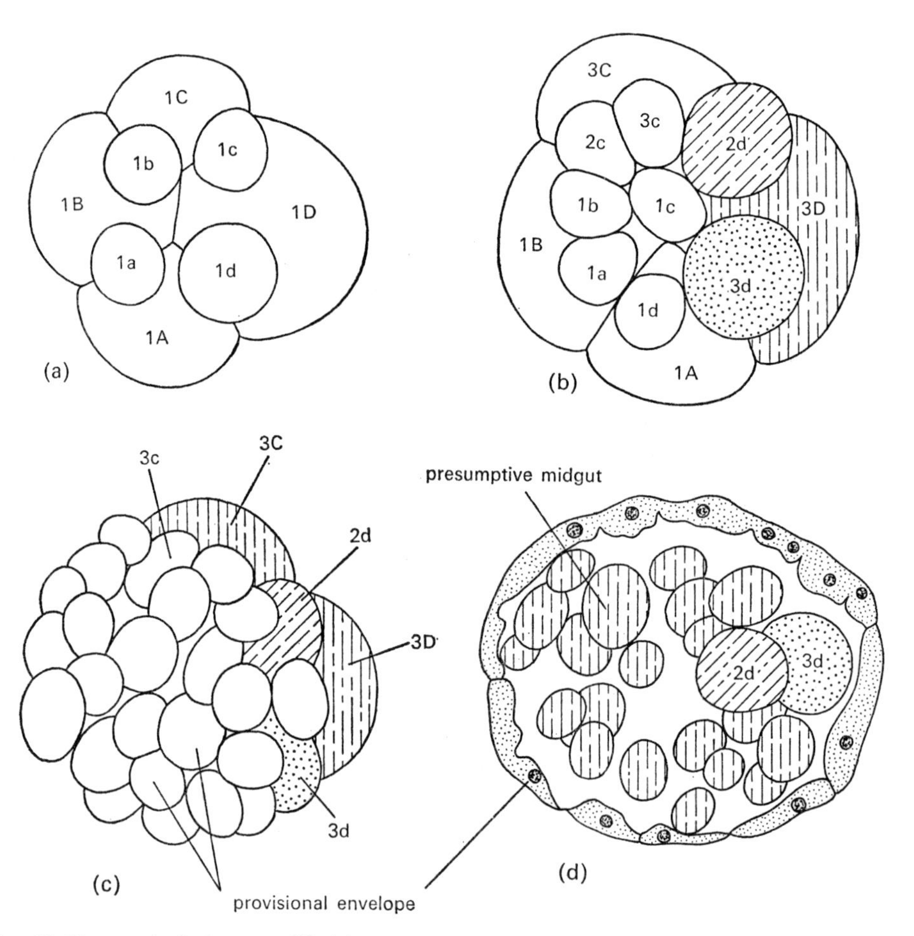

FIG. 25. Cleavage in *Stylaria*, modified from Anderson (1971) after Dawydoff (1942). (a) 8-cell, dorsal view; (b) 12-cell dorsal view; (c) early stage of overgrowth by provisional envelope, dorsal view; (d) early cell divisions within provisional envelope, dorsal view.

while the yolky stem cells 3A–3C and 4D are relatively small. 2d occupies a posterodorsal position and 4d is posteroventral. It is therefore immediately clear that the major modification associated with the reduction of yolk in *Bdellodrilus* is a reduction in size of the yolky stem cells.

The naidids *Stylaria* und *Chaetogaster*, although they retain eggs of considerable diameter (Stylaria, 350–400 μ; *Chaetogaster*, 400–500 μ), show more extreme modifications (Fig. 25). In *Stylaria*, the first three cleavage divisions remain typical, yielding a dorsal quartet of micromeres, 1a–1d, above four, large stem cells, 1A–1D, of which 1D is the largest. Then, during the completion of the blastula, the first quartet 1a–1d and the stem cells 1A and 1B remain undivided, while small 2c and 3c and larger 2d and 3d cells are cut off in the usual way, leaving stem cells 3C and 3D still large. The blastula stage thus retains the typical configuration but is attained through fewer cell divisions than in *Tubifex* and has a large 3d cell instead of a large 4d cell. How development proceeds after this stage will be discussed below. *Chaetogaster* is only slightly different from *Stylaria*. The 8-cell stage is reached in the same way and 1D cuts off 2d and 3d as relatively large cells, but 1C does not divide.

Among the oligochaetes, only the earthworms have small eggs exhibiting a marked secondary reduction of yolk. The early students of earthworm cleavage up to and including Swetloff (1923a) were unable to discern any evidence of basic spirality. Anderson (1966) reinterpreted Swetloff's results in accordance with the principle of formation of presumptive areas of the blastula, and found that earthworm cleavage is clearly a modification of the *Tubifex* pattern of cleavage, given secondary loss of yolk. Subsequently, Devries (1968), in a careful cell lineage study of the cleavage of *Eisenia*, has carried this analysis an important step further. We can now see that the D-quadrant in earthworms is emphasized to an even greater degree than in naidids (Fig. 26). The egg of *Eisenia foetida* is about 100–120 μ in diameter. At the first cleavage division, a small AB cell separates in the usual way from a larger CD cell. The second division is approximately equal in the AB cell, but again highly unequal in the CD cell, so that D remains large. The left, anterior and right cells of the 4-cell stage, A, B and C, now become specialized (see below) and take no further part in cleavage, but the D cell undergoes some of the further divisions of the clitellate spiral sequence. 1d is cut off first as a micromere on the upper left side. The large 1D then divides equally into a posterodorsal 2d and posteroventral 2D. The posterodorsal cell, 2d, cuts off a micromere to the upper right, $2d^2$, a micromere to the upper left, $2d^{12}$ and a micromere in front of itself, $2d^{112}$. These micromeres, together with 1d, form the usual dorsal micromere arc around the anterior face of the persistent, large, posterodorsal cell, $2d^{111}$. Meanwhile, the posteroventral cell 2D cuts off a small cell 3d on the left, then subdivides into a large 4d cell posteroventrally behind a ventral cell 4D. In the completion of the blastula, both 3d and 4D divide again, and are joined by a ventral division product of $2d^2$ to give a group of five, small, ventral cells, which contain most of the yolk of the original egg.

The similarity of this stage to the characteristic blastula of yolky oligochaete eggs is unmistakable, in spite of the reduction in relative size of the ventral yolky cells (even more marked than in *Bdellodrilus*) and the specialization of the three anterior cells A, B and C. A group of yolky cells, now 2d, 3d and 4D products, lies ventrally. The usual large 4d cell is present posteroventrally. The usual large 2d cell lies posterodorsally, surrounded anteriorly

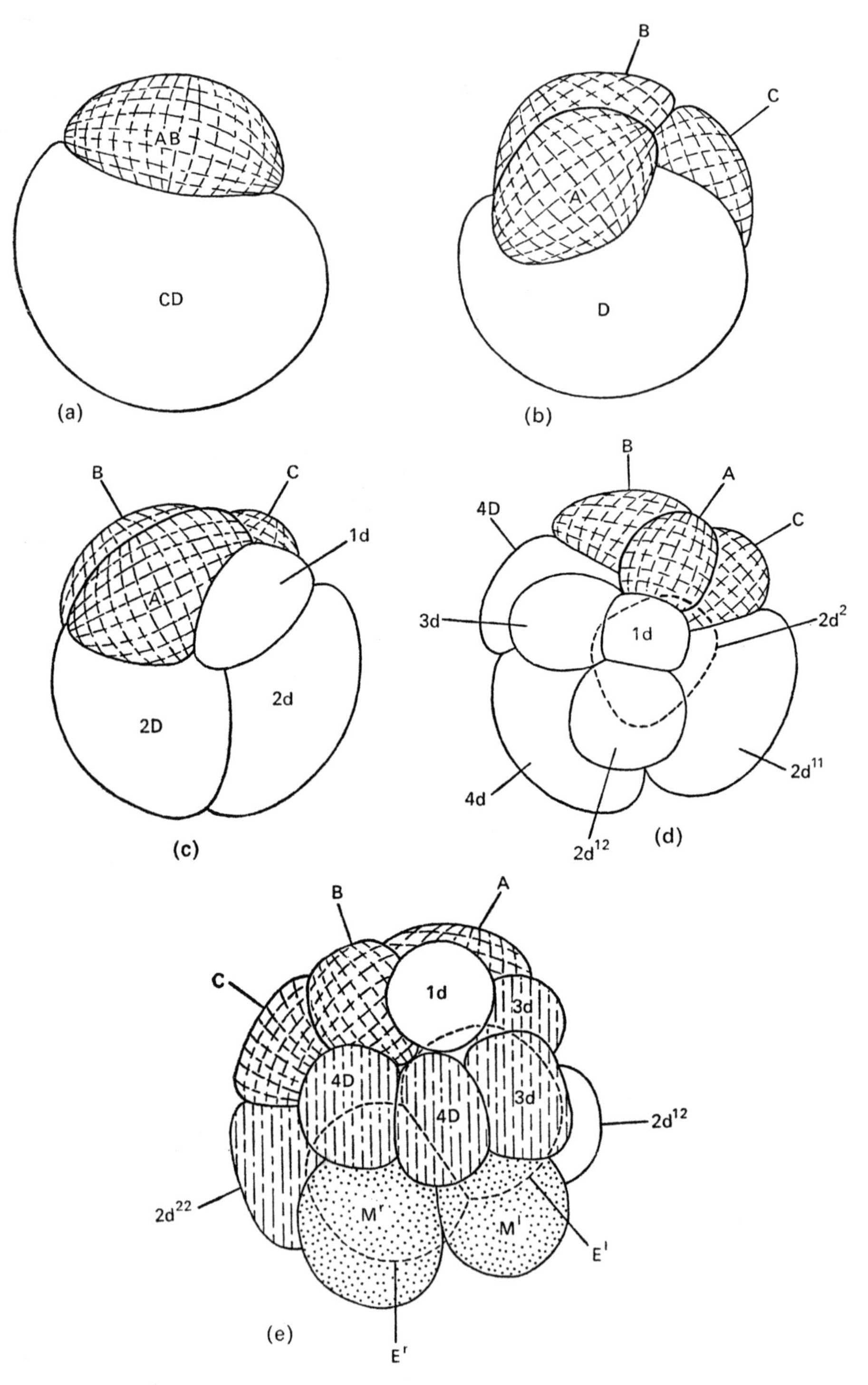

FIG. 26. Cleavage in *Eisenia*, after Devries (1968). (a) 2-cell, right lateral view; (b) 4-cell, left lateral view; (c) 6-cell, left lateral view; (d) 10-cell, left lateral view; (e) 17-cell early blastula, ventral view.

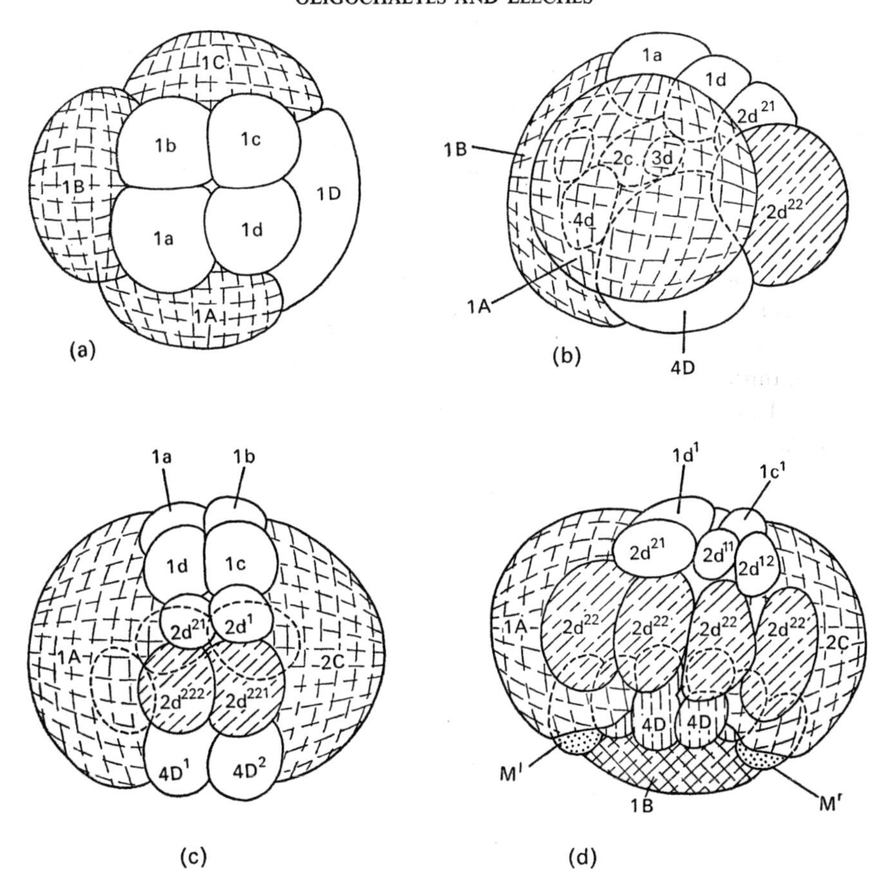

FIG. 27. Cleavage in *Erpobdella*. (a) 8-cell, dorsal view; (b) 14-cell, left lateral view; (c) 16-cell, postero-dorsal view; (d) 27-cell, posterior view. [(a) after Sukatschoff, 1903; (b–d) after Dimpker, 1917; modified from Anderson, 1966.]

and laterally by the usual arc of micromeres (now 1d and 2d products). We shall see below that this similarity becomes even more marked when functional considerations are entertained.

Returning to the leeches, both the gnathobdellids and pharyngobdellids on the one hand and the piscicolids on the other display cleavage modifications in their secondarily small eggs. Remarkably, the modifications of gnathobdellid and pharyngobdellid leeches are convergently similar to those of earthworm cleavage, while the modifications of piscicolids are convergently similar to those of naidids. Taking the gnathobdellids (e.g. *Hirudo*) and pharyngobdellids (e.g. *Erpobdella*) first (Fig. 27), in spite of a great reduction in egg size, the first three cleavage divisions proceed in the normal manner and, in this repect, are less specialized than those of earthworms. We now see, however, that the anterior and lateral stem cells 1A–1C persist without further division, except that 1C cuts off a small cell, 2c, into the interior and becomes 2C. The three cells 1A, 1B and 2C subsequently become specialized in the same manner as the three anterior cells A, B and C of earthworms (see below). The posterior cell, 1D, continues initially along the normal pathway, cutting off a large,

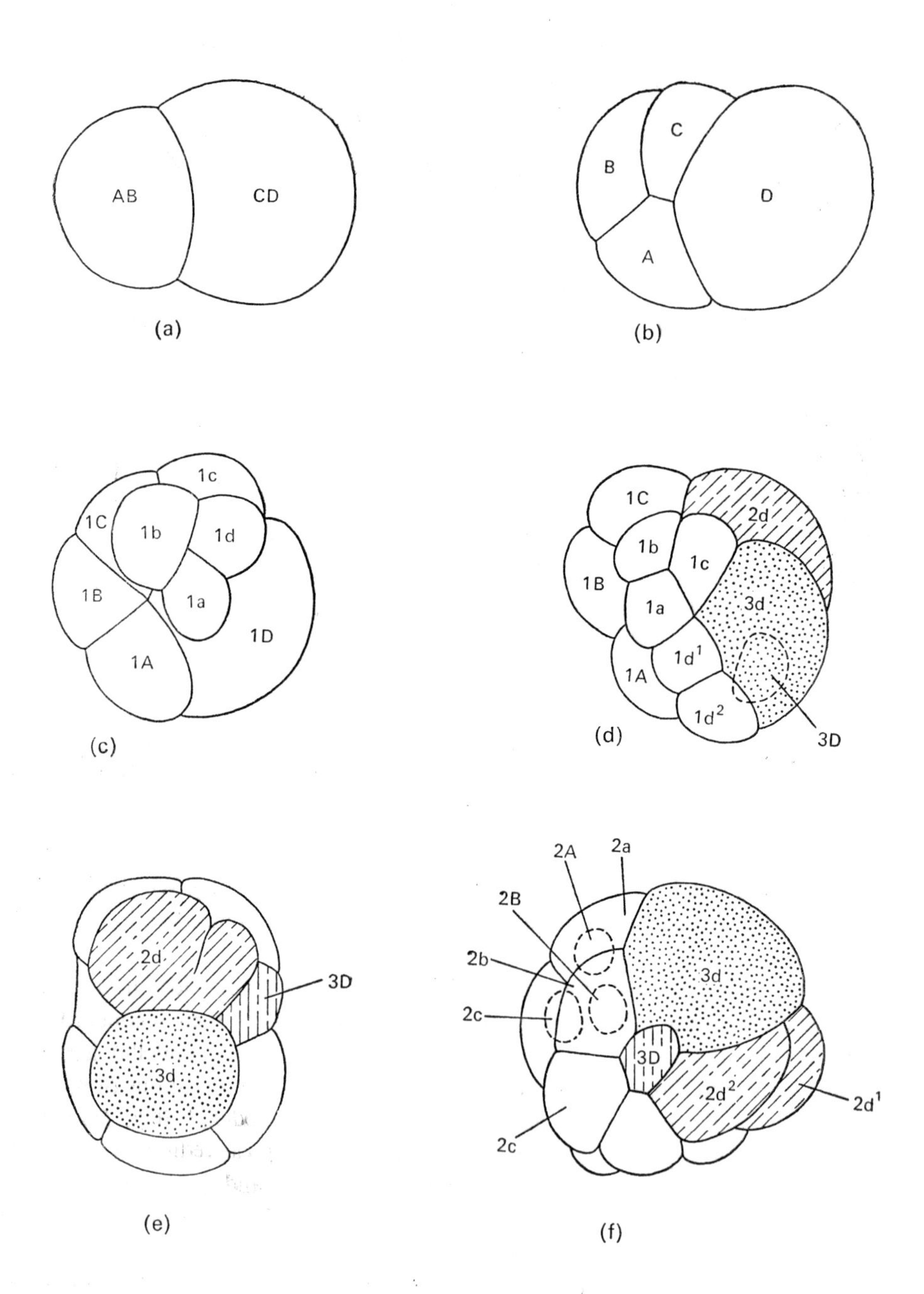

FIG. 28. Cleavage in *Piscicola*, after Schmidt (1925b). (a) 2-cell, dorsal view; (b) 4-cell, dorsal view; (c) 8-cell, left dorsolateral view; (d) 11-cell, dorsal view; (e) onset of division of the 2d cell, posterior view; (f) completion of formation of the second quartet, ventral view.

posterodorsal 2d cell which itself adds two small cells in the usual way to the micromere arc in front of it. 2D, however, then enters into a series of specialized divisions. It first cuts off two small cells, 3d and 4d, into the interior to join 2c and then divides into several ventral cells (products of 4D), of which only the two most lateral are the functional equivalent of 3D in glossiphoniids, the remainder being small, yolky cells. A typical clitellate blastula is thus attained but, as in earthworms, it includes three specialized cells anteroventrally and has several relatively small cells as ventral, yolky cells.

Cleavage in the piscicolids (Fig. 28) proceeds in a different manner. In spite of an egg diameter of only 40–60 μ, the first three cleavage divisions are normal in orientation and relative blastomere size, with 1D being the largest cell. The latter now divides equally, producing a large 2d cell on the right of an equally large 2D cell. 2D then cuts off a small 3D cell in a ventral position and persists at the surface as a large, posterior 3d cell. The remaining stem cells 1A–1C, like those of naidids, show a partial suppression of further divisions at this stage. They simply cut off small stem cells, 2A–2C, into the interior to join 3D and persist at the surface as three larger cells, 2a–2c, anteriorly and laterally adjacent to the first quartette cells. The convergent similarity with *Stylaria* and *Chaetogaster* is not yet obvious, but soon manifests itself in convergent functional specializations as development proceeds. The significance of these and other functional modifications in clitellate cleavage is made clear by a comparison of fate maps of the blastula (Anderson, 1966, 1971; Cather, 1971), to which we shall now turn.

Presumptive Areas of the Blastula

Yolky or not, the outcome of clitellate cleavage is a blastula of constant basic pattern, with a large 2d cell or its equivalent dorsally, a similar 4d cell or its equivalent behind the 2d cell, the stem cells ventrally and an arc of micromeres dorsally in front of 2d. By the time this stage is reached, the presumptive areas of the embryo are fully segregated from one another in the blastula wall. The divisions which segregate the areas from one another vary markedly in different species, especially when yolk is reduced, but the areas themselves are surprisingly constant in position and general form. They comprise, in the basic arrangement (Fig. 29):

1. a dorsal area of presumptive ectoderm;
2. at the anterior edge of this area, in the midline, an area of presumptive stomodaeum;
3. a posterior area of presumptive mesoderm;
4. a large ventral area of presumptive midgut.

This pattern is clear in *Tubifex*, *Peloscolex*, *Rhychelmis*, *Glossiphonia* and *Theromyzon*. It is also retained (Fig. 30), with minor functional modifications, in the secondarily less yolky blastulae of *Bdellodrilus*, naidids, earthworms, piscicolids, gnathobdellids and pharyngobdellids. In comparison with polychaetes (Fig. 6), the presumptive prototroch and ectomesoderm are absent, the presumptive anterior ectoderm is merged with the presumptive posterior ectoderm and the presumptive stomodaeum is shifted from a ventral to an anterior position, sitting on top of the enlarged presumptive midgut. All of these differences between

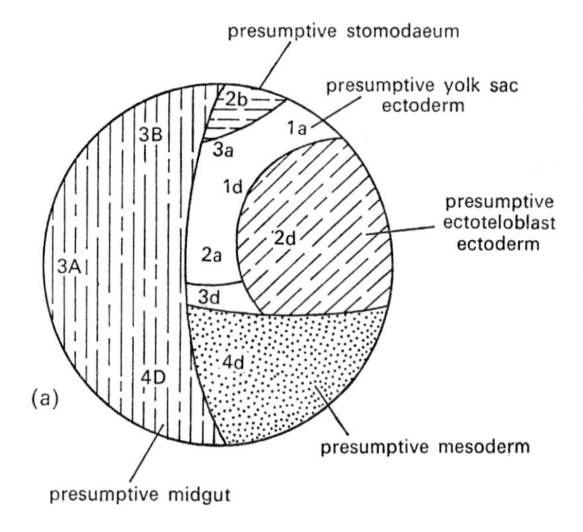

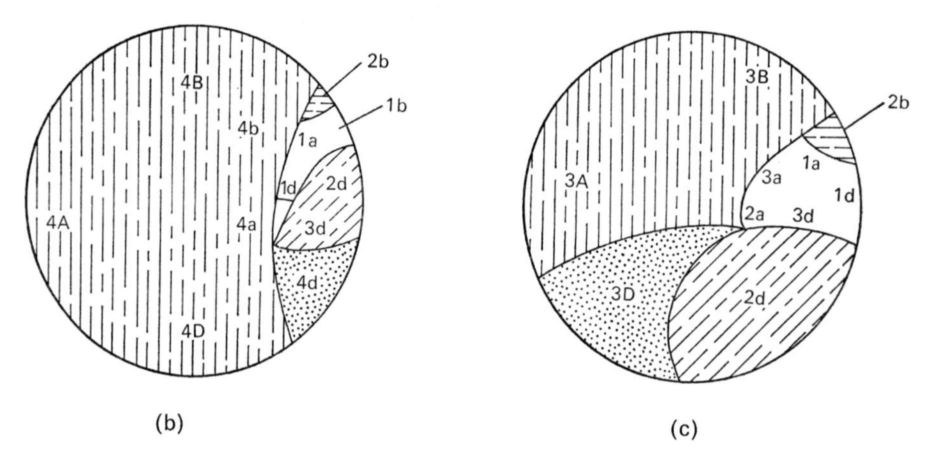

FIG. 29. Fate maps of yolky clitellate blastulae, drawn in left lateral view. (a) *Tubifex*; (b) *Rhynchelmis*; (c) *Glossiphonia* and *Theromyzon* [modified from Anderson, 1966.]

clitellates and polychaetes can be given a functional explanation in terms of the type of egg laid and the mode of development followed. We shall return to this topic at the end of the present chapter, after taking account of the functional variations in presumptive areas within the clitellates and the manner in which the areas undergo their subsequent development.

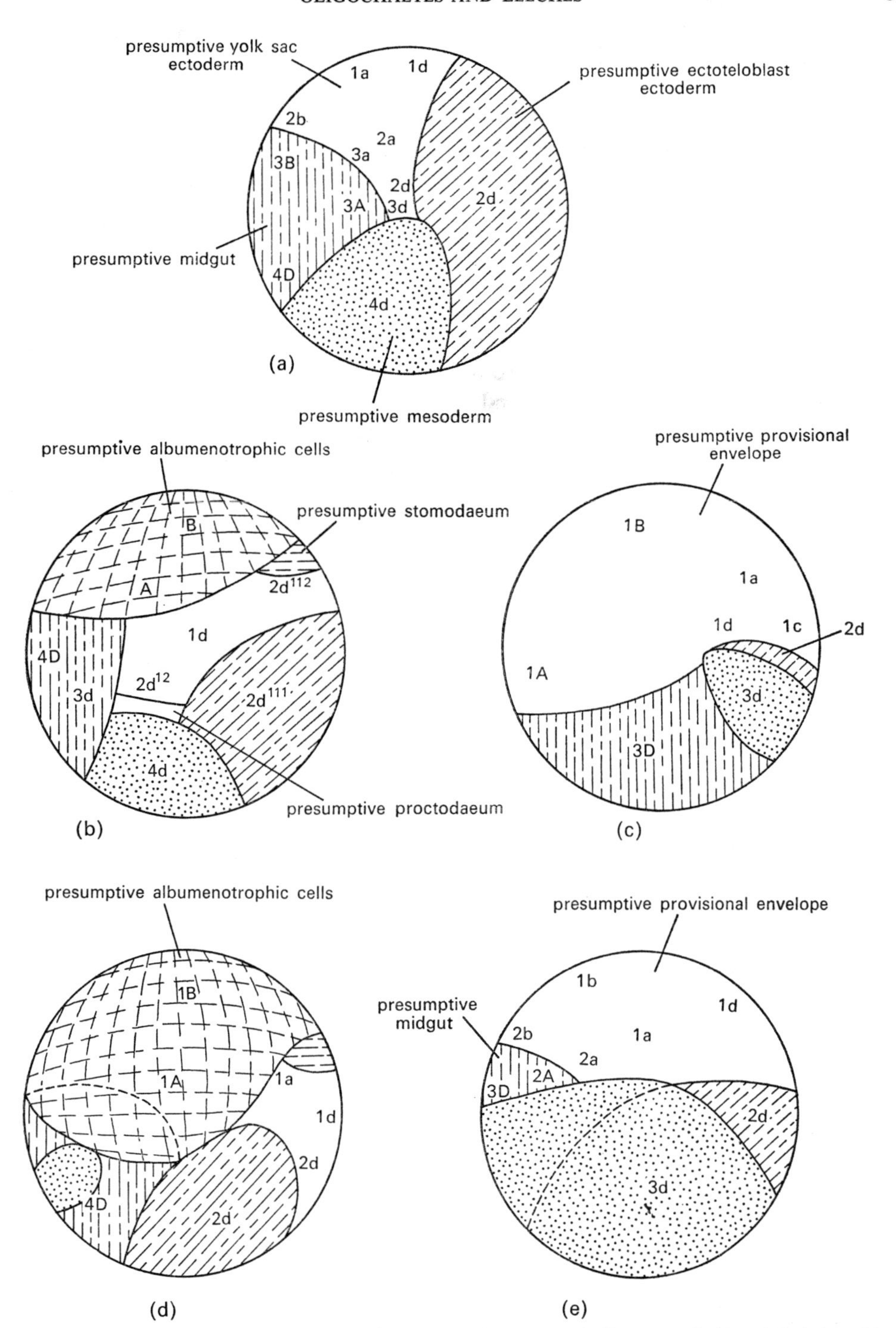

FIG. 30. Fate maps of clitellate blastulae in which yolk is secondarily reduced. (a) *Bdellodrilus*; (b) *Eisenia*; (c) *Stylaria*; (d) *Erpobdella*; (e) *Piscicola*. Drawn in left lateral view. [(a–d) modified from Anderson, 1966, 1971; (e) based on data of Schmidt, 1925b.]

The presumptive mesoderm

The presumptive mesoderm is little modified throughout the clitellates. It always comprises a pair of posterodorsal to posterior M-cells, almost always (gnathobdellid and pharyngobdellid leeches being slightly aberrant in this respect) formed by equal bilateral division of a single mother cell. In the oligochaetes *Tubifex*, *Peloscolex*, *Rhynchelmis*, *Bdellodrilus* and *Eisenia*, as in polychaetes, the mother cell is 4d. In the glossiphoniid leeches *Glossiphonia* and *Theromyzon* it is 3D, the stem cell from which 4d is cut off in oligochaetes at the next division. In the naidid oligochaetes and piscicolid leeches, the presumptive mesoderm is segregated one division earlier, as 3d. Gnathobdellid and pharyngobdellid leeches show direct formation of a pair of M-cells as posteroventral division products of 2D, in a specialized manner.

Whatever the manner in which the clitellate M-cells are segregated, the two cells subsequently become internal and give rise teloblastically to a pair of ventrolateral mesodermal bands. In yolky clitellate embryos, the M-cells are relatively even larger than in yolky polychaete embryos and contribute numerous somites before feeding begins. In albumenotrophic clitellate embryos, the M-cells are smaller but equally productive. There is no good evidence that the clitellate M-cells make any preliminary contribution to the presumptive midgut, as they do in polychaetes, but it has been well substantiated for *Tubifex* that 4d buds off a pair of primordial germ cells before dividing to form the M-cells (see below).

The presumptive midgut

The basic type of presumptive midgut in clitellates, like the presumptive mesoderm, has the same cellular composition as that of polychaetes, namely the yolky stem cells 3A, 3B, 3C and 4D. This mode of segregation is displayed in *Tubifex*, *Peloscolex*, *Rhynchelmis* and *Bdellodrilus*. The cells are disproportionately large except in *Bdellodrilus*, with its secondarily small egg. *Glossiphonia* and *Theromyzon* differ from *Tubifex* only in the exclusion of the D stem cell from the presumptive midgut and its devotion exclusively to presumptive mesoderm as 3D. The presumptive midgut of glossiphoniid leeches is thus anteroventral rather than ventral.

In cases of more extreme modification associated with reduced yolk, the presumptive midgut is more specialized. Convergently in earthworms and in gnatho- and pharyngobdellid leeches, the presumptive midgut comprises a group of small ventral cells containing little yolk. These cells, in the leeches, are all division products of 2D, except for one additional small cell, 2c. The corresponding cells of the earthworm blastula are all D-quadrant cells ($2d^{22}$, 3d and 4D) and the convergent similarity is further emphasized by an associated specialization in parallel in the two types of embryo. The three anterior stem cells, 1A, 1B and 2B, of gnathobdellids and pharyngobdellids, and the three equivalent anterior cells A, B and C of earthworms, no longer form part of the presumptive midgut. Their fate is curious and specialized and can conveniently be dealt with at this point. Throughout cleavage, the three cells persist, making up most of the anterior half of the embryo. During gastrulation, the three cells are overgrown by other cells and come to lie internally (Figs. 33 and 38), where they eventually degenerate. Although the function of these cells has never been established, it is known for earthworms that they develop a complex series of intracellular canals and show pulsatile activity. It seems likely (Anderson, 1966) that the three cells act as precocious absorbers and assimilators of albumen until the embryonic gut cells assume this function. The three cells can thus be interpreted as specialized derivatives of the presumptive midgut, with a new, temporary, albumenotrophic function.

The A, B and C stem cells of naidids and piscicolids are also largely or entirely specialized for a temporary albumenotrophic function and are excluded from the definitive presumptive midgut, but the nature of their specialization is different and is best discussed in connection with the presumptive ectoderm (see below). It is sufficient to notice at this point that the small presumptive midgut of the piscicolids is identifiable as the small cells 2A, 2B, 2C and 3D, while the presumptive midgut of the naidids is reduced to two relatively large cells, 1C and 3D in *Stylaria*, 2d and 3D in *Chaetogaster*. Both specializations are clearly derived from the more basic condition of four yolky cells.

For clitellates in general, therefore, the basic presumptive midgut is identical with that of polychaetes, save that the four cells are proportionately larger and contain more yolk. When yolk is secondarily reduced, the presumptive midgut rudiment is smaller and some of the ancestral midgut cells have assumed a new function as temporary albumenotrophic cells.

The presumptive stomodaeum

Although, in association with the relative enlargement of the basic presumptive midgut, the presumptive stomodaeum of clitellates is displaced to an anterodorsal location (Figs. 29 and 30), the cells comprising this area are essentially the same cells as in polychaetes. In yolky embryos, they comprise a transverse arc of small cells descended mainly from 2b and 3b, e.g. *Tubifex*, *Rhynchelmis*, *Glossiphonia*. In secondarily less yolky embryos with modified cleavage, the lineage of the presumptive stomodaeal cells is different, e. g. 1a and 1b in *Erpobdella*, or cannot be traced, but the cells themselves always occupy the same relative position and have the same fate.

The presumptive ectoderm

The presumptive ectoderm of the clitellates is the most highly modified presumptive area as compared with those of polychaetes, but all of its basic modifications are related to two functional innovations, loss of the prototroch and increase in bulk of the presumptive midgut. The components of the presumptive ectoderm common to all yolky clitellate embryos (Fig. 29) are a large dorsal cell descended from 2d (exceptionally in *Rhychelmis*, augmented by a second large cell, 3d) and an anterior group of micromeres in front of the large cell. The micromeres are normally segregated as 1a–1d, 2a and 2c, 3a and 3d, and one or two minor products of 2d. These, together with the large 2d cell, are the same cells that form the presumptive anterior and posterior ectoderm, the prototroch and the presumptive ectomesoderm of polychaetes. Allowing for the fact that clitellates have no prototroch or ectomesoderm, the identity of formation of the presumptive ectoderm in the two groups is plain. As we have already seen, the same presumptive area in related animals can be established through a variety of cleavage sequences. Furthermore, the large 2d cell of yolky clitellate embyros (2d and 3d in *Rhynchelmis*) corresponds to the large division products of 2d in yolky polychaete embryos as the source of the ectoteloblasts of the trunk segments (see below). The essential modifications of clitellate presumptive ectoderm are twofold. Firstly, the large ectoteloblast mother cell derived from 2d is not only the source of the trunk segment ectoderm, but also the source of the prostomial and peristomial ectoderm. In the absence of a prototroch, there is no separate formation during cleavage of an anterior, prostomium-forming, presumptive ectoderm. Secondly, the remainder of the presumptive ectoderm no longer develops directly as definitive ectoderm, but passes first through a phase of proliferation and spread as a temporary, yolk-sac ectoderm which rapidly encloses the grossly enlarged, yolky presumptive midgut before reverting to its fate as definitive ectoderm. The first expression of these modifications is the formation of the presumptive ectoderm as a large dorsal cell (the ectoteloblast mother cell) surrounded anteriorly and laterally by a simple arc of micromeres (the temporary yolk-sac ectoderm). The complex of sub-areas seen in the polychaete presumptive ectoderm has been eliminated.

With secondary loss of yolk, these basic modifications are further exaggerated (Fig. 30). The large ectoteloblast mother cell is retained as a product of 2d in *Bdellodrilus*, *Stylaria*, earthworms, gnatho- and pharyngobdellid leeches and piscicolid leeches, and as 1d (i.e. cut off one division earlier) in *Chaetogaster*; but the formation and fate of the micromeres is variously modified. In *Bdellodrilus*, the micromere cap is formed in the same manner as in *Tubifex* but, as will be seen below, its spread during gastrulation to form an external, temporary yolk-sac ectoderm is rapid and precocious, in spite of the fact that the embryo has reduced yolk. We can infer from this the introduction of a new albumenotrophic function for these cells, balancing the reduction of yolk. In naidids and piscicolids, convergently, the same adaptation is even more precocious and specialized. In naidids, not only the micromere cap (1a–1d in *Stylaria*, 1a–1c in *Chaetogaster*), but also the anterior stem cells (1A and 1B in *Stylaria*, 1A–1C in *Chaetogaster*) are now part of the presumptive ectoderm. Similarly, in piscicolids, not only the micromere cap, 1a–1d, but also the large descendants 2a–2c of the stem cells 1A–1C form part of the presumptive ectoderm. Proliferation and spread of these cells begins at a very early stage of cleavage, forming an enclosing provisional envelope which rapidly covers the other presumptive rudiments and then persists until the end of development (Fig. 25). The provisional envelope makes no contribution to the definitive ectoderm of the embryo and is obviously albumenotrophic. Equally obviously, the provisional envelope is a specialization of the temporary, yolk-sac ectoderm of yolky clitellate embryos. In contrast, in the earthworms and the gnatho- and pharyngobdellid leeches, which have convergently evolved the three, large, anterior stem cells as specialized, albumenotrophic cells deriving from the presumptive midgut, the micromere cap is reduced and does not undergo such precocious proliferation, though it still participates in the formation of a provisional ectoderm as development proceeds.

Two generalizations can usefully be made about the presumptive areas of the blastula in clitellate annelids. Firstly, there is a basic configuration of these areas, common to all species of oligochaetes and leeches with large yolky eggs. When the egg is small, the same areas are functionally modified in association with secondary yolk loss and albumenotrophic nutrition of the embryo. Secondly, the presumptive areas of the yolky clitellate blastula are a functional modification of those of polychaetes, displaying an integrated combination of adaptations to increased yolk and prolonged lecithotrophic development in the cocoon. We can now follow out the fates of these areas in yolky clitellate embryos, with passing reference to attendant modifications associated with the secondary reduction of yolk.

Gastrulation

In view of the fact that the two presumptive areas whose gastrulation movements are the most conspicuous in polychaete embryos, namely, the presumptive ectoderm and presumptive midgut, are the most modified areas in clitellates, it is not surprising that clitellate gastrulation is also highly specialized. A clue to the nature of this specialization is provided by the functional modification of gastrulation displayed in the polychaete *Neanthes*. As we have seen in the previous chapter (Figs. 7 and 9), this species has four relatively large and densely yolky presumptive midgut cells, which become internal partly as ectoderm cells are cut off from them anteriorly and dorsally during cleavage, and partly as the ectoderm cells then spread posteriorly and ventrally over the surface. The yolky midgut cells undergo no autonomous gastrulation movements. In clitellate gastrulation, except when secondary yolk reduction has occurred, the same functional condition is displayed in a more exaggerated form. The presumptive midgut cells are even larger, so that very little coverage of their surface results from cleavage, but without undergoing any autonomous gastrulation movement, the cells become covered as a result of ectodermal spread. This process is prolonged and overlaps the onset of proliferation of the trunk segments.

With few exceptions, the workers who have described clitellate gastrulation are not those who were successful in the study of cleavage. The classical workers on *Rhynchelmis* (Kowalevsky, 1871; Vedjovsky, 1888–92) and *Glossiphonia* (Whitman, 1878, 1887; Nusbaum, 1886; Bürger, 1902), even though they failed to interpret cleavage satisfactorily, described gastrulation well. Schmidt (1917), Bychowsky (1921) and Penners (1924) added further details on the gastrulation of yolky clitellate embryos and their work was later augmented in a most useful way by Iwanoff (1928) and Meyer (1929). The results of these studies have been summarized by Anderson (1966, 1971).

As mentioned above, the midgut rudiment of *Tubifex*, *Rhynchelmis*, *Glossiphonia* and *Theromyzon* becomes internal as a result of overgrowth by other cells (Fig. 31). The midgut cells begin to undergo equal divisions while still exposed at the surface, yielding a solid mass of yolky, polygonal cells. The presumptive ectoderm gradually spreads over this mass eventually meeting in the ventral midline, but complete enclosure is not achieved until most, of the segments of the embryo have already been formed.

In contrast, the M-cells or presumptive mesoderm are overgrown by the ectoderm quite rapidly and retain some intrinsic gastrulation movement into the interior which aids this process. As in yolky polychaetes, the teloblastic activity of the M-cells begins while they are still at the surface of the embryo.

The presumptive stomodaeum does not exhibit a gastrulation movement. Instead, it penetrates the interior by ingrowth as a solid mass of cells (Fig. 34), and the formation of a stomodaeal lumen and mouth is much delayed.

Obviously, spread of the presumptive ectoderm is the major gastrulation process in yolky clitellate embryos. The cells of the micromere cap proliferate and gradually extend as a

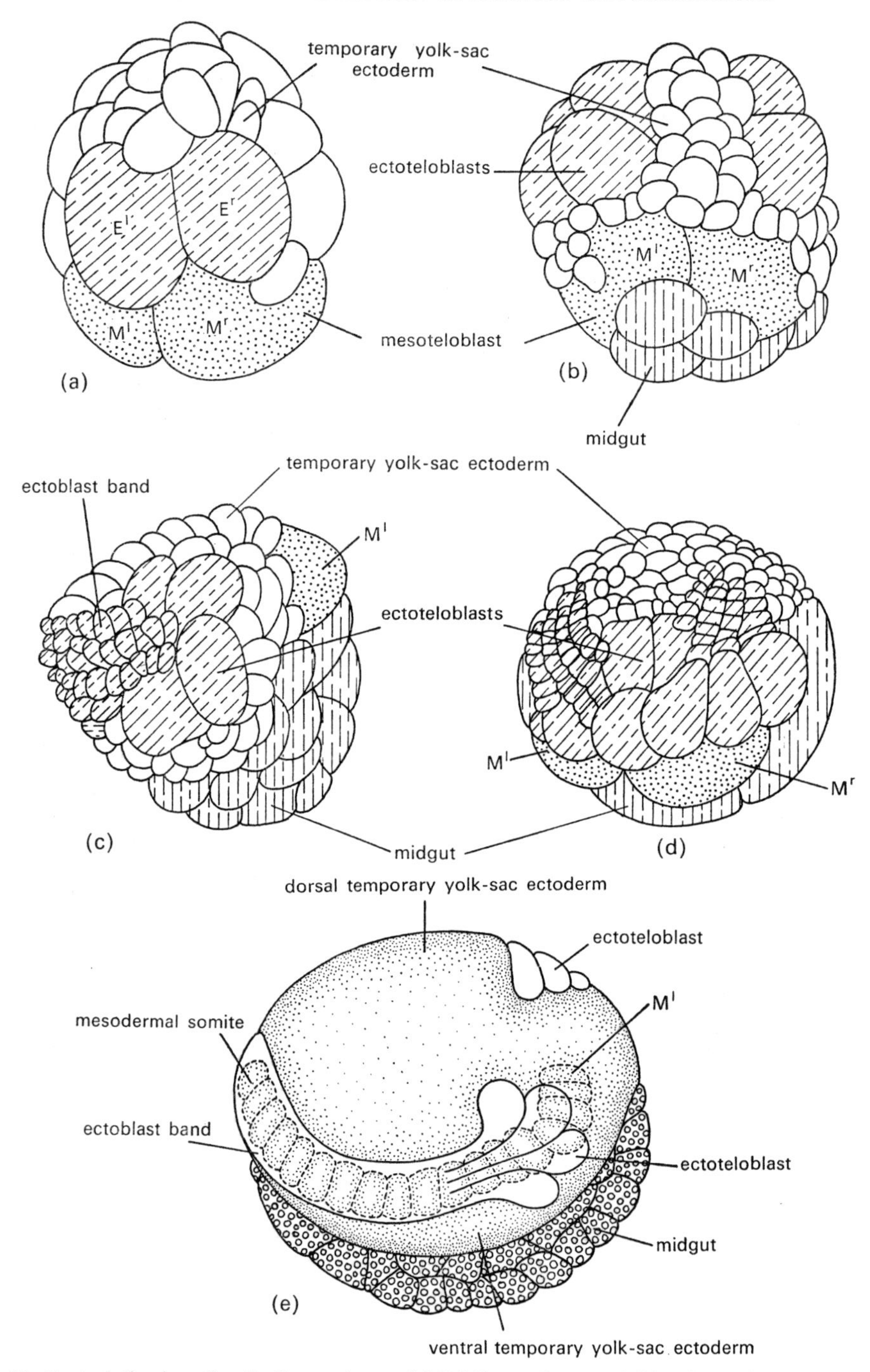

FIG. 31. Gastrulation in yolky clitellate embryos. (a) *Tubifex*, early ectoteloblast formation, posterodorsal view; (b) *Tubifex*, completion of ectoteloblast formation, posterodorsal view; (c) *Tubifex*, early development of ectoblast bands, left lateral view; (d) *Glossiphonia*, early development of ectoblast bands, posterodorsal view; (e) *Tubifex*, gastrulating embryo with 17 pairs of somites, left lateral view. [(a–c) after Penners, 1924; (d) after Schleip, 1914; (e) after Meyer, 1929; modified from Anderson, 1966, 1971.]

sheet over the surface of the yolky midgut rudiment, firstly in all directions, later mainly as two lateral sheets which spread towards the ventral midline (Fig. 31). Ectoteloblast cells are also formed and begin to proliferate during this time (see below), so that part of the spread of ectoderm at the surface is due to the formation of segmental ectoderm as ectoblast bands, but the major factor in enclosure is proliferation and spread of the micromere cap as flattened, temporary yolk-sac ectoderm.

The modifications of gastrulation associated with the secondary reduction of yolk in clitellates have been less well documented, since they are difficult to interpret unless one appreciates the significance of presumptive areas; but pertinent observations are to be found in the work of Bergh (1891), Bürger (1891, 1894), Sukatschoff (1903) and Schoumkine (1953) on gnathobdellid and pharyngobdellid leeches, Wilson (1889), Vedjovsky (1888–92), Hoffman (1899), Swetloff (1928) and Devries (1971) on earthworms, Schmidt (1921, 1924, 1925b, 1930, 1939, 1941, 1944) on piscicolids, Tannreuther (1915) on *Bdellodrilus* and Swetloff (1926) and Dawydoff (1942) on naidids. The last three of these papers are discussed in detail by Anderson (1971).

Some of the modifications displayed have already been hinted at in discussing the clitellate presumptive ectoderm. In *Bdellodrilus*, with a reduced presumptive midgut as the only peculiarity, overgrowth by the specialized, temporary yolk-sac ectoderm proceeds rapidly and is completed when proliferation by the ectoteloblasts and mesoteloblasts has only just begun (Fig. 32). Associatedly, it seems probable that the temporary yolk-sac ectoderm at first has an albumenotrophic function, though it later participates in the formation of the definitive ectoderm. In the naidids and also in the piscicolids, the same modification is carried to an extreme in which gastrulation is wholly eliminated (Figs. 25 and 33). The overgrowth by temporary yolk-sac ectoderm is even more precocious and rapid, but can no longer be regarded as a part of gastrulation, since the resulting provisional envelope does not contribute structurally to the embryonic ectoderm. The presumptive midgut, M-cells and ectoteloblast cells attain their definitive positions directly as a result of cleavage and proceed directly into organogenetic activity.

In the earthworms and the gnathobdellid and pharyngobdellid leeches, in contrast, new patterns of gastrulation movements have evolved. That of earthworms has interesting analogies with gastrulation in polychaetes. The basic event of gastrulation in the small hollow blastula is an invagination of the presumptive midgut, albumenotrophic cells and stomodaeum into the interior (Fig. 33). Accompanying this invagination, the M-cells slip into the interior at the posterior end of the embryo and the temporary yolk-sac ectoderm begins to spread towards the ventral midline. Closure of the invagination opening proceeds from behind forwards, the anterior end of the opening persisting as the mouth. At the same time, however, the ectoteloblasts at the posterior end of the embryo and the M-cells which lie internal to them begin to proliferate typical clitellate germ bands (see below). This, together with the evidence of cleavage discussed in the previous section, shows that invaginate gastrulation in the earthworms is a secondary phenomenon, derived from a prior condition of yolky clitellate gastrulation. The specializations of gastrulation in the gnathobdellid and pharyngobdellid leeches are different. Here, as in the piscicolids and naidids, rapid overgrowth by the temporary yolk-sac ectoderm is again the major gastrulation movement, though the process is assisted to some extent by an inward migration of the presump-

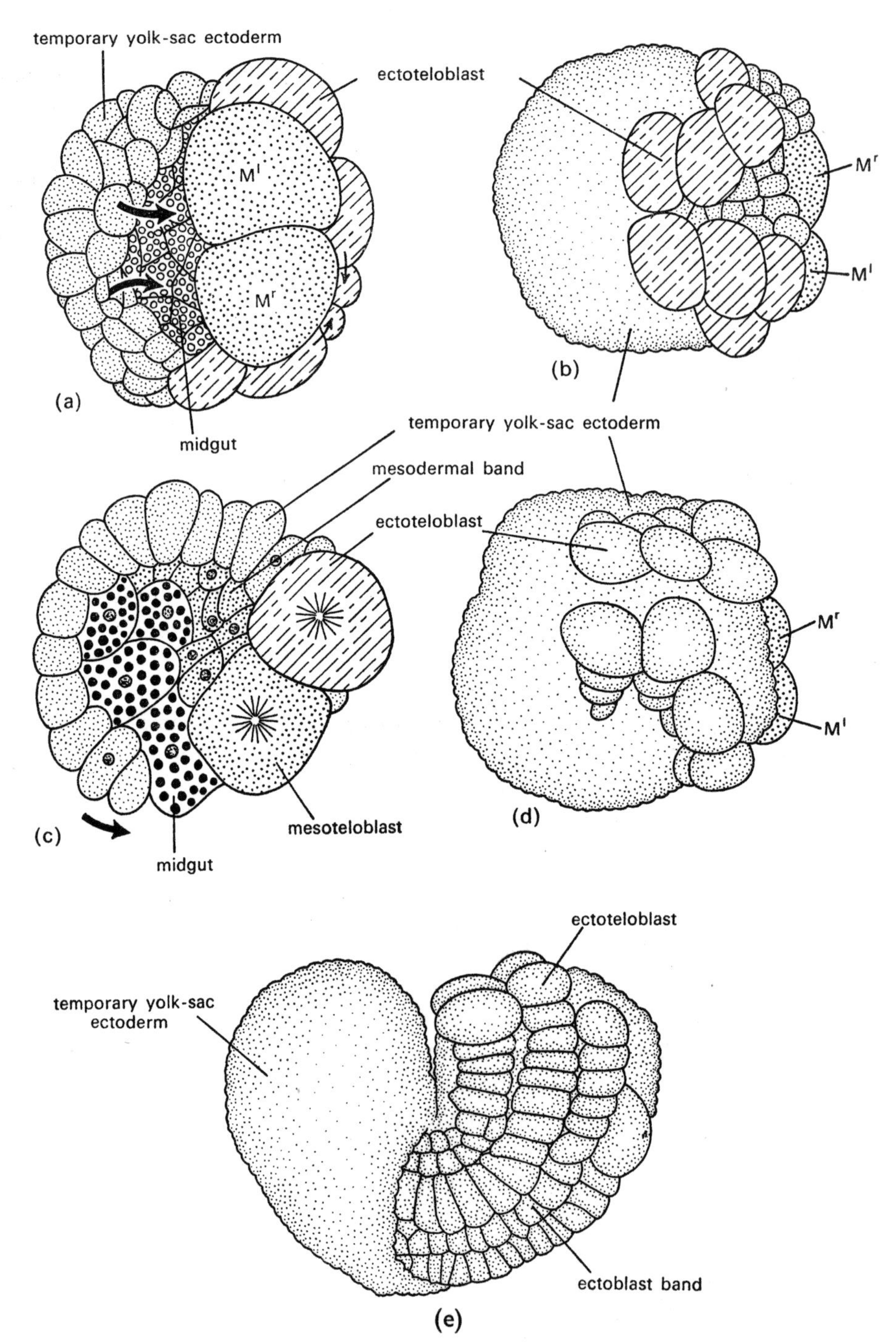

FIG. 32. Gastrulation and development of the ectoblast bands in *Bdellodrilus*, modified from Anderson (1971) after Tannreuther (1915). (a) Gastrulation, ventral view; (b) completion of ectoteloblast formation, dorsal view; (c) sagittal section through gastrulating embryo; (d) early formation of ectoblast bands, dorsal view; (e) later development of ectoblast bands, lateral view.

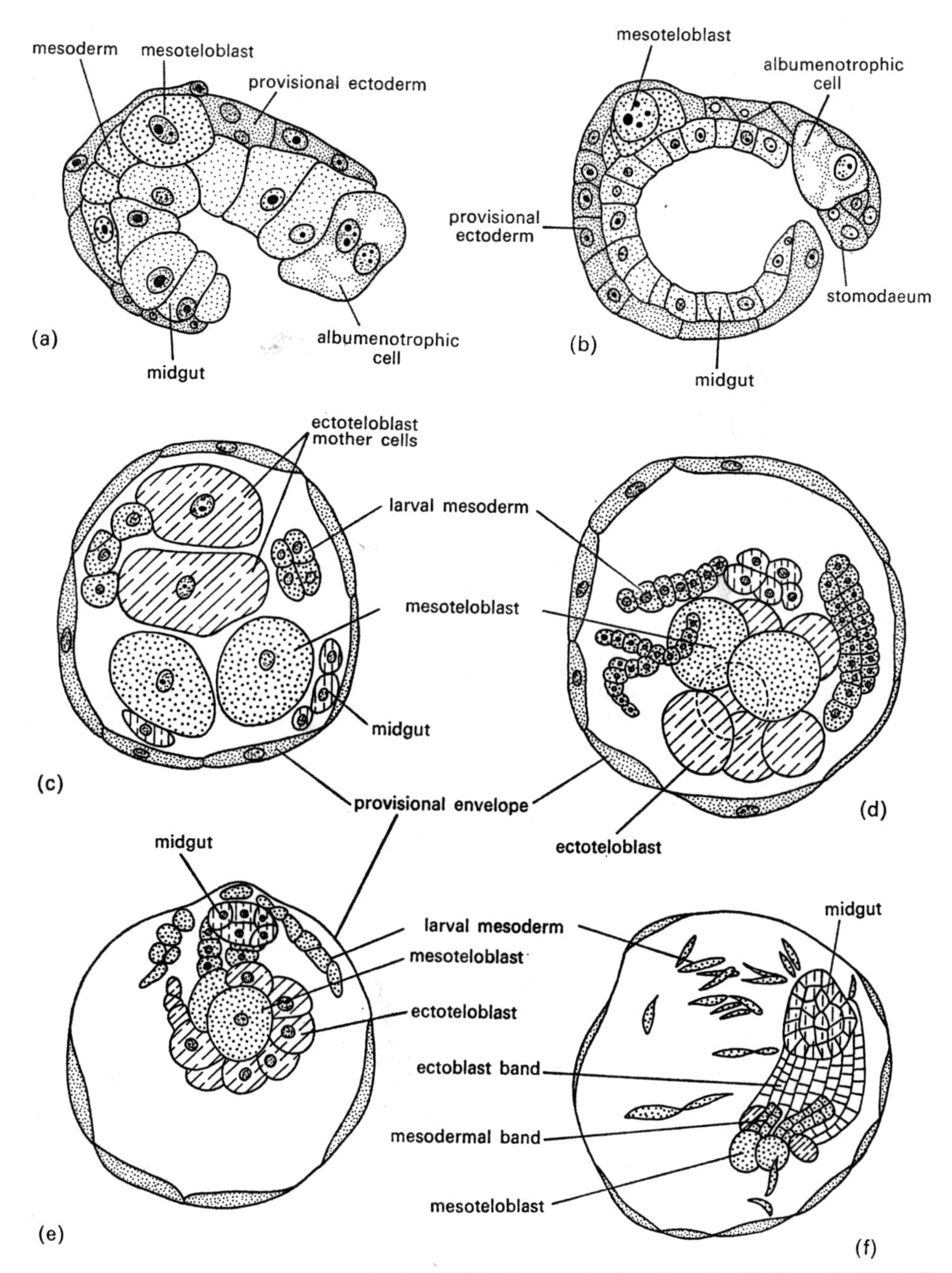

FIG. 33. (a and b) Stages of gastrulation in *Allolobophora*, sagittal section after Hoffman (1899); (c–f) stages during the formation of the ectoblast and mesoblast bands of *Piscicola*, after Schmidt (1925b).

tive midgut and mesoderm cells. The overgrowth process effects enclosure of the ectoteloblast mother cell and also of the three anterior, albumenotrophic cells 1A, 1B and 2B.

In sum, there seems to be no doubt that the basic mode of clitellate gastrulation is exemplified by such species as *Tubifex* and *Glossiphonia* and that all gastrulation processes in small-egged clitellates are highly specialized. We shall see below that this conclusion is borne out by the further development of the clitellate midgut, mesoderm and ectoderm.

Further Development of the Gut

The difficulties presented by the further development of the large, yolky midgut of primitive clitellate embryos were not surmounted by the classical workers of the late nineteenth century, but considerable success was obtained with the later stages of earthworm development by Vedjovsky (1888–92), Wilson (1889), Bergh (1890), Beddard (1892) and Hoffman (1899) and with gnathobdellid and pharyngobdellid leeches by Bergh (1885, 1890, 1891), Nusbaum (1884, 1885, 1886), Apáthy (1891), Bürger (1891, 1894, 1902), Filatoff (1898) and Sukatschoff (1903). The further development of the gut of yolky embryos was established in a later era, by Penners (1924, 1929, 1934) for tubificids, Schmidt (1922) and Iwanoff (1928) for *Rhynchelmis* and Schmidt (1917, 1925a) and Bychowsky (1921) for glossiphoniids. The development of the gut of branchiobdellids and naidid oligochaetes is still not well understood (Tannreuther, 1915; Dawydoff, 1942; Anderson, 1971) but that of the piscicolid leeches was treated in some detail by Schmidt (1921, 1924, 1925b, 1930, 1936). The functional differentiation of the midgut and that of the stomodaeum are closely linked in all clitellates.

In yolky clitellate embryos, the presumptive stomodaeum begins to bud off a mass of cells which pushes inwards against the yolky midgut rudiment (Fig. 34). This part of the stomodaeum eventually grows back through the first few trunk segments, develops a lumen and differentiates as the lining epithelium of the pharynx. Meanwhile, the stomodaeal cells that remain at the surface of the embryo thicken and invaginate as a plate which subsequently forms the lining of the buccal cavity. Continuity between the buccal and pharyngeal lumina is established later by a breakthrough at the base of the buccal invagination.

Albumenotrophic clitellate embryos of all types, in contrast, develop a precociously functional embryonic pharynx lined by cilia (Figs. 34, 38). The earthworm provisional pharynx is formed during gastrulation, while that of naidids, gnatho- and pharyngobdellid leeches and piscicolid leeches arises as an independent invagination after the completion of the provisional ectoderm at the surface of the embryo. The provisional pharynx acts in the ingestion of the ambient albumen and is later transformed into or replaced by the definitive pharynx. Replacement is observed in earthworms and in gnathobdellid and pharyngobdellid leeches. The source of the adult pharyngeal cells of the latter is not clear, but the metamorphosis of the earthworm pharynx proceeds in a manner analogous to that of certain planktotrophic polychaete larvae (see Chapter 2). While the lining cells of the embryonic pharynx become ciliated and provisionally functional, several layers of small cells develop external to them. Late in development, the ciliated lining cells are shed and the outer layers of cells give rise to the definitive pharynx.

In typical contrast, the further development of the midgut is more complicated in the

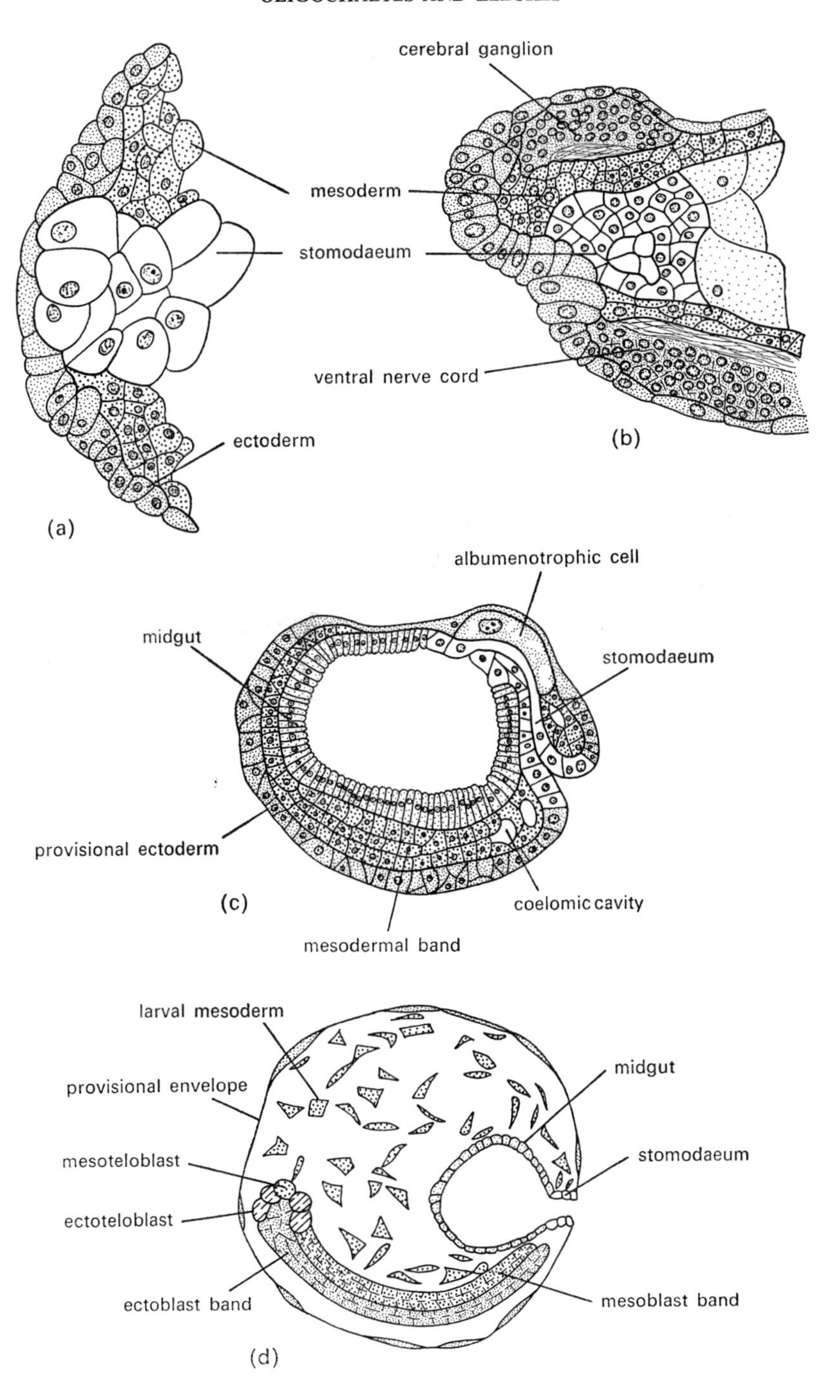

FIG. 34. (a) *Tubifex*, frontal section showing early ingrowth of the stomodaeum; (b) *Tubifex*, sagittal section through the anterior end of a stage similar to 35c; (c) *Allolobophora*, sagittal section through a young post-gastrula; (d) *Piscicola*, sagittal section through a young albumenotrophic embryo. [(a and b) after Penners 1924; modified from Anderson, 1971; (c) after Hoffman, 1899; (d) after Schmidt, 1925b.]

large, yolky embryos of tubificids, lumbriculids and glossiphoniids than in small, yolkless clitellate embryos. In *Tubifex*, for example, although the large mass of yolky cells filling the interior of the embryo (Fig. 36) develops more or less directly into a midgut epithelium, there is some evidence of vitellophage specilization. Basically, a central split develops in the cell mass, around which the cells gradually become arranged as an epithelial layer. Small cells at the surface of the cell mass act as vitellophages and are later resorbed without becoming part of the midgut epithelium. The same vitellophage specialization is more strongly manifested in *Rhynchelmis*. Initially, the yolky midgut cells fuse to form a syncitium. A number of yolk-free cells persists at the outer surface of the syncitium, acting as vitellophages which are later resorbed. Other nuclei within the syncitium collect around a small space which arises near the anterior end and give rise to a small epithelial vesicle enclosing the space. As the yolk is resorbed, the epithelial vesicle expands to form the lining epithelium of the midgut. Analogous specializations in which the definitive midgut cells cluster at the centre of the yolk mass instead of enclosing it as a peripheral layer are seen in the polychaete *Capitella* (Chapter 2), in many myriapods (Chapter 5) and in some Crustacea (Chapter 8). The glossiphoniid midgut shows an intermediate level of specialization. The yolky cell mass becomes syncitial, as in *Rhynchelmis*, but the nuclei proliferate and move to the periphery of the syncitium. Cellularization at the periphery then follows and the midgut lumen develops as a central split, as in *Tubifex*.

The midgut rudiment of secondarily yolkless clitellate embryos develops in a variety of specialized ways. In *Bdellodrilus*, simplicity is the keynote. The small midgut cells (Fig. 32) mass together in the interior of the embryo after gastrulation and gain a central lumen. The peripheral cells of the mass become arranged as an epithelium, while the more central cells around the lumen degenerate and are gradually resorbed. The midgut sac now becomes connected with the provisional pharynx and fulfils a temporary role as an albumenotrophic sac before becoming the definitive midgut. Piscicolid leeches display the same pattern of development of the midgut except that there is no degeneration of cells. The midgut sac is initially small (Figs. 33 and 34) in relation to the volume of the provisional albumenotrophic within which the piscicolid embryo develops, but soon enlarges to line the inner surface of the envelope (Fig. 38). With elongation of the embryo and formation of the definitive body wall, the midgut sac transforms directly into the definitive midgut epithelium.

In naidids, the development of the midgut retains more of the pattern observed in yolky clitellate embryos. A mass of midgut cells is formed within the provisional albumenotrophic envelope (Fig. 25). A peripheral cell layer is then segregated at the surface of the mass and seems likely to have a vitellophage role. Within this layer, the central mass of midgut cells hollows out to form an epithelial sac which becomes connected with the provisional pharynx. Presumably, the precocious midgut sac acts temporarily in feeding on the ambient albumen. Later, the epithelial walls of the sac merge to form a syncitial mass in which the nuclei gradually migrate to the periphery and become the focus for differentiation of the definitive midgut epithelium. The central part of the syncitial mass is gradually resorbed. The naidids appear to display a functional condition of midgut development intermediate between that of yolky clitellates and that in which the entire midgut rudiment is specialized as a provisional midgut sac.

The latter specialization, already discussed for *Bdellodrilus* and piscicolids, is also found in the earthworms and the gnathobdellid and pharyngobdellid leeches. Both of these groups, as we have seen in previous sections of the present chapter, have three large albumenotrophic cells which appear to play an absorptive role during cleavage, become internal during gastrulation (Figs. 33, 34 and 38) and are gradually resorbed. The albumenotrophic role is taken over after gastrulation by the midgut itself. The midgut sac formed in earthworms as a result of invaginate gastrulation is already connected with the provisional pharynx (Fig. 34). As development continues, the midgut sac elongates and later transforms directly into the lining epithelium of the oesophagus, crop, gizzard and intestine. Midgut development in the gnatho- and pharyngobdellid leeches takes a slightly different course. Initially the provisional pharynx opens directly into the blastocoelic body cavity of the embryo and the midgut cells are scattered large cells, presumably with an albumenotrophic function (Fig. 38). The scattered cells then fuse to form a syncitium, which hollows out as a syncitial midgut epithelium connected to the provisional pharynx. Cellularization of the syncitial epithelium as definitive midgut epithelium takes place a few days after hatching. In general, there seems little doubt that the development of a yolky midgut rudiment represents a primary condition in clitellates and that the functional variations observed in albumenotrophic embryos are secondary.

The clitellate proctodaeum is only a minor component of the functional gut, developing at a late stage of embryonic development as a short invagination on the pygidium (Fig. 36). In all species in which proctodaeal development has been observed, the cells which give rise to the proctodaeum, like the pygidial cells which surround them, have a prior temporary role as part of the temporary yolk-sac ectoderm. The formation of the proctodaeum in naidid and piscicolid embryos, in which the provisional ectoderm does not contribute to the definitive ectoderm, needs further study.

Further Development of the Mesoderm

The definitive mesoderm of clitellate embryos originates as a pair of mesodermal bands formed through the proliferative activity of the paired M-cells (Fig. 31). The precociously functional, albumenotrophic embryos of earthworms, naidids and non-glossiphoniid leeches develop a variety of temporary mesodermal organs, the source of origin of which is usually not known, but none of these structures persists into definitive mesodermal organogeny. In yolky species, the proliferation of the mesodermal bands begins before the M-cells become internal (e.g. Fig. 32) and continues during and after the prolonged gastrulation enclosure of the large midgut rudiment. The early growth of the bands follows a paired dorsolateral pathway, but the bands are gradually displaced to a ventrolateral position as gastrulation overgrowth of the midgut rudiment proceeds (Figs. 31 and 35). In albumenotrophic species, the onset of proliferation of the mesodermal bands is usually delayed until the M-cells have become internal and the bands then form by forward proliferation directly in a ventrolateral position (Figs. 34 and 38).

Somite formation and somite development in yolky clitellate embryos were studied first by Kowalevsky (1871), using the oligochaete *Rhynchelmis*. Vedjovsky (1888–92) and Iwanoff (1928) made further investigations on *Rhynchelmis*, but the most informative studies are those of Penners (1924) and Meyer (1929) on *Tubifex*. Mesodermal development in glossi-

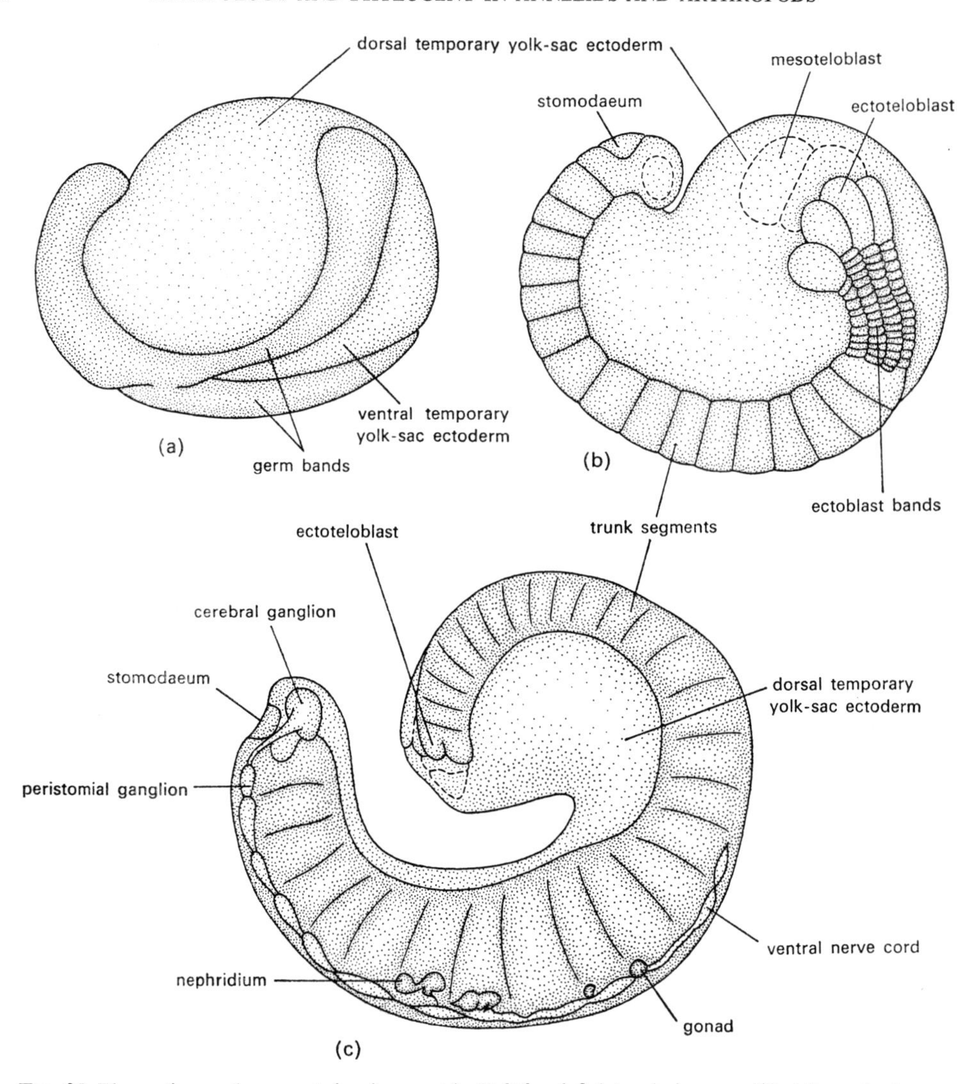

FIG. 35. Elongation and segment development in *Tubifex*, left lateral view, modified from Anderson (1971) after Penners (1924) and Meyer (1929). (a) Diagrammatic view showing midventral fusion of germ bands; (b) stage with 24 pairs of somites; (c) stage with 36 pairs of somites.

phoniid leeches is less well known, the most useful accounts still being those of Whitman (1878, 1887) and Bürger (1891, 1902). Among albumenotrophic embryos, the development of the mesoderm of earthworm embryos was investigated in some detail by Kleinenberg (1879), Wilson (1889), Bergh (1890), Vedjovsky (1888–92), Staff (1910), Bahl (1922), Iwanoff (1928) and Swetloff (1928), though it still awaits a detailed modern investigation. Fragments of information are also to be found in the work of Tannreuther (1915) on *Bdellodrilus* and Dawydoff (1942) on naidids and in the several papers by Schmidt on piscicolids. The details of mesoderm development in the tubificid, lumbriculid, bdellodrilid and naidid oligochaetes were summarized by Anderson (1971).

In *Tubifex* and *Rhynchelmis*, each somite originates as a single large cell cut off from the

M-cell of its side. The cell first undergoes a series of equal divisions, forming a group of cells which then becomes hollow by internal splitting. A similar mode of somite formation is evident in glossiphoniid leeches. The somites develop in strict anteroposterior succession and gradually extend upwards to enclose the yolky midgut. The walls of the somites develop in the usual way as somatic, splanchnic and septal mesoderm.

A similar separation of somite blocks which become hollow in anteroposterior succession is found in naidid embryos. In other clitellates, somite formation has been less carefully studied, but the present knowledge of earthworm and bdellodrilid development indicates that the first sign of somite delineation is the formation of paired coelomic cavities in already well-developed mesoderm bands (Figs. 34 and 36). The cells around each cavity then individuate as the walls of a distinct somite. In spite of this difference, which is the reverse of the polychaete condition (Chapter 2), the strict anteroposterior succession of somite formation appears to be retained in all clitellates.

Iwanoff (1928) attempted to demonstrate for *Rhynchelmis* and, especially, for the lumbricid *Eisenia*, that the anterior segments of the trunk each contain a pair of vestigial larval somites developed from ectomesoderm, as well as the normal pair of somites developed from the mesodermal bands. The evidence for this interpretation cannot now be seriously entertained. It is probable that renewed investigation would reveal some peculiarities in the development of the mesoderm of the anterior end of the oligochaete trunk, associated with specializations of the gut, blood system and musculature of this region, but it seems unlikely that these anterior segments are fundamentally different from those behind than in the heteronomous sense implied by Iwanoff.

As the paired somites enlarge and extend towards the dorsal midline, their somatic walls exhibit a differentiation of myoblasts towards the external surface (Fig. 36). The muscle fibrils of the longitudinal muscles of the body wall then develop within the myoblasts, first ventrolaterally, then dorsolaterally (Kowalevsky, 1871; Kleinenberg, 1879; Wilson, 1889; Bergh, 1890, 1891; Bürger, 1891, 1902; Vedjovsky, 1892; Staff, 1910; Penners, 1924; Swetloff, 1928). When the chaetal sacs develop as ectodermal invaginations, some of the somatic myoblasts cluster around them and differentiate as chaetal sac musculature. The origin and development of the circular muscle of the clitellate body wall is unresolved. We have already seen in Chapter 2 that a resolution of the same problem in polychaetes, assigning the circular muscles to a somatic mesodermal origin, has only recently been attained. In clitellates, the circular muscles arise from cells lying laterally between cells definitely assignable to the somatic wall of the somite and cells definitely descended from the ectoderm (Fig. 34). Most authors (e.g. Vedjovsky, 1888–92; Bergh, 1891; Bürger, 1902; Bychowsky, 1921; Penners, 1922, 1924; Meyer, 1929) have interpreted these cells as derivatives of the ectoderm, but their evidence for this is inconclusive. Early interpretations of the circular muscle cells as derivatives of the somatic mesoderm are even more poorly documented. The recent results for polychaetes argues in favour of the latter view, but a final solution awaits fresh evidence. If judgement is ultimately passed in favour of an ectodermal origin of circular muscle in clitellates, the fate map of the clitellate blastula will require some modification.

The development of blood vessels in oligochaete embryos proceeds in the same general manner as in polychaetes (Hanson, 1949). The leeches present additional complications in

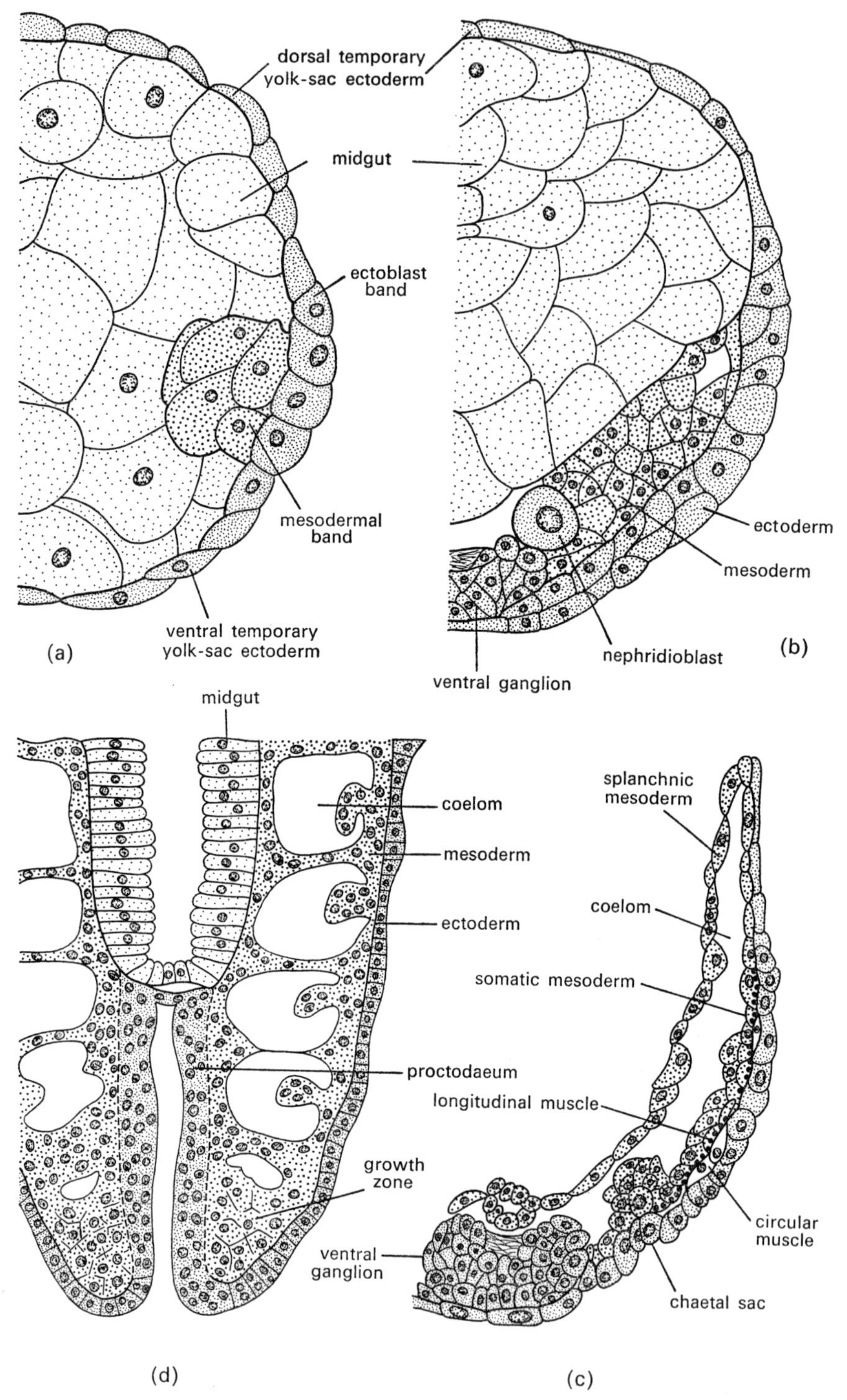

FIG. 36. (a–c) *Tubifex*, transverse sections showing somite development; (a) is through the middle region of a stage similar to Fig. 31e; (b) is through the middle region of a stage similar to Fig. 35b; (c) is through a stage similar to Fig. 35c; (d) *Allolobophora*, longitudinal section through the posterior end showing proctodaeum formation and segment proliferation. [(a–c) after Penners, 1924; modified from Anderson, 1971; (d) after Hoffman, 1899.]

the later incorporation of parts of their coelomic cavities into the blood system (see below).

The main ventral blood vessel of *Tubifex* is formed by separation of the apposed walls of the ventral mesentery (Penners, 1924; Meyer, 1929). Thus, as in polychaetes, the blood vascular system occupies the site of the former blastocoel. In *Criodrilus* and *Eisenia*, the ventral vessel forms in a different way. It is first seen as a space between the ventral mesentery and the floor of the midgut and later closes off dorsally by apposition of the mesenterial edges (Wilson, 1889; Sterling, 1909). The dorsal vessel develops precociously, as a pair of lateral dorsal vessels between the upper margins of the somites and the lateral walls of the midgut. When the margins of the somites later come together in the dorsal midline, the two half vessels are combined into a single mid-dorsal vessel in the resulting dorsal mesentery.

Little attention has been paid to the formation of minor vessels in oligochaete embryos, though in *Tubifex* the segmental commissural vessels arise, as in polychaetes, by separation of the apposed walls of the intersegmental septa (Penners, 1924; Meyer, 1929). Iwanoff (1928) described the development of blood vessels in several oligochaete embryos from a mesenchyme distinct from the mesodermal somites, but his results cannot be accepted without corroboration. On the other hand, the existence of mesenchyme in leeches is well known. Budded off mainly from the somatic walls of the somites, but also from the splanchnic walls (Bürger, 1891, 1902), the mesenchyme gradually fills the coelomic spaces of the somites. The intersegmental septa also break down, so that the coelom is reduced to a series of longitudinal channels (basically, one dorsal, one ventral and two lateral) connected by a complex network of transverse channels. The coelomic and blood-vascular spaces become contiguous during this process, but dorsal and ventral longitudinal blood vessels persist in the dorsal and ventral coelomic channels of glossiphoniid leeches.

As in polychaetes, the splanchnic walls of the somites of clitellate embryos become applied to the outer surfaces of the stomodaeum and midgut as splanchnic mesoderm (Fig. 36). The musculature of the midgut wall presumably develops from these cells, but its development has not yet been described satisfactorily. The stomodaeal musculature is augmented in a complex manner by cells of the walls of the anterior somites (e.g. Wilson, 1889; Bürger, 1891, 1902; Iwanoff, 1928), but further details of its formation require further study.

Albumenotrophic clitellate embryos with a precociously functional gut develop a temporary gut musculature in the form of long-stranded muscles running from the larval pharynx and midgut to the provisional body wall (e.g. Bergh, 1885 a, b; Wilson, 1889; Filatoff, 1898; Schmidt, 1921, 1924, 1925a). The cells from which these muscles form are uncertain in origin. According to Wilson (1889), they are proliferated in earthworms from the anterior ends of the mesodermal bands. Schmidt (1925b), in contrast, claimed that the larval mesenchyme of piscicolid leeches (Fig. 34) is proliferated from the ectoteloblasts. In all cases, when the somites spread around the gut, the larval muscles are resorbed.

Gonoducts

In all clitellate embryos in which the development of the gonoducts has been adequately investigated, they have been identified as coelomoducts (Goodrich, 1946). The oligochaetes retain a relatively generalized condition of the development of coelomoduct gonoducts (earthworms, Vedjovsky, 1884; Bergh, 1885 a, b; Lehmann, 1887; *Tubifex*, Gatenby, 1916; Iwanoff, 1928; Meyer, 1929). Each gonoduct originates as a thickening of the coelomic epithelium opposite a gonad. The thickening develops into a funnel, while an outgrowth from the base of the thickening forms the duct. At the end of each male duct, a small ectodermal invagination establishes the opening to the exterior. The receptacula seminales of oligochaetes also develop as simple invaginations of the ventrolateral ectoderm (Vedjovsky, 1884; Bergh, 1885 a, b; Gatenby, 1916; Mehra, 1924).

The development of the gonoducts of leeches is a more specialized process, since it stems from a prior development of walled gonosacs (ovisacs and testis sacs) within the segmental coelomic cavities of the leech (see below). Bürger (1891, 1894, 1902) studied gonoduct development in a wide range of representative leeches (*Glossiphonia*, *Haemopsis*, *Hirudo* and *Erpobdella*). The paired vasa deferentia of the male system are formed by the union of successive coelomoduct outgrowths from the testis sacs. The most anterior pair meet to form the median genital atrium. The paired oviducts arise as outgrowths from the single pair of ovisacs lying just behind the genital atrium. The ducts meet to open by a median vagina, which Bürger at first (1891, 1894) interpreted as an ectodermal ingrowth, but later (1902) regarded as the conjoined ends of the mesodermal oviduct. Neither in oligochaetes nor leeches is there any association between the gonoducts and the nephridia of the genital segments.

Gonads

The timing of the first visible differentiation of primordial germ cells varies considerably among clitellate embryos, but is better documented than it has been in polychaetes. Although the gonoducts of oligochaete embryos arise relatively late in development, for example, there is now good evidence that the primordial germ cells are set aside at a very early stage. The work of Iwanoff (1928), Penners (1929, 1930), Meyer (1929), Penners and Stablein (1931) and Herlant-Meewis (1946) on tubificids and enchytraeids showed that a pair of small cells, which Penners and Stablein found were descendants of 4d, is closely associated with the presumptive mesoderm at the posterior end of the blastula. These small cells proliferate primordial germ cells, which spread forwards through the mass of yolky midgut cells as gastrulation proceeds and settle in clumps in the vicinity of the genital segments. Most of the cells in the clumps, together with the few that remain scattered among the midgut cells, appear to degenerate, since their fate cannot be traced further. Two pairs of cells persist in each genital segment and proliferate to form the testes and ovaries, projecting into the segmental coelom. Each gonad is covered by a thin layer of somatic peritoneum.

Among the earthworms, several early workers were able to trace the development of the gonads from the stage at which primordial germ cells can be identified in the genital segments (Bergh, 1885 a, b; Lehmann, 1887; Beddard, 1892), but could not elucidate the history of the germ cells before this stage. Devries (1971) has recently re-examined the problem both histologically and experimentally, using *Eisenia foetida*, with unexpected results. *Eisenia* buds off a pair of small cells as the first division products of the mesoteloblasts during gastrulation, much in the same manner as *Tubifex*. Unlike the latter, however, the small cells in *Eisenia* are not primordial germ cells. They undergo no further division, gradually lose their stainability and finally, late in gastrulation, disappear. The primordial germ cells of *Eisenia*, like those of other earthworms, first become histologically distinguishable (by the usual criteria of large nuclei, small nucleoli and finely granular cytoplasm) in the walls of the genital somites. In an attempt to trace the origins of these cells, Devries experimentally deleted the mesoteloblasts of young embryos before the somites of the genital segments had been formed. He found that primordial germ cells were then differentiated in segments anterior to the normal genital segments, indicating that the primordial germ cells are segregated earlier than the somites of the genital segments. On the other hand, there is no evidence in *Eisenia* of a direct lineal connection between the primordial germ cells and the pair of cells cut off as the first products of the mesoteloblasts.

Leech embryos present the same problem, in that the primordial germ cells have not been traced earlier than their first visible differentiation in the genital somites (Bürger, 1891, 1894, 1902). The problem is further compounded by the complexity of development of the hirudinean gonads. Each gonad develops as a sac enclosed within the coelomic space of its somite. The early development of the testis sacs is not yet clear, but the single pair of ovisacs originates from a single pair of gonocyte in the walls of the female genital somites (e.g. segment 11 of *Glossiphonia*). The first division of each gonocyte yields two cells, one of which divides again many times and surrounds the other cell as a capsule. The capsule then develops as the wall of the ovisac, which subsequently also gives rise to the oviduct. The enclosed cell divides repeatedly to form the ovarian germinal tissue within the sac. In leeches, therefore, the clear distinction between gonads and coelomoducts formed in oligochaetes is no longer maintained.

Further Development of the Ectoderm

In an earlier section of this chapter, discussing gastrulation in clitellate embryos, a brief mention was made of the formation of paired ectoblast bands through the proliferative activity of ectoteloblasts. In order to establish a basis for the description of the further development of the ectoderm, we need first to examine the ectoteloblasts and their activities in more details.

The ectoteloblasts of yolky clitellate embryos arise by the further division of the large 2d cell which makes up the bulk of the presumptive ectoderm. In *Tubifex* (Fig. 31), this cell divides equally bilaterally into a pair of cells, each of which gives rise by further divisions to a transverse row of four ectoteloblasts (Penners, 1922, 1924). *Rhynchelmis* is slightly modified, since the 2d cells is here accompanied by a similar 3d cell (Swetloff, 1923b). The 2d cell divides into a pair of cells which migrate away from one another in a lateral direction. The 3d cell also divides into a pair of cells, but each cell then divides to form a row of three cells with the large 2d cell at the lateral end of the row. The same two rows of four ectoteloblasts are thus attained by a different sequence of divisions. Formation of the bilateral rows of four ectoteloblasts in the glossiphoniid leeches *Glossiphonia* and *Theromyzon* (Figs. 23 and 31) in contrast, proceeds in the same manner as in *Tubifex* (Schleip, 1914; Schmidt, 1917, 1925a).

Among albumenotrophic embryos, the events of *Tubifex* are again usually retained in their entirety, even though the cells are smaller. This has been determined for earthworms (Swetloff, 1923b; Anderson, 1966; Devries, 1968), *Bdellodrilus* (Figs. 24 and 32; Tannreuther,

1915), piscicolid leeches (Figs. 28, 33 and 34; Schmidt, 1921, 1924, 1925b, 1930, 1939, 1941, 1944) and gnathobdellid and pharyngobdellid leeches (Fig. 27; Sukatschoff, 1903; Dimpker, 1917; Schoumkine, 1953). Minor modifications are reported only in the naidid oligochaetes (Dawydoff, 1942; see also Anderson, 1971). 2d is the source of the segmental ectoderm of *Stylaria* (Fig. 25), but ectoteloblast formation is omitted in favour of a more generalized proliferation of ectoderm. *Chaetogaster* shows the same specialization, but omits one of the preliminary divisions leading to the formation of the ectoderm mother cell, which is thus 1d.

As soon as the bilateral rows of ectoteloblasts are formed each ectoteloblast cell begins to proliferate a row of small cells forward from itself. The four rows of each side make up an ectoblast band, positioned external to an underlying mesodermal band. In yolky clitellate embryos (Fig. 31), a complex relationship exists between the forward growth of the ectoblast bands and the proliferation and spread of the temporary yolk-sac ectoderm. As the ectoteloblasts are formed, the micromeres anterior to them begin to proliferate. The main direction of spread of the micromeres as temporary yolk-sac ectoderm is lateral to ventral, effecting gastrulation enclosure of the bulky midgut rudiment. The micromeres also spread back along the dorsal midline, however, separating the two groups of ectoteloblasts. Simultaneously, the ectoteloblasts begin to proliferate the ectoblast bands. These push aside the yolk-sac ectoderm as they grow forwards, separating a broad dorsal area of yolk-sac ectoderm from two narrow band of ventral yolk-sac ectoderm. The ectoblast bands are at first lateral (*Tubifex*, Penners, 1924; Meyer, 1929) or dorsolateral (*Rhynchelmis*, Kowalevsky, 1871; Vedjovsky, 1888–92; Iwanoff, 1928; glossiphoniids, Whitman, 1878, 1887; Bergh, 1891; Bürger, 1902; Bychowsky, 1921) in position, but gradually curve down to meet in the ventral midline as the bands themselves become longer and the dorsal yolk-sac ectoderm becomes more attenuated (Fig. 35). The two narrow bands of ventral yolk-sac ectoderm come together in the ventral midline as ventral closure proceeds.

In albumenotrophic clitellate embryos, in which a precocious coverage of the surface of the embryo is always attained by rapid spread of the temporary yolk-sac ectoderm, the ectoblast bands proliferated by the ectoteloblasts lie within the provisional ectoderm and grow forwards in their definitive ventrolateral positions, apposed in the ventral midline (Figs. 32, 33 and 34). The fate of the temporary yolk-sac ectoderm varies. That of *Tubifex*, *Rhynchelmis* and the glossiphoniids is gradually incorporated into the segmental surface epithelium as the embryo elongates and becomes tubular (Vedjovsky, 1888–92; Bürger, 1902; Bychowsky, 1921; Penners, 1924). In *Bdellodrilus*, the provisional ectoderm is said by Tannreuther to differentiate later as the surface epithelium of the worm. The provisional ectoderm of the albumenotrophic embryos of earthworms (Vedjovsky, 1888–92; Wilson, 1889; Bergh, 1890; Beddard, 1892) and gnathobdellid and pharyngobdellid leeches (Bergh, 1885 a, b; Filatoff, 1898) develops a temporary ciliation along the ventral midline, through the action of which the embryo rotates in the albumen. Subsequently, the cilia are resorbed and the provisional ectoderm is incorporated into the surface epithelium of the worm. In contrast, the provisional envelope of naidid and piscicolid embryos remains external to the embryo throughout development and is finally shed (Schmidt, 1921, 1925b, 1944; Dawydoff, 1942).

As the ectoblast bands elongate, the cells of the four cell rows of each band undergo repeated divisions. The division products of the most ventral row on each side remain distinct, forming a pair of ventrolateral bands. The division products of the three lateral rows on each side merge to form a single broad band of small cells on that side (*Tubifex*, Penners, 1924; Meyer, 1929; *Rhynchelmis*, Vedjovsky, 1888–92; earthworms, Wilson, 1889; Bergh, 1890; Staff, 1910; Swetloff, 1928; leeches, Bergh, 1891; Filatoff, 1898; Bürger, 1902). Segment delineation occurs in the ectoblast bands by the formation of segmental annuli in anteroposterior succession (Fig. 35), subsequent to the formation of somites. In *Tubifex*, for example, when twenty-four pairs of somites are present, sixteen segments are delineated externally (Penners, 1924). The evocation of ectodermal segmentation by mesodermal somites is experimentally well established for clitellate annelids (Devries, 1969; Reverberi, 1971).

Although, as described above, the temporary yolk-sac ectoderm usually contributes to the definitive surface epithelium of the embryo as development proceeds, all other ectodermal structures are derived wholly from the ectoblast bands. The two ventral rows of cells derived from the most ventral ectoteloblasts give rise, for example, to the segmental ganglia of the ventral nerve cord (e.g. Whitman, 1887; Wilson, 1889; Bergh, 1890, 1891; Filatoff, 1898; Bürger, 1902; Schmidt, 1921; Penners, 1924; Meyer, 1929). The two rows of cells come to lie on either side of the ventral midline, beneath a thin layer of superficial cells. The latter has a disputed origin. For *Tubifex*, it is not clear whether the superficial ventral ectoderm arises from the ventral yolk-sac ectoderm or is cut off from the neuroblasts. The corresponding ventral ectoderm of earthworms seems to be a product of the provisional (=ventral yolk-sac) ectoderm. The two rows of ganglion cells proliferate the paired segmental ganglia, which then sink inwards and fuse together in the ventral midline before neuropile formation begins (Fig. 36).

The contribution of the more lateral parts of the ectoblast bands derived from the three more lateral ectoteloblasts on each side of the embryo is varied and, to some extent, controversial. It is from these cells, for example, that the circular muscle cells of the body wall have been said to be derived in several clitellate species (Vedjovsky, 1888–92; Bergh, 1891; Bürger, 1902; Bychowsky, 1921; Penners, 1922, 1924; Meyer, 1929). After the segmental delineation of the ectoblast bands, there is no doubt that groups of ectoblast cells in each segment proliferate in oligochaetes as ingrowths into the interior, giving rise to the chaetal sacs (Fig. 35; Kowalevsky, 1871; Vedjovsky, 1886, 1888–92; Wilson, 1889; Bergh, 1890; Bourne, 1894; Staff, 1910; Penners, 1924; Swetloff, 1928; Meyer, 1929; Vanderbroek, 1936). In general, the segmental units of ectoblast also contribute to the surface epithelium of the body wall, usually in consort with the temporary yolk-sac ectoderm, but independently of the temporary yolk-sac ectoderm when the latter is specialized as a temporary provisional envelope, as in naidids and piscicolids. In being the source of proliferation of cells which give rise to the ventral ganglia and surface epithelium of the segments, and to the chaetal sacs in oligochaetes, the 2d ectoteloblasts of clitellates display their homology with the 2d ectoteloblast ring of the prepygidial growth zone of polychaetes. The question of the descent of clitellate circular muscle from the same source deserves further study.

The role of the ectoblast bands of clitellates in the development of segmental nephridia also remains undecided. Early studies of nephridial development in oligochaetes (Hatscheck, 1878; Kleinenberg, 1879; Boutchinsky, 1881; Vedjovsky, 1884, 1886, 1887, 1888–92, 1900; Bergh, 1888, 1890, 1891; Wilson, 1887, 1889; Staff, 1910; Tannreuther, 1915) were

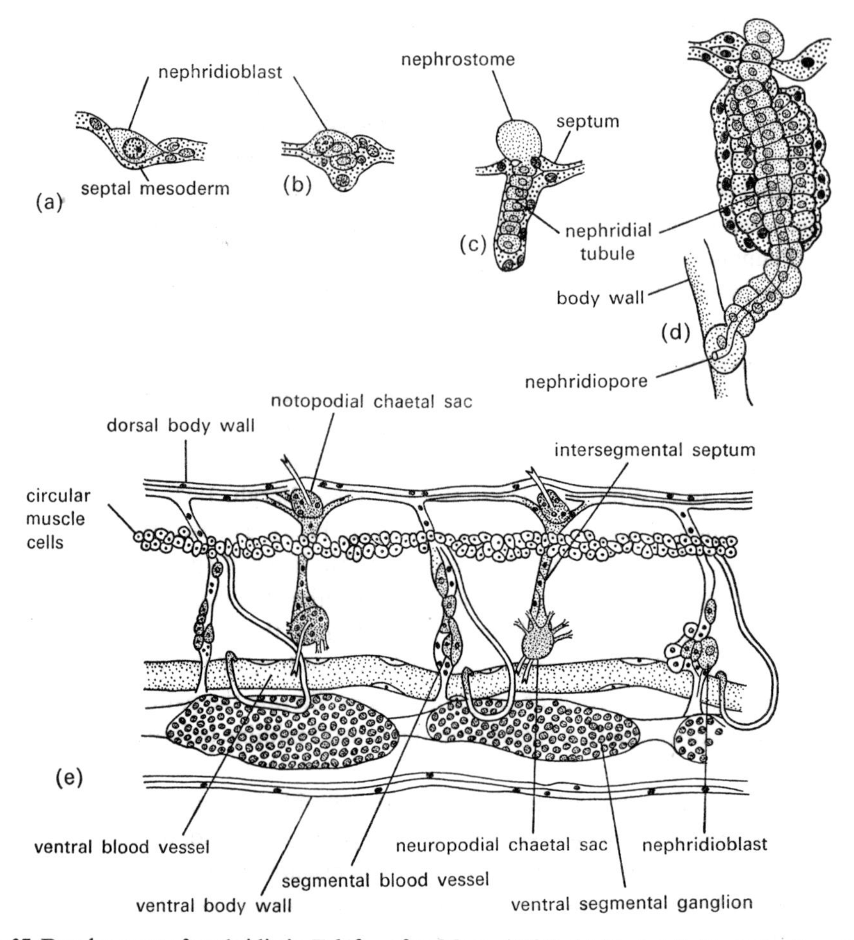

FIG. 37. Development of nephridia in *Tubifex*, after Meyer (1929). (a) Nephridioblast; (b–d) proliferation of nephridium and formation of its mesodermal covering; (e) diagram showing intersegmental location of nephridioblasts.

inconclusive, but more recent investigators are agreed that each nephridium develops from a single, large nephridioblast cell (Bahl, 1922; Penners, 1924; Iwanoff, 1928; Meyer, 1929; Goodrich, 1932, 1946; Vanderbroek, 1932, 1934). The cell (Figs. 36 and 37) lies in the intersegmental septum. It buds off a row of small cells which develops as the nephridial tubule, and then divides into a group of cells which forms the nephridiostome. The septal location of the nephridioblast has generally been taken as evidence that the cell has a mesodermal origin, but the possibility of an earlier derivation from ectoderm cannot be disregarded. The same problem of nephridioblast origin exists in polychaetes (Chapter 2) and recurs in leeches. The development of the nephridia of leeches from intersegmental nephridioblasts in the same manner as oligochaetes was demonstrated by Bürger (1891, 1894, 1902) and confirmed by Bychowsky (1921), but the origin of the nephridioblasts is still uncertain. The terminal bladder and exit duct of the hirudinean nephridium are formed as an independent ectodermal invagination.

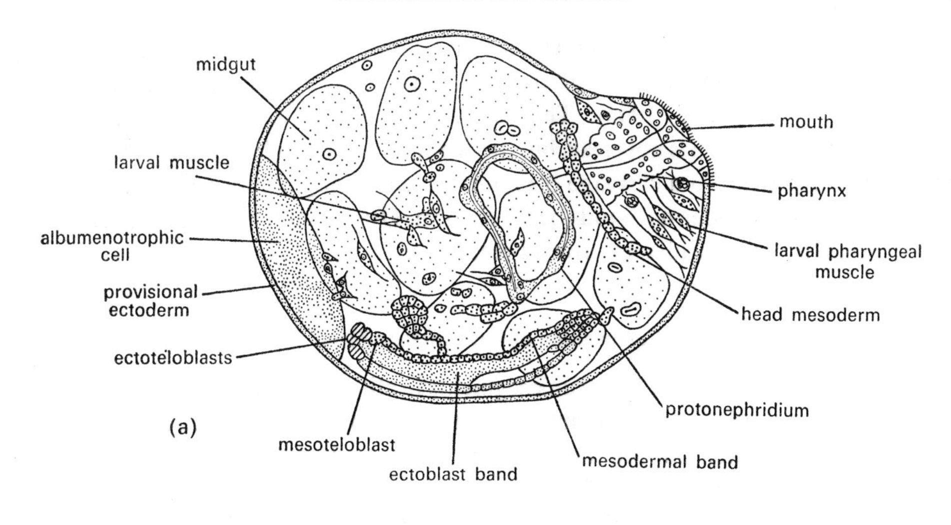

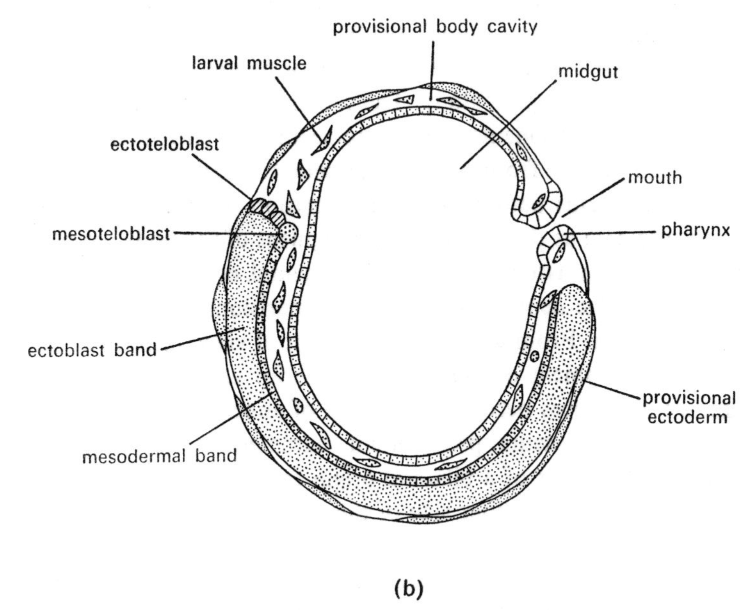

FIG. 38. Albumenotropic clitellate embryos. (a) *Erpobdella*, after Bergh (1885); (b) *Piscicola*, after Schmidt (1925b).

Prior to the formation of the definitive segmental nephridia, many clitellate embryos develop provisional protonephridia (Fig. 38), which differentiate precociously during gastrulation. The only notable groups in which provisional protonephridia seem to be absent are the naidid oligochaetes and piscicolid leeches. In *Tubifex*, *Rhynchelmis*, and the albumenotrophic embryos of earthworms, a single pair of protonephridia is developed (Kleinenberg, 1879; Vedjovsky, 1886, 1887, 1888–92; Lehmann, 1887; Bergh, 1888; Wilson, 1889; Hoffman, 1899; Iwanoff, 1928; Swetloff, 1928). The protonephridia lie at the anterior end of the embryo, bilaterally between the ectoderm and the gut. Each protonephridium comprises a solenocyte, a ciliated intracellular duct and a terminal opening to the surface laterally or ventrolaterally. The albumenotrophic embryos of gnathobdellid and pharyngobdellid leeches develop two to four pairs of similar protonephridia (Bergh, 1885 a, b; Filatoff, 1898; Sukatschoff, 1903). The origin and early development of the clitellate provisional protonephridia is not clear, but the work of Iwanoff (1928) and Swetloff (1928) suggests that they arise from cleavage blastomeres which migrate into the interior of the embryo before the mesodermal bands begin to form. During later embryonic development, the provisional protonephridia are always resorbed.

Composition of the Clitellate Head

Although the development of the trunk segments in clitellate embryos is quite well understood and bears a fundamental resemblance to that of polychaetes, the development of the clitellate head is less well known. There seems to be no doubt that the prostomium and peristomium have a bilateral origin from the anterior ends of the ectoblast bands and their underlying mesodermal bands (Wilson, 1889; Bergh, 1891; Filatoff, 1898; Bürger, 1902; Tannreuther, 1915; Penners, 1924; Iwanoff, 1928), but the details of the development of the cerebral ganglia and prostomial mesoderm and the formation and segmental significance of the peristomium all require further study. The cerebral ganglia of oligochaetes develop from the ventral neuroblast components at the extreme anterior end of each ectoblast band. Those of leeches are said to arise from similar components which separate off from the remainder of the ectoblast bands before ganglion formation begins. The oligochaete peristomium appears to originate as the first definitive segment of the body, comprising a pair of segmental somites and an associated pair of segmental ectodermal areas just behind the prostomial rudiment. The neuroblast rows of the peristomial ectoderm give rise to a single pair of ventral ganglia. From a comparative point of view, in spite of the poor state of knowledge of the development of the clitellate anterior end, considerable significance attaches to the known differences between this group of annelids and the polychaetes. The apical plate origin of a presegmental prostomium characteristic of polychaetes has been abandoned in clitellates in favour of a bilateral origin from cells budded from a posterior growth zone, moving forwards on either side of the stomodaeum and fusing together in a preoral location. Associatedly, a mouth region is absent in clitellates and the peristomium is formed by segmental components which also move forwards bilaterally from a postoral to a para-oral location, though they do not migrate in front of the mouth. These differences between polychaetes and clitellates, which are a functional corollary of the new pattern of presumptive areas associated with clitellate lecithotrophy, have important implications for comparisons with the Onychophora, as we shall see in Chapters 4 and 10.

The Basic Pattern of Development in Clitellates

Comparative considerations make it clear that the tubificid and lumbriculid oligochaetes and the glossiphoniid leeches share a common pattern of cleavage and presumptive area formation (Figs. 21, 22, 23 and 29). Minor specializations are evident in the expression of this pattern in the leeches, and the most generalized expression of clitellate spiral cleavage and presumptive area formation is undoubtedly that of *Tubifex*. If we then take into account the fact that tubificid development is modified in association with increased yolk and the elimination of a larval phase, it becomes immediately clear that the basic pattern of clitellate development exemplified by *Tubifex* is a functional modification of the pattern basic to polychaetes. The argument in support of this conclusion was set out in detail by Anderson (1966) and will now be briefly reiterated.

The conclusion rests on a comparison of the formation and fates of the presumptive areas of the blastula. The most conspicuous feature of the clitellate fate map (Fig. 29) as compared with that of polychaetes (Fig. 6) is simplification. Presumptive prototroch and presumptive ectomesoderm are missing and the presumptive ectoderm is not divided into anterior and posterior components. Of the four presumptive areas shared in common by clitellates and polychaetes, two are fundamentally similar in both. The presumptive mesoderm is segregated into a single, posterior cell 4d. This cell then divides equally bilaterally into the two mesoteloblasts M^l and M^r. Before assuming their definitive fate, the mesoteloblasts bud off a pair of small primordial germ cells. They then proliferate the paired, ventrolateral mesodermal bands, which give rise to the paired somites. The presumptive midgut of clitellates, like that of polychaetes, comprises the four stem cells 3A–3C and 4D. Like the presumptive mesoderm cell, the presumptive midgut cells are larger in clitellates than in polychaetes, providing the basis for the direct development of a large embryo with numerous segments (Figs. 31 and 35).

A close correspondence can also be recognized between the presumptive stomodaeal area of clitellates and that of polychaetes. The micromeres which make up this area in clitellates stem mainly from 2b and 3b, with perhaps some contribution from 3a. In polychaetes, as discussed in Chapter 2, the presumptive stomodaeum also arises from ventral cells of the second and third quartets, variously involving 2a, 2b, 2c, 3a and 3b. The displacement of the clitellate presumptive stomodaeum to an anterodorsal location is a functional corollary of the enlargement and forward displacement of the presumptive midgut.

Comparison between the presumptive ectoderm in clitellates and polychaetes is less simple. The composition of the polychaete presumptive ectoderm (Fig. 6) reflects a dual role in subsequent development, larval and definitive. The presumptive anterior ectoderm (mainly $1a^1$–$1d^1$), which gives rise to the prostomium, cerebral ganglion and larval episphere, is separated by the presumptive prototroch (mainly $1a^2$–$1d^2$) from the presumptive posterior ectoderm. The latter includes the presumptive ectoderm of the larval hyposphere (mainly 2a, 2c, 2d, 3c and 3d), the ectoteloblast ring of the segmental growth zone (mainly 2d) and the pygidium (also 2d). Of these components, the ectoteloblast ring makes the greatest contribution to the polychaete body, since it proliferates the ectoderm of the trunk segment.

The clitellate presumptive ectoderm contrasts with this arrangement in comprising a single, large 2d cell surrounded anteriorly and laterally by an arc of micromeres, formed mainly by 1a–1d, 2a, 2c, 3c and 3d. The simplification, however, is entirely functional. Firstly, the large 2d cell, as in polychaetes, is the ectoteloblast mother cell, from which all of the trunk ectoderm arises. In the absence of presumptive episphere ectoderm, prototroch and hyposphere ectoderm, the presumptive anterior ectoderm is also incorporated into the large 2d cell. The remainder of the presumptive ectoderm has a new presumptive fate as temporary yolk-sac ectoderm, although it retains a capacity for forming definitive trunk ectoderm and pygidial ectoderm. This functional modification of part of the presumptive ectoderm as temporary yolk-sac ectoderm is directly correlated with the need for rapid enclosure of the large, yolky presumptive midgut and with the associated redundancy of larval ectodermal differentiation. The role of 2d as the source of proliferation of segmental ectoderm, established in polychaetes, is enhanced in clitellates as part of the complex of adaptations to the direct formation of numerous segments at the expense of yolk.

The basic pattern of presumptive areas in the clitellate blastula is thus functionally derivable from that of polychaetes as a further adaptation to the direct lecithotrophic development of a yolky egg without the loss of total spiral cleavage. As will be expanded in Chapter 10, the clitellate modification of annelid development points the way to yet further modifications associated with lecithotrophy, providing a basis for the interpretation of the embryonic development of the Onychophora (see also Anderson, 1966). In the clitellates, much of the presumptive ectoderm is simplified to a proliferative rudiment which will form a temporary yolk-sac epithelium over the surface of the massive midgut rudiment. In arthropods, it is customary to refer to such temporary, yolk-enclosing epithelia as extra-embryonic ectoderm. This terminology will be employed in the following chapters. We can turn now to the variety of modes of embryonic development displayed among the arthropods, beginning with the Onychophora.

References

ANDERSON, D. T. (1966) The comparative early embryology of the Oligochaeta, Hirudinea and Onychophora. *Proc. Linn. Soc. N.S.W.* **91,** 10–43.

ANDERSON, D. T. (1971) The embryology of aquatic oligochaetes. In BRINKHURST, R. O. and JAMIESON, B. G. (eds.), *Aquatic Oligochaeta of the World*, Oliver & Boyd, Edinburgh.

APÁTHY, S. (1891) Keimstreifen und Mesoblaststreifen bei Hirudineen. *Zool. Anz.* **14,** 388–93.

BAHL, K. N. (1922) On the development of the "entonephric" type of nephridial system found in Indian earthworms of the genus *Pheretima*. *Q. Jl. microsc. Sci.* **66,** 49–103.

BEDDARD, F. E. (1892) Researches into the embryology of the Oligochaeta. No. 1. On certain points in the development of *Acanthodrilus multiporus*. *Q. Jl. microsc. Sci.* **33,** 497–540.

BERGH, R. S. (1885a) Ueber die Metamorphose von *Nephilis*. *Z. wiss. Zool.* **41,** 284–301.

BERGH, R. S. (1885b) Die Metamorphose von *Aulostomum gulo*. *Arb. zool.-zootom. Inst. Würzburg*, **7,** 231–91.

BERGH, R. S. (1890) Neue Beiträge zur Embryologie der Anneliden. I. Zur Entwicklung und Differenzierung des Keimstreifens von *Lumbricus*. *Z. wiss. Zool.* **50,** 469–526.

BERGH, R. S. (1891) Neue Beiträge zur Embryologie der Anneliden. II. Die Schichtenbildung im Keimstreifen der Hirudineen. *Z. wiss. Zool.* **52,** 1–17.

BOURNE, A. G. (1894) On certain points in the development and anatomy of some earthworms. *Q. Jl. microsc. Sci.* **36,** 11–34.

BOUTCHINSKY, P. (1881) Zur Frage über die Entwicklung des Regenwurms. *Trav. Soc. nat. nouv. Russie, Odessa*, **7.**

BÜRGER, O. (1891) Beiträge zur Entwicklungsgeschichte der Hirudineen. Zur Embryologie von *Nephilis*. *Zool. Jb. Anat. Ont.* **4,** 697–783.

BÜRGER, O. (1894) Neue Beiträge zur Entwicklungsgeschichte der Hirudineen. Zur Embryologie von *Hirudo medicinalis* und *Aulostomum gulo*. *Z. wiss. Zool.* **58,** 440–59.

BÜRGER, O. (1902) Weitere Beiträge zur Entwicklungsgeschichte der Hirudineen. Zur Embryologie von *Clepsine*. *Z. wiss. Zool.* **72,** 525–44.

BYCHOWSKY, A. (1921) Über die Entwicklung der Nephridien von *Clepsine sexoculata* Bergmann (*complanata* Savigny). *Revue suisse Zool.* **29,** 41–131.

CATHER, J. N. (1971) Cellular interactions in the regulation of development in annelids and molluscs. *Adv. Morphog.* **9,** 67–125.

DAWYDOFF, C. (1942) Étude sur l'embryologie des Naididae indochinois. *Archs. Zool. exp. gén. Notes et Revue*, **81**, 173–94.

DEVRIES, J. (1968) Les premières étapes de la segmentation (formation de la jeune blastule) chez le lombricien *Eisenia foetida*. *Bull. Soc. zool. Fr.* **93**, 87–97.

DEVRIES, J. (1969) Le développement des embryons d'*Eisenia foetida* aprés la destruction unilatérale des mesoteloblastes. *Bull. Soc. zool. Fr.* **94**, 663–71.

DEVRIES, J. (1971) Origine de la lignée germinale chez le lombricien *Eisenia foetida*. *Ann. Emb. Morph.* **4**, 37–43.

DIMPKER, A. M. (1917) Die Eifurchung von *Herpobdella atomaria* Carena. (*Nephilis vulgare* Moq. Tand.). *Zool. Jb. Anat. Ont.* **40**, 249–90.

FILATOFF, D. (1898) Einige Beobachtungen über die Entwicklungsvorgänge bei *Nephilis vulgaris* M.T. *Zool. Anz.* **21**, 645–7.

GATENBY, J. B. (1916) The development of the sperm duct, oviduct and spermatheca in *Tubifex rivulorum*. *Q. Jl. microsc. Sci.* **61**, 317–36.

GOODRICH, E. S. (1932) On the nephridiostome of *Lumbricus*. *Q. Jl. microsc. Sci.* **75**, 165–79.

GOODRICH, E. S. (1946) The study of nephridia and genital duct since 1895. *Q. Jl. microsc. Sci.* **86**, 113–301.

HANSON, J. (1949) The histology of the blood system in oligochaetes and polychaetes. *Biol. Rev.* **24**, 127–73.

HATSCHEK, B. (1878) Studien über Entwicklungsgeschichte der Anneliden. *Arb. zool. Inst. Univ. Wien* **3**, 277–404.

HERLANT-MEEWIS, H. (1946) Contribution à l'étude de la régénération chez les Oligochetes. II. Reconstitution du germen chez *Lumbricillus lineatus* (Enchytraeidés). *Archs. Biol.* **57**, 197–306.

HOFFMAN, R. W. (1899) Beiträge zur Entwicklungsgeschichte der Oligochäten. *Z. wiss. Zool.* **66**, 335–57.

INASE, M. (1967) Behaviour of the pole plasm in the early development of the aquatic worm, *Tubifex hattai*. *Sci. Rep. Tohóku Univ.*, Ser. IV, *Bio.* **33**, 223–31.

IWANOFF, P. P. (1928) Die Entwicklung der Larvalsegmente bei den Anneliden. *Z. Morph. Ökol. Tiere* **10**, 62–161.

KLEINENBERG, N. (1879) The development of the earthworm. *Q. Jl. microsc. Sci.* **19**.

KOWALEVSKY, A. (1871) Embryologische Studien an Würmen und Anthropoden, *Zap. imp. Akad. Nauk.* **16**, No. 12, 1–70.

LEHMANN, O. (1887) Beiträge zur Frage von der Homologie der Segmental-organe und Ausfuhrgänge der Geschlechtsprodukte bei den Oligochaeten. *Jena Z. naturwiss* **21**, 322–60.

MEHRA, H. R. (1924) The genital organs of *Stylaria lacustris*, with an account of their development. *Q. Jl. microsc. Sci.* **68**, 147–86.

MEYER, A. (1929) Die Entwicklung der Nephridien und Gonoblasten bei *Tubifex rivulorum* Lam. nebst Bemerkungen zum natürlich System der Oligochäten. *Z. wiss. Zool.* **133**, 517–62.

NUSBAUM, J. (1884) Zur Entwicklungsgeschichte der Hirudineen (*Clepsine*). *Zool. Anz.* **7**, 609–15.

NUSBAUM, J. (1885) Zur Entwicklungsgeschichte der Geschlechtsorgane der Hirudineen (*Clepsine complanata* Sav.). *Zool. Anz.* **8**, 181–4.

NUSBAUM, J. (1886) Recherches sur l'organo-genèse des Hirudinées (*Clepsine complanata* Sav.). *Arch. Slav. Biol.* **1**, 320–40 and 539–56.

PENNERS, A. (1922) Die Furchung von *Tubifex rivulorum* Lam. *Zool. Jb. Anat. Ont.* **43**, 323–68.

PENNERS, A. (1924) Die Entwicklung des Keimstreifs und die Organbildung bei *Tubifex rivulorum* Lam. *Zool. Jb. Anat. Ont.* **45**, 251–308.

PENNERS, A. (1929) Entwicklungsgeschichte Untersuchungen an marinen Oligochäten. I. Furchung, Keimstreif, Vorderdarm und Urkeimzellen von *Peloscolex benedini* Udekem. *Z. wiss. Zool.* **134**, 307–44.

PENNERS, A. (1934) Die Ontogenese des entodermalen Darmepithel bei limicolen Oligochäten. *Z. wiss. Zool.* **145**, 497–507.

PENNERS, A. and STABLEIN, A. (1931) Über die Urkeimzellen bei Tubificiden (*Tubifex rivulorum* Lam. und *Limnodrilus udekemianus* Claparéde). *Z. wiss. Zool.* **137**, 606–26.

REVERBERI, G. (1971) Annelids. In REVERBERI, G. (ed.), *Experimental Embryology of Marine and Freshwater Invertebrates*, North Holland, Amsterdam.

SCHLIEP, W. (1913) Die Furchung des Eies von *Clepsine* und ihre Beziehung zur Furchung des Polychaeteneies. *Ber. Naturforsch. Ges. Freiburg i. Br.* **20**, 1–12.

SCHLIEP, W. (1914) Die Furchung des Eies der Russelegel. *Zool. Jb. Anat. Ont.* **37**, 313–68.

SCHMIDT, G. A. (1917) Zur Entwicklung des Entoderms bei *Protoclepsis tesselata*. *Ann. zool. Abt. Ges. Freunde N.A. und E.* **4**, (N.S.) *Mosc.* **1**, 22.

SCHMIDT, G. A. (1921) Die Embryonalentwicklung von *Piscicola geometra* Blainv. *Zool. Anz.* **53**, 123–7.

SCHMIDT, G. A. (1922) Zur Frage über die Entwicklung des Entoderms bei der *Rhynchelmis limosella* Hoffm. *Rusk. zool. Zh.* **3**, 74–93.

SCHMIDT, G. A. (1924) Untersuchungen über die Embryologie der Anneliden. 2. Die Besonderheiten der Embryonalentwicklung der Ichthyobdellidae und ihre Entstehung. *Zool. Jb. Anat. Ont.* **46**, 199–244.

SCHMIDT, G. A. (1925a) Die polaren plasmatischen Massen Ei von *Protoclepsis tesselata. Rusk. zool. Zh.* **5**, 138–64.

SCHMIDT, G. A. (1925b) Untersuchungen über die Embryologie der Anneliden. I. Die Embryonalentwicklung von *Piscicola geometra* Blainv. *Zool. Jb. Anat. Ont.* **47**, 319–428.

SCHMIDT, G. A. (1930) Der Keimentwicklungstypus der *Crangonobdella murmicanica* W.D. Selenk. *Rusk. zool. Zh.* **10**, 5–15.

SCHMIDT, G. A. (1936) Gesetzmassigkeiten des Wechsels der Embryonalanpassungen. *Biol. Zh. Mosc.* **5**, 633–56.

SCHMIDT, G. A. (1939) Dégénérescence phylogénétique des modes de développement des organes. *Arch. Zool. exp. gén.* **81**, 317–70.

SCHMIDT, G. A. (1941) Researches on the comparative embryology of leeches. **10.** Early stages of development of the embryo in Ichthyobdellidae. *A la memoire A. L. Sewertzoff, Moscow* **2**, 357–489.

SCHMIDT, G. A. (1944) Adaptive significance of the peculiarities of the cleavage process in leeches. *Sh. obshch. Biol.* **5**, 284–303.

SCHOUMKINE, O. B. (1953) Embryonic development of *Hirudo. Trudy Inst. Morf. Zhivot.* **8**, 216–79.

STAFF, F. (1910) Organogenetische Untersuchungen über *Criodrilus lacuum. Arb. zool. Inst. Univ. Wien,* **18**, 227–56.

STERLING, S. (1909) Das Blutgefässsystem der Oligochäten. *Jena Z. naturwiss.* **44**, 253–352.

SUKATSCHOFF, B. (1903) Beiträge zur Entwicklungsgeschichte der Hirudineen. II. Über die Furchung und Bildung der Embryonalen Anlagen bei *Nephilis vulgaris* Moq.-Tand. (*Herpobdella atomaria*). *Z. wiss. Zool.* **73**, 321–67.

SWETLOFF, P. (1923a) Sur la segmentation de l'œuf chez le *Bimastus constrictus* R. (fam. Lumbricidae). *Izv. biol. nauchno-issled. Inst. biol. Sta. Perm. gosud. Univ.* **1**, 101–10.

SWETLOFF, P. (1923b) Sur la segmentation de l'œuf de *Rhynchelmis limosella* Hoffmstr. *Izv. biol. nauchno-issled. Inst. biol. Sta. Perm. gosud. Univ.* **1**, 141–52.

SWETLOFF, P. (1926) Über die Embryonalentwicklung bei den Naididen. *Izv. biol. nauchno-issled. Inst. biol. Sta. Perm. gosud. Univ.* **4**, 359–72.

SWETLOFF, P. (1928) Untersuchungen über die Entwicklungsgeschichte der Regenwurmer. *Trudy osob. zool. Lab. Sebastop. biol. Sta.* **13**, 95–329.

TANNREUTHER, G. W. (1915) The early embryology of *Bdellodrilus philadelphicus. J. Morph.* **26**, 143–216.

VANDERBROEK, G. (1932) Origine et développement des saccules mesodermiques et des néphridies chez une oligochète tubicole: *Allolobophora foetida* Sav. *C.r. Ass. f. Advanc. Sci.* **56**, 292–6.

VANDERBROEK, G. (1934) Organogénèse du systeme nephridien chez les oligochètes et plus specialement chez *Eisenia foetida* Sav. *Recl. Inst. zool. Torley-Rousseau* **5**, 1–72.

VANDERBROEK, G. (1936) Organogénèse des follicules setigères chez *Eisenia foetida* Sav. *Mem. Mus. r. Hist. nat. Belg.* **3**, 559–68.

VEDJOVSKY, F. (1884) *System und Morphologie der Oligochaeten*, Prague.

VEDJOVSKY, F. (1886) Die Embryonalentwicklung von *Rhychelmis. Sber. K. Bohm. Ges. Wiss.* **2**, 227–39.

VEDJOVSKY, F. (1887) Das larval und definitiv Excretionssytem. *Zool. Anz.* **10**, 681–5.

VEDJOVSKY, F. (1888–92) *Entwicklungsgeschichtliche-Untersuchungen*, Prague.

VEDJOVSKY, F. (1900) Noch ein Wort über die Entwicklung der Nephridien. *Z. wiss. Zool.* **67**, 247–54.

WHITMAN, C. O. (1878) The embryology of *Clepsine. Q. Jl. microsc. Sci.* **18**, 215–315.

WHITMAN, C. O. (1887) A contribution to the history of the germ layers of *Clepsine. J. Morph.* **1**, 105–82.

WILSON, E. B. (1887) The germ bands of *Lumbricus. J. Morph.* **1**, 183–92.

WILSON, E. B. (1889) The embryology of the earthworm. *J. Morph.* **3**, 387–462.

CHAPTER 4

ONYCHOPHORANS

AMONG the modern arthropods, the Onychophora have long been recognized as a relict group whose morphology displays many primitive features. It is only to be expected, therefore, that the embryonic development of onychophorans should be highly pertinent to any comparison of annelid and arthropod embryology, and such it is. At the same time, the facts of onychophoran development have always been elusive, and it is also an unfortunate accident of history that no worker on onychophoran embryos had previously had any direct experience of annelid embryology. As it transpires, the most penetrating interpretation of onychophoran embryology is one which first takes account of how annelids have responded to yolk (Anderson, 1966). We shall return to this in a later chapter. In the meantime, what are the facts of onychophoran development?

Most Onychophora have a highly specialized mode of development, associated with a small, secondarily yolkless egg and viviparity. Only a few Australasian species retain a yolky egg and of these, only one or two are oviparous, the remainder being ovoviviparous. The oviparous species, *Symperipatus* and *Ooperipatus*, are also rare, have an astonishingly prolonged development (17 months, according to Dendy, 1902) and are consequently unknown embryologically. Their ovoviviparous relatives are a little better known. As we shall see, they undoubtedly represent a primitive mode of development for Onychophora, in which the egg is large and development proceeds to an advanced stage, with the full complement of segments, at the expense of the yolk of the egg. The viviparous species fall into two groups, those of southern Africa, which develop at the expense of maternal nutrients but lack a placenta, and those of Central and South America, which have a placental attachment to the oviducal wall.

The fact that onychophorans live mainly where descriptive embryologists do not, namely, in the Southern Hemisphere, is strongly reflected in the history of investigation of onychophoran development. As part of the upsurge of classical descriptive embryology in the later nineteenth century, three important studies were made whose results, especially for the later phases of development, are still largely unsupplemented. The first was by the German embryologist Kennel (1884, 1885), working at the University of Würzburg on two species of placental Onychophora obtained from Trinidad, *Epiperipatus trinidadensis* and *Macroperipatus torquatus*. Almost at the same time, the well-known English zoologist Sedgwick (1885, 1886, 1887, 1888) published a series of papers on the non-placental, viviparous species *Peripatopsis capensis*, obtained from South Africa. Sheldon (1888, 1889 a, b), working with Sedgwick at Cambridge, then described the development of the yolky egg of the ovoviviparous *Peripatoides novae-zealandiae* from New Zealand. For all these workers, the early stages of development, especially cleavage and gastrulation, were extremely difficult to interpret. Those of the viviparous species are highly aberrant, while the yolky eggs of *Peripatoides* proved far from amenable to nineteenth-century histological technique.

Three lesser contributions, with similar attendant problems, were made by other workers at the same period. Sclater (1888) studied a viviparous South American species, Willey (1898) a viviparous species in New Britain and Evans (1902) an ovoviviparous species from Malaya. Then followed a long period of neglect, ending only with the publication by Manton (1949) of the first and sole modern account of onychophoran embryonic development. Manton, one of the great invertebrate zoologists of this century, had previously made revolutionary contributions to crustacean embryology, as will be seen in Chapter 8. Her paper on the embryology of four species of the viviparous *Peripatopsis* from South Africa was the outcome of intensive work over several years, and was carried out mainly at the University of Cambridge. It remains the crucial source of information on cleavage, gastrulation, segment formation and the modifications associated with secondary yolk loss in the Onychophora. Since 1949 the only additional information of which I am aware is a number of unpublished observations by Manton and myself on placental species from Jamaica, and by myself on the ovoviviparous *Peripatoides orientalis* from New South Wales, referred to in Anderson (1966). A new interpretation of onychophoran cleavage and gastrulation was offered, however, by Anderson (1966) and is used here as the basis of the present discussion of the theme and variations of onychophoran development.

Cleavage

The yolky eggs of oviparous and ovoviviparous Onychophora, ovoid in shape, combine dimensions of more than a millimetre (*Ooperipatus*, 1·9 mm, Dendy, 1902; *Peripatoides novae zealandiae*, 1·5 mm, Sheldon, 1888; *Peripatoides orientalis*, 1·3 mm, Anderson, 1966; *Eoperipatus weldoni*, 1·3 mm, Evans, 1902) with what can be called a primitive centrolecithal structure. The egg cytoplasm is dense with yolk and devoid of a periplasm, and the zygote nucleus lies in a central position, surrounded by a small halo of yolk-free cytoplasm. Two membranes enclose the egg, a thin vitelline membrane and an external chorion which is thick in the oviparous *Ooperipatus*, thin in the ovoviviparous *Peripatoides* and *Eoperipatus*. The accounts of cleavage in these eggs given by Sheldon (1888, 1889a) and Evans (1902) are incomplete, but it is clear that cleavage is of the intralecithal kind found in many arthropods (Fig. 39). The yolk mass remains undivided while the zygote nucleus and its daughters divide and spread, with accompanying divisions of their cytoplasmic haloes. Some of the nuclei and their haloes rise to the surface of the yolk mass to form a small disc of blastomeres. With further division and spread, these cells then give rise to a low, cuboidal blastoderm at the surface. Simultaneously, the yolk mass divides into a number of yolk spheres. It is not clear whether any of these retain cleavage nuclei.

As might be expected, the spherical eggs of the non-placental viviparous Onychophora are much smaller than those of the lecithotrophic species when released from the ovary. Manton (1949) confirmed this fact with especial clarity for four species of *Peripatopsis*,

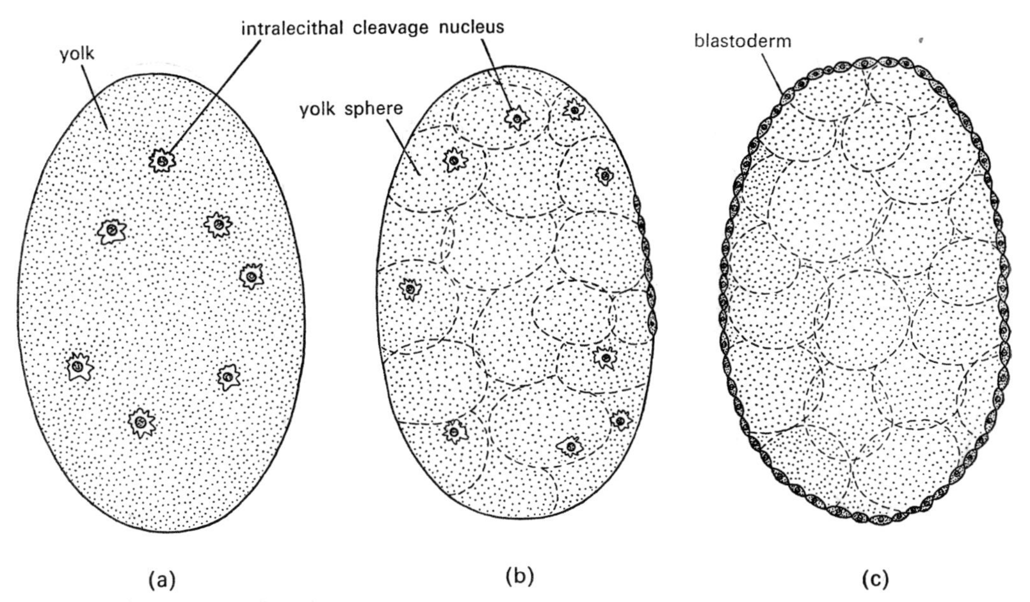

FIG. 39. Cleavage and blastoderm formation in a yolky onychophoran egg, after Anderson (1966). (a) Early intralecithal cleavage; (b) initiation of blastoderm formation; (c) blastoderm.

whose initial egg dimensions are 380 μ (*P. balfouri*), 260 μ (*P. capensis*), 150–170 μ (*P. moseleyi*) and 65–80 μ (*P. sedgwicki*). On entry into the oviduct, the egg swells and becomes ellipsoidal (*P. balfouri*) or cylindrical, with hemispherical ends. Anderson (1966) suggested that the swelling is a delayed partial recapitulation of the enlargement which occurs in a yolky egg during vitellogenesis. The dimensions of the egg become, in *P. capensis* 600×145 μ, in *P. moseleyi* 520×160 μ, in. *P. balfouri* 480×220 μ and in *P. sedgwicki*,

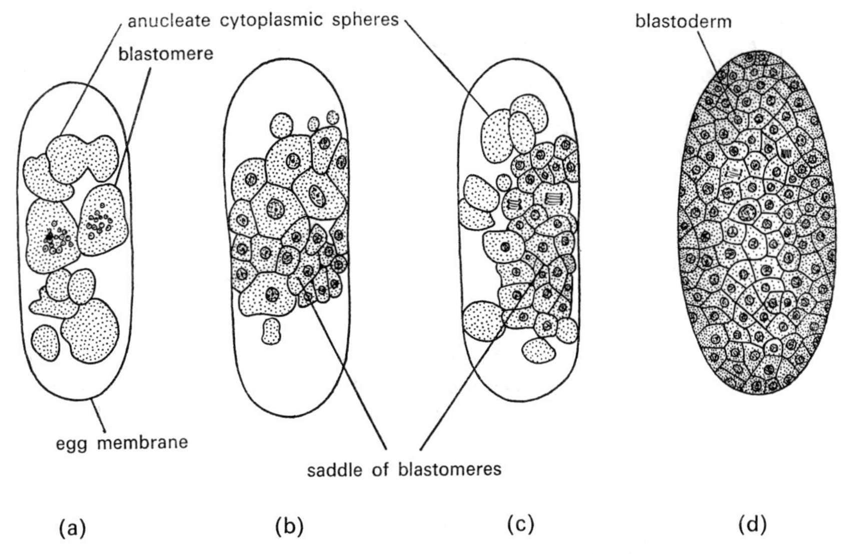

FIG. 40. Early cleavage in *Peripatopsis moseleyi*, after Manton (1949). (a) 2-cell stage; (b) early saddle of blastomeres; (c) later saddle of blastomeres; (d) blastoderm, before dilatation (see text).

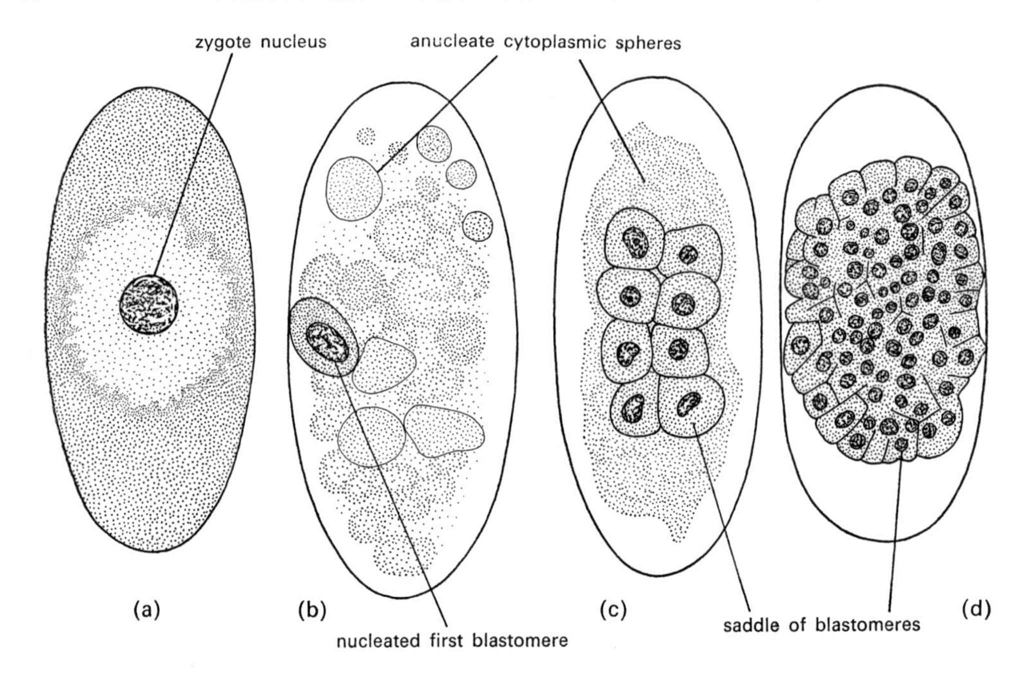

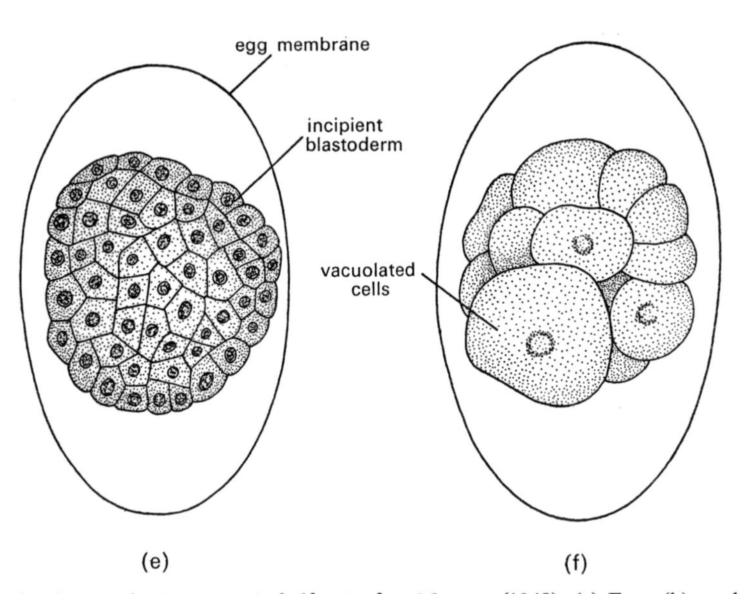

FIG. 41. Early cleavage in *Peripatopsis balfouri*, after Manton (1949). (a) Egg; (b) nucleated first blastomere and anucleate cytoplasmic spheres; (c) 8-cell stage; (d) saddle of blastomeres, with marginal cells migrating beneath the saddle; (e) contracted saddle of blastomeres, external surface; (f) contracted saddle of blastomeres, internal surface.

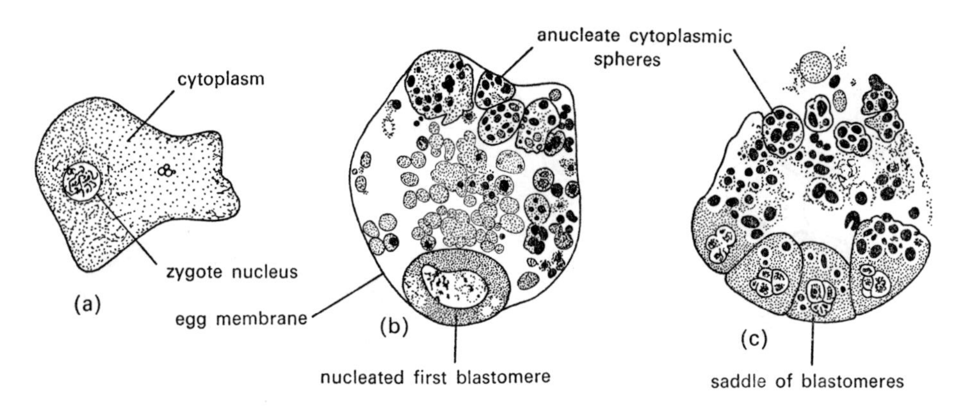

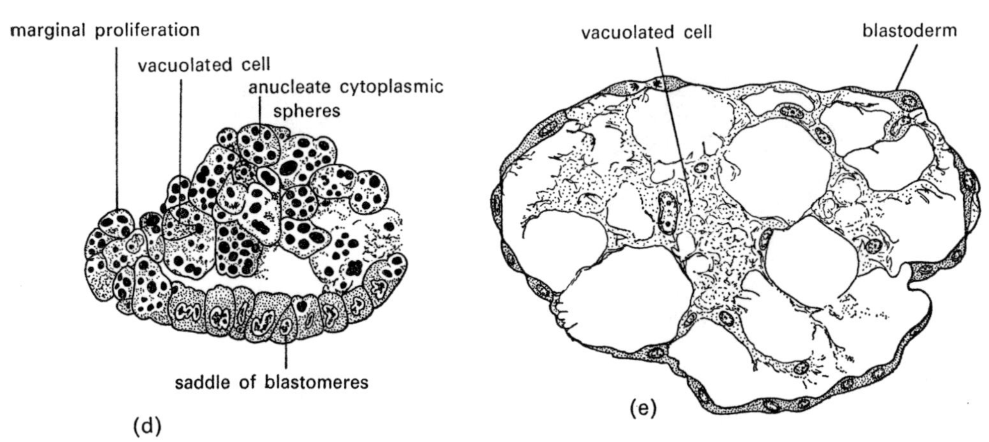

FIG. 42. Cleavage in *Peripatopsis balfouri*, after Manton (1949). (a) Transverse section through egg; (b) transverse section through first blastomere and anucleate cytoplasmic spheres; (c) transverse section through early saddle of blastomeres; (d) transverse section through later saddle of blastomeres with immigrating marginal cells; (e) section through early blastoderm.

260×80 μ. After swelling has taken place, the zygote nucleus lies in an irregular, granular mass of cytoplasm at one side of the egg, and the egg becomes enclosed in two membranes which Manton (1949) interpreted as the vitelline membrane and chorion.

Early cleavage is similar in the four species (Figs. 40–42). The cytoplasm of the egg breaks up into a number of spheres, recalling the yolk spheres of *Peripatoides* and *Eoperipatus*. One sphere is a nucleated blastomere, while the remainder are anucleate pseudoblastomeres. When mitosis begins, the blastomere divides into a disc of blastomeres apposed to the inner surface of the egg membranes on one side of the egg, recalling the blastomere disc of the yolky-egged species. The blastomeres absorb the disintegrating pseudoblastomeres. By the time 64 cells are present, the blastomere disc extends as a saddle around two-thirds of the circumference of the egg, leaving one side and both ends free of cells.

In *P. moseleyi* and *P. sedgwicki* subsequent events reveal clearly that cleavage is a simple modification of its counterpart in yolky onychophoran eggs. Cell division continues (Fig. 40) and the saddle of blastomeres becomes a continuous blastoderm, which then swells by fluid uptake, stretching the enclosing membranes. The final dimensions of the blastoderm,

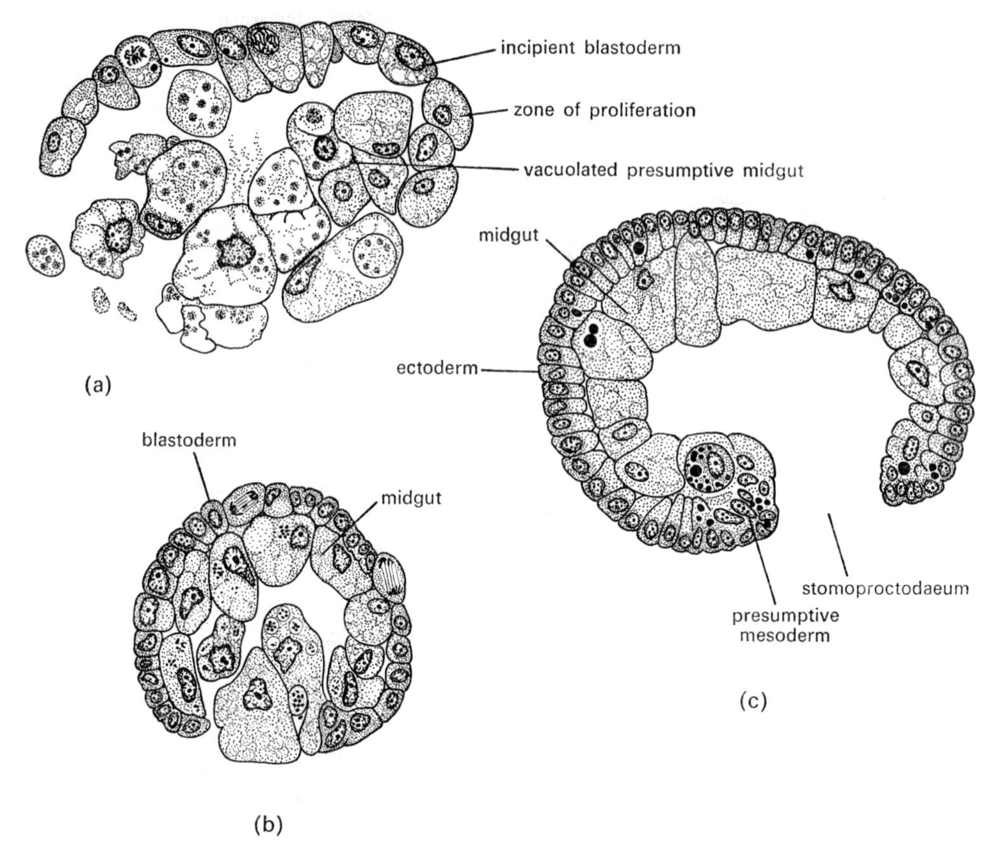

FIG. 43. Late cleavage and early gastrulation in *Peripatopsis capensis*, after Manton (1949). (a) Section through saddle of blastomeres filled with vacuolated midgut cells; (b) section through almost completed blastoderm; (c) longitudinal section through gastrulating embryo (see text).

about 900×500 μ in *P. moseleyi* and 1000×500 μ in *P. sedgwicki* approach those of the blastoderm of ovoviviparous species, and the only difference between the two types of embryo at this stage is the replacement of the yolk mass by fluid in the viviparous species.

P. capensis and *P. balfouri* are more modified. In *P. balfouri* (Figs. 41 and 42) the marginal cells of the saddle of blastomeres swell and move into the interior as the saddle spreads to form a blastoderm. Then, as the blastoderm dilates, the mass of vacuolated cells in the interior gradually disintegrates. *P. capensis* also undergoes swelling and ingression of cells from the margin of the saddle of blastomeres, but in this species the vacuolated cells line the interior of the blastoderm as an inner epithelium around a central cavity (Fig. 43). After the edges of the external blastodermal epithelium have converged as a ventral longitudinal slit and then fused, the blastoderm in this species does not dilate, but the surrounding membranes swell in the usual way. The significance of the ingression of marginal cells during blastoderm formation in *P. capensis* and *P. balfouri* is best explained by reference to the presumptive areas of the onychophoran blastoderm, and will be examined below.

The placental viviparous Onychophora have even smaller eggs than those of the non-placental species and lack the recapitulatory swelling of the egg after release from the ovary.

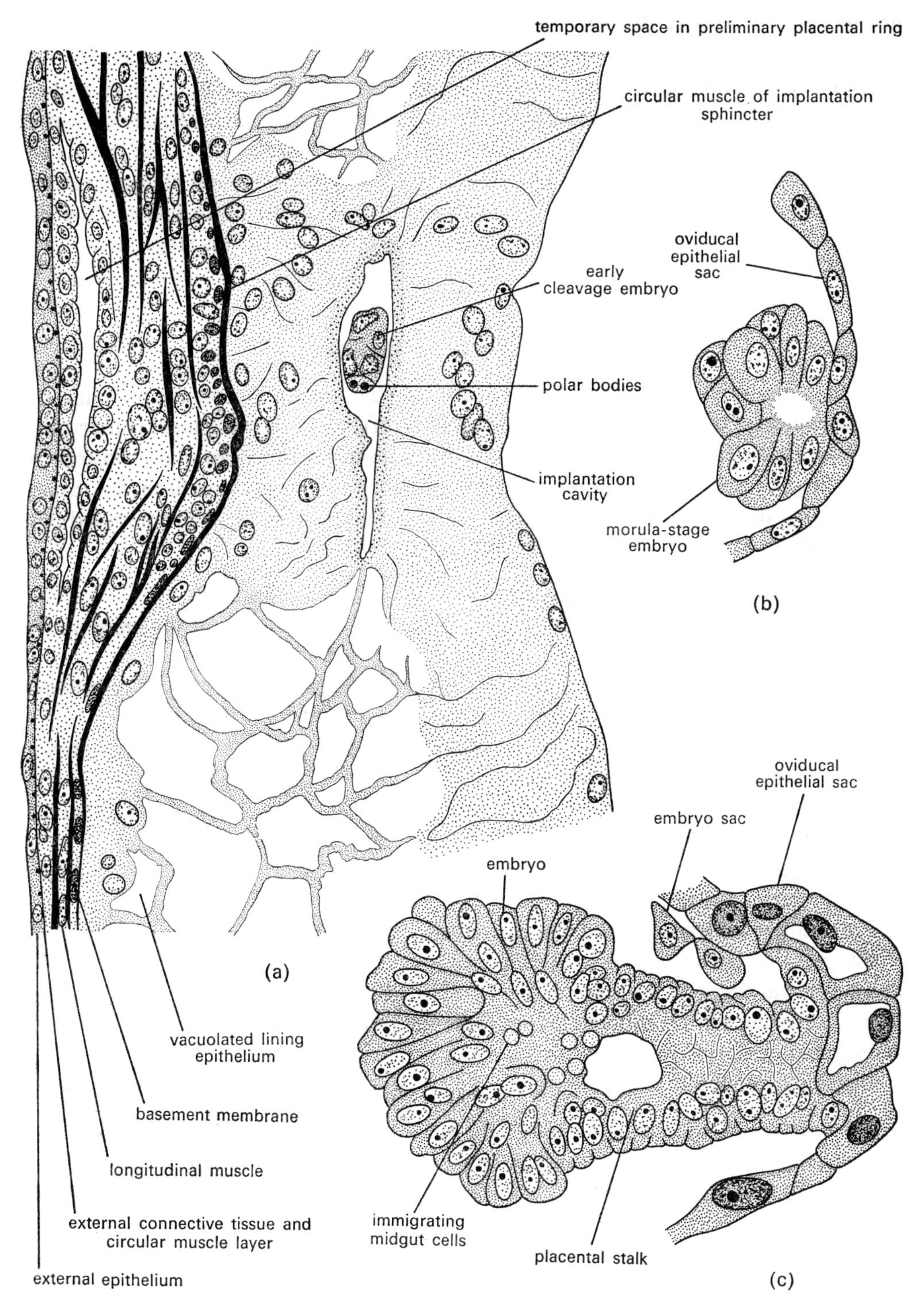

FIG. 44. Cleavage, leading to formation of the stalked vesicle, in a viviparous neotropical onychophoran. (a) early cleavage embryo becoming implanted in oviduct; (b) morula stage; (c) stalked vesicle stage.

The eggs are 25–40 μ in diameter and enclosing membranes are absent (Kennel, 1884, 1885; Sclater, 1888). Cleavage is total and equal, yielding a small morula which lies free in the lumen of the oviduct. With further division (Fig. 44), the morula becomes hollow, with a wall one cell thick, and attaches to the wall of the oviduct. The vesicle now enters directly into a further specialized phase of development, in which the cells of the attachment region proliferate as a hollow stalk, spread out basally as a flat placental plate against the oviducal wall, while the remaining cells undergo further divisions which increase the size and cell number of the hollow vesicle at the free end of the stalk. This stage, as we shall see, is the modified blastoderm.

Presumptive Areas of the Blastoderm

Although the results of cleavage are highly variable among the onychophorans, the resulting pattern of presumptive areas of the blastoderm (Fig. 45) is much more conservative (Anderson, 1966). The basic pattern for the group is displayed in species whose blastoderm covers a mass of yolk spheres. Anteroposterior and dorsoventral axes are fixed, the former corresponding to the long axis of the egg, and the presumptive areas, each composed of a sheet of small cells at the surface, of the yolk, are as follows:

1. Presumptive anterior midgut, a long narrow band of midventral cells.
2. Presumptive stomodaeum, an arc of cells around the anterior end of the presumptive anterior midgut.
3. Ventral presumptive extra-embryonic ectoderm, a pair of narrow bands of cells flanking the presumptive anterior midgut.
4. Presumptive proctodaeum, a pair of small groups of cells on either side of the posterior end of the presumptive anterior midgut.
5. Presumptive posterior midgut, a small group of cells just behind the posterior end of the presumptive anterior midgut.
6. Presumptive mesoderm, a small posteroventral group of cells just behind the presumptive posterior midgut.
7. Presumptive embryonic ectoderm, extending as two broad, ventrolateral bands from the presumptive mesoderm to the anterior end of the blastoderm, where they meet. The ventral edges of the presumptive ectoderm flank, in succession, the presumptive proctodaeum, ventral extra-embryonic ectoderm and stomodaeum.
8. Presumptive dorsal extra-embryonic ectoderm, making up the dorsal half of the blastoderm covering the large yolk mass.

The delineation of these areas is still in need of full confirmation through new studies on yolky onychophoran embryos, since it is based mainly on the results of Sheldon (1888, 1889a) and Evans (1902). It seems likely to be a reasonably accurate assessment, however, because the presumptive areas of viviparous species are so obviously modifications of this pattern. For example, in *Peripatopsis moseleyi* and *P. sedgwicki*, the only differences in the pattern of presumptive areas of the blastoderm are absence of the presumptive anterior midgut, which in yolky species is the source of cells acting as temporary vitellophages digesting the yolk, and a relative increase in the area of extra-embryonic ectoderm as

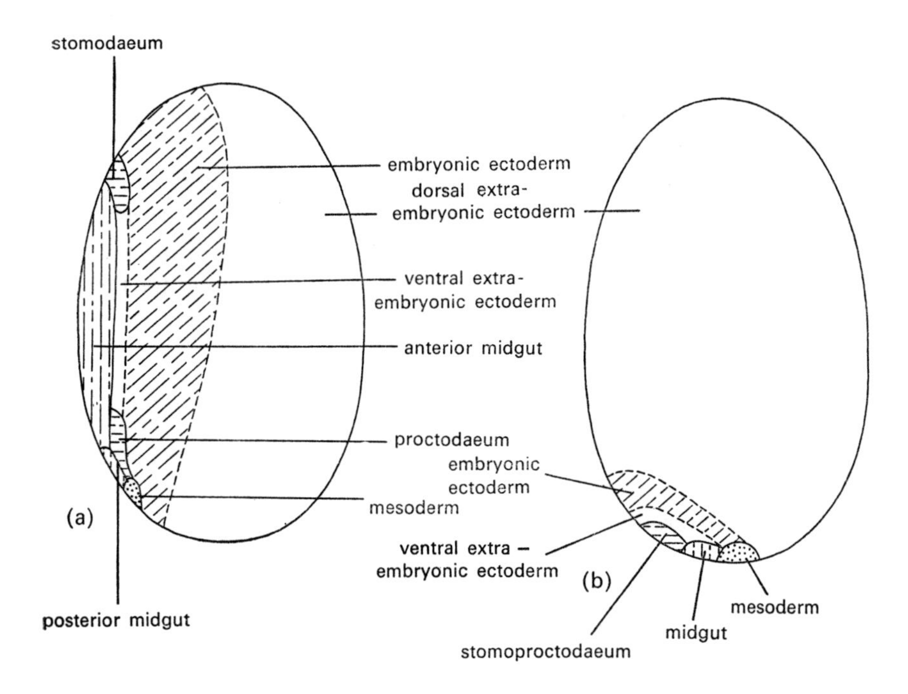

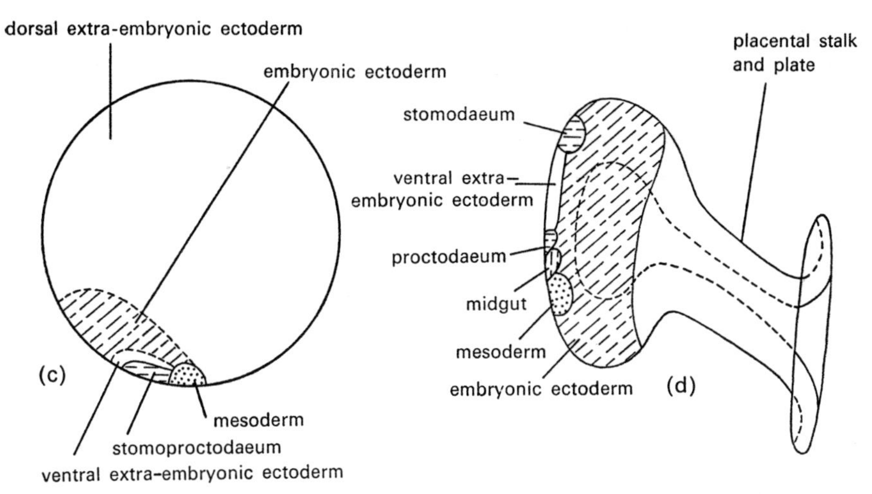

FIG. 45. Fate maps in Onychophora (after Anderson 1966). (a) Yolky species, based on *Peripatoides*; (b) *Peripatopsis sedgwicki*; (c) *Peripatopsis capensis*; (d) viviparous placental species.

compared with embryonic ectoderm. The latter difference no doubt reflects a new functional role for the extra-embryonic ectoderm as an absorptive epithelium, convergently similar to the albumenotrophic yolk-sac ectoderm of secondarily yolkless clitellate embryos. *P. balfouri* has a similar pattern of blastodermal presumptive areas, but retains a vestige of the presumptive anterior midgut in the form of the temporary vacuolated cells which roll into the interior from the edge of the saddle of blastomeres during blastoderm formation. *P. capensis* is specialized in a uniquely different way. The same anterior midgut cells form a precocious lining to the blastoderm, which probably has an absorptive function. In association with this, the presumptive posterior midgut is lost and the extra-embryonic ectoderm does not dilate. It was Sedgewick's misfortune to select the highly specialized *P. capensis* for his pioneering study of onychophoran development.

In viviparous placental Onychophora, a consideration of presumptive areas shows that the placental stalk and plate are modified dorsal extra-embryonic ectoderm and that the remaining areas are established in the usual way in the vesicular part of the embryo, except for complete loss of the presumptive anterior midgut. Thus, throughout the Onychophora, in spite of great variations of the mode of cleavage, the presumptive areas formed as a result of cleavage show a remarkably stable configuration, only the presumptive midgut and the extra-embryonic ectoderm, the two areas intimately involved with yolk, being much affected by secondary loss of yolk. It has already been pointed out that the same generalization can be made for annelid embryos in respect both of increase in yolk and of secondary loss of yolk, and we shall see in due course that all arthropods express the same functional correlation, though in several different ways.

Gastrulation

Gastrulation in the Onychophora has to be different from that of any annelid, simply because the presumptive areas entering into gastrulation are sheets of small, yolk-free cells at the surface of the yolk (or its equivalent fluid-filled cavity in secondarily yolkless embryos), and include no large cells whatsoever. Only Manton (1949) has studied onychophoran gastrulation in detail, but it is possible to discern in the results of the several classical workers that the events in *Peripatopsis*, although happening in secondarily yolkless embryos, are generally typical of all species (Anderson, 1966).

For annelids, gastrulation is the movement of presumptive areas of the blastula into their organ-forming positions. In the Onychophora, the extent to which the small-celled presumptive areas move as whole areas is extremely limited. With the exception of the precocious movement of the presumptive anterior midgut in *Peripatopsis capensis*, becoming internal during cleavage, and of its vestigial equivalent in *P. balfouri*, the presumptive areas in the Onychophora tend not to perform gastrulation movements. Rather, they enter into proliferative organogeny in their primary superficial positions, so that the cells which invade the interior of the embryo pass in as small cells through and around the yolk mass, and immediately begin to form organ rudiments.

This being the case, it is particularly significant that the presumptive anterior midgut in yolky onychophoran embryos, although a midventral sheet of small cells at the surface of the yolk, retains a definite and quite remarkable gastrulation movement (Fig. 46). The sheet of cells sinks inwards at the onset of gastrulation, then separates along the midventral line to leave the yolk broadly exposed along the ventral surface. This seems to be the only case among arthropods in which the yolk mass or yolk spheres are directly exposed to the exterior at any stage during development. The paired bands of sunken presumptive anterior midgut cells bordering the ventral slit now become mitotically active. Cells proliferated from these bands migrate upwards through the peripheral yolk, taking up yolk and acting as temporary vitellophages as they move. Eventually the temporary vitellophage cells form

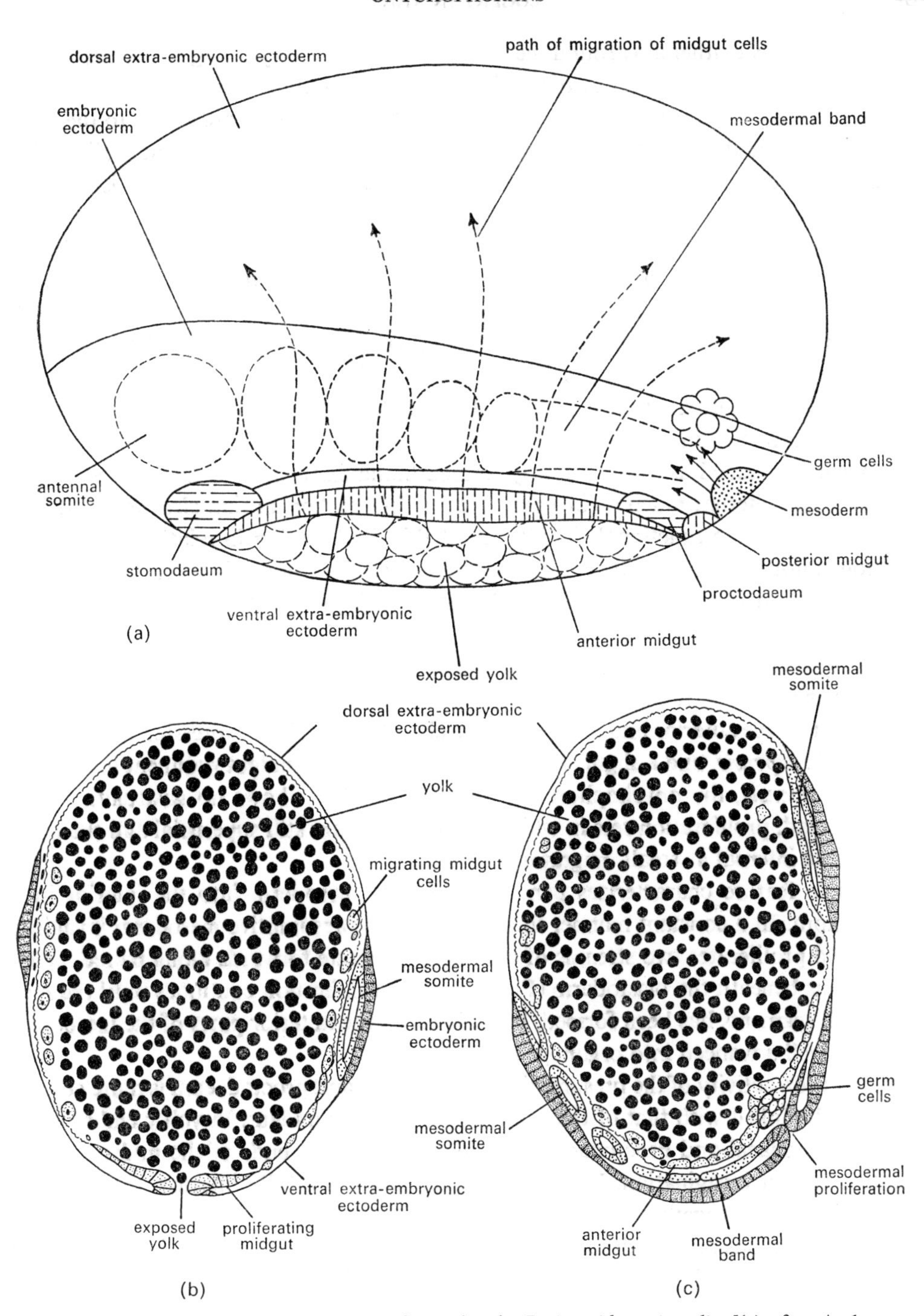

FIG. 46. Gastrulation and early segment formation in *Peripatoides orientalis*. [(a) after Anderson (1966).] (a) Diagrammatic lateral view; (b) section showing immigrating midgut; (c) section showing immigrating mesoderm and somite formation.

a complete, yolky epithelium around the yolk mass, subsequently giving rise to the definitive epithelium of the anterior part of the midgut. At the same time, the presumptive posterior midgut sinks slightly inwards and begins to proliferate cells which are added to the midgut epithelium behind the level of the yolk mass. It is likely that posterior midgut proliferation is prolonged, extending the midgut along the growing trunk (see below), but is completed before the full number of trunk segments is formed.

In the secondarily yolkless embryos of *Peripatopsis sedgwicki* and *P. moseleyi*, the presumptive anterior midgut and its temporary vitellophage function are lost, and in *P. balfouri* only vestigially retained. The entire midgut in these species is formed by cells proliferated from a slightly invaginated posterior midgut rudiment (Figs. 47, 48). The cells spread forwards and upwards beneath the blastoderm to enclose a central, fluid-filled space, and continue to be proliferated until almost all of the trunk segments have been formed. In *Peripatopsis capensis*, in contrast, as pointed out above, the presumptive anterior midgut component is retained as a precociously formed lining epithelium of the blastoderm (Fig. 43), becoming internal much as in yolky species, but earlier, during cleavage, and in the absence of yolk. The functional significance of this modification in relation to embryonic nutrition is not yet established. The presumptive posterior midgut is absent in *P. capensis*.

The midgut in the viviparous placental species again arises by prolonged proliferation from the presumptive posterior midgut, which sinks slightly inwards before proliferation begins (Fig. 49). The proliferated cells give rise to an epithelial sac. No trace of the presumptive anterior midgut is retained. Thus, apart from the unique invagination and separation of the presumptive anterior midgut in yolky species, its precocious expression in the absence of yolk in *P. capensis*, and a slight sinking in of the posterior midgut rudiment in all species except *P. capensis*, the onychophoran presumptive midgut shows no gastrulation movements. Instead, proliferation sets in while the small cells are still superficial and the midgut cells become internal as small, migratory cells.

The same feature characterizes penetration of the mesoderm into the interior in the Onychophora. In a constant manner in all species, the presumptive mesoderm invaginates slightly and proliferates two streams of cells which migrate forwards along paired, ventrolateral paths beneath the blastoderm (Figs. 46–49). Proliferation continues until the mesoderm of all segments has been formed, by which time the anterior parts of the mesodermal bands are well advanced into organogeny.

Gastrulation activity of the presumptive stomodaeum and proctodaeum is more complex, being closely linked with that of the presumptive anterior midgut. In yolky embryos (Fig. 46), the stomodaeal and proctodaeal areas invaginate at the ends of the midventral, presumptive midgut invagination. When the midgut bands are fully internal, the lips of the invagination merge together along the ventral midline, except at the ends. Internally, the paired midgut bands meet midventrally. Externally, as a result of gastrulation spread of the ventral extra-embryonic ectoderm, the superficial layer of cells also meets midventrally (see below). At the ends of the original invagination, the stomodaeal and proctodaeal rudiments are closed off as short tubes, opening by the mouth and anus respectively.

In *Peripatopsis*, in spite of the absence of a midventral anterior midgut component in the blastoderm, much of this sequence is retained. The confluent presumptive stomodaeal and proctodaeal areas invaginate together, producing a short midventral slit (Fig. 48). The slit

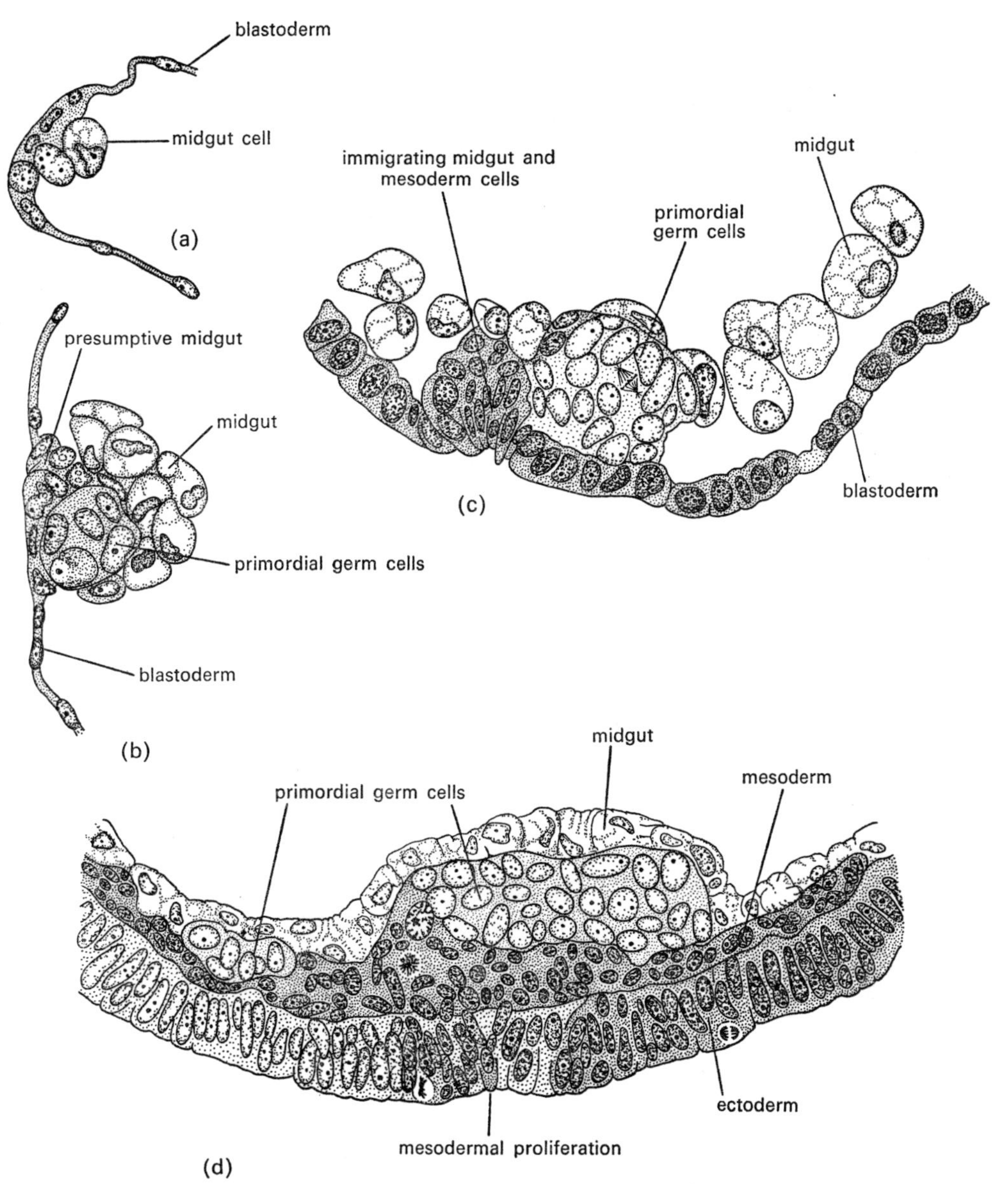

FIG. 47. Gastrulation in *Peripatopsis moseleyi*, after Manton (1949). (a) Section showing onset of proliferation of presumptive midgut; (b) sagittal section showing later midgut proliferation; (c) sagittal section at onset of proliferation of presumptive mesoderm; (d) transverse section showing later mesodermal proliferation.

then elongates as a result of cell proliferation in its lateral lips, pushing the stomodaeal and proctodaeal rudiments apart at the ends of the slit. By this time, the midgut epithelium is present as a sac whose ventral wall lies within the ventral slit. Now, in what appears to be a reminiscence of early midgut formation in yolky onychophoran embryos, the ventral wall of the midgut sac opens along the ventral midline and the margins of the opening fuse with those of the superficial midventral slit. Just as in yolky species, therefore, the interior

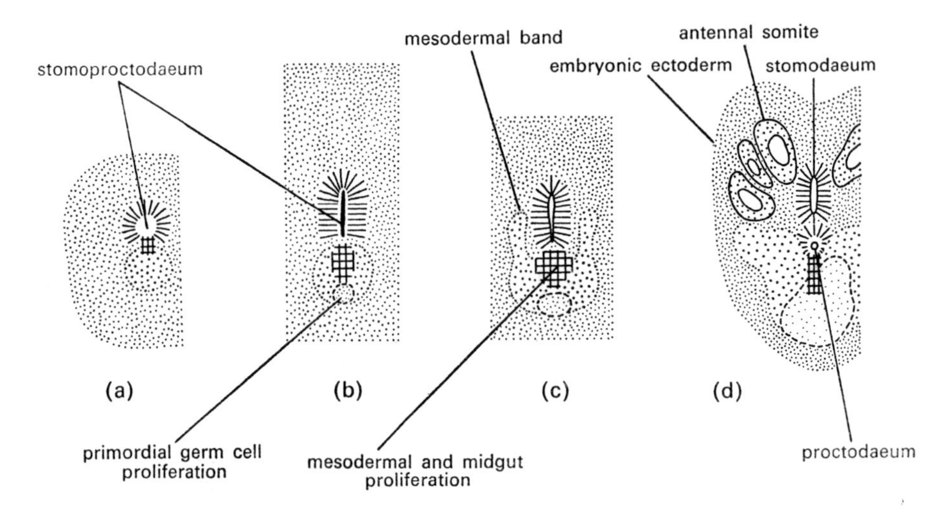

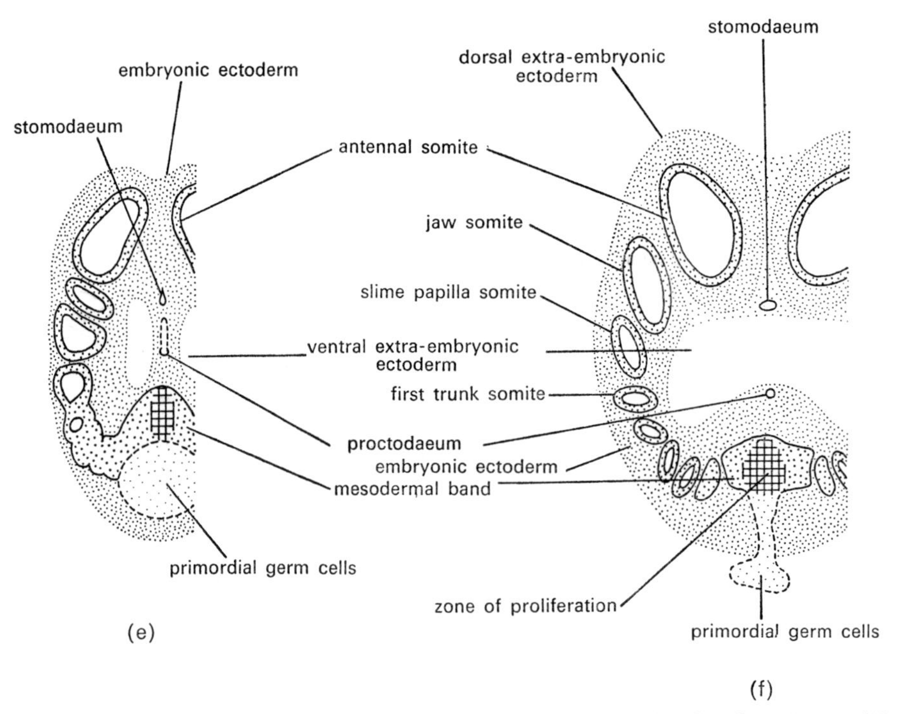

FIG. 48. Gastrulation and early somite formation in *Peripatopsis sedgwicki*, after Manton (1949). (a) Presumptive areas; (b) formation of stomoproctodaeum; (c) early mesoderm proliferation; (d) separation of mouth and anus, early somite formation; (e) and (f) continued mesoderm proliferation and somite formation.

of the midgut in *Peripatopsis* becomes temporarily open to the exterior. The functional significance of retention of this phenomenon in the absence of yolk is not understood. Midventral closure of the slit now proceeds as in yolky species. The midgut cells separate from the superficial cells and reunite, restoring the complete ventral wall of the midgut sac.

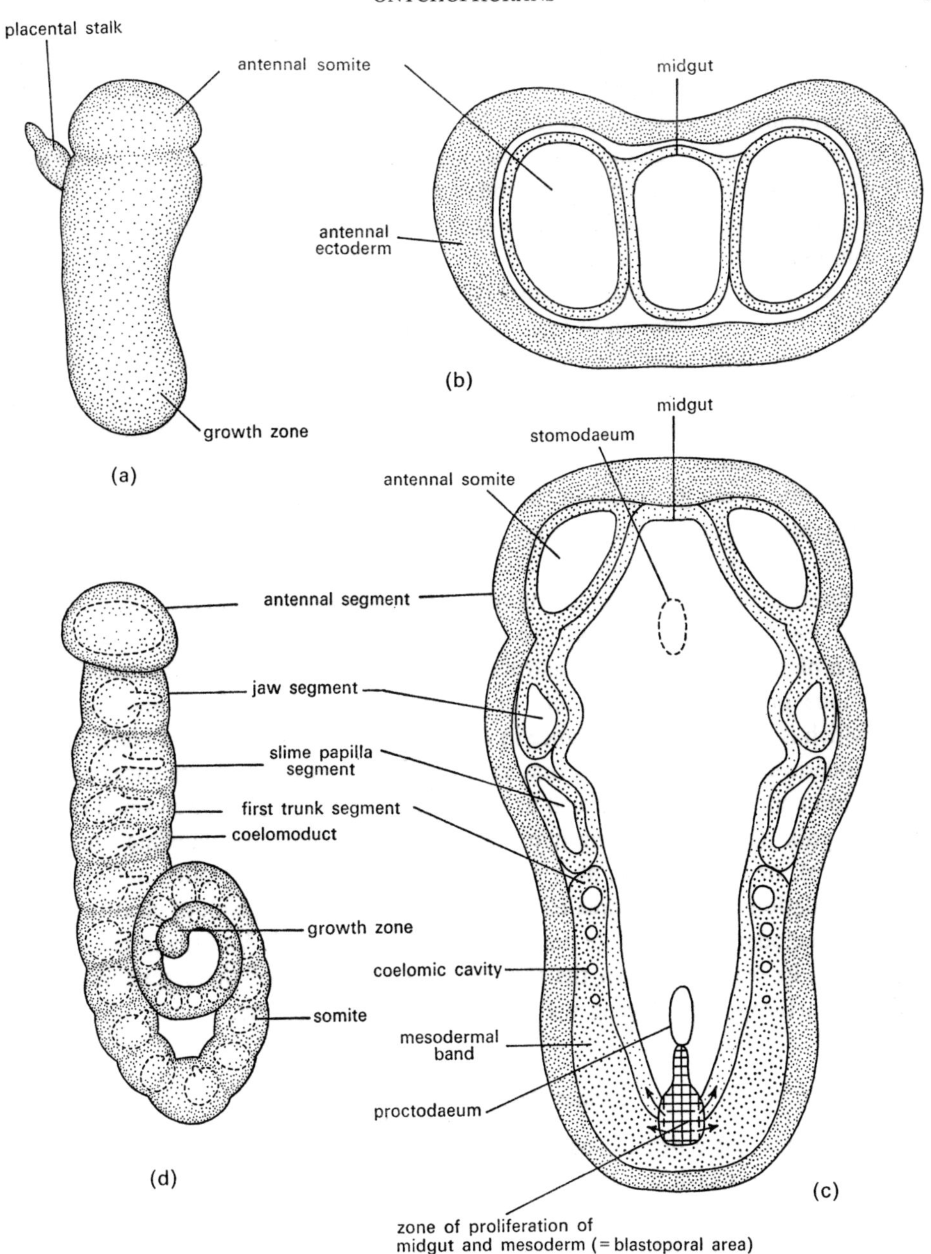

FIG. 49. Gastrulation and somite formation in viviparous placental Onychophora. (a) Early elongation of stalked vesicle and onset of segment delineation; (b) transverse section through the antennal segment of the same embryo; (c) diagrammatic frontal section of the same embryo; (d) coiled, segmenting embryo in lateral view [(d) after Manton, 1960.]

The superficial cells also unite along the ventral midline, at the same time completing the tubulation of the stomodaeum and proctodaeum. One possible explanation of the retention of this curious sequence in the absence of yolk in *Peripatopsis* is that it is an essential phase in the tubulation of the stomodaeum and proctodaeum. In placental, viviparous species, however, the midventral slit is entirely eliminated and the presumptive stomodaeum and proctodaeum invaginate independently as short tubes which break through secondarily to establish continuity with the midgut (Fig. 49).

We are left now with the changes in presumptive ectoderm during onychophoran gastrulation (Figs. 46, 48 and 49). The presumptive embryonic ectoderm is formed directly as paired ventrolateral bands of blastoderm cells and shows no gastrulation movement. The dorsal component of the extra-embryonic ectoderm also lacks this activity, being formed initially as a broad area of blastodermal cells covering the yolk. In contrast, the paired ventrolateral bands of extra-embryonic ectoderm in the yolky and the viviparous non-placental species come together in the ventral midline by a gastrulation spreading movement which completes the ectodermal covering of the embryo. In the placental species, the same component is narrow and unpaired and forms directly in its definitive position, a secondary simplification.

Consideration of the gastrulation movements in onychophoran embryos bears out the fact, already apparent from the cleavage processes and presumptive area patterns in these animals, that when yolk is absent, it is absent secondarily. The movements which occur are obviously basically related to the setting out of a blastoderm around a large, internal yolk mass. Furthermore, we can arrive at two useful further generalizations at this point. One is that when a blastoderm is formed, gastrulation movements are largely eliminated in favour of precocious proliferation, especially of the presumptive mesoderm and midgut. The second, which is of especial importance in comparing the development of blastula-forming and blastoderm-forming species, is that presumptive areas of the same location, relative juxtaposition and fate may be formed as large cells containing yolk or as small cells at the surface of yolk (e.g. the presumptive mesoderm and midgut of clitellate annelids and Onychophora), or alternatively, as a small group of cells which proliferates and spreads during gastrulation or a broad area of cells formed directly as a result of cleavage (e.g. the presumptive ectoderm of clitellate annelids and Onychophora). Various intermediates are also possible, as we shall see in the Crustacea. The bearing of these generalizations on annelid and arthropod phylogeny will be discussed in due course.

Further Development of the Gut

Apart from the early development of the embryonic ectoderm and mesoderm during segment formation, which was described in detail by Manton (1949), the further development of onychophoran embryos after gastrulation is still known mainly from the classical works of Sedgewick, Sheldon and Kennel; but Manton also made a number of interesting observations on the development of the gut in *Peripatopsis*.

In all species, the stomodaeum and proctodaeum develop directly, through cell division and differentiation, as the foregut and hindgut epithelia. The further development of the midgut is more diverse. In yolky embryos, in which the anterior part of the midgut is filled with yolk and has a wall of temporary vitellophage function, the further development of the midgut proceeds partly by resorption of yolk and partly by the addition of further cells from the posterior midgut rudiment. Both the temporary vitellophage portion and the posterior portion then differentiate directly, after further cell divisions, as midgut epithelium. Viviparous non-placental species show various modifications of this condition. *Peripatopsis sedgwicki* and *P. moseleyi* retain the large area of dorsal, extra-embryonic ectoderm as a swollen, anterodorsal sac during a long period of their later development, in spite of the absence of yolk (see below, Fig. 51, and development of external form). In yolky species, of course, the yolk-filled anterior midgut is mainly covered by this ectoderm. *P. sedgwicki* and *P. moseleyi* form a small midgut sac entirely as a result of proliferation from the presumptive posterior midgut, but the midgut sac then expands, with temporary attenuation and vacuolation of its cells, to line the inside of the swollen sac of dorsal extra-embryonic ectoderm. As elongation of the embryo proceeds, the midgut cells enter into mitosis and become more compactly arranged once more, accompanied by shrinkage of the antero-dorsal sac, to form the definitive midgut epithelium. The temporary vacuolated phase is obviously equivalent to the temporary vitellophage phase of the midgut cells of yolky species. Its functional role in embryonic nutrition is not known, but seems likely to justify the name of trophic vesicle for the swollen sac as a whole. The problem is compounded, however, by the fact that the temporary vacuolated phase is eliminated in *P. balfouri* (Fig. 52) and *P. capensis*. The extra-embryonic ectoderm of *P. balfouri* shrinks before the midgut cells have spread beneath it and the midgut epithelium develops directly. A surplus of cells is produced dorsally, followed by the degeneration of many of these cells, presumably as a vestige of the swollen phase of the midgut in *P. sedgwicki* and *P. moseleyi*. In *Peripatopsis capensis*, finally, the blastoderm does not dilate and the midgut epithelium develops directly, without even a vestige of temporary swelling.

The viviparous placental Onychophora also show a direct and simplified development of the midgut epithelium proliferated from the presumptive area of posterior midgut (Fig. 49). No specialization of the midgut is associated in these species with the modification of the dorsal extra-embryonic ectoderm as a placenta.

Development of External Form

All onychophorans are epimorphic, attaining the full complement of segments and adult organ systems before they hatch or are born as juveniles. In spite of the frequency of yolklessness and viviparity among species, the development of the external form of the embryo is not greatly variable. The external features of development have been described and illustrated by Sheldon (1889) for the yolky, ovoviviparous *Peripatoides novae-zealandiae*, by Manton (1949) for *Peripatopsis* and by Kennel (1884, 1885) for placental species, as well as in a more fragmentary manner by a number of other workers.

Development always proceeds in strict anteroposterior succession. As the mouth and anus are being formed, the paired lateral halves of the embryo begin to develop as thickened bands on the surface of the blastoderm and soon exhibit paired segmental swellings demarcated by intersegmental annuli. The first pair of swellings develops as the two halves of the antennal segment, in a preoral position in front of the mouth. The two halves of the antennal segment are always in contact in the anterior midline. Behind them, on either side of the mouth, the two halves of the jaw segment arise, followed by the slime-papilla segment and a series of trunk segments. Trunk segment formation continues at the posterior end at a relatively rapid rate, so that limb buds do not begin to appear on the anterior segments until most of the trunk segments have been formed. Due to acceleration of segment formation relative to the further development of the segments, the full complement of segments is formed before the anterior segments have progressed far along their pathway of further development. A similar developmental balance between segment formation and segment elaboration is observed in some yolky polychaetes (e.g. *Scoloplos*, Chapter 2) and clitellate annelids (e.g. *Tubifex*, Chapter 3) as a corollary of the direct development of a long-bodied, metameric animal from a yolky egg.

In the yolky embryo of *P. novae-zealandiae*, the segment halves of the anterior part of the embryo are widely separated on either side of the yolk mass (Fig. 50). As segment proliferation continues, the posterior part of the embryo is prolonged beyond the yolk mass as a tubular caudal papilla, in which the two halves of each segment are only slightly separated

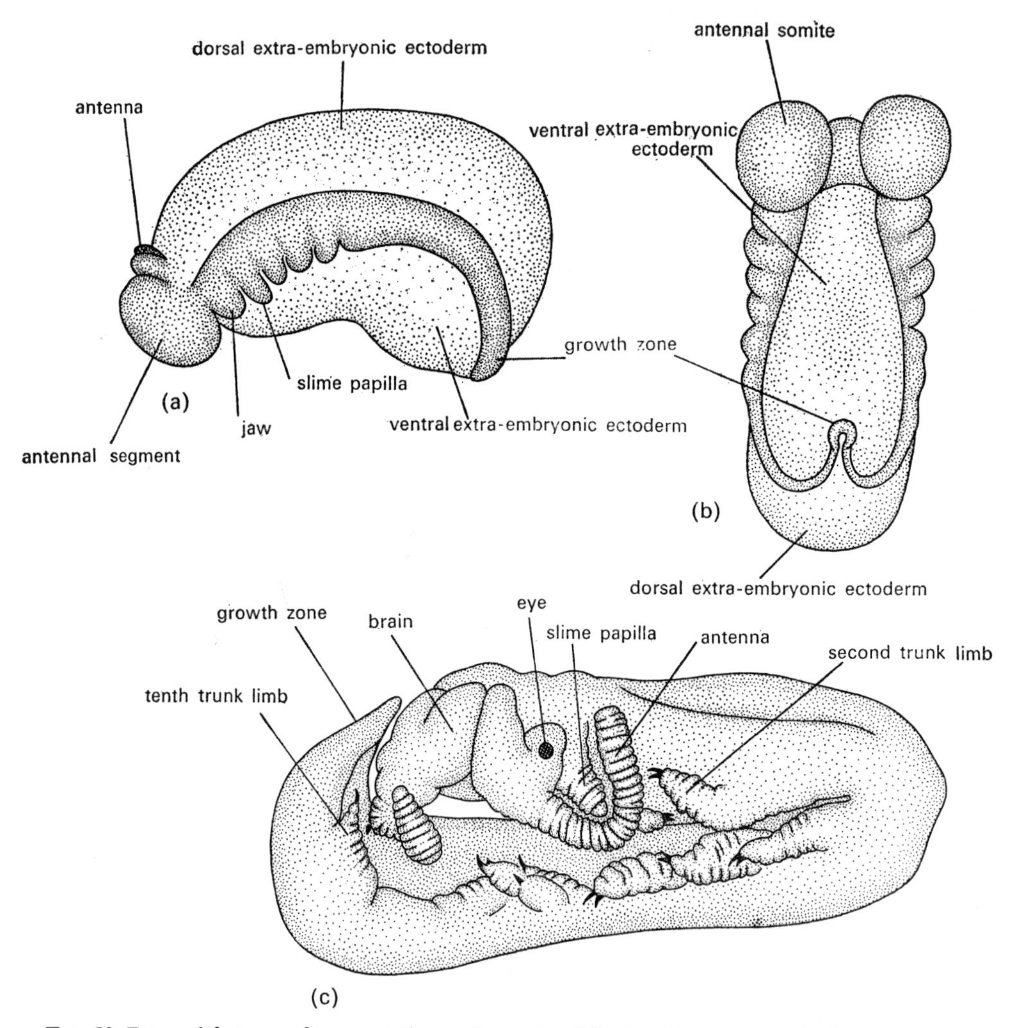

FIG. 50. External features of segmentation and growth of *Peripatoides novae-zealandiae* [(a) and (b) after Sheldon, 1889a; (c) original]. (a) Early segment delineation, lateral view; (b) the same stage in ventral view; (c) flexed embryo, lateral view.

by narrow mid-dorsal and mid-ventral prolongations of the extra-embryonic ectoderm. Associated with development in a confined space enclosed by ovoid membranes, the elongating caudal papilla is flexed forwards ventrally beneath the yolk-filled anterior part of the embryo. As the development of limb buds continues and the full number of trunk segments is gradually attained, the yolk within the anterior part of the embryo diminishes in volume. This part, therefore, also becomes tubular, with the two halves of each segment coming together in the midline, first ventrally, later dorsally. The final, flexed position of the embryo within the egg membranes results from this change.

In those species of *Peripatopsis* which retain a swollen sac of dorsal extra-embryonic ectoderm and still develop within enclosing egg membranes (*P. moseleyi*, *P. sedgwicki*; Figs. 51, 52), this sequence of development is modified only in minor ways. The ventral separation of the two halves of the anterior segments is slight, but the dorsal separation resulting from the swollen trophic vesicle is extensive. It is notable, however, that the forward flexure of the embryo takes place when only a few segments have been formed, so that the dorsal separation due to the trophic vesicle is confined to a few anterior segments only. In *P. moseleyi*, the trophic vesicle is gradually resorbed without further specialization, but in *P. sedgwicki*, the attachment of the trophic vesicle to the embryo soon becomes restricted to a narrow neck inserted at the dorsal surface of the antennal and jaw region. Ventral, forward flexure of the embryo persists during the completion of segment formation, and the increasing size of the embryo is accommodated by progressive shrinkage of the trophic vesicle in both species. The final position of the embryo within the egg membranes is the same as that of *P. novae-zealandiae*, developing from a yolky egg.

When, with further modification, the trophic vesicle is no longer a feature of development in *Peripatopsis*, as in *P. balfouri* (Fig. 53) and *P. capensis*, dorsal separation of the two halves of the anterior segments is almost eliminated from development. The embryo develops in a simple direct manner, by proliferation of segments accompanied by forward flexure of the posterior part beneath the anterior part in the usual way. The same simple pattern of growth and segment formation, accompanied by ventral forward flexure of the posterior part of the body, is retained in the viviparous placental species (Fig. 49). The fact that the insertion of the placental stalk in these species is located in the same position as the insertion of the trophic vesicle in *Peripatopsis* lends further weight to the view that the placenta in neotropical Onychophora is a modification of the dorsal, extra-embryonic ectoderm.

Before passing on to the details of segment development in onychophoran embryos, mention can be made of a special problem peculiar to the development of placental Onychophora, the problem of having an embryo which is attached to the oviducal wall yet moves progressively down the oviduct as it develops. Anderson and Manton (1972) have recently examined this question in the two species of neotropical Onychophora whose embryology was studied by Kennel (1884, 1885), namely, *Epiperipatus trinidadensis* and *Macroperipatus torquatus*. The relationship between the embryo and the oviduct wall is one of remarkable complexity. The embryo first becomes implanted in a newly proliferated section of proximal oviduct leading from the ovary. It remains within this section of oviduct as it develops. Meanwhile, further sections of oviduct containing a succession of younger embryos are added proximally, while older embryos are born distally. The sections of oviduct that had contained the older embryos are resorbed in association with each birth.

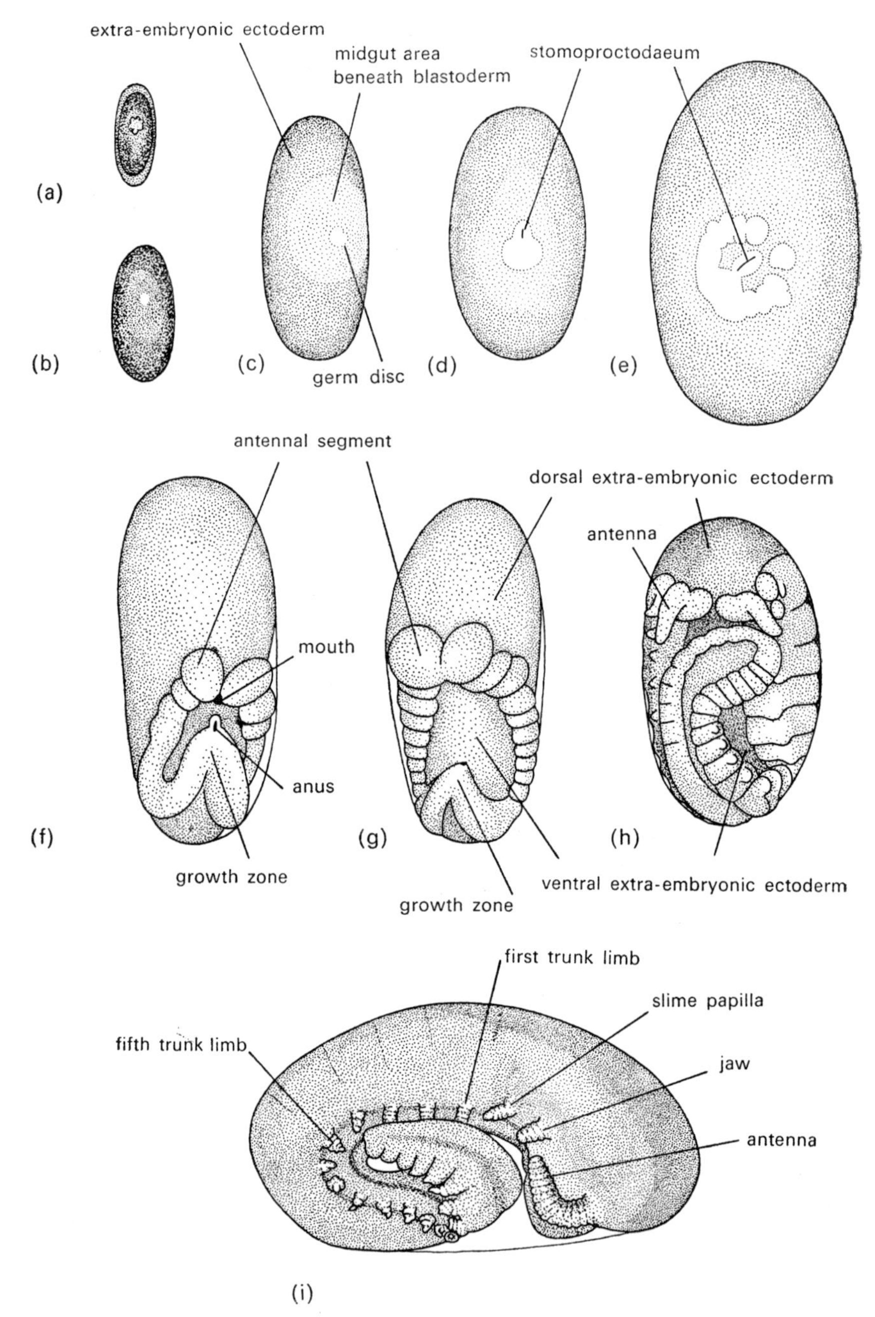

FIG. 51. External features of segmentation and growth of *Peripatopsis mosleyi*, after Manton (1949). (a) Blastoderm before dilatation; (b) and (c) early dilatation and formation of the embryonic primordium; (d) further enlargement of the embryonic primordium; (e) initial segment delineation; (f) and (g) continued segment formation and ventral flexure; (h) fully segmented embryo; (i) older flexed embryo after resorption of dorsal extra-embryonic ectoderm.

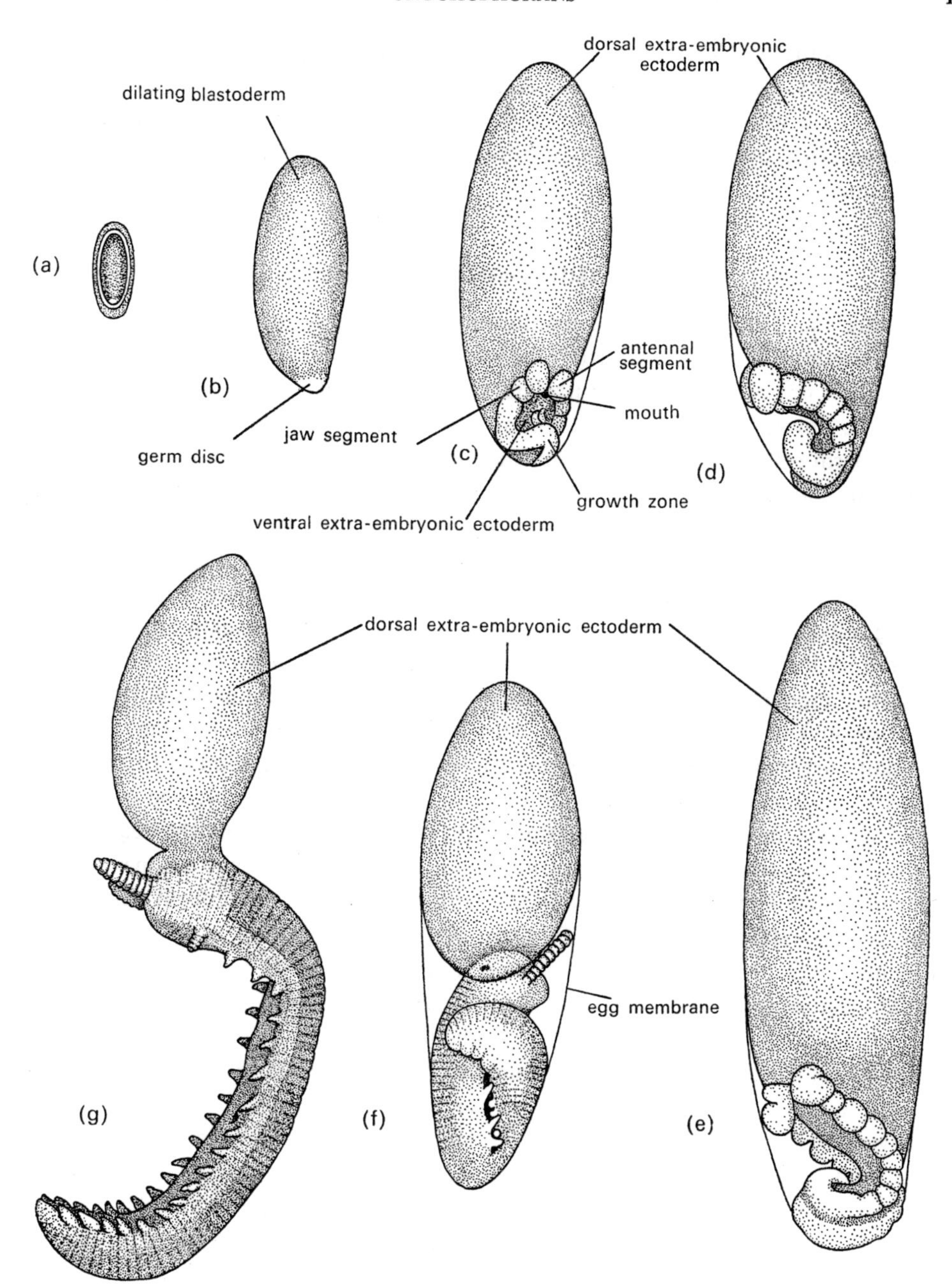

FIG. 52. External features of segmentation and growth of *Peripatopsis sedgwicki*, after Manton (1949). (a) Blastoderm before dilatation; (b) dilated blastoderm with embryonic primordium; (c), (d) and (e) stages in segment formation and ventral flexure; (f) older embryo with shrinking trophic vesicle; (g) slightly older embryo, unfolded to show details of head and trunk.

The oviduct is thus in effect a conveyor belt, in which each fixed embryo is gradually transported towards the genital opening as a result of proximal growth and distal resorption of the oviduct as a whole. The oviducal epithelium forms a closed sac around the embryo, while the external layers of the wall contribute to a complex placentation zone external to the attachment of the placental stalk.

Further Development of the Mesoderm

Preceding the onset of external manifestation of segmentation in the Onychophora, copious mesoderm is proliferated from the small posteroventral area of presumptive mesoderm, in the form of two broad ventrolateral bands of small cells which spread forwards between the developing midgut epithelium and the overlying blastoderm (Figs. 46–49). The mesoderm cells push forwards until they meet in the anterior midline in front of the stomodaeum. Somite formation, which is similar in yolky and viviparous embryos and has been described in detail by Manton (1949), begins as the ends of the two mesodermal bands approach the level of the stomodaeum. Each pair of somites originates by coelomic cavity formation in the mesoderm, followed by the individuation of somite walls from the surrounding cells, much as in yolky annelid embryos. The somites develop in anteroposterior succession until the last segmental pair has been individuated, when proliferation from the presumptive mesoderm ceases. As a result of continued forward migration, the first three pairs of somites come to lie in front of, at, and just behind the level of the stomodaeum. These pairs are subsequently incorporated into the head as the somites of the antennal, jaw and slime papilla segments and will be discussed separately below. The swelling of the somites initiates the superficial delineation of the segments throughout the head and trunk.

Although the paired somites of the trunk segments form in a simple manner reminiscent of that of annelids and enlarge as swollen coelomic sacs, their subsequent development is entirely arthropodan (Figs. 54 and 55). Only Kennel (1884, 1885), Sedgwick (1887, 1888) and Evans (1902) have so far described what takes place, but there seems little doubt that the initial change following swelling of a somite is partial subdivision into three compartments, dorsolateral, medioventral and appendicular. Only the appendicular compartment contributes cells which differentiate as somatic musculature. As this compartment pushes outwards, the overlying ectoderm protrudes as a segmental limb bud. The wall of the somite in this vicinity then disaggregates and spreads as a mesenchyme which differentiates into intrinsic and extrinsic limb musculature, circular muscle and longitudinal muscle of its half-segment. In contrast to the varied accounts of the origin of annelid circular muscle (see Chapter 3), the origin of circular muscle from somatic mesoderm in Onychophora seems to be well established.

With the transformation of somite walls partially into mesenchyme, the coelomic cavities become confluent with spaces which develop between the mesenchyme cells. An extensive system of blood spaces thereby results, forming the haemocoelic body cavity of the developing animal. The only definite vessel is the mid-dorsal heart, formed by cells of the upper margins of the dorsolateral compartments of the somites. Anteriorly, if the somites are widely separated, the walls of these compartments initially become mesenchymatous and push up towards the dorsal midline over the surface of the swollen anterior midgut. The

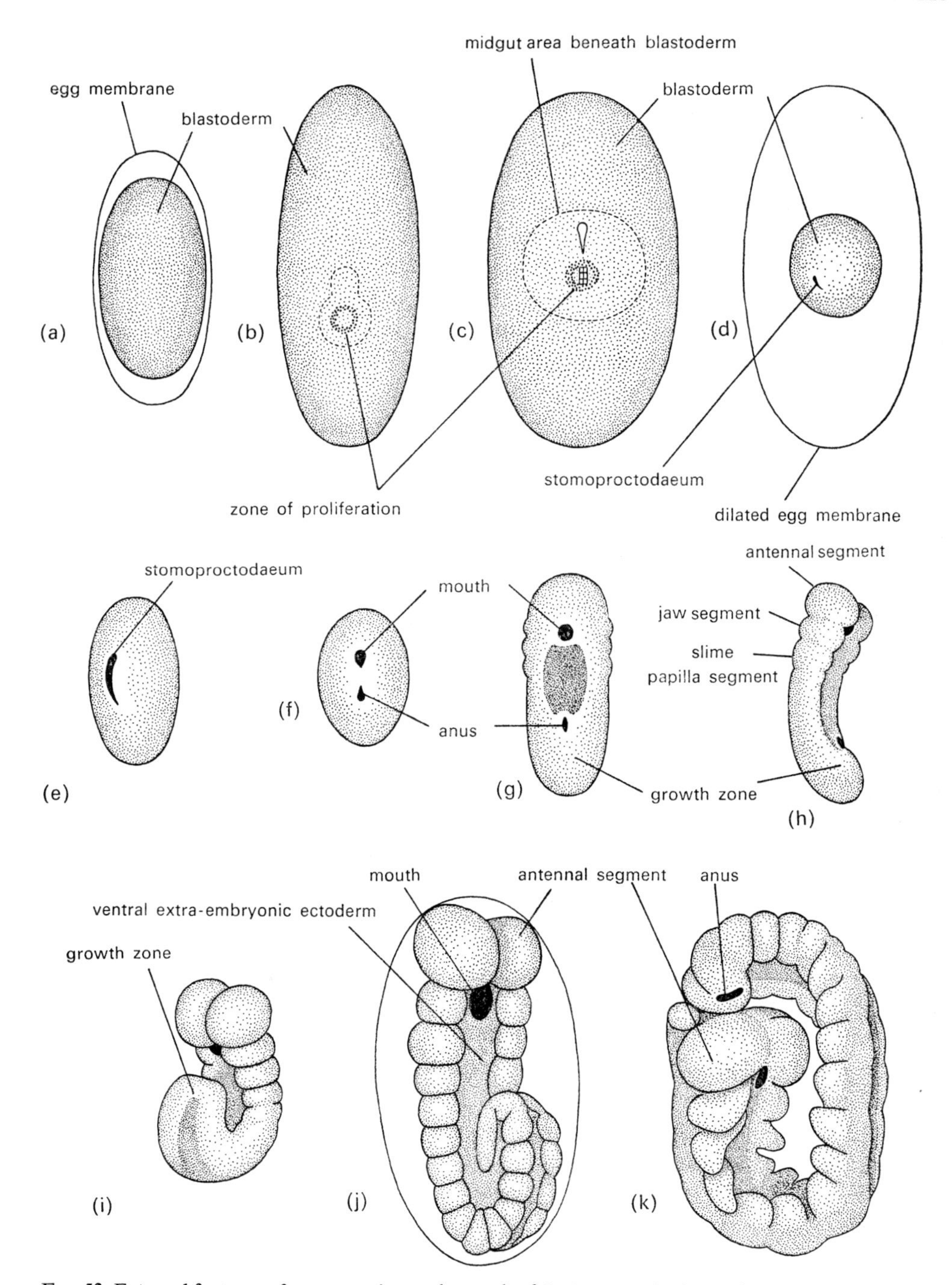

FIG. 53. External features of segmentation and growth of *Peripatopsis balfouri*, after Manton (1949). (a) Blastoderm before dilatation; (b) dilated blastoderm with embryonic primordium; (c) enlargement of embryonic primordium and formation of stomoproctodaeum; (d) embryo soon after secondary shrinkage; (e), (f), (g) and (h) early elongation and initiation of segment delineation; (i) and (j) coiled segmenting embryo; (k) older, flexed embryo.

marginal cells then differentiate as heart walls around a mid-dorsal haemocoelic space, while the more lateral cells separate into a pericardial floor, supporting the heart, and a layer of splanchnic mesoderm on the surface of the gut. More posteriorly, where the midgut is not swollen, the dorsolateral compartments of the somites close off as hollow sacs which then extend over the gut to the dorsal midline. The dorsal margins of the sacs continue the progressive formation of the heart. The somatic walls of the sacs form continuations of the pericardial floor. Laterally, the splanchnic walls merge into this floor with obliteration of the coelom, but more medially, the splanchnic walls become cut off around persistent parts of the coelom as small sacs below the pericardial floor. These later give rise to the gonads. Finally, the median edges of the splanchnic mesoderm, adjacent to the heart, spread downwards as a splanchnic layer on the surface of the midgut epithelium. From this layer arises the splanchnic musculature of the midgut. The origin and development of the stomodaeal and proctodaeal musculature of the Onychophora is still not satisfactorily resolved.

The major contribution of the medioventral compartments of the coelomic sacs in further development is a pair of ventrolateral segmental organs in every trunk segment. The compartments first become cut off as separate ventrolateral sacs adjacent to the limb bases. Parts of these sacs then persist as the end sacs of the segmental organs. According to Sedgwick (1887, 1888) and others, the ducts of the segmental organs arise as coelomoducts growing from the mesodermal sacs to the exterior, but both Kennel (1884, 1885) and Glen (1918) interpreted the ducts as ectodermal ingrowths equivalent to nephridia. The point requires further study.

In the preanal segment, the ducts of the segmental organs are modified as gonoducts. Losing their end sacs, the ducts elongate and become attached to the posterior ends of the paired gonads. Proximally, the two ducts unite and joint a short mid-ventral tube invaginated from the ectoderm and opening at a single genital aperture. The gonoducts develop in the male as vasa deferentia, in the female as oviducts. The segmental organs of the anal segment also develop in the male as accessory genital structures. In females these organs are lost.

As in oligochaetes and leeches, the development of the gonads of Onychophora is best treated with development of the mesoderm. The primordial germ cells are segregated during early gastrulation, as a group of cells which migrates into the interior with the early mesoderm or posterior midgut cells and congregates beneath the proliferating presumptive mesoderm (Figs. 46–48). The same genital rudiment is formed whether yolk is present in the embryo or not. From this position, Manton (1949) has traced the migration of the germ cells in *Peripatopsis moseleyi* to the dorsal edges of the seventh and succeeding pairs of trunk somites. Many of the primordial germ cells degenerate before reaching the somites, but why this should be so is still unexplained.

The actual formation of the gonads after the primordial germ cells have reached the somites was accounted for by Sedgwick (1887, 1888) and Kennel (1884, 1885). The region of somite wall in which the germ cells lie is the splanchnic wall of the dorsolateral compartment of each somite. As the lateral part of this wall merges into the pericardial floor, and the median part spreads over the surface of the gut as incipient splanchnic muscle, the part containing the germ cells separates off as a closed vesicle beneath the pericardial floor, the cavity of the vesicle being a pinched-off piece of coelom. The paired vesicles of successive segments fuse to form two longitudinal, dorsolateral tubes. The germ cells now migrate into the coelomic lumen of the tube and begin to proliferate as oogonia or spermatogonia. The posterior ends of the gonads unite with the paired genital ducts growing forwards and upwards from the pre-anal segment. The testes remain separate, but in the female, the ovaries join together posteriorly before uniting with the genital ducts.

Further Development of the Ectoderm

Unlike the mesoderm, which is represented in the blastoderm by only a small group of cells, the presumptive ectoderm of the Onychophora, especially in yolky embryos and those which retain a swollen blastoderm in the absence of yolk, is already extensive in area before gastrulation and segment formation begins. As we have seen, the presumptive ectoderm comprises three distinct components, paired bands of presumptive embryonic ectoderm ventrolaterally, a broad area of presumptive extra-embryonic ectoderm dorsally and a paired band of extra-embryonic ectoderm, becoming a single united band, ventrally (Figs. 45 and 46).

In the further development of the embryonic ectoderm, the majority of the cells of the presumptive bands of embryonic ectoderm give rise to ectoderm of the head. Most of the trunk ectoderm is proliferated from cells at the posterior ends of these bands, which form an ectodermal growth zone on either side of the presumptive mesoderm, budding off cells in a forward direction. In the small embryos of viviparous placental species and of non-placental species lacking a swollen yolk sac (*Peripatopsis balfouri*, *P. capensis*, Manton (1949)) it seems likely that the ectoderm of all the trunk segments is formed in this way.

As the paired bands of mesoderm push forwards beneath the blastoderm, the blastoderm cells outside them begin to divide and become cuboidal (Fig. 46). At the same time, adjacent blastoderm cells begin to concentrate into the same region, so that the two bands of embryonic ectoderm become dense, while the temporary yolk sac ectoderm above and below them becomes attenuated. Manton (1949) has shown that in *Peripatopsis*, the antero-posterior axis of the embryo does not become fixed until the migrating mesoderm is set on a particular path under the blastoderm, and it seems that the initial divergence of the presumptive ectoderm into either a definitive embryonic or an extra-embryonic pathway of development depends on an influence of mesoderm on ectoderm cells. Formation and swelling of somites also precedes the division of the embryonic ectoderm into segmental units outside the somites (Figs. 46 and 49). The intersegmental grooves that mark these units traverse only the embryonic ectoderm, without extending onto the extra-embryonic ectoderm. It should be noted that the ectodermal segmental units do not become distinct until the anterior somites have completed their forward migration on either side of the stomodaeum, so that the head ectoderm develops from blastoderm cells that are already in the appropriate position.

When the embryo is large, whether yolky (Sheldon, 1888, 1889a; Evans, 1902; unpublished personal observations) or non-yolky (Manton, 1949), the posterior parts of the bands of presumptive embryonic ectoderm give rise directly to the ectoderm of several trunk segments and as we have seen, each half-segment is widely separated from its partner by broad dorsal and ventral areas of extra-embryonic ectoderm. More posteriorly, in the

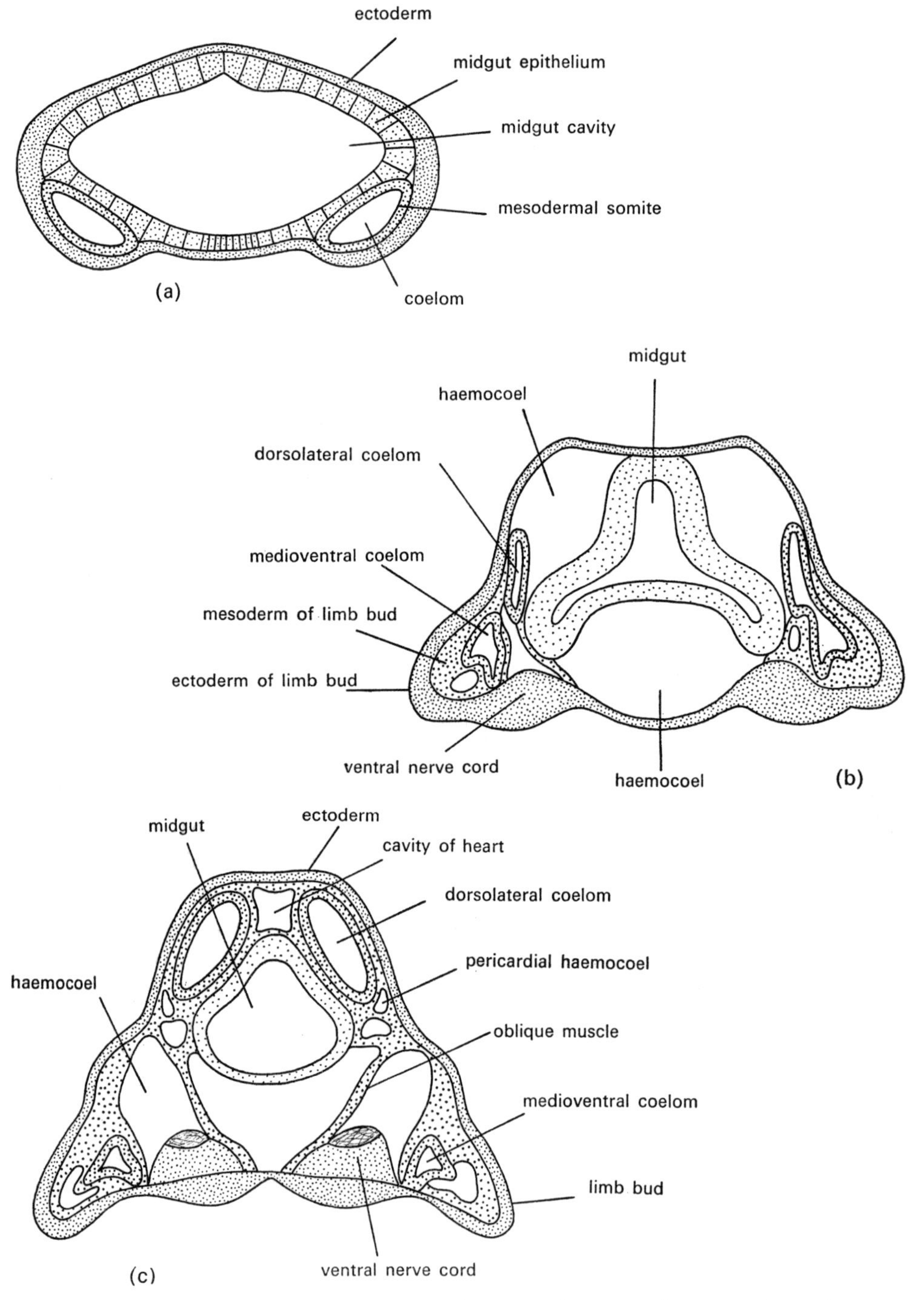

FIG. 54. Diagrammatic transverse sections showing the early stages of segment differentiation in *Peripatopsis capensis*, after Sedgwick (1888).

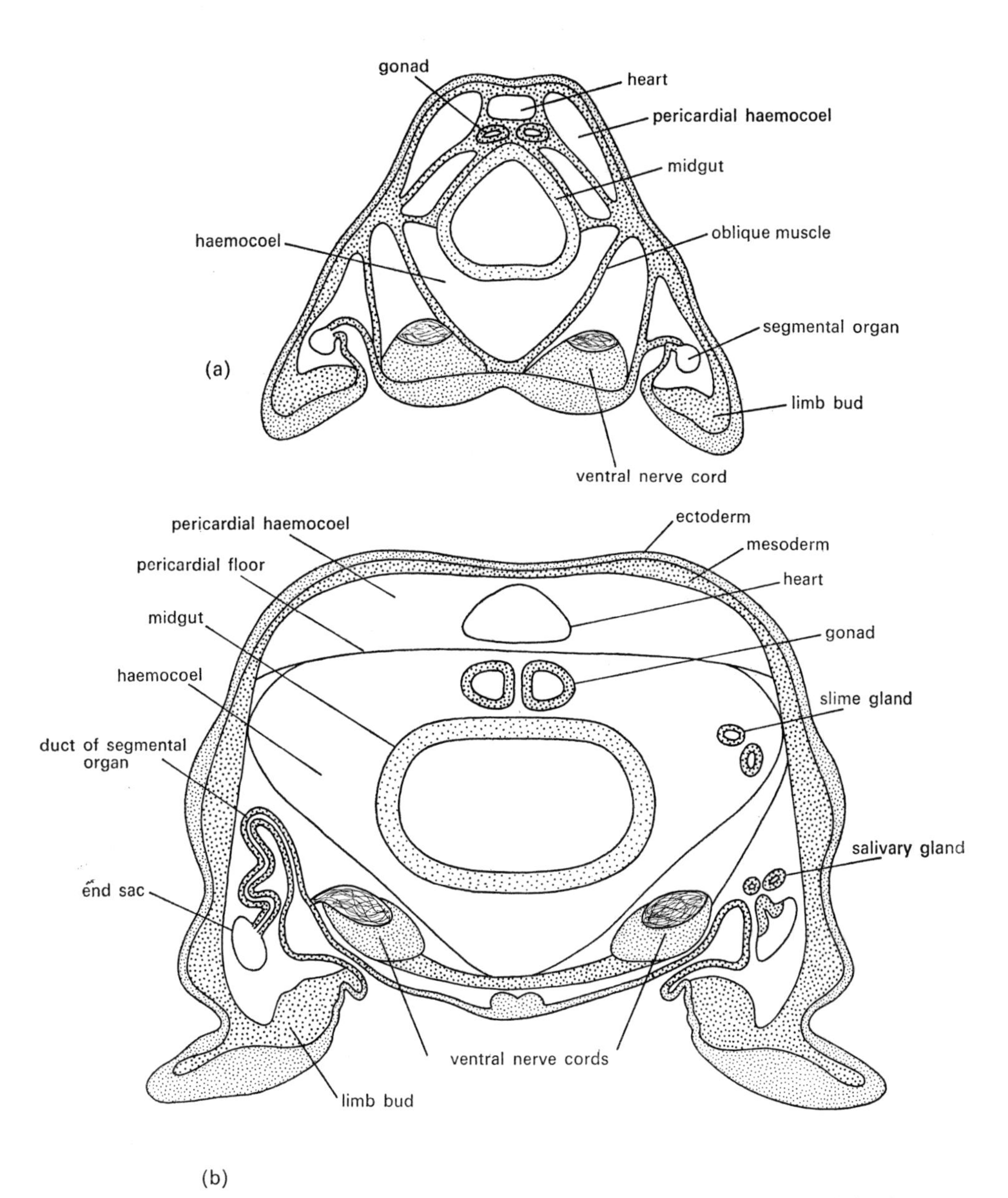

FIG. 55. Diagrammatic transverse sections showing the later stages of segment differentiation in *Peripatopsis capensis*, after Sedgwick (1888).

segments proliferated from the growth zone, the two halves of each segment are separated by only narrow strips of extra-embryonic ectoderm. The extra-embryonic ectoderm is later mostly transformed into definitive segmental ectoderm. Before discussing this, we can examine the further development of the embryonic ectoderm outside each pair of trunk somites (Figs. 54 and 55). The products of this layer are the surface epithelium, ventral nerve cord, tracheae, coxal glands and exit ducts of the segmental organs. The tracheae and coxal glands do not develop until after birth or hatching, and need further study. The ventral nerve cords arise in association with so-called ventral organs. At an early stage in the development of each trunk segment, the ventrolateral embryonic ectoderm of each side of the segment begins to bud off small cells into the interior. A thickening is formed which soon divides into two parts. The lateral part, continuous from segment to segment, sinks into the interior to give rise to the ventral nerve cord of its side. The median part becomes segmentally distinct and then partially invaginates, to form a so-called ventral organ on each side of the segment. There is no suggestion in the work of either Kennel (1885) or Sedgwick (1887) that the ventral organs contribute cells to the ventral nerve cords. As the ventral epithelium of each segment is completed, the ventral organs fuse in the ventral midline, then gradually shrink and disappear. Their function is still unknown. Similar ventral organs in Symphyla persist and develop as exsertile vesicles (Tiegs, 1947; see Chapter 5).

The remainder of the embryonic ectoderm gives rise to the surface epithelium of the segments, pouching out ventrolaterally as the surface epithelium of the limb buds. As the segment halves spread around the gut, the extra-embryonic ectoderm is for the most part transformed into surface epithelium along the dorsal and ventral midlines. Only when the dorsal component of this ectoderm is highly specialized and persistent, as in the swollen but empty trophic vesicle of some viviparous non-placental species (*Paraperipatus novae-britanniae*, Willey, 1898; *Peripatopsis sedgwicki* and *P. moseleyi*, Manton, 1949) and the placental stalk of viviparous placental species, is there any evidence that extra-embryonic ectoderm is subsequently resorbed and thus genuinely temporary and specialized.

Composition of the Onychophoran Head

With the onychophoran head, we enter into the difficult question of the incorporation into the pre-oral part of the head of structures which appear to be formed initially as post-oral segments. Whether these components of the head are segments or not has been the subject of argument and counter-argument by many authors. For the Onychophora, admittedly, much of the controversy on head composition preceded the detailed investigations of Manton (1949), but the old prejudices die hard, as witnessed in the relatively recent theoretical discussion by Butt (1960), rebutted by Manton (1960). The present position is still unsatisfactory, because the later development of the head and especially the brain has not yet been properly investigated, but we shall see that a reasonable interpretation can be given of the composition of the pre-oral part of the head and an exact analysis is available of the composition of the post-oral part.

As we have already seen, three pairs of somites lie in front of the first trunk segment. The first pair are large and lie anterior to the stomodaeum. The second and third pairs, somewhat smaller, lie on either side of the stomodaeum and behind the stomodaeum respectively. These three pairs of somites (Fig. 48) originate posteriorly and migrate forwards (Kennel, 1884, 1886; Evans, 1902; Manton, 1949) to comprise the entire mesoderm of the head. Pflugfelder (1948) suggested that there are three pairs of somites in front of the stomodaeum in *Paraperipatus amboinensis*, but his evidence is unconvincing (Tiegs and Manton, 1958).

External to the three pairs of head somites, the anterior parts of the bands of presumptive embryonic ectoderm become marked off as the ectoderm of three segments, a large pre-oral antennal segment, a jaw segment flanking the mouth and a slime-papilla segment behind the mouth (Fig. 56). The two halves of the antennal segment already meet in the midline when first formed, but the jaw segment and slime-papilla segment are divided dorsally and ventrally by extra-embryonic ectoderm, just like the trunk segments which follow them.

The course of further development confirms that the jaw segment and slime-papilla segment are undoubtedly cephalized segments belonging to the same metameric series as the trunk segments (Kennel, 1884, 1885; Sedgwick, 1887, 1888; Evans, 1902; Glen, 1918). The somites become subdivided into the usual three compartments, of which the lateral, appendicular compartments grow out with the ectoderm as limb buds, the jaws and slime papillae respectively. The paired medioventral compartments in the slime-papilla segment develop coelomoducts, which then becomes greatly elongated, retaining their coelomic end sacs, to form salivary glands (Sedgwick, 1888; Evans, 1902). The proximal ends of the two glands later come together in the ventral midline to open by a common pore into the preoral cavity. Kennel (1885) and Glen (1918) interpreted the salivary glands, like the segmental organs of the trunk segments, as ectodermal ingrowths, but did not deny their homology. The jaw segment lacks coelomoducts. Its small somites are mainly concerned in the formation of specialized jaw muscles. In fact, the major products of the two pairs of somites we are discussing are the muscles of the foregut, jaws and slime-papillae, but the details of their development, including the fate of the dorsolateral components, require further study. These somites do not contribute to the heart.

The ectoderm of the jaw segment and slime-papilla segment also follows essentially the same course of development as that of the trunk segments. Surface epithelium and ventral nerve cords result, the latter being associated with paired ventral organs that are later resorbed. The ventral organs of the jaw segment lie just behind the posterior lip of the mouth. The only ectodermal specializations observed in these segments is the invagination of the ectoderm of the tips of the slime-papillae as long slime glands growing back into the trunk.

Whether the pre-oral part of the onychophoran head constitutes more than a single cephalized segment that has migrated forwards in front of the mouth is a more controversial question. The later development of this part of the head has not been redescribed since the work of Kennel, Sedgwick, Sheldon and Evans. The paired antennal somites are obviously serially homologous with those behind them, becoming subdivided into three compartments, developing a pair of limb buds in association with the appendicular compartments and forming a pair of transient coelomoducts ventrally (Manton, 1960). Beyond this, they develop as antennal musculature, foregut muscles and other muscles of the head, all of which require further study. The preoral ectoderm also shares many features in common with the postoral segmental ectoderm. Much of it develops as surface epithelium of the head and antennae. Over a broad ventrolateral area on each side, the ectoderm thickens as

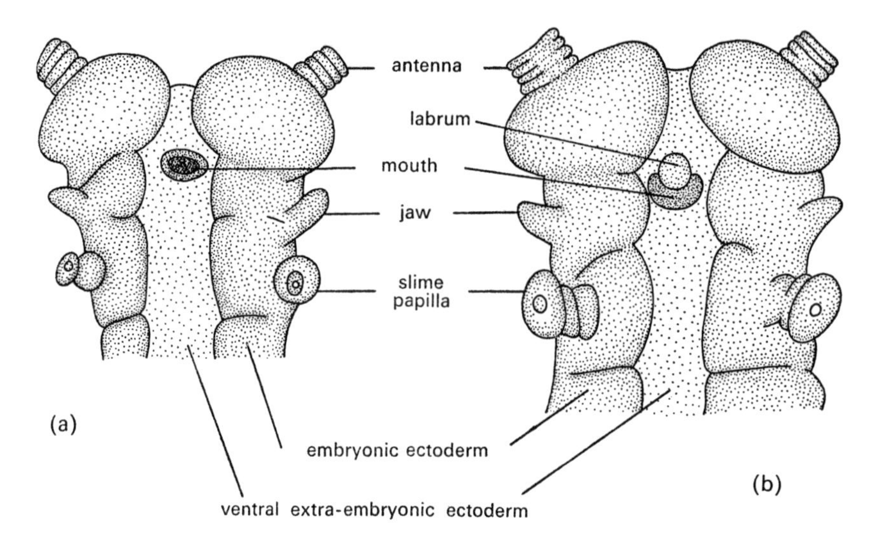

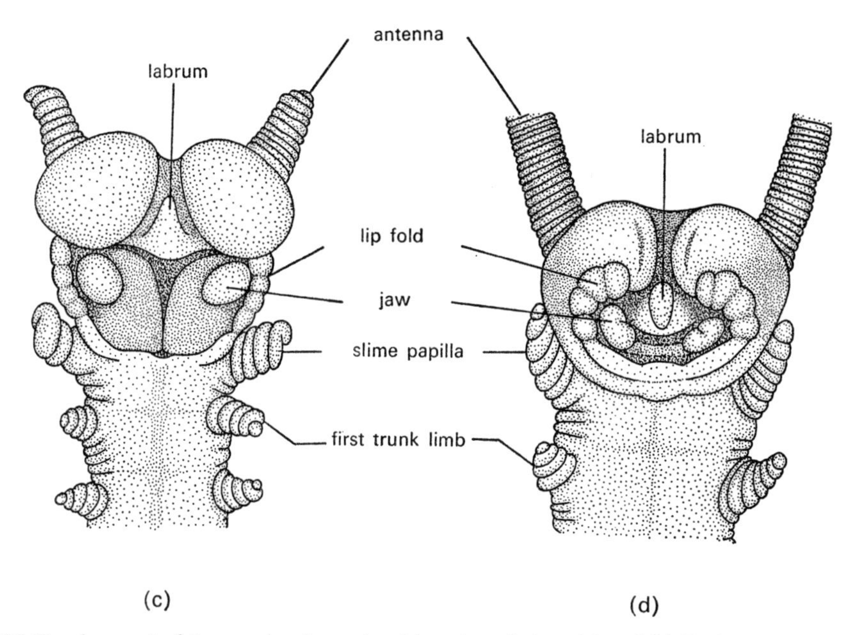

FIG. 56. Development of the onychophoran head in external view. (a) and (b) Early stages in formation of the head of *Peripatopsis capensis*, modified from Butt (1959) after Sedgwick (1888); (c) and (d) later stages in formation of the head of *Epiperipatus trinidadensis*, modified from Manton (1960) after Kennel (1885).

antennal ganglionic rudiments, continuous with the ventral nerve cords, from which arise the paired ganglia of the brain and the circumpharyngeal connectives (Kennel, 1885; Sedgwick, 1887; Evans, 1902). Part of each thickening also remains at the surface and develops as a temporary ventral organ. Although Evans (1902) and others have interpreted the dorsal lobes of the antennal ganglia as distinct, pre-segmental, protocerebral ganglia, no evidence has ever been offered that these ganglia originate separately from the antennal ganglia.

We cannot, therefore, escape the conclusion that the major part of the pre-oral region of the head in Onychophora comprises a cephalized segment that has moved forwards in front of the mouth as the antennal segment (Manton, 1949, 1960). All the necessary segmental criteria of somites, limbs, coelomoducts, paired ganglia and associated ventral organs, are fulfilled. The only question that cannot be answered at the present time is whether the pre-oral part of the head in Onychophora retains any pre-segmental components. A median acron, equivalent to the annelid prostomium, appears to be absent (Goodrich, 1897; Manton, 1949), but whether the brain incorporates anterior components, perhaps intimately associated in origin and structure with the antennal ganglia, that might be equated with a pair of protocerebral ganglia, has yet to be decided. That a pre-segmental component of the brain was present in the ancestors of onychophorans seems highly probable, since it is retained as the entire brain in annelids and is identifiable as large protocerebral ganglia in all myriapods and hexapods.

A Pygidium in Onychophora?

Just as a pre-segmental acron cannot be identified in Onychophora, so too is there no external sign of a post-segmental pygidium or telson. In this case, however, the absence of such a structure finds a reasonable explanation in terms of the total mode of development. The growth zone from which the trunk segments arise is post-anal in position, having ectoderm on either side of posteroventral mesoderm. When the last mesoderm is proliferated, the growth zone ectoderm develops directly into the ectoderm of the anal segment and no post-segmental ectoderm remains. In annelids, as we have seen above, and in most other arthropods, as we shall see below, the ectodermal growth zone lies in front of the anus and the ectoderm around the anus forms the pygidium or telson.

The Basic Pattern of Development in Onychophora

Obviously, although no such egg has yet been studied embryologically, we can say with confidence that the basic pattern of development in the Onychophora proceeds from a large yolky egg, upwards of 2 mm in length, deposited by an oviparous female and protected by two membranes, an outer chorion and an inner vitelline membrane. Development is epimorphic, all segments of the body being developed before hatching takes place, but involves the anteroposterior addition of several trunk segments beyond the length of the original egg, formed by proliferation from a terminal growth zone. The increasing length is mainly added while the yolk mass within the embryo is still voluminous, and is accommodated within the limited egg space by forward flexure of the growing region beneath the ventral surface of the anterior region.

The egg undergoes intralecithal cleavage, forming a blastoderm around a mass of yolk spheres. The blastoderm has a characteristic distribution of presumptive areas, each composed of small cells at the surface of the yolk. The presumptive midgut, a long midventral band of cells, is in two parts, a long anterior midgut and a compact posterior midgut behind it. The presumptive anterior midgut splits along the ventral midline, exposing the yolk, and invaginates slightly as paired cell bands. These now proliferate cells which migrate around the yolk, taking up part of the yolk material, and form a diffuse vitellophage epithelium at the surface of the yolk mass. The ventral split subsequently closes so that the vitellophage epithelium becomes continuous. After completion of its vitellophage function this layer of cells transforms into the definitive epithelium of the anterior part of the midgut. The posterior midgut rudiment remains superficial and proliferates cells which form a tubular continuation of the midgut epithelium in the growing trunk, behind the yolk-filled anterior region.

The presumptive stomodaeum and proctodaeum lie at the ends of the presumptive anterior midgut. The former is a short arc of cells, the latter a paired rudiment. When invagination of the anterior midgut takes place, the stomodaeal and proctodaeal rudiments form the anterior and posterior margins of the ventral opening. As the lateral margins of this opening subsequently come together in the ventral midline, the stomodaeum and proctodaeum are closed off as short tubes, which subsequently gain openings into the midgut. The presumptive posterior midgut lies behind the anus at the end of this process.

Behind the presumptive posterior midgut is a second small group of cells, posteroventral in position, the presumptive mesoderm. The cells of this area sink slightly below the surface and proliferate two streams of mesoderm cells, which pass forwards as paired lateral bands on either side of the midgut and eventually meet in the anterior midline. During forward growth, the mesodermal bands begin to segment into paired somites. Each somite is formed by individuation of a layer of mesoderm cells around a coelomic space and undergoes considerable swelling, developing a large coelomic cavity. Forward penetration of the mesodermal bands carries the first and largest pair of somites into a pre-oral position. The second pair comes to lie on either side of the mouth, the third pair just behind the mouth. These are the three paired somites of the head: antennal, jaw and slime-papilla respectively. The remaining pairs are somites of the trunk. The somites are formed in anteroposterior succession. The last mesodermal cells produced by the presumptive mesoderm enter into the last pair of somites and there is no residual mesoderm when hatching takes place.

The trunk somites become partially subdivided into three compartments, dorsolateral, medioventral and appendicular. The appendicular compartment pushes out as a lobe and then becomes mesenchymatous, giving rise to the somatic musculature (circular, dorsal longitudinal, ventral longitudinal, oblique, extrinsic limb and intrinsic limb muscles). The dorsolateral compartment extends towards the dorsal midline. Its somatic wall gives rise to the heart and pericardial floor. The splanchnic wall merges into the pericardial floor and, medially below the heart, gives rise to splanchnic mesoderm spreading downwards around the midgut. The medioventral compartment develops a coelomoduct and becomes the coelomic end sac of a segmental organ, subsequently lost in the first three segments. The coelom is otherwise obliterated or gains continuity with the haemocoel, except where it persists within the gonads. Primordial germ cells are segregated as early products of the proliferating presumptive mesoderm. The germ cells initially form a compact group of cells in a posteroventral position just below the surface, but then migrate outwards and upwards to enter the splanchnic walls of the more posterior trunk somites. Subsequently, these splanchnic areas close off a small portion of the dorsolateral coelom on each side of the body as the cavity of a tubular gonad, and the germ cells migrate into the gonad cavities. The gonoducts are specialized coelomoducts of the pre-anal segment.

The somites of the head develop in a slightly modified manner but are serially homologous with the trunk somites. Antennal coelomoducts are only transient and the jaw segment lacks coelomoduct vestiges, but those of the slime-papilla segment become specialized as salivary glands.

The presumptive ectoderm comprises all except the ventral part of the blastoderm. External to the mesodermal bands, it concentrates as dense bands of embryonic ectoderm. Dorsally and ventrally, it becomes attenuated as extra-embryonic ectoderm, later incorporated into the segmental ectoderm. The ventral component of the extra-embryonic ectoderm is paired at first and borders the ventral presumptive areas (stomodaeum, anterior midgut, proctodaeum and posterior midgut) but with closure of the ventral opening, the two halves come together as a midventral band.

Segmental subdivision of the embryonic ectoderm follows swelling of the paired somites in anteroposterior succession. The intersegmental grooves do not extend onto the extra-embryonic ectoderm. Each segmental unit of ectoderm thickens ventrally, the lateral part of the thickening forming ventral nerve cord, the median part a temporary ventral organ. The remainder of the ectoderm gives rise to surface epithelium, including ventrolateral outpouchings as limb buds, and later invaginations as coxal glands and tracheae. The surface epithelium spreads dorsally and ventrally and incorporates the extra-embryonic ectoderm.

The posterior ends of the paired bands of presumptive embryonic ectoderm, lying on either side of the presumptive mesoderm, constitute a growth zone from which the ectoderm of the more posterior trunk segments is proliferated. The growth zone ectoderm finally forms part of the ectoderm of the anal segment and there is no post-segmental pygidium. Since the growth zone lies behind the anus, which is formed before proliferation of new segments begins, the ectoderm and mesoderm of new segments must pass forwards on either side of the anus as the trunk grows longer.

The ectoderm of the jaw segment and slime-papilla segment develops in essentially the same way as that of the trunk segments. The slime glands are probably modified coxal glands. As far as can be seen, the pre-oral ectoderm external to the antennal somites also develops in the same manner, though its ganglionic components are disproportionately enlarged. Whether the pre-oral region of the head includes any presegmental components of ectoderm, anterior to the cephalized antennal segment, cannot be answered satisfactorily on present evidence.

Superficially, the basic pattern of embryonic development in Onychophora appears to have almost nothing in common with that of polychaete annelids beyond paired somites and metameric segmentation. Again superficially, it appears to be quite different from that of clitellate annelids, i.e. annelids which show a particular response of basic annelid (polychaete) development to increased yolk and a protective environment for development. As I have recently argued, however (Anderson, 1966), if one speculates on the possible further modification of clitellate development in response to yet additional yolk, a remarkably onychophoran-like picture begins to emerge. I will return to this possibility in the final chapter.

References

ANDERSON, D. T. (1966) The comparative early embryology of the Oligochaeta, Hirudinea and Onychophora. *Proc. Linn. Soc. N.S.W.* **91,** 10–43.

ANDERSON, D. T. and MANTON, S. M. (1972) Studies on the Onychophora VIII: The relationship between the embryos and the oviduct in the viviparous onychophorans *Epiperipatus trinidadensis* (Bouvier) and *Macroperipatus torquatus* (Kennel) from Trinidad. *Phil. Trans. R. Soc.* B, **264,** 161–89.

BUTT, F. H. (1959) The structure and some aspects of the development of the onychophoran head. *Smiths. misc. Coll.* **137,** 43–60.

BUTT, F. H. (1960) Head development in the arthropods. *Biol. Rev.* **35,** 43–91.

DENDY, A. (1902) On the oviparous species of Onychophora. *Q. Jl. microsc. Sci.* **45,** 363–416.

EVANS, R. (1902) On the Malayan species of Onychophora. Part II. The development of *Eoperipatus weldoni. Q. Jl. microsc. Sci.* **45,** 41–86.

GLEN, E. (1918) Certain points in the early development of *Peripatopsis capensis. Q. Jl. microsc. Sci.* **63,** 283–92.

GOODRICH, E.S. (1897) On the relation of the arthropod head to the annelid prostomium. *Q Jl. microsc. Sci.* **40,** 259–68.

KENNEL, J. (1884) Entwicklungsgeschichte von *Peripatus Edwardsi* Blanch. und *Peripatus torquatus* n.sp. *Arb. zool. zootom. Inst. Würzburg* **7,** 95–229.

KENNEL, J. (1885) Entwicklungsgeschichte von *Peripatus Edwardsi* Blanch. und *Peripatus torquatus* n.sp. *Arb. zool. zootom. Inst. Würzburg* **8,** 1–93.

MANTON, S. M. (1949) Studies on the Onychophora VII. The early embryonic stages of *Peripatopsis* and some general considerations concerning the morphology and phylogeny of the Arthorpoda. *Phil. Trans. R. Soc.* B, **233,** 483–580.

MANTON, S. M. (1960) Concerning head development in the arthropods. *Biol. Rev.* **35,** 265–82.

PFLUGFELDER, O. (1948) Entwicklung von *Paraperipatus amboinensis* n.sp. *Zool. Jb. Anat. Ont.* **69,** 443–92.

SCLATER, W. L. (1888) On the early stages of the development of a South American species of *Peripatus. Q. Jl. microsc. Sci.* **28,** 343–63.

SEDGWICK, A. (1885) The development of *Peripatus capensis.* Part 1. *Q. Jl. microsc. Sci.* **25,** 449–68.

SEDGWICK, A. (1886) The development of the Cape species of *Peripatus.* Part 2. *Q. Jl. microsc. Sci.* **26,** 175–212.

SEDGWICK, A. (1887) The development of the Cape species of *Peripatus.* Part 3. *Q. Jl. microsc. Sci.* **27,** 467–550.

SEDGWICK, A. (1888) The development of the Cape species of *Peripatus.* Part 4. *Q. Jl. microsc. Sci.* **28,** 373–98.

SHELDON, L. (1888) On the development of *Peripatus Novae-Zealandiae. Q. Jl. microsc.* **28,** 205–37.

SHELDON, L. (1889a) On the development of *Peripatus Novae-Zealandiae. Q. Jl. microsc. Sci.* **29,** 283–94.

SHELDON, L. (1889b) The maturation of the ovum in the Cape and New Zealand species of *Peripatus. Q. Jl. microsc. Sci.* **30,** 1–29.

TIEGS, O. W. (1947) The development and affinities of the Pauropoda, based on a study of *Pauropus sylvaticus. Q. Jl. microsc. Sci.* **88,** 165–267 and 275–336.

TIEGS, O. W. and MANTON, S. M. (1958) The evolution of the Arthropoda. *Biol. Rev.* **33,** 255–337.

WILLEY, A. (1898) Anatomy and development of *Peripatus novae-brittaniae. Zoological Results, Cambridge* **1,** 1–52.

CHAPTER 5

MYRIAPODS

SINCE the myriapodous arthropods are common and well-known members of the leaf litter fauna and lack the complications of viviparity prevalent amongst the Onychophora, it might be expected that a substantial understanding of their embryonic development has been attained. Unfortunately this is not the case. The myriapods yield up their embryological secrets only as the reward of great patience, ingenuity and histological skill. Myriapod habits of oviposition are secretive, and a combination of good fortune and high resolve is essential to the procurement of eggs for laboratory study. The eggs are yolky, the early embryonic rudiments are small and diffuse and the external membranes are highly resistant to penetration by fixatives and other histological reagents. The fact that we have any detailed understanding of myriapod embryonic development is due almost entirely to the efforts of the late Professor O. W. Tiegs of the University of Melbourne. In two papers, published in 1940 and 1947, Tiegs set out in meticulous detail the embryology of the symphylan *Hanseniella agilis* and the pauropod *Pauropus sylvaticus*. Neither group had previously been known embryologically and both offer in full measure the difficulties that myriapod embryos can provide. It is a continuing tribute to Tiegs that, with such unfavourable material, he succeeded in writing two of the great classics of arthropod descriptive embryology.

The Chilopoda and Diplopoda have been less well served. Chilopods, indeed, have attracted embryological attention on only four occasions. The first investigator, whose paper now has little to offer, was Zograff (1883), working on *Geophilus*. The second, more fortunately, was the outstanding insect embryologist, R. Heymons. In a rare excursion into myriapod embryology, Heymons (1901) wrote an account of the development of *Scolopendra cingulata* and *S. dalmatica* which still remains our only adequate source of information on chilopod embryos. Although his description of the later stages of development is very detailed, however, his analysis of the early stages was more fragmentary and less convincing. Dawydoff (1956) subsequently attempted a rectification of this situation, but his comments on the embryos of a variety of centipedes from Indo-China are brief and, unfortunately, not illustrated. More recently, Dohle (1970) has given a brief description of the embryos of *Scutigera coleoptrata*.

The history of diplopod embryology has involved a greater number of workers than have essayed the chilopods, but with scarcely more satisfactory results. Metschnikoff (1874) made a number of preliminary observations of the eggs of several species (*Julus moreleti*, *Strongylosoma Guerinii*, *Polyxenus lagurus*, *Polydesmus complanatus*). More detailed studies were published by Heathcote (1886, 1888) on *Julus terrestris* and Cholodkowsky (1895) on *Julus subulosus*, in which the main features of diplopod organogeny were reasonably firmly established, but such problems as the origin of mesoderm and the segmental composition of the diplopod head remained unresolved. Later investigations by Silvestri (1903, 1933, 1949) on several species, Hennings (1904) on *Glomeris*, Robinson (1907) on *Archispirostreptus*, Lignau (1911) on *Polydesmus*, Pflugfelder (1932) on *Platyrrhacus* and Seifert (1960) on *Polyxenus* did little to improve understanding, and it is only quite recently that careful and detailed investigations of *Glomeris marginata* by Dohle (1964) and *Narceus annularis* by Bodine (1970) have provided better descriptions of the segmental composition of the diplopod head and trunk and resolved a long-standing argument about the significance of the gnathochilarium. The origin of the mesoderm in diplopod embryos is still uncertain.

The new facts provided by Dohle and Bodine in supplementation of those already given by Heymons and Tiegs, and the new recognition that the Myriapoda are a respectable systematic unit (Manton, 1964, 1970, 1972) now permit the embryonic development of the myriapods to be reappraised independently of that of the hexapods. We shall see that much of the evolutionary modification of embryonic development within the Myriapoda has been concerned with the secondary reduction of yolk and associated precocious hatching with only a few segments present. This condition is obvious in the Diplopoda, Pauropoda and Symphyla and can also be discerned in the anamorphic chilopods. The scolopendromorph centipedes, in contrast, show an extreme of epimorphic development and hatch as juveniles with the full complement of trunk segments.

Cleavage

All myriapods eggs exhibit the same structural organization. The egg is ovoid or spherical, with a central nucleus in a small halo of yolk-free cytoplasm, surrounded by a dense accumulation of yolk granules enmeshed in a sparse reticulum of cytoplasm. A single membrane, usually termed the chorion, encloses the egg. At the same time, myriapod eggs exhibit a wide range of sizes. The ovoid egg of *Scolopendra cingulata* is 3 mm long (Heymons, 1901), while that of *Scutigera coleoptrata* has a diameter of about 1–2 mm (Dohle, 1970). Diplopod eggs are generally smaller, ranging from 1·3 mm in *Strongylosoma guerinii* (Metschnikoff, 1874), through 0·96 mm *Platyrrhacus amauros* and 0·70 to 0·80 mm in *Glomeris marginata* (Pflugfelder, 1932; Dohle, 1964), to 0·35 mm in *Polyxenus lagurus* and 0·30 mm in *Polydesmus abchasius* (Metschnikoff, 1874; Lignau, 1911). The symphylan *Hanseniella agilis* produces spherical eggs 0·37 mm in diameter (Tiegs, 1940), while the spherical egg of *Pauropus sylvaticus* is the smallest of all, 0·11 mm in diameter (Tiegs, 1947).

In spite of these differences in size, cleavage always leads to the formation of a blastoderm. For centipedes, there is still some uncertainty about the sequence of events through which the blastoderm stage is attained. There is no doubt (Zograff, 1883; Heymons, 1901; Dawydoff, 1956) that repeated nuclear divisions take place within the yolk mass in scolopendromorphs and geophilomorphs and that most of the resulting cleavage energids migrate to the surface of the yolk mass and emerge in small groups of cells uniformly scattered over the surface (Fig. 57). With further divisions and spread, the groups of cells merge to form a uniform low blastoderm. The remaining energids persist within the yolk mass as vitello-

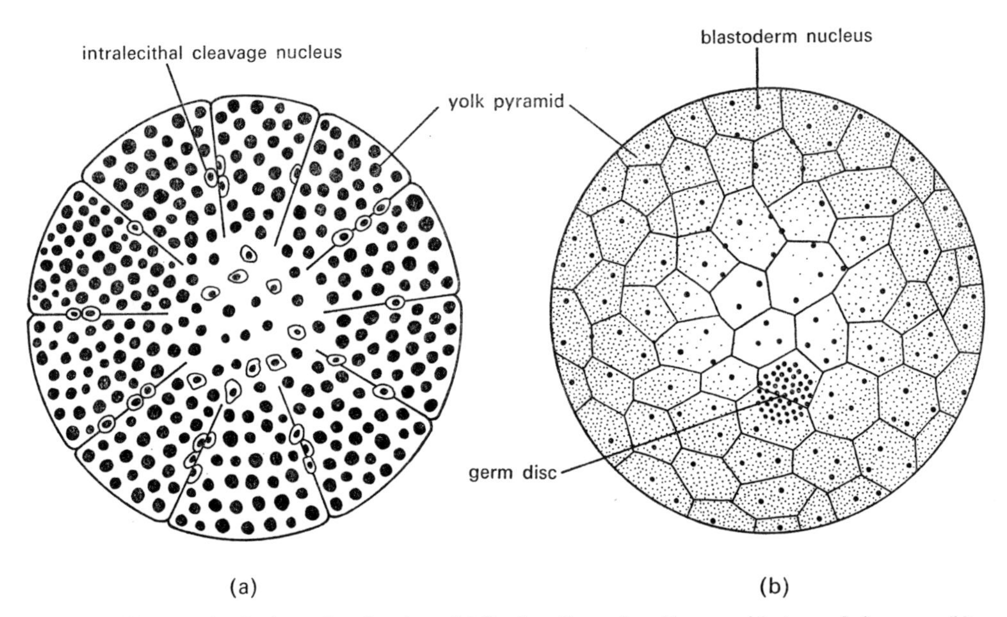

FIG. 57. Cleavage in *Scolopendra cingulata.* (a) Section through yolk pyramid stage of cleavage; (b) blastoderm. [After Heymons, 1901.]

phages. Heymons suggests, however, and Dawydoff tends to confirm, that the yolk mass divides during early cleavage into a number of radial yolk pyramids converging on an undivided central mass of yolk. It is also inferred that the yolk pyramids are anucleate and that the cleavage energids which migrate to the yolk surface do so by passing along the interfaces between pyramids. If so, the chilopods are unique in this respect. Yolk pyramids are formed during the cleavage of other myriapods (see below) and also in many Crustacea (see Chapter 7), but each pyramid always contains a single cleavage energid. A more precise description of scolopendromorph cleavage is still required, but on present evidence, cleavage can be regarded as essentially of the intralecithal type. As the blastoderm is completed, the yolk pyramids fuse once more to form a unitary nucleated yolk mass.

In the diplopods, symphylans and pauropods, in contrast, total cleavage (Figs. 58–60) precedes blastoderm formation (Tiegs, 1940, 1947; Dohle, 1964). After each nuclear division, a superficial furrowing initiates the interposed cleavage plane, but the deeper parts of the plane are formed as a thin sheet of cytoplasm which segregates within the yolk mass and then splits between the two cells. In consequence of this mode of subdivision, the spherical outline of the egg is maintained throughout cleavage and the division sequence exhibits no trace of an ancestral spiral pattern.

The first division divides the egg into two equal blastomeres. The second, slightly less equal, is perpendicular to the first. As division now continues equally but become more irregular in timing, pyramidal yolky blastomeres begin to be formed around a small, central blastocoel. Each pyramidal blastomere contains a nucleus and associated cytoplasmic halo (energid) and, at each division, the energids move closer to the exposed outer surfaces of the pyramids.

In the continuation of its cleavage sequence, *Glomeris marginata* can be taken as repre-

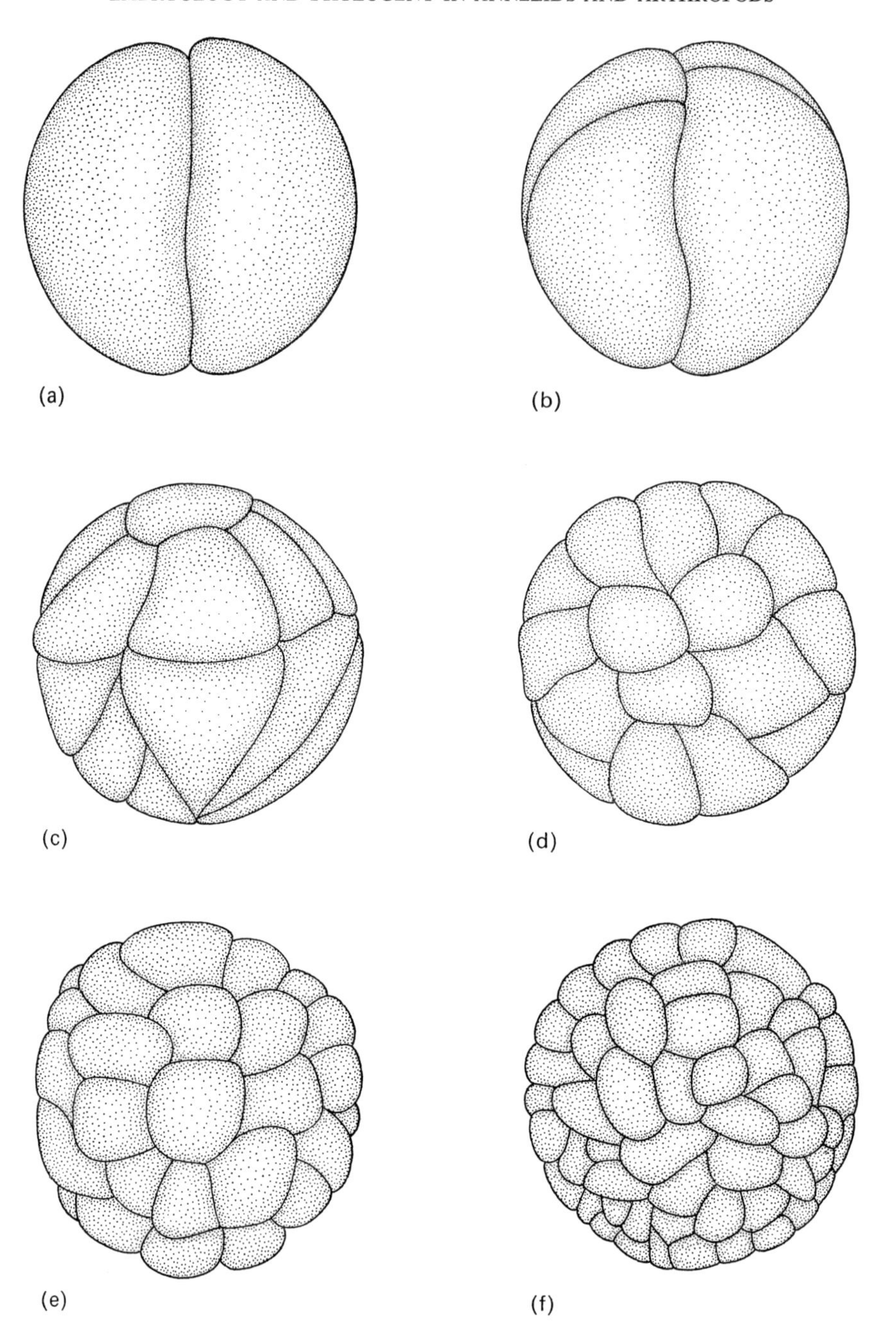

FIG. 58. Cleavage in *Hanseniella agilis*. (a) 2-cell; (b) 4-cell; (c) 17-cell; (d) 30-cell; (e) 50-cell; (f) 100-cell. [After Tiegs, 1940.]

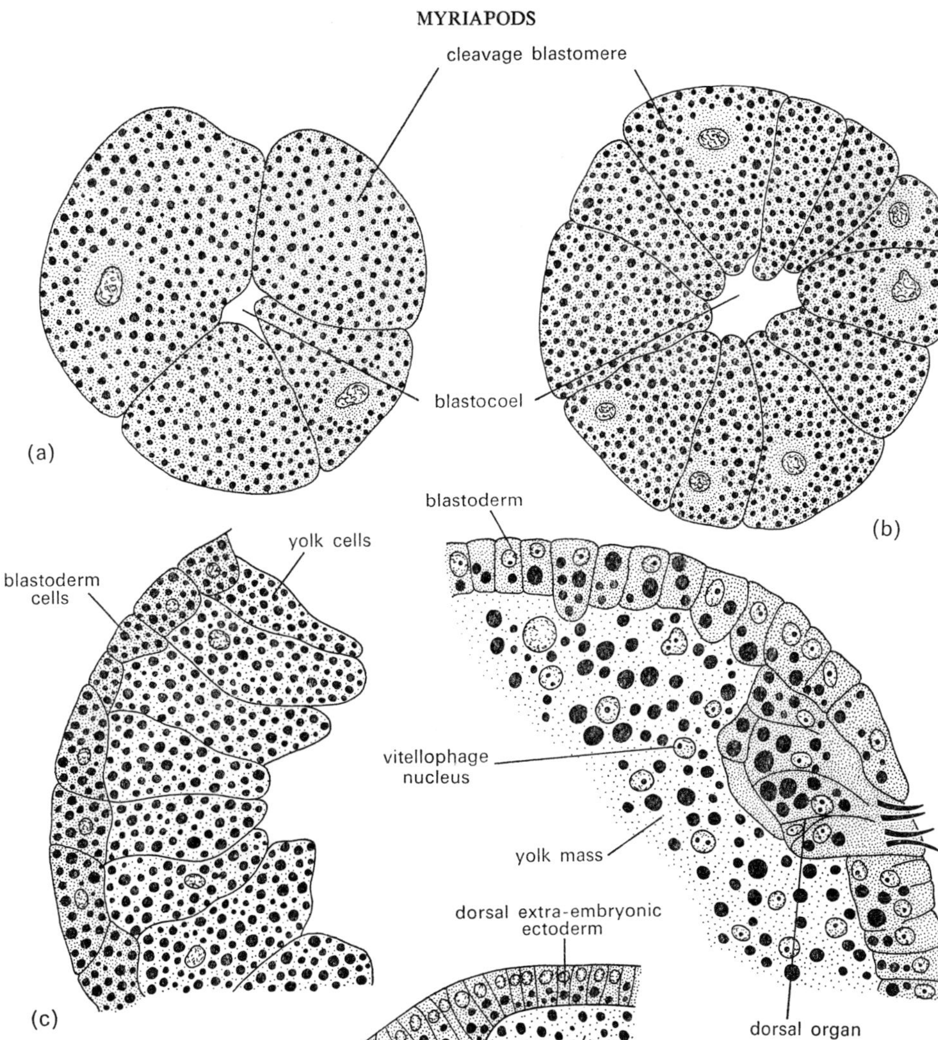

FIG. 59. *Hanseniella agilis*. (a) Transverse section through 4-cell stage; (b) section through an egg with 58 blastomeres; (c) section through an egg with more than 120 cells; (d) section through completed blastoderm with dorsal organ; (e) transverse section through fourth trunk segment at the stage of somite formation. [After Tiegs, 1940.]

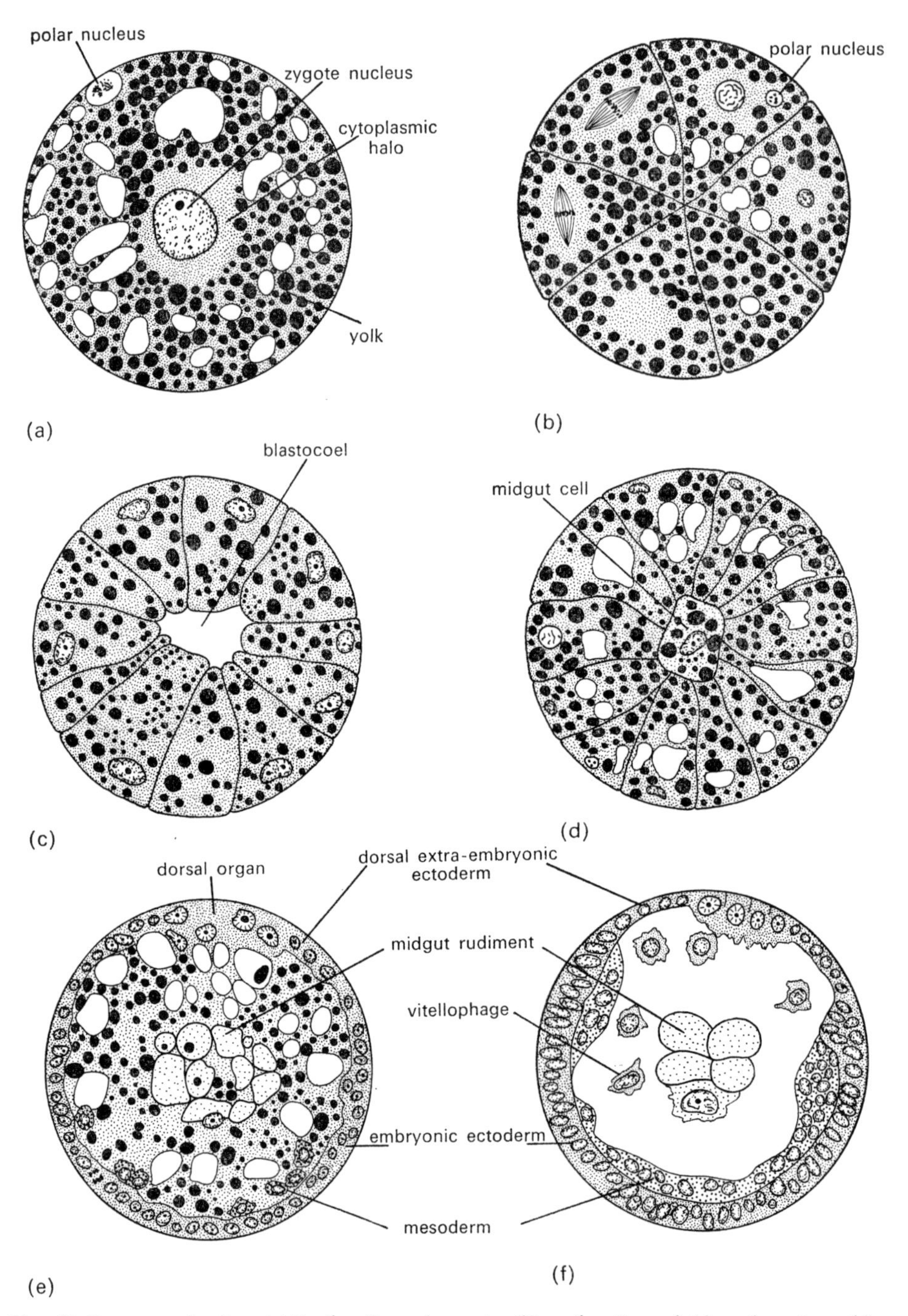

FIG. 60. *Pauropus sylvaticus.* (a) Section through zygote; (b) section through 11-nucleus stage; (e) section through 45-nucleus stage; (d) section through 80-cell stage; (e) sagittal section through mature blastoderm; (f) sagittal section through germ band, shown with anterior end on the right. [After Tiegs, 1947.]

sentative of the diplopods (Dohle, 1964). At approximately the 16-cell stage, some of the nuclei begin to undergo radial mitoses, cutting off polygonal yolky cells towards the centre of the egg and leaving truncated pyramids peripherally. The latter also continue to divide, with tangential mitoses. When about 130 cells have been formed, the cell boundaries begin to disappear and, as cleavage continues, a unitary yolk mass is restored, with numerous energids at its periphery and others more centrally placed. Finally, the peripheral energids are cut off uniformly at the surface of the yolk mass as small, cuboidal blastoderm cells. The final product, a uniform low blastoderm around a nucleated yolk mass, is similar to that of chilopods.

The later events of cleavage in the Symphyla (Figs. 58 and 59) are somewhat more specialized (Tiegs, 1940). Radial mitoses do not begin until about 100 pyramidal blastomeres have been formed. A superficial layer of small, cuboidal cells, still containing some yolk, is then cut off from an inner mass of large, polygonal yolky cells surrounding the small blastocoel. The peripheral cells resume tangential divisions but, unlike those of diplopods, they retain their cell boundaries and give rise directly to a columnar blastoderm. Furthermore, the divisions proceed in such a way that the cells of the ventral half of the blastoderm are about twice as deep as those of the dorsal half, indicating a precocious differentiation of the embryonic primordium and a dorsal, extra-embryonic ectoderm.

The central cells, meanwhile, also differ from those of diplopods in that they undergo further radial mitoses, cutting off a roughly double layer of small irregular cells beneath the blastoderm, from a central group of anucleate yolk masses. But finally, as blastoderm formation is completed, the yolk masses and the irregular cells around them fuse to form a nucleated yolk mass within the blastoderm, establishing a blastodermal organization similar to that of diplopods in all respects save in the precocity of differentiation of the embryonic primordium.

As might be expected, the later cleavage of the small egg of *Pauropus* is yet more specialized again (Fig. 60). The early blastomeres are, of course, small, and the nuclei are almost peripheral by the time forty to forty-five pyramidal blastomeres have been formed. Generalized radial divisions do not ensue at this stage, however. Instead, only two yolky cells are cut off into the centre of the egg, to fill the blastocoel. The site of origin of these cells relative to the axes of development of the future embryos cannot be established, an unfortunate circumstance in view of the fact that, as discussed below, the central cells develop as midgut.

Meanwhile, the yolk pyramids of *Pauropus* continue their tangential mitotic divisions until 160–200 cells are present and each pyramid is more or less columnar. During this phase, divisions are more numerous ventrally than dorsally, indicating a highly precocious differentiation of the embryonic primordium and the dorsal extra-embryonic ectoderm. Additional cytoplasm begins to accumulate around the peripheral nuclei, leaving only the central parts of the cells yolky. Further divisions continue wholly with tangential spindle so that no cells are cut off into the interior at this stage as they are in the diplopods and Symphyla. Instead, the yolky inner parts of the cells separate inwards as anucleate yolk masses beneath a superficial layer of blastoderm cells and fuse together to form an anucleate yolk shell between the blastoderm and the two central, yolky cells. Thus, in addition to the greater distinction that exists between the embryonic primordium and the dorsal extra-embryonic ectoderm in the blastoderm of *Pauropus* than in Symphyla, the pauropod blastoderm encloses, not a nucleated yolk mass as in other myriapods, but an anucleate shell of yolk around two central, yolky cells. As we shall see in due course, this difference is one of degree rather than of kind. The outer yolk subsequently becomes nucleated in *Pauropus* and the central yolk subsequently becomes different from the peripheral yolk in diplopods and Symphyla, though not in the chilopods.

When complete, the blastoderm of diplopods (Metschnikoff, 1874; Dohle, 1964), Symphyla (Tiegs, 1940) and pauropods (Tiegs, 1947) secretes a thin but highly resistant blastodermal cuticle beneath the chorion. The presence of this cuticle is a major barrier to the penetration of reagents into the egg during histological treatment. The chilopod blastoderm, as far as is known, does not secrete a blastodermal cuticle.

Presumptive Areas of the Blastoderm

Having established the formation and morphology of the blastoderm stage for the major groups of myriapods, we can now examine the presumptive fates of the components of the embryo at this stage. The construction of fate maps for myriapod blastoderms is more

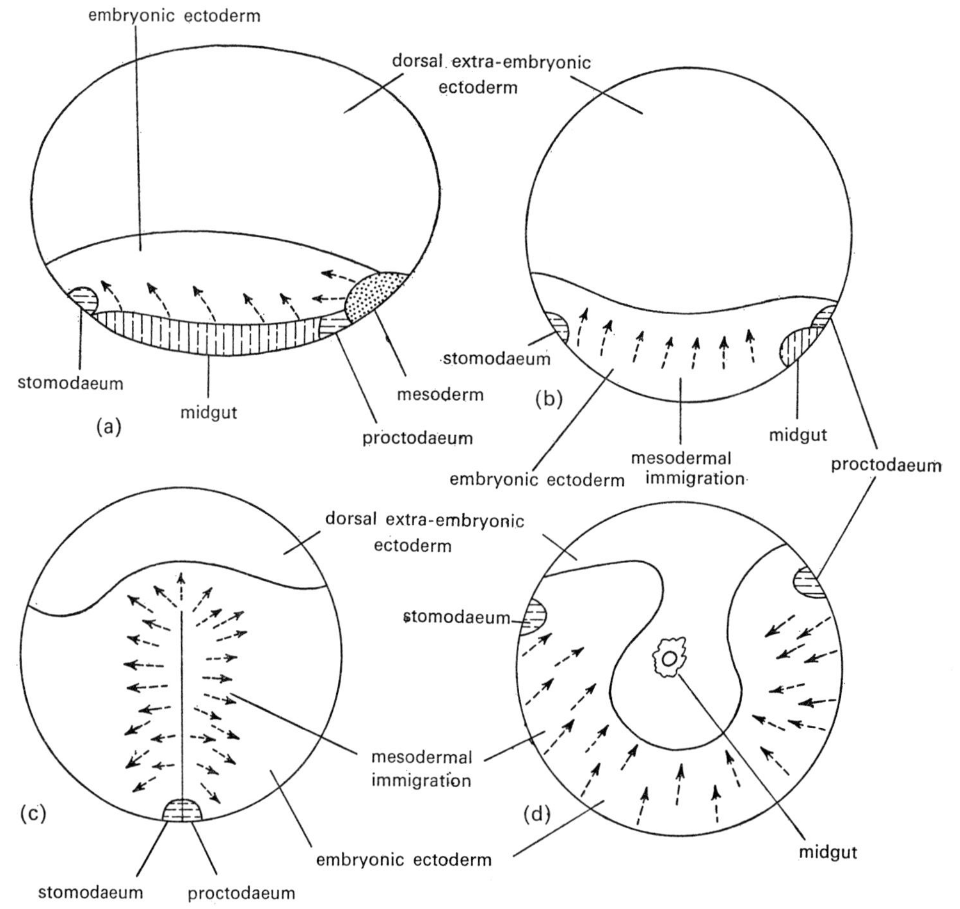

FIG. 61. Fate maps of the myriapod blastoderm. (a) Chilopoda; (b) Diplopoda; (c) Symphyla; (d) Pauropoda. (Based on the data summarized in the text.)

difficult than for any other group of arthropods. The difficulty arises partly because many myriapod embryos already have cells internal to the blastoderm at the blastoderm stage, but is also compounded by the fact that myriapod presumptive rudiments segregate from one another in a diffuse manner during subsequent development. The sharp boundaries between the presumptive areas of the annelid blastula, sufficiently retained to permit a reasonable degree of exactitude in constructing fate maps of the blastoderm of the Onychophora, and reconvened in pterygote embryos with commendable clarity (see Chapter 7), are missing in the myriapods. The elucidation of myriapodan fate maps (Fig. 61) therefore necessitates extended discussion.

The presumptive midgut

According to Heymons (1901), the midgut epithelial cells of *Scolopendra* arise as cells which separate inwards in scattered array throughout the ventral surface of the blastoderm, leaving a continuous blastodermal epithelium at the surface. The midgut cells spread around the yolk mass to form an enclosing midgut epithelium, but do not enter the yolk mass as vitellophages. The fact that the presumptive midgut cells enclose but do not enter the yolk mass has been confirmed for a number of scolopendromorph and geophilomorph centipedes by Dawydoff (1956), but the latter author offers a different interpretation of the origin of these cells. According to Dawydoff, the midgut cells are proliferated from the centre of a small posteroventral concentration of blastoderm cells, the lateral parts of which proliferate mesoderm cells. Heymons also observed that cells were released into the yolk from a posteroventral concentration of blastoderm cells which subsequently gives rise to mesodermal bands, but interpreted the immigrating cells as vitellophages, supplementary to those left in the yolk mass during cleavage. Since Heymons' description of the origin of the midgut in *Scolopendra* is not wholly convincing and Dawydoff's expression of opinion is brief and not illustrated, we are left with a dilemma which only renewed investigation can resolve. Two things, however, are already clear. Firstly, the newly segregated midgut cells of *Scolopendra* surround the yolk mass. Secondly, these cells are distinct from the vitellophages within the yolk mass, which subsequently play no part in formation of the midgut epithelium. If we provisionally follow Tiegs (1940) in accepting that Heymons presented a correct interpretation, then the presumptive midgut of *Scolopendra* occupies the ventral surface of the blastoderm, and a localized area of presumptive vitellophages lies posteroventrally. Gastrulation with respect to midgut takes the form of a scattered inward migration of the presumptive midgut cells over a wide area, leaving other cells as an ectodermal epithelium at the surface of this area. On present evidence, the limits of the area cannot be accurately defined.

In spite of the doubtful nature of the evidence, I have discussed what is known of the presumptive midgut in chilopods at some length in order to emphasize that it is unquestionably different from the presumptive midgut of other myriapods. Unfortunately, early midgut development is again not well understood in diplopods, although the recent account for *Glomeris* by Dohle (1964) is more informative than any previous account; but the formation of midgut is accurately known for pauropods and symphylans (Tiegs, 1940, 1947) and has a common ground in all three groups.

Several authors (Metschnikoff, 1874; Cholodkowsky, 1895; Lignau, 1911; Pflugfelder, 1932; Dohle, 1964) have agreed that the midgut rudiment of diplopods is a strand of cells which runs through the centre of the yolk mass. Dohle (1964) has now shown that the strand is proliferated from a localized posteroventral area of the blastoderm, which must therefore be deemed the presumptive midgut. This area, as will be seen below, lies immediately in front of the presumptive proctodaeum. The yolk mass itself is occupied by nuclei resulting from radial mitoses during early cleavage. These act as vitellophages and then develop, as the yolk is gradually resorbed, into fat-body cells.

The Symphyla are further specialized along the same lines (Tiegs, 1940). Yolk nuclei originate as a result of radial divisions during cleavage, as in the diplopods, but only the more laterally placed nuclei become the nuclei of fat-body cells. Those lying along the central core of the yolk mass become the nuclei of midgut cells. Presumptive midgut in the Symphyla is thus segregated precociously in the form of yolk cells during cleavage and retains no representation in the blastoderm.

Finally, the Pauropoda show an extreme of specialization (Tiegs, 1947), in which the presumptive midgut is cut off during early cleavage as the two distinct yolk cells in the centre of the yolk mass and the more peripheral yolk is populated at a later stage by vitellophages released by radial divisions throughout the early blastoderm. The vitellophages later develop as fat-body cells, as in the diplopods and Symphyla. Obviously, in pauropods, the presumptive midgut again has no blastodermal representation.

When one takes into account that the chilopod fat body is a product of the mesoderm (Heymons 1901; see below), it is clear that the chilopods and the remaining myriapods exhibit two alternative conditions of the presumptive midgut and fat-body. In the chilopods, the midgut originate through a broadly spread, radial immigration of ventral blastoderm cells which migrate around the yolk mass. Vitellophages arise as persistent cleavage energids and as invasive cells proliferated from a small posteroventral area of blastoderm in the midline behind the presumptive proctodaeum. Fat-body arises from mesoderm. In the remaining myriapods, the presumptive midgut either originates by proliferation of a strand of cells through the centre of the yolk from a small posteroventral area of blastoderm in front of the proctodaeum (diplopods), or is cut off in this internal position during cleavage (Symphyla and Pauropoda). Vitellophages arise throughout the forming blastoderm as a result of radial divisions and later develop as fat-body. We shall see later that this difference between the Chilopoda on the one hand and the Diplopoda, Symphyla and Pauropoda on the other, is reflected in many aspects of myriapod embryonic development.

The presumptive stomodaeum and proctodaeum

In contrast to the vagaries of midgut development, the stomodaeum and proctodaeum retain a uniformly simple development throughout the myriapods. Each originates as a cellular ingrowth in the ventral midline, the stomodaeum anteriorly between the incipient procephalic lobes, the proctodaeum posteriorly. The ingrowth hollows out as a short blind-ending tube, the further development of which will be discussed in a subsequent section.

Localization of the presumptive stomodaeum in the blastoderm of chilopods requires further study, but the evidence of Heymons (1901) indicates a likely occupation of the anterior margin of the area from which midgut cells arise. For the Diplopoda, Symphyla and Pauropoda, the presumptive stomodaeum obviously cannot be identified in the same way, since there is no long ventral component of presumptive midgut in the blastoderm, but it can be placed quite easily relative to presumptive embryonic ectoderm and in this respect occupies the same location as in chilopods (see below).

The presumptive proctodaeum in the Chilopoda lies just anterior to the posterior proliferative zone from which the secondary vitellophages and mesoderm arise, and is thus at the posterior margin of the presumptive midgut area. For the Diplopoda, Dohle (1964) has shown that the presumptive proctodaeum again lies immediately behind the small, posteroventral area of presumptive midgut. In the absence of a posterior area of presumptive mesoderm in diplopods, however (see below), the presumptive proctodaeum occupies the extreme posterior end of the embryonic primordium. The same posterior location of the presumptive proctodaeum is observed in the Symphyla and Pauropoda (Tiegs, 1940, 1947), even though the presumptive midgut is no longer represented in the blastoderm.

The presumptive mesoderm

Chilopods retain the small posteroventral area of presumptive mesoderm already seen in Onychophora. From this area, paired ventrolateral bands of mesoderm are proliferated forwards on either side of the proctodaeum, passing between the midgut epithelium on the yolk and the blastoderm cells at the surface. We have already seen that the same proliferative area acts previously as a source of secondary vitellophages. In addition, however, the mesodermal bands are supplemented along their length by further cells which move inwards from the overlying blastoderm. Other cells migrate inwards from the ventral blastoderm between the two mesodermal bands, to lie beneath the surface as scattered median mesoderm. Thus the presumptive mesoderm of chilopods has in part a localized distribution posteroventrally in the blastoderm, but in part is also diffusely spread bilaterally along the presumptive embryonic ectoderm.

The embryos of the Diplopoda, Symphyla and Pauropoda show this diffuse pattern in a more modified form. As already mentioned, the origin of the mesoderm in diplopod embryos is uncertain (Dohle, 1964), but the work of Tiegs (1940, 1947) provides a precise account for Symphyla and pauropods and the limited observations on diplopods point to a similar interpretation. The localized posteroventral area of presumptive mesoderm is absent. Mesoderm cells separate into the interior diffusely over the entire length of the embryonic primordium. In the Symphyla (Tiegs, 1940) and probably also in diplopods (Cholodkowsky, 1895; Lignau, 1911; Dohle, 1964), the majority of the mesoderm cells pass in along the lateral margins of the embryonic primordium and pile up as paired lateral cords. Immigration nearer the midline also produces a thin, irregular sheet of median mesoderm between the two lateral cords. *Pauropus* is slightly more modified (Tiegs, 1947) in that cells separate into the interior over the entire surface of the embryonic primordium to form an irregular layer beneath the columnar outer layer and only begin to aggregate as lateral cords at a later stage. The cells remaining at the surface after the presumptive mesoderm cells have passed inwards are components of the embryonic ectoderm. Presumptive mesoderm in the

Diplopoda, Symphyla and Pauropoda can thus be represented only as a diffuse distribution of cells throughout the presumptive ectoderm, normally more concentrated ventrolaterally than ventrally.

The presumptive ectoderm

The dorsal part of the blastoderm, occupying more than half of the surface in the Chilopoda and Diplopoda, less than half in the Symphyla and Pauropoda, develops directly as extra-embryonic ectoderm and is thus presumptive for this fate. The presumptive embryonic ectoderm arises from the remainder of the blastoderm. Broadly bilaterally distributed on either side of the ventral midline, this area includes the presumptive ectoderm of the acron (in front of the presumptive stomodaeum), the cephalic segments and a number of trunk segments, together with a posterior growth zone on either side of the presumptive proctodaeum. We shall see that further development usually proceeds in such a way that the mid-ventral presumptive ectoderm between the stomodaeum and proctodaeum assumes a temporary fate as ventral extra-embryonic ectoderm before reverting to definite ectoderm. This temporary deviation is conspicuous in the large embryo of *Scolopendra*, vestigial in the smaller embryos of *Geophilus*, *Scutigera* and the Diplopoda and Symphyla, and absent in the small *Pauropus* embryo.

Before attaining its strict condition as presumptive ectoderm, the blastodermal area which has this subsequent fate releases many cells into the interior. As we have seen, in chilopods, presumptive midgut cells are released first throughout the area, and presumptive mesoderm cells later, mainly along paired bilateral zones. In the diplopods and Symphyla, there is again a mainly bilateral release of mesoderm cells and in the small embryo of *Pauropus*, a generalized release of mesoderm cells. The causal processes underlying this mode of formation and segregation of presumptive areas in myriapod embryos have not yet been analysed.

Gastrulation

Associated with the diffuse nature of the presumptive areas of myriapod embryos and with modifications of cleavage in the small eggs of symphylans and pauropods, gastrulation is scarcely represented as a phase in myriapod development. If we retain the definition of gastrulation already applied to annelids and onychophorans, namely the movement of presumptive areas into their organ-forming positions, we can summarize the position as follows:

1. The presumptive midgut displays a gastrulation movement in the Chilopoda, as a diffuse migration of cells into the interior over a wide area. The diplopod presumptive midgut enters directly into organogeny in its superficial position. The presumptive midgut of Symphyla and Pauropoda is cut off precociously into the interior during cleavage and exhibits no gastrulation movement.
2. The presumptive stomodaeum and proctodaeum of all myriapods undergo independent invagination as short tubes which then enter directly into organogeny.
3. The posterior area of presumptive mesoderm in chilopods shows a slight insinking, but

then proceeds directly into proliferative organogeny without moving into the interior as a whole. The bilateral supplementary mesoderm of the chilopods enters the interior by diffuse migration, as does the entire presumptive mesoderm of the remaining myriapods.

4. The presumptive embryonic ectoderm and the dorsal extra-embryonic ectoderm enter directly into organogeny in the positions in which they are formed as components of the blastoderm.
5. The presumptive fat body, in the form of vitellophages, is cut off into the interior during cleavage in diplopods, symphylans and pauropods. The chilopod fat-body has a later origin from mesoderm.

Myriapods, then, have all but abandoned gastrulation as a significant phase in development and attain the arrival of cells in their organ-forming positions through a variety of specialized segregations during and immediately following blastoderm formation. We shall return to the possibility of a basic theme of early development for myriapods when the further development of the various components of the embryo during organogeny has been expounded.

Further Development of the Gut

In strong contrast to the complex manner in which myriapods segregate and redistribute their presumptive embryonic rudiments during early development, the further development of each rudiment during organogeny presents a pleasingly simple picture. Simplest of all is the further development of the gut. The stomodaeal and proctodaeal rudiments grow into the interior as blind-ending tubes in the usual way and differentiate as the epithelia of the foregut and hindgut respectively. The free end of the proctodaeum also pouches out as pair of evaginations which form the Malpighian tubules. According to Tiegs (1947), the end of the proctodaeum distal to the Malpighian tubules in *Pauropus* differentiate into the posterior termination of the midgut, a specialization presumably associated with the small size of the basic midgut rudiment in this animal (see below). The stomatogastric ganglia originate as dorsal thickenings of the wall of the stomodaeum (Heymons, 1901; Tiegs 1940, 1947; Dohle, 1964).

The midgut of *Scolopendra* (Fig. 62) develops directly from the epithelial layer of small midgut cells formed around the yolk mass. The vitellophages within the yolk mass play no part in midgut formation and are digested with the yolk. In diplopods, in contrast, the midgut epithelium develops from the strand of cells which runs through the centre of the yolk mass from the stomodaeum to the proctodaeum (Cholodkowsky, 1895; Lignau, 1911; Pflugfelder, 1932; Dohle, 1964). The strand hollows out and the cells increase in number and form an epithelium around the central lumen. The nucleated yolk mass lies entirely external to the developing midgut and transforms gradually into the cells of the fat body.

Midgut development in the Symphyla and Pauropoda (Fig. 63) proceeds in an essentially similar way (Tiegs, 1940, 1947). In *Hanseniella*, the mesoderm, as will be described below, develops so as to separate a central strand of nucleated yolk from paired lateral masses. The central strand differentiates into a midgut tube joining the stomodaeum to the proctodaeum. The lateral masses develop into fat-body cells. *Pauropus*, of course, already has a

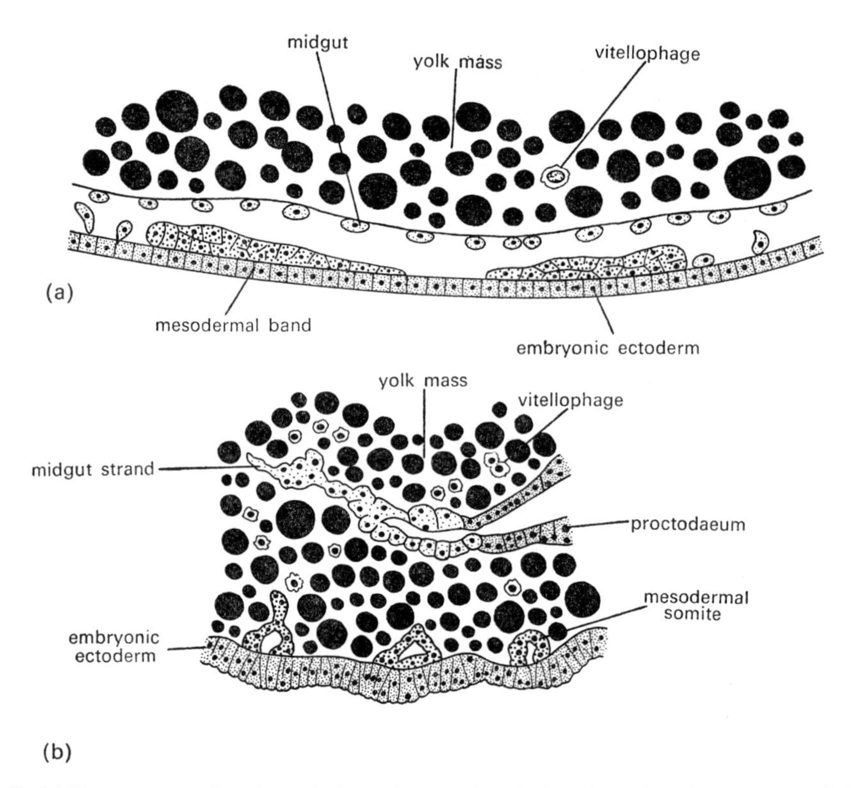

FIG. 62. (a) Transverse section through the early germ band of *Scolopendra*, after Heymons (1901); (b) sagittal section showing early development of the midgut strand of *Glomeris*, after Dohle (1964).

distinct midgut rudiment in the form of two yolky cells in the centre of the embryo. These cells slowly divide, increasing the number of cells, and give rise to the midgut tube joining the stomodaeum to the proctodaeum (Fig. 63). The surrounding nucleated yolk develops in the usual way as fat-body cells.

Plainly, the Diplopoda, Symphyla and Pauropoda share a mode of midgut and fat-body development in common, in spite of variations in the mode of origin of the midgut rudiment. Clearly, also, the Chilopoda achieve midgut development in a different way. The indications of two alternative modes of myriapod embryonic development observed during cleavage, blastoderm formation and the completion of gastrulation thus begin to receive much stronger support from a consideration of the further development of the midgut. This dichotomy is equally evident, as we shall see, in the further development of the mesoderm.

Development of External Form

Before discussing in detail the contributions made by the mesoderm and ectoderm during myriapod organogeny, a brief excursion must be made into the gross development of the embryo as observed in external view (Figs. 64–67). Following the blastoderm phase in *Scolopendra*, a small, circular, posteroventral swelling develops. This thickening, which Heymons (1901) called the primitive cumulus in accordance with the terminology of his

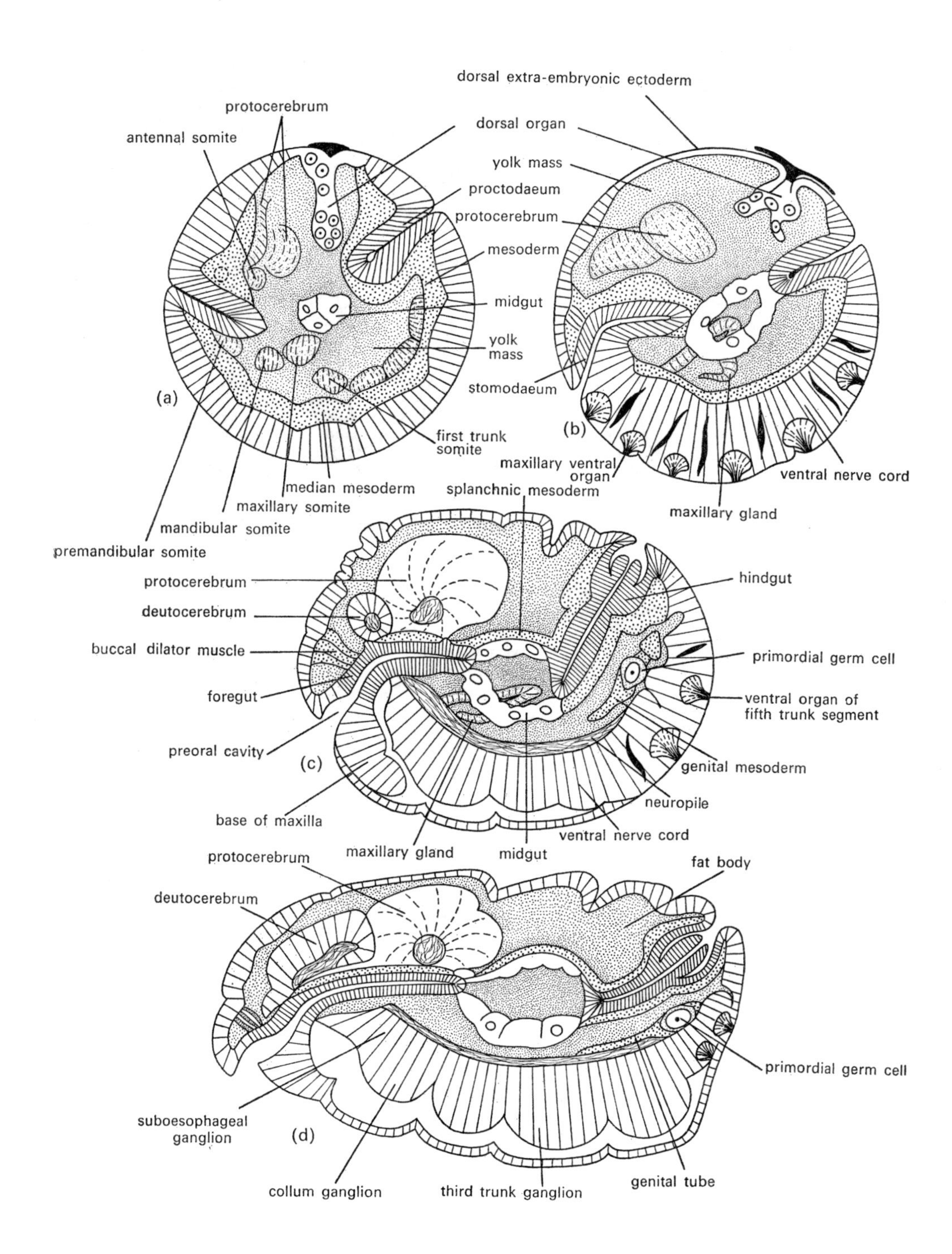

FIG. 63. Diagrammatic sagittal sections through the embryo of *Pauropus sylvaticus*. (a) 7-day stage; (b) 9-day stage; (c) advanced embryo, 11-day stage; (d) pupoid stage. (After Tiegs, 1947.)

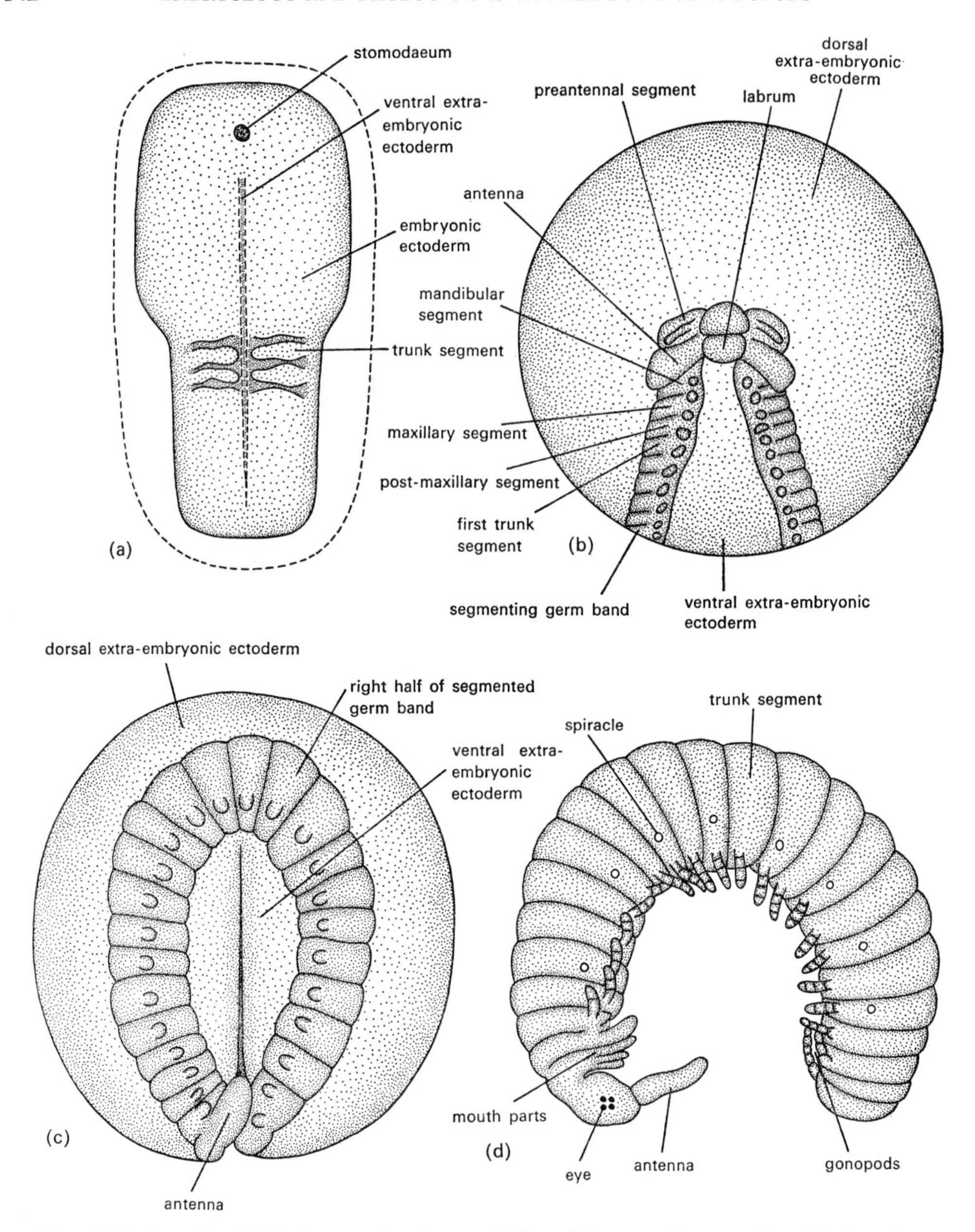

FIG. 64. *Scolopendra.* (a) Early germ band, ventral view; (b) segmenting germ band before dorso-ventral flexure, anterior view; (c) segmental germ band after flexure; (d) embryo approaching hatching stage. (After Heymons, 1901.)

day, results from an onset of proliferative activity of the presumptive mesoderm. As the paired mesodermal bands are proliferated forwards ventrolaterally from this site, the cells of the blastoderm increase in number over an area bounded posteriorly by the site of mesoderm proliferation, laterally by the mesodermal bands and anteriorly by a transverse line anterior to the presumptive stomodaeum. As the cells become more numerous, the area can be distinguished externally as the germ band. The remainder of the blastoderm remains attenuated, forming the dorsal extra-embryonic ectoderm. The germ band is broad throughout its length, but is slightly broader anteriorly than posteriorly.

The onset of external delineation of segments begins before any further increase takes place in the length of the germ band, and is accompanied by another significant change. Along the ventral midline between the sites of formation of the stomodaeum and proctodaeum, the germ band becomes less dense, while bilaterally on either side of this line, the density of cells continues to increase. Thus, when the first segment rudiments are delineated, they take the form of bilateral paired swellings on either side of a midventral groove.

The first segments to become obvious lie in the middle region of the germ band as three pairs of short, wide swellings, the rudiments of trunk segments 11–13. Further segment delineation quickly follows. The cephalic and the first ten trunk segments are delineated simultaneously. A pair of small pre-antennal segment rudiments flanks the stomodaeal opening, leaving a short, broad acron anteriorly. Paired, larger rudiments of the antennal segment lie behind the stomodaeal opening, followed by small mandibular, larger maxillary (=first maxillary) and post-maxillary (=second maxillary) segments and a sequence of ten trunk segment rudiments. The pre-mandibular segment has not been identified externally, though it has the usual coelomic sacs and ganglia (see below). Behind the thirteenth trunk segment, the next three pairs of segment rudiments are delineated in rapid succession, leaving a bilateral posterior growth zone on either side of the proctodaeal opening. The growth zone converges posteriorly on the median site of mesoderm proliferation.

Since *Scolopendra* has twenty-four trunk segments and a telson, the growth zone produces only segments 17–24 before persisting as the telson rudiment around the anus. This proliferation is completed before limb buds have begun to appear on the more anterior segments, so that the entire complement of head and trunk segments, together with the terminal acron and telson, are formed as rudiments in the germ band. The cephalic and first sixteen trunk segment rudiments, however, increase considerably in length as the last few trunk segments are proliferated by the growth zone. The increase in total length is accommodated by an upward flexure of each half of the segmenting germ band on its own side of the yolk mass. The two halves remain attached only at the anterior end by the acron and at the posterior end by the telson. The midventral area between the two halves of the germ band expands as ventral extra-embryonic ectoderm covering the yolk mass between the ventral edges of the two halves of the germ band. The dorsal extra-embryonic ectoderm becomes narrower dorsally and is stretched further over the anterior and posterior ends of the yolk mass. It must be remembered, of course, that throughout this phase of development the yolk mass is already enclosed in the midgut epithelium, which lies just beneath the dorsal and ventral extra-embryonic ectoderm, as well as beneath the halves of the segmented germ bands.

As the elongation and dorsoventral flexure of each half of the germ band is completed, the anterior and posterior ends, the acron and telson respectively, move towards one another over the ventral surface of the yolk mass and come into close proximity in a midventral position. This posture is retained during further development. The yolk mass shrinks, together with the extra-embryonic ectoderm, and the two halves of the body merge in the dorsal and ventral midline to complete the tubular body wall. Thus the long-bodied,

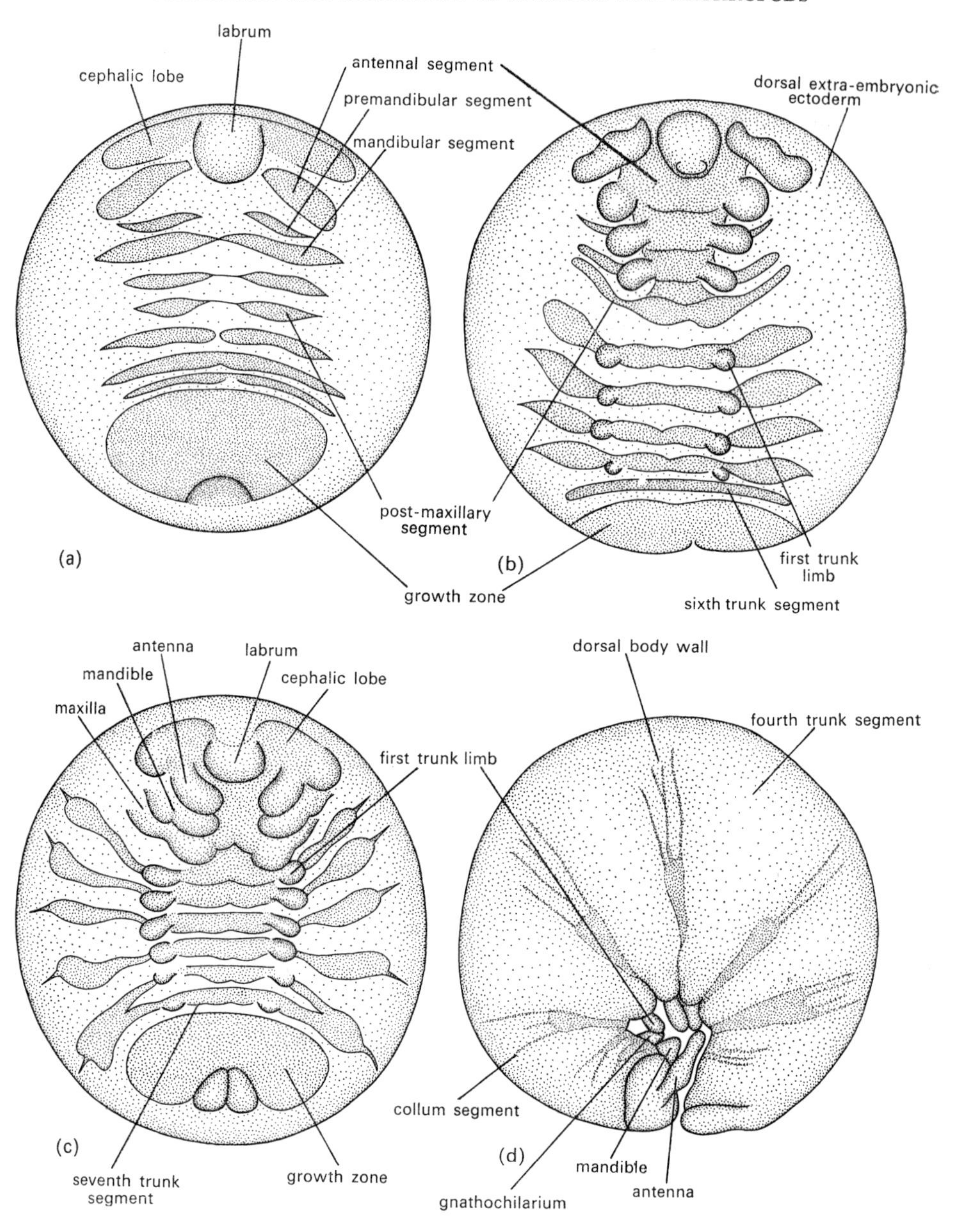

FIG. 65. *Glomeris*. (a) Segmenting germ band, ventral view; (b) limb bud formation; (c) continued segment proliferation; (d) embryo after dorsoventral flexure, lateral view. (After Dohle, 1964.)

tubular embryo is accommodated within the ovoid egg space by being tightly dorsoventrally flexed about its midpoint.

The Diplopoda (Dohle, 1964; Bodine, 1970), Symphyla and Pauropoda (Tiegs, 1941, 1947) and the anamorphic Chilopoda (Dohle, 1970) contrast with the epimorphic Chilopoda in having a smaller yolk mass relative to the size of the germ band and in hatching with only a few trunk segments completed. Associated with these differences they lack the early elongation, rapid completion of segment formation and wide bilateral separation of the two halves of the germ band observed in the scolopendromorphs and do not develop an extensive, ventral extra-embryonic ectoderm. We can take the Diplopoda first (Fig. 65). As described by Dohle (1964) and Bodine (1970), the ventral half of the blastoderm develops into a broad germ band which segments simultaneously into a series of short, broad segment rudiments. Initially, the rudiments are paired and flank a narrow midventral band of thin tissue as they do in the chilopods. Anteriorly, an acron is demarcated, consisting of a median clypeolabral lobe and paired proctocerebral lobes. Behind this, flanking the site of formation of the stomodaeum, are the halves of the antennal segment, followed by the pre-mandibular, mandibular, maxillary and post-maxillary (= first trunk) segments, then the second to fourth trunk segments. The pre-antennal segment is not delineated externally in diplopods. Behind the fourth trunk segment, the remainder of the germ band soon proliferates the fifth and sixth trunk segments, leaving a posterior growth zone in front of the terminal anus. As development proceeds further, the two halves of each segment concentrate towards the ventral midline and the thin midventral area is eliminated. It can be regarded as a vestigial representation of ventral extra-embryonic ectoderm. At the same time, the proliferative zone cuts off the short, broad rudiments of the seventh and eighth trunk segments. In *Glomeris*, no further segments are delineated before hatching takes place but in *Narceus* a further fourteen segment rudiments are established during later embryonic development. The cephalic and first eight trunk segments increase in length and breadth in the usual way. The increase in length is accommodated by a dorsoventral flexure which develops gradually and brings the anterior and posterior ends into close proximity (Metschnikoff, 1874; Lignau, 1911; Pflugfelder, 1932; Siefert, 1960; Dohle, 1964; Bodine, 1970). The increase in breadth carries the lateral walls of the segments around the yolk mass towards the dorsal midline, replacing the dorsal extra-embryonic ectoderm. The dorsal wall of the post-maxillary or first trunk segment forms the collum. The dorsal walls of the fifth and sixth trunk segments, and of the seventh and eighth trunk segments, merge to form the tergites of the first two diplosegments.

Gross development of the Symphyla (Fig. 66) shows further modification in relation to the reduction of the yolk mass (Tiegs, 1940). The embryonic primordium is at first a hemispherical cup of cells formed directly as blastoderm cells and covering slightly more than the ventral half of the yolk mass. Precociously, before segment delineation begins, the primordium undergoes a transverse dorsoventral flexure which pulls the anterior and posterior ends downwards to meet each other in the midventral position and flexes the entire ventral surface between them up into the yolk mass. At the same time, the dorsal extra-embryonic ectoderm is stretched over the anterior and posterior poles of the yolk. Segmentation of the germ band now takes place. The cephalic segments are delineated first. A median clypeolabral and paired protocerebral lobes lying in front of the stomodaeal opening are followed by postoral antennal, pre-mandibular, mandibular, maxillary and post-maxillary (= labial) segments. The pre-antennal segment is not delineated externally but can be identified by other criterial that will be discussed in later sections. Behind the point of flexure, the first three trunk segments are delineated in rapid succession. They take up most of the posterior half of the germ band, leaving a segment-forming growth zone in front of the anus. The growth zone then gives rise to four further trunk segments and then

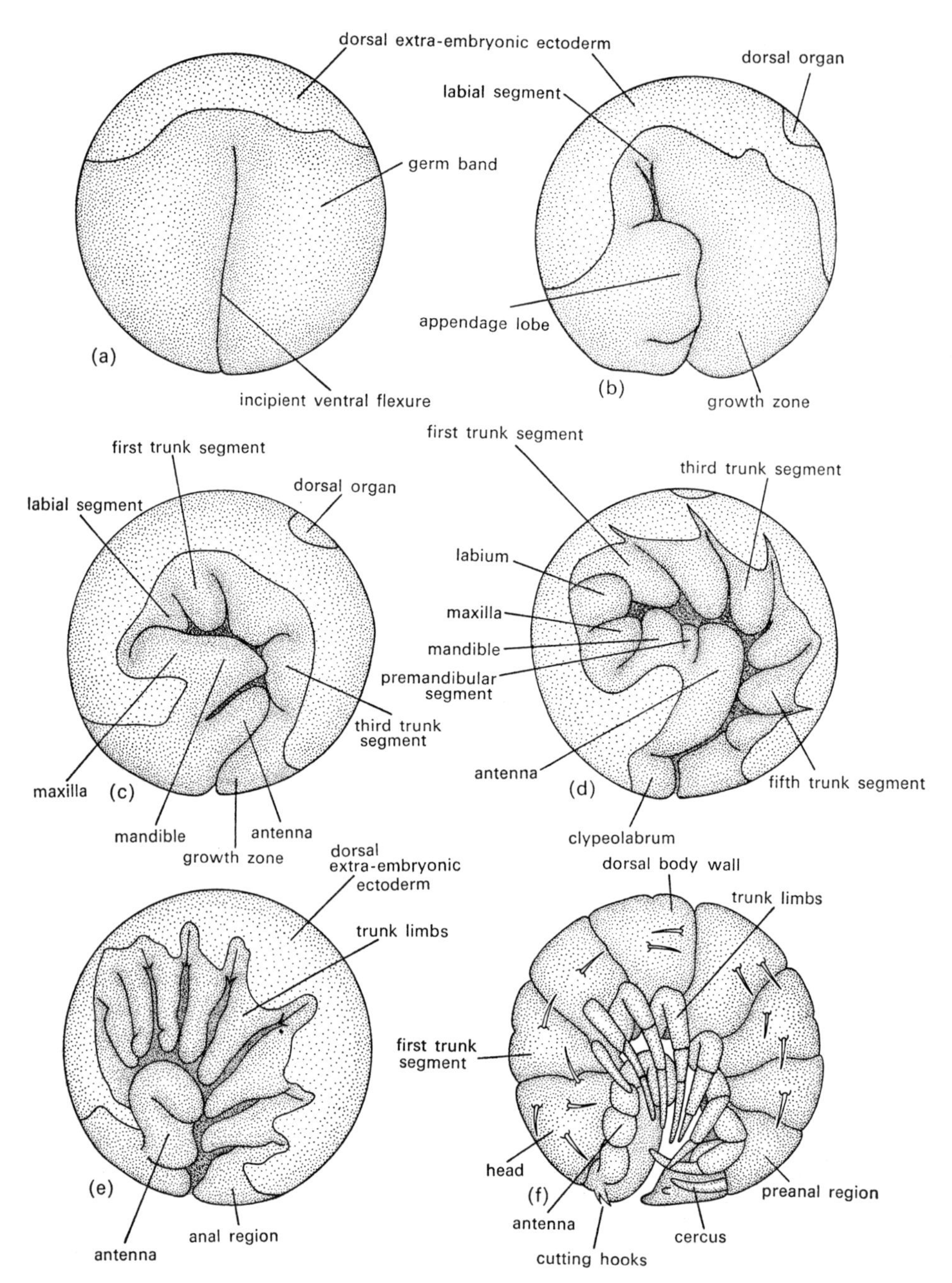

FIG. 66. *Hanseniella.* (a) Germ band showing formation of precocious ventral flexure; (b) 5-day stage; (c) early 6-day stage; (d) late 6-day stage; (c) 8-day stage; (f) approaching hatching. (After Tiegs, 1940.)

divides into a large pre-anal "segment" and a small anal segment on which the anus lies. The pre-anal "segment" is more than a single segment, since it includes the growth zone from which all subsequent segments arise during anamorphic post-embryonic growth. The terminal unit, the anal segment, presumably incorporates a vestigial telson. As the fourth to

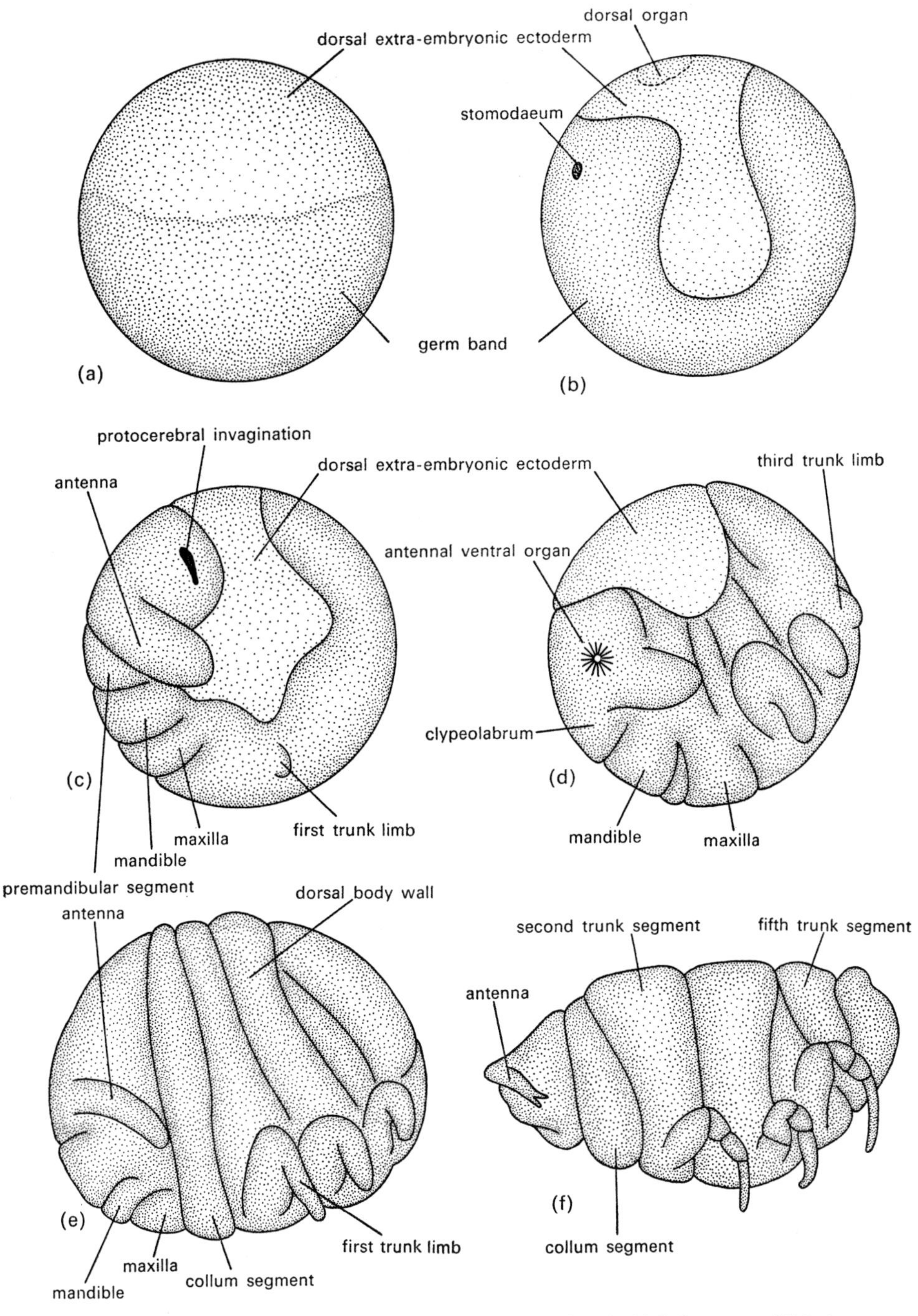

FIG. 67. *Pauropus*. (a) Differentiation of blastoderm; (b) germ band; (c) 7-day stage; (d) 9-day stage; (e) 10-day stage; (f) early pupoid stage with sheath removed. (After Tiegs, 1947.)

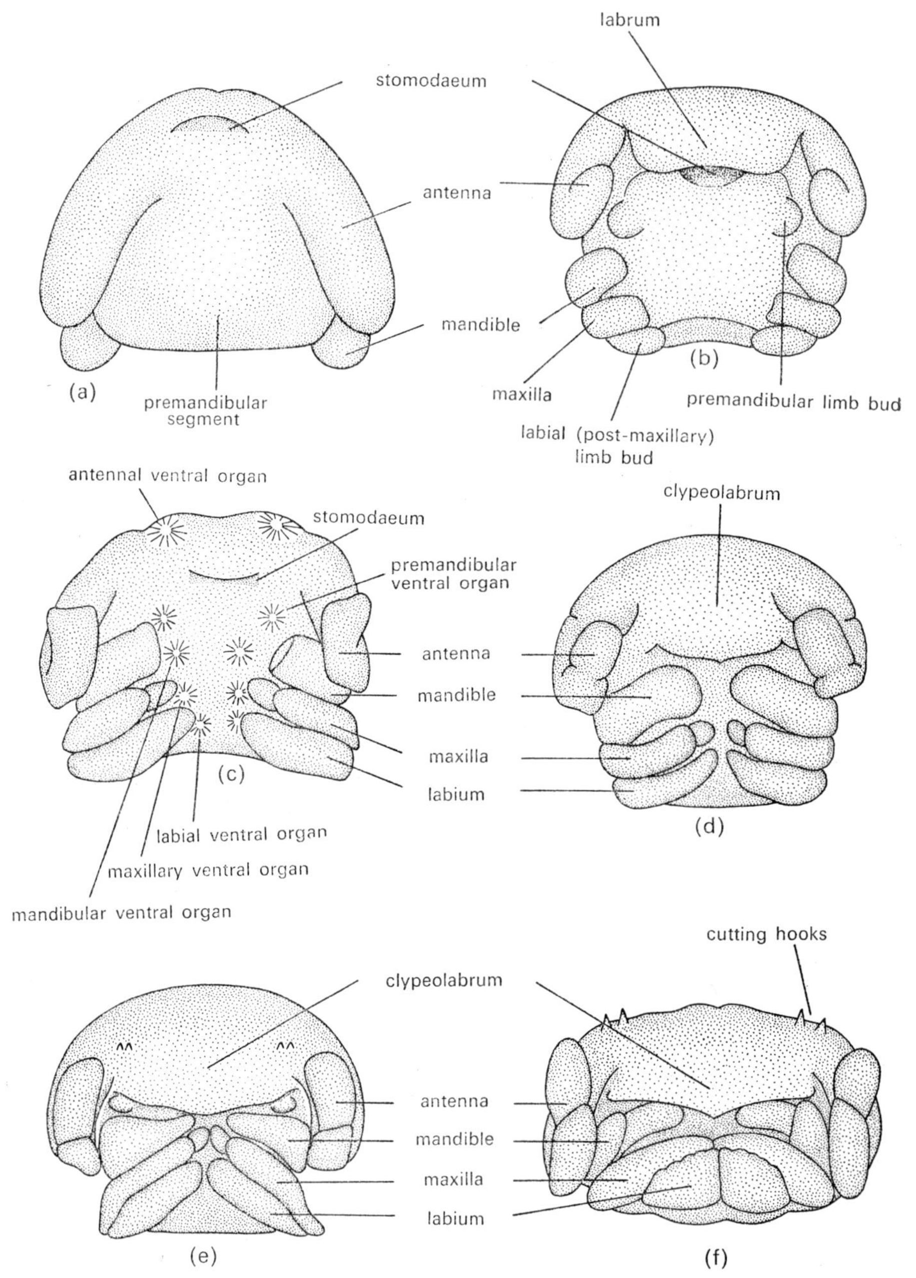

FIG. 68. External development of the head of *Hanseniella*, ventral view. (a) Early 7-day; (b) late 7-day; (c) early 8-day; (d) late 8-day; (c) early 9-day; (f) late 9-day. (After Tiegs, 1940.)

seventh trunk segments are formed, the dorsoventral flexure of the elongating embryo increases and the embryo becomes tightly coiled within the spherical egg space. Lateral upgrowth of the body wall then completes dorsal closure, replacing the dorsal extra-embryonic ectoderm.

Finally, the gross development of the embryonic primordium in the very small egg of *Pauropus* (Tiegs, 1947) omits ventral flexure entirely (Fig. 67). The embryonic primordium formed at the blastoderm stage occupies all except the dorsal part of the spherical yolk surface. As it narrows and elongates into a germ band, the primordium remains curved around the yolk surface and the anterior and posterior ends approach one another in the mid-dorsal position. Associated with this, the small area of dorsal extra-embryonic ectoderm is stretched down the sides of the yolk. The anterior end of the germ band, anterior to the stomodaeal opening, develops as a median clypeolabral area and paired protocerebral lobes. There is no external expression of a pre-antennal segment, but a large antennal segment can now be recognized just behind the level of the stomodaeum. Behind this again is a small pre-mandibular segment, followed in temporal succession by mandibular and maxillary segments, then the post-maxillary and second to fourth trunk segments. As in diplopods, the dorsal wall of the post-maxillary or first trunk segment forms the collum. The fourth trunk segment is delineated near the posterior end of the germ band, leaving a terminal growth zone in front of and around the anus. While upgrowth of the lateral walls of the segments effects dorsal closure, the curved embryo straightens out. Short fifth and sixth trunk segment rudiments are then demarcated in front of the growth zone, but at hatching, only the first four trunk segments have attained the functional condition, as compared with the first seven in symphylans and eight in diplopods.

The very small egg of *Pauropus* thus develops with a considerably greater degree of anamorphosis than the larger eggs of diplopods and Symphyla. It is notable, however, that in the embryos of all three classes, the germ band includes the rudiments of the first nine serial segments (pre-antennal, antennal, pre-mandibular, mandibular, maxillary, post-maxillary and three trunk segments) and that all additional trunk segments stem from a posterior growth zone. Thus the varying degrees of anamorphosis in these myriapods, attaining an extreme in the small embryo of *Pauropus*, are associated with different degrees of suppression of the activity of the growth zone in the embryo, but in no case involve inhibition of development of any of the nine anterior, serial segments. As we shall see below, the mode of development of the mesoderm gives further support to this conclusion. The anamorphosis of *Scutigera*, in contrast, is based on an initial development of twelve serial segments (Dohle, 1970).

Development of limbs

All myriapod embryos develop a pair of large, antennal limb buds, which are post-oral in position when first formed. Behind them lie a pair of mandibular limb buds and a pair of maxillary limb buds. Only *Scolopendra* has been found to have a pair of transitory, pre-antennal limb vestiges in front of the antennae, on either side of the mouth (Heymons, 1901; Dawydoff, 1956). These limbs are not formed in other myriapods (Tiegs, 1940, 1947; Dohle, 1964). Similarly, only the symphylan *Hanseniella* has been found to develop a pair of transient, pre-mandibular limb buds. Other myriapods lack these vestiges.

During the further development of the head (Fig. 68), the median clypeolabral lobe forms the anteroventral surface of the head, including the labrum. The lateral protocerebral lobes meet in the dorsal midline to form the dorsal wall of the head and the pre-antennal, antennal and anterior margins of the pre-mandibular segment halves migrate forwards on either side of the mouth, to contribute to the lateral walls of the head and the undersurface of the labrum. As part of this forward migration, the antennae are carried into a pre-oral position and the mandibles come to flank the mouth, with the maxillae just behind them ventrolaterally. The maxillae remain paired in the Chilopoda, Symphyla and Pauropoda, but unite to form the gnathochilarium in the Diplopoda (see below).

The developmental role of the post-maxillary segment differs in each group of myriapods. In the chilopods (Heymons, 1901) this segment is incorporated into the head as the second maxillary segment, but its limb buds develop in a generalized way and remain quite leg-like. The post-maxillary segment of the Symphyla is also incorporated into the head, but shows a greater specialization of its limb buds, which unite and develop as the symphylan labium (Tiegs, 1940). The same segment in the Pauropoda, in contrast, is the limbless first trunk segment, which gives rise to the collum (Tiegs, 1947). A careful recent investigation by Bodine (1970) has now shown that the post-maxillary segment of the diplopods is similarly a limbless first trunk segment, developing as the collum. In those millipedes in which the collum carries a pair of limbs, the limbs of the second to fifth trunk segments undergo a forward shift of one segment during later development. Bodine also confirms the view put forward earlier by Silvestri (1903, 1933, 1949) and Dohle (1964) that the gnathochilarium of diplopods is formed by fusion of the maxillae alone.

The myriapods, therefore, share a common pre-oral and mandibular–maxillary head development, but exhibit three conditions of the post-maxillary segment, in the Chilopoda, Symphyla and Pauropoda–Diplopoda respectively. It seems likely that in protomyriapods, the head ended with the maxillae and the post-maxillary segment was the limb-bearing first trunk segment. We shall return to the implications of this possibility later in the present chapter.

Development of limb buds on the trunk segments of myriapod embryos varies with the degree of anamorphosis. *Scolopendra* develops a pair of limb buds on each trunk segment. The first pair, on the seventh serial segment, transform into the specialized poisonjaws, the next twenty-one develop as locomotory limbs and the last two pairs are rudimentary gonopods. In the diplopods, in contrast, only seven pairs of trunk limbs develop in the embryo, the last four pairs usually becoming grouped together as the pairs of the first two diplo-segments. The Symphyla also develop seven pairs of functional trunk limbs in the embryo, but display no diplopody. Finally, in *Pauropus*, only three pairs of trunk limbs are developed in the embryo, the remainder being added anamorphically after hatching.

Further Development of the Mesoderm

Having established an understanding of the gross pattern of development in the four main types of myriapod embryos, it now becomes possible to consider in more detail the underlying role of the mesodermal bands and the median mesoderm in this sequence of events. Fortunately, careful and detailed accounts of the development of the mesoderm have been given for one species in each major group of myriapods, by Heymons (1901), Tiegs (1940, 1947) and Dohle (1964) respectively.

Mesodermal segmentation always precedes ectodermal segmentation. *Scolopendra* shows

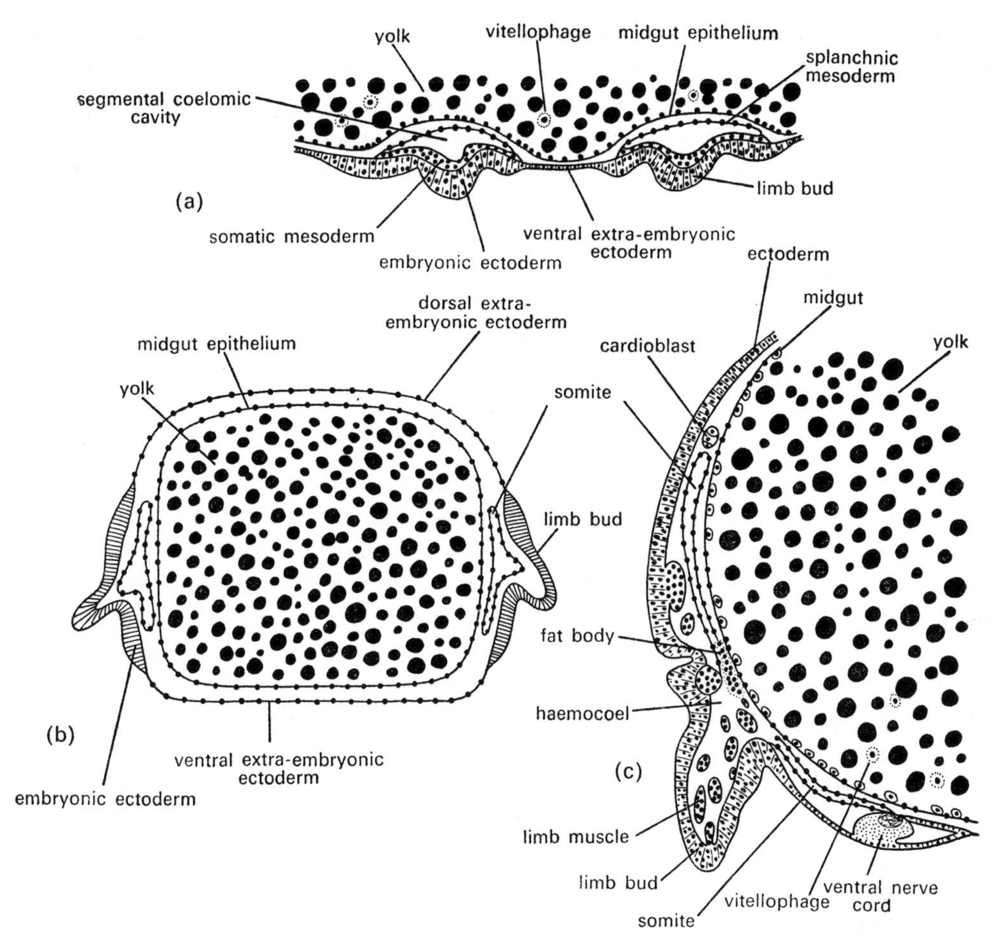

FIG. 69. Transverse sections through stages of segment development in *Scolopendra*, after Heymons (1901). (a) Before dorsoventral flexure; (b) after flexure; (c) during dorsal closure.

the most generalized mode of mesoderm development among the myriapods (Fig. 69). In contrast to the pattern of external definition of the segments, the cells of the mesodermal bands group into paired somites in an anteroposterior sequence. The most anterior parts of the bands form a small pair of pre-antennal somites on either side of the site of presumptive stomodaeum, the acron being devoid of mesodermal components. The pre-antennal somites are followed by a large antennal pair, a small pre-mandibular pair and then a sequence of somites of more uniform intermediate size, those of the mandibular, first maxillary, second maxillary and trunk segments. The last few pairs of trunk somites are proliferated from the growth zone after the proctodaeum has been formed and move forwards on either side of the proctodaeum to complete the sequence of twenty-two pairs of trunk somites and two pairs of genital somites. The telson has no somites.

During the elongation and dorsoventral flexure of each half of the germ band, the paired somites become widely separated from one another on either side of the yolk mass, especially in the middle region of the trunk. Each somite hollows out during this phase and develops a trilobed form. The somatic wall pouches out into the overlying limb bud as an appendicular lobe, while the remainder of the somite expands dorsally and ventrally to form dorsolateral and medioventral lobes beneath the embryonic ectoderm of the segment. The splanchnic walls of the dorsolateral and medioventral lobes are continuous and lie against the underlying midgut epithelium at the surface of the yolk mass.

During subsequent development, the pre-antennal and antennal somite pairs migrate forwards on either side of the stomodaeum to attain a preoral position and the pre-mandibular somites come to lie on either side just behind the mouth. All pairs of cephalic and trunk somites, however, undergo the same major changes. As the yolk mass shrinks, the dorsolateral and medioventral lobes extend around the midgut towards the dorsal and ventral midlines. The appendicular lobes of the somites develop into limb musculature, except in the pre-antennal segment, whose limbs are transient, and the pre-mandibular segment, which lacks limb vestiges. The basal regions of the appendicular lobes release cells which develop as the fat body. The dorsolateral lobes give off cell masses from their somatic walls which develop as dorsal longitudinal muscles. The dorsal edges of the dorsolateral lobes then come together in the dorsal midline to form the heart and anterior aorta. All pairs of somites, including the pre-antennal and pre-mandibular pairs, contribute to the dorsal blood vessel. In the trunk, the somatic walls of the dorsolateral lobes also stretch out to form the pericardial septum, while beneath the septum, part of the coelomic cavity of each dorsolateral lobe is enclosed and becomes confluent with the others of its side, forming a pair of genital tubes. The splanchnic walls, applied to the external surfaces of the midgut, develop as midgut musculature.

Development of the medioventral lobes begins in a similar way. The lobes extend towards the ventral midline, except for those anterior to the mandibular segment, and their edges come together to form the ventral blood vessel. The splanchnic walls develop as gut musculature and the somatic walls give off cell masses which develop as ventral longitudinal muscle. The remainder of the development of the medioventral lobes differs from that of the dorsolateral lobes, however, in that the coelomic cavities are lost and the somatic cells differentiate as transverse ventral muscles. The last two pairs of medioventral lobes, in the two genital segments, develop ventrolateral outgrowths in the form of coelomoducts, retaining coelomic internal cavities. The coelomoducts subsequently develop as the genital ducts.

Certain peculiarities attend the development of the anterior cephalic somites. Obviously, none of them contributes to the pericardial septum, gonads or ventral blood vessel. The somatic walls of the pre-antennal somites form, in addition to the anterior ends of the ventral longitudinal muscles, only clypeolabral musculature. The somatic walls of the antennal somites contribute, apart from antennal muscles and components of the anterior aorta, only ventral longitudinal muscle and somatic musculature of the head capsule. The somatic walls of the pre-mandibular somites are specialized in another way. Apart from their contribution to the anterior aorta and longitudinal muscles, the walls of these somites develop mainly as paired masses of lymphoid tissue in the head. For reasons that will become apparent below, the lymphoid masses are usually interpreted as the vestiges of a pair of pre-mandibular segmental organs.

The three pairs of postoral cephalic segments are, of course, much more normal in development, especially the second maxillary pair. As compared with the trunk somites, they lack only the pericardial and genital components. None of these three somite pairs develops any trace of segmental organs.

When the mesodermal bands are first formed in the early embryo, a scattering of mesoderm cells lies between them as median mesoderm. The median mesoderm is subsequently left in the ventral position while the somites form and separate. According to Heymons (1901), the median mesoderm of chilopods has a twofold fate. Anteriorly and posteriorly, the cells are carried inwards as a covering sheath on the ingrowing stomodaeum and proctodaeum and develops as stomodaeal and proctodaeal musculature respectively. The intervening median mesoderm gives rise to blood cells.

In discussing the further development of the mesoderm of the remaining classes of myriapods, it is advantageous to give consideration first to the Symphyla (Fig. 70). As pointed out by Tiegs (1940), their mesodermal development has remained generalized in many respects, but at the same time, exhibits a number of significant differences from the Chilopoda.

We have already seen that the early germ band of *Hanseniella* develops a pair of lateral cords of mesoderm extending from the stomodaeal to the proctodaeal level, together with a sheet of median mesoderm between the lateral cords. Before external segmentation becomes apparent, the lateral cords pile up into paired somite masses, linked by narrow intersegmental bridges. Somite formation begins at the mandibular level, but spreads rapidly forwards and backwards. As in chilopods, a pair of small pre-antennal somites is formed anterolateral to the stomodaeum, followed by a pair of large antennal somites behind the stomodaeum, then pre-mandibular, mandibular, maxillary and labial pairs and three pairs of trunk somites, formed in rapid succession. Mesodermal proliferation at the posterior end of the segmenting germ band soon results in the formation of four more pairs of trunk somites. The remaining leg-bearing segments of the trunk in *Hanseniella* are not developed until after hatching, but a mass of mesoderm is left posterior to the seventh pair of trunk somites. The further development of this mass in the embryo will be discussed below.

Associated with the secondary reduction in length and precocious flexure of the germ band in Symphyla, the two halves of the germ band do not arch away from one another over the sides of the yolk mass as they do in the scolopendromorph chilopods. The paired somites remain near to the ventral midline during their further development, with median mesoderm between them as a sheet of cells extending from the mandibular to the seventh trunk segment. In spite of this difference, the early development of the trunk somites in *Hanseniella* follows a similar course to that in *Scolopendra*. Each pair of somites hollows out and develops a coelomic cavity, a thick somatic wall and a thin splanchnic wall. An appendicular lobe then pushes out into the adjacent limb bud, leaving a large dorsolateral lobe, which soon develops a medioventral lobe. The only difference from *Scolopendra* up to this stage is that the paired coelomic cavities are linked from segment to segment by narrow channels through the persistent intersegmental bridges. As development continues, however, a more fundamental difference supervenes. Instead of remaining immediately beneath the lateral ectoderm of the body wall, the dorsolateral lobes of the somites shift towards the midline of the embryo. As they now grow upwards towards the dorsal surface in the usual way, these lobes cut vertically through the yolk mass, separating a central

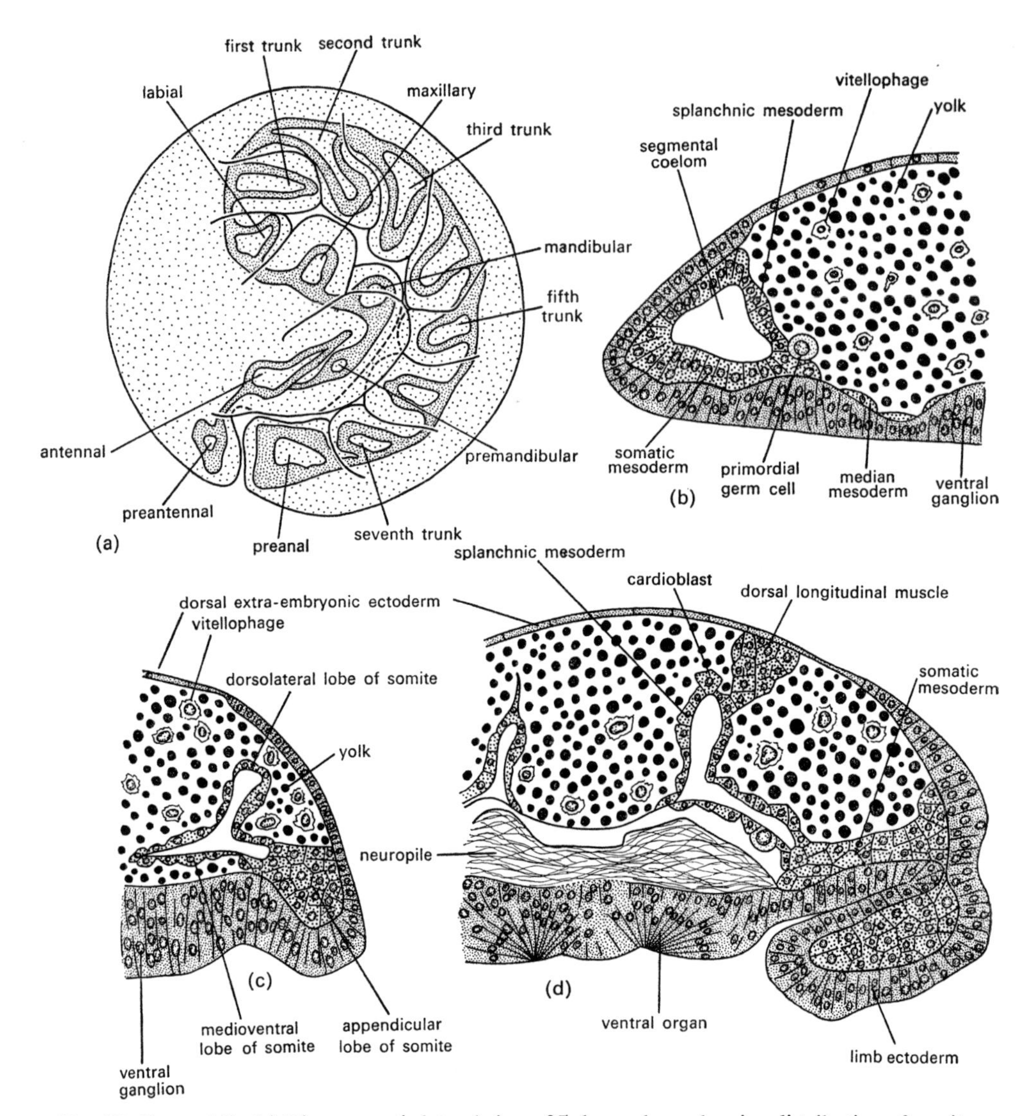

FIG. 70. *Hanseniella.* (a) Diagrammatic lateral view of 7-day embryo showing distribution of somites; (b) transverse section through fifth trunk segment of 7-day embryo; (c) transverse section through fifth trunk segment of early 8-day embryo; (d) transverse section through sixth trunk segment of late 8-day embryo. (After Tiegs, 1940.)

column of nucleated yolk from two lateral yolk masses. The central column of yolk develops into the midgut epithelium, while the lateral masses develop as fat body. The splanchnic walls of the dorsolateral and medioventral lobes are adpressed to the central column of yolk as splanchnic mesoderm. In due course, the splanchnic mesoderm spreads ventrally beneath the developing midgut epithelium and develops as splanchnic muscle in the usual way.

Even though they shift inwards at an early stage towards the midline, the somatic walls of the dorsolateral lobes of the trunk somites of *Hanseniella* complete their development in essentially the same way as in chilopods. At their dorsal edges, these walls develop cardioblasts which come together from either side above the midgut and give rise to the heart. Dorsolaterally, cell masses are given off which differentiate as dorsal longitudinal muscles and tergosternal muscles. The remainder of the wall in each segment thins out, with obliteration of the coelomic cavity, and forms the pericardial septum. The genital tubes, however, are not formed from the dorsolateral lobes (see below).

The appendicular lobes of the trunk somites of *Hanseniella* also develop in the same manner as in chilopods, giving rise to the musculature of the limb and limb base. A major difference is observed, however, in the origin of fat body, not from these cells, but from the lateral yolk masses above the pericardial septum.

Finally, the medioventral lobes show some similarities and some differences as compared with those of chilopods. We have already seen that they contribute to the splanchnic mesoderm. The somatic walls of the medioventral lobes also give off cell masses which develop as ventral longitudinal muscles and sternal muscles, as they do in *Scolopendra*. In addition, however, the lateral parts of the medioventral lobes retain their coelomic cavities and give rise to the paired, ventrolateral gonads. Germ cells differentiate in those parts of the genital tubes derived from the fifth to seventh pairs of trunk somites. Adjacent to their gonad-forming regions, the somatic walls of the medioventral lobes of the third to seventh pairs of trunk somites grow out towards the limb bases as ventrolateral coelomoducts, but these are only transient vestiges. Mesodermal genital ducts opening to a posterior, opisthogoneate gonopore are not developed in the Symphyla. Instead, secondary ectodermal gonoducts are developed in an anterior, progoneate position during post-embryonic development. We will return to the significance of this difference later in the present chapter.

It will be noted that the medioventral lobes of the somites do not extend towards the midline beneath the yolk mass in the way that they do in *Scolopendra*. Formation of the ventral blood vessel has been divorced from ventral spread of the splanchnic mesoderm and now takes place directly from cells of the median mesoderm. At the same time, the median mesoderm does not give rise to blood cells, as it does in *Scolopendra*. The blood cells of *Hanseniella* appear to be proliferated from the pericardial septum, and the other main product of the symphylan median mesoderm, apart from the ventral blood vessel, is the neurilemma of the ventral nerve cord. We shall see in due course that the chilopod neurilemma develops from the nerve cord itself.

What, now, of the cephalic somites of *Hanseniella?* As in *Scolopendra*, the pre-antennal and antennal pairs migrate forwards in front of the mouth, the premandibular pair comes to lie on either side of the mouth, and all pairs retain much of the generalized pattern of development observed in the trunk segments. Each pair of cephalic somites also has its own specializations and requires brief individual treatment.

The labial somites are the least modified. They undergo the typical trilobed outpouching and develop the typical splanchnic mesoderm, cardioblasts, ventral longitudinal muscle and sternal muscles, and extrinsic and intrinsic limb muscles previously seen in the trunk. Vestigial coelomoducts are absent, as are dorsal longitudinal muscle, pericardial septum and genital tube components. None of the last three structures extends into the head.

The maxillary somites are more specialized. The splanchnic wall separates off and grows up through the yolk in the usual way with cardioblasts at its dorsal edge, to contribute to the splanchnic musculature and heart. The somatic wall mainly breaks up directly into maxillary muscles and ventral longitudinal muscles; but in addition, a small portion of the coelom median to the base of the limb remains enclosed in an epithelial sac. From these rudiments, a pair of coiled, maxillary salivary glands with small coelomic end sacs is developed. They acquire openings directly into the pre-oral cavity, with no ectodermal exit ducts, and are interpreted as coelomoduct segmental organs of the maxillary segment.

The mandibular somites, although large, with a spacious coelom, display a greater specialization than the maxillary somites. The splanchnic wall separates off in the usual way and contributes to the splanchnic musculature and heart, but the remainder of the somite becomes mesenchymatous and develops almost entirely as mandibular and ventral longitudinal muscle. The only other product is a few cells which separate from the main mass as a pair of paracardial glands of uncertain function and homologies.

The pre-mandibular somites are the most specialized of all. The segment is mainly vestigial and has only transient limb buds. The somites are small and develop almost exclusively into a pair of coiled, pre-mandibular segmental organs with small end sacs. These organs, unlike those of the maxillary segment, acquire short, ectodermal exit ducts just lateral to the pre-mandibular ganglia. The premandibular segmental organs are large and conspicuous in the embryo, but degenerate and disappear shortly after hatching has taken place. Tiegs (1940) was unable to ascribe any function to them. Some of the cells of the pre-mandibular somites spread back into the first three trunk segments and differentiate as glandular nephrocytic organs. Unlike the pre-mandibular segmental orders, the nephrocytic organs persist into the adult.

The antennal somites, in contrast to the pre-mandibular pair, are remarkably unspecialized, being typically large and trilobed, but their later products are somewhat aberrant. The appendicular lobe develops, of course, as antennal muscle, but the medioventral lobe spreads over the dorsal wall of the stomodaeum and develops as the major part of the anterior aorta and the musculature of the stomodaeum. It is to be noted that the pre-mandibular somites are the only ones that do not contribute to the dorsal blood vessel in *Hanseniella*. The dorsolateral lobes are more specialized again, since they spread thinly beneath the large protocerebral ganglia above them and develop as the neurilemma of the brain.

Finally, the pre-antennal somites, lying in front of the stomodaeum in the developing clypeolabrum, are small and vestigial. Although transiently hollow, they soon break up into small clumps of mesenchyme, from which arise the clypeolabral muscles above the anterior end of the stomodaeum and also the anterior end of the anterior aorta.

Thus, in addition to the delay in development of the more posterior segments in the Symphyla associated with a secondarily small egg and post-embryonic anamorphosis, there are a number of other differences in the development of the trunk somites between the Chilopoda and the Symphyla. The somites form the ventral blood vessel in chilopods, but not the blood cells, which develop from median mesoderm. The ventral blood vessel in Symphyla develops from median mesoderm, and the blood cells from somite mesoderm. Chilopods retain coelomoducts only posteriorly, but develop definitive gonoducts from them. Symphylans develop transient coelomoducts in most of the anterior trunk segments, but retain none of them, the gonoducts being formed as secondary ectodermal ingrowths in an anterior, progoneate position. Finally, and perhaps most significantly, chilopod somites give off fat-body cells, while the symphylan fat-body develops from yolk cells.

In the head, the differences are more pronounced. Comparing the gnathal segments, the symphylan maxillary segment develops a pair of large segmental organs, not found in the chilopods. The pre-mandibular segment also develops a pair of large, though transient, segmental organs in Symphyla, and also some nephrocytic tissue. Only the latter is observed in chilopods. On the other hand, the chilopod pre-mandibular somites contribute to the dorsal blood vessel and splanchnic mesoderm. In Symphyla they do not. The antennal somites also remain generalized in chilopods, but are specialized in Symphyla in that they contribute stomodaeal mesoderm medioventrally and neurilemma of the brain dorsolaterally. The chilopod stomodaeal mesoderm arises from median mesoderm and the neurilemma from ectoderm. We are thus faced with patterns of somite development in chilopods and Symphyla, differing in ways which indicate that each is an alternative specialization derived from a more basic generalized pattern common to both. Thus, the indication of alternative specializations suggested by fat-body development alone is borne out in the Chilopoda and Symphyla by other aspects of mesoderm development, and the question arises—does the development of the mesoderm in Diplopoda and Pauropoda accord with the symphylan pattern of specialization, as the mode of fat-body development would suggest?

Knowledge of somite development in the Diplopoda is still deficient on many respects, but some valuable pointers can be gained from the work of Dohle (1964) on *Glomeris marginata*. The ventrolateral mesodermal bands segregate into paired hollow somites joined by median mesoderm in the usual way, but the somites are small and have rather thin walls. Eight pairs of somites are formed in the trunk, the first three being formed simultaneously with the cephalic somites, the remainder being segregated in succession from the proliferating growth zone mesoderm. As in Symphyla, they do not move away from one another up the sides of the yolk mass. In the head, paired antennal, small pre-mandibular, mandibular and maxillary somites are formed, but there is no satisfactory evidence for the formation of pre-antennal somites. According to Dohle (1964), the mesoderm in front of the antennal somites remains continuous across the head lobes and intervening clypeolabral rudiment. As we have seen previously, there is also no external indication of a pre-antennal segment in diplopods.

The first pair of trunk somites, belonging to the apodous collum segment, is vestigial and contributes only to the musculature of the gnathochilarium. The second and succeeding trunk somites develop typical appendicular lobes, together with medioventral lobes, but it is generally reported that dorsolateral lobes do not form. The mode of development of the splanchnic mesoderm, heart, pericardial septum and dorsal longitudinal muscles is still not clear. Presumably, these structures arise from the dorsal walls of the so-called medioventral lobes. The gonads definitely develop from the medioventral lobes, with germ cells first becoming apparent in the eighth and ninth pairs of somites, but no coelomoduct gonoducts are formed. Indeed, there is no evidence of the development of any, even transient, coelomoducts in the trunk segments of diplopods. The genital ducts arise, as in the Symphyla, as secondary ectodermal ingrowths in the progoneate position. The development of the ventral longitudinal muscles, ventral blood vessel, blood cells, stomodaeal mesoderm and proctodaeal mesoderm requires further investigation.

Plainly, apart from knowing that the splanchnic mesoderm grows up between the central midgut and the lateral yolk masses developing as fat body, we are still too ill-informed about diplopod trunk somites to make any satisfactory comparisons with the chilopods or Symphyla. Fortunately, the somites of the head are slightly better understood. The maxillary pair, in addition to playing a major part in the formation of gnathochilarial muscles, give rise to a pair of large segmental organs which differentiate as salivary glands, as they do in symphylans. The mandibular pair develops as mandibular musculature and the antennal pair as antennal musculature. The fate of the small premandibular pair is not clear, but there is no indication that they give rise to segmental organs as in Symphyla. Pre-antennal somites are not formed and the labral musculature develops from the generalized mesoderm in front of the antennal somites. Thus for the head, as for the trunk, an impression is gained that somite development in diplopods is generally reduced, but that significant similarities are retained with the Symphyla (anamorphosis; yolk-cell fat-body; progoneate ectodermal gonoducts; maxillary, segmental organ, salivary glands) in which both differ from the scolopendromorph Chilopoda. There is no doubt, however, that the embryology of diplopods is still in need of the intensive level of investigation devoted by Tiegs to the Symphyla and Pauropoda.

Finally, then, we can consider the Pauropoda themselves, as exemplified by *Pauropus sylvaticus* (Fig. 71). The work of Tiegs (1947) shows that in the small embryo of *Pauropus*, although the somites are relatively vestigial, a distinct pair is formed in each of the six trunk segments developed in the embryo and also in the pre-antennal, antennal, pre-mandibular, mandibular and maxillary segments of the head. Furthermore, the somites arise in the usual way, by the break up of the paired ventrolateral bands of mesoderm formed in the germ band. The cephalic and first four trunk somites are formed directly. The fifth and sixth trunk somites are proliferated from posterior, growth-zone mesoderm. As in the Symphyla and Diplopoda, the somites remain near the ventral midline during their subsequent development and do not become separated from one another bilaterally as they do in *Scolopendra*.

At the same time, the further development of the mesoderm in *Pauropus* is specialized in various ways and much of the mesoderm has escaped from the somite route of development. In addition to the median mesoderm lying between the somites, continuous cords of mesoderm are left dorsolateral to the somites. From these cords, the dorsal longitudinal muscle is formed by direct development. The median mesoderm gives rise, not only to the neurilemma of the nerve cord as in Symphyla, but also to the stomodaeal muscle layer, the splanchnic mesoderm of the midgut and the genital tube. When we take into account that the Pauropoda lack blood vessels and blood cells, so that no formation of a heart, pericardial septum or ventral vessel takes place in the embryo, it can be seen that very little of the normal generalized development of the somites takes place in *Pauropus*. The only mesodermal development retaining the somite route in the trunk, in fact, is that of the limb musculature and ventral longitudinal muscles. In the limbless collum segment, even this contribution is severely reduced and the somites are very small and lacking in coelomic cavities. Finally, none of the trunk somites develops coelomoducts, the gonoducts being formed as ectodermal ingrowths in the progoneate position.

Similar reductions obtain in the development of the cephalic somites of *Pauropus*. Obviously, these somites cannot contribute to a dorsal blood vessel, splanchnic mesoderm or dorsal longitudinal muscles. However, the maxillary somites release cells which develop as limb muscles and ventral longitudinal muscles at the maxillary level, as in the trunk, and

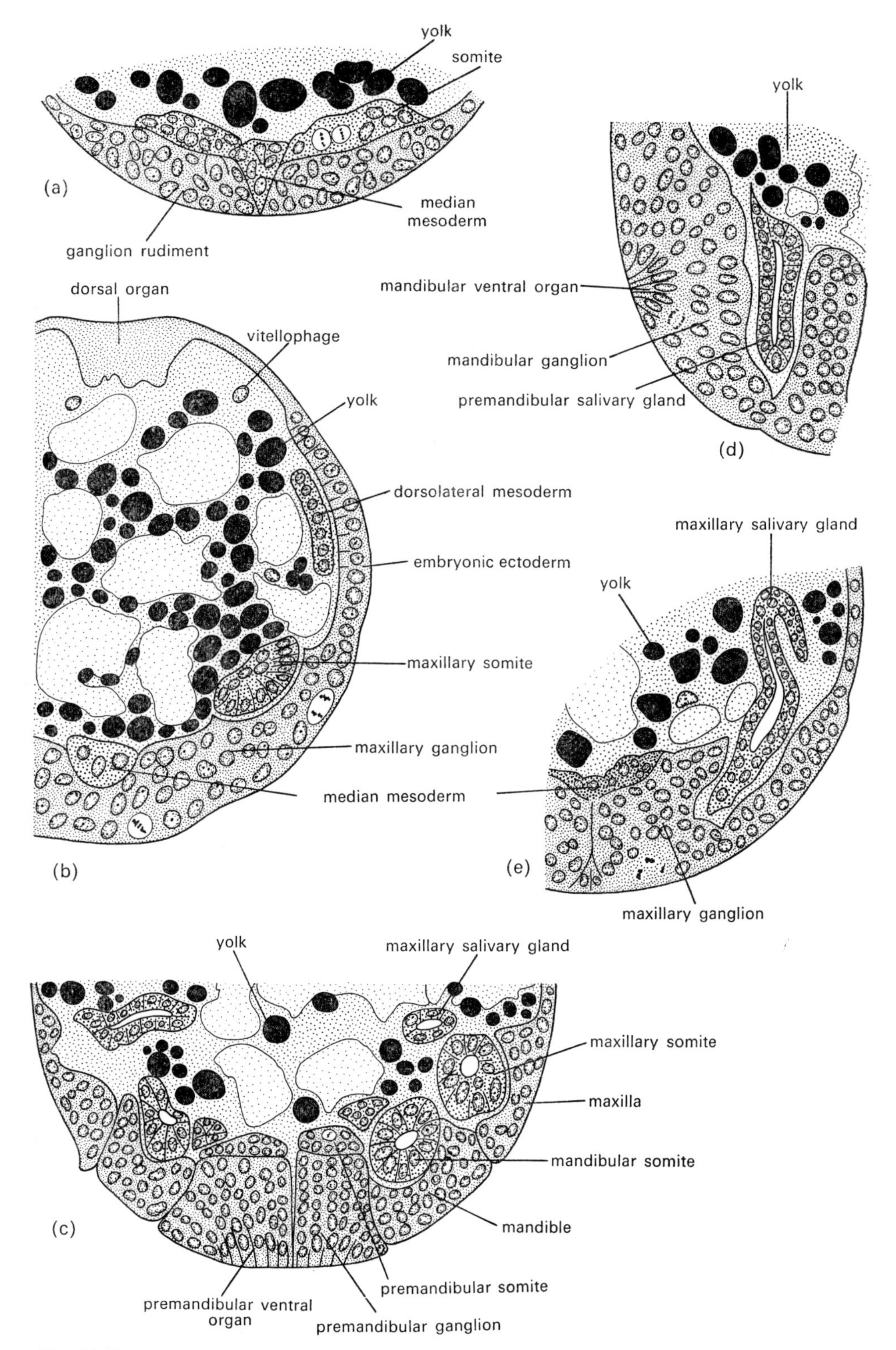

FIG. 71. Transverse sections showing segment development in *Pauropus*. (a) Maxillary segment of early 7-day embryo; (b) maxillary segment of late 7-day embryo; (c) mandibular segment of early 8-day embryo; (d) mandibular segment of late 8-day embryo; (e) maxillary segment of late 8-day embryo. (After Tiegs, 1947.)

then elongate and differentiate as a pair of maxillary segmental organs which become maxillary salivary glands, as in the Symphyla and Diplopoda. The mandibular somites break up and develop as mandibular and ventral longitudinal muscle. The pre-mandibular somites, as in the Symphyla, elongate and differentiate into a pair of pre-mandibular segmental organs with short ectodermal exit ducts. Unlike those of *Hanseniella*, the pre-mandibular segmental organs of *Pauropus* become functional as pre-mandibular salivary glands. The antennal somites disrupt and give rise to the antennal musculature and other dorsolateral muscles in the head, while the very small pair of pre-antennal somites forms the clypeolabral musculature. If allowance is made for the relative vestigiality of the cephalic somites of *Pauropus*, it is clear that they develop along the same general lines as the corresponding somites in the Symphyla and Diplopoda.

Since the pauropod stomodaeal mesoderm, midgut mesoderm and genital tube have specialized direct development from median mesoderm, we need to look at this in relation to what happens in other myriapods. The development of stomodaeal mesoderm from median mesoderm at the antennal level is observed in the Chilopoda and Diplopoda but not in the Symphyla, where the antennal somites themselves contribute this component. However, the difference is minor. A more striking difference exists in the formation of the splanchnic mesoderm of the pauropod midgut. The normal source of midgut mesoderm is the splanchnic walls of the somites, with contributions from all somites in the Chilopoda and from the mandibular and subsequent somites in the Symphyla. The pauropod midgut, in contrast, becomes invested in a sheath of cells which spreads back from the dorsal mesoderm on the distal end of the stomodaeum. No useful speculation can be made concerning the origin of such a difference. The same extreme of specialization characterize the pauropod genital tube. In contrast to the paired segmental origin of the genital tubes of other myriapods, that of *Pauropus* arises directly from the median mesoderm of the second to sixth trunk segments and has a single primordial germ cell embedded in its posterior end. However, the paired ventrolateral genital tubes developed from the medioventral lobes of the trunk somites of diplopods and symphylans, with a few germ cells near their posterior ends, could have been reduced to a single median rudiment in association with a general reduction in size in the pauropod embryo.

The final point to be made concerns the mesoderm at the posterior end of the anamorphic embryos of the pauropods, diplopods and Symphyla. In both pauropods and diplopods, a residual mass of mesoderm is left posterior to the last definitive pair of somites in the embryonic trunk. From this mass, further pairs of somites are proliferated during post-embryonic development. The median mesoderm of the same region is in part carried inwards as an investing layer on the ingrowing proctodaeum and develops as the proctodaeal splanchnic musculature. The posterior end of Symphyla, however, has a more elaborate morphology. After the seventh pair of trunk somites has been formed, the remaining mesoderm at the posterior end subdivides between a pair of large "pre-anal somites" in front of the proctodaeum and a pair of small anal somites on either side of the proctodaeum. The anal somites contribute splanchnic mesoderm to the proximal part of the proctodaem and minor musculature in the terminal anal segment on which the anus lies in Symphyla. The "pre-anal somites" are more complicated. They become hollow in the usual way, but then give rise to a pair of clumps of cells, which act as the growth zone for post-embryonic somite formation, in front of a set of more typical somite derivatives. The splanchnic mesoderm invests the proctodaeum and the posterior end of the midgut. The dorsal somatic wall becomes a part of the dorsal longitudinal muscles. An appendicular lobe grows out into the cercus of its side, though it subsequently differentiates, not as muscle, but as the cells of the cercal spinning gland. The important comparative point is that the specializations of posterior mesoderm associated with anamorphosis in *Hanseniella* are fundamentally different from those of *Pauropus* and *Glomeris*, even though the Symphyla, Diplopoda and Pauropoda share the progoneate condition and a number of other basic features of mesoderm development in which they differ from the opisthogoneate Chilopoda. We shall pursue these tentative conclusions further after an examination of the further development of the ectoderm in the same animals.

Further Development of the Ectoderm

When the two halves of the germ band are externally obvious and enter into segment delineation and limb bud formation, the ectoderm of each half of the germ band develops as epidermis, ganglia of the central nervous system, tracheal invaginations, apodemes and a variety of glandular structures. Once again it is simplest to give consideration, first to the trunk, then to the head.

As the halves of the trunk segments of *Scolopendra* become separated from one another during dorsoventral flexure of the germ band, the ectoderm of each half-segment pushes out as a limb bud (Fig. 69). Dorsal to the limb bud, the ectoderm remains epithelial. Ventral to the limb bud, a thickening develops as a result of cell proliferation, forming a ventral ganglion. Each ganglion arises from a superficial focus of cell proliferation which invaginates slightly during ganglion formation, but does not persist at the surface after the ganglion rudiment has separated into the interior. The paired rudiments link up to form a pair of widely separated nerve cords, which then gradually come together along the ventral midline as the yolk mass shrinks and the tubular body wall is completed. During the latter process, the dorsal epithelium of each half of the segment proliferates and spreads dorsally, taking in the cells of the dorsal extra-embryonic ectoderm, while the ventral epithelium left outside the ventral ganglion spreads ventrally, taking in the cells of the ventral extra-embryonic ectoderm.

Apart from the vestigiality or absence of the ventral extra-embryonic ectoderm, the general development of the ectoderm of the trunk segments of other myriapods follows the pattern described above (Figs. 70 and 71). The only exceptions are the absence of a pair of limb buds on the first trunk segment in pauropods and diplopods and the development of ventral organs at the sites of ganglionic proliferation in the Pauropoda and Symphyla. The Diplopoda have pit-like sites of ingress of the ganglion cells, similar to those of the Chilopoda (Heathcote, 1888; Robinson, 1907; Pflugfelder, 1932; Dohle, 1964), but in the Pauropoda (Tiegs, 1947) the cells around these pits are arranged in characteristic rosettes, forming pairs of temporary ventral organs like those of Onychophora in each segment. Mitoses are numerous both in the cells of the ventral organs and in the cells of the thickening ganglion rudiments beneath them. The ventral organs are gradually incorporated into the ganglia, except in the collum segment. Here, vestiges of the ventral organs remain at the surface after the segmental ganglion has become internal, and later develop into a pair of exsertile vesicles. The Symphyla (Tiegs, 1940) similarly develop ventral organs in association with each pair of ganglia, but retain them as distinct structures in each segment after the ganglia have separated into the interior. Each pair of ventral organs then develops into a pair of exsertile vesicles. It will be recalled that the ventral organs of the Onychophora remain at the surface after separation of the nerve cords into the interior, as in Symphyla, but then subsequently degenerate without giving rise to any definitive structures.

A pair of ganglia develops in each of the twenty-two trunk segments and two genital segments of *Scolopendra*. The eight trunk segments formed in the *Glomeris* embryo each develop a pair of ganglia, but those of segments 5 and 6 and segments 7 and 8 respectively unite to form the ventral ganglia of the first two diplosegments. In *Pauropus*, the six trunk segments of the embryo, including the collum segment, develop paired ganglia, but the ganglia of segments 5 and 6 are still at an early stage of development when hatching takes place. Finally, in *Hanseniella*, in which the specialization of the posterior end of the embryo for anamorphosis is uniquely different from that of other myriapods, paired ganglia and exsertile vesicles are developed on the first seven trunk segments and the preanal segment, and a pair of ganglia is also developed in the anal segment. The anal ganglia are small and lack associated ventral organs.

Ectoderm cells along the midventral line also play a part in the formation of the ventral nerve cord in all myriapod embryos except *Pauropus*. The two halves of the nerve cord are close together during their early formation in the small embryo of *Pauropus*, and fuse into a single median cord during their movement into the interior (Tiegs, 1947). In *Scolopendra*, in contrast, where the two halves of the nerve cord are at first widely separated, a strand of cells is proliferated along the ventral midline of the extra-embryonic ectoderm and separates inwards to form a median strand (Heymons, 1901). When the two halves of the nerve cord come together, the median strand becomes wedged between them. Some of its cells possibly become incorporated into the nerve cord. Others spread over the surface of the nerve cord and give rise to the neurilemma.

A similar median cord is formed along the midventral line of the vestigial extra-embryonic ectoderm of diplopods (Lignau, 1911; Pflugfelder, 1932; Dohle, 1964) and Symphyla (Tiegs, 1940). Here, however, it merges entirely into the nerve cord, the neurilemma being formed from median mesoderm.

Behind the last trunk segment of *Scolopendra*, the embryonic ectoderm around the anus develops as epidermis of the telson. The ectoderm in the corresponding location in Symphyla is ectoderm of the anal segment, no definitive telson rudiment being retained. Growth zone ectoderm here lies at the anterior end of the pre-anal "segment". The Pauropoda and Diplopoda, in contrast, retain growth zone ectoderm as a ring around the anus, behind the last distinct segment formed in the embryo. No telson ectoderm can be discerned at hatching, but a telson becomes defined during the final phase of post-embryonic growth.

Symphylans and pauropods lack tracheal invaginations on the trunk segments, but the tracheae of the chilopod and diplopod trunk originate as paired invaginations of ectoderm (Heathcote, 1888; Heymons, 1901; Dohle, 1964). Similarly, the apodemes of the trunk segments of all myriapod embryos arise as ectodermal invaginations. However, characteristic differences mark the development of the ectodermal components of gonoducts. In *Scolopendra*, the gonopore is formed in front of the anus as a short median invagination of ventral ectoderm with which the mesodermal gonoducts become connected. In *Pauropus* and the Symphyla, in contrast, the gonoducts develop entirely from a pair of ventral ectodermal invaginations, on the third trunk segment in *Pauropus* and the fourth in Symphyla. Presumably they develop in the same way in diplopods, in which the genital opening again lies on the third trunk segment.

The ectoderm of the head

As we have seen, the embryonic head in myriapods comprise a median clypeolabral lobe flanked by broad protocerebral lobes in front of the mouth, a pair of pre-antennal segment rudiments just behind the mouth (except in the Diplopoda, in which this segment is no longer represented in the embryo as a distinct unit), and a sequence of paired antennal, pre-mandibular, mandibular, maxillary and post-maxillary segment rudiments. The post-maxillary segment of *Pauropus* and the Diplopoda is the limbless collum segment, the first segment of the trunk. In the Chilopoda, the post-maxillary is the second maxillary segment of the head, and in the Symphyla it is the labial segment of the head. Limb buds are developed as transient vestiges on the pre-mandibular segment only in the Symphyla (Tiegs, 1940) and as transient vestiges on the pre-antennal segment only in the Chilopoda (Heymons, 1901). The antenna, mandibles and maxillae are well developed and homologous in all myriapods (Manton, 1964, 1970).

We have also seen that this interpretation of the segmental composition of the myriapod head is sustained by the development of the mesoderm. Except for the absence of a pair of pre-antennal somites in the Diplopoda, each segment as defined above develops a pair of somites, and the anterior protocerebral region or acron lacks paired somites. It now becomes possible to discuss the further development of the cephalic ectoderm in segmental terms.

For the gnathal segments, such discussion is easily dismissed. The development of their ectoderm follows the same basic pattern as that of the trunk segments. The ventral ganglia fuse to form the suboesophageal ganglion. The suboesophageal ganglion of *Pauropus* includes only the mandibular and maxillary ganglia. In other myriapods, the post-maxillary ganglion is also added to it. The general ectoderm of the segments develops as epithelium of the posterior part of the head capsule and the mouth-parts. There are, however, a number of special features which require emphasis. Firstly, the gnathal segments develop a variety of ectodermal invaginations. The salivary glands of chilopods are formed as ectodermal invaginations on the second maxillary segment (Heymons, 1901). Those of other myriapods are mesodermal segmental organs of the maxillary segment. All myriapods develop their tentorium from a single pair of ectodermal invaginations on the mandibular segment (Tiegs, 1940, 1947). Symphyla develop their only pair of tracheae as ectodermal invaginations on the mandibular segment (Tiegs, 1940).

Secondly, the superlinguae of the Symphyla develop directly from the ventral organs of the mandibular segment, while the ventral organs of the maxillary and labial segments make a major contribution to the formation of the hypopharynx (Tiegs, 1940). These structures are thus entirely independent of the segmental limbs in their origin and development.

The development of the ectoderm in front of the gnathal segments is understandably more complicated, since it involves the pre-oral migration of cephalized segments to contribute to the brain and the anterior part of the head. As shown especially by Tiegs (1940, 1947), the ventral organs of these segments in *Pauropus* and *Hanseniella* are particularly useful as indicators of forward migration. The process is most easily comprehended by taking account first of the further development of the acron.

The ectoderm of the median clypeolabral region pouches out posteroventrally over the mouth to form the labral rudiment. In front of the labral rudiment, the same ectoderm persists as clypeal ectoderm. In *Scolopendra*, though not in other myriapods, a cellular thickening is proliferated inwards from the clypeal ectoderm and separates off as a small, median, "archicerebral" ganglion (Heymons, 1901). The large, paired, protocerebral lobes of the acron spread upwards and backwards towards the dorsal midline, forming the dorsolateral regions of the head and extending back above the gnathal segments. As they do so, cellular thickenings are proliferated into the interior to form the paired frontal, lateral and posterior lobes of the protocerebral ganglia. The sites of proliferation of the three lobes form three temporary invaginations in *Scolopendra*. In *Pauropus*, in contrast, only the posterior lobe invaginates, the median and frontal lobes being centred on "ventral organs". *Hanseniella* develops a single thickening, which then displays invaginations at the lateral and posterior lobes as the initial thickening subdivides into three (Fig. 72). Finally, in diplopods, the initial thickening arises from a single site of proliferation and then subdivides into the three lobes without invagination. All three pairs of ganglia eventually separate into the interior, leaving cephalic epithelium at the surface. The frontal lobes become united in the anterodorsal midline, above the stomodaeum.

As the protocerebral lobes extend upwards and backwards, the pre-antennal, antennal and pre-mandibular segments move fowards on either side of the mouth to occupy the sides of the pre-oral region. The small pre-antennal segment develops, except in the Diplopoda, a pair of distinct pre-antennal ganglia, with associated ventral organs in the Pauropoda. As the pre-antennal ganglia become internal, they merge with the frontal lobes of the protocerebral ganglia, but in *Pauropus*, the pre-antennal components can still be distinguished in the fully developed brain. The large antennal segment, which makes up most of the lateral walls of the anterior part of the head, buds out typical limb buds before forward migration. Ventral to the limb buds, a pair of antennal ganglia is proliferated in the usual way, with associated ventral organs in the Pauropoda and Symphyla. The ganglia become internal after they have moved forwards on either side of the mouth and lie on either side beneath the lateral lobes of the protocerebral ganglia. They become connected with the ipsilateral frontal lobes of the protocerebrum and are also joined transversely with one another by a preoral commissure. Once internal, the antennal ganglia become the deutocerebral ganglia.

The pre-mandibular areas of segmental ectoderm can be identified in the embryo mainly as the site of origin of the pre-mandibular ganglia. Ventral organs are again associated with these ganglia in the Symphyla and Pauropoda and provide important evidence of the distribution of the pre-mandibular ectoderm after the ganglia of the segment have become internal. The pre-mandibular ganglia are joined anteriorly to the antennal ganglia and come to lie ventrolaterally, just anterior to the mouth, as the tritocerebral ganglia of the brain. They retain a post-oral transverse connection, however, as a tritocerebral commissure, as

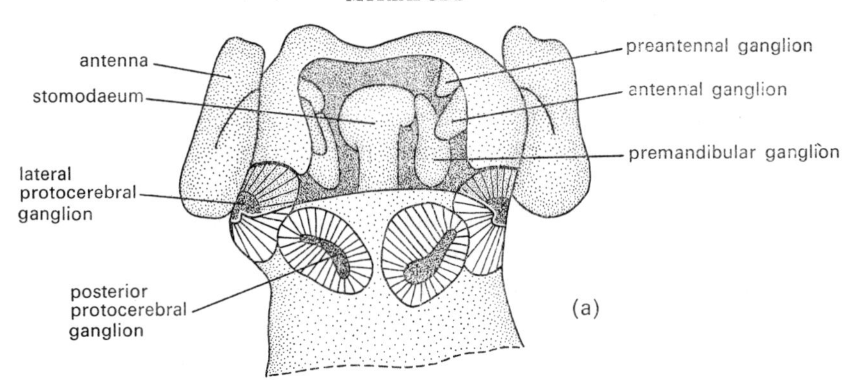

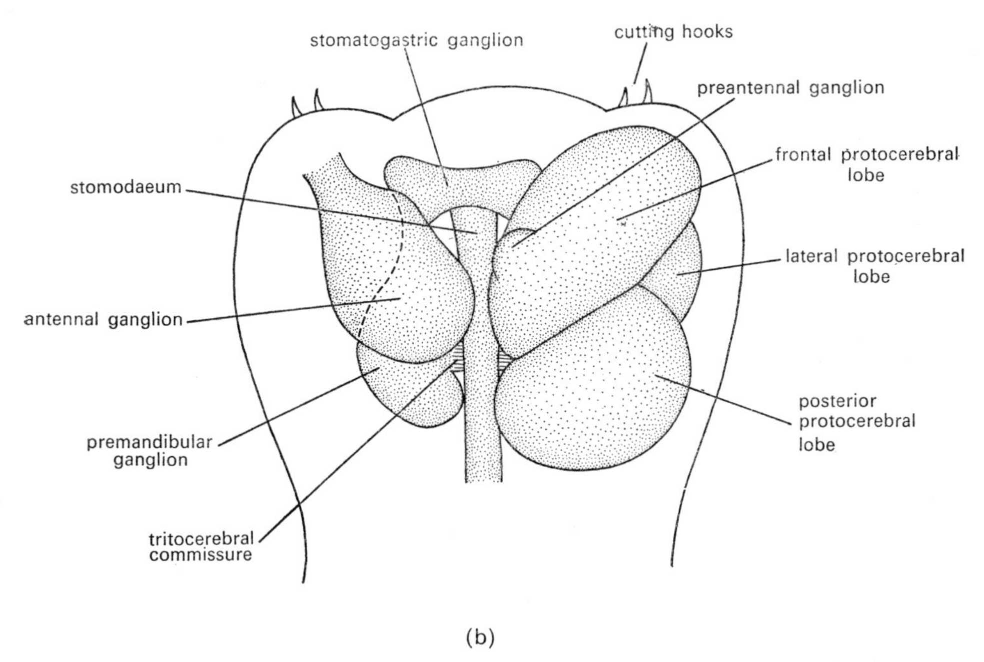

FIG. 72. Development of the head of Hanseniella, diagrammatic dorsal views. (a) 8-day stage; (b) 10-day stage. (After Tiegs, 1940.)

well as being connected longitudinally with the suboesophageal ganglion. The forward migration of the superficial, pre-mandibular ectoderm continues as the pre-mandibular ganglia become internal. Much of the superficial tissue spreads medially in front of the mouth and contributes to the floor of the pre-oral cavity.

Unlike the ventral organs of the gnathal segments, those of the pre-antennal, antennal and pre-mandibular segments merge without trace into the surface epithelium of the head. None of these segments produces any major ectodermal invaginations, though the protocerebral ectoderm develops a pair of organs of Tömösváry adjacent to the lateral lobes of the protocerebrum, just behind the bases of the antennae. A similar formation of these characteristic myriapod sense organs has been observed in the Chilopoda (Heymons, 1901), Diplopoda (Hennings, 1904; Dohle, 1964), Symphyla (Tiegs, 1940) and Pauropoda (Tiegs, 1947).

The dorsal organ

While the broadly attenuated, dorsal extra-embryonic ectoderm of the chilopod embryo has a uniform structure and fate, being gradually incorporated into the definitive dorsal epithelium of the body wall, that of other myriapods includes a localized mid-dorsal region which has a different form and fate. This is the dorsal organ, first developed when the dorsal extra-embryonic ectoderm becomes distinct from the embryonic primordium. Diplopod embryos were long thought not to have a dorsal organ, but recent work by Siefert (1960) on *Polyxenus* and Dohle (1964) on *Glomeris* has revealed a localized area of cells which become deeper than the attenuated cells around them and project downwards into the yolk mass. The subsequent fate of the cells is not clear.

Hanseniella develops an elaborate dorsal organ (Tiegs, 1940). A circular patch of 20–30 dorsal blastoderm cells enters into cell elongation. The cells penetrate deeply into the underlying yolk mass and the external ends of the cells become narrow, so that the cells take on a flask-shaped form. External to the narrow ends of the cells, an aperture is left in the newly secreted blastodermal cuticle.

The necks of the dorsal organ cells now become fibrillated. As segmentation of the germ band takes place, thin filaments begin to grow out through the narrow necks of the cells and radiate into the space between the blastodermal cuticle and the chorion. In 2 days, the filaments have reached the ventral pole of the egg. The cells of the dorsal organ shrink and flatten as the threads are produced. Subsequently, the cells gradually degenerate, but the threads remain in position until the embryo hatches. No satisfactory explanation of the functional significance of this remarkable sequence of events has yet been attained. As we shall see in the next chapter, a similar dorsal organ is developed in collembolan embryos.

Tiegs (1947) subsequently showed that *Pauropus* develops a dorsal organ of essentially the same type, but does not produce the radiating threads of secretion. The organ invaginates slightly, then secretes a material which fills the invagination cavity and spreads out locally beneath the adjacent blastodermal cuticle. Once again, the functional significance of the secretion is obscure. As development proceeds, the dorsal organ sinks into the underlying yolk mass and gradually disintegrates.

Hatching

Symphylan embryos hatch directly, through rupture of the enclosing chorion and blastodermal cuticle, and become immediately active (Tiegs, 1940), but other myriapods hatch in two stages. As the embryo develops, an embryonic cuticle is secreted. In centipedes (Heymons, 1901; Dohle, 1970) the chorion ruptures before development is complete and the late embryo is released as a motionless "pupoid" phase, enclosed in the embryonic cuticle. The active first instar larva escapes from this after a further period of development. Similar preliminary hatching as a "pupoid" phase is well known for the diplopods (Metschnikoff, 1874; Heathcote, 1886; Robinson, 1907; Siefert, 1960; Dohle, 1964) and pauropods (Tiegs, 1947).

The Basic Pattern of the Development in Myriapods

We have seen in the preceding pages that the evidence of embryology consistently places the modern myriapods in two discrete groups, the Chilopoda on the one hand and the Symphyla, Pauropoda and Diplopoda on the other. Furthermore, there are also indications that the Pauropoda and Diplopoda together differ in certain ways from the Symphyla. In order to test the possibility of a basic theme in myriapod development, it is necessary to summarize these resemblances and differences more succinctly, in terms of the formation and fates of presumptive areas of the blastoderm.

The basic mode of development in the Chilopoda is epimorphic, stemming from a large, yolky egg, up to 3 mm or more in diameter, laid by an oviparous female. Cleavage is intralecithal and results in the formation of a uniform blastoderm. Some of the cleavage energids remain within the yolk mass as vitellophages which are subsequently resorbed with the yolk.

The Symphyla, Pauropoda and Diplopoda are also oviparous, but lay eggs of smaller dimensions than those of most epimorphic chilopods and have anamorphic development. Cleavage is total at first, proceeding in a specialized manner through the formation of yolk pyramids. Radial mitotic divisions then segregate an inner mass of yolky cells from a superficial layer which forms a blastoderm. In the Symphyla and Pauropoda, the blastoderm exhibits precocious differentiation of an embryonic primordium. The inner mass of cells fuses to form a yolk mass containing vitellophages. The fat body develops from vitellophages during later development, whereas in the Chilopoda the fat-body originates from the mesodermal somites.

The chilopod presumptive midgut is broadly spread throughout the ventral blastoderm. Numerous cells immigrate beneath the blastoderm, spread around the yolk mass and develop as the midgut epithelium. Additional vitellophages invade the yolk mass from a small posteroventral area behind the presumptive proctodaeum. This area subsequently becomes the site of proliferation of the mesodermal bands.

The diplopod presumptive midgut is localized as a small blastodermal area in front of the presumptive proctodaeum. From this site, cells are proliferated as a strand through the centre of the yolk mass and develop directly into the midgut epithelium. *Pauropus* exhibits a precocious segregation of the same cells in the centre of the embryo during cleavage and has no blastodermal representation of presumptive midgut. The Symphyla also develop their central midgut strands from cells which are cut off into the interior during cleavage, but in a different way. Initially, all of the internal cells cut off during cleavage become vitellophages within the yolk mass. A subsequent separation of the central vitellophages as midgut and lateral vitellophages as fat-body is then brought about by the mesoderm.

The stomodaeum, in contrast, develops in a constant manner in all myriapods, originating from a small presumptive area situated in the ventral midline of the blastoderm, between the presumptive protocerebral ectoderm and pre-antennal ectoderm. The stomodaeal rudiment invaginates independently and develops as foregut. The proctodaeum, originating from a small presumptive area situated immediately in front of the presumptive mesoderm in chilopods and in an equivalent position in other myriapods, also invaginates independently and develops into the hindgut and Malpighian tubules.

The mesodermal bands of chilopods are, for the most part, proliferated forwards from a small, posteroventral area of presumptive mesoderm behind the presumptive proctodaeum. Additional cells migrate into the mesodermal bands from the overlying blastoderm. Ventrally between the mesodermal bands, similar immigrating cells form scattered median mesoderm. The mesodermal bands extend anteriorly as far as the presumptive stomodaeum and include the mesoderm of twenty-two serial segments and a posterior growth zone.

The symphylan, pauropod and diplopod mesoderm arises entirely through a diffuse immigration of ventral blastoderm cells into the interior, mainly ventrolaterally as paired bands extending from the stomodaeal to the proctodaeal level, but also as median mesoderm between them. The mesodermal bands include the mesoderm of nine serial segments and a posterior growth zone.

In the further development of the mesoderm of all myriapods, the pre-antennal and antennal somites migrate to a pre-oral position and the pre-mandibular somites come to lie on either side of the mouth. The mandibular and maxillary somites are always cephalized post-orally, but the post-maxillary somites vary. In chilopods and symphylans, they become part of the head but retain a generalized form and development. In pauropods and diplopods, the post-maxillary somites are reduced in association with specialization of the post-maxillary segment as the limbless collum at the anterior end of the trunk.

The chilopod trunk somites become partially subdivided into three compartments, dorsolateral, medioventral and appendicular. The appendicular compartment pushes out as a lobe which gives rise to limb musculature and, basally, to fat body cells. The dorsolateral compartment extends towards the dorsal midline. Its somatic wall gives rise to dorsal longitudinal muscle, pericardial floor, gonad and heart components, while its splanchnic wall forms midgut musculature. The medioventral compartment extends towards the ventral midline. Its somatic wall gives rise to ventral longitudinal and transverse muscle and, marginally, to ventral blood vessel, while its splanchnic wall again forms midgut musculature. The gonoducts arise as coelomoduct outgrowths of the medioventral compartments of the last two trunk segments. Coelomoducts are absent from the other trunk somites.

The chilopod median mesoderm gives rise to blood cells and, anteriorly and posteriorly, to the stomodaeal and proctodaeal mesoderm.

The somites of the chilopod head are variously modified. The three post-oral pairs lack only the gonad and pericardial septum components, but the pre-oral pairs are further reduced. They make no contribution to the splanchnic musculature, ventral blood vessel or fat-body. The antennal somites retain appendicular lobes but the small pre-antennal and pre-mandibular somites lack these components. Like the trunk somites, the cephalic somites of chilopods develop no coelomoducts, but the pre-mandibular somites develop in part as lymphoid tissue.

The further development of the somites of symphylan embryos concurs in general with that of the chilopods, but there are certain fundamental differences. The dorsolateral lobes shift towards the midline and cut upwards through the yolk mass, separating central midgut from lateral fat body. Concomitantly, the appendicular lobes of the symphylan somites have no fat-body component. The gonads in Symphyla develop from the medioventral lobes of the trunk somites and these lobes, in several somites, also develop transient coelomoducts. The definitive gonoducts, however, are secondary ectodermal ingrowths. The ventral blood vessel in Symphyla develops from median mesoderm, while the blood cells are given off by the dorsolateral lobes of the somites. The stomodaeal and proctodaeal mesoderm also have a somitic origin, antennal anteriorly, pre-anal and anal posteriorly.

The symphylan median mesoderm, in addition to forming the ventral blood vessel, also forms the neurilemma of the ventral nerve cord, which in chilopods develops from ectoderm.

In the symphylan head, a major difference from chilopods is seen in the development of the pre-mandibular and maxillary somites in part as segmental organs. The pre-mandibular pair are transient but the maxillary pair persist and develop as salivary glands. The salivary glands of chilopods develop from ectoderm of the post-maxillary (= second maxillary) segment.

The further development of the diplopod mesoderm is not well understood and that of the pauropod mesoderm is specialized in many ways, but several significant points can be made. Diplopods are said to lack the dorsolateral lobes of the somites, but it can be inferred from the available evidence that mesoderm grows dorsally from the trunk somites, between the central midgut strand and lateral masses of fat body cells, to form the splanchnic mesoderm of the midgut, the heart and the dorsal longitudinal muscles. The gonads develop from the medioventral lobes of the trunk somites, although no coelomoducts have been reported, even as vestiges, in the trunk. The gonoducts arise as secondary ectodermal ingrowth. Development of the ventral blood vessel and formation of the blood cells have not yet been described, but the stomodaeal and proctodaeal mesoderm arise from median mesoderm. The origin of the neurilemma in diplopods is not yet known. In the head, the fate of the pre-mandibular somites is still unknown, but it is clear that they do not form segmental organs. The maxillary somites, however, give rise to a pair of segmental organs which develop as salivary glands.

Pauropod trunk somites lack dorsolateral and medioventral lobes and give rise only to muscles. The dorsal longitudinal muscles have a specialized, extra-somitic origin. The heart and pericardial floor are absent and the splanchnic mesoderm of the midgut is proliferated from the terminal splanchnic mesoderm of the stomodaeum, clearly a specialization. The single, medioventral gonad develops from median mesoderm, a second specialization. Coelomoducts are absent and the gonoducts arise as secondary ectodermal ingrowths. There is no ventral blood vessel and the origin of the few blood cells is unknown. The stomodaeal and proctodaeal mesoderm, and the neurilemma of the ventral nerve cord arise from median mesoderm. In the head, both the pre-mandibular and maxillary somites give rise to segmental organs, and both pairs develop as salivary glands.

Thus, on present evidence, the further development of the mesoderm of the Symphyla, Diplopoda and Pauropoda can be seen to have a common ground, retained in its most generalized form in the Symphyla. A minor difference between the Symphyla on the one hand and the Diplopoda and Pauropoda on the other is seen in the mode of formation of the stomodaeal and proctodaeal mesoderm. The further development of the mesoderm of the Chilopoda, on the other hand, proceeds in a manner fundamentally different in many ways from that of other myriapods. In addition, of course, the epimorphic Chilopoda retain no growth zone mesoderm at the end of embryonic development, whereas a posterior growth zone persists in the mesoderm of the anamorphic Symphyla, Diplopoda and Pauropoda. It is probably significant that the specialization of the posterior mesoderm for anamorphosis differs in the Symphyla from that shared in common by the Diplopoda and Pauropoda. The latter have a residual mass of mesoderm in front of the anus, behind the last pair of segmental somites formed in the embryo (the thirteenth serial segment in *Glomeris*, the eleventh in *Pauropus*). The Symphyla have a subterminal pair of large preanal somites in front of the anus and a pair of small anal somites on either side of the anus. The growth zone mesoderm separates off in a forward direction from the preanal somites, as a pair of mesodermal masses interposed between these and the thirteenth serial segment.

Since the further development of both the midgut and the mesoderm support the idea that the Symphyla, Pauropoda and Diplopoda are related more closely to one another than to the Chilopoda, and hint at an even closer affinity between the Pauropoda and Diplopoda, can any additional evidence bearing on this interpretation be gained from the further development of the ectoderm?

In all classes, the presumptive ectoderm comprises a broad band of ventral blastoderm cells, left at the surface after the immigration of mesoderm (and, in chilopods, midgut) into the interior. The remainder of the blastoderm, dorsally to laterally, makes up the dorsal extra-embryonic ectoderm. A midventral band of ventral extra-embryonic ectoderm develops between the presumptive stomodaeum and proctodaeum and becomes extensive during the later elongation and separation of the two halves of the germ band in scolopendromorphs, though not in other myriapods.

The presumptive ectoderm of the epimorphic chilopods includes the rudiments of the cephalic and first sixteen trunk segments, as well as a posterior growth zone. The latter proliferates the ectoderm of the remaining trunk segments in the embryo and then gives rise to the telson. Following segment delineation, the segmental ectoderm pouches out ventrolaterally as limb buds and thickens ventrally as segmental ganglia. The chilopod ventral ganglia develop by proliferation from invaginated pits of ectoderm but lack associated ventral organs. A median strand of ectoderm developed between the ganglia contributes the neurilemma of the ventral nerve cord. A short, midventral invagination of ectoderm in front of the anus gives rise to the opisthogoneate gonopore. The salivary glands are formed as ectodermal invaginations on the second maxillary segment, the tentorial arms as ectodermal invaginations on the mandibular segment. Apodemes and tracheae are also ectodermal invaginations. The dorsal extra-embryonic ectoderm of chilopods does not develop a dorsal organ.

The presumptive ectoderm of the Symphyla, Pauropoda and Diplopoda includes the rudiments of the cephalic and first four trunk segments (three in Symphyla, in which the post-maxillary segment is cephalic), together with a posterior growth zone. The symphylan growth zone adds the rudiments of the next four trunk segments, the preanal "segment" and the anal segment in the embryo. The diplopod growth zone adds the ectoderm of at least four further segments in the embryo and persists behind the last of these, in front of the anus. The pauropod growth zone acts in a similar way to that of diplopods, but adds the rudiments of only two further segments in the embryo.

In the further development of the segmental ectoderm, the ventral ganglia of the Symphyla and Pauropoda develop in association with ventral organs and there are hints of a similar mode in the Diplopoda. The ventral organs of the trunk segments in the Symphyla, and perhaps the Diplopoda, subsequently transform into exsertile vesicles on the trunk segments. *Pauropus* develops exsertile vesicles in this manner on the collum alone. The median strand, absent in the Pauropoda, is incorporated in the Symphyla and Diplopoda into the ventral nerve cord. The paired gonoducts arise as ventral invaginations of ectoderm in a progoneate position, on the tenth serial segment in the Symphyla (fourth trunk segment) but on the eighth serial segment in the Pauropoda and Diplopoda (third trunk segment). The salivary glands in these three myriapod classes are maxillary and mesodermal, though the tentorial arms, apodemes and tracheae are as in chilopods. The dorsal extra-embryonic ectoderm develops a temporary dorsal organ.

Thus the ectodermal development of the Symphyla, Pauropoda and Diplopoda shares in common a basic pattern of differences from that of the Chilopoda, but also differs in certain ways between the Symphyla on the one hand and the Pauropoda and Diplopoda on the other. The position of the presistent growth zone ectoderm and the position of the progoneate gonopores support the evidence of the mesoderm in suggesting that the Symphyla display one solution, and the Pauropoda and Diplopoda an alternative solution, to the problem of anamorphic development based on an initial formation of only nine serial segments in an embryo with a secondarily reduced yolk mass.

Taking the entire course of embryonic development into account, there is evidence at every level, from cleavage through presumptive area formation to the further development of the midgut, mesoderm and ectoderm, that the Chilopoda have arisen from one line of myriapod evolution and the Symphyla, Pauropoda and Diplopoda from another. The same evidence further suggests that the Pauropoda and Diplopoda are more closely related to one another than either is to the Symphyla but does not give a clear indication of the relative timing of their divergences from a common ancestry. A possible solution to this problem, however, emerges from a reconsideration of the comparative development of the myriapod head. All myriapods share the same basic construction of the head as far back as the maxillary segment, but differ in the functional expression of the post-maxillary segment. In the Pauropoda and Diplopoda, this segment remains part of the trunk, but is specialized as the limbless collum. In the Symphyla, as in the Chilopoda, the post-maxillary segment is incorporated into the head, but retains limbs modified as jaws. It follows that, if the Symphyla, Pauropoda and Diplopoda had a common dignathan ancestry, the post-maxillary or first trunk segment of the ancestral forms must have been a limb-bearing segment. Dignathous ancestors with a collum could not have given rise to the trignathous Symphyla. Hence, the specialization of the post-maxillary segment as a limbless collum in the Pauropoda and Diplopoda must have occurred after divergence of the dignathous ancestors of the Symphyla from those of the Pauropoda and Diplopoda. This, in turn, accounts for the fact that the Symphyla, although trignathous and labiate, retain many primitive features in their embryonic development. The Chilopoda must similarly have arisen from dignathous ancestors that had not lost the first pair of trunk limbs, but the evidence of the head supports the extensive evidence from other aspects of embryology that these were a different group of dignathan myriapods from those which gave rise to the Symphyla, Pauropoda and Diplopoda. These conclusions can be expressed diagrammatically as follows:

SUPERCLASS MYRIAPODA

CLASS	CHILOPODA	SYMPHYLA	PAUROPODA	DIPLOPODA
	Trignathy	Trignathy		
	Epimorphic	Anamorphic		
	Dignathous proto myriapods			

Obviously, as pointed out by Manton (1949, 1964, 1970) on the basis of comparative morphology, dignathous and trignathous are grades in myriapod evolution. A taxon Dignatha in the sense used by Tiegs (1940, 1947) for the Pauropoda and Diplopoda cannot be sustained, since the modern dignathous myriapods are not the ancestors of the modern trignathous myriapods, and are in all probability phylogenetically "younger" than the dignathous ancestors of the modern trignathous forms. Similarly, a taxon Trignatha is artificial, since it includes two fundamentally different groups of myriapods that have evolved the trignathous condition independently.

In view of the deep cleft between the two basic modes exhibited in myriapod embryonic development, it is not surprising that a common theme of development for modern myriapods is impossible to envisage. At the same time, the two alternatives retain a sufficient number of fundamental features in common to indicate that they are both ultimately derived from a common, more generalized, developmental mode. Further discussion of this topic will be deferred to Chapter 10, but it can be pointed out in anticipation that the requisite mode, of which both myriapod alternatives are modifications, is to be found in the basic theme of development of the Onychophora.

References

BODINE, M. W. (1970) The segmental origin of the appendages of the head and anterior body segments of a spiroboloid millipede, *Narceus annularis*. *J. Morph.* **132,** 47–67.

CHOLODKOWSKY, N. (1895) Zur Embryologie der Diplopoden. *St. Petersburg Soc. Imp. Nat.* **2,** 10–12 and 17–18.

DAWYDOFF, C. (1956) Quelques observations sur l'embryogénèse des myriapodes scolopendromorphes et geophilomorphes indochinois. *C.R. Acad. Sci., Paris* **242,** 2265–7.

DOHLE, W. (1964) Die Embryonalentwicklung von *Glomeris marginata* (Villers) im Vergleich zur Entwicklung anderer Diplopoden. *Zool. Jb. Anat. Ont.* **81,** 241–310.

DOHLE, W. (1970) Über Eiablage und Entwicklung von *Scutigera coleoptrata* (Chilopoda). *Bull. Mus. Nat. d'Hist. Nat.*, sér. 2, **41,** 53–57.

HEATHCOTE, F. G. (1886) The early development of *Julus terrestris*. *Q. Jl. microsc. Sci.* **26,** 449–70.

HEATHCOTE, F. G. (1888) The post-embryonic development of *Julus terrestris*. *Phil. Trans. R. Soc.* B, **179,** 157–79.

HENNINGS, C. (1904) Das Tömöswarische Organ der Myriapoden. *Z. wiss. Zool.* **76,** 26–52.

HEYMONS, R. (1901) Die Entwicklungsgeschichte der Scolopender. *Zoologica, Stuttgart,* **13,** 1–224.

LIGNAU, N. (1911) Embryonalentwicklung des *Polydesmus abchasius*. *Zool. Anz.* **37,** 144–53.

MANTON, S. M. (1949) Studies on the Onychophora VII. The early embryonic stages of *Peripatopsis* and some general considerations concerning the morphology and phylogeny of the Arthropoda. *Phil. Trans. R. Soc.* B, **233,** 483–580.

MANTON, S. M. (1964) Mandibular mechanisms and the evolution of the arthropods. *Phil. Trans. R. Soc.* B, **247,** 1–183.

MANTON, S. M. (1970) Arthropoda: Introduction. *Chemical Zoology* **5,** 1–34.

MANTON, S. M. (1972) The evolution of arthropodan locomotory mechanisms, Part 10. *J. Linn. Soc. Zool.* **51,** 203–400.

METSCHNIKOFF, E. (1874) Embryologie der doppeltfüssigen Myriapoden (Chilognatha). *Z. wiss. Zool.* **24,** 253–82.

PFLUGFELDER, O. (1932) Über den Mechanismus der Segmentbildung bei der Embryonalentwicklung und Anamorphose von *Platyrrhacus amauros* Attems. *Z. wiss. Zool.* **140**, 650–723.

ROBINSON, J. (1907) On the segmentation of the head of Diplopoda. *Q. Jl. microsc. Sci.* **51**, 607–24.

SEIFERT, G. (1960) Die Entwicklung von *Polyxenus lagurus* L. (Diplopoda, Pselaphognatha). *Zool. Jb. Anat. Ont.* **78**, 257–312.

SILVESTRI, F. (1903) Classis Diplopoda. In: *Acari, Myriapoda et Scorpiones huiusque in Italia reperta* **1**, 1–272. Premiato stab. tip. Vesuviano Portici.

SILVESTRI, F. (1933) Sulli appendici del capo degli "Japygidae" (Thysanura entotropha) e rispettivo confronto con quelle dei Chilopodi, dei Diplopodi e dei Crostacei. *5th Intern. Congr. Ent. Paris*, pp. 329–43.

SILVESTRI, F. (1949) Segmentazione del corpo dei Colobognati (Diplopodi). *Boll. Lab. Ent. agr. Portici* **9**, 115–21.

TIEGS, O. W. (1940) The embryology and affinities of the Symphyla, based on a study of *Hanseniella agilis*. *Q. Jl. microsc. Sci.* **82**, 1–225.

TIEGS, O. W. (1947) The development and affinities of the Pauropoda, based on a study of *Pauropus sylvaticus*. *Q.Jl. microsc. Sci.* **88**, 165–267 and 275–336.

ZOGRAFF, N. (1883) Materialy k poznaniju embrional'nago razvitija *Geophilus ferrugineus* L. k *Geophilus proximus* L. *Nachr. Ges. Freunde Naturk., Anthrop. und Ethnogr., Moscow.*

CHAPTER 6

APTERYGOTE INSECTS

LIKE the myriapods, the apterygote hexapods fall into four classes, whose interrelationships are still far from clear (Manton, 1970, 1972). Of these four, two have been severely neglected by embryologists. The embryonic development of the Protura remains entirely unknown, while that of the Diplura (e.g. *Campodea, Japyx*) has been the subject of only a few superficial investigations. Grassi (1885) reported on the embryos of *Japyx*, limiting his observations essentially to the fact that *J. solifugus* has a dorsal organ. The development of *Campodea staphylinus* was examined briefly by Heymons (1897a) and then in more detail by Uzel (1897 a, b, 1898). Uzel's work is informative on the general course of embryonic development, but can no longer be regarded as satisfactory in respect of questions of internal development, especially of the mesoderm. More recently, Tiegs (1942b) made valuable observations on the germ band and dorsal organ of *Campodea fragilis*, but his intimation of further work to come was, unfortunately, not to be fulfilled. The classic paper on dipluran embryology, therefore, still remains to be written.

The third class of apterygotes, the Collembola, as befits their greater diversity of species and more common occurrence than the Diplura, has been studied by a greater number of embryologists, although knowledge of the internal development of collembolan embryos is still deficient in many ways. Uljanin (1875, 1876) was the first worker to give a reasonably detailed account of the external features of collembolan development, treating four species. His work was soon followed by a similar paper by Lemoine (1883) and later by a brief comment from Wheeler (1893), but the first substantial accounts of collembolan embryos were those of Uzel (1898) on *Tomocerus* and *Achorutes* and Claypole (1898) on *Anurida maritima* Guér. Once again, however, the development of the mesoderm eluded satisfactory analysis. Folsom (1900) and Hoffman (1911) gave further attention to the development of the head of *Anurida* and *Tomocerus* respectively and then, in the last of the classical papers on collembolan embryology, Philiptschenko (1912) described the development of *Isotoma cinerea*. Although this paper has been much quoted in subsequent discussions (e.g. Tiegs, 1940, 1947; Johannsen and Butt, 1941), there is little doubt that its interpretation of internal details is highly questionable.

After a long period of neglect, during which the only further investigation was a study by Tiegs (1942a) of the dorsal organs of a number of species, the Collembola have re-emerged on the embryological scene in the last few years. The Polish worker C. Jura (1965, 1966, 1967a, 1967b) has described several aspects of the development of the primitive species

Tetradontophora bielanensis (Waga), extending and correcting an earlier account of this species by Lemoine (1883). The same worker (Jura, 1972) has also reviewed apterygote embryology, pointing out the need for renewed studies. Parallel with this work, M. Garaudy-Tamarelle, working in Professor A. Haget's laboratory in Bordeaux (Haget and Garaudy, 1964; Garaudy, 1967; Garaudy-Tamarelle, 1969 a, b, 1970), has re-examined certain features of the development of *Anurida maritima*. Bruckmoser (1965) has made a detailed and informative study of the development of the head of *Orchesella villosa* L. and Zakhvatkin (1969) has described cleavage in *Sinella cubiseta* Brooks. Zakhvatkin's attempt to interpret collembolan cleavage as spiral cleavage is, in my opinion, unconvincing.

The fourth class of apterygotes is the Thysanura. Although thysanuran embryonic development has been examined by only a few workers, it is better known than in the other classes, probably because the eggs of Thysanura are larger than those of Diplura or Collembola and develop in a manner generally similar to that of pterygote insects. During the classical period of comparative embryology, however, only R. Heymons (1896, 1897 a, b), in the midst of his immensely productive engagement in pterygote embryology (see Chapter 7 and Anderson, 1972), paid any detailed attention to the Thysanura. Heymons' description of the development of *Lepisma saccharina* L., which included an account of the formation and segmentation of the germ band, the development of embryonic membranes, blastokinesis and the development of the gut, was later supplemented by an account he gave with H. Heymons (R. and H. Heymons, 1905) of blastokinesis in *Trigoniophthalmus* (-*Machilis*) *alternatus*. Recent studies by Sharov (1953, 1966) and Larink (1970) on *Lepisma* and by Woodland (1957) on *Thermobia domestica* have confirmed and extended Heymons' work. The development of the mesoderm of the Thysanura is not well understood, even following the important study by Larink (1969) of the embryology of the littoral thysanuran *Petrobius brevistylis* Carpenter, but enough information is now available to show that the development of the Thysanura shares much in common with that of pterygotes and is fundamentally different from the more myriapod-like development of the Collembola and Diplura. We shall find confirmation of this difference, first noted by Heymons (1896), at every phase of development.

Cleavage

All apterygote eggs have a central nucleus in a small cytoplasmic halo, surrounded by uniformly distributed yolk granules in a sparse cytoplasmic reticulum. A few workers have reported the presence of a thin layer of yolk-free cytoplasm, the periplasm, at the surface of the egg (e.g. Uzel, 1898, *Campodea*; Claypole, 1898, *Anurida*; Larink, 1969, *Petrobius*), but others have denied that a periplasm is present (Heymons, 1897b; Philiptschenko, 1912; Jura, 1965). The egg is enclosed in an external chorion secreted by the follicle cells of the parent ovariole, and a few Collembola are also said to have a thin vitelline membrane beneath the chorion (Lemoine, 1883; Claypole, 1898; Jura, 1965).

Marked differences are displayed between classes in the size, shape and mode of cleavage of the egg. Thysanuran eggs are ovoid, about 1·0 mm long and 0·8 mm in diameter (Heymons, 1897b; R. and H. Heymons, 1905; Woodland, 1957; Sharov, 1966; Larink, 1969). Cleavage is intralecithal (Fig. 73), with repeated mitoses leading to the formation of numerous cleavage energids within the yolk mass. The first three cleavage divisions of *Lepisma* and the first five of *Ctenolepisma* are synchronous, but thereafter synchrony is lost. The energids spread through the yolk mass and many of them emerge at the surface and divide further to form a uniform blastoderm of low cuboidal cells. Other energids persist within the yolk mass and assume the role of vitellophages. The uniform blastoderm then differentiates, through a differential concentration of cell divisions towards the posterior end and a gradual aggregation of most of the cells into a small posterior or posteroventral disc, to

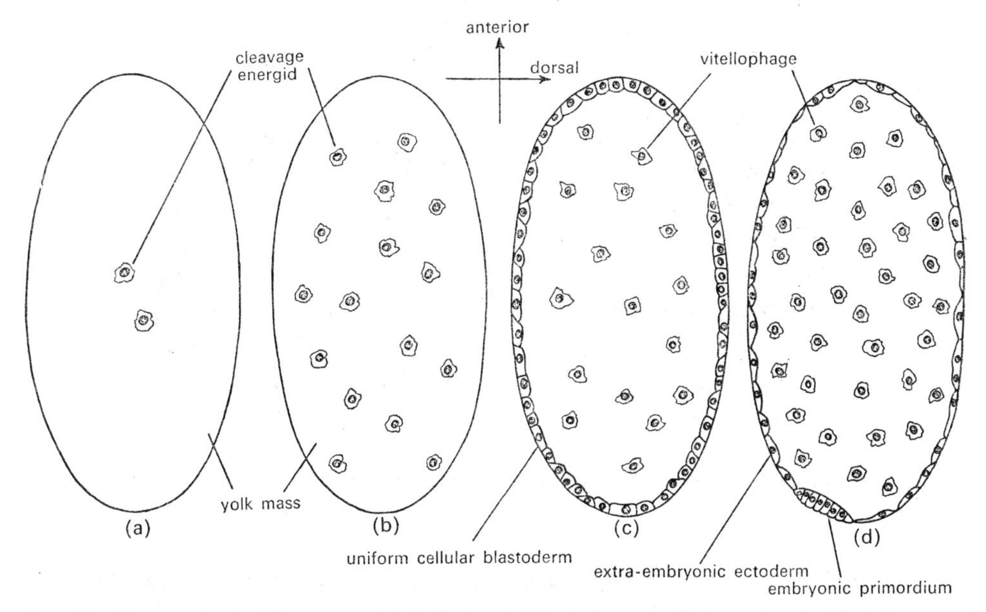

FIG. 73. Cleavage in the Thysanura. (a) Early intralecithal cleavage; (b) later intralecithal cleavage; (c) uniform blastoderm; (d) differentiated blastoderm.

form a compact embryonic primordium and an attenuated dorsal, extra-embryonic ectoderm. The cells of the embryonic primordium are columnar and closely packed. The extra-embryonic ectoderm consists of the usual flattened cells spread out as a thin layer over most of the surface of the yolk mass.

The differentiation of the uniform blastoderm in this way to form a compact embryonic primordium is confined to Thysanura among the apterygotes and is otherwise seen only in the pterygote insects, but the Diplura resemble the Thysanura in retaining intralecithal cleavage. Dipluran eggs are spherical, those of *Campodea* being 0·40 mm in diameter (Uzel, 1898; Tiegs, 1942b), while those of the unusually large Australian species *Heterojapyx gallardi* are 1·6 mm in diameter (Tiegs, 1942b). According to Uzel (1898), when the uniform blastoderm of low cuboidal cells has formed in *Campodea*, no energids remain as vitellophages within the yolk mass. Differentiation of the germ band and dorsal, extra-embryonic ectoderm (Fig. 77) takes place, as in myriapods, after the blastoderm has begun to release cells into the interior as mesoderm and midgut cells. Furthermore, the germ band, when it forms, occupies a large part of the surface of the yolk mass. These differences are an outward expression of a more fundamental difference between the Diplura and Thysanura in the distribution of the presumptive areas of the blastoderm (see below).

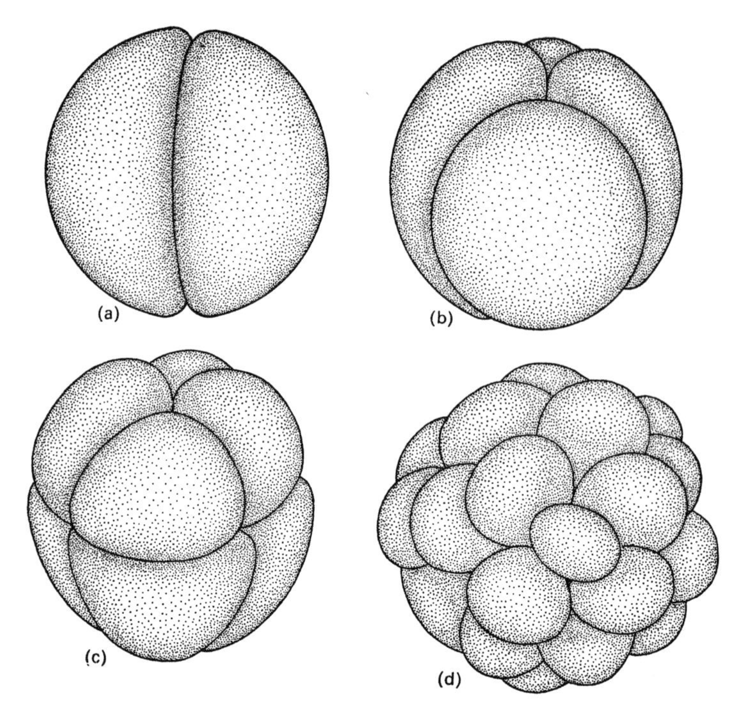

FIG. 74. Cleavage in the collembolan *Anurida maritima*, after Claypole (1898). (a) 2-cell; (b) 4-cell; (c) 8-cell; (d) 16–32 cell.

The Collembola parallel the Diplura in the production of spherical eggs, but have a different range of egg sizes. The largest collembolan egg whose embryonic development has been studied is that of the poduromorph *Tetradontophora bielanensis*, with a diameter of 0·50 mm (Jura, 1965). Other poduromorphs produce eggs approaching this size, e.g. *Achorutes hirtellus*, 0·42 mm (Tiegs, 1942a), *Anurida maritima*, 0·27 mm (Claypole, 1898), as do some entomobryomorphs, e.g. *Tomocerus plumbeus*, 0·37 mm (Uzel, 1898), *Entomobrya marginata*, 0·27 mm (Tiegs, 1942a) and sminthurids, e.g. *Sminthurus viridis*, 0·27 mm (Tiegs, 1942a), *Sminthurinus mime*, 0·25 mm (Ashraf, 1969), but the eggs of many species are between 0·12 and 0·20 mm in diameter (Uzel, 1898; Philiptschenko, 1912; Tiegs, 1942a; Bruckmoser, 1965; Zakhvatkin, 1969). Throughout the entire range of size, collembolan eggs exhibit a phase of total cleavage (Figs. 74 and 75) preliminary to the formation of a blastoderm (Uljanin, 1875b; Lemoine, 1883; Claypole, 1898; Uzel, 1898c; Philiptschenko, 1912; Jura, 1965; Zakhvatkin, 1969; Garaudy-Tamarelle, 1970). The first three divisions are synchronous, but synchrony is then lost. In the large egg of *Tetradentophora bielanensis*, the first two nuclear divisions are completed within an undivided yolk mass, each quadrant of which becomes nucleated, and the egg then divides directly into four equal blastomeres. In other species, total, almost equal cleavage of the yolk mass accompanies the first two divisions. The third synchronous cleavage division, perpendicular to the previous two, is equal in large eggs and slightly unequal in smaller eggs. On the grounds that the 8-cell stage of *Sinella cubiseta* has four micromeres set slightly obliquely above four macromeres, Zakhvatkin has suggested that collembolan cleavage is a form of spiral cleavage, but since

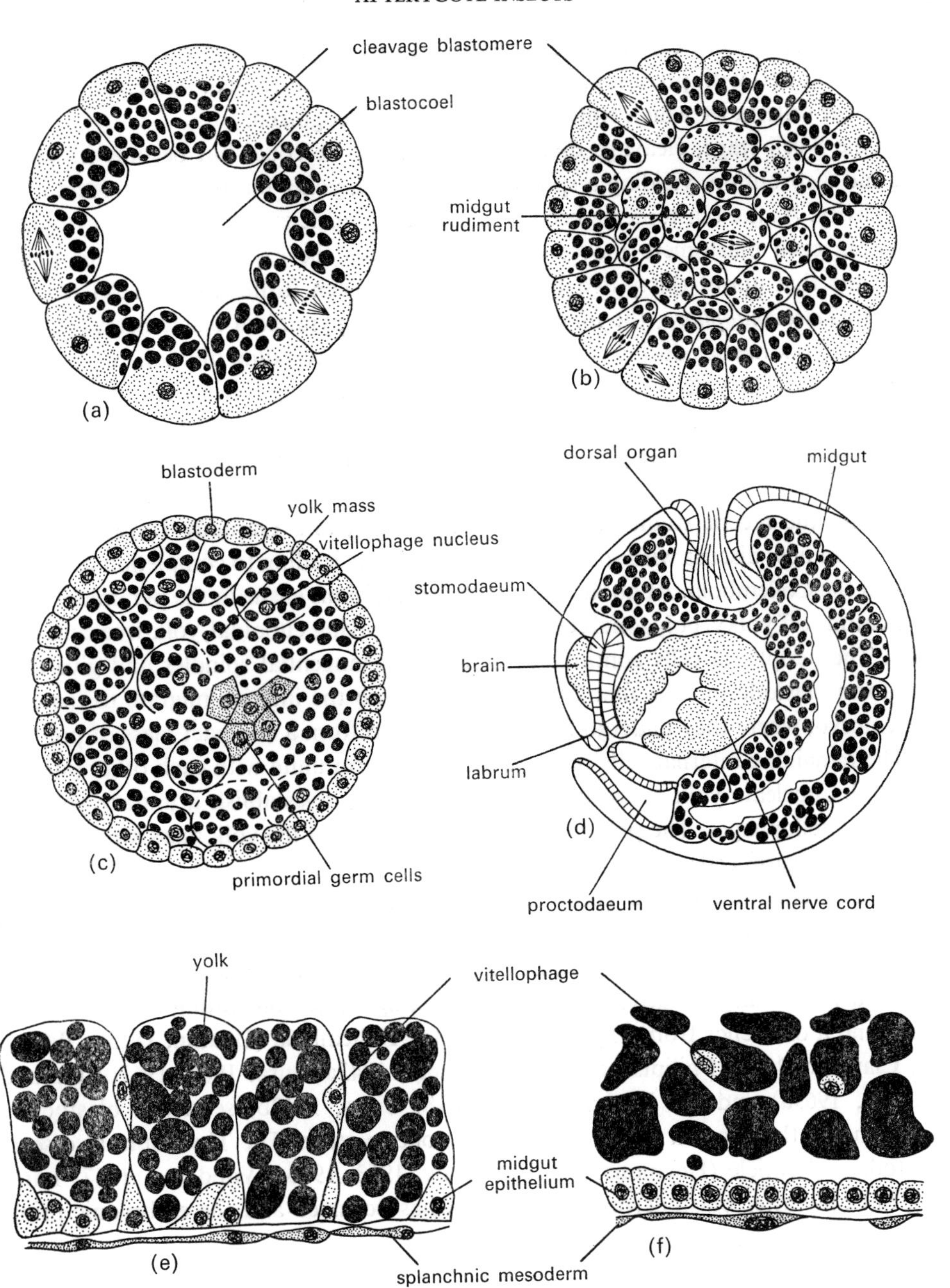

FIG. 75. *Tetradontophora bielanensis* (Collembola). (a) Section through late stage of total cleavage; (b) section showing blastoderm formation; (c) section through blastoderm stage; (d) sagittal section through flexed embryo; (e) transverse section through early midgut wall; (f) transverse section through late midgut wall. (After Jura, 1966.)

the axes of development cannot be established at this stage and the subsequent divisions cannot be followed with precision, such a conclusion is unwarranted. Anderson (1969) has discussed the criteria by which spiral cleavage can be recognized.

After the 8-cell stage, cleavage divisions continue to be equal but become irregular in timing and distribution, yielding a sphere of pyramidal, yolky blastomeres around a small central cavity. As the blastomeres become narrow, the nucleus and associated cytoplasmic halo within each blastomere moves outwards towards the exposed surface of the cell. By the time that about thirty-two blastomeres have been formed, the nuclei reach this surface. In *Tetradontophora*, the innermost parts of the pyramidal blastomeres are cut off as anucleate yolk masses during these divisions and the succeeding divisions which lead to the 64-cell stage. The yolk masses fuse into a single mass, filling the interior of the embryo, leaving a layer of truncated pyramidal blastomeres at the surface. In the smaller egged species, however, it seems likely that the yolk pyramids remain intact and the central blastocoel persists.

Following the recent studies by Jura (1965, 1967a), and Garaudy-Tamarelle (1969a, 1970), we can now discuss with some certainty the events leading to completion of the blastoderm in the Collembola. At about 64 cells, radial divisions ensue in the pyramidal blastomeres, cutting off yolky, polygonal cells inwards and leaving cuboidal, yolk-free cells at the surface. The peripheral cells continue to divide with tangential spindles and give rise to the uniform blastoderm. The first 2–5 cells cut off into the interior migrate to the centre of the embryo, where they undergo further divisions to form a compact mass of polygonal cells. The remaining internal cells beneath the blastoderm unite with each other (and with the central, anucleate yolk mass in *Tetradontophora*) to re-establish a unitary yolk mass containing vitellophages. The polygonal cells at the centre of the yolk mass are, as we shall see, primordial germ cells.

When differentiation of the germ band and dorsal extra-embryonic ectoderm takes place in the Collembola, the events concerned with the internal placement of mesoderm and midgut cells have already begun, as they have in the Diplura at the same stage. The germ band is also large relative to the yolk mass, again demonstrating a similarity with Diplura and a difference from Thysanura. Both the Collembola (Uljanin, 1875b; Wheeler, 1893; Claypole, 1898; Philiptschenko, 1912; Tiegs, 1942a; Jura, 1965, 1967b; Garaudy-Tamarelle, 1969a) and the Diplura (Uzel, 1898; Tiegs, 1942b) secrete a blastodermal cuticle beneath the chorion at the uniform blastoderm stage. The Thysanura also secrete a blastodermal cuticle (Heymons, 1897b; Sharov, 1953, 1966; Woodland, 1957; Larink, 1969), but only after blastoderm differentiation has occurred and the embryonic primordium has begun to transform into a germ band.

Presumptive Areas of the Blastoderm

As described above, the blastoderm stage of development varies in its organization in the three main classes of apterygotes. The dipluran blastoderm is a uniform layer of cuboidal cells around an anucleate yolk mass. The collembolan blastoderm is also uniform, but surrounds a yolk mass containing vitellophages and a central group of primordial germ cells. The thysanuran blastoderm, in contrast, becomes differentiated into a small, posterior

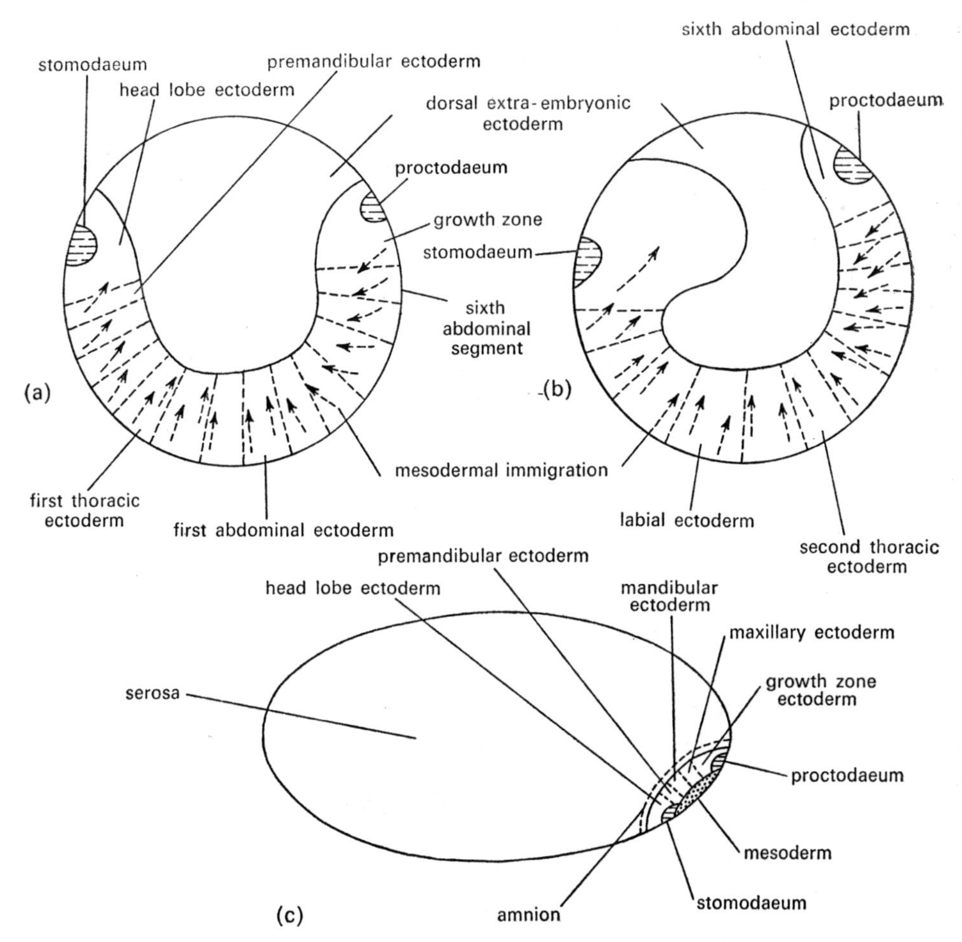

FIG. 76. Fate maps of the apterygote blastoderm. (a) Diplura; (b) Collembola; (c) Thysanura. (Based on the data summarized in the text.)

embryonic primordium and an attenuated extra-embryonic ectoderm and surrounds a yolk mass whose vitellophages are persistent cleavage energids.

Not surprisingly, the fate maps of the three types of blastoderm vary considerably (Fig. 76). The lack of knowledge of internal events during later development and the uncertainty about the accuracy of some of the details in the older accounts, together with the fact that the mesoderm and midgut components of apterygotes often originate diffusely, create considerable problems in the construction and comparison of fate maps for these animals. Even so, the exercise is quite rewarding.

The presumptive midgut

According to Uzel (1898), the blastoderm of the dipluran *Campodea* releases scattered cells into the yolk mass and these cells, after acting as vitellophages, subsequently give rise to the midgut epithelium. If this is so, and it requires confirmation, the presumptive midgut has no localized representation in the dipluran blastoderm and retains a gastrulation movement only in the form of a diffuse migration of blastoderm cells into the interior.

On much more firmly established evidence, the collembolan midgut epithelium also develops from the vitellophages within and around the yolk mass (Jura, 1966). As we have seen, these cells are cut off into the interior by radial divisions during cleavage. Presumptive midgut in the Collembola thus has no representation in the blastoderm. Again, the presumptive midgut of the Thysanura is now known to be the vitellophages left within the yolk mass during cleavage and formation of the blastoderm (Sharov, 1953, 1966; Woodland, 1957) and lacks blastodermal representation. A generalization can thus be made for the apterygote hexapods, that the presumptive midgut is constituted by vitellophages which attain an internal position during blastoderm formation or earlier in cleavage and persist as midgut cells after a period of temporary vitellophage activity.

The presumptive stomodaeum and proctodaeum

When the presumptive midgut has no localized distribution in the blastoderm, the locations of the presumptive stomodaeum and proctodaeum can be established only in relation to the presumptive ectoderm. The events of later development show that in all apterygotes, the stomodaeum invaginates in the ventral midline between the head lobes (see below) and the proctodaeum invaginates at the posterior end of the germ band, behind the growth zone (Uljanin, 1875b; Heymons, 1897 a, b; Claypole, 1898; Uzel, 1898; Sharov, 1953, 1966; Woodland, 1957; Jura, 1966; Larink, 1969). The presumptive areas of the stomodaeum and proctodaeum can thus be localized as shown in Fig. 76.

The presumptive mesoderm

Such knowledge as is available indicates that the mesoderm of the Diplura and Collembola has a diffuse origin over the entire area of the presumptive ectoderm, as cells which move inwards to form an inner layer beneath the ectoderm (Uzel, 1898; Philiptschenko, 1912; Jura, 1966). In this respect, the Diplura and Collembola convergently resemble the Pauropoda. A strong contrast is observed in the Thysanura. Here, the mesoderm migrates into the interior from the midline of the disc-shaped embryonic primordium and spreads out from this site to form an inner layer of polygonal cells beneath the columnar outer layer (Sharov, 1953; Woodland, 1957). Thus, the presumptive mesoderm of the Thysanura has a precise location as a narrow, short, midventral area of cells in the embryonic primordium, between the presumptive stomodaeum and presumptive proctodaeum.

The presumptive ectoderm

The presumptive ectoderm of all apterygote embryos is a bilateral area of blastoderm cells extending on either side of the ventral midline. In the Diplura and Collembola, the area is both broad and long, and is also the site of the diffuse presumptive mesoderm. Presumptive midgut cells may also have been released from the same area in the Diplura.

It can be deduced from the work of Uzel (1898) that the presumptive ectoderm of the

Diplura consists of a pair of presumptive head lobes, extending in front of the presumptive stomodaeum and including the presumptive antennal ectoderm at their posterior margins, followed by presumptive premandibular, mandibular, maxillary, labial and thoracic ectoderm, the ectoderm of the first six abdominal segments and an ectodermal growth zone in front of the presumptive proctodaeum. The several descriptions of the early germ band of collembolan embryos indicate that the presumptive ectoderm of the Collembola is similar, with the exception that the posterior growth zone is not present in these short-bodied hexapods.

In strong contrast, the presumptive ectoderm of the Thysanura is separate from the presumptive mesoderm and is extremely short by the time the embryonic primordium has differentiated (Woodland, 1957; Larink, 1969). Presumptive head lobes are present, followed by presumptive premandibular, mandibular and perhaps maxillary ectoderm, but then, only a short posterior growth zone. Although development is epimorphic in Thysanura in the sense that all segments are present and functional at hatching, most of the segments are proliferated in the embryo in an anteroposterior sequence.

Gastrulation

Obviously, in apterygotes, gastrulation with respect of midgut has all but been abandoned. The presumptive midgut cells of the Thysanura and Collembola become internal, in different ways, during cleavage, while those of Diplura are released into the interior from the early blastoderm, retaining a slight gastrulation movement in the form of a diffuse immigration. Gastrulation with respect to the mesoderm is also a diffuse immigration throughout the length of the presumptive germ band in the Diplura and Collembola, but has gained a new precision in the Thysanura, as a rapid immigration of presumptive mesoderm cells in the ventral midline of the embryonic primordium and spread of these cells to form an inner layer. We shall see in the next chapter that the same localization and mode of gastrulation of the mesoderm is characteristic of pterygotes. The stomodaeum and proctodaeum of apterygote embryos, like those of myriapods, penetrate the interior by independent invaginations.

Further Development of the Gut

Throughout the apterygotes, the stomodaeum and proctodaeum display the simple mode of direct development already seen in the myriapods. The stomodaeum always invaginates earlier in development than the proctodaeum. Once invaginated, the stomodaeum and proctodaeum (Fig. 75) continue to grow into the interior of the embryo as blind-ending tubes, the walls of which differentiate as the epithelium of the foregut or hindgut respectively (Uljanin, 1875b; Heymons, 1897 a, b; Claypole, 1898; Uzel, 1898; Sharov, 1953, 1966; Woodland, 1957; Jura, 1966; Larink, 1969). The stomatogastric ganglia arise as outgrowths of the dorsal wall of the stomodaeum. The Malpighian tubules develop as evaginations of the wall of the distal end of the proctodaeum.

The apterygote midgut (Fig. 75) has a simpler mode of further development than the variety of modes observed among myriapods. As development proceeds (Heymons 1897 a, b; Claypole, 1898; Woodland, 1957; Jura, 1966), many of the vitellophages within the yolk mass migrate towards the surface of the yolk and gather beneath the splanchnic mesoderm. In Collembola, the further development of the midgut epithelium proceeds in the embryo. In the Thysanura, this process is delayed until after hatching, but the sequence of events is similar in both. The central part of the yolk mass tends to break down, but the peripheral yolk subdivides into cuboidal blocks, each containing one or more peripheral vitellophages and a few vitellophages placed more deeply within the block. The peripheral vitellophages proliferate, forming small nests of cells beneath the splanchnic mesoderm. The cell nests spread out, beginning to do so earlier in the vicinity of the stomodaeum and proctodaeum than in the middle region of the embryo, and gradually appropriate the remaining yolk into a columnar epithelium which becomes the midgut epithelium. A few workers (Philiptschenko, 1912; Sahrhage, 1953; Wellhouse, 1953) have claimed that midgut epithelium is proliferated from the distal ends of the stomodaeum and proctodaeum, but the work of Woodland (1957) and Jura (1966) provides convincing evidence that the vitellophages within the yolk mass constitute the midgut rudiment, and merely begin their differentiation into definitive epithelium earlier at the ends than in the middle region. The fat-body of apterygotes develops from mesoderm (see below).

Development of External Form

A consideration of the development of the external form in apterygote embryos is particularly valuable, since it emphasizes more than any other feature the fundamental difference between the Collembola and Diplura on the one hand and the Thysanura on the other. The formation and further development of the germ band in the Collembola and Diplura is convergently reminiscent of that of the Symphyla, Pauropoda and Diplopoda among the myriapods. The germ band of the Thysanura, in contrast, exhibits an alternative pattern of further development which is also found in the pterygote insects.

We may begin with the Diplura, as exemplified by *Campodea* (Uzel, 1898; Tiegs, 1942b). It has already been described above how the embryo of *Campodea* develops a uniform blastoderm about an anucleate yolk mass, which is then reinvaded by cells from the blastoderm, acting as vitellophages. Differential cell division and aggregation now ensue in the ventral third of the spherical blastoderm, resulting in the formation of a hemispherical germ band in this region and an attenuated layer of dorsal, extra-embryonic ectoderm over the remainder of the surface of the yolk mass. During proliferation and aggregation of the cells of the germ band, many of the cells slip beneath the surface to establish an inner layer, the mesoderm.

The germ band now enters into anteroposterior elongation and becomes narrower at the same time (Fig. 77). The anterior and posterior ends of the germ band extend on to the dorsal surface of the yolk mass and the extra-embryonic ectoderm is redistributed down the sides of the yolk mass. The anterior region of the germ band is relatively wide, forming incipient head lobes. The middle region, occupying the ventral surface of the yolk mass, is somewhat narrower, with a slight widening towards the posterior end.

Segmentation arises simultaneously in all but the posterior third of the germ band. At the anterior end, the head lobes broaden, the labrum and stomodaeal invagination develop in the midline between the head lobes and the antennal limb rudiments pouch out at the

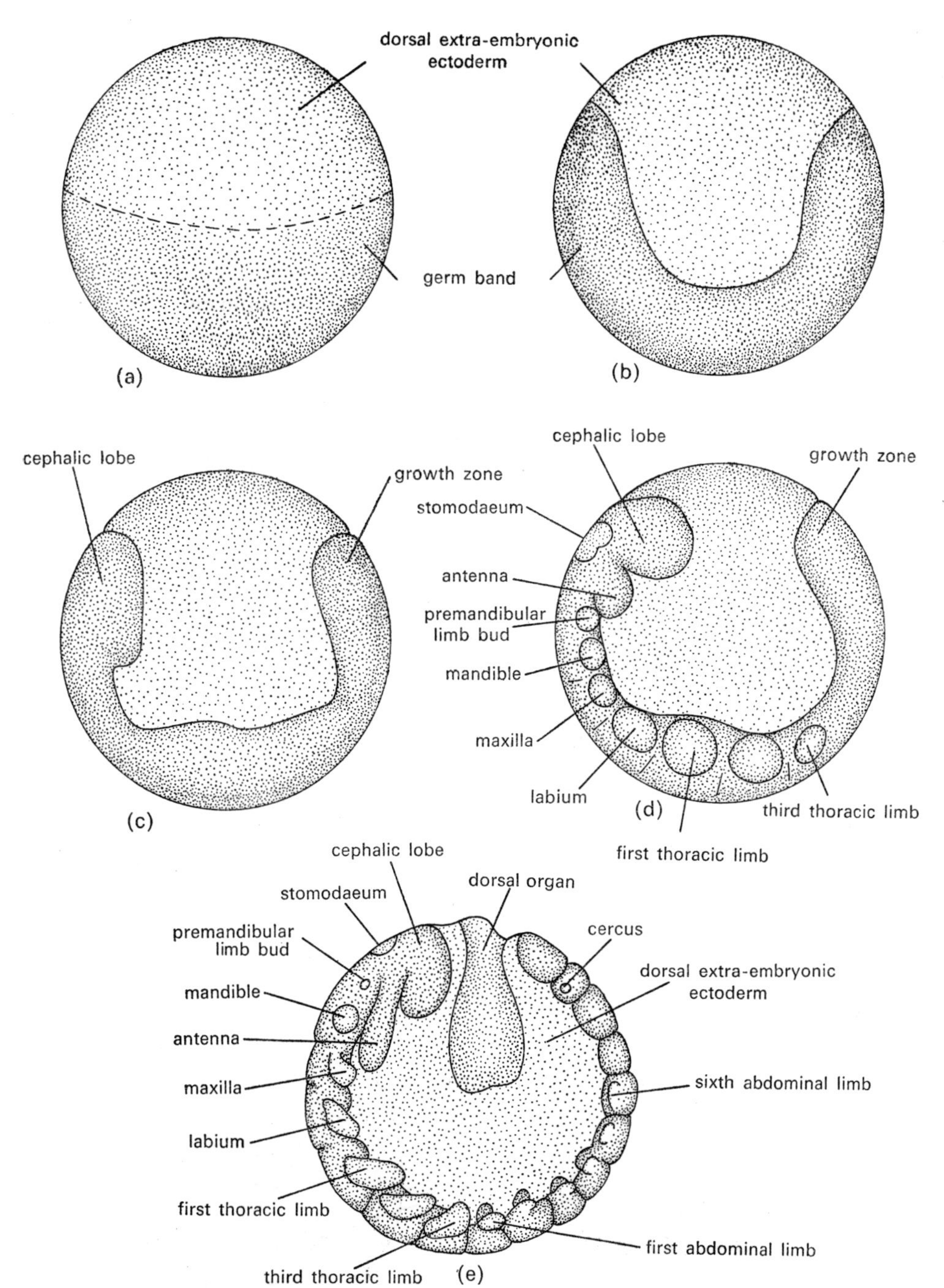

FIG. 77. Formation and segmentation of the germ band of *Campodea*, after Uzel (1898). (a) Differentiation of the germ band; (b) and (c) elongation of the germ band; (d) segmenting germ band; (e) segmented germ band.

posterior margins of the head lobes, behind the stomodaeum. The first part of the post-antennal region of the germ band remains limbless at this stage, but later develops a pair of transient pre-mandibular limb buds, indicating that it is the pre-mandibular segment of the germ band. The remainder of the anterior third becomes demarcated as the mandibular, maxillary and labial segments, while the middle third becomes demarcated as the three thoracic segments and the first abdominal segment. All of these segments develop limb buds.

By the time that this stage of development is reached, the posterior third of the germ band has begun to grow longer. It seems likely that the presumptive rudiments of the second to fifth abdominal segments are already present, followed by the segment-forming growth zone. As elongation continues, the end of the abdomen approaches the anterior margin of the head on the dorsal surface of the yolk mass, though the two ends always remain separated by the dorsal organ (see below). The second to ninth abdominal segments are delineated in succession, leaving a terminal region which carries the anus and cerci. Since the second to ninth abdominal segments develop limb buds in sequence with those of the first abdominal

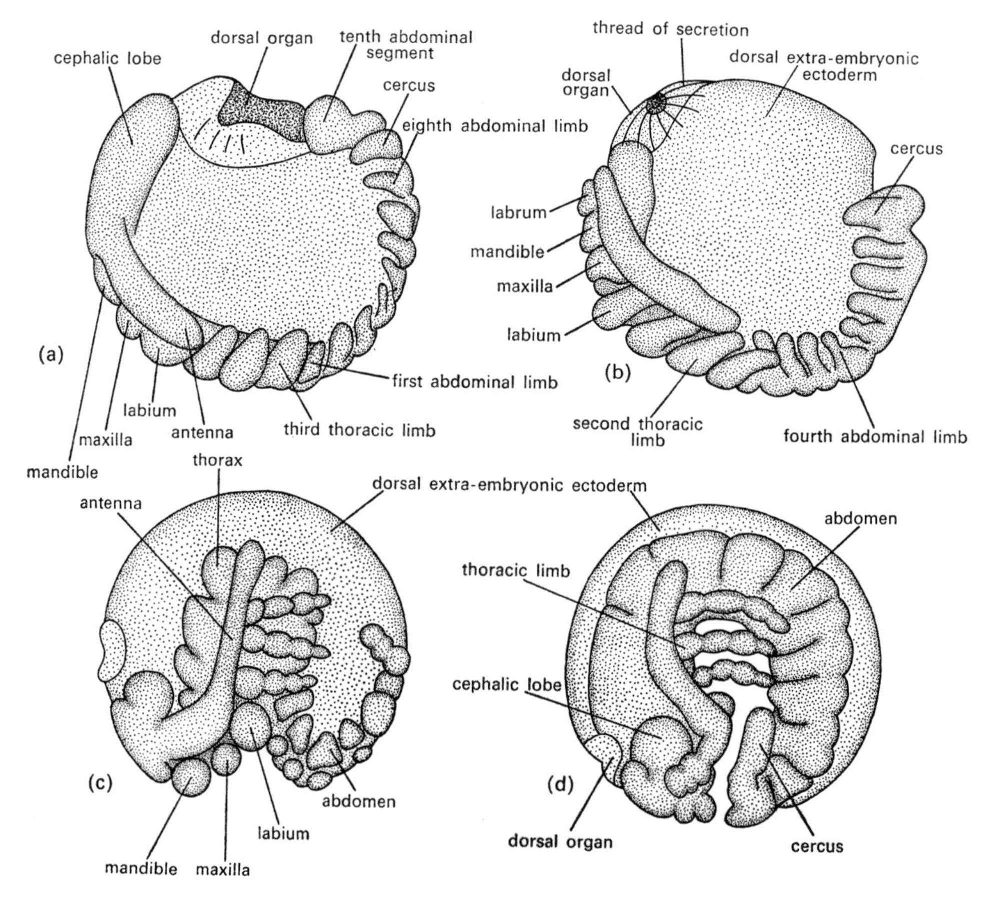

FIG. 78. Later development of *Campodea*. (a) Formation of limb buds and dorsal organ; (b) shortening of the segmented germ band; (c) ventral flexure; (d) dorsal closure. [(a) and (b) after Tiegs, 1942b; (c) and (d) after Uzel, 1898.]

segment, and the cercal rudiments are the next pair of limb buds in the same sequence, it is possible that the dipluran cerci are the limbs of the tenth abdominal segment and that the terminal abdominal region on which they lie constitutes this segment together with a terminal telson. Quite obviously, however, a more detailed investigation is required to test this interpretation. The eighth and ninth pairs of abdominal limb buds are transient, but the pairs on segments 1–7 persist and develop as styles.

Once segmentation of the germ band and formation of the limb buds is completed, a deep dorsoventral flexure begins to develop at the midventral point of the germ band, between the third thoracic and first abdominal segments (Fig. 78). As the flexure deepens, the germ band gradually folds inwards about this point, bringing the ventral surface of the thorax and head to face the ventral surface of the abdomen, and eventually bringing the anterior end of the head and the posterior end of the abdomen together at the mid-ventral point of the egg space. At the same time, the antennae become pre-oral and the mouth-parts crowd forward behind the mouth. The hypopharynx arises from ventral maxillary ectoderm and the super-linguae from ventral mandibular ectoderm.

The final posture within the egg space is similar to that observed in all myriapods except *Pauropus*. As a result of dorsoventral flexure of the germ band, the spherical yolk mass is stretched out into an elongate form and bent into a convex hoop above the germ band. The dorsal, extra-embryonic ectoderm is simultaneously returned to the convex dorsal surface of the yolk. The antennae, thoracic limbs and cerci increase considerably in length as the embryo undergoes dorsoventral flexure and are accommodated with the space created by the flexure itself. Finally, upgrowth of the lateral walls of the segments effects dorsal closure in the flexed position. Hatching takes place directly, without the intervention of a pupal phase.

The further development of the external form of the Collembola (Fig. 79) has been studied in many species and always follows the same course, whether in the large embryo of *Tetradontophora*, the moderate-sized embryo of *Anurida* or the small embryo of *Orchesella* (Uljanin, 1875b; Lemoine, 1883; Wheeler, 1893; Claypole, 1898; Uzel, 1898; Folsom, 1900; Hoffmann, 1911; Tiegs, 1942a; Haget and Garaudy, 1964; Bruckmoser, 1965; Jura 1965; Garaudy-Tamarelle, 1971). From the beginning, it is obvious that the embryonic primordium takes in a greater proportion of the blastoderm in the Collembola than it does in *Campodea*. When differential cell division makes the germ band externally distinct from the dorsal, extra-embryonic ectoderm, the latter can be seen to be confined to the dorsal surface of the yolk mass. The germ band changes shape and becomes U-shaped, as it does in *Campodea*, but the anterior and posterior ends of the germ band are already close to one another on the dorsal surface before external segmentation becomes apparent.

The first phase of external segmentation of the germ band in the Collembola takes place, as it does in *Campodea*. The head lobes broaden and develop antennal buds at their postero-medial corners, a labral lobe between their anterior margins and a stomodaeal invagination behind the labral lobe. At the same time, the gnathal and thoracic limb buds develop, together with a conspicuous pair of transient premandibular limb buds on the most anterior part of the post-antennal region (Claypole, 1898; Folsom, 1900; Hoffmann, 1911; Haget and Garaudy, 1964; Bruckmoser, 1965). In contrast to *Campodea*, the outgrowth of the first pair of abdominal limbs does not begin until after the onset of dorsoventral flexure

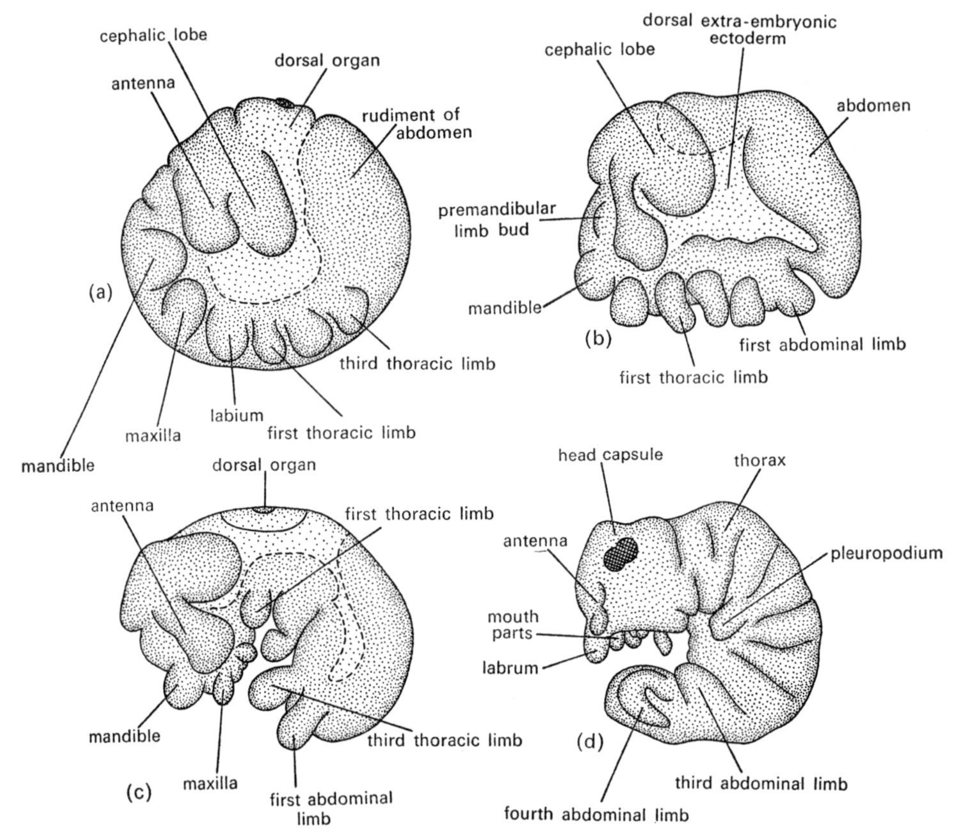

FIG. 79. Later embryonic development of the collembolan *Orchesella*, after Bruckmoser (1965). (a) Segmenting germ band; (b) early ventral flexure; (c) later ventral flexure; (d) dorsal closure.

of the germ band. The unsegmented rudiment of the abdomen, making up the posterior third of the U-shaped germ band, probably already consists of the rudiments of the five abdominal segments and sixth or terminal unit. There is no indication, on present evidence, that a segmented growth zone is retained in the Collembola.

Dorsoventral flexure itself is also more gradual in the Collembola than in the Diplura, and takes place in a different way. Flexure begins with a flattening of the middle region of the germ band, from the labial segment to the third thoracic segment, which gradually deepens into an inflexion about the first and second thoracic segments. The posterior part of the head is slightly inflected, but the major part of the head retains its anterior location within the egg space and simply undergoes a slight downward rotation. By far the major movement of dorsoventral flexure in Collembola is a shortening and forward flexure of the abdomen, which brings the posterior end of the abdomen forwards beneath the head. As this happens, the abdomen becomes segmented into five segments and a terminal unit which bears the anus. A pair of limb buds develops on the first abdominal segment at an early stage of dorsoventral flexure. As development proceeds, these limb buds come together to form the ventral tube. Limb buds are also formed on the second and third abdominal segments. The second pair is transistory, but the third pair usually develops into the hamula.

A large pair of limb buds develops on the fourth abdominal segment as the posterior end of the abdomen turns forwards beneath the head. These limbs develop into the characteristic collembolan furcula or spring. The sixth or terminal unit of the abdomen, which carries the anus, may be the telson or may incorporate a sixth abdominal segment. Further embryological evidence is needed on this point.

During dorsoventral flexure, the yolk mass is extended and curved within the germ band in the usual way. Dorsal closure, however, proceeds simultaneously with flexure and is not delayed until after dorsoventral flexure has been completed, as it is in *Campodea*. The development of the head takes place in the usual hexapod manner (Fig. 80). Careful recent studies by Haget and Garaudy (1964) and Bruckmoser (1965) have shown that the antennae migrate forward to a pre-oral position, the pre-mandibular region forms folds at the sides of the labrum and the mouth-parts crowd forward to form the sides and posterior wall of the pre-oral cavity. The superlinguae arise as ventral outgrowths on the mandibular segment and the hypopharynx as a ventral outgrowth on the maxillary and labial segments (Uzel, 1898; Hoffmann, 1911; Philiptschenko, 1912; Bruckmoser, 1965).

Clearly, the external development of the embryo is generally similar in the Collembola and Diplura. The external development of the Thysanura, in contrast, is distinct from that of other apterygotes (Heymon, 1897b; Sharov, 1953; Woodland, 1957; Larink, 1969, 1970), continuing and furthering a difference already apparent at the blastoderm stage. We have already seen that the thysanuran egg is relatively large and ovoid and that the blastoderm undergoes differentiation into a small embryonic primordium and extensive dorsal, extra-embryonic ectoderm before the embryonic primordium begins to release mesoderm into the interior. The embryonic primordium is circular in outline when first formed, but after the inner layer has developed, the primordium begins to elongate as a germ band (Fig. 81) and soon develops small head lobes and a short post-antennal region which then divides into the mandibular and maxillary segments and a posterior growth zone. As the growth zone becomes active, antennal limb buds arise at the posterior margins of the head lobes, together with mandibular and maxillary limb buds on their respective segments. At this stage, the growing germ band is curved around the posteroventral surface of the yolk mass, but as increase in length continues and the labial limb buds develop, a precocious dorsoventral flexure of the short germ band takes place, flexing the germ band upwards as an arc into the yolk (Fig. 82). By this time, the yolk mass has become temporarily cellularized into nucleated yolk spheres, which are displaced as the germ band pushes between them. Early growth and flexure of the germ band is accompanied by another precocious event, outgrowth of the labrum, combined with invagination of the stomodaeum and forward migration of the antennae to their pre-oral position.

As growth in length of the germ band now continues, accompanied by the serial delineation of further segments, the flexed germ band sinks more deeply into the yolk mass. Simultaneously, the margin of the germ band proliferates a sheet of small cells between itself and the margin of the superficial, extra-embryonic ectoderm. This sheet of cells forms the wall of a cavity which arises as the flexed germ band sinks into the yolk mass. The opening of the cavity to the exterior becomes temporarily closed in *Petrobius* and *Ctenolepisma* but remains as a small, superficial pore in *Lepisma*. The extra-embryonic ectoderm of the Thysanura is now customarily called the serosa. The newly formed cavity, roofed by the

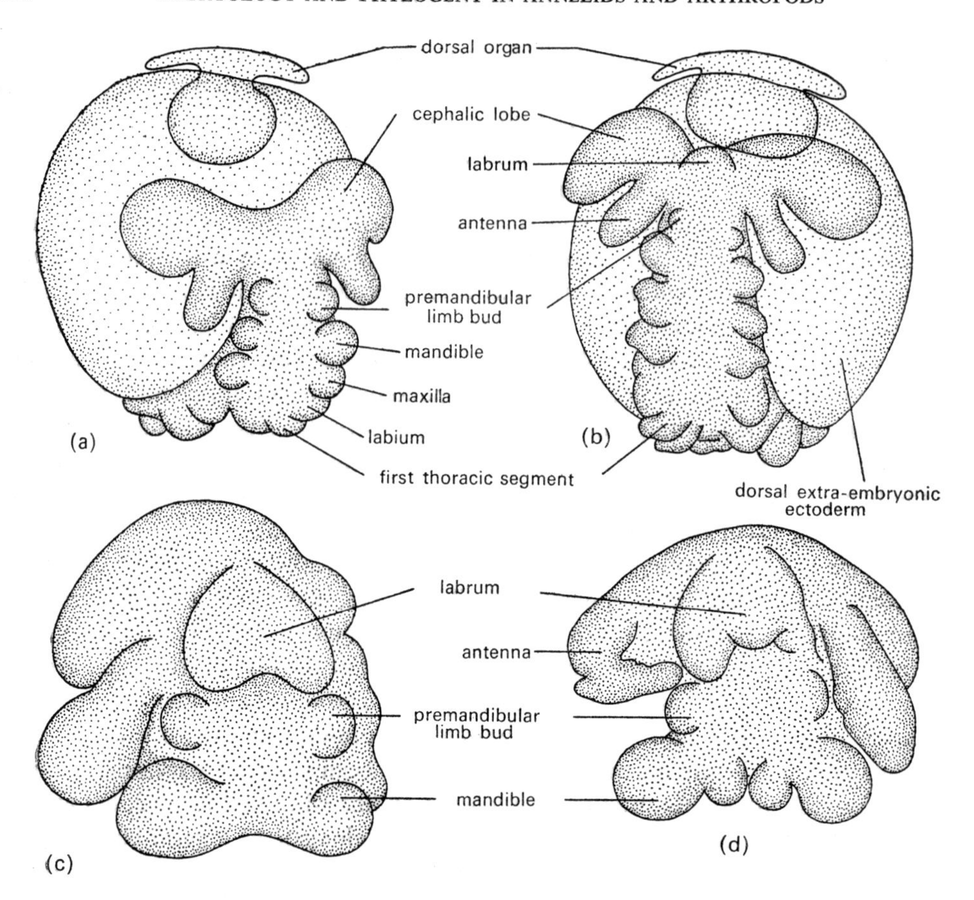

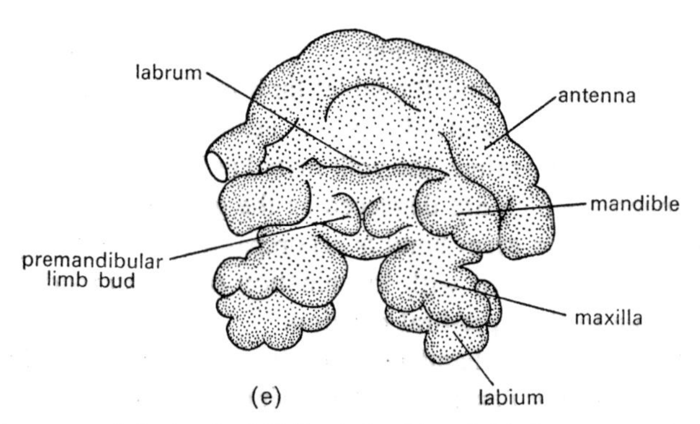

FIG. 80. Development of the head of Collembola. (a) and (b) Anteroventral views of the anterior end of the segmenting germ band of *Anurida*, showing early development of the head, after Haget and Garaudy (1964); (c)–(e) stages in the later development of the head of *Orchesella*, in ventral view, after Bruckmoser (1965).

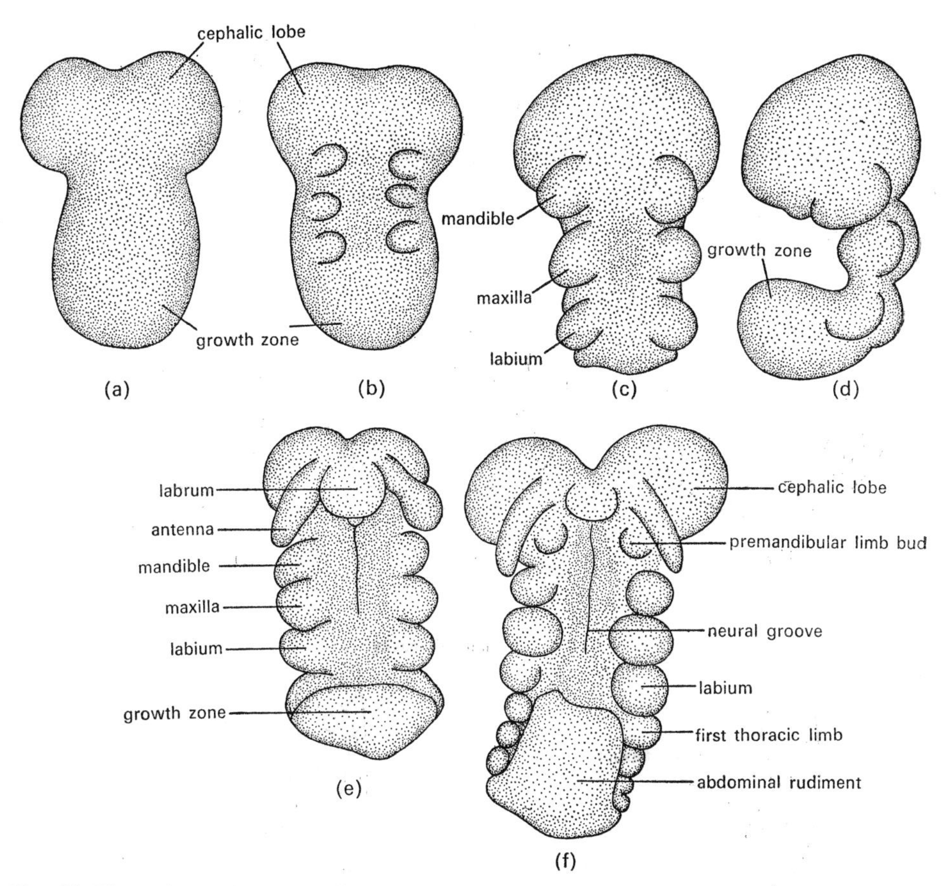

FIG. 81. Elongation and segmentation of the germ band of the thysanuran *Petrobius*, after Larink (1969). (a) Embryonic primordium; (b) onset of segmentation; (c) and (d) initial formation of limb buds; (e) early elongation; (f) further elongation and limb bud formation.

flexed germ band, is the amniotic cavity and the newly proliferated wall of the cavity constitutes the amnion. It is thus clear that the thysanuran amnion is a temporary outgrowth of the embryonic ectoderm, associated with precocious dorsoventral flexure of the short germ band into the yolk mass. The functional significance of the precocious dorsoventral flexure and accompanying amnion formation are still not clear, either for the Thysanura or for corresponding events in various Pterygota (see Chapter 7).

Once immersed in the yolk mass, the thysanuran germ band continues to increase in length through the activity of its posterior growth zone. The labial, three thoracic and ten abdominal segments are delineated in succession, leaving an eleventh abdominal unit at the posterior end of the body, bearing the anus. Thoracic limb buds develop, the antennae grow longer and a pair of transient pre-mandibular limb buds develops in *Petrobius* and *Ctenolepisma*, in the region between the mandibular segment and the mouth. The development of limb buds on the abdominal segments, however, is delayed until the onset of the next phase of development. This phase, unrepresented in the development of the Collembola or Diplura, consists in the re-emergence of the immersed germ band at the surface

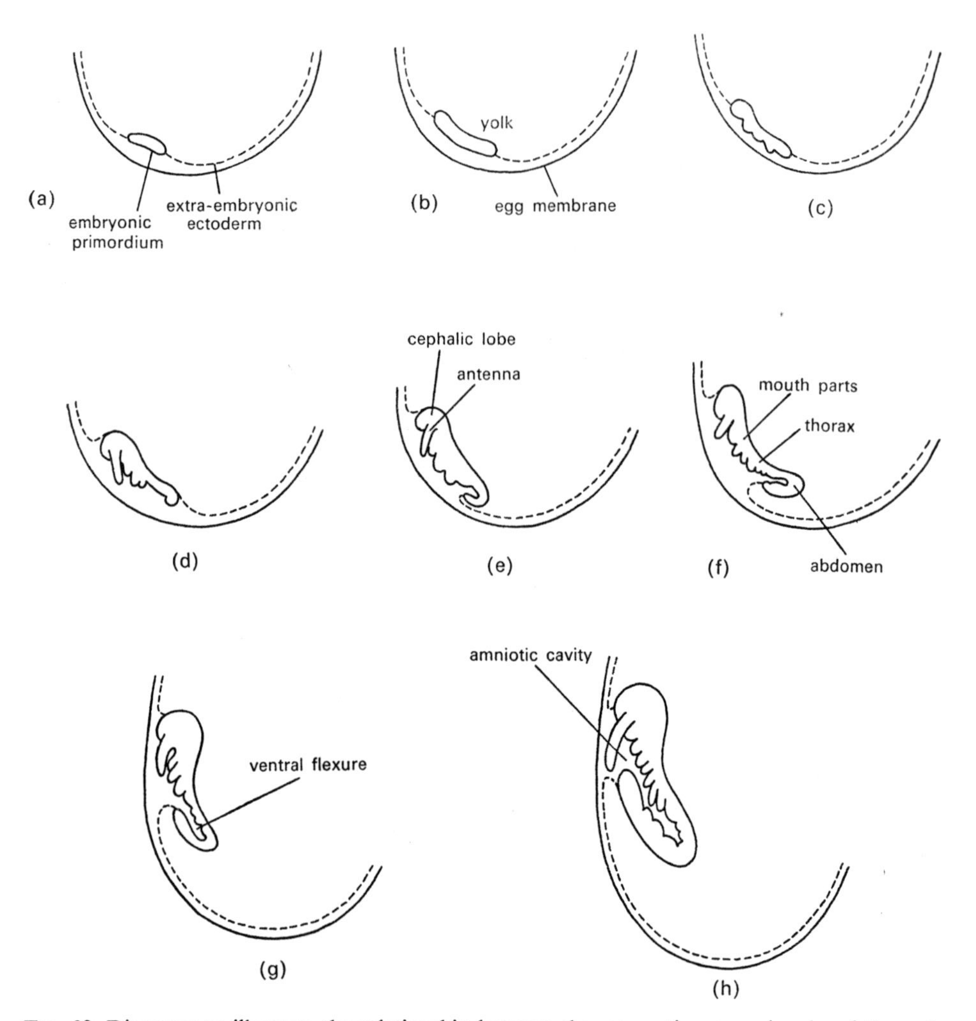

FIG. 82. Diagrams to illustrate the relationship between the segmenting germ band and the yolk mass of *Petrobius*, after Larink (1969).

of the yolk mass, by a process usually spoken off as blastokinesis (Fig. 83). We shall see in the next chapter that a variety of blastokinetic movements occurs amongst pterygote embryos and that the most universal of these, katatrepsis, is equivalent to blastokinesis in the Thysanura.

The work of Sharov (1953), Woodland (1957) and Larink (1969) has expanded considerably the knowledge of thysanuran blastokinesis first layed down by R. Heymons (1897b) and R. and H. Heymons (1905).

The process begins when the germ band is fully segmented and proceeds slowly while the segments and limbs of the germ band undergo further development. The first sign of blastokinesis is a reopening and widening of the amniotic cavity and an eversion of the amnion around the anterior margin of this entrance. Gradually, as eversion of the amnion continues, the amniotic cavity gapes at its anterior end and the head of the germ band,

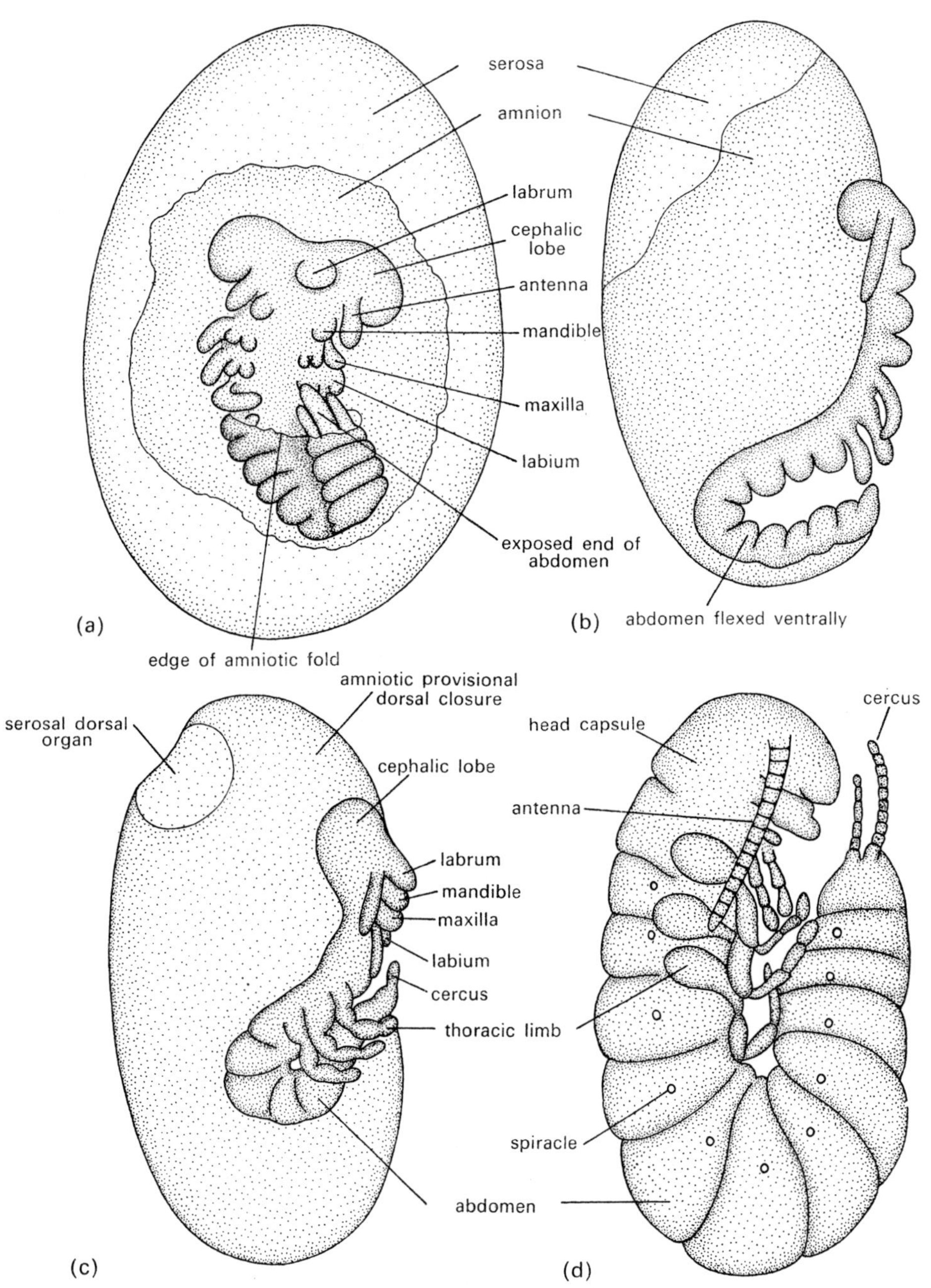

FIG. 83. Blastokinesis and dorsal closure in *Lepisma*, after Heymons (1897b). (a) Onset of blastokinesis; (b) late blastokinesis; (c) blastokinesis completed; (d) embryo after completion of dorsal closure.

followed by the thorax, emerges on to the ventral surface of the yolk mass. The anterior half of the germ band straightens out and stretches forward along the ventral surface of the yolk mass, though the abdomen remains flexed within the posterior part of the amniotic cavity.

By this stage it is beginning to become apparent how the process of blastokinesis comes about. The everting amnion emerges onto the surface of the yolk and the anterior end of the germ band follows it, in consequence of a progressive contraction of the serosa towards an anterodorsal focus. As serosal contraction continues, the amnion stretches more and more to replace the serosa at the surface of the yolk mass, establishing a new, provisional dorsal closure. Gradually, the amnion is peeled back over the flexed posterior end of the germ band, so that the germ band is entirely exposed at the surface of the yolk. The abdomen retains the original dorsoventral flexure, however, and subsequently increases this flexure, so that the posterior end of the abdomen comes to lie beneath the head. The serosa contracts into a tight knot of cells at the anterodorsal focus and degenerates in this position. The contracted serosa is usually called a dorsal organ, though it differs in many ways from the dorsal organs of the Collembola and Diplura (see below). There is no justification for distinguishing a primary dorsal organ in the latter from a secondary dorsal organ in the Thysanura (and by implication, in the Pterygota, see Chapter 7). The two are alternative and different types of dorsal organ associated with different developmental functions. The thysanuran dorsal organ is a disposal centre for the contracted serosa or dorsal extra-embryonic ectoderm. The possible functions of the collembolan and dipluran dorsal organs will be discussed in a later section. As blastokinesis proceeds, the thysanuran germ band continues its further development. The pre-mandibular limb buds are resorbed. The mouthparts crowd forwards behind the mouth and develop their various component lobes. The hypopharynx and superlingae arise midventrally between the maxillae and mandibles respectively (Larink, 1969). The antennae and thoracic limb buds increase in length. Limb buds also develop on the first and eleventh abdominal segments, then on the intervening segments. The second to tenth abdominal limb buds of *Petrobius* each divide into a median and a lateral part. The median part swells, forming a lobe which later develops a pair of eversible sacs. The lateral part of the limb bud elongates, forming a style (Larink, 1969). In the lepismatids, styles persist on segments 7–9 or 8–9, but the second to sixth (or seventh) and the tenth pairs of abdominal limb buds are resorbed.

The first pair of abdominal limb buds swells to form a pair of rounded protuberances, the pleuropodia. The pleuropodia are glandular structures and it seems likely that they secrete an enzyme which digests the serosal cuticle, as they are thought to do in those pterygote embryos which develop pleuropodia (see Chapter 7). Shortly before hatching takes place, the pleuropodia invaginate into the body and are resorbed (Larink, 1969). The eleventh pair of abdominal limbs persists and develops into the cerci (Heymons, 1896a, 1897b). The caudal filament is developed as a median protuberance of the posterior termination of the body, behind the anus.

After the completion of blastokinesis and the tightening of the abdominal flexure, the head, thoracic segments and abdominal segments begin to increase in length and to grow dorsally around the yolk mass. Dorsal closure progresses from behind forwards. The amnion is gradually replaced at the surface of the yolk by the definitive body wall and is eventually incorporated into the dorsal part of the body wall (Woodland, 1957). When dorsal closure is completed, the flexed, tubular embryo occupies the entire space within the egg membranes.

Embryonic development in the Thysanura differs greatly from that of the Collembola and Diplura in the relationship between the germ band and the yolk. We shall see in the next chapter that the peculiarities of thysanuran development are also much in evidence in the Pterygota. Before attempting to assess the implications of developmental variation within the apterygote hexapods, we need to examine as far as possible the further development of the mesoderm and ectoderm of these animals. Information on internal development is scanty at present for the Collembola and virtually unavailable for the Diplura, but has been given useful expression by Woodland (1957) and especially Larink (1969, 1970) for the Thysanura.

Further Development of the Mesoderm

The further development of the mesoderm of the Diplura has not yet been described, so that no comment can be made at the present time on the formation, distribution and further development of the somites of this class of apterygotes. It is also true to say that, with the exception of the development of the gonads, the further development of the mesoderm of collembolan embryos is rather poorly known. Philiptschenko (1912) and, more recently, Jura (1966) have described how the mesodermal inner layer of the germ band piles up laterally into paired mesodermal bands, united by a thin sheet of median mesoderm. The mesodermal bands then subdivide into paired somites which develop small coelomic cavities. The antennal, mandibular, maxillary, labial, three thoracic and five abdominal pairs of somites have been identified and Garaudy-Tamarelle (1969b) has recently described a pair of small but typical pre-mandibular somites in *Anurida*, but there is at present no good evidence for distinct pre-antennal or sixth abdominal somites in the Collembola. The absence of pre-antennal somites may be a matter of inadequate observation, but the absence of a sixth abdominal pair seems likely to be correct. In their further development, the somites become trilobed and follow the expected pattern of general development, with the appendicular lobes giving rise to limb muscles, the dorsal edges of the dorsolateral lobes to the heart, the somatic walls of the dorsolateral lobes to dorsal and lateral longitudinal muscles and fat-body cells, the somatic walls of the medioventral lobes to ventral longitudinal and other muscles, and the splanchnic walls of the somites to the splanchnic musculature the midgut and to the gonads and gonoducts. According to Jura (1966), the stomodaeal and proctodaeal splanchnic mesoderm arises from the median mesoderm that coats the stomodaeal and proctodaeal invaginations. It is obvious, however, that the fine details of mesoderm development in the Collembola still await elucidation. Only in one respect has the mesoderm of Collembola received careful attention, namely, in the development of the gonads.

We have already seen that the primordial germ cells of the Collembola are segregated during cleavage and occupy the centre of the yolk at the blastoderm stage. A recent investigation of *Anurida maritima* by Garaudy-Tamarelle (1970) has suggested that the cytoplasm of these cells may incorporate an oösome shifted from the posterior pole to the centre of the egg during total cleavage. The same author (Garaudy-Tamarelle, 1969a) has also confirmed and extended the evidence of Philiptschenko (1912) on *Isotoma* and Jura (1967a) on *Tetradontophora* that the primordial germ cells move away from their central position during segmentation of the germ band and come to rest as a bilateral pair of strands beneath the splanchnic mesoderm of the first to fifth abdominal segments. As dorsoventral flexure

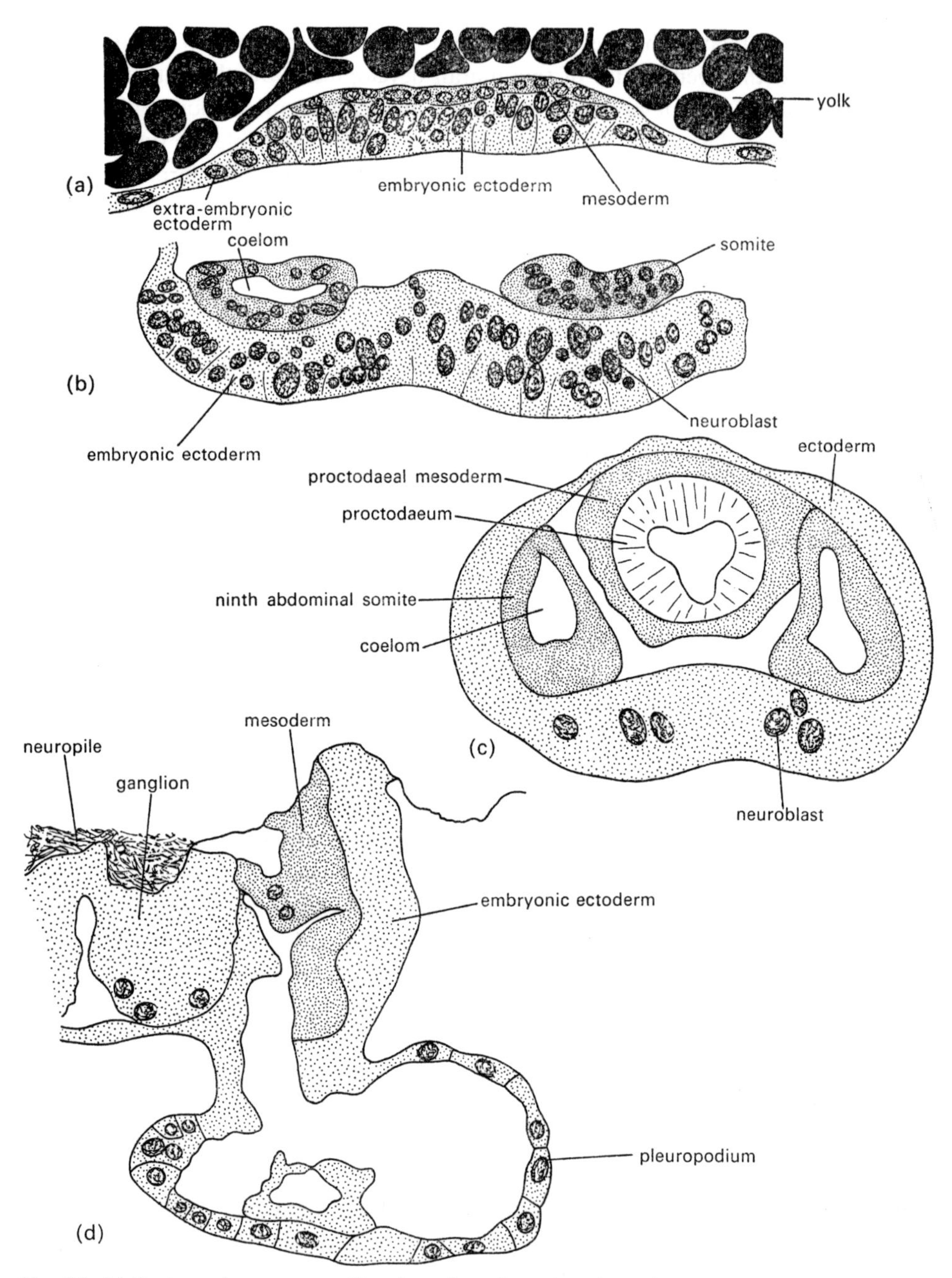

FIG. 84. (a) *Lepisma*, transverse section through embryonic primordium; (b) *Petrobius*, transverse section through early segmenting germ band; (c) *Petrobius*, transverse section through developing abdomen; (d) *Petrobius*, transverse section through first abdominal ganglion and pleuropodium. [(a) after Woodland, 1957; (b)–(d) after Larink, 1969.]

proceeds and the abdominal segments continue to develop, the two strands of germ cells shorten, forming a pair of compact groups of cells in the third to fourth abdominal segments. Each group of germ cells is covered with a thin layer of splanchnic mesoderm. The mesodermal gonad sheaths proliferate extensions which form the paired mesodermal gonoducts. The midventral genital opening arises as an ectodermal invagination (see below).

The association of primordial germ cells with the abdominal splanchnic mesoderm during formation of the gonads is a distinctive feature of hexapods as compared with the myriapods and Onychophora. As we shall see, the gonoducts of other hexapods also originate as segmental outgrowths of the splanchnic mesoderm.

Following the recent investigations by Sharov (1953, 1966), Woodland (1957) and Larink (1969, 1970) on *Lepisma*, *Thermobia* and *Petrobius*, the further development of the mesoderm of the Thysanura is now quite well understood. Proliferation of the mesodermal inner layer continues as the germ band increases in length, accompanied by anteroposterior segregation of the mesoderm into paired somite rudiments (Fig. 84). No evidence has been obtained for the persistence of median mesoderm. The paired somites are flat sheets of cells when first formed, but soon thicken and develop coelomic cavities. In *Petrobius* the coelomic cavities develop as splits within the thickened somites, but those of *Thermobia*, according to Woodland (1957), arise through the median overfolding of the lateral edges of the somite sheets, as they do in some pterygotes (see Chapter 7). A pair of large antennal somites is formed, followed by pre-mandibular somite masses which do not become hollow, small mandibular somites, large maxillary, labial and thoracic and first to ninth abdominal somites and a pair of small, tenth abdominal somites flanking the proctodaeum. Behind the tenth abdominal somites, a mass of mesoderm cells persists at the posterior end of the abdomen. In front of the antennal somites, a sheet of mesoderm extends forwards beneath the head lobes and develops paired thickenings, but does not break up into paired pre-antennal somites. Thus, on present evidence, distinct pre-antennal somites have not been observed in any apterygote hexapod.

In their further development, the somites develop the typical trilobation, with appendicular, large dorsolateral and small medioventral lobes. Woodland (1957) and Larink (1969) have shown that the appendicular lobes develop in the usual way as limb musculature, the somatic walls of the dorsolateral and medioventral lobes as longitudinal muscle and fat body, and the splanchnic walls as splanchnic mesoderm of the midgut. The modifications of this general fate that might be expected for the cephalic somites have not yet been studied in detail. The dorsal edges of the somatic walls of the dorsolateral lobes give rise to the heart during dorsal closure, but there are no distinct cardioblasts. Adjacent somatic mesoderm develops as the pericardial septum. Anteriorly, the median edges of the antennal somites come together above the stomodaeum to form the anterior aorta (Woodland, 1957). The small masses of pre-mandibular mesoderm develop into the glandular suboesophageal body (Larink, 1969, 1970), as they do in some pterygotes. The stomodaeal and labral mesoderm arise from the sheet of mesoderm in front of the antennal somites, while the proctodaeal mesoderm is formed by the posterior mesoderm which remains behind the tenth abdominal somites.

The origin of primordial germ cells in the Thysanura is not clear. Heymons (1897b), who gave the first account of this process, claimed that a posterior group of primordial germ cells can be distinguished in the embryonic primordium of *Lepisma*. This observation was supported by Sharov (1953), but neither Woodland (1957, *Thermobia* and *Ctenolepisma*) nor Larink (1969, *Petrobius*) were able to observe these cells. Since the migration of the ostensible primordial germ cells of *Lepisma* to their next-observed site in the splanchnic walls of certain somites has not been followed, it cannot be said with certainty that an early segregation of primordial germ cells occurs in any thysanuran. All authors are agreed, however, that these cells can be identified at the segmented germ-band stage, as large, pale-staining cells in the splanchnic walls of the dorsolateral lobes of several trunk somites. Larink (1969) found them in the second and third thoracic somites and first abdominal somites of *Petrobius*, but did not pursue the further development of the gonads. In *Thermobia*, *Ctenolepisma* and *Lepisma*, the distribution of the primordial germ cells in the somites is more precise, since they lie in abdominal somites 2–6 in females and 4–6 in males (Heymons, 1897b; Sharov, 1953, 1966; Woodland, 1957). In these positions, each group of germ cells is carried dorsally during dorsal closure and becomes individually sheathed by splanchnic mesoderm to form an ovariole or a testicular follicle. The splanchnic mesoderm of abdominal segments 1–10 in males and 1–7 in females also produces lateral segmental outgrowths or ampullae which join up to form the paired mesodermal gonoducts. These outgrowths can be interpreted as modified coelomoducts. The tenth pair in the male abdomen migrate forwards into the ninth abdominal segment to form the terminal parts of the male gonoducts leading to the gonopore. As in the Collembola, the gonopore in both sexes is formed as an ectodermal invagination. The similarity between thysanuran and pterygote genital development (see Chapter 7) is most striking.

Further Development of the Ectoderm

The embryonic ectoderm of apterygote embryos has not been well studied in any species. Although it seems likely that the same general pattern of ectodermal development is expressed in all apterygotes, we have no information bearing directly on this matter in the Diplura and very little on any collembolan. The recent workers on collembolan embryology have had little to say about the ectoderm, except for the point made by Garaudy-Tamarelle (1969a) that the gonopore of *Anurida maritima* forms as a short ectodermal invagination at the midventral, posterior margin of the fifth abdominal segment. Any further information on the further development of the ectoderm of Collembola must be gleaned from the studies of Claypole (1898) and Philiptschenko (1912), neither of which can now be regarded as wholly reliable on internal development. Both workers demonstrated, however, that the ventral ganglia are formed by neuroblasts (see below) and Claypole was also able to show that the segmental ganglia of the ventral nerve cord of *Anurida maritima* are the mandibular, maxillary and labial ganglia, which fuse to form the suboesophageal ganglion, the three thoracic ganglia, and six abdominal ganglia which fuse together in later development. Thus the terminal abdominal unit in Collembola, the sixth, has a distinct pair of ganglia even though it retains no paired somites. Claypole also demonstrated that the collembolan brain develops in the usual hexapod manner, from large paired protocerebral lobes, a pair of antennal deutocerebral ganglia and a pair of pre-mandibular tritocerebral ganglia which retain a post-oral commissure. The antennal and pre-mandibular ganglia develop from neuroblasts, but there is no suggestion that neuroblasts are involved in the development of the protocerebral ganglia arising from the head-lobe ectoderm. No ventral organs are associated with the developing ganglia in any apterygote hexapod.

In the collembolan labial segment, a pair of ventral ectodermal invaginations gives rise to the salivary glands. Philiptschenko (1912) interpreted these glands as mesodermal segmental organs, but it seems likely that this interpretation is incorrect. The development of mesodermal segmental organs in the head has not been observed by any other worker on apterygote or pterygote embryos, whereas many workers have described the development of ectodermal, labial salivary glands in the hexapods.

For the Thysanura, the work of Heymons (1897b), Woodland (1957), and Larink (1969, 1970) has given a much clearer understanding of the further development of the ectoderm (Fig. 84). The general product of the ectoderm is, of course, the hypodermis of the body wall, but at a very early stage of growth and segmentation of the germ band, segmental groups of neuroblasts differentiate to form the primordial cells of the ventral segmental ganglia. Four to five rows of neuroblasts differentiate by cell enlargement beneath the outer surface of the ectoderm on each side of the ventral midline, leaving small cells at the surface which subsequently form ventral hypodermal cells. Each neuroblast buds off small cells which build up as a radial row beneath the neuroblasts. With further proliferation, the rows of ganglion cells merge to form the paired ganglia of the segment and the neuroblasts eventually become indistinguishable. Along the midventral line between the ganglion rudiments, the ectoderm thickens to form a median strand, which is later incorporated into the ganglia as they move into the interior, away from the hypodermis. The mode of development of the neurilemma is unknown. Three gnathal, three thoracic and eleven abdominal ganglia are developed, of which the three gnathal ganglia fuse together in the usual way to form the suboesophageal ganglion, and the last three abdominal ganglia also fuse together. As in the Collembola, the terminal unit of the abdomen in the Thysanura thus has a distinct pair of ganglia, although it lacks distinct mesodermal somites.

In the head (Fig. 85), the antennal and pre-mandibular ganglia arise from neuroblasts and shift forwards relative to the mouth to give rise to the deutocerebrum and tritocerebrum of the brain. At the same time, each head lobe gives rise to two thickenings, one anteriorly, one more laterally. The thickenings separate into the interior, leaving hypodermis at the surface, and each lateral thickening subdivides into a lateral and a posterior ganglion. Together with the anterior thickenings, which form a pair of frontal ganglia, the lateral and posterior ganglia make up the protocerebral lobes of the brain. Neither superficial invaginations nor neuroblasts are involved in the development of the protocerebral ganglia. External to the posterior lobes of the latter, the ectoderm at the surface thickens and develops as the eyes. At the present time, no worker has distinguished a pair of pre-antennal ganglia during the development of the apterygote brain, but the level of analysis has not yet been sufficient to discount the possibility that pre-antennal ganglia exist in these animals.

Like the Collembola, the Thysanura develop a pair of salivary glands as ectodermal invaginations on the labial segment. In addition a pair of ectodermal invaginations develops at the anterior margin of the labium and forms the posterior tentorial arms, while a similar pair of invaginations in front of the median edges of the mandibles gives rise to the anterior

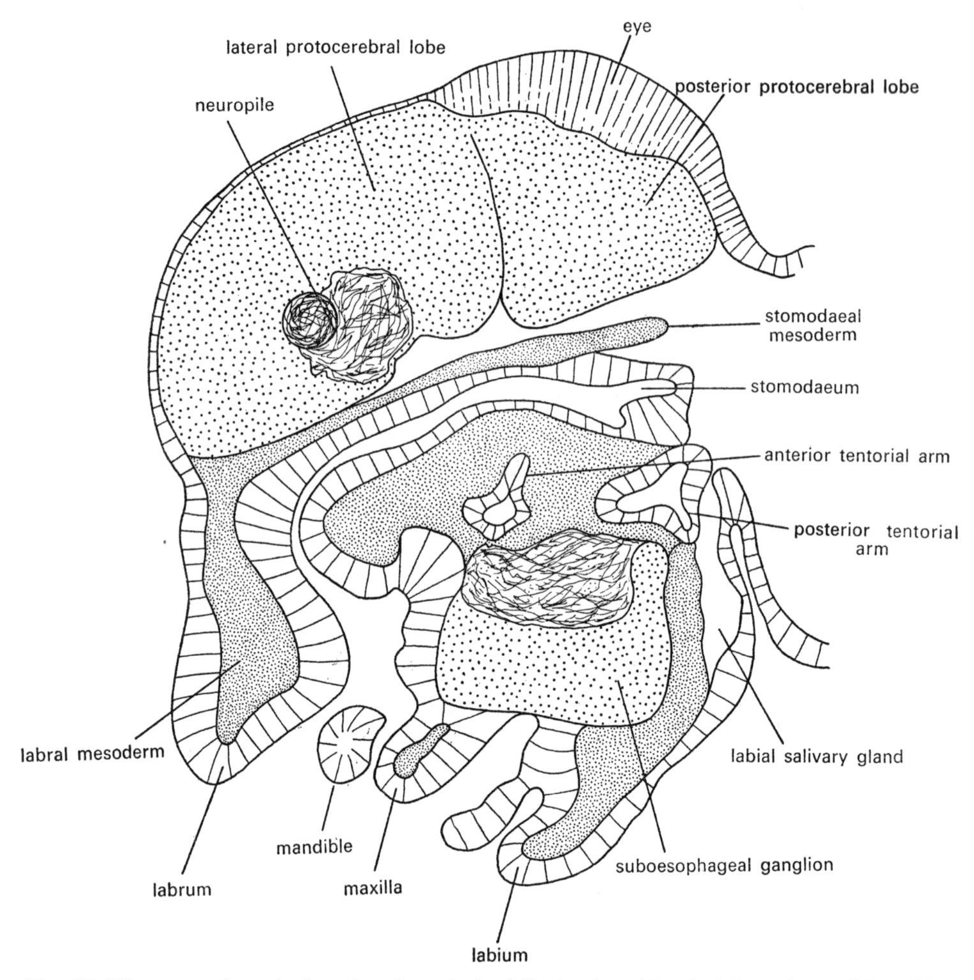

FIG. 85. Diagrammatic sagittal section through the fully developed head of the embryo of *Petrobius*, after Larink (1969).

tentorial arms (Larink, 1969, 1970). The two pairs of arms remain separate from one another in thysanurans, becoming fused only in the pterygotes (see Chapter 7).

A number of ectodermal invaginations also develops on the trunk. The mesothoracic, metathoracic and first eight abdominal segments each develop a pair of lateral invaginations which give rise to the tracheal system. Transient tracheal invaginations were also reported on the ninth and tenth abdominal segments of the embryo of *Lepisma* by Heymons (1897b) but were not found by Woodland (1957) in *Thermobia*. An opisthogoneate gonopore is formed as a short midventral invagination of ectoderm on the seventh (female) or ninth (male) abdominal segment.

Dorsal organs

We have already discussed, in describing the blastokinesis of the thysanuran embryo, how the serosa contracts to form a serosal dorsal organ in an anterodorsal position and is replaced by the amnion as a provisional dorsal closure. The process of blastokinesis does not occur in the Diplura or Collembola, which show, instead, a gradual dorsoventral flexure of the embryo. In these animals, an alternative type of dorsal organ, reminiscent of that of the Pauropoda and Symphyla, is developed (Figs. 75, 77, 78, 79, 80). The dipluran dorsal organ was first noticed by Grassi (1886) and Uzel (1898) and was described in detail by Tiegs (1942b) in *Campodea.* When the germ band has become distinct from the dorsal extra-embryonic ectoderm and has acquired cephalic and thoracic limb buds, a sheet of columnar, closely packed cells differentiates in the dorsal, extra-embryonic ectoderm on the dorsal surface of the yolk mass. The remainder of the extra-embryonic ectoderm simultaneously becomes much more attenuated. The large dorsal organ is closely associated with the anterior margins of the head lobes. Before dorsoventral flexure of the germ band begins, the sheet of cells invaginates and pushes down into the yolk, becoming a sphere with a central cavity opening to the dorsal surface by a small orifice. As invagination proceeds, the invagination cavity becomes filled with fluid and also with fine filaments secreted by the deeply columnar cells in the floor of the organ. The bundle of filaments penetrates through the opening of the dorsal organ and radiates out between the surface of the embryo and the blastodermal cuticle. The threads are fine and delicate and do not reach the equator of the egg.

The dorsal organ of *Campodea* persists until dorsoventral flexure of the germ band has taken place. During this process, it remains attached to the margin of the head lobes and is carried over the surface of the yolk mass to an eventual anteroventral position. Here the dorsal organ begins to degenerate, passing inwards during dorsal closure and eventually disintegrating within the haemocoel. The function of the dipluran dorsal organ is still unknown, but may be similar to the function of the more specialized dorsal organ of Collembola, which is now understood to some extent as a result of the work of Jura (1967b).

The dorsal organ of collembolan embryos is so conspicuous in the early embryo that it has been reported on by almost all workers on collembolan embryology (Uljanin, 1875b; Lemoine, 1883; Wheeler, 1893; Heymons, 1896; Claypole, 1898; Uzel, 1898; Folsom, 1900; Philiptschenko, 1912), but for a detailed description and investigation one has to turn to the recent studies by Tiegs (1942a) and Jura (1967b). Tiegs examined the dorsal organs of several species, confirming the indications from earlier workers that all collembolan dorsal organs are much alike. He found that the dorsal organ arises through the enlargement of a circular patch of cells on the dorsal surface of the blastoderm, before the germ band can be distinguished. The cells elongate and the dorsal organ gradually invaginates as the germ band becomes distinct, with progressive constriction of the invagination open. During invagination the cell boundaries disappear, the nuclei migrate to the deepest parts of the cells and the entire organ becomes mushroom-shaped, with fibrillar cytoplasm in the neck region.

By this time, in the small collembolan embryos studied by Tiegs, the chorion has ruptured and the blastodermal cuticle has expanded. The dorsal organ now begins to secrete radiating filaments which fan out beneath the blastodermal cuticle and soon reach the opposite pole of the egg. Fluid is secreted at the same time. When dorsoventral flexure begins, the secreted fluid fills the space created beneath the embryo by this flexure. The dorsal organ has already begun to regress before dorsoventral flexure begins, and is gradually withdrawn into the yolk mass during the process of dorsal closure. Eventually, it degenerates within the yolk.

In the unusually large embryo of *Tetradontophora bielanensis*, Jura (1967b) has found that the dorsal organ develops in the manner previously described by Tiegs and others, but reaches its maximum activity in fluid and filament secretion when dorsoventral flexure of the germ band is taking place. In this species, which may represent a primitive type of development among Collembola, the rupture of the chorion and swelling of the embryo within the blastodermal cuticle accompanies dorsoventral flexure. Jura gives experimental evidence to show that, in the absence of a functional dorsal organ, the embryo of *T. bielanensis* fails to undergo swelling or dorsoventral flexure. He has also shown that fluid is transferred to the exterior through the dorsal organ as a result of yolk digestion within the dorsal organ. It seems clear that the dorsal organ plays a part in the change of shape essential to flexure and elongation of the developing embryo and it can also be suggested that the radiating threads act in strengthening the blastodermal cuticle, which stretches to accommodate the change of shape. In small collembolan eggs, the swelling and the dorsal organ action are precocious. In the Diplura, swelling does not occur and the dorsal organ secretes only vestigial threads, though its action in fluid secretion is still important. In the Thysanura, with a different mode of development, this type of dorsal organ is absent.

Hatching

All apterygote embryos eventually hatch directly from their enclosing membranes, without the intervention of a pupoid phase within an embryonic cuticle such as occurs in many myriapods. The Collembola must rupture only the blastodermal cuticle in order to escape to a free life, but the Diplura still have to dispose of the chorion as well as the blastodermal cuticle at the end of development. In the Thysanura, the chorion again persists throughout development, but it seems likely that the blastodermal cuticle is digested by enzymes secreted by the pleuropodia during the later part of embryonic development, leaving only the chorion to be ruptured by the escaping first instar.

The duration of development is highly variable among apterygotes. *Tetradontophora bielanensis* does not hatch for 6 months (Jura, 1965), but *Isotoma cinerea* hatches after 20 days (Philiptschenko, 1912), *Campodea staphylinus* after 14 days (Uzel, 1898) and *Onychiurus bhattii* after only 6 days (Ashraf, 1969). Similarly, among the Thysanura, *Petrobius brevistylis* takes a full year over its embryonic development (Larink, 1969), but *Lepisma saccharina* hatches after about 8–9 weeks (Sharov, 1966) and *Thermobia domestica* completes its embryonic development in 10 days (Woodland, 1957).

The Basic Pattern of Development in Apterygotes

In order to bring embryological evidence to bear on the question of the relationships between the myriapods, the apterygote hexapods and the pterygotes, we need to establish whether or not there is a common theme of embryonic development within the aptergygotes. Superficially, this seems unlikely, since the Diplura and Collembola appear to be remarkably myriapod-like in their mode of development while the Thysanura share many of the specializations of pterygote embryos. Furthermore, the limitations of the available evidence make it difficult to propound a comprehensive statement of development variation within the apterygotes. The discussion of developmental variation in the myriapods presented in

Chapter 6 emphasized the great contribution made by a detailed knowledge of the mesoderm and nervous system to the elucidation of resemblances and differences between animals of this type, and such knowledge is not available for the apterygotes at the present time. Yet there are certain points which merit consideration and permit the arrival at useful tentative conclusions. The simplest mode of entry into the question of an underlying developmental unity amongst apterygotes is through a consideration of the development of the germ band.

Each class of apterygotes (excluding the Protura, which remain unknown embryologically) presents a variation on a fundamentally similar germ band. The germ band is a single unitary structure, lacking any vestige of ventral, extra-embryonic ectoderm. The head lobes are sharply rounded in outline, project slightly in a posterolateral direction and develop antennae at their posteromedian corners. The narrow anterior end of the post-antennal region behind the head lobes is the pre-mandibular segment, which develops transient limb buds. This is followed by mandibular, maxillary and labial segments with short, forwardly directed limb buds, three thoracic segments with longer, posteriorly directed limb buds, and a series of abdominal segments (probably basically eleven) with short limb buds which never develop as locomotory limbs in the manner of the thoracic limb buds, though some of them may later be involved in specialized jumping locomotion (e.g. in *Petrobius*, Collembola; see Manton, 1972).

Associated with this type of germ band, which also occurs in pterygotes, are a number of fundamental developmental features shared amongst the apterygotes and not found in myriapods. The segmental ventral ganglia develop from neuroblasts, differentiated in the ectoderm on either side of the ventral midline from the antennal to the last abdominal segment. Posterior tentorial arms arise as ectodermal invaginations on the labial segment, and a second pair of labial ectodermal invaginations gives rise to the salivary glands. The pre-mandibular mesodermal somites give rise to the suboesophageal body. The gonads develop through the segmental association of primordial germ cells with splanchnic mesoderm and the gonoducts arise as segmental outgrowths of abdominal splanchnic mesoderm (coelomoduct segmental organs) which link up to form paired ducts. Finally, the midgut epithelium in apterygote embryos develops from vitellophages within the yolk mass, while the fat-body cells arise from segmental somatic mesoderm.

If we assume that this pattern of development indicates a common archihexapod ancestry for extant apterygotes, then the large differences between the embryos of the different modern classes must be explainable as functional variations on a common theme. Here, as in the case of other groups of animals discussed in earlier chapters, the formation and fate of presumptive areas provides the key. All apterygotes develop a blastoderm around a yolk mass which contains vitellophages whose future fate is as midgut. The temporary intervention of total cleavage in the Collembola, as compared with intralecithal cleavage in the Diplura and Thysanura, does not change the outcome at the blastula stage. Furthermore, the modes of internal segregation of the vitellophages, varying from reinvading blastoderm cells (Diplura) to cells cut off by radial divisions in the forming blastoderm (Collembola) to energids left within the yolk mass during cleavage (Thysanura), are all acceptable as functional variations, once an origin of presumptive midgut as diffuse vitellophages has been established.

Within the blastoderm, a clear distinction can be made between dorsal, presumptive extra-embryonic ectoderm and ventral presumptive embryonic ectoderm, mesoderm, stomodaeum and proctodaeum. There is no need to be concerned with the precocious segregation of primordial germ cells in the centre of the yolk mass in the Collembola. The timing of the first visible differentiation of primordial germ cells is highly variable within groups of related animals, as seen, for instance, in pterygotes and crustaceans, and the cells can undergo extensive migrations without disturbing the functional morphological configuration of the embryo as a whole. An important problem arises, however, in comparing the fate maps oe the Diplura, Collembola and Thysanura, since the thysanuran map, when drawn for the differentiated blastoderm, appears to differ from those of the other two classes. There are no essential differences in the presumptive stomodaeum and proctodaeum, but the presumptive ectoderm and mesoderm appear to be remarkably different. Firstly, the presumptive ectoderm of the Diplura and Collembola is long and broad and already includes much of the ectoderm of the germ band. In contrast, that of the Thysanura is small in area and includes only the ectoderm of the anterior part of the head and a growth zone. Secondly, the presumptive mesoderm of the Diplura and Collembola is diffusely spread throughout the presumptive ectoderm, while that of the Thysanura is compact, distinct from the ectoderm and situated in the ventral midline between the presumptive stomodaeum and proctodaeum. A proper approach, however, can reduce these differences to the status of acceptable functional variations.

The presumptive ectoderm of the Diplura includes the ectoderm of the head, thorax and first six abdominal segments, followed by a segment-forming growth zone. The presumptive ectoderm of the Collembola differs from this only in the absence of the growth zone. In both types of embryo, once the mesoderm has migrated inwards to form an inner layer, the ectoderm thickens by generalized cell division, without preliminary aggregation and decrease in area. The Thysanura, in contrast, like the primitive pterygotes (see Chapter 7), establish their presumptive ectoderm first in a thinly spread blastoderm and undergo a preliminary period of aggregation of the cells into a compact embryonic primordium occupying a much smaller area on the surface of the yolk mass. If the comparison of presumptive ectoderm is made before the thysanuran blastoderm has undergone this preliminary differentiation, an impression is gained immediately of a much closer resemblance to the pattern detectable in the Diplura and Collembola. The only difference, in fact, is the concentration of all post-antennal ectoderm into a proliferative growth zone. Why the thysanuran should have evolved this pattern of specialization of their ectodermal development is not clear. Neither is it obvious why aggregation of the presumptive areas into a compact embryonic primordium has replaced direct thickening in the initial extended distribution. Since the pterygotes also share these features, we can be sure that they offer some functional advantage in later development. One possibility is that the are functional precursors of amnion formation, which itself is advantageous; but since we do not yet know the functional roles of the embryonic membranes of pterygotes, no useful opinion can be given.

The difference in distribution of the presumptive mesoderm of the Thysanura as compared with the Diplura and Collembola requires much briefer discussion. It is not difficult to envisage functional advantages in a precise segregation of midventral presumptive mesoderm between bilateral ectoderm, rather than a diffuse release of mesoderm cells throughout the area of presumptive ectoderm, and the former can easily be interpreted as a functional modification of the latter. The same specialization is also observed in the pterygote hexapods (Anderson, 1972 a, b, and Chapter 7). We are thus led to the conclusion that the mode of formation and the pattern of relative juxtaposition of presumptive areas are not incompatible with the idea of a common basis for all apterygote development. Further support for this view is given by a consideration of the development of the head, which will be treated below. It also becomes clear that the Diplura retain the most generalized mode of development of the three classes under discussion, with the Collembola exhibiting a few specializations (e.g. total cleavage, precocious segregation of primordial germ cells, absence of a segment-forming growth zone) and the Thysanura displaying a different and much greater range of specializations, which they share with the Pterygota.

It now remains to discuss the later phases of development and the relationship between the germ band and the yolk mass. The Collembola differ little from the Diplura in this respect. Dorsoventral flexure sets in earlier and is attained in a slightly different way, probably reflecting the secondary abbreviation of the abdomen. Associatedly, the dorsal organ, which appears to have a functional role in dorsoventral flexure, is larger and more complex in the Collembola than in the Diplura. The development of limbs on the abdomen follows a more specialized course, with the second pair being resorbed in association with specialization of the first pair as a ventral tube, the third pair usually persisting as the hamula and the fourth pair as the furcula. In other respects, development is the same, suggesting that the Collembola are a neotenic stock with an ancestry close to that of the Diplura among the ancestral hexapods. The Thysanura, in contrast, combine their extreme shortening of the embryonic primordium and prolonged activity of a segmental growth zone with a highly precocious dorsoventral flexure and the associated proliferation of an amnion. In the continuation of this specialization, the extra-embryonic yolk-sac ectoderm or serosa later contracts and is replaced by the amnion as a provisional dorsal closure, accompanied by restoration of the flexed, segmented germ band to a normal superficial position relative to the yolk mass. Lacking evidence of intermediates, there is little point in speculating how the specialized mode of development in the Thysanura might have evolved from that of ancestral hexapods. All that can be said with a reasonable degree of certainty is that there is no reason to assume a dipluran ancestry for the Thysanura and there is strong evidence of an early divergence of the Thysanura from the ancestors of the Diplura and Collembola. One primitive feature retained by the Thysanura in common with the Diplura is the persistence of abdominal limbs as styles. Only the first pair, which develop as temporary pleuropodia, show unusual specialization.

Development of the Head

The general indication of divergence from a common ancestral hexapod stock, evident throughout the embryology of the apterygotes, is borne out further by the mode of development of the head. Although the functional configuration of the mouth-parts differs greatly in the four classes of apterygotes (Manton, 1964, 1970, 1972), the segmental composition and mode of development of the basic components of the head are identical in the Diplura, Collembola and Thysanura (unknown in Protura), despite the remarkable variations in other aspects of their embryology. Externally, as we have seen, the head rudiments comprise a pair of rounded head lobes with posteromedial antennal rudiments, a pre-mandibular segment region and three gnathal segments. Internally lie paired antennal, pre-mandibular and gnathal somites. No distinct pre-antennal somites have yet been observed in the pre-antennal mesoderm. The further development of the cephalic mesoderm is not well known, but there is evidence that the antennal somites contribute the anterior aorta and the pre-mandibular somites develop into the suboesophageal body.

Invagination of the stomodaeum, overgrowth of the labrum and forward migration of the antennal rudiments to a pre-oral position all take place while the segments behind the antennal segment are still well spread out along the ventral midline. Mandibular, maxillary and labial limb buds develop early, simultaneously with or shortly after the antennae, and conspicuous, transient pre-mandibular limb buds arise while the antennae perform their pre-oral migration. Only after this stage do the mouth-parts begin to crowd forwards around the mouth and exhibit their characteristic class specializations. Meanwhile, of course, the head lobes have developed the paired protocerebral ganglia, in the same way in all classes (compare the variations in mode of protocerebral ganglion formation among the myriapods), the deutocerebral and tritocerebral ganglia have developed from neuroblasts, and the gnathal ganglia have developed from neuroblasts and fused to form the suboesophageal ganglia. As previously noted, no pre-antennal ganglia have been found, so that there is no direct evidence of a pre-antennal segment in the apterygote head, but it should be borne in mind that the pterygote head, which develops in precisely the same manner, does show slight evidence of a vestigial pre-antennal segment (see Chapter 7), and similar evidence may yet come to light in more detailed studies of the apterygotes.

When one adds the facts of posterior tentorial arms and salivary glands developing as ectodermal invaginations on the labial segment of all apterygotes, it is reasonable to conclude that the precise and complex sequence of events described above as the common mode of development of the apterygote head is an indication of common ancestry. Taken together with the other evidence of a fundamental commonality of development, discussed above, this development of the head provides further support for the view, based on embryology, that all extant apterygotes diverged from an ancestral lobopod stock which was already labiate, with the Diplura and Collembola being more closely related to one another than either is to the Thysanura. In the following chapter we shall see that the Thysanura are related to the Pterygota, which also diverged from the same labiate stock.

References

ANDERSON, D. T. (1969) On the embryology of the cirripide crustaceans *Tetraclita rosea* (Krauss), *Tetraclita purpurascens* (Wood), *Chthamalus antennatus* Darwin and *Chamaesipho columna* (Spengler) and some considerations of crustacean phylogenetic relationships. *Phil. Trans. Roy. Soc.* B, **256,** 183–235.

ANDERSON, D. T. (1972a) The development of hemimetabolous insects. In COUNCE, S. J. (ed.), *Developmental Systems—Insects,* Academic Press, New York.

ANDERSON, D. T. (1972b) The development of holometabolous insects. In COUNCE, S. J. (ed.), *Developmental Systems—Insects,* Academic Press, New York.

ASHRAF, M. (1969) Studies on the biology of Collembola. *Revue d'écol. biol. sol* **6,** 361–72.

BRUCKMOSER, P. (1965), Untersuchungen über den Kopfbau der Collembole *Orchesella villosa* L. *Zool. Jb. Anat. Ont.* **82,** 299–364.

CLAYPOLE, A. M. (1898) The embryology and oogenesis of *Anurida maritima*. *J. Morph.* **14,** 219–300.

FOLSOM, J. W. (1900) The development of the mouthparts of *Anurida maritima*. *Bull. Mus. comp. Zool., Harvard* **36,** 87–158.

GARAUDY, M. (1967) Quelques observations sur l'apparition et le développement des ébauches appendiculaires de l'abdomen chez les embryons du collembole *Anurida maritima* Guér. *Actes Soc. Linn. Bordeaux* **104,** ser. A (2), 1–13.

GARAUDY-TAMARELLE, M. (1969a) Quelques observations sur le développement embryonnaire de l'ébauche genitale chez le collembole *Anurida maritima* Guérin. *C.R. Acad. Sci., Paris* **268,** 945–7.

GARAUDY-TAMARELLE, M. (1969b) Les vesicules coelomiques du segment intercalaire (= premandibulaire) chez les embryons du collembole *Anurida maritima* Guér. *C.R. Acad. Sci., Paris* **269,** 198–200.

GARAUDY-TAMARELLE, M. (1970) Observations sur la ségregation de la ligne germinale chez le collembole *Anurida maritima* Guérin. Explication de son caractère intravitellin. *C.R. Acad. Sci., Paris* **270,** 1149–51.

GARAUDY-TAMARELLE, M. (1971) Principales étapes du développement embryonnaire chez le collembole *Anurida maritima*. *Revue d'écol. biol. sol* **8,** 159–62.

GRASSI, B. (1885) I progenitori degli insetti e dei miriapodi, l'*Japyx* e la *Campodea*. *Atti accad. Gioenia di Sci. Nat. Catania* **19,** 1–83.

HAGET, A. and GARAUDY, M. (1964) Quelques precisions sur les "appendices intercalaires" de l'embryon du collembole *Anurida maritima* Guér. *C.R. Acad. Sci., Paris* **258,** 3364–6.

HEYMONS, R. (1896) Ein Beitrag zur Entwicklungsgeschichte der Insecta apterygota. *Sitzb. K. Preuss. Akad. Wiss., Berlin* **51,** 1385–9.

HEYMONS, R. (1897a) Ueber die Bildung und den Bau des Darmkanals bei niederen Insekten. *Sitzb. d. Ges. Naturf. Freunde, Berlin* **1897,** 111–19.

HEYMONS, R. (1897b) Entwicklungsgeschichteliche Untersuchungen an *Lepisma saccharina* L. *Z. wiss. Zool.* **62,** 583–631.

HEYMONS, R. and HEYMONS, H. (1905) Die Entwicklungsgeschichte von *Machilis*. *Verh. deut. Zool. Ges.* **15,** 123–35.

HOFFMAN, R. W. (1911) Zur Kenntnis der Entwicklungsgeschichte der Collembolen. *Zool. Anz.* **37,** 353–77.

JOHANNSEN, O.A. and BUTT, F.H. (1941) *Embryology of Insects and Myriapods*, McGraw-Hill, New York.

JURA, C. (1965) Embryonic development of *Tetradontophora bielanensis* (Waga) (Collembola) from oviposition till germ band formation. *Acta Biol. Cracow Zool.* **8,** 141–57.

JURA, C. (1966) Origin of the endoderm and embryogenesis of the alimentary system in *Tetradontophora bielanensis* (Waga) (Collembola). *Acta Biol. Cracow Zool.* **9,** 95–102.

JURA, C. (1967a) Origin of germ cells and gonad formation in embryogenesis of *Tetradontophora bielanensis* (Waga) (Collembola). *Acta Biol. Cracow Zool.* **10,** 97–103.

JURA, C. (1967b) The significance and function of the primary dorsal organ in embryonic development of *Tetradontophora bielanensis* (Waga) (Collembola). *Acta Biol. Cracow Zool.* **10,** 301–11.

JURA, C. (1972) The development of apterygote insects. In COUNCE, S. J. (ed.), *Developmental Systems—Insects*, Academic Press, New York.

LARINK, O. (1969) Zur Entwicklungsgeschichte von *Petrobius brevistylis* (Thysanura, Insecta). *Helgolander wiss. Meeresunters* **19,** 111–55.

LARINK, O. (1970) Die Kopfentwicklung von *Lepisma saccharina* L. (Insecta, Thysanura). *Z. Morph. Tiere* **67,** 1–15.

LEMOINE, M. (1883) Recherches sur le développement des podurelles, *C.R. Assoc. France l'Avanc. Sci.* **11.** 483–520.

MANTON, S. M. (1964) Mandibular mechanisms and the evolution of the arthropods. *Phil. Trans. Roy. Soc.* B, **247,** 1–183.

MANTON, S. M. (1970) Arthropods: Introduction. *Chemical Zoology* **5,** 1–34.

MANTON, S. M. (1972) The evolution of arthropodan locomotory Mechanisms, Part 10. *J. Linn. Soc. Zool.* **51,** 203–400.

PHILIPTSCHENKO, J. (1912) Beiträge zur Kenntnis der Apterygoten. III. Die Embryonalentwicklung von *Isotoma cinerea*. *Z. wiss. Zool.* **103,** 519–660.

SAHRHAGE, D. (1953) Oekologische Untersuchungen an *Thermobia domestica* (Packard) und *Lepisma saccharina* L. *Z. wiss. Zool.* **157,** 77–168.

SHAROV, A. G. (1953) Development of bristle tails (Thysanura, Apterygota) in connection with the problem of insect phylogeny. *Trud. inst. morf. zhivot.* **8,** 63–127.

SHAROV, A. G. (1966) *Basic Arthropod Stock*, Pergamon Press, Oxford.

TIEGS, O. W. (1940) The embryology and affinities of the Symphyla, based on a study of *Hanseniella agilis*. *Q. Jl. microsc. Sci.* **82,** 1–225.

TIEGS, O. W. (1942a) The "dorsal organ" of collembolan embryos. *Q. Jl. microsc. Sci.* **83,** 153–70.

TIEGS, O. W. (1942b) The "dorsal organ" of the embryo of *Campodea*. *Q. Jl. microsc. Sci.* **84**, 35–47.

TIEGS, O. W. (1947) The development and affinities of the Pauropoda, based on a study of *Pauropus sylvaticus*. *Q. Jl. microsc. Sci.* **88**, 165–267 and 275–336.

ULJANIN, M. (1875) Beobachtung über die Entwicklung der Poduren. *Arch. Zool. éxp. gen., Notes et Revue* **4**, 39–4.

ULJANIN, M. (1876) Beobachtung über die Entwicklung der Poduren. *Arch. Zool. éxp. gen., Notes et Revue* **5**, 17–19.

UZEL, H. (1897a) Vorlaufige Mittheilungen über die Entwicklung der Thysanuren. *Zool. Anz.* **20**, 125–32.

UZEL, H. (1897b) Beiträge zur Entwicklungsgeschichte von *Campodea staphylinus*. *Zool. Anz.* **20**, 232–7.

UZEL, H. (1898) *Studien über die Entwicklung der apterygoten Insekten*, Friedlander, Berlin.

WELLHOUSE, W. T. (1953) The embryology of *Thermobia domestica* Packard (Thysanura). *Iowa State Coll. J. Sci.* **28**, 416–17.

WHEELER, W. M. (1893) A contribution to insect embryology. *J. Morph.* **8**, 1–160.

WOODLAND, J. T. (1957) A contribution to our knowledge of lepismatid development. *J. Morph.* **101**, 523–77.

ZAKHVATKIN, Y. A. (1969) The morphology of cleavage of the Collembola egg. *Zool. Zhur., Moscow* **48**, 1029-40.

CHAPTER 7

PTERYGOTE INSECTS

IN CONTRAST to the Myriapoda and the apterygote insects, the pterygote insects have been the focus of a large amount of embryological research for almost a century. During this time, more than 400 hundred significant papers have been published on the descriptive embryology of the group. The reasons for this are not difficult to discern. Many pterygote species are of economic significance in human affairs and their embryology has been closely studied as part of the investigations underlying control methods. The easy availability of pterygotes, their often short life-cycles and their favorable response to laboratory culture have also made them attractive to embryologists. Furthermore, the pterygotes display a fascinating diversity in their embryonic development, ranging from the large yolky embryos of the more primitive orders to instances of specialized polyembryony and viviparity in orders of more recent origin. Comprehension of this diversity is a continuing challenge.

In the other chapters of the present book, I have endeavoured to provide a comprehensive survey of the variations in modes of development of the animals to which each chapter is devoted, before elucidating a basic theme of development for the group. Fortunately, from the aspect of brevity, this approach is not required for the pterygotes. Anderson (1972 a, b) and Ivanova-Kazas (1972) have recently reviewed the descriptive embryology of pterygotes in detail, giving full place and prominence to all of the remarkable specializations of embryonic development that occur among the pterygote orders. Among other things, their work shows that beneath the striking differences in development of the various orders, a characteristic theme of embryonic development is strongly maintained. I shall therefore concentrate in the present context mainly on the evidence bearing on this theme, with little reference to the more modified modes of development manifested in such orders as the Hymenoptera, Diptera and Lepidoptera. For the sake of completeness, however, the papers bearing on the more specialized orders are included in the following lines, so that their place in the historical development of pterygote embryology can be appreciated.

In spite of easy availability, pterygote embryos present a number of technical problems to the investigator. Their tough external membranes, the outer chorion and inner vitelline membrane, are a barrier to many fixatives. Their yolk, voluminous in many species, creates difficulties in the preparation of histological sections. The cells and tissues of the embryonic rudiments also tend to be small, especially during the middle stages of embryonic development, so that very careful observation is necessary to distinguish one part from another. The early students of insect embryology in the eighteen-seventies and eighteen-eighties were much troubled by these refractory properties of insect eggs and the yolky embryos of the hemimetabolous insects only began to yield satisfactorily to investigation at the hands of two outstanding workers of the time, Wheeler (1889a, 1893) in America and Heymons (1895, 1896 a, b, 1897, 1899, 1912) in Germany. Although many attempts were made by other investigators, the work of Wheeler and Heymons was not augmented satisfactorily for hemimetabolous insects until the nineteen-twenties. The embryology of the holometabolous species fared a little better, in that the foundations of understanding were laid for several orders during this period (Coleoptera, Lecaillon, 1897 a, b, c, 1898; Hirschler, 1909; Hegner, 1909, 1910, 1911, 1912; Korschelt, 1912; Strindberg, 1913; Blunck, 1914; Hymeoptera, Carrière and Bürger, 1897; Nelson, 1915; Strindberg, 1914, 1915c, 1917; Hegner, 1914, 1915; Diptera, Escherich, 1900 a, b, 1901 a, b, 1902; Noack, 1901; Lepidoptera, Toyama, 1902), but no progress was made with the more generalized holometabolan embryos of the Neuroptera, Megaloptera, Mecoptera and Trichoptera. It is not surprising, therefore, that controversy was rife in classical pterygote embryology (see Johannsen and Butt, 1941), since diversity was dominant and unity elusive.

Following the improvements in histological and microscopical techniques made after the First World War, pterygote embryology entered a new phase with the work of Seidel (1924) on the bug *Pyrrhocoris* and Leuzinger and Wiesmann (1926) on the phasmid *Carausius*. Like the early papers by Wheeler and Heymons, these were outstanding contributions that still have an important place in the literature of insect embryology. With gradually increasing impetus during the nineteen-thirties, new information was then accumulated by many workers. Slifer (1932 a, b, 1937), Nelsen (1931, 1934) and especially Roonwal (1936, 1937) and Krause (1938 a, b, 1939) did much to clarify orthopteran development. Searching studies were also carried out by Mellanby (1936, 1937) on the heteropteran *Rhodnius* and by Miller (1939, 1940) on the plecopteran *Pteronarcys*. The holometabolous insects were even more intensively attacked. Paterson (1931, 1932, 1936), Smreczyński (1932, 1934, 1938), Inkmann (1933), Mansour (1934), Butt (1963), Wray (1937) and Tiegs and Murray (1938) described the development of a wide variety of Coleoptera. Eastham (1927, 1930), Johannsen (1929), Sehl (1931), Saito (1934, 1937) and Mueller (1938) made excellent progress with the peculiar embryos of the Lepidoptera. The small, highly specialized embryos of the Diptera were studied by Hardenburg (1929), Du Bois (1932), Gambrell (1933), Butt (1934), Lassman (1936) and Poulson (1937), those of the Siphonaptera by Kessel (1939) and those of the Hymenoptera by Henschen (1928), Schnetter (1934) and Speicher (1936). Towards the end of this decade, two important papers were also published on holometabolan orders with a generalized mode of embryonic development, the Megaloptera (Du Bois, 1938) and the Neuroptera (Bock, 1939). The advances made during this productive period of pterygote embryology provided the material for the well-known textbook by Johannsen and Butt (1941), much used as a standard reference in the ensuing thirty years.

At the same time, it must be realized that the years since the Second World War have seen further great advances in our understanding of insect embryos. Certain orders, the Ephemeroptera, Plecoptera, Embioptera, Dermaptera, Mecoptera and Trichoptera, remain little known, since their embryos are difficult to obtain and slow to develop. Their neglect must be regarded as a matter of some regret, for the fragmentary information available on each hints that these orders are the repositories of relatively generalized modes of development within the hemimetabolous and holometabolous insects. In the case of the more accessible orders, however, the picture is entirely different. Among hemimetabolous orders, the orthopterans and cheleutopterans have been thoroughly investigated (Moscona, 1950; Kanellis, 1952; Ibrahim, 1957; Bergerard, 1958; Mahr, 1960; Louvet, 1964; Scholl, 1965, 1969; Fournier, 1967; Cavallin, 1969, 1970; Malzacher, 1968; Vignau-Rogueda, 1969; Bedford, 1970; Stringer, 1969). Hemipteran development has also been further clarified by Butt (1949), Sander (1956, 1959), Behrendt (1963), Voegele (1967), Cobben (1968) and Mori (1969, 1970), while detailed single contributions have transformed our knowledge of the embryos of mantids (Görg, 1959), Isoptera (Striebel, 1960), Odonata (Ando, 1962), Psocoptera (Goss, 1952, 1953), Anoplura (Piotrowski, 1953) and Thysanoptera (Bournier, 1960). The major holometabolous orders have been even more intensively studied. The development of the Coleoptera is now extremely well known, due to the work of Brauer (1949), Krause and Ryan (1953), Luginbill (1953), Haget (1953, 1955, 1957), Jura (1957), Surowiak (1958), Dobrowski (1959), Such and Haget (1962), Such (1963), Ullman (1964, 1967), Jung (1966 a, b), Küthe (1966), Zakhvatkin (1967 a, b, 1968), Rempel and Church (1965, 1969, 1971 a, b) and Ressouches (1969). Studies on the Lepidoptera, although confined mainly to the important pest family Tortricidae, have provided answers to many of the problems still unsolved in the nineteen-thirties (Presser and Rutschky, 1957; Okada, 1960; Stairs, 1960; Bassand, 1965; Guénnelon, 1966; Reed and Day, 1966; Srivastava, 1966; Anderson and Wood, 1968). The Diptera have been widely investigated by such workers as Sonnenblick (1950), Poulson (1950), Ede and Counce (1956), Anderson (1962, 1963, 1964, 1966), Schoeller (1964), Craig (1967), Davis (1967) and Guichard (1971), and the Hymenoptera by Baerends and Baerends-von Roon (1950), Shafiq (1954), Ivanova-Kazas (1954, 1958, 1959, 1960), Ando and Okada (1958), Bronskill (1959, 1964), Ochiai (1960), Amy (1961), Farooqui (1963) and Dupraw (1967). With so many new facts to hand for comparison, it is not surprising that the theme and variations of pterygote embryonic development are clearer now than at any previous time.

Cleavage

Since the hemimetabolous insects are a diverse assemblage (e.g. Smart, 1963), I shall simplify the present discussion by employing the term primitive Hemimetabola for the palaeopteran orders (Odonata and Ephemeroptera) and polyneopteran orders (Dictyoptera, Isoptera, Plecoptera, Cheleutoptera, Orthoptera, Embioptera and Dermaptera), but retain the term Paraneoptera for those orders recognized under this heading (Psocoptera, Mallophaga, Anoplura, Thysanoptera, Homoptera and Heteroptera). On the other hand, I shall employ the term Holometabola, as the more familiar term than its synonym Oligoneoptera, for the endopterygote orders (Coleoptera, Megaloptera, Neuroptera, Mecoptera, Trichoptera, Lepidoptera, Diptera, Siphonaptera and Hymenoptera). The value of this simplification becomes immediately apparent in a consideration of the generalizations that can be made about pterygote eggs.

In spite of wide variations in shape, the eggs of the primitive Hemimetabola always have at least one dimension close to or greater than 1 mm. The generalized shape is ovoid, with distinct anteroposterior and dorsoventral axes, and it is not uncommon for the ventral

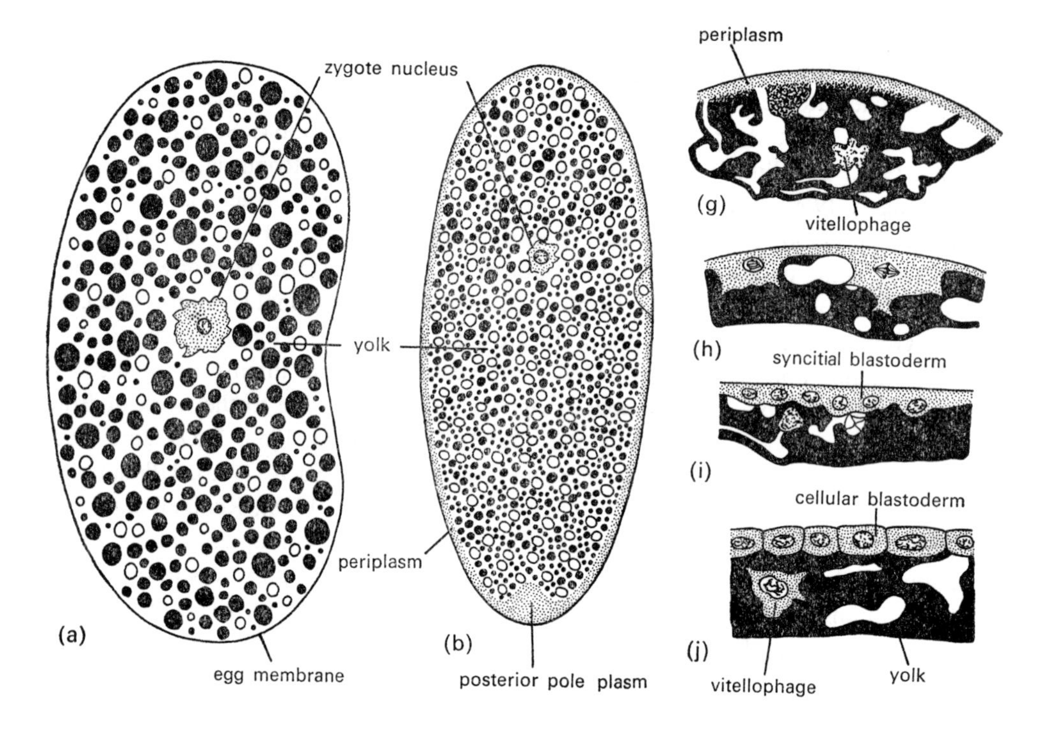

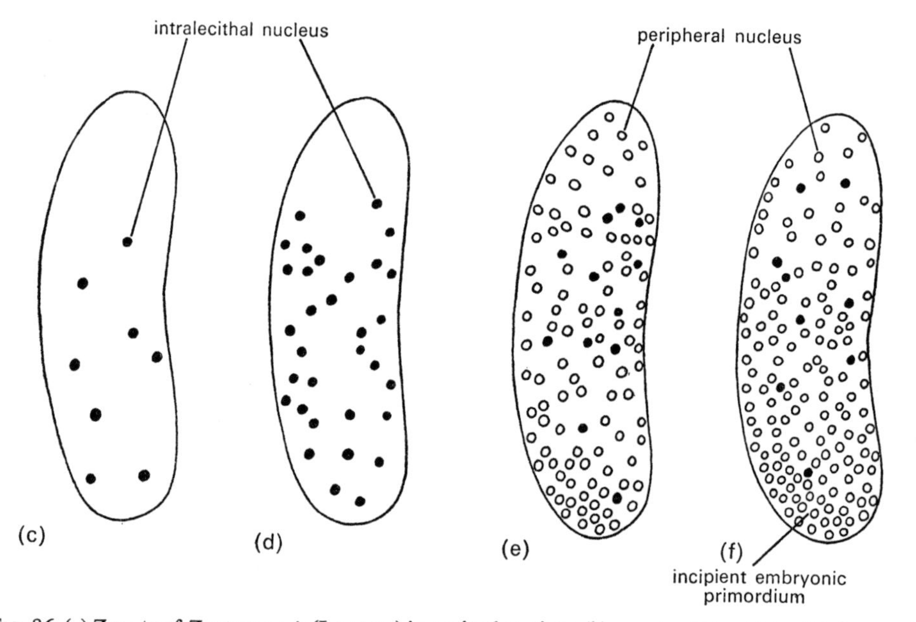

FIG. 86. (a) Zygote of *Zootermopsis* (Isoptera) in sagittal section; (b) zygote of *Bruchidius* (Coleoptera) in sagittal section; (c)–(f) stages in cleavage of *Kalotermes* (Isoptera) in diagrammatic left lateral view [(c) 8-nucleus stage; (d) 32-nucleus stage; (e) 140-nucleus stage; (f) 194-nucleus stage); (g)–(j) stage in blastoderm formation of *Rhodnius* (Heteroptera) in transverse section. [(a) and (c)–(f) after Striebel, 1960; (b) after Jung, 1966a; (g)–(j) after Mellanby, 1936; modified from Anderson, 1972a.]

surface to be convex and the dorsal surface concave (Fig. 86). Some representative egg lengths in this group are: *Platycnemis pennipes* (Odonata, Seidel, 1929), 1·00 mm; *Kalotermes flavicollis* (Isoptera, Striebel, 1960), 1·22 mm; *Tachycines asynomorus* (Orthoptera, Mahr, 1960), 2·6 mm; *Carausius morosus* (Cheleutoptera, Thomas, 1936), 5·0 mm. The egg contains relatively little cytoplasm, the bulk of its substance being made up of closely packed yolk spheres. The nucleus is usually centrally placed in a small cytoplasmic halo. Eggs of this type have been described by many workers (e.g. Wheeler, 1889a, 1893; Heymons, 1895, 1896 a, b; Giardina, 1897; Tschuproff, 1903; Strindberg, 1913; Kershaw, 1914; Hagan, 1917; Leuzinger and Wiesmann, 1926; Roonwal, 1936; Thomas, 1936; Krause, 1938 a, b, 1939; Miller, 1939; Johannsen and Butt, 1941; Brookes, 1952; Kanellis, 1952; Ibrahim, 1957; Bergerard, 1958; Görg, 1959; Mahr, 1960; Striebel, 1960; Stefani, 1960; Ando, 1962; Bhatnagar and Singh, 1965; Sauer, 1966; Fournier, 1967; Illies, 1968; Khoo, 1968; Bedford, 1970). Varying reports have been given of the existence of a periplasm at the surface of the egg. A thin periplasm has been identified in certain odonatan, orthopteran and dermapteran eggs (Heymons, 1895; Roonwal, 1936; Johannsen and Butt, 1941; Bhatnagar and Singh, 1965), but is said to be absent in other species of these orders (Seidel, 1929; Mahr, 1960) and in representative Dictyoptera, Isoptera and Plecoptera (Seidel, 1929; Görg, 1959; Striebel, 1960).

The eggs of the more recently evolved group of hemimetabolous pterygotes, the Paraneoptera, retain most of the characteristics described above (Shinji, 1919; Seidel, 1924; Mellanby, 1936; Schölzel, 1937; Krause, 1939; Böhmel and Jancke, 1942; Butt, 1949; Goss, 1952, 1953; Piotrowski, 1953; Sander, 1959; Bournier, 1960; Behrendt, 1963; Cobben, 1968), although the smaller insects of this group tend to lay eggs with dimensions less than 1 mm (e.g. the psocopteran *Liposcelis*, 340 μ; the homopteran *Aphis*, 550 μ). In one respect, however, paraneopteran eggs are more highly organized than those of the primitive Hemimetabola. A cytoplasmic reticulum enmeshes their yolk spheres and merges at the surface with a distinct, yolk-free periplasm. The same specialization is convergently expressed in the eggs of the Holometabola. Here, too, the dimensions of the egg are generally small, being frequently less than 1 mm. The egg is moderately yolky in generalized orders, such as the Coleoptera (Fig. 86), Megaloptera and Neuroptera (Hirschler, 1909; Strindberg, 1915a; Paterson, 1931; Inkmann, 1933; Butt, 1936; Du Bois, 1936, 1938; Tiegs and Murray, 1938; Bock, 1939; Kessel, 1939; Ando, 1960; Ullman, 1964; Rempel and Church, 1965; Jung, 1966a), but the more specialized holometabolans such as the Lepidoptera, Diptera and Hymanoptera show a relative reduction of yolk, a preponderance of the cytoplasmic reticulum and a conspicuous periplasm (Nelson, 1915; Huie, 1918; Eastham, 1927, 1930; Pauli, 1927; Johannsen, 1929; Du Bois, 1932; Lautenschlager, 1932; Schnetter, 1934; Saito, 1934, 1937; Christensen, 1943 a, b; Sonnenblick, 1950; Rempel, 1951; Formigoni, 1954; Müller, 1957; Presser and Rutschky, 1957; Ando and Okada, 1958; Ivanova-Kazas, 1958, 1959; Bronskill 1959, 1964; Geyer-Duzynska, 1959; Idris, 1960; Reinhardt, 1960; Yajima, 1960; Amy, 1961; Anderson, 1962; Bassand, 1965; Guénnelon, 1966; Davis, 1967; Dupraw, 1967). There are also, of course, instances of more extreme secondary yolk loss in those pterygotes whose development is viviparous or parasitic (e.g. Hagan, 1951; Ivanova-Kazas, 1965, 1972; Counce, 1966), but the peculiarities of these forms need not concern us in the present context.

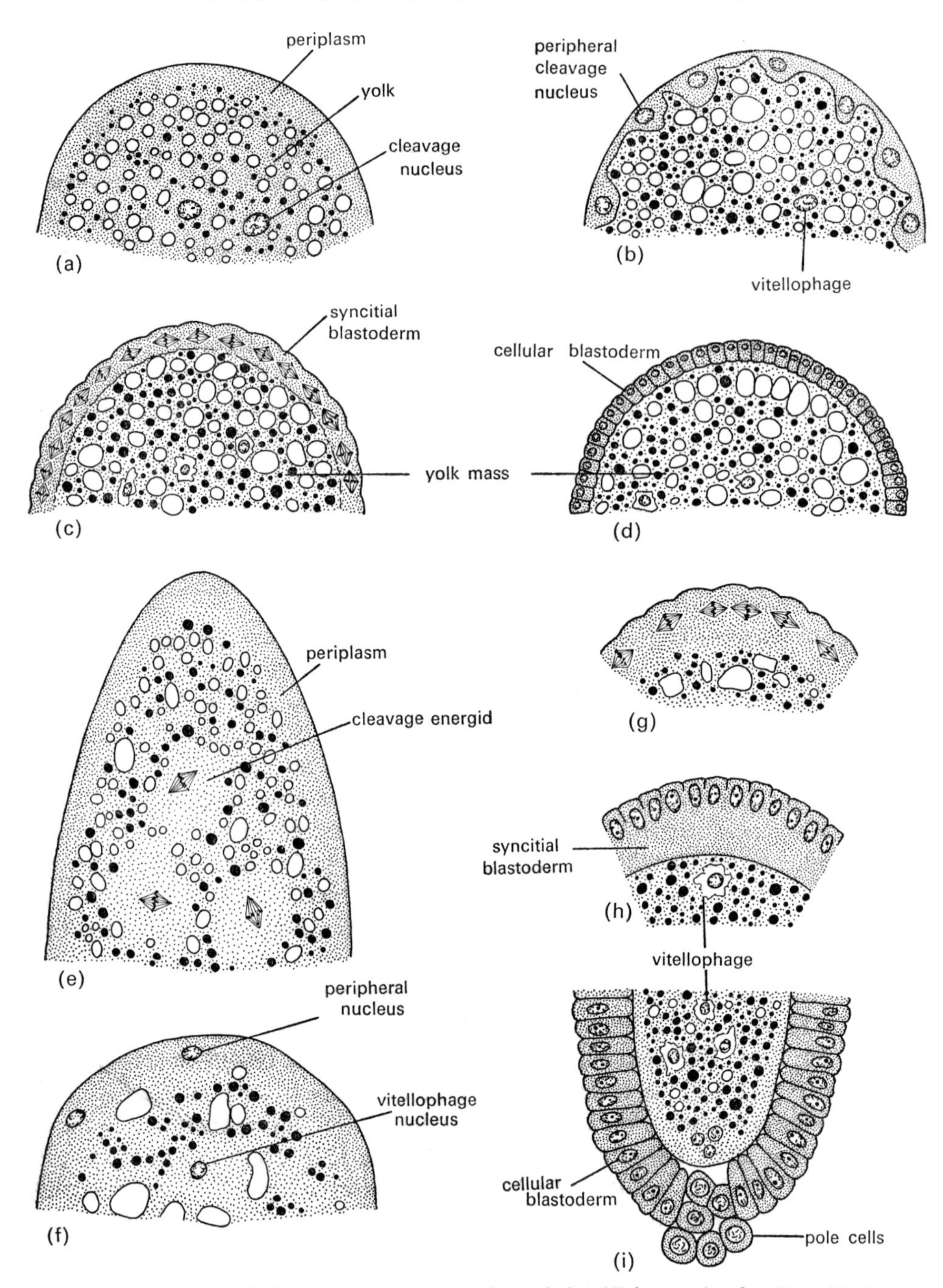

FIG. 87. (a)–(d) Stages in blastoderm formation of *Bruchidius* (Coleoptera), after Jung (1966a), in transverse section [(a) 2-nucleus stage; (b) 256-nucleus stage; (c) approximately 4000-nucleus stage; (d) blastoderm]; (e)–(i) stages in blastoderm formation of *Dacus* (Diptera), after Anderson (1962), in section [(e) 16-nucleus stage, longitudinal; (f) 512-nucleus stage, transverse; (g) approximately 4000-nucleus stage, transverse; (h) approximately 8000-nucleus stage or syncitial blastoderm, transverse; (i) posterior end of cellular blastoderm and pole cells, longitudinal. Modified from Anderson, 1972b].

Even though the eggs of pterygotes vary markedly in size and structural configuration, their basic and almost universal mode of cleavage is the same intralecithal mode described in the preceding chapter for the Thysanura (Figs. 86, 87). Nuclear mitoses, always synchronous at first and often synchronous throghout cleavage, proceed without corresponding division of the cytoplasm. Cytoplasmic haloes accumulate around the daughter nuclei, forming cleavage energids. The energids spread through the yolk mass and gradually approach the periphery. In most pterygote eggs, some of the cleavage energids remain within the yolk mass as vitellophages, though many examples have been described of the secondary formation of vitellophages by cells reinvading the yolk mass after reaching the surface (see Anderson, 1972 a, b, for details).

In many primitive Hemimetabola and in all Paraneoptera and Holometabola, the cleavage energids arrive uniformly at the surface of the yolk mass. The next event is formation of a blastoderm. We have already seen in the preceding chapter that cleavage in the Thysanura first results in the formation of a uniform blastoderm and is then succeeded by a process of blastodermal differentiation, yielding a differentiated blastoderm, composed of a small but dense embryonic primordium and an extensive, attenuated extra-embryonic ectoderm. Among the pterygotes, many species follow this mode, though others show a specialized condition in which blastodermal differentiation begins before the blastoderm is completed (see below). We can give consideration firstly to the primitive Hemimetabola. In these orders, blastoderm formation results from the division and spread of the peripheral cleavage energids, which are separated from one another when they first emerge. An attenuated, uniform blastoderm results from this process in the Odonata, Ephemeroptera, Plecoptera, Embioptera, Dermaptera and some Orthoptera (Wheeler, 1893; Heymons, 1895; Seidel, 1929; Miller, 1939; Johannsen and Butt, 1941; Kanellis, 1952; Mahr, 1960; Stefani, 1961; Bhatnagar and Singh, 1965; Sauer, 1966; Illies, 1968). Blastodermal differentiation then ensues, resulting in the formation of an embryonic primordium as a compact disc or band of columnar cells, ventrally or posteroventrally placed on the surface of the yolk mass.

A more precocious differentiation of the embryonic primordium during blastoderm formation is seen in the Dictyoptera, Isoptera, Cheleutoptera and many Orthoptera (Korotneff, 1885; Wheeler, 1889a; Heymons, 1895; Nusbaum and Fulinski, 1909; Leuzinger and Wiesmann, 1926; Roonwal, 1936; Thomas, 1936; Krause, 1939; Kanellis, 1952; Ibrahim, 1957; Bergerard, 1958; Görg, 1959; Mahr, 1960; Striebel, 1960; Sauer, 1966; Bedford, 1970). In these orders, cellular concentration to form the embryonic primordium proceeds while the cleavage energids are still dividing and spreading over the surface of the yolk mass. In the Isoptera and Cheleutoptera, the embryonic primordium has already begun to enter into further development at the posterior pole of the egg by the time that blastoderm formation is completed at the anterior pole, but this can be regarded as an extreme specialization. In general, the completion of the differentiated blastoderm marks the end of cleavage in the primitive Hemimetabola and precedes the onset of gastrulation.

The Paraneoptera and Holometabola, with their distinct periplasm at the surface of the egg, proceed towards blastoderm formation by a more direct route. The formation of the blastoderm has been described by several workers on paraneopteran embryos (Seidel, 1924; Mellanby, 1936; Schölzel, 1937; Böhmel and Jancke, 1942; Goss, 1952; Piotrowski, 1953; Sander, 1956; Jura, 1959) and on numerous occasions for the Holometabola (see Anderson,

1962, 1966; Ullman, 1964; Bassand, 1965; Rempel and Church, 1965, 1969; Guénnelon, 1966; Jung, 1966a; Davis, 1967; Dupraw, 1967; Davis *et al.*, 1968; West *et al.*, 1969; Wolf, 1969, for recent accounts). The cleavage energids arrive uniformly at the periphery of the yolk mass and merge with the periplasm. Further mitoses then increase the surface population of nuclei, giving a syncitial blastoderm. In the more specialized Holometabola, cytoplasm from the reticulum within the yolk mass also moves out to thicken the syncitial blastoderm. When the nuclear divisions are complete, a simultaneous formation of cell membranes throughout the syncitial blastoderm transforms the latter into a cuboidal or columnar uniform, cellular blastoderm. The columnar condition is characteristic of the Diptera and Hymenoptera, the two orders exhibiting the greatest preponderance of cytoplasm relative to yolk in the egg. Differentiation of the uniform cellular blastoderm to a differentiated blastoderm then proceeds. The formation of the embryonic primordium and attenuation of the remainder of the blastoderm as extra-embryonic ectoderm takes place in the Paraneoptera in much the same manner as in the primitive Holometabola, producing a relatively small embryonic primordium (e.g. Mellanby, 1936; Krause, 1939; Butt, 1949; Goss, 1952; Sander, 1959). The same process of cellular aggregation in the Holometabola, on the other hand, yields a larger embryonic primordium, occupying relatively more of the surface of the yolk mass (Patten, 1884; Hegner, 1909; Hirschler, 1909; Blunck, 1914; Strindberg, 1915a; Mansour, 1927; Paterson, 1931, 1936; Du Bois, 1932, 1936, 1938; Gambrell, 1933; Inkmann, 1933; Butt, 1934, 1936; Smreczyński, 1934; Wray, 1938; Tiegs and Murray, 1938; Bock, 1939; Kessel, 1939; Mulnard, 1947; Krause and Ryan, 1953; Luginbill, 1953; Shafiq, 1954; Jura, 1957; Ando and Okada, 1958; Ivanova-Kazas, 1959; Ando, 1960; Ede, 1964; Ullman, 1964; Jung, 1966 a, b; Küthe, 1966; Craig, 1967; Zakhvatkin, 1967 a, b, 1968; Kalthoff and Sander, 1968; Rempel and Church, 1969; Wolf, 1969). The significance of this difference will become apparent when we come to consider the question of presumptive areas in the pterygote blastoderm. The higher Diptera and Hymenoptera, in particular, develop a relatively very large embryonic primordium from their columnar blastoderm and show some delay in the differentiation of their small, dorsal area of extra-embryonic ectoderm (see, for example, Anderson, 1966; Davis, 1967; Dupraw, 1967). In the present context, the important point to emphasize is that differentiation of the blastoderm and formation of an embryonic primordium is a basic characteristic of pterygote embryos, which they share with the Thysanura.

Before leaving cleavage in the pterygotes, brief mention must be made of the precocious formation of germ cells during the cleavage of certain types of pterygote eggs. In the Dermaptera (Heymons, 1895; Singh, 1967), Psocoptera (Goss, 1953), Homoptera (Gerwel, 1950) and certain Coleoptera (Hegner, 1909, 1914; Inkmann, 1933); Mulnard, 1947; Jung, 1966 a, b), a group of rounded cells is separated off at the posterior pole of the embryonic primordium during differentiation of the blastoderm. These pole cells, as they are called, are the primordial germ cells. As we shall see later, the primordial germ cells do not become distinct at this early stage in most pterygotes, but a few holometabolan orders emulate the Coleoptera in this respect. The Diptera (Fig. 87) show a precocious segregation of pole cells during cleavage and blastoderm formation (see Anderson, 1962, 1966, 1972b; Davis, 1967; Mahowald, 1968, 1971; West *et al.*, 1968; Wolf, 1969), as do the Siphonaptera (Kessel, 1939) and the parasitic apocritan Hymenoptera (Henschen, 1928; Ivanova-Kazas, 1958;

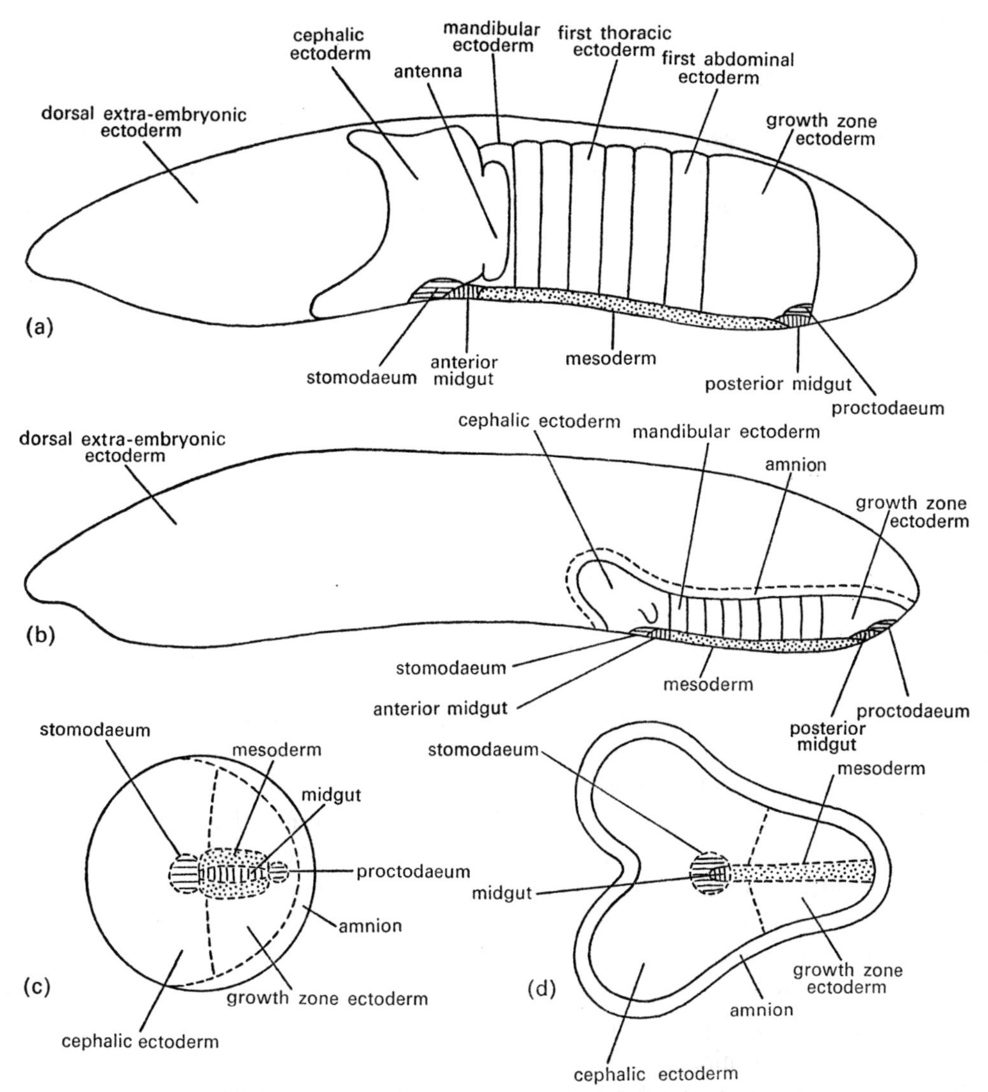

FIG. 88. Fate maps of the pterygote blastoderm, after Anderson (1972a). (a) The blastoderm of of *Platycnemis* (Odonata) in left lateral view at the uniform blastoderm stage; (b) the differentiated blastoderm of *Platycnemis* in left lateral view; (c) the embryonic primordium of *Kalotermes* (Isoptera), ventral view; (d) the embryonic primordium of *Carausius* (Cheleutoptera), ventral view. (Originals based on data of Seidel, 1935; Striebel, 1960; and Louvet, 1964.)

Bronskill, 1959, 1964; Amy, 1961; Meng, 1968). We shall discuss the further development of the gonads in due course.

Presumptive Areas of the Blastoderm

Since the differentiation of the blastoderm intervenes between cleavage and the onset of gastrulation in pterygote embryos, it is simplest to construct fate maps for the pterygote blastoderm at the differentiated blastoderm stage. By this time, all of the presumptive areas except the attenuated, extra-embryonic ectoderm are concentrated within the embryonic primordium. As pointed out by Anderson (1972 a, b), however, there is also much to be gained from giving consideration to the prior distribution of the presumptive areas at the uniform blastoderm stage, before the concentration and aggregation of these areas as an embryonic primordium has begun. In the primitive Holometabola and the Paraneoptera, the cells of the presumptive areas of the uniform blastoderm are thinly spread over the surface of the yolk mass. As blastodermal differentiation proceeds, all except one of these areas lessens greatly in area as the cells become tightly packed and columnar. The exception of course, is the dorsal, extra-embryonic ectoderm, which becomes more attenuated and spread as blastodermal differentiation proceeds. The majority of Holometabola, in contrast, having a more deeply cuboidal blastoderm, undergo less concentration of cells during formation of the embryonic primordium. Their presumptive areas therefore remain relatively large in area, even at the differentiated blastoderm stage. In the cyclorrhaphous Diptera, this specialization is carried to an extreme in which the deeply columnar presumptive areas of the uniform blastoderm undergo no concentration and reduction in area before gastrulation begins. As shown by Anderson (1972 a, b), if one compares the fate map of the blastoderm of the dragonfly *Platycnemis* at the uniform blastoderm stage with that of a cyclorrhaphous dipteran at the same stage, a striking similarity emerges. The presumptive areas are distributed over the surface of the yolk mass in the same way in both, but those of *Platycnemis* are made up of flattened, thinly spread cells, while those of the Cyclorrhapha are made up of closely packed, columnar cells. We thus have no difficulty in perceiving the general trend of specialization in pterygote embryos. The blastodermal fate map has changed little, but specializations of cleavage have led in the Holometabola, and to a lesser extent in the Paraneoptera, to a thicker blastoderm, in which each presumptive area has more and larger cells. A more substantial basis is thus laid down for further development, and much of the cell proliferation that occurs during the further development of the small embryonic primordium of primitive hemimetabolans is obviated. At the same time, it is still not obvious why the basic development of pterygotes, like that of the Thysanura among the apterygotes, includes the process of blastodermal differentiation or, as we can now put it, aggregation of the blastodermal presumptive areas to form an embryonic primordium. Cellular proliferation and thickening of these areas could equally well occur directly, following the uniform blastoderm stage, as it does in the Onychophora, myriapods and other apterygotes. The functional significance of blastodermal differentiation in the Thysanura and Pterygota still awaits elucidation.

We are now in a position to consider how the presumptive areas of the pterygote blastoderm are arranged (Figs. 88, 89). The resemblance between the Odonata and Diptera in this respect is indicative of a constant basic pattern and this is borne out by studies on other orders. Allowing for certain functional variations which will be mentioned below, the survey of recent evidence made by Anderson (1972 a, b) has revealed that the blastodermal fate map is remarkably stable throughout the pterygotes. The cyclorrhaphous Diptera provide a valuable guide to this map. Their large presumptive areas are easily distinguished at the onset of gastrulation (Poulson, 1950; Anderson, 1962, 1966; Davis, 1967). In most pterygote embryos, the cells of the embryonic rudiments are difficult to distinguish from one another until gastrulation is well advanced, so that the interpretation of their prior arrangement in

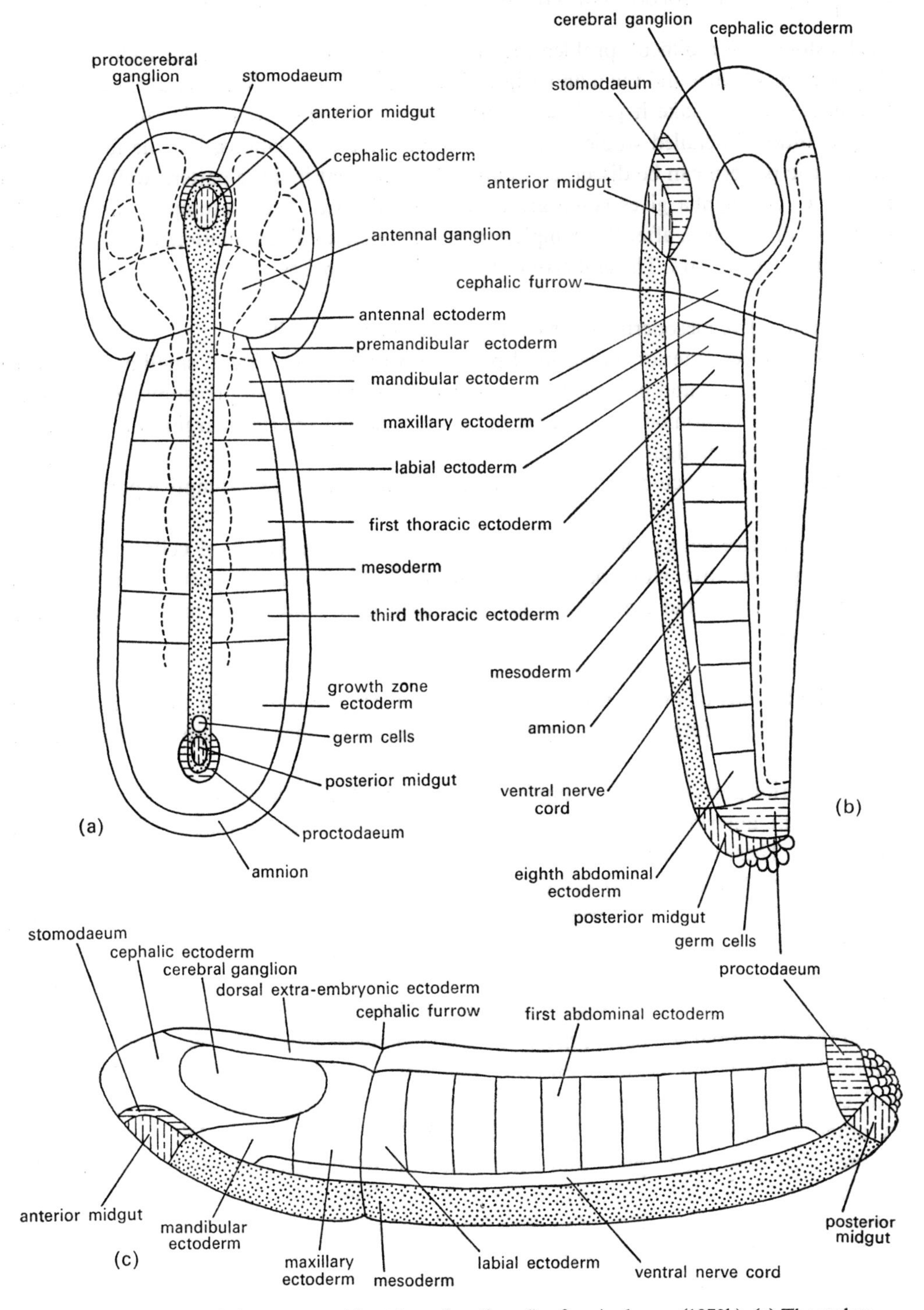

FIG. 89. Fate maps of the pterygote blastoderm (continued), after Anderson (1972b). (a) The embryonic primordium of *Tenebrio* (Coleoptera), ventral view; (b) the blastoderm of *Culex* (Diptera) in left lateral view; (c) the blastoderm of *Dacus* (Diptera) in left lateral view. (Originals based on data of Anderson, 1962, 1963, 1964; Ullman, 1964; and Davis, 1967.)

the blastoderm is a difficult problem. Even so, the relative positions of the rudiments when they become distinct and the manner in which the blastoderm cells arrive at these positions (see next section) make it possible to postulate where the cells lay prior to gastrulation. Only additional detailed studies can substantiate these postulations, but the fact that the same basic pattern can be discerned in all orders and that it is the pattern well recognized for the Cyclorrhapha gives some grounds for confidence. The presumptive areas of the pterygote blastoderm, after the completion of blastodermal differentiation, can therefore be summarized as follows. The embryonic primordium comprises:

1. Presumptive mesoderm, a narrow mid-ventral band of cells.
2. Presumptive anterior midgut and presumptive posterior midgut, small areas of cells at the anterior and posterior ends respectively of the band of presumptive mesoderm.
3. Presumptive stomodaeum, an arc of cells around the anterior margin of the presumptive anterior midgut.
4. Presumptive proctodaeum, a small area of cells behind the presumptive posterior midgut.
5. Presumptive embryonic ectoderm, paired ventrolateral bands of cells which border the presumptive stomodaeum, anterior midgut, mesoderm, posterior midgut and proctodaeum, and meet in the anterior midline in front of the presumptive stomodaeum.

The paired bands of presumptive embryonic ectoderm are broader in the Holometabola, especially in the Diptera, Hymenoptera and Lepidoptera, than they are in the hemimetabolous insects.

The anterior parts of the presumptive ectoderm, in front of and around the presumptive stomodaeum, anterior midgut and anterior part of the mesoderm, take the form of broad head lobes. These are made up of the presumptive ectoderm of the pre-antennal part of the head, together with antennal ectoderm at the posteromedian corners of the head lobes. The remainder of each ectoderm band is narrower. In the blattid Dictyoptera, Odonata, Ephemeroptera and Dermaptera among the primitive Hemimetabola, the narrow part of the band on each side is relatively long and comprises a sequence of presumptive pre-mandibular, gnathal and thoracic segmental zones, followed by a terminal growth zone from which the abdominal ectoderm arises. The same elongate bands of presumptive ectoderm are retained in the Paraneoptera and the Holometabola, with some of the anterior abdominal ectoderm already segmentally zoned within them in the latter group. In contrast, in other orders of primitive Hemimetabola, including the mantid Dictyoptera, Cheleutoptera, Orthoptera, Isoptera and Plecoptera, all of the presumptive ectoderm behind the antennal level is concentrated into a growth zone from which the postantennal ectoderm subsequently arises by proliferation. The functional significance of this secondary specialization of development is not clear.

Around the grouped presumptive areas of the embryonic primordium, the remainder of the surface of the yolk mass is occupied by presumptive extra-embryonic ectoderm, broadly attenuated in hemimetabolous species, less so in the Holometabola and much reduced in the cyclorrhaphous Diptera. It should also be borne in mind that the yolk mass contains vitellophages additional to the presence of the anterior and posterior midgut rudiments in the embryonic primordium. We can now discuss gastrulation in pterygote embryos in terms of the movements of presumptive areas prior to organogeny.

Gastrulation

Commensurate with the strict localization of the presumptive mesoderm along the ventral midline of the embryonic primordium, the gastrulation of pterygote embryos centres around the entry of midventral cells into the interior, accompanied by the formation of a temporary midventral groove, the gastral groove (Fig. 90). This groove may be long or short, deep or shallow and transient or relatively persistent, depending on the species, but is always closed before the external delineation of the segments of the embryo begins.

Gastrulation in the primitive Hemimetabola takes place while the embryonic primordium is still small. This fact led to some confusion of interpretation in the early papers on the subject. More recent papers, however, particularly those of Roonwal (1936), Krause (1938a) Görg (1959), Striebel (1960) and Louvet (1964), have confirmed that a narrow gastral groove is formed in the primitive Hemimetabola as the presumptive mesoderm cells sink into the interior. The edges of the paired bands of presumptive ectoderm then come together in the ventral midline, closing the gastral groove. The mesoderm spreads out as a layer of cells beneath the embryonic ectoderm.

The same mode of entry of presumptive mesoderm into the interior is seen in the Paraneoptera (Seidel, 1924; Mellanby, 1936; Böhmel and Jancke, 1942; Butt, 1949; Goss, 1952, 1953; Piotrowski, 1953; Jura, 1959) and in the relatively long embryonic primordia of most Holometabola (Fig. 101). The papers of Paterson (1931), Eastham (1927), Tiegs and Murray (1938), Bock (1939) and Ullman (1964) provide especially clear accounts of generalized gastrulation in the Holometabola. Only in the Hymenoptera and Diptera, in which the presumptive mesoderm is a much broader band of ventral cells, is the mode of entry into the interior substantially modified (Fig. 91). In the Hymenoptera, as shown for example by Schnetter (1934), Ivanova-Kazas (1958), Amy (1967) and Dupraw (1967), the presumptive mesoderm first separates from the edges of the presumptive ectoderm and begins to sink inwards as a broad plate of cells. Rapid spreading of the presumptive ectoderm towards the ventral midline then completes the process of coverage of the mesoderm. In the Diptera, in contrast, recent studies by Poulson (1950), Anderson (1962), Schoeller (1964), Davis (1967) and others have fully confirmed the classical observation that the mesoderm becomes internal by invagination. The broad plate of presumptive mesoderm folds into the interior along the ventral midline and closes off as a tube. The edges of the presumptive ectoderm come together in the usual way and the walls of the tube then dissociate and spread out to form the usual single layer of mesoderm beneath the ectoderm. Clearly, the events of mesodermal gastrulation in the Hymenoptera and Diptera are specializations of the basic mode of mesodermal gastrulation in the Holometabola, associated with a broadening of the presumptive mesoderm in both orders.

In discussing gastrulation in pterygote embryos, one cannot easily separate the gastrulation movements of the mesoderm from those of the presumptive midgut rudiments. Indeed, in most species, the midgut cells, which lie at the two ends of the mesoderm, cannot

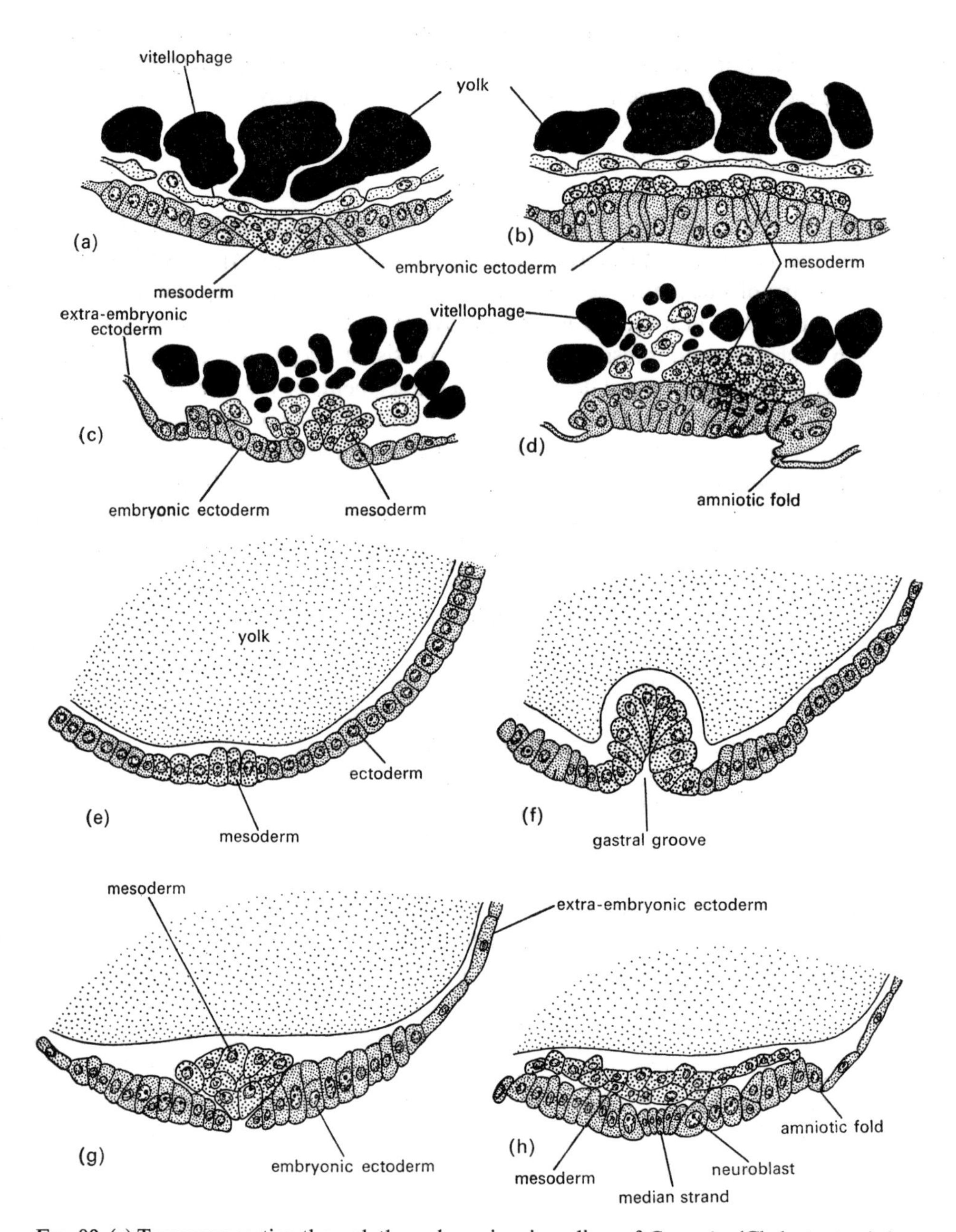

FIG. 90. (a) Transverse section through the embryonic primordium of *Carausius* (Cheleutoptera) during gastrulation; (b) transverse section through the early germ band of *Carausius*; (c) sagittal section through the embryonic primordium of *Kalotermes* (Isoptera) during gastrulation; (d) sagittal section through the early germ band of *Kalotermes*; (e–h) stages in gastrulation of *Chrysopa* (Neuroptera) in transverse section. [(a) and (b) after Louvet, 1964; (c) and (d) after Striebel, 1960; (e–h) after Bock, 1939; modified from Anderson, 1972a, b.]

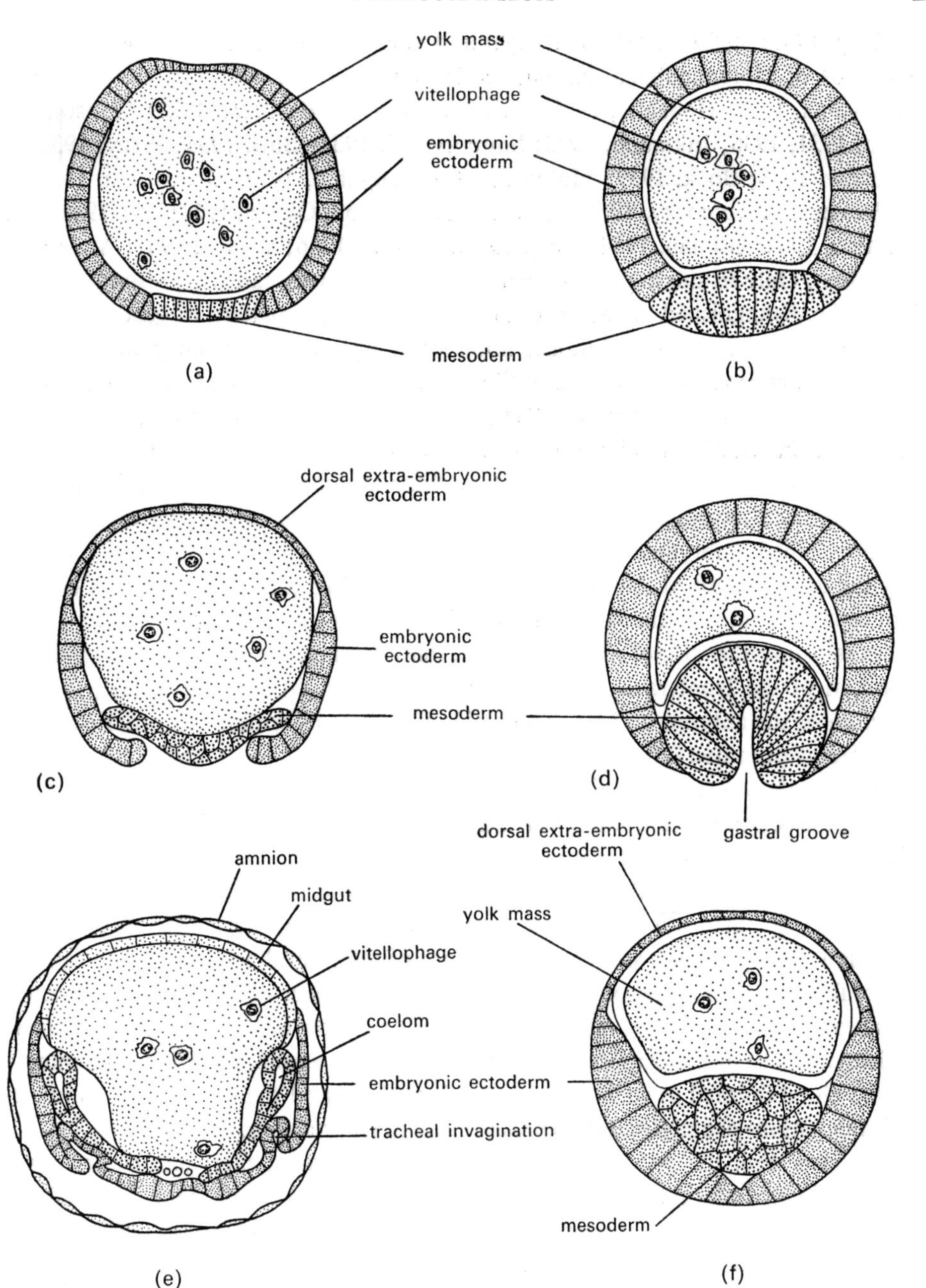

FIG. 91. Transverse sections showing stages in gastrulation of *Apis* (Hymenoptera) [(a), (c) and (e)] and *Dacus* (Diptera) [(b), (d) and (f)], after Schnetter (1934) and Anderson (1962); modified from Anderson, 1972b.

usually be distinguished from the mesoderm histologically until gastrulation is complete. This is especially true of hemimetabolous embryos (Heymons, 1895, 1896b, 1912; Rabito, 1898; Tschuproff, 1903; Seidel, 1924; Ries, 1931; Nelsen, 1934; Mellanby, 1936, 1937; Roonwal, 1936, 1937; Thomas, 1936; Schölzel, 1937; Krause, 1938b; Butt, 1949; Goss,

1952; Görg, 1959; Stefani, 1961; Bhatnagar and Singh, 1965), but is also true in general for the embryos of Holometabola (Patten, 1884; Heider, 1889; Wheeler, 1889a; Deegener, 1900; Toyama, 1902; Hirschler, 1909; Strindberg, 1915a; Eastham, 1927; Mansour, 1927; Paterson, 1931, 1936; Inkmann, 1933; Smreczyński, 1934; Butt, 1936; Wray, 1937; Tiegs and Murray, 1938; Bock, 1939; Mulnard, 1947; Luginbill, 1953; Deobahkta, 1957; Jura, 1957; Presser and Rutchsky, 1958; Ivanova-Kazas, 1959; Okada, 1960; Ede, 1964; Krause and Krause, 1964; Ullman, 1964; Jung, 1966a; Küthe, 1966; Vaidya, 1967; Anderson and Wood, 1968). Only in the higher Hymenoptera and Diptera can the presumptive midgut rudiments be distinguished at the onset of gastrulation. The pattern of gastrulation movements and relative cell juxtapositions, however, suggests very strongly that the midgut gastrulation movements observed so easily in the Hymenoptera and Diptera are, in fact, standard processes in all pterygotes. The presumptive anterior and posterior midgut cells move into the interior by immigration at the ends of the gastral groove and are overgrown by marginal cells at the surface. The Isoptera alone are known to be exceptional (Striebel, 1960), exhibiting a secondary modification associated with precocious gastrulation in a very short, disc-shaped embryonic primordium. The presumptive midgut cells of the Isoptera occupy the entire length of the midline of the immigrating, midventral band of cells. They are not separated into an anterior and posterior rudiment and the presumptive mesoderm lies bilaterally on either side of the presumptive midgut. Since the extremely short embryonic primordium of the Isoptera is a secondary modification (Anderson, 1972a), the unique conjunction of the anterior and posterior midgut rudiments within this order can scarcely be regarded as primitive for pterygotes.

As the gastral groove closes midventrally over the now internal mesodermal and midgut rudiments, the presumptive stomodaeal and presumptive proctodaeal cells lie in the ventral midline at the surface of the embryo, at the two ends of the groove. Without exception, these groups of cells now make independent invaginations into the interior as short tubes, the superficial openings of which are the mouth and anus respectively (Fig. 92). Stomodaeal invagination usually follows shortly after closure of the gastral groove, in association with the general lead taken in development by the anterior part of the embryo. Proctodaeal invagination is almost always delayed until the formation of the abdominal segments is well advanced. The cyclorrhaphous Diptera are again an exception. Their abdominal segments are all established as rudiments in the blastoderm and their stomodaeal and proctodaeal invaginations proceed simultaneously, early in gastrulation.

As we have seen, the presumptive embryonic ectoderm of pterygote embryos always plays a part in gastrulation, through a ventral spreading movement from either side which closes the gastral groove. Apart from this specialized movement, the presumptive embryonic ectoderm of pterygotes normally resembles that of myriapods and apterygotes in making no gastrulation movements. It already lies in its definitive, organ-forming position when formed in the differentiated blastoderm. The extra-embryonic ectoderm also has the usual exclusion from gastrulation movements, although we shall see below that it later becomes involved in an elaborate sequence of movements of the germ band relative to the yolk mass.

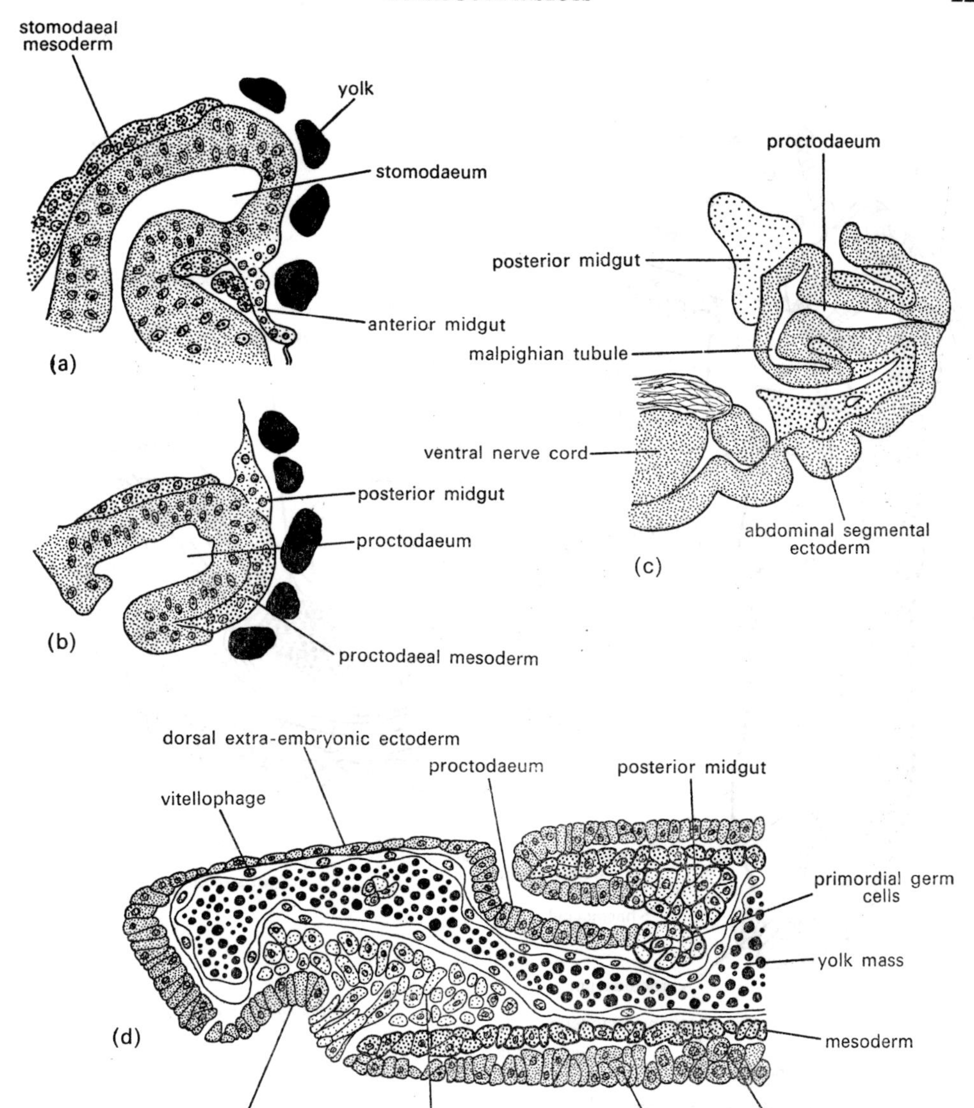

FIG. 92. (a) The stomodaeum and anterior midgut rudiment of *Carausius* (Cheleutoptera) in sagittal section; (b) the proctodaeum and posterior midgut rudiment of *Carausius*; (c) the proctodaeum and posterior midgut of *Rhodnius* (Heteroptera) in sagittal section; (d) the anterior half of the gastrula of *Dacus* (Diptera) in sagittal section. [(a) and (b) after Thomas, 1936, (c) after Mellanby, 1937, (d) after Anderson, 1962.]

Further Development of the Gut

The pterygote stomodaeum and proctodaeum continue their development in a manner which varies little among species (Fig. 93). The numerous descriptions of these processes have almost always been associated with accounts of development of the midgut and can

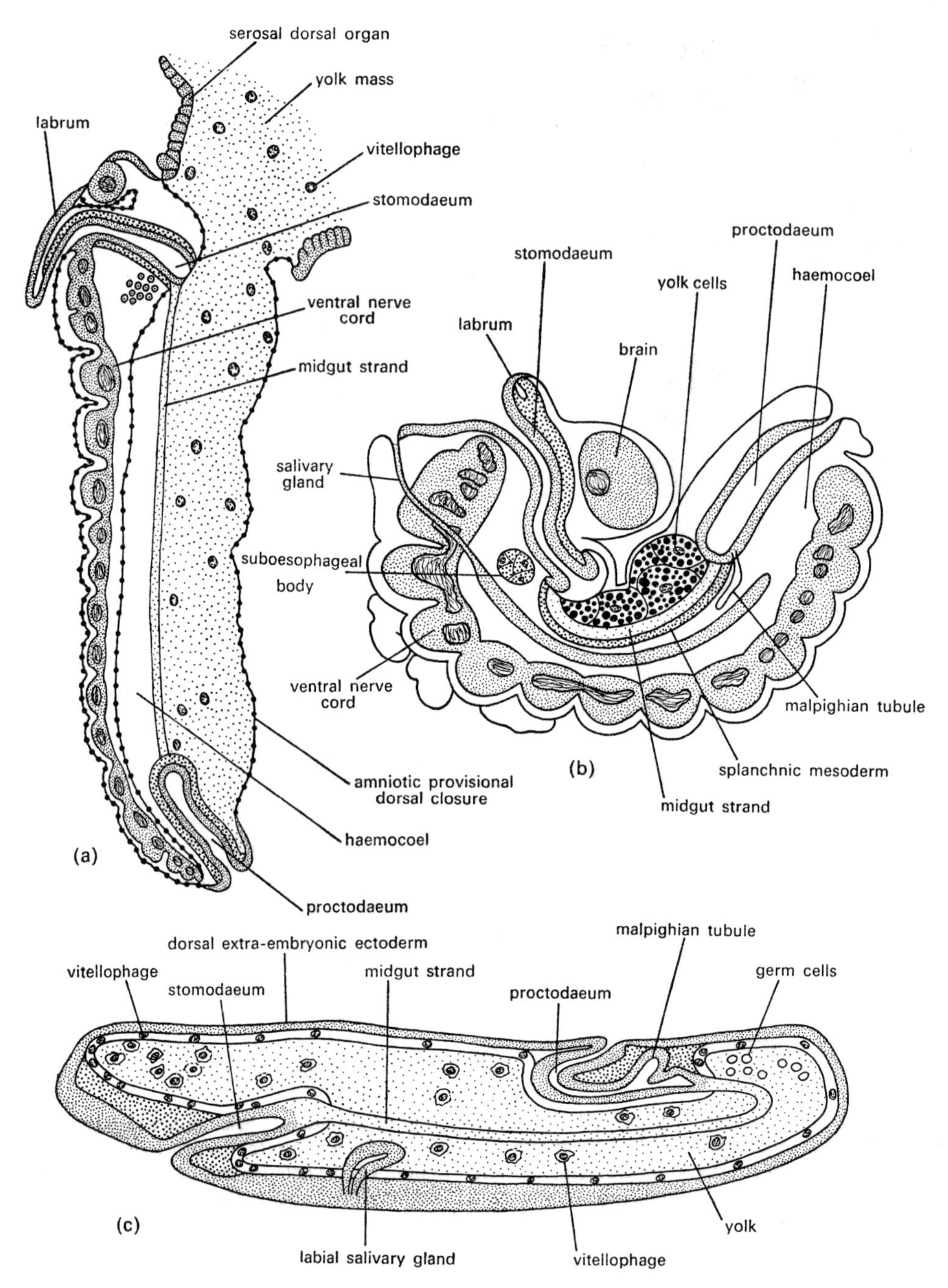

FIG. 93. Sagittal sections showing the development of the midgut strands of pterygote embryos. (a) *Locusta* (Orthoptera), after Roonwal (1937); (b) *Epiphyas* (Lepidoptera), after Anderson and Wood (1968); (c) *Dacus* (Diptera), after Anderson (1962).

therefore be cited in connection with the latter topic, which is discussed below. The stomodaeum grows inwards and backwards as a simple epithelial tube, reaching as far as the thorax. The distal end of the tube becomes reflexed on itself, forming the central part of the proventriculus. The remainder of the stomodaeal wall differentiates as the epithelium of the oesophagus. Three median outgrowths arise from the dorsal wall of the stomodaeum, either as thickenings or evaginations, and give rise to the ganglia of the stomatogastric nervous system. The proctodaeum also grows inwards as an epithelial tube and develops as the lining epithelium of the hindgut. The distal end of the proctodaeum becomes pouched out as the rudiments of the Malpighian tubules, which grow as blind-ending tubes into the haemocoel.

Although the development of the stomodaeum and proctodaeum in pterygote embryos is similar to that of the myriapods and apterygotes, important differences occur in the development of the midgut. We have already seen that the precursors of the midgut in pterygote embryos can be identified at the end of gastrulation as an anterior midgut rudiment attached to the end of the stomodaeum and a posterior midgut rudiment attached to the end of the proctodaeum (Fig. 92). In addition, vitellophages are present in the yolk mass. The mode of subsequent development of the midgut from these components has been the subject of much controversy in pterygote embryology, especially in relation to the contribution made by the vitellophages. The weight of recent evidence, however, favours a complete functional separation between vitellophage function and midgut formation. There is no good evidence that the vitellophages contribute structurally to the wall of the midgut, though it is perhaps too soon to exclude this possibility entirely. A large volume of well-substantiated evidence, on the other hand, demonstrates that the formation of the midgut epithelium takes place from the cells of the anterior and posterior midgut rudiments. In several species of primitive Hemimetabola (Wheeler, 1889a; Heymons, 1895; Rabito, 1898; Nusbaum and Fulinski, 1909; Strindberg, 1915b; Nelsen, 1934; Thomas, 1936; Roonwal, 1936, 1937; Görg, 1959; Stefani, 1961) and Paraneoptera (Hirschler, 1912; Seidel, 1924; Ries, 1931; Mellanby, 1937; Schölzel, 1937; Böhmel and Jancke, 1942; Butt, 1949; Goss, 1952), the anterior and posterior midgut rudiments have been shown to enter into proliferation as the growth and segmentation of the germ band nears completion, producing paired ventrolateral strands of cells which grow along the surface of the yolk mass and meet in the middle region of the embryo. After the two midgut strands have become continuous, they begin to spread, first ventrally, then dorsally, over the surface of the shrinking yolk mass. In this way the midgut strands give rise to an epithelial midgut sac, filled with yolk, joining the stomodaeum to the proctodaeum. The yolk and vitellophages are gradually resorbed as the midgut sac differentiates into the definitive midgut epithelium. The same mode of development of the midgut (Figs. 93, 105) has also been found in numerous holometabolans (Noack, 1901; Strindberg, 1915a; Mansour, 1927; Du Bois, 1932, 1938; Smreczyński, 1932; Gambrell, 1933; Schienert, 1933; Butt, 1934, 1936; Lassman, 1936; Wray, 1937; Tiegs and Murray, 1938; Bock, 1939; Ivanova-Kazas, 1949; Poulson, 1950; Luginbill, 1953; Formigoni, 1954; Deobahkta, 1957; Christophers, 1960; Anderson, 1962, 1966; Ullman, 1964; Davis, 1967; Rempel and Church, 1969, 1971a) with minor functional modifications being noted in the Hymenoptera and Lepidoptera (see Anderson, 1972b, for references). In spite of continuing controversy, we can now be certain that midgut development in the pterygotes proceeds in a fundamentally different manner from that of apterygotes, in spite of the many similarities between pterygote and thysanuran embryos. The significance of this difference will be discussed in Chapter 10. In passing, it should be noted that the pterygote fat-body, like that of apterygotes, develops from mesoderm cells.

Development of External Form

Although in general the development of the external form of pterygote embryos follows the lines already described in the previous chapter for the Thysanura, there are manifold variations and complications in the different orders. Some are related to the functional specializations of the newly hatched young, e.g. a cyclorrhaphous dipteran larva as compared with a blattid numph, but others are temporary, embryonic peculiarities, most of which defy functional explanation at the present time. The details of these are fully discussed by Anderson (1972 a, b). For the sake of completeness we will touch briefly on them in the present context, while emphasizing the basic features that are important for comparative purposes. The starting point for discussion of this aspect of pterygote development is the differentiated blastoderm. It is also useful to note that the development of all pterygote embryos is epimorphic, so that the full complement of segments is formed and functional in all species when hatching takes place. Among hemimetabolous species the location of the small embryonic primordium on the surface of the yolk mass varies considerably. Most frequently, the embryonic primordium lies in a posteroventral portion (Fig. 94), as in the Odonata, Dictyoptera, Isoptera, Embioptera, some Orthoptera and all Paraneoptera, but it may be posterior, as in the Ephemeroptera, Cheleutoptera and certain Orthoptera, or midventral, as in the Plecoptera, Dermaptera and a few Orthoptera. The shape of the primordium also varies. Most hemimetabolous orders display the precocious formation of incipient head lobes as paired enlargements of the anterior part of the primordium. As indicated in the earlier discussion of presumptive areas, the head lobes comprise the preantennal and antennal regions of the embryo. The remainder of the primordium, or post-antennal region, is relatively long in the Ephemeroptera, Odonata, blattid Dictyoptera, Embioptera, Dermaptera and the paraneopterans (Wheeler, 1893; Heymons, 1895, 1896 a, b; Kershaw, 1914; Seidel, 1924, 1929, 1935; Mellanby, 1936; Schölzel, 1937; Johannsen and Butt, 1941; Böhmel and Jancke, 1942; Butt, 1949; Goss, 1952; Sander, 1959; Behrendt, 1963; Illies, 1968; Mori, 1969; Böhle, 1969) and includes the rudiments of the gnathal and thoracic segments as well as the posterior growth zone (Figs. 88, 94 and 97). For reasons which will become apparent below, this form probably constitutes the basic type of pterygote embryonic primordium. In other Dictyoptera and in the Cheleutoptera and Orthoptera, the post-antennal region is abbreviated to a pointed growth zone only (Fig. 95), so that the primordium as a whole is heart-shaped (Wheeler, 1893; Heymons, 1895; Hagan, 1917; Leuzinger and Wiesmann, 1926; Nelsen, 1931, 1934; Slifer, 1932b; Else, 1934; Roonwal, 1936; Thomas, 1936; Krause, 1938 a, b, 1939; Steele, 1941; Jhingran, 1947; Salt, 1949; Bodenheimer and Shulov, 1951; Matthée, 1951; Kanellis, 1952; Ibrahim, 1957; Bergerard, 1958; Görg, 1959; Shulov and Pener, 1959, 1963; Mahr, 1960; Reigert, 1961; Rakshpal, 1962; Koch, 1964; Louvet, 1964; Van Horn, 1966; Fournier, 1967; Bedford, 1970); while in the Isoptera and Plecoptera, even the incipient head lobes are reduced (Fig. 97) and the primordium is disc-shaped (Miller, 1939; Striebel, 1960). In contrast, in the Holometabola,

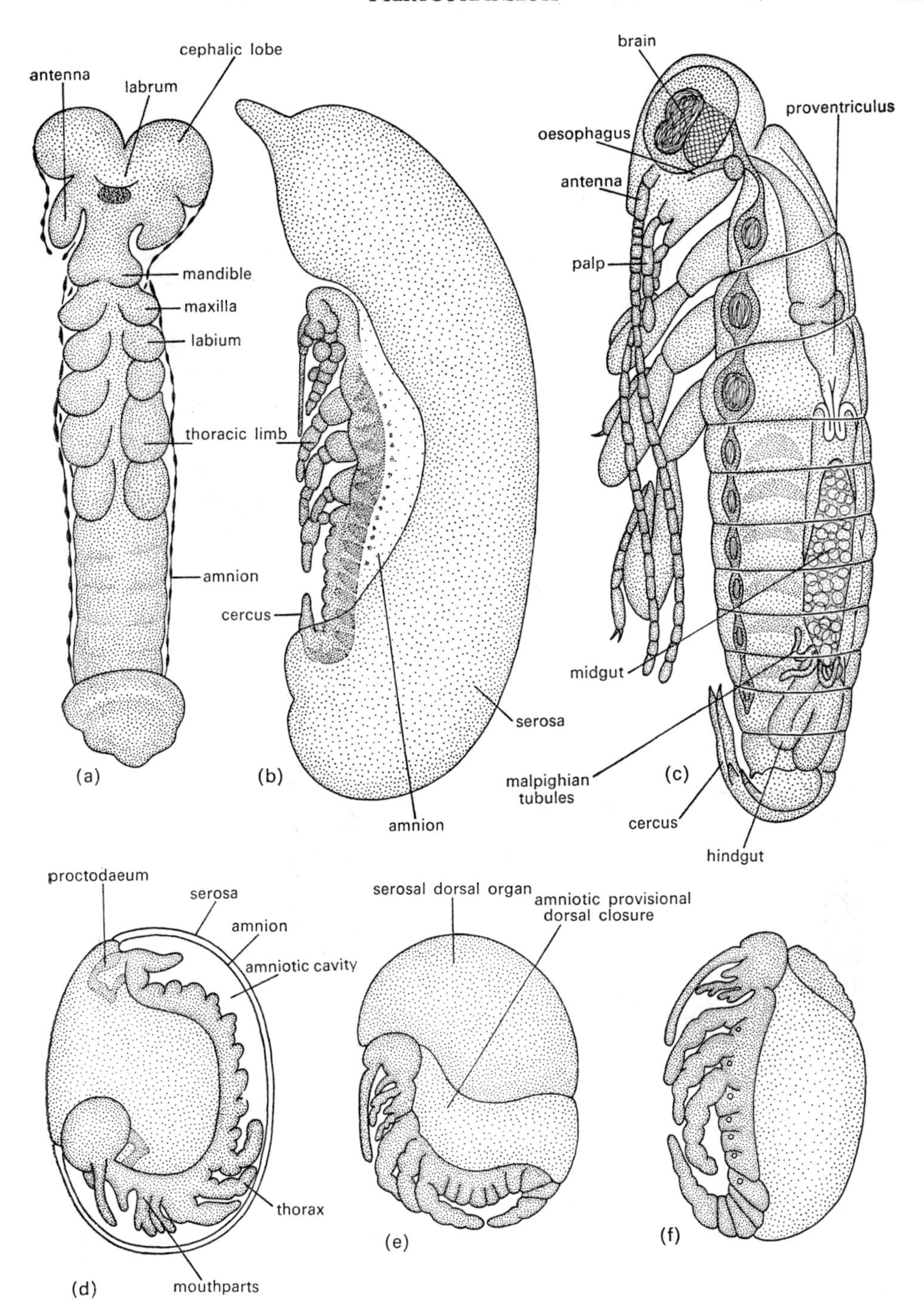

FIG. 94. (a)–(c) The external features of development of *Blatta* (Dictyoptera), after Wheeler (1889) [(a) segmenting germ band, ventral view; (b) embryo during katatrepsis, left lateral view; (c) embryo approaching hatching). (d)–(f) The external features of development of *Forficula* (Dermaptera), after Heymons (1895), in left lateral view [(d) segmented germ band; (e) early katatrepsis; (f) katatrepsis completed].

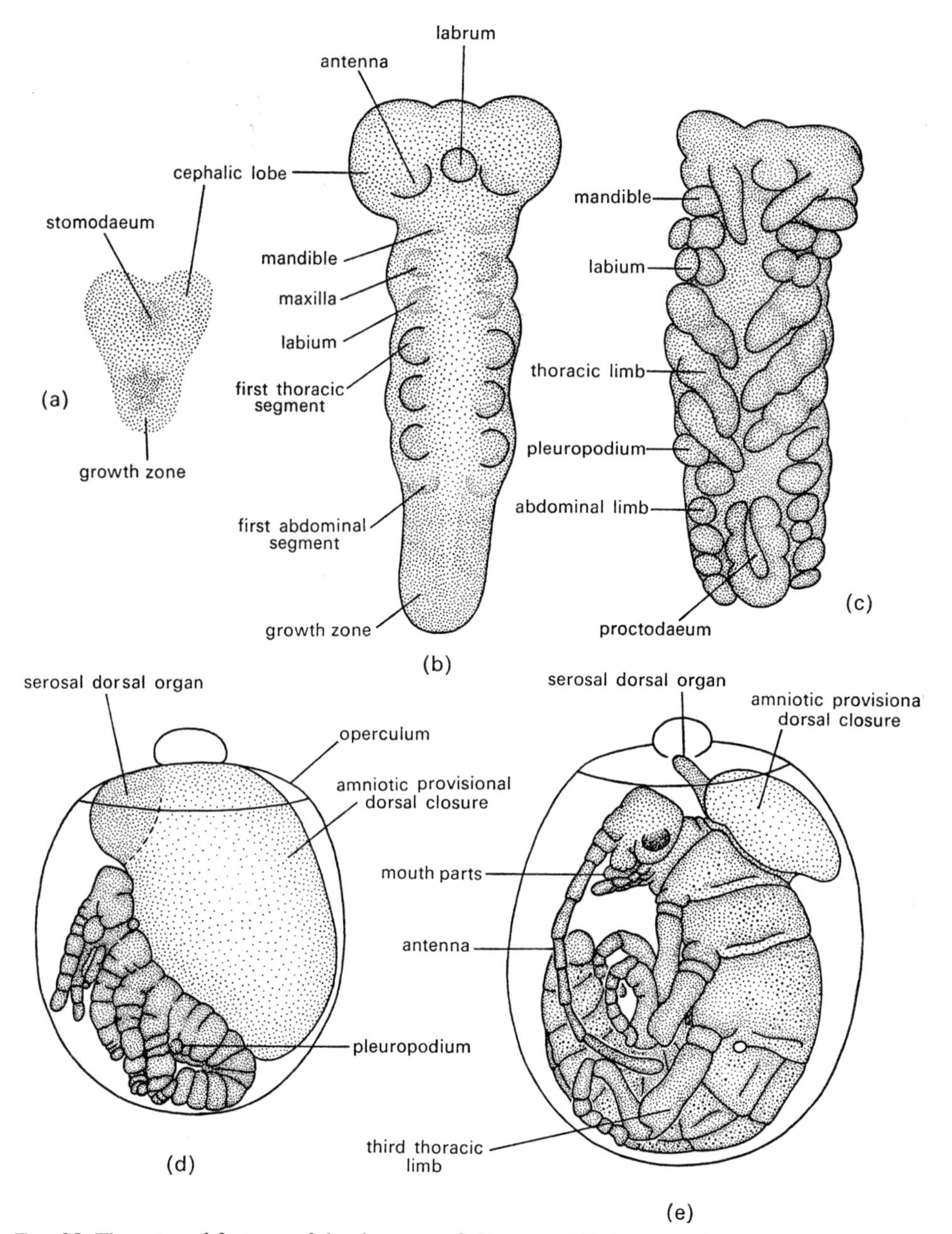

FIG. 95. The external features of development of *Carausius* (Cheleutoptera), after Fournier (1967). (a) Embryonic primordium, ventral view; (b) segmenting germ band, ventral view; (c) segmented germ band, ventral view; (d) katatrepsis completed, left lateral view; (e) dorsal closure almost completed.

the embryonic primordium is both long and broad, with large head lobes (Figs. 98 and 99), although the composition of the head lobes as incipient preantennal and antennal components, and of the post-antennal region as gnathal and thoracic segments and a growth zone (Fig. 89), is usually similar to the generalized hemimetabolan condition described above

(Patten, 1884; Hegner, 1909; Hirschler, 1909; Blunck, 1914; Strindberg, 1915a; Mansour, 1927; Paterson, 1931, 1936; Du Bois, 1932, 1936, 1938; Gambrell, 1933; Inkmann, 1933; Butt, 1934, 1936; Smreczyński, 1934; Wray, 1937; Tiegs and Murray, 1938; Bock, 1939; Kessel, 1939; Mulnard, 1947; Krause and Ryan, 1953; Luginbill, 1953; Shafiq, 1954; Jura, 1957; Ando and Okada, 1958; Ivanova-Kazas, 1959; Ando, 1960; Ede, 1964; Ullman, 1964; Jung, 1966 a, b; Küthe, 1966; Craig, 1967; Zakhvatkin, 1967 a, b, 1968; Kalthoff and Sander, 1968; Rempel and Church, 1969, 1971a; Wolf, 1970). The higher Hymenoptera and Diptera and the Lepidoptera exhibit further specializations of this condition, the nature of which will be briefly mentioned in later paragraphs.

Accompanying and following the onset of gastrulation, the embryonic primordium begins to increase in length and shortly after this, the external delineation of segments begins. The segments are usually delineated in anteroposterior succession, though some primitive Hemimetabola begin this process at the third thoracic segment and proceed both anteriorly and posteriorly from this point. The simplest pattern of growth of the segmenting germ band (Fig. 94) is seen in the Dictyoptera (Wheeler, 1889a; Hagan, 1917; Görg, 1959). Growth proceeds in a posterior direction along the ventral surface of the yolk mass, accompanied by a slight forward shift of the anterior end of the germ band and a downward flexure of the posterior end when it reaches the posterior pole of the yolk mass. The Dermaptera, Cheleutoptera and gryllotalpid Orthoptera also retain a superficial germ band, but in these examples (Figs. 94 and 95) the segmenting germ band curves around the posterior pole and forwards along the dorsal surface of the yolk mass towards the anterior pole (Heymons, 1895; Koch, 1964; Bhatnagar and Singh, 1965; Fournier, 1967; Bedford, 1970). A similar extension of the posterior part of the growing germ band along the dorsal surface of the yolk mass is found (Figs. 98 and 99) in the more generalized Holometabola (Patten, 1884; Hegner, 1909; Hirschler, 1909; Blunck, 1914; Mansour, 1927; Paterson, 1932, 1936; Inkmann, 1933; Smreczyński, 1934; Butt, 1936; Wray, 1937; Tiegs and Murray, 1938; Bock, 1939; Mulnard, 1947; Krause and Ryan, 1953; Luginbill, 1953; Jura, 1957; Krause and Sander, 1962; Ede, 1964; Ullman, 1964; Miya, 1965; Jung, 1966 a, b; Küthe, 1966; Zakhvatkin, 1967 a, b, 1968). The hemimetabolans in this category retain the forward flexure of the posterior part of the abdomen, but the holometabolans have lost this feature.

In other primitive Hemimetabola, the dorsoventral curvature of the growing germ band becomes more pronounced, resulting in partial or complete immersion of the germ band in the yolk mass. The Isoptera (Fig. 97) and Embioptera exhibit partial immersion (Kershaw, 1914; Striebel, 1960), while the Ephemeroptera, Odonata, most Orthoptera and almost all paraneopterans achieve total immersion of the germ band within the yolk mass (Figs. 94, 96 and 97). The specialized combination of growth and movement which brings the germ band into this position is called anatrepsis (Wheeler, 1893) (see Anderson, 1972a, for further discussion and a full list of references). Being a secondary phenomenon confined to certain pterygote orders, anatrepsis need not concern us further here, beyond making the point that the dorsoventral curvature of the pterygote germ band is in the opposite direction to the dorsoventral flexure which carries the thysanuran germ band deep into the yolk mass.

During the growth and segmentation of the germ band, limb buds develop on the head and thorax, usually in anteroposterior succession. The antennal limb buds arise at the posteromedian corners of the head lobes, behind the level of the invaginating stomodaeum,

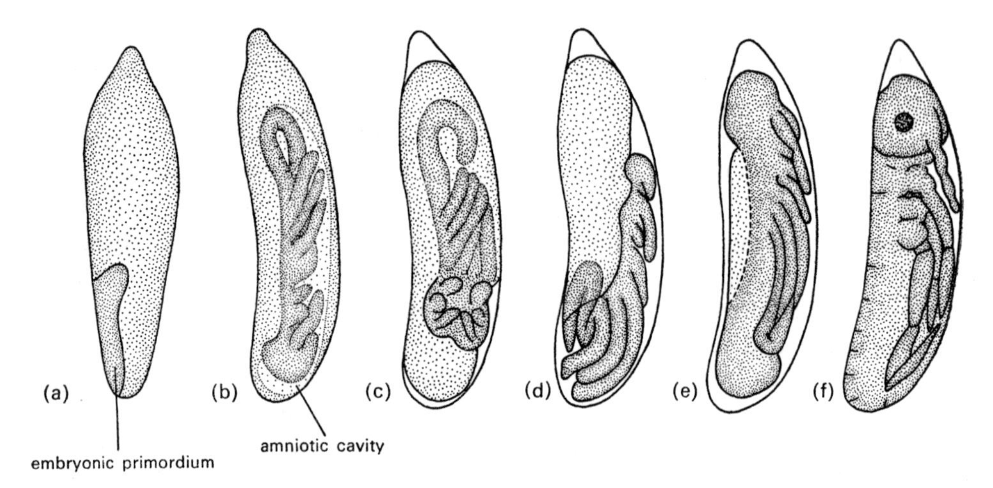

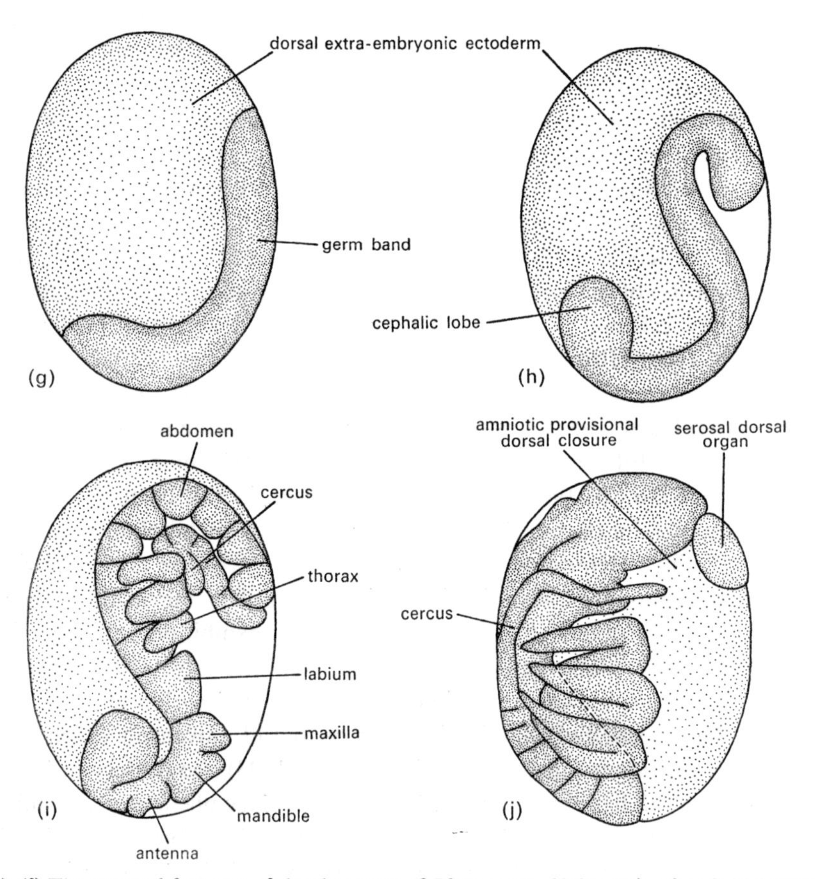

FIG. 96. (a)–(f) The external features of development of *Platycnemis* (Odonata), after Seidel (1935), in left lateral view [(a) embryonic primordium; (b) segmented germ band after anatrepsis; (c) onset of katatrepsis; (d) and (e) stages in katatrepsis; (e) dorsal closure completed). (g)–(j) External features of development of *Baetis* (Ephemeroptera), after Böhle (1969), in left lateral view [(g) embryonic primordium; (h) segmenting germ band; (i) segmented germ band before katatrepsis; (j) embryo after katatrepsis].

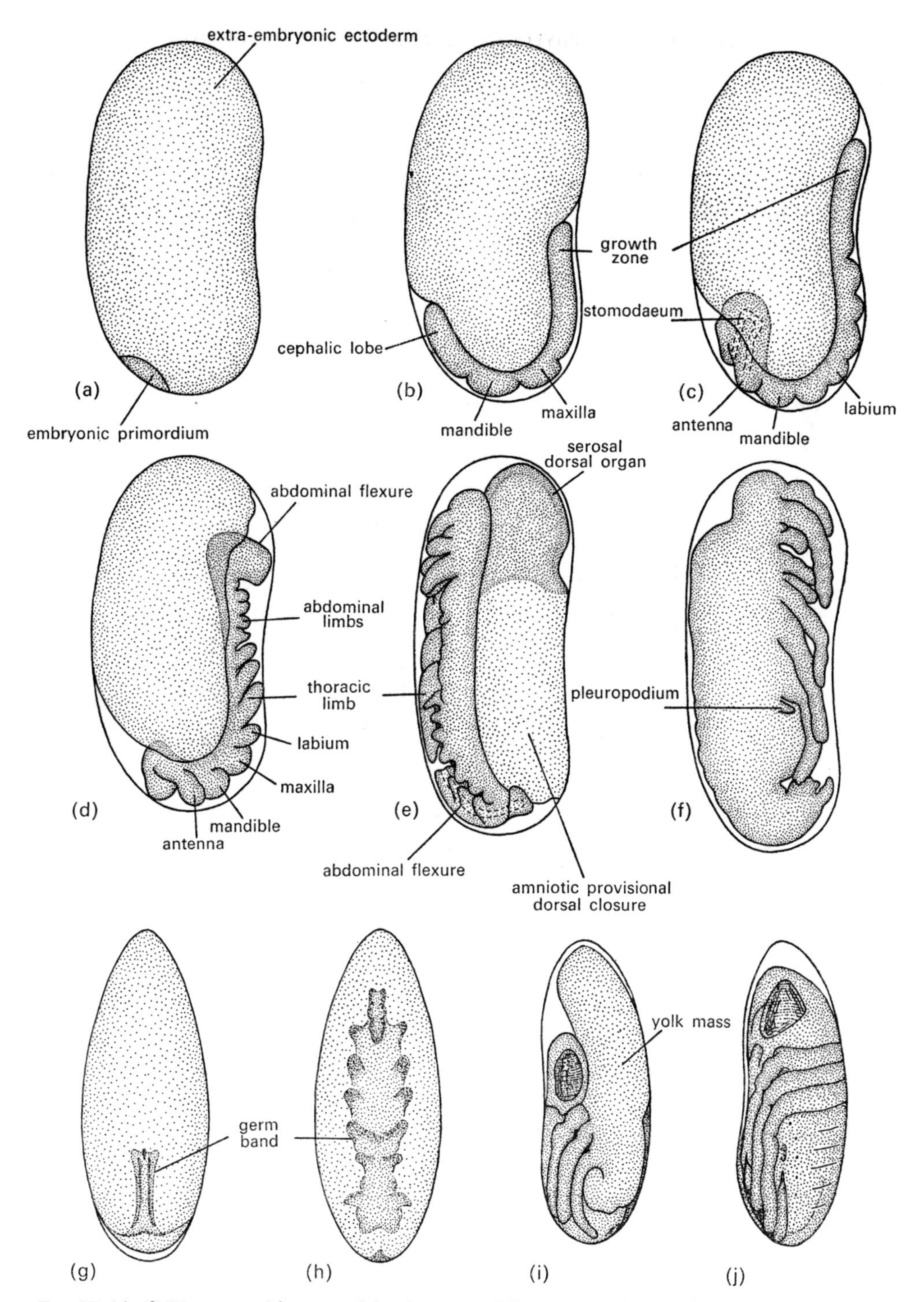

FIG. 97. (a)–(f) The external features of development of *Zootermopsis* (Isoptera), after Striebel (1960), in left lateral view [(a) embryonic primordium; (b) and (c) segmenting germ band; (d) segmented germ band; (e) embryo after katatrepsis; (f) embryo after dorsal closure and 180° rotation on long axis]. (g)–(j) the external features of development of *Saldula* (Heteroptera), after Cobben (1968) [(g) early growth of the germ band and anatrepsis, ventral view; (h) segmented germ band immersed in yolk, ventral view; (i) embryo during katatrepsis, left lateral view; (j) embryo after dorsal closure, left lateral view].

and are followed by the gnathal and thoracic limb buds in their respective segments. The labrum pouches out midventrally between the head lobes, usually as a single rudiment, but sometimes as a pair of rudiments which later fuse (Leuzinger and Wiesmann, 1926; Mellanby, 1937; Roonwal, 1937, for primitive Holometabola and Paraneoptera; Toyama, 1902; Blunck, 1914; Williams, 1916; Johannsen, 1929; Eastham, 1930; Paterson, 1932, 1936; Tiegs and Murray, 1938; Kessel, 1939; Haget, 1955; Bronskill, 1959, 1964; Okada, 1960; Ryan, 1963; Farooqui, 1963; Ullman, 1964; Dupraw, 1967; Anderson and Wood, 1968;

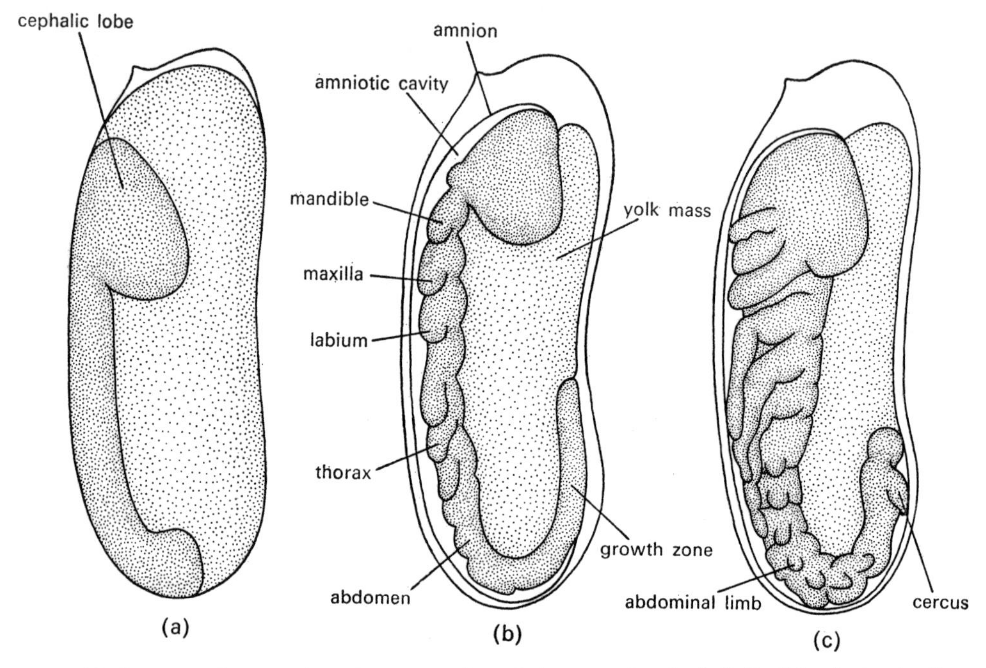

FIG. 98. Stages in the growth and segmentation of the germ band of *Sialis* (Megaloptera), after Du Bois (1938).

Rempel and Church, 1971 a, b, for Holometabola). The paired origin of the labrum is thought to be secondary (Matsuda, 1965). Somewhat later than the other cephalic limb buds, a transient pair of small, premandibular limb buds develops in front of the mandibles in certain Orthoptera (Viallanes, 1891; Roonwal, 1937), Heteroptera (Mellanby, 1937) and many Holometabola (see Anderson, 1972b).

During the completion of abdominal segmentation, the antennae grow longer and move forwards to a preoral position, the pre-mandibular limb buds are resorbed and the gnathal limb buds crowd forwards behind the mouth (Fig. 95). The mandibular and maxillary pairs remain distinct, but the post-maxillary pair, as is well known, fuses to form the labium. In the ventral midline in front of the labium, a median ventral protuberance derived from the sternal ectoderm of the pre-mandibular and gnathal segments develops as the hypopharynx (Heymons, 1895, 1899; Leuzinger and Wiesmann, 1926; Roonwal, 1937; Young, 1953; Matsuda, 1965; Wada, 1966; Scholl, 1969). The thoracic limbs, of course, develop directly as three pairs of locomotory limbs, except in those specialized cases among the

Holometabola (some Coleoptera; Diptera, Siphonaptera; some Hymenoptera) in which they are temporarily reduced and invaginated during the development of an apodous larva.

As segment delineation nears completion along the abdomen, limb buds also begin to develop on the abdominal segments (Figs. 94, 95 and 98). Eleven pairs of abdominal limb buds are formed in the primitive Hemimetabola, the first and last pairs being formed before the intervening pairs develop. The first and last pairs only are developed in the Homoptera and Heteroptera, while the other paraneopterans lack abdominal limb buds. Most orders of Holometabola, in contrast, develop ten pairs of abdominal limb buds. The further development of these limb buds varies. The eleventh pair, where present, form the cerci. The first pair is resorbed in the Ephemeroptera, Odonata, Isoptera, Plecoptera, Embioptera, Dermaptera and most holometabolans, but develops in the Dictyoptera, Cheleutoptera, Orthoptera, Homoptera, Heteroptera, some Coleoptera and some Lepidoptera as temporary pleuropodia, lost before hatching takes place (Graber, 1888 a, b, c, 1890; Wheeler, 1889 b, c, 1890; Heymons, 1895, 1899; Blunck, 1914; Hagan, 1917; Hussey, 1926; Slifer, 1932 b, 1937; Mellanby, 1937; Roonwal, 1937; Salt, 1949; Brookes, 1952; Jones, 1956; Ibrahim, 1957; Görg, 1959; Shulov and Pener, 1959; Mahr, 1960; Rakshpal, 1962; Bedford, 1970; Rempel and Church, 1971 a, b). The most favoured interpretation of pleuropodial function is the secretion of hatching enzymes. The second to seventh pairs of abdominal limbs are resorbed, except where they persist in some Holometabola as larval prolegs. The eighth to tenth pairs contribute to the formation of the external genitalia, and to proleg formation in some Holometabola.

Extra-embryonic membranes

So far, in our discussion of the growth and segmentation of the pterygote germ band and the formation and further development of its limb buds, we have set aside the role of the extra-embryonic ectoderm. Special attention must now be directed to this matter, since the extra-embryonic ectoderm of pterygote embryos becomes elaborated in a manner not shown by any other type of arthropod embryo save, to a degree, the Thysanura. The basic mode of elaboration is found in those primitive Hemimetabola whose germ band remains at the surface of the yolk mass as it grows (the Dictyoptera, Cheleutoptera, Embioptera, Dermaptera, Isoptera) and in certain others whose germ band subsequently becomes immersed (the Ephemeroptera, Orthoptera and Plecoptera). The margin of the embryonic primordium is composed, not of embryonic cells, but of presumptive amnion. As soon as gastrulation and elongation begin, this marginal tissue folds ventrally over the embryonic ectoderm, carrying the edge of the attenuated, extra-embryonic ectoderm with it (Fig. 101). As the folds meet and merge in the ventral midline, the germ band becomes covered by a double layer of extra-embryonic ectoderm (Wheeler, 1889a, 1893; Heymons, 1895; Kershaw, 1914; Hagan, 1917; Leuzinger and Wiesmann, 1926; Roonwal, 1936; Thomas, 1936; Krause, 1938a; Miller, 1939, 1940; Kanellis, 1952; Görg, 1959; Mahr, 1960; Striebel, 1960; Stefani, 1961; Rakshpal, 1962; Bhatnagar and Singh, 1965). The resulting inner layer is the amnion, enclosing an amniotic cavity outside the ventral surface of the germ band. The outer layer, continuous over the yolk mass, is the serosa. In many hemimetabolous pterygotes, the serosa secretes a serosal cuticle beneath the vitelline membrane soon after amnio-serosal separation has occurred (Kershaw, 1914; Slifer, 1937; Miller, 1940; Ibrahim, 1957; Striebel, 1960; Cobben, 1968). Although the mode of formation of the amnion and serosa is more specialized and direct in the primitive Hemimetabola than in the Thysanura discussed in the preceding chapter and does not involve dorsoventral flexure of the germ band, the general similarity of the two groups in this respect is most striking. We shall see below that this similarity becomes even more pronounced during the further development of the amnion and serosa. For the moment, however, we must give brief accord to the development of the extra-embryonic membranes of other pterygotes.

A similar mode of formation of the amnion and serosa is retained (Fig. 101) in the majority of the Holometabola (Patten, 1884; Heider, 1889; Wheeler, 1889a; Hirschler, 1909; Blunck, 1914; Strindberg, 1915a; Paterson, 1931, 1936; Inkmann, 1933; Butt, 1936; Wray, 1937; Du Bois, 1938; Tiegs and Murray, 1938; Bock, 1939; Kessel, 1939; Mulnard, 1947; Luginbill, 1953; Shafiq, 1954; Weglarska, 1955; Ivanova-Kazas, 1959; Krzystofowicz, 1960; Ede, 1964; Ullman, 1964; Anderson, 1966; Jung, 1966a; Küthe, 1966; Craig, 1967; Davis 1967; Zakhvatkin, 1967 a, b; Rempel and Church, 1969). Only the Lepidoptera and the higher Hymenoptera and Diptera show certain functional specializations, beyond the scope of the present discussion (see Anderson, 1972 b). Among the hemimetabolous insects, the Odonata and the paraneopterous orders also show a modified mode of membrane formation, but here it is associated with anatrepsis (Shinji, 1919; Seidel, 1924, 1929; Mellanby, 1936; Schölzel, 1937; Johannsen and Butt, 1941; Böhmel and Jancke, 1942; Butt, 1949; Gerwel, 1950; Goss, 1952; Piotrowski, 1953; Sander, 1959; Ando, 1962; Behrendt, 1963; Cobben, 1968; Mori, 1969, 1970). As the growing germ band turns into the yolk, the presumptive amnion at the posterior and lateral margins of the germ band folds over in the usual way to enclose an amniotic cavity, but the cavity remains open to the exterior at the point of inturning. The extra-embryonic ectoderm which will give rise to the serosa does not contribute to the amniotic folds. A delayed formation of anterior amniotic folds finally takes place around the margins of the head lobes and the amniotic cavity is then closed off.

Katatrepsis

Following the completion of elongation and segment formation, the hemimetabolan germ band undergoes further growth and development, including elongation of the cephalic and thoracic limbs and formation of the abdominal limb buds, as it lies within the amniotic cavity. This period of development then comes to an abrupt end with the fusion and rupture of the extra-embryonic membranes in the vicinity of the head of the embryo. The extra-embryonic membranes now roll back over the yolk mass, exposing the germ band at the surface. The serosa becomes concentrated to form a dorsal organ on the anterodorsal surface of the yolk, mass, and the amnion replaces the serosa as a covering epithelium on the yolk mass, just as it does during blastokinesis in the Thysanura.

In the Dictyoptera (Fig. 94), little or no movement of the germ band takes place as the membranes rupture and roll back (Bruce, 1887; Wheeler, 1889 a; Heymons, 1895; Hagan, 1917; Görg, 1959). The germ band retains its definitive position along the ventral surface of the yolk mass, with the posterior end of the abdomen flexed downwards and forwards.

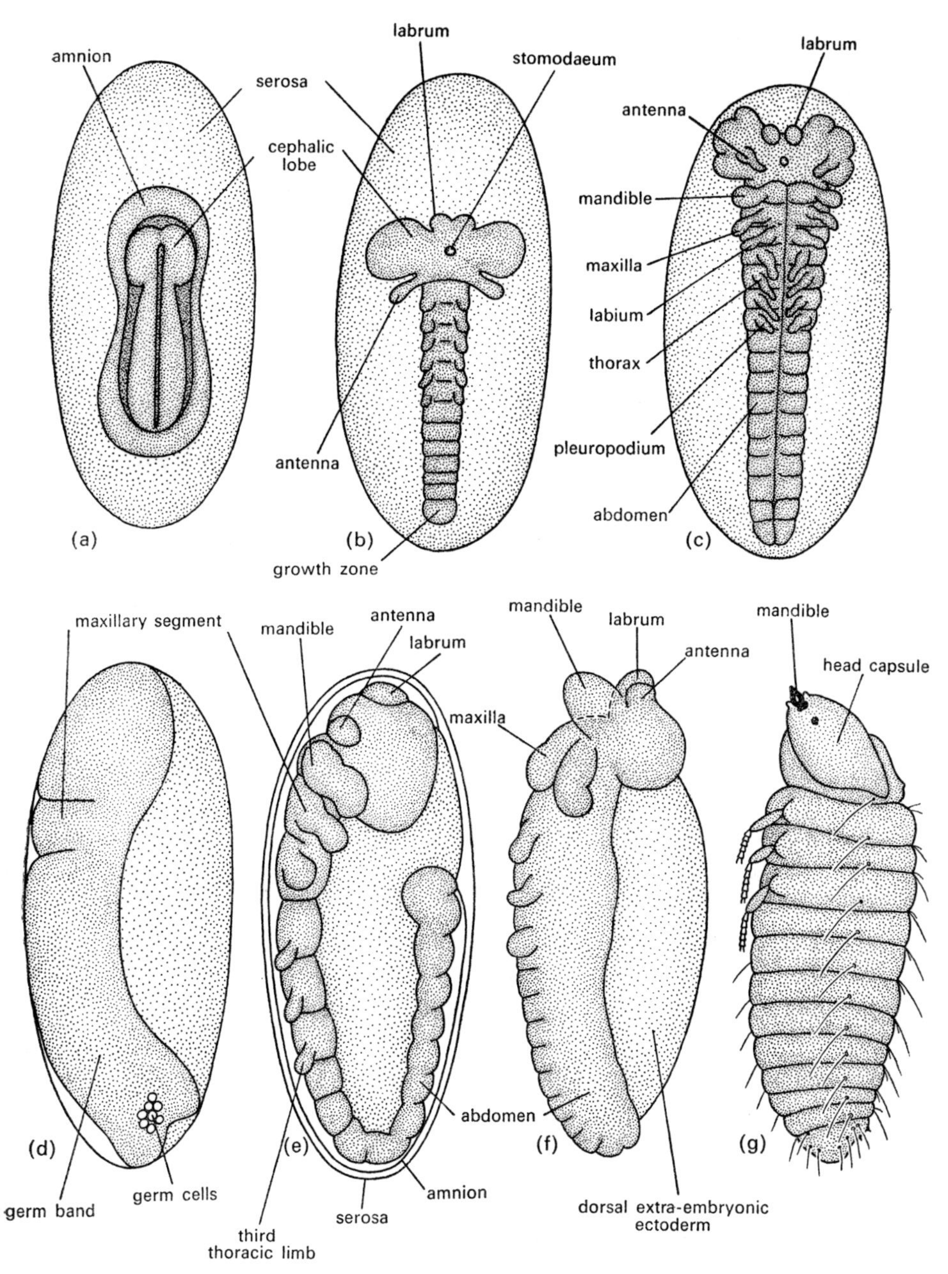

FIG. 99. (a)–(c) Stages in the growth and segmentation of the germ band of *Tenebrio* (Coleoptera), after Ullman (1964); (d)–(g) the external features of development of *Bruchidius*, after Jung (1966a), in left lateral view [(d) early segmenting germ band; (e) segmented germ band; (f) during shortening of the germ band; (g) about to hatch].

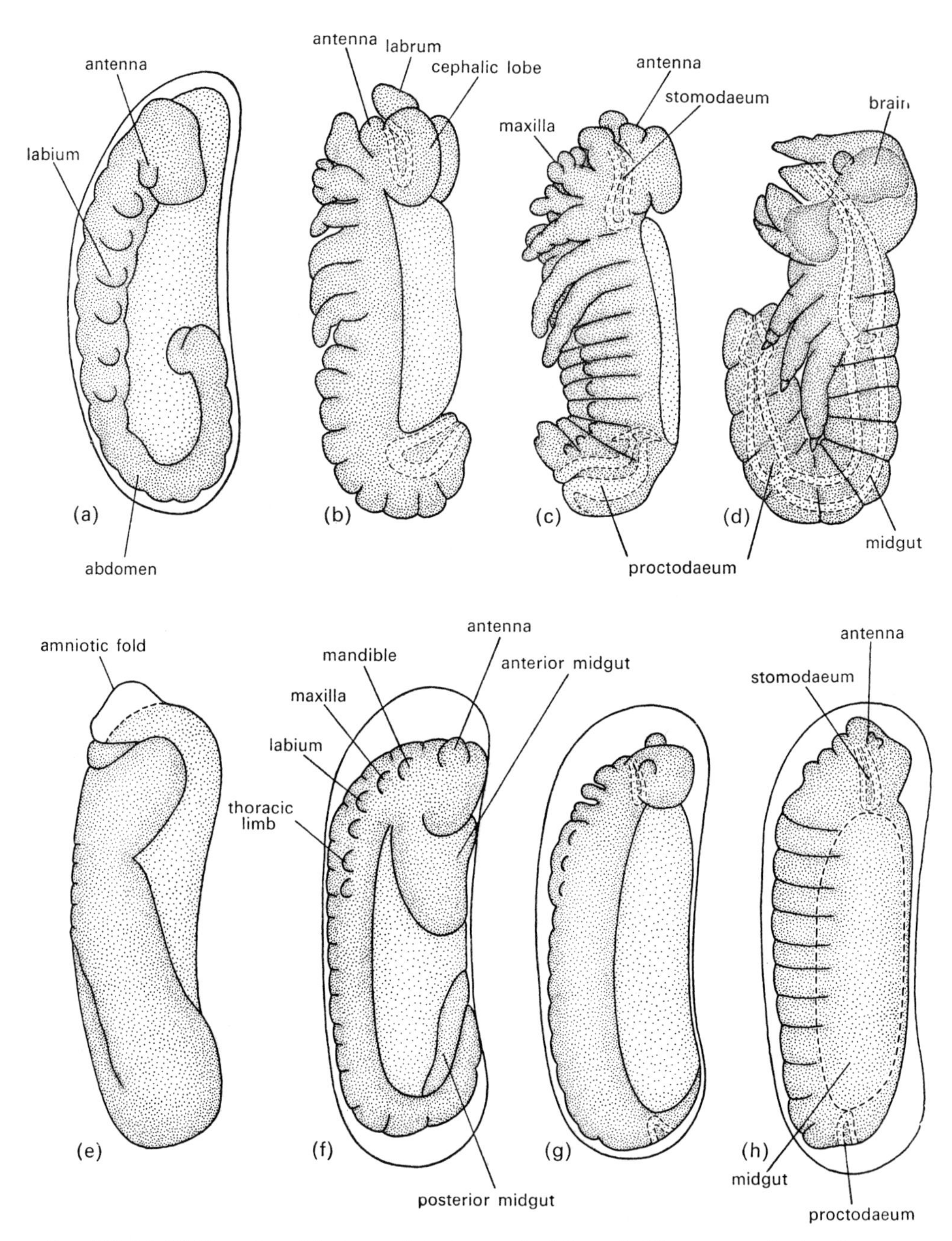

FIG. 100. (a)–(d) The external features of development of *Pontania* (Hymenoptera), after Ivanova-Kazas (1959), in left lateral view [(a) the segmented germ band; (b) embryo during shortening of the germ band; (c) embryo during dorsal closure; (d) embryo approaching hatching)]. (e)–(h) The external features of development of *Apis* (Hymenoptera), after Schnetter (1934), in left lateral view [(e) segmenting germ band; (f) segmented germ band; (g) shortening and dorsal closure; (h) approaching hatching].

All other hemimetabolous embryos, however, reach this stage with the germ band on the dorsal side of the yolk mass and the head facing the posterior pole. Several orders also have the additional complication that the germ band is immersed in the yolk as a result of anatrepsis (Figs. 94–96). Rupture of the extra-embryonic membranes in these species is always followed by a rapid migratory movement which brings the germ band into its definitive position relative to the yolk mass. This movement is called katatrepsis (Wheeler, 1893). Usually, katatrepsis is a straight migratory movement of the germ band over the posterior pole of the yolk mass and forwards along the ventral surface, though some species exhibit additional complications (see Anderson, 1972a, for further discussion and references). Contraction of the serosa to form a dorsal organ and reflexion of the amnion over the surface of the yolk mass accompany the movement. The general result is to establish the germ band on the ventral surface of the yolk, with the anterior end at the anterior pole of the yolk mass and the posterior end of the abdomen flexed downwards and forwards, in the same position as the thysanuran and dictyopteran germ band at the same stage. Katatrepsis is clearly a specialized process evolved within the pterygotes in association with the earlier growth of the segmenting germ band in the "wrong" direction relative to the anteroposterior axis of the egg, but the functional reasons for such specialized procedures remains obscure.

Not unexpectedly, these complications are avoided in the Holometabola, since the anterior end of the germ band retains its ventral, anteriorly directed location on the surface of the yolk mass during growth and segmentation. As discussed earlier, however, it is usual for the posterior end of the germ band to be extended around the posterior pole of the yolk mass onto the dorsal surface by the time growth and segmentation are completed. When the embryonic membranes rupture and roll back, the germ band does not move as a whole, but contracts strongly in length, so that the posterior end withdraws from the dorsal surface to the posterior pole of the yolk mass (Figs. 99 and 100). This process of shortening, characteristic of holometabolan embryos as an accompaniment to membrane rupture, can be regarded as a specialization consequent on the earlier formation of a large germ band relative to the yolk mass. Rupture and reflection of the embryonic membranes and formation of a serosal dorsal organ follow the same course as in the Dictyoptera in many generalized Holometabola (Patten, 1884; Heider, 1889; Hirschler, 1889; Blunck, 1914; Strindberg, 1915a; Paterson, 1932; Du Bois, 1938; Bock, 1939; Deobahkta, 1957; Ullman, 1964; Jung, 1966a; Küthe, 1966; Rempel and Church, 1971a), though a variety of specializations is found in the Coleoptera, Lepidoptera, Siphonaptera, Diptera and Hymenoptera (Anderson, 1972b).

Dorsal closure

When the extra-embryonic membranes have ruptured and rolled back, the serosal dorsal organ has formed and katatrepsis has occurred, the majority of pterygote embryos have the same general configuration. The segmented germ band, with its well-developed cephalic and thoracic limbs and its developing abdominal limbs, occupies the ventral surface of the yolk mass. The lateral and dorsal surfaces of the yolk mass are covered by the attenuated amnion and the serosa is reduced to an anterodorsal vestige. The major events in the completion of external development are now dorsal closure, further development of the head and limbs and secretion of the exoskeleton (Figs. 94, 95, 99 and 100). Dorsal closure takes place by an upgrowth of the lateral walls of the segments towards the dorsal midline. The dorsal organ is simultaneously resorbed into the yolk mass, followed by the shrinking vestige of the amnion. In some pterygote orders, notably the Orthoptera and Cheleutoptera, an embryonic cuticle is formed and shed before the first definitive nymphal cuticle is secreted (Jones, 1956; Bedford, 1970).

Throughout the external development of the more generalized pterygote embryos, the sequence of events is highly reminiscent of the external development of the Thysanura. The only marked difference is that the pterygotes no longer retain the dorsoventral flexure of the germ band that the Thysanura share in common with the other apterygotes. We shall see below that the same close similarity is evident in the details of development of the mesoderm and embryonic ectoderm after gastrulation.

Further Development of the Mesoderm

As soon as the gastrulation movement of the mesoderm into the interior is completed in a pterygote embryo and the mesoderm cells have spread out as a single layer beneath the ectoderm of the germ band, further changes begin. Anteriorly, the mesoderm cells begin to express the first stages of organogeny. Posteriorly, proliferation of new cells usually continues, giving rise to the mesoderm of further segments. In the Dictyoptera, Dermaptera and Odonata, the Paraneoptera and the majority of Holometabola, orders in which the blastodermal presumptive areas are relatively elongated, only the abdominal mesoderm is formed by posterior proliferation after gastrulation. In orders in which the embryonic primordium is very short, as in the Orthoptera, Cheleutoptera and Isoptera, posterior proliferation is the source of all of the mesoderm behind the head lobes. In contrast, in the exceptionally long primordia of the higher Diptera and Hymenoptera, posterior proliferation of the mesoderm after gastrulation does not take place.

The first stage in the further development of the mesoderm (Fig. 101) is a bilateral concentration of the mesoderm cells towards the sides of the germ band, giving two, thick lateral bands and a thin, median strip. The lateral bands now typically separate into segmental blocks, forming paired somites. It is usual in the Hemimetabola, for example, in the Isoptera and Orthoptera (Roonwal, 1937; Striebel, 1960), for the gnathal and thoracic segments to develop more or less simultaneously, followed by those in front of the mandibular segment, than those along the abdomen. An anteroposterior sequence of formation of the somites, in contrast, is more common among the Holometabola (e.g. Ullman, 1964).

Among hemimetabolous pterygotes, the somites become hollow in two ways. Either an internal split develops in each mesodermal block, as in the Cheleutoptera, some Orthoptera, most Paraneoptera and the abdominal somites of the Isoptera and certain Orthoptera (Heymons, 1895; Leuzinger and Wiesmann, 1926; Thomas, 1936; Mellanby, 1937; Roonwal, 1937; Schölzel, 1937; Böhmel and Jancke, 1942; Goss, 1952; Striebel, 1960), or the lateral part of each block folds over the median block to enclose a coelomic cavity, as in the Odonata, the Dictyoptera and the gnathal and thoracic segments of the Isoptera and certain Orthoptera (Wheeler, 1889a; Faussek, 1911; Roonwal, 1937; Johannsen and Butt, 1941; Ibrahim, 1957; Görg, 1959; Striebel, 1960). Internal splitting is more usual in the Holometabola (Hirschler 1909; Eastham, 1927; Paterson, 1932; Butt, 1936; Mansour, 1937; Tiegs and Murray, 1938; Bock, 1939; Kessel, 1939; Luginbill, 1953; Deobahkta,

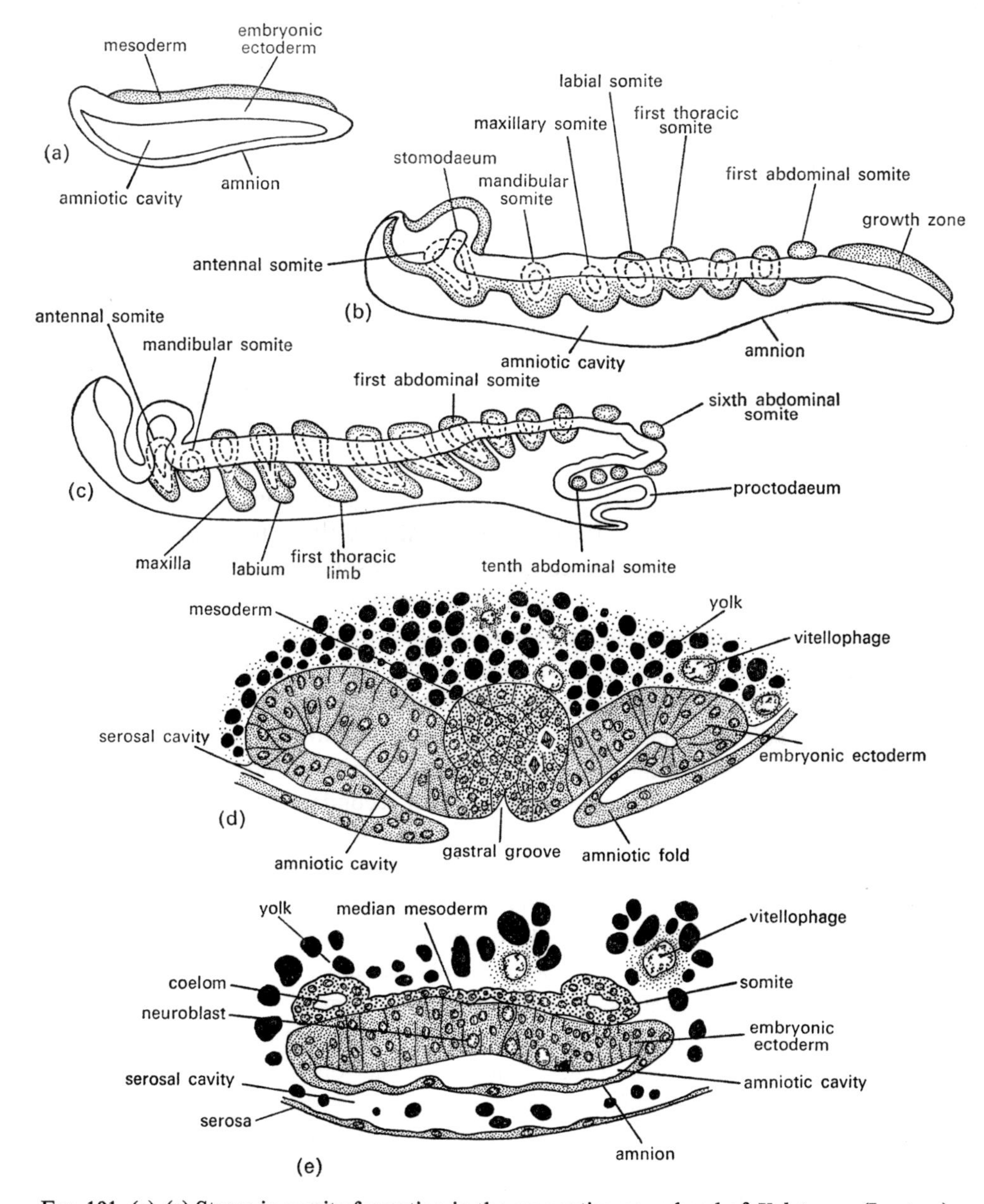

FIG. 101. (a)–(c) Stages in somite formation in the segmenting germ band of *Kalotermes* (Isoptera), drawn diagrammatically in left lateral view; (d) transverse section through the embryonic primordium of *Tenebrio* (Coleoptera) during gastrulation and formation of the amniotic folds; (e) transverse section through the segmenting germ band of *Tenebrio*. [(a)–(c) after Striebel, 1960; (d) and (e) after Ullman, 1964; modified from Anderson, 1972a, b.]

1957; Jura, 1957; Ullman, 1964; Rempel and Church, 1969), with only occasional reports of the enclosure of coelomic cavities by folding. In many holometabolous embryos, somite formation is partially or wholly suppressed (see Anderson, 1972b).

In the primitive Hemimetabola, the labial, thoracic and first abdominal somites are large, while the mandibular, maxillary and second to tenth abdominal pairs are relatively small (Fig. 101). Behind the tenth pair of abdominal somites, a small mass of mesoderm persists around the proctodaeum. In the Dictyoptera and Orthoptera, a pair of temporary cavities develops in this mass (Wheeler, 1889a; Heymons, 1895; Roonwal, 1937). The generalized Holometabola have a similar pattern of somite development, except that the first abdominal somites are small and the residual posterior mass of mesoderm behind the tenth abdominal somites never develops cavities.

In front of the mandibular somites, the mesoderm continues forward on either side of the stomodaeum, beneath the head lobes, and meets anteriorly beneath the labrum. Many primitive Hemimetabola develop a distinctive series of somites in this region. A large antennal pair is formed just behind the level of the stomodaeum (Leuzinger and Wiesmann, 1926; Roonwal, 1937; Striebel, 1960; Scholl, 1965). Then, in the mesoderm linking the antennal to the mandibular somites, a pair of vestigial premandibular somites is developed (Wheeler, 1889a, 1893; Heymons, 1895; Leuzinger and Wiesmann, 1926; Roonwal, 1937; Jhingran, 1947; Striebel, 1960; Scholl, 1965). In front of the antennal somites, some of the mesoderm also aggregates as a pair of vestigial pre-antennal somites (Heymons, 1895; Leuzinger and Wiesmann, 1926; Mellanby, 1937; Roonwal, 1937; Miller, 1940; Jhingran, 1947; Striebel, 1960; Scholl, 1969). It is generally agreed that six pairs of somites is the basic number in the developing pterygote head. The generalized Holometabola confirm this (Fig. 102), since they too have a small pre-antennal, large antennal and small pre-mandibular pair in front of the mandibular pair (Eastham, 1927, 1930; Bock, 1939; Ullman, 1964; Rempel and Church, 1971a). In most Holometabola, cephalic somite formation is variously suppressed in favour of a more direct development of cephalic mesodermal structures.

The further development of the somites is remarkably uniform throughout the pterygotes (Fig. 102), except in those holometabolans in which secondary specialization has clearly occurred. The labial and trunk somites all develop in a similar manner (primitive Hemimetabola, Heymons, 1895; Leuzinger and Wiesmann, 1926; Roonwal, 1937; Miller, 1939; Ibrahim, 1957; Görg, 1959; Striebel, 1960; Paraneoptera, Mellanby, 1937; Butt, 1949; Holometabola, Bock, 1939; Ullman, 1964). Each somite pouches out ventrolaterally as an appendicular lobe and expands dorsally as a dorsolateral lobe, although the ventral part of the somite does not extend towards the ventral midline as a medioventral lobe. The appendicular lobe gives rise to the intrinsic musculature of the attendant limb bud, except in those abdominal segments that later lose their limb buds. The somatic wall of the dorsolateral lobe develops as dorsal longitudinal muscle, fat body, pericardial septum and, at its dorsal margin, cardioblasts. During the upward spread of mesoderm that accompanies dorsal closure, the cardioblasts come together in the dorsal midline in the usual way to form the heart. The somatic wall of the ventral part of the somite also gives rise to fat-body cells, as well as to ventral longitudinal muscles and extrinsic limb muscles. The splanchnic walls of the somites separate off and become applied to the midgut strands as splanchnic mesoderm. In addition to giving rise to the splanchnic musculature and additional fat body cells, the

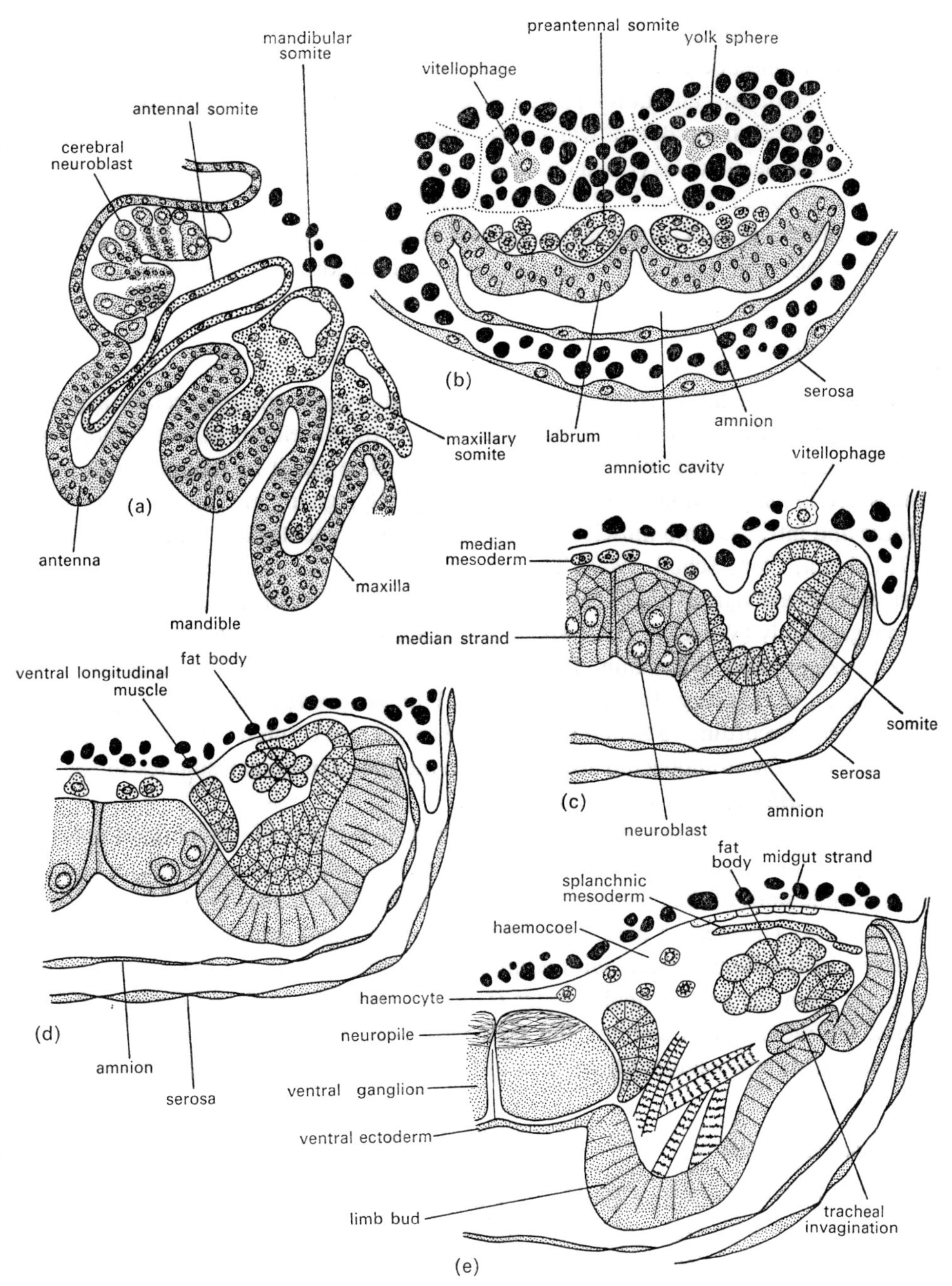

FIG. 102. (a) Parasagittal section through the antennal and gnathal segments of *Tenebrio* (Coleoptera) at a stage when the gnathal limb buds are beginning to crowd forwards around the mouth; (b) transverse section through the labral region of the segmented germ band of *Tenebrio*; (c)–(e) stages in the further development of a thoracic somite of *Chrysopa* (Neuroptera) in transverse section. [(a) and (b) after Ullman, 1964; (c)–(e) after Bock, 1939; modified from Anderson, 1972b].

splanchnic mesoderm of pterygote embryos is the site of development of the gonads and genital ducts (see below).

The residual mesoderm behind the tenth pair of abdominal somites becomes associated with the proctodaeum and develops as the splanchnic mesoderm of the proctodaeum. The thin sheet of median mesoderm which lies between the paired segmental somites normally gives rise to the haemocytes (Heymons, 1895; Hirschler, 1909; Nelson, 1915; Leuzinger and Wiesmann, 1926; Eastham, 1930; Smreczyński, 1932; Butt, 1936; Mellanby, 1937; Roonwal, 1937; Bock, 1939; Görg, 1959; Wigglesworth, 1959; Amy, 1961), although the origin of haemocytes is often more diffuse in embryos in which somite formation is suppressed.

Cephalic somites

Even though the labial somites of generalized pterygote embryos retain a pattern of development similar to that of the thoracic somites, the development of the more anterior cephalic somites displays the usual specializations. The small mandibular and maxillary somites retain only a limited developmental capacity, giving rise to no more than the extrinsic and intrinsic limb muscles of their respective segments (Roonwal, 1937; Jhingran, 1947; Ullman, 1964; Scholl, 1969). The pre-mandibular somites are even more specialized. Most authors are agreed that they give rise, as in the Thysanura, to the glandular suboesophageal body (Wheeler, 1893; Heymons, 1895, 1899; Strindberg, 1913, 1915a, 1915d, 1916; Leuzinger and Wiesmann, 1926; Eastham, 1930; Wray, 1937; Tiegs and Murray, 1938; Miller, 1939; Görg, 1959; Kessel, 1961; Ullman, 1964; Anderson and Wood, 1968; Scholl, 1969). According to Scholl (1969) and Rempel and Church (1971a), certain muscles associated with the stomodaeum also arise from the pre-mandibular somites in the cheleutopteran *Carausius* and the coleopteran *Lytta*.

The large antennal somites retain more of the basic features of somite development. The appendicular lobe is typical. The somatic wall of the rest of the somite forms fat body, while the splanchnic walls of the paired somites come together above the stomodaeum to form the anterior aorta. The median mesoderm between the antennal somites forms the splanchnic mesoderm of the stomodaeum, with additional contributions in the coleopterans *Tenebrio* and *Lytta* from the preantennal somites (Ullman, 1964; Rempel and Church, 1971a). In general, the pre-antennal somites and their associated median mesoderm develop as the labral and other musculature at the front of the head (Ullman, 1964; Scholl, 1969).

Gonads and gonoducts

In the majority of primitive Hemimetabola and in the Heteroptera, the primordial germ cells either first become recognizable in the splanchnic walls of certain abdominal somites (Orthoptera, Wheeler, 1893; Roonwal, 1937) or are segregated from the posterior end of the mesoderm during gastrulation and migrate subsequently to the splanchnic walls of the abdominal somites (Dictyoptera, Cheleutoptera, Embioptera, Heteroptera; Heymons, 1895; Seidel, 1924; Leuzinger and Wiesmann, 1926; Mellanby, 1936, 1937; Stefani, 1961; Cavallin, 1969). Once in this location (Fig. 103), the segmental groups of germ cells proliferate and merge to form two continuous strands, each covered by a thin splanchnic epithelium. The gonads arise from these strands. In other pterygotes (the Dermaptera, Pscocoptera, Homoptera and Holometabola), the early segregation of primordial germ cells is followed by a more direct development of the gonads. The germ cells become distinct either during gastrulation or, in exceptional cases, during blastoderm formation (see p. 216), again in the vicinity of the posterior end of the presumptive mesoderm or the adjacent presumptive posterior midgut and proctodaeum. Once internal, they segregate as two groups which become embedded in the splanchnic mesoderm strands. A temporary distribution as segmental

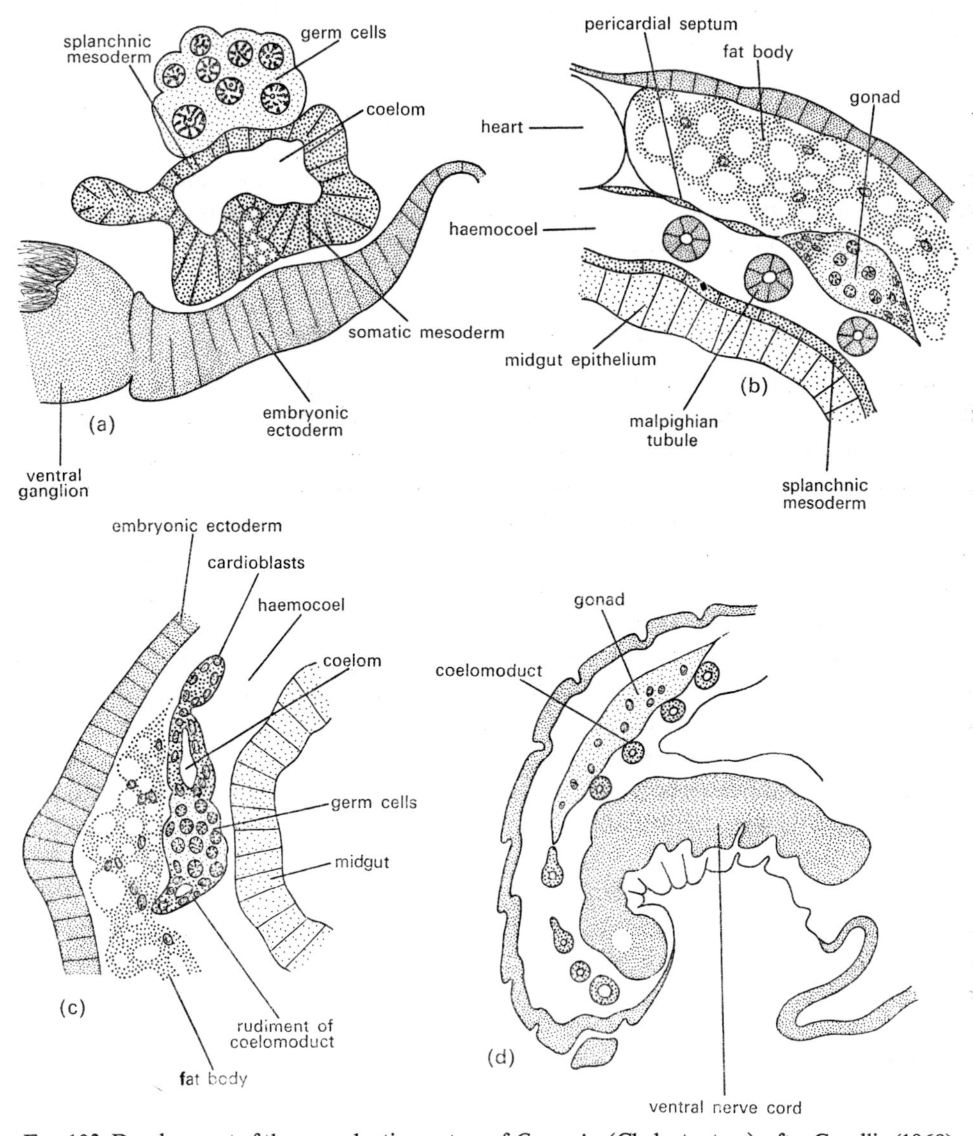

FIG. 103. Development of the reproductive system of *Carausius* (Cheleutoptera), after Cavallin (1968). (a) Early stage in development of abdominal segment, in transverse section; (b) transverse section through abdominal segment after formation of gonad; (c) transverse section through fourth abdominal segment, showing rudiment of coelomoduct; (d) parasagittal section through the abdomen of an embryo at the same stage as Fig. 95d, showing arrangement of gonad and coelomoducts.

groups in the abdominal mesoderm has been observed in the Coleoptera (Stanley and Grundman, 1970; Rempel and Church, 1971a). Mesoderm cells cluster around them, completing a pair of compact gonads (see Anderson, 1972 a, b, for references).

Since the association of germ cells with abdominal somites is obviously the more generalized condition, we can confine our account of gonoduct development to species of this type. Typically, as in the Dictyoptera, Isoptera, Cheleutoptera, Orthoptera and Heteroptera (Heymons, 1891, 1895; Wheeler, 1893; Strindberg, 1913; Seidel, 1924; Leuzinger and Wiesmann, 1926; Roonwal, 1937; Jhingran, 1947; Cavallin, 1969, 1970), the splanchnic walls of some of the abdominal somites give rise to paired coelomoducts, of which one or more posterior pairs enlarge and persist as gonoducts while the remainder disappear. In the cheleutopterans, Cavallin (1969, 1970) has found that the seventh and eighth abdominal somites contribute to the female gonoducts, the eighth and ninth to the male gonoducts. The opisthogoneate gonopore, which develops on the seventh or eighth abdominal segment in females and the eighth or ninth segment in males, is formed by a median ventral invagination of ectoderm.

Further Development of the Ectoderm

Like the embryonic ectoderm of the Thysanura, the pterygote embryonic ectoderm exhibits an early differentiation of neuroblasts as a preliminary to the proliferation of segmental ganglia (Figs. 101 and 102). As the activity of the growth zone adds progressively to the posterior end of the germ band, the mandibular and succeeding ectoderm develops a longitudinal thickening on either side of the ventral midline, producing paired neural ridges bordering a neural groove. The thickening is due to the proliferative activity of three to five rows of neuroblasts formed by the enlargement of ventral ectoderm cells (primitive Hemimetabola, Wheeler, 1893; Heymons, 1895; Baden, 1936; Thomas, 1936; Roonwal, 1937; Ibrahim, 1957; Görg, 1959; Striebel, 1960; Paraneoptera, Seidel, 1924; Mellanby, 1937; Goss, 1952, 1953; Holometabola, Ullman, 1967; Rempel and Church, 1969). Through teloblastic budding, the neuroblasts proliferate radial rows of small cells inwards as ganglion cells. The neuroblasts sink slightly into the interior as they bud, leaving other ectoderm cells as dermatoblasts at the surface. Between the neural ridges, a median strand of tall ectoderm cells forms the floor of the neural groove, with occasional neuroblasts differentiated along its length.

As the number of cells proliferated by the neuroblasts increases, the radial cell rows merge to form ganglionic cell masses associated with the paired mesodermal somites. In front of the mandibular segment, the neuroblast rows diverge on either side of the stomodaeum and are much broader over the anterior parts of the head lobes. Ullman (1967), Malzacher (1968), Scholl (1969) and Rempel and Church (1971a) have recently made detailed studies of the development of the brain in *Tenebrio*, *Carausius*, *Periplaneta* and *Lytta*. The neuroblasts give rise to paired pre-mandibular or tritocerebral ganglia behind the stomodaeum, paired antennal or deutocerebral ganglia on either side of the stomodaeum, and large, paired, protocerebral masses in front of the stomodaeum (Fig. 104).

The early development of the brain and ventral nerve cord from neuroblasts takes place before the extra-embryonic membranes rupture and roll back. During this time, the more

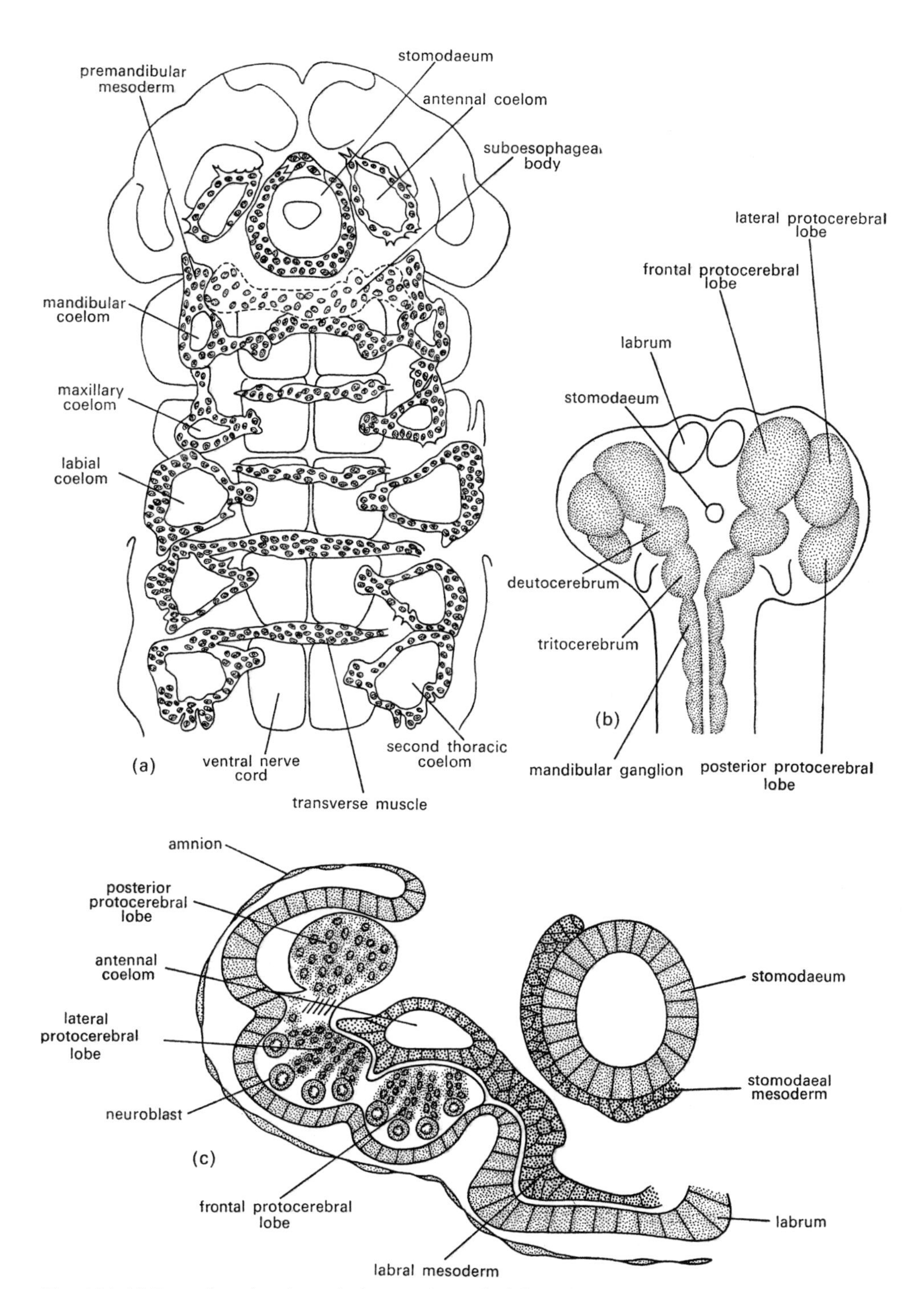

FIG. 104. (a) Frontal section through the anterior end of the germ band of *Carausius* (Cheleutoptera) at a stage intermediate between Figs. 95b and 95c; (b) diagrammatic reconstruction of the cephalic ganglia of the segmented germ band of *Tenebrio* (Coleoptera); (c) frontal section through developing cephalic lobes of *Locusta* (Orthoptera). [(a) after Scholl, 1969; (b) after Ullman, 1967; (c) after Roonwal, 1937.]

lateral embryonic ectoderm pouches out as limb buds and produces a variety of ectodermal invaginations. In due course, after membrane rupture has occurred, the lateral embryonic ectoderm spreads dorsally towards the midline, effecting dorsal closure.

Ectodermal invaginations on the trunk segments gives rise to the thoracic and abdominal apodemes, the tracheal system and the genital opening (see Anderson, 1972 a, b, for references). Invaginations on the head develop as cephalic apodemes and glands. The largest ectodermal ingrowths on the pterygote head are the salivary glands (Fig. 93), which have been found by many workers to originate near the bases of the labial limb buds. When the labial limb buds fuse, the paired openings of the salivary glands come together in the ventral midline to form a median opening on to the hypopharynx. Many holometabolans also have mandibular glands and maxillary glands of similar ectodermal origin (Eastham, 1930; Smreczyński, 1932; Wray, 1937; Tiegs and Murray, 1938).

Other ectodermal invaginations on the head (Fig. 104) form the arms of the tentorium. The anterior tentorial arms originate in the ventrolateral ectoderm between the antennal and mandibular segments. The posterior tentorial arms invaginate from the ventrolateral ectoderm between the maxillary and labial segments. At the same time, between the anterior and posterior pairs of tentorial invaginations, a pair of invaginations at the bases of the mandibles gives rise to the mandibular apodemes (primitive Hemimetabola, Heymons, 1895; Riley, 1904; Strindberg, 1913; Roonwal, 1937; Striebel, 1960; Ando, 1962; Matsuda, 1965; Wada, 1966; Scholl, 1969; Vignau-Rogueda, 1969; Paraneoptera, Mellanby, 1937; Sander, 1959; Holometabola, Heider, 1889; Carrière and Bürger, 1897; Strindberg, 1913, 1915a, 1917; Nelson, 1915; Eastham, 1930; Paterson, 1932, 1936; Smreczyński, 1932; Wray, 1937; Tiegs and Murray, 1938; Kessel, 1939; Baerends and Baerends-von Roon, 1950; Deobahkta, 1957; Presser and Rutchsky, 1957; Ivanova-Kazas, 1959; Okada, 1960; Amy, 1961; Schoeller, 1964; Ullman, 1967; Rempel and Church, 1971b).

The remaining important structures arising as invaginations of the cephalic ectoderm in the pterygote embryo are the corpora allata. Their origin is closely associated with that of the posterior tentorial arms and has been traced in several species to the region of the mandibular–maxillary intersegment (Heymons, 1895, 1897; Carrière and Bürger, 1897; Janet 1899; Strindberg, 1913; Nelson, 1915; Leuzinger and Wiesmann, 1926; Smreczyński, 1932; Paterson, 1936; Roonwal, 1937).

The further development of the brain and ventral nerve cord proceeds in the same general manner throughout the pterygotes. The many descriptions of this process are listed by Anderson (1972 a, b), but for detailed accounts of individual species the reader is referred to the work of Roonwal (1937), Tiegs and Murray (1938), Poulson (1950), Striebel (1960), Ullman (1967), Malzacher (1968), Rempel and Church (1969, 1971a) and Scholl (1969). As the segmental ganglionic masses become more compact and move into the interior of the embryo, they become linked together through the development of neuropile to form the ventral nerve cord. Three pairs of gnathal ganglia, three thoracic pairs and ten abdominal pairs always develop. An eleventh pair of abdominal ganglia has been identified in certain primitive Hemimetabola (Graber, 1890; Heymons, 1895, 1896a; Roonwal, 1937). In later development, the gnathal ganglia fuse to form the suboesophageal ganglion, and the posterior abdominal ganglia also fuse. In many paraneopterans and some Holometabola, a much greater condensation and fusion of the ventral ganglia takes place.

The median strand has been interpreted in various ways by different authors. The intra-segmental regions of the median strand are known to be incorporated into the segmental ganglia (Baden, 1937; Ullman, 1967; Springer, 1967; Springer and Rutschky, 1969). The intersegmental regions are said by some workers on holometabolous embryos to contribute to the ganglia, by others to the neurilemma of the nerve cord. For many species, however, the neurilemma has been found to originate from peripheral ganglion cells (Mazur, 1960; Ashhurst, 1965; Springer, 1967; Springer and Rutschky, 1969) and the fate of the inter-segmental regions of the median strand remains uncertain.

The development of the antennal and pre-mandibular ganglia follows the same course as that of the ventral ganglia, except that the antennal ganglia move forward in front of the mouth and do not develop a transverse commissure. The commissure connecting the pre-mandibular ganglia (the tritocerebral commissure) remains post-oral. There has been considerable controversy in pterygote embryology over the existence of a pair of distinct pre-antennal ganglia in the head. Malzacher (1968) and Scholl (1969) have now provided convincing evidence that pre-antennal ganglia are present in the embryos of *Carausius* and *Periplaneta*. At the front of the head, the neuroblasts on either side of the labrum give rise to the paired lobes of the protocerebrum (Fig. 104), connected by the supraoesophageal commissure. The lateral, optic ganglia of the protocerebrum develop as separate prolifera-tions of the anterolateral ectoderm of the head lobes, without the intervention of neuro-blasts (Viallanes, 1891; Cholodkowsky, 1891; Wheeler, 1893; Heymons, 1895; Baden, 1936; Roonwal, 1937; Ando, 1962; Ullman, 1967; Malzacher, 1968).

Hatching

Like the apterygotes, pterygote embryos hatch directly from their enclosing membranes. Hatching thus involves the rupture of the chorion and vitelline membrane. In species in which a serosal cuticle is secreted, the cuticle appears to be digested by enzymes secreted by the pleuropodia during the later part of embryonic development.

The duration of development varies greatly among species, but as a general rule is more prolonged in hemimetabolous species than in the Holometabola. Typical examples of the duration of development in the primitive Hemimetabola include *Carausius morosus*, 85 days (Fournier, 1967), *Kalotermes flavicollis*, 54 days (Striebel, 1960) and the ephemeropteran *Baetis rhodani*, 45 days (Illies, 1968). On the other hand, *Locusta migratoria* hatches after 13 days (Salzen, 1960). The latter duration approximates to that of various Paraneoptera, but most holometabolous embryos hatch in 10 days or less. Extremely rapid development is observed among the cyclorrhaphous Diptera. *Drosophila melanogaster* hatches after 22 hours (Ede and Counce, 1956) and *Lucilia sericata* after 20–24 hours (Davis, 1967).

The Basic Pattern of Development in Pterygotes

In spite of their immense variability, the pterygotes reveal their standing as a unitary group by the ease with which all of their modes of embryonic development can be related to a single common theme. Much of this theme is further shared in common with the aptery-gotes and, in particular, with the Thysanura, although there are some significant differences, a matter which will be further discussed in Chapter 10.

Pterygote development begins with the formation of a uniform blastoderm covering a yolk mass which contains vitellophages. Within the blastoderm, the presumptive areas centre on a narrow band of presumptive mesoderm cells in the ventral midline. The presumptive embryonic areas are broadly spread at this stage, but become concentrated together to form an embryonic primordium, with associated attenuation of the presumptive extra-embryonic ectoderm, before gastrulation begins.

The presumptive midgut comprises two small areas, also in the ventral midline; an anterior midgut area in front of the presumptive mesoderm and a posterior midgut area behind the presumptive mesoderm. These areas are distinct from the vitellophages within the yolk, which arise during cleavage and blastoderm formation but do not subsequently contribute to the formation of the midgut epithelium. The presumptive stomodaeum lies as an arc around the anterior face of the presumptive anterior midgut. The presumptive proctodaeum forms a small area in the midline behind the presumptive posterior midgut.

Lateral to the midventral areas, and more broadly spread around the anterior end of the latter, is a bilateral band of presumptive embryonic ectoderm, with a marginal zone of presumptive amnion. The broad anterior parts are the presumptive pre-antennal and antennal ectoderm. The narrower, post-antennal bands consist of the presumptive pre-mandibular, gnathal and thoracic segmental ectoderm, followed by growth zone ectoderm. When blastodermal differentiation has occurred, the embryonic primordium occupies the ventral midline of the surface of the yolk mass and is nearer to the posterior pole than to the anterior pole.

Gastrulation takes place by a midventral immigration of the mesoderm and midgut rudiments, accompanied by the temporary formation of a gastral groove and followed by closure of the groove through the median approximation of the embryonic ectoderm. The midgut rudiments now lie internal to the presumptive stomodaeum and proctodaeum respectively. The mesoderm spreads out as a layer beneath the embryonic ectoderm. Within this layer, the cells of the cephalic and thoracic mesoderm are already present, but the abdominal mesoderm is only represented by a posterior growth zone associated with the ectodermal growth zone.

As growth zone activity proceeds, the embryonic primordium elongates and undergoes segmentation, then limb bud formation. Increase in length is accompanied by a forward shift of the anterior end of the germ band along the ventral midline of the yolk mass, and later by a downward flexure of the posterior end of the abdomen at the posterior pole of the yolk mass. This is the only vestige of dorsoventral flexure. Early in elongation and segment formation the presumptive amnion folds ventrally over the surface of the germ band, carrying the edge of the extra-embryonic ectoderm with it, and merges in the midline to complete the amnion and external serosa. The amnion stretches as elongation of the germ band continues.

The segmenting germ band develops distinct mandibular, maxillary and post-maxillary (= labial) segments behind the head lobes, followed by three thoracic segments and eleven abdominal segments. A small pre-mandibular segment later becomes transiently distinct between the head lobes and mandibular segment. The head lobes do not develop any intersegmental grooves, but the labrum arises in front of the mouth and the antennae grow out at the posteromedian corners of the head lobes, behind the mouth. Mandibular, maxillary and labial limb buds arise, followed by three thoracic pairs, eleven abdominal pairs and a small, transient, pre-mandibular pair. In later development, the antennae become pre-oral, the gnathal limb buds crowd forwards behind the mouth, the post-maxillary pair fuse to form the labium and the thoracic pairs develop as locomotory limbs. The first seven and the tenth pair of abdominal limb buds are resorbed before hatching, but the eighth and ninth pairs contribute to the external genitalia and the eleventh pair develops as the cerci.

The amnion and serosa remain intact until segmentation and limb bud formation along the germ band are complete, but then fuse, rupture and roll back to expose the embryo once more at the surface of the yolk mass. The serosa contracts into a small mass, the serosal dorsal organ, on the anterodorsal surface of the yolk mass, and is subsequently resorbed into the yolk. The amnion replaced the serosa as a provisional membrane at the yolk surface, but is gradually contracted and resorbed into the yolk as the definitive body wall grows dorsally to replace it. During later growth and development after rupture of the extra-embryonic membranes, the germ band extends forwards to reach the anterior pole of the yolk mass, as well as upward towards the dorsal midline. The forward flexure of the posterior end of the abdomen persists until hatching.

Underlying the changes summarized above, the further development of the mesoderm begins with posterior proliferation at the growth zone, accompanied by a bilateral segregation of the mesoderm into paired bands separated by a thin layer of median mesoderm in the ventral midline. The median mesoderm gives rise to haemocytes, except in the antennal segment and pre-antennal part of the head, where it develops as stomodaeal musculature and as labral and other cephalic muscles. The mesodermal bands break up into paired somites which develop coelomic cavities. The labial, thoracic and first abdominal somites are large, the second to tenth abdominal somites small, but all follow the same mode of development. An appendicular lobe grows into the associated limb bud and gives rise to intrinsic limb musculature, except in those abdominal segments in which the limbs are resorbed. A dorsolateral lobe grows towards the dorsal midline. Cardioblasts at the dorsal margin of the dorsolateral lobe contribute to the heart. The somatic wall of the same lobe develops as the dorsal longitudinal muscle, fat-body, pericardial floor and associated structures. The somatic wall of the ventral part of the somite develops as ventral longitudinal muscle, extrinsic limb muscle and fat-body. The splanchnic walls of the somites separate off and join up to form paired splanchnic mesodermal strands associated with the midgut strands. The musculature of the midgut arises from these strands. Germ cells become associated with the splanchnic mesoderm of certain abdominal somits and give rise to the gonads. The gonoducts are formed from coelomoduct outgrowths of the splanchnic mesoderm of certain posterior abdominal somites. Behind the tenth abdominal segment, a posterior mass of mesoderm associated with the proctodaeum develops a pair of temporary coelomic cavities, but later spreads over the proctodaeum and gives rise to proctodaeal musculature.

In front of the labial segment, somite development is more specialized. The maxillary and mandibular pairs are small and develop as limb musculature only. The small pre-mandibular pair gives rise to the suboesophageal body and perhaps contributes to the stomodaeal musculature. The antennal pair, which originate post-orally but subsequently become pre-oral, are large and retain a typical appendicular lobe, but otherwise develop as fat-body and, at their dorsal margins, anterior aorta. A vestigial pre-antennal pair of somites is formed and contributes to the labral musculature.

The embryonic ectoderm proliferates posteriorly and becomes segmentally delineated in association with the mesoderm. Ventrally on either side of the midline from the mandibular to the eleventh abdominal segment, the ectoderm differentiates in part as neuroblasts from which the ganglia of the ventral nerve cord are proliferated. The neurilemma arises from ganglion cells. Along the ventral midline, the ectoderm forms a median strand of tall cells. The intrasegmental parts of the median strand are incorporated into the ganglia, but the fate of the intersegmental parts is still not clear. Ventrolaterally, the segmental ectoderm pouches out as limb buds. Laterally, the ectoderm gives rise mainly to hypodermis, gradually spreading towards the dorsal midline as dorsal closure proceeds.

In front of the mandibular segment, paired pre-mandibular (tritocerebral) and antennal (deutocerebral) ganglia develop from neuroblasts. The pre-mandibular ganglia retain a post-oral tritocerebral commissure but the antennal ganglia become pre-oral and are not connected by a transverse commissure. The pre-oral protocerebral ganglia develop in part from neuroblasts, but the lateral, optic lobes of the protocerebrum arise as proliferations of the lateral ectoderm of the head lobes, without the intervention of neuroblasts.

Paired, segmental, ectodermal invaginations on the trunk give rise to the tracheal system and apodemes. Ectodermal invaginations on the head give rise to the labial salivary glands, anterior and posterior tentorial arms, mandibular apodemes and corpora allata. The anterior tentorial arms originate ventrolaterally in the region of the antennal–mandibular intersegment. The posterior tentorial arms and corpora allata arise in the region of the maxillary–labial intersegment.

In the further development of the gut, the stomodaeum develops as the epithelial lining of the foregut and also buds off the ganglia of the stomatogastric system from its dorsal midline. The proctodaeum develops as the epithelial lining of the hind gut and proliferates Malpighian tubules at its free end. The anterior and posterior midgut rudiments, attached to the free ends of the stomodaeum and proctodaeum respectively, proliferate paired midgut strands which flank the yolk mass and meet in the midregion of the embryo. The paired midgut strands then spread to enclose the yolk mass and form the lining epithelium of the midgut. The yolk and vitellophages are digested.

Development of the Head

Although the development of the pterygote head is touched upon in various parts of the summary given above, an integrated summation of head development is also required, in order to emphasize the similarity with other hexapods. Externally, as we have seen, the head rudiments comprise a pair of rounded head lobes with posteromedian antennal rudiments, a pre-mandibular segment region and three gnathal segments. Internally lie paired pre-antennal, antennal, pre-mandibular and gnathal somites. As in the Thysanura, the antennal somites contribute the anterior aorta and the pre-mandibular somites develop as the suboesophageal body.

Invagination of the stomodaeum, overgrowth of the labrum and pre-oral migration of the antennal rudiments take place, as they do in the apterygotes, while the segments behind the antennal segment are still well spread out along the ventral midline. Mandibular, maxillary and labial limb buds develop and conspicuous pre-mandibular limb buds are also transiently developed during the forward migration of the antennae. The mouth-parts then crowd forwards around the mouth and the labial limb buds unite to form the labium. Posterior tentorial arms develop as ectodermal invaginations associated with the labium, while labial salivary glands also arise as ectodermal invaginations on the labial segment. Meanwhile, the head lobes have developed the protocerebral ganglia, partly from neuroblasts. The antennal, pre-mandibular and gnathal ganglia also develop from neuroblasts and the gnathal ganglia merge to form the suboesophageal ganglion. Recent evidence that a pair of segmental pre-antennal ganglia can be distinguished within the protocerebral complex supports the argument that a vestigial pre-antennal segment is incorporated into the pre-antennal part of the pterygote head, though most of this region is still best interpreted as a large, pre-segmental acron.

The basic pattern of development of the pterygote head is thus similar in all respects to that of the apterygote hexapods and finds its closest comparison among the Thysanura. These findings lend further weight to the view that the Pterygota are more closely related to the Thysanura than to the other apterygote classes. They also indicate that all extant hexapods, including the Pterygota, diverged from an ancestral stock which was already labiate. It should therefore be possible to elucidate a basic theme for hexapod development, through which the hexapods can be compared with the myriapods and Onychophora. This task will be attempted in Chapter 10.

References

AMY, R. J. (1961) The embryology of *Habrobracon juglandis* (Ashmead). *J. Morph.* **109**, 199–228.

ANDERSON, D. T. (1962) The embryology of *Dacus tryoni* (Frogg.) (Diptera, Trypetidae (= Tephritidae)), the Queensland fruit fly. *J. Embryol. exp. Morph.* **10**, 248–92.

ANDERSON, D. T. (1963) The embryology of *Dacus tryoni*. 2. Development of imaginal discs in the embryo. *J. Embryol. exp. Morph.* **11**, 339–51.

ANDERSON, D. T. (1964) The embryology of *Dacus tryoni* (Diptera). 3. Origins of imaginal rudiments other than the principal discs. *J. Embryol. exp. Morph.* **12**, 64–75.

ANDERSON, D. T. (1966) The comparative embryology of the Diptera. *Ann. Rev. Entom.* **11**, 23–64.

ANDERSON, D. T. (1972a) The development of hemimetabolous insects. In COUNCE, S. J. (ed.), *Developmental Systems—Insects*, Academic Press, New York.

ANDERSON, D. T. (1972b) The development of holometabolous insects. In COUNCE, S. J. (ed.), *Developmental Systems—Insects*, Academic Press, New York.

ANDERSON, D. T. and WOOD, E. C. (1968) The morphological basis of embryonic movements in the light brown apple moth, *Epiphyas postvittana* Walk. (Lepidoptera, Tortricidae). *Australian J. Zool.* **16**, 763–93.

ANDO, H. (1960) Studies on the early embryonic development of a scorpion fly, *Panorpa pryeri* MacLachlan (Mecoptera, Panorpidae). *Sci. Rep. Tokyo Kyoiku Daig.* B **9**, 141–3.

ANDO, H. (1962), *The Comparative Embryology of Odonata with Special Reference to a Relict Dragonfly* Epiophlebia superstes *Selys*, Japanese Society for the Promotion of Science, Tokyo.

ANDO, H. and OKADA, J. (1958) Embryology of the butterbur stem sawfly *Aglaeostigma occipitosa* (Malaise) as studied by external observation (Tenthredinidae, Hymenoptera). *Acta hymenop.* **1**, 55–62.

ASHHURST, D. E. (1965) The connective tissue sheath of the locust nervous system: its development in the embryo. *Q. Jl. microsc. Sci.* **106**, 61–73.

BADEN, V. (1936) Embryology of the nervous system in the grasshopper, *Melanoplus differentialis. J. Morph.* **60**, 156–90.

BADEN, V. (1937) Origin and fate of the median cord in the grasshopper, *Melanoplus differentialis. J. Morph.* **63**, 219–23.

BAERENDS, G. P. and BAERENDS-von ROON, J. M. (1950) Embryological and ecological investigations on the development of the egg of *Ammophila campestris* Jur. (Hym.). *Tijdschr. Ent.* **92**, 53–112.

BASSAND, D. (1965) Contribution a l'étude de la diapause embryonnaire et de l'embryogénèse de *Zeiraphera griseana* Hubner (= *Z. dianana* Guénée) (Lepidoptera: Tortricidae). *Revue suisse Zool.* **72,** 431–542.

BEDFORD, G. O. (1970) The development of the egg of *Didymuria violescens* (Phasmatodea: Phasmatidae: Podacanthinae)—embryology and determination of the stage at which first diapause occurs. *Australian J. Zool.* **18,** 155–69.

BEHRENDT, K. (1963) Über die Eidiapause von *Aphis fabae* Scop. (Homoptera, Aphididae). *Zool. Jb. Physiol.* **70,** 309–98.

BERGERARD, J. (1958) Étude de la parthogénèse facultative de *Clitumnus extradentatus* Br. (Phasmidae). *Bull. biol. Fr. Belg.* **92,** 87–182.

BHATNAGAR, R. D. S. and SINGH, J. P. (1965) Studies on the embryonic development of the earwig *Labidura riparia* (Pallas) (Labiduridae, Dermaptera). *Res. Bull. Panjab Univ., Hoshiapur* (Sci.) **16,** 19–30.

BLUNCK, H. (1914) Die Entwicklung der *Dytiscus marginalis* L. vom Ei bis zur Imago. I Th. Das Embryonalleben. *Z. wiss. Zool.* **111,** 76–151.

BOCK, E. (1939) Bildung und Differenzierung der Keimblätter bei *Chrysopa perla* (L.). *Z. Morph. Ökol. Tiere* **35,** 615–702.

BODENHEIMER, F. S. and SHULOV, A. (1951) Egg development and diapause in the Moroccan locust (*Dociostaurus maroccanus* Thnb. *Bull. Res. Coun. Israel* **1,** 58–75.

BÖHLE, H. W. (1969) Untersuchungen über die Embryonalentwicklung und die embryonale Diapause bei *Baëtis vernus* Curtis und *Baëtis rhodani* (Pictet) (Baetidae, Ephemeroptera). *Zool. Jb. Anat. Ont.* **86,** 493–575.

BÖHMEL, W. and JANCKE, O. (1942) Beitrag zur Embryonalentwicklung der Wintereies von Aphiden. *Z. angew. Ent.* **29,** 636–58.

BOURNIER, A. (1960) Sur l'existence et l'évolution du corps dorsal secondaire dans l'ontogénèse de *Caudothrips buffai* Karny (Thysanoptera, Tubulifera). *C.R. Acad. Sci., Paris* **250,** 1347–8.

BRAUER, A. (1949) Localization of presumptive areas in the blastoderm of the pea beetle *Callosobruchus maculatus* Fabr. as determined by ultraviolet (2537 Å) irradiation injury. *J. exp. Zool.* **112,** 165–93.

BRONSKILL, J. F. (1959) The embryology of *Pimpla turionellae* (L.) (Hymenoptera, Ichneumonidae). *Can. J. Zool.* **34,** 655–88.

BRONSKILL, J. F. (1964) Embryogenesis of *Mesoleius tenthredinis* Morl. (Hymenoptera, Ichneumonidae). *Can. J. Zool.* **42,** 439–53.

BROOKES, H. M. (1952) The morphological development of the embryo of *Gryllus commodus* Walker (Orthoptera: Gryllidae). *Trans. Roy. Soc. S. Australia* **75,** 150–9.

BRUCE, A. T. (1887) *Observations on the Embryology of Insects and Arachnids*, Johns Hopkins University, Baltimore.

BUTT, F. H. (1934) Embryology of *Sciara*. *Ann. ent. Soc. Amer.* **27,** 565–79.

BUTT, F. H. (1936) The early embryological development of the parthenogenetic alfalfa snout beetle, *Brachyrhinus ligustici* L. *Ann. ent. Soc. Amer.* **29,** 1–13.

BUTT, F. H. (1949) Embryology of the milkweed bug, *Oncopeltus fasciatus* (Hemiptera). *Mem. Cornell Univ. agric. Exp. Stn.* **283,** 1–43.

CARRIÈRE, J. and BÜRGER, O. (1897) Die Entwicklungsgeschichte der Mauerbeine (*Chalicodoma muraria* Fabr.) im Ei. *Nova Acta Caesar Leop. Carol.* **69,** 255–420.

CAVALLIN, M. (1969) Étude descriptive du développement et de la differenciation sexuelle de l'appareil génital chez les embryons des phasmes *Clitumnus extradentatus* Br. et *Carausius morosus* Br. *C.R. Acad. Sci., Paris* **268,** 2189–92.

CAVALLIN, M. (1970) Développement embryonnaire de l'appareil génital chez le phasme *Carausius morosus* Br. *Bull. Biol. Fr. Belg.* **104,** 343–66.

CHOLODKOWSKY, N. (1891) Ueber die Entwicklung des centralen Nervensystems bei *Blatta germanica*. *Zool. Anz.* **14,** 115–16.

CHRISTENSEN, P. J. H. (1943a) Embryologische und zytologische Studien über die erste und frühe Eientwicklung bei *Orgyia antiqua* Linné (Fam. Lymantridae, Lepidoptera). *Vidensk. Meddr. dansk. natuh. Foren.* **106,** 1–223.

CHRISTENSEN, P. J. H. (1943b) Serosa und Amnionbildung der Lepidopteren. *Ent. Meddr.* **23,** 206–23.

CHRISTOPHERS, Sir S. R. (1960) "*Aedes aegypti L.*", *the Yellow Fever Mosquito. Its Life History, Bionomics and Structure*, University Press, Cambridge.

COBBEN, R. H. (1968) *Evolutionary Trends in Heteroptera*. Pt. 1. *Eggs, Architecture of the Shell, Gross Embryology and Eclosion*, Centre for Agricultural Publishing and Documentation, Wageningen.

COUNCE, S. J. (1966) Culture of insect embryos *in vitro*. *Annls. N.Y. Acad. Sci.* **139,** 65–78.

CRAIG, D. A. (1967) The eggs and embryology of some New Zealand Blepharoceridae (Diptera, Nematocera) with reference to the embryology of other Nematocera. *Trans.Roy. Soc. N.Z. Zool.* **8,** 191–206.
DAVIS, C. W. (1967) A comparative study of larval embryogenesis in the mosquito *Culex fatigans* Wiedermann (Diptera: Culicidae) and the sheep fly *Lucilia sericata* Meigen (Diptera: Calliphoridae). *Australian J. Zool.* **15,** 547–79.
DAVIS, C. W. C., KRAUSE, J. and KRAUSE, G. (1968) Morphogenetic movements and segmentation of posterior egg fragments *in vitro* of *Calliphora erythrocephala* Meigen (Diptera). *Arch. EntwMech. Org.* **161,** 209–40.
DEEGENER, P. (1900) Entwicklung der Mundwerkzeuge und des Darmkanals von *Hydrophilus. Z. wiss. Zool.* **68,** 113–68.
DEOBAHKTA, S. R. (1957) Embryonic development of *Mylabria pustulata* Thunb. *Agra. Univ. J. Res.* **6,** 92–172.
DOBROWSKI, Z. (1959) The early stages of embryonic development of *Blastophagus piniperda* L. (Coleoptera, Scolytidae). *Zesz. nauk. Univ. jagiellonsk. Zool.* **4,** 5–29.
DU BOIS, A. M. (1932) A contribution to the embryology of *Sciari. J. Morph.* **54,** 161–92.
DU BOIS, A. M. (1936) Recherches expérimentales sur la détermination de l'embryon dans l'œuf de *Sialis lutaria*, Neuroptera. *Revue suisse Zool.* **43,** 519–23.
DU BOIS, A. M. (1938) La determination de l'ébauche embryonnaire chez *Sialis lutaria* (Megaloptera). *Revue suisse Zool.* **45,** 1–92.
DUPRAW, E. J. (1967) The honey bee embryo. In WILT, F. H. and WESSELS, N. K. (eds.), *Methods in Developmental Biology*, Crowell, New York.
EASTHAM, L. E. S. (1927), A contribution to the embryology of *Pieris rapae. Q. Jl. micros. Sci.* **71,** 353–94.
EASTHAM, L. E. S. (1930) The embryology of *Pieris rapae*. Organogeny. *Phil. Trans. Roy. Soc.* B, **219,** 1–50.
EDE, D. A. (1964) An inherited abnormality affecting the development of the yolk plasmodium and endoderm in *Dermestes maculatus* (Coleoptera). *J. Embryol. exp. Morph.* **12,** 551–62.
EDE, D. A. and COUNCE, S. J. (1956) A cinematographic study of the embryology of *Drosophila melanogaster. Arch. EntwMech. Org.* **148,** 259–66.
ELSE, F. L. (1934) The developmental anatomy of male genitalia in *Melanoplus differentialis* (Locustidae (Acrididae) Orthoptera). *J. Morph.* **55,** 577–610.
ESCHERICH, K. (1900a) Keimblatterbildung bei den Musciden. *Verh. dt. zool. Ges.* **10,** 130–4.
ESCHERICH, K. (1900b) Ueber die Bildung der Keimblätter bei den Musciden. *Nova Acta Acad. Caesar Leop. Carol.* **77,** 299–357.
ESCHERICH, K. (1901a) Ueber die Bildung der Keimblätter bei den Insekten. *Allg. Z. Ent.* **6,** 79.
ESCHERICH, K. (1901b) Das Insekten—Entoderm. *Biol. Zbl.* **21,** 416–31.
ESCHERICH, K. (1902) Zur Entwicklung des Nervensystems der Musciden, mit besonderer Berücksichtigung des sog. Mittelstranges. *Z. wiss. Zool.* **71,** 525–49.
FAROOQUI, M. M. (1963) The embryology of the mustard sawfly *Athalia proxima* Klug. (Tenthredinidae, Hymenoptera). *Aligarh Musl. Univ. Publs.* **6,** 1–68.
FAUSSEK, V. (1911) Vergleichend—embryologisch Studien. *Z. wiss. Zool.* **98,** 529–625.
FORMIGONI, A. (1954) Studio sullo sviluppo embrionale di *Musca domestica* L. *Boll. zool. agrar. Bachic.* **20,** 111–54.
FOURNIER, B. (1967) Echelle résumée des stades du développement embryonnaire du Phasme *Carausius morosus* Br. *Actes Soc. Linn. Bordeaux* **104**A, 1–30.
GAMBRELL, F. L. (1933) The embryology of the black fly *Simulium pinctipes* Hagen. *Ann. ent. Soc. Amer.* **26,** 641–71.
GERWEL, C. (1950) The embryonic development of *Porphyrophora polonica* Ckll. (Coccidae) Part I. *Pr. Kom. biol., Poznan* **12** (10), 1–33.
GEYER-DUZINSKA, I. (1959) Experimental research on chromosome elimination in Cecidomyidae. *J. exp. Zool.* **141,** 391–449.
GIARDINA, A. (1897) Prima stadi embrionali della *Mantis religiosa. Monitore zool. ital.* **8,** 275–80.
GÖRG, I. (1959) Untersuchungen am Keim von *Hierodula (Rhombodera) crassa* Giglio Tos, ein Beitrag zur Embryologie des Mantiden (Mantodea). *Deutsch. ent. Z.* **6,** 390–450.
GOSS, R. J. (1952) The early embryology of the book louse, *Liposcelis divergens* Badonnel (Psocoptera: Liposcelidae). *J. Morph.* **91,** 135–67.
GOSS, R. J. (1953) The advanced embryology of the book louse, *Liposcelis divergens* Badonnel (Psocoptera: Liposcelidae). *J. Morph.* **92,** 157–91.
GRABER, V. (1888a) Ueber die Polypodie bei den Insekten Embryonen. *Morph. Jb.* **13,** 586–613.

GRABER, V. (1888b) Vergleichenden Studien über die Keimhüllen und die Rückenbildung der Insekten. *Denkschr. Akad. Wiss. Wien* **55**, 109–62.

GRABER, V. (1888c) Ueber die primäre Segmentierung des Keimstriefs der Insekten. *Morph. Jb.* **14**, 345–68.

GRABER, V. (1890) Vergleichenden Studien eim Keimstreifen der Insekten. *Denkschr. Akad. Wiss. Wien* **57**, 621–734.

GUÉNNELON, G. (1966) Contribution a l'étude de la diapause embryonnaire chez *Archips rosana* L. (Lepidoptera–Tortricidae). *Annls. Epiphyt.* **17**, 3–135.

GUICHARD, M. (1971) Étude *in vivo* du développement embryonnaire de *Culex pipiens*. Comparison avec *Calliphora enythrocephala* (Diptera). *Ann. Soc. ent. France* N.S. **7**, 325–42.

HAGAN, H. R. (1917) Observations on the development of the mantid *Paratenodera sinesis*. *J. Morph.* **30**, 223–43.

HAGAN, H. R. (1951) *Embryology of the Viviparous Insects*, Ronald Press, New York.

HAGET, A. (1953) Analyse experimentale des facteurs de la morphogenèse embryonnaire chez le coléoptère *Leptinotarsa*. *Bull. biol. Fr. Belg.* **87**, 123–217.

HAGET, A. (1955) Experiences mettant en évidence l'origine paire du labre chez l'embryon du coléoptère *Leptinotarsa*. *C.R. Acad. Biol.* **149**, 690–2.

HAGET, A. (1957) Recherches expérimentale sur l'origine embryonnaire du crâne d'un coléoptère, *Leptinotarsa decemlineata* Say. *Bull. Soc. zool. Fr.* **82**, 269–95.

HARDENBERG, J. D. F. (1929) Beiträge zur Kenntnis der Pupiparen. *Zool. Jahrb. Anat. Ont.* **50**, 497–570.

HEGNER, R. W. (1909) The origin and early history of the germ cells of some chrysomelid beetles. *J. Morph.* **20**, 231–96.

HEGNER, R. W. (1910) Experiments with chrysomelid beetles. *Biol. Bull., Woods Hole* **19**, 18–30.

HEGNER, R. W. (1911) Experiments with chrysomelid beetles. *Biol. Bull., Woods Hole* **20**, 237–51.

HEGNER, R. W. (1912) The history of the germ cells in the paedogenetic larva of *Miastor*. *Science* **36**, 124–6.

HEGNER, R. W. (1914) Studies on germ cells I. *J. Morph.* **25**, 375–509.

HEGNER, R. W. (1915) Studies on germ cells II. *J. Morph.* **26**, 495–561.

HEIDER, K. (1889) *Die Embryonalentwicklung von* Hydrophilus piceus *L.*, Gustav Fischer, Jena.

HENSCHEN, W. (1928) Über die Entwicklung der Geschlechts-drüsen von *Habrobracon juglandis* Ash. *Z. Morph. Ökol. Tiere* **13**, 144–79.

HEYMONS, R. (1895) *Die Embryonalentwicklung von Dermapteren und Orthopteren unter besonderer Berücksichtigung der Keimblatterbildung*, Gustav Fischer, Jena.

HEYMONS, R. (1896a) Grundzüge der Entwicklung und des Körperbaues von Odonaten und Ephemeriden, *Dtsch. Akad. Wiss. Abh.* **1896**, 1–66.

HEYMONS, R. (1896b) Ueber die Lebensweise und Entwicklung von *Ephemera vulgata* L. *Sitzb. Ges. naturf. Freunde Berl.* **1896**, 82–96.

HEYMONS, R. (1897) Ueber die Organisation und Entwicklung von *Bacillus rossii* Fabr. *Sitzb. preuss. Akad. Wiss.* **1897**, 363–73.

HEYMONS, R. (1899) Beiträge zur Morphologie und Entwicklungsgeschichte der Rhynchoten. *Nova Acta Acad. Caesar Leop. Carol.* **74**, 351–456.

HEYMONS, R. (1912) Ueber den Genitalapparat und die Entwicklung von *Hemimerus talpodes* Walker. *Zool. Jb. Suppl.* **15**, 141–84.

HIRSCHLER, J. (1905) Embryologische Untersuchungen an *Calocala rupta* L. *Bull. int. Acad. Sci. Lett. Cracovie* B, **1905**, 802–10.

HIRSCHLER, J. (1909) Die Embryonalentwicklung von *Donacia crassipes* L. *Z. wiss. Zool.* **92**, 627–744.

HIRSCHLER, J. (1912) Embryologische Studien an Aphiden. *Z. wiss. Zool.* **100**, 393–446.

HUIE, L. H. (1918) The formation of the germ band in the egg of the Holly Tortrix Moth. *Proc. Roy. Soc. Edinb.* **38**, 154–65.

HUSSEY, P. B. (1926) Studies on the pleuropodia of *Belostoma flumineum* and *Ranatra fusca*, with a discussion of these organs in other insects. *Entomologica Amer.* **7**, 1–80.

IBRAHIM, M. M. (1957) Grundzüge der Organbildung im Embryo von *Tachycines* (Insecta Saltatoria). *Zool. Jb. Anat. Ont.* **76**, 541–94.

IDRIS, B. E. M. (1960) Die Entwicklung im normalen Ei von *Culex pipiens* L. (Diptera). *Z. Morph. Ökol. Tiere* **49**, 387–429.

ILLIES, J. (1968) Ephemeroptera. In *Handbuch der Zoologie*, 4(2), 2/5 (M. BIEIER ed.), pp. 1–63, Gruyter, Berlin.

INKMANN, F. (1933) Beiträge zur Entwicklungsgeschichte des Kornkafers (*Calandra granaria* L.) Die Anfangsstadien der Embryogenese. *Zool. Jahrb. Anat. Ont.* **56**, 521–58.

IVANOVA-KAZAS, O. M. (1949) Embryological development of *Anopheles maculipennis* Mg. *Izv. Akad. Nauk. SSSR. Ser. Biol.* **2,** 140–70.

IVANOVA-KAZAS, O. M. (1954) The effects of parasitism on the embryonal development of *Caraphractus reductus* R.-Kors. (Hymenoptera). *Trudy leningr. Obshch. Estest.* **72,** 53–73.

IVANOVA-KAZAS, O. M. (1958) Biology and embryonic development of *Eurytoma aciculator* Ritz. (Hymenoptera Eurytomidae). *Ent. Obozr.* **37,** 1–18.

IVANOVA-KAZAS, O. M. (1959) Die embryonale Entwicklung der Blattwaspe *Pontania capreae* L. (Hymenoptera, Tenthredinidae). *Zool. Jb. Anat. Ont.* **77,** 193–228.

IVANOVA-KAZAS, O. M. (1960) Embryologische Entwicklung von *Angita vestigialis* Ratz. (Hymenoptera, Ichneumonidae)—des Entoparasiten von *Pontania capreae* L. (Hymenoptera, Tenthredinidae). *Ent. Obozr.* **39,** 284–95.

IVANOVA-KAZAS, O. M. (1965) Trophic connections between the maternal organism and the embryo in paedogenetic Diptera (Cecidomyidae). *Acta Biol., Szeged* **16,** 1–24.

IVANOVA-KAZAS, O. M. (1972) Polyembryony in insects. In COUNCE, S. J. (ed.), *Developmental Systems—Insects*, Academic Press, New York.

JANET, C. (1899) Sur les nerfs céphaliques, les corpora allata et le tentorium de la fourmi *(Myrmica rubra)*. *Mem. Soc. zool. Fr.* **12,** 295–335.

JHINGRAN, V. G. (1947) Early embryology of the desert locust *Schistocerca gregaria* (Förskal) (Orthoptera, Acrididae). *Rec. Indian Mus.* **45,** 181–200.

JOHANNSEN, O. A. (1929) Some phases in the embryonic development of *Diacrisia virginica* Fabr. (Lepidoptera). *J. Morph.* **48,** 493–541.

JOHANNSEN, O. A. and BUTT, F. H. (1941) *Embryology of Insects and Myriapods*, McGraw-Hill, New York.

JONES, B. M. (1956) Endocrine activity during insect embryogenesis. Control of events in development following the embryonic moult (*Locusta migratoria* and *Locusta pardalina*, Orthoptera). *J. exp. Biol.* **33,** 685–96.

JUNG, E. (1966a) Untersuchungen am Ei des Speisebohnenkäfers *Bruchidius obtectus* Say (Coleoptera). I. Entwicklungsgeschichtliche Ergebnisse zur Keimzeichnung des Eitypus. *Z. Morph. Ökol. Tiere* **56,** 444–80.

JUNG, E. (1966b) Untersuchungen am Ei des Speisebohenkäfers *Bruchidius obtectus* Say (Coleoptera). II. Entwicklungsphysiologische Ergebnisse der Schnurungs-experimente. *Arch. EntwMech. Org.* **157,** 320–92.

JURA, C. Z. (1957) Experimental studies on the embryonic development of the *Melasoma populi* L. (Chysomelidae, Coleoptera). *Zoologica Pol.* **8,** 177–99.

JURA, C. Z. (1959) The early developmental stages of the ovoviviparous scale insect, *Quadraspidiotus ostreaeformis* (Curt.) (Homoptera, Coccidae, Aspidiotini). *Zoologica Pol.* **9,** 17–34.

KALTHOFF, K. and SANDER, K. (1968) Der Entwicklungsgang der Missbildung "Doppelabdomen" im partiell UV-bestrahlten Ei von *Smittia parthenogenetica* (Dipt. Chironomidae). *Arch. EntwMech. Org.* **161,** 128–46.

KANNELIS, A. (1952) Anlagenplan und Regulationserscheinungen in der Keimanlage des Eies von *Gryllus domesticus*. *Arch. EntwMech. Org.* **145,** 417–61.

KERSHAW, J. C. (1914) Development of an embiid. *J. Roy. micr. Soc.* **1914,** 24–27.

KESSEL, E. L. (1939) The embryology of fleas. *Smithson misc. Collns.* **98,** 1–78.

KESSEL, R. G. (1961) Cytological studies on the sub-oesophageal body cells and pericardial cells in the embryo of the grasshopper *Melanoplus differentialis differentialis* (Thomas). *J. Morph.* **109,** 289–322.

KHOO, S. G. (1968) Experimental studies on diapause in stone flies. III Eggs of *Brachyptera risi* (Morton). *Proc. Roy. ent. Soc. Lond.* **43**A, 141–6.

KOCH, P. (1964) In vitro-Kultur und entwicklungsphysiologische Ergebnisse an Embryonen der Stabheuschrecke *Carausius morosus* Br. *Arch. EntwMech. Org.* **155,** 549–93.

KOROTNEFF, A. (1885) Die Embryologie der *Gryllotalpa*. *Z. wiss. Zool.* **41,** 507–604.

KORSCHELT, E. (1912) Zur Embryonalentwicklung des *Dytiscus marginalis*. *Zool. Jahrb. suppl.* **15,** 499–532.

KRAUSE, G. (1938a) Einzelbeobachtungen und typische Gesamtbilder der Entwicklung von Blastoderm und Keimanlage im Ei der Gewachschausschrecke, *Tachycines asynamorus*. *Z. Morph. Ökol. Tiere* **34,** 1–78.

KRAUSE, G. (1938b) Die Ausbildung der Körpergrundgestalt im Ei der Gewachshausschrecke, *Tachycines asynamorus*. *Z. Morph. Ökol. Tiere* **34,** 499–564.

KRAUSE, G. (1939) Die Eitypen der Insekten. *Biol. Zbl.* **59,** 495–536.

KRAUSE, G. and KRAUSE, J. (1964) Schichtenbau und Segmentierung junger Keimanlagen von *Bombyx mori* L. (Lepidoptera) in vitro ohne Dottersystem. *Arch. EntwMech. Org.* **155,** 451–510.

KRAUSE, G. and SANDER, K. (1962) Ooplasmic reaction systems in insect embryogenesis. *Adv. Morphogenesis* **2,** 259–303.

KRAUSE, J. B. and RYAN, M. T. (1953) The stages of development in the embryology of the horned passalus beetle. *Popilius disjunctus* Illiger. *Ann. ent. Soc. Amer.* **46,** 1–20.

KRZYSZTOFOWICZ, A. (1960) Comparative investigations on the embryonic development of the weevils (Coleoptera, Curculionidae), and an attempt to apply them to the systematics of this group. *Zoologica Pol.* **10,** 3–27.

KÜTHE, H. W. (1966) Das Differenzierungszentrum als selbst regulier-endes Faktorensystem für die Aufbau der Keimanlage im Ei von *Dermestes frischi* (Coleoptera). *Arch. EntwMech. Org.* **157,** 212–302.

LASSMAN, G. W. P. (1936) The early embryological development of *Melophagus ovinus*, with special reference to the development of the germ cells. *Ann. Ent. Soc. Amer.* **29,** 397–413.

LAUTENSCHLAGER, F. (1932) Die Embryonalentwicklung der weiblichen Keimdrüse bei der Psychide *Solenobia triquetella. Zool. Jahrb. Anat. Ont.* **56,** 121–62.

LECAILLON, A. (1897a) Sur les feuillets germinatifs des coléoptères. *C.R. Acad. Sci., Paris* **125,** 876–9.

LECAILLON, A. (1897b) Note preliminaire relative aux feuillets germinatifs des coléoptères. *C.R. Soc. Biol. Paris* **4,** 1014–16.

LECAILLON, A. (1897c) Contribution a l'étude des premiers phénomènes due développement embryonnaire chez les insectes particulièrement chez les coléoptères. *Archs. Anat. Microsc.* **1,** 205–24.

LECAILLON, A. (1898) Recherches sur le développement embryonnaire de quelques chrysomelides. *Archs. Anat. Microsc.* **2,** 118–76.

LEUZINGER, H., WIESMANN, R. and LEHMANN, F. E. (1926) *Zur Kenntnis der Anatomie und Entwicklungsgeschichte der Stabheuschrecke* Carausius morosus *Br.* Gustav Fischer, Jena.

LOUVET, J. P. (1964) La ségrégation du mesoderme chez l'embryon du Phasme *Carausius morosus* Br. *Bull. Soc. zool. Fr.* **89,** 688–701.

LUGINBILL, P., Jr. (1953) A contribution to the embryology of the may beetle. *Ann. ent. Soc. Amer.* **46,** 505–28.

MAHOWALD, A. P. (1968) Polar granules of *Drosophila*. II. Ultrastructural changes during early embryogenesis. *J. exp. Zool.* **167,** 237–62.

MAHOWALD, A. P. (1971) Polar granules of Drosophila. III. The continuity of polar granules during the life cycle of *Drosophila. J. exp. Zool.* **176,** 329–44.

MAHR, E. (1960) Normale Entwicklung, Pseudofurchung und die Bedeutung des Furchungszentrum im Ei des Heimchens *(Gryllus domesticus). Z. Morph. Ökol. Tiere* **49,** 263–311.

MALZACHER, P. (1968) Die Embryogenese des Gehirns paurometaboler Insekten; Untersuchungen an *Carausius morosus* und *Periplaneta americana. Z. Morph. Ökol. Tiere* **62,** 103–61.

MANSOUR, K. (1927) The development of the larval and adult midgut of *Calandra oryzae* (Lim.): the rice weevil. *Q. Jl. microsc. Sci.* **71,** 313–52.

MANSOUR, K. (1934) The development of the adult midgut of coleopterous insects and its bearing on systematics and embryology. *Bull. Fac. Sci. Egypt Univ.* **3,** 1–34.

MANSOUR, K. (1937) The endoderm problem in insects. *C.R. Congr. int. Zool., Lisbon* **12,** 567–70.

MATSUDA, R. (1965) Morphology and evolution of the insect head. *Mem. Amer. Inst. Entom.* **4,** 1–334.

MATTHÉE, J. J. (1951) The structure and physiology of the egg of *Locustana pardalina. Sci. Bull. Dep. Agric. Forest. Union S. Afr.* **316,** 1–83.

MAZUR, Z. T. (1960) The embryogenesis of the central nervous system of *Agelastica alni* L. (Coleoptera, Chrysomelidae). *Zesz. nauk. Univ. jagiellonsk. Zool.* **5,** 205–29.

MELLANBY, H. (1936) The early embryonic development of *Rhodnius prolixus. Q. Jl. microsc. Sci.* **78,** 71–90.

MELLANBY, H. (1937) The later development of *Rhodnius prolixus. Q. Jl. microsc. Sci.* **79,** 1–42.

MENG, C. (1968) Strukturwandel und histochemische Befunde inbesondere am Oösom während der Oogenese und nach der Ablage des Eies von *Pimpla turrionella* L. (Hymenoptera, Ichneumonidae). *Arch. EntwMech. Org.* **161,** 162–208.

MILLER, A. (1939) The egg and early development of the stone fly *Pteronarcys proteus. J. Morph.* **64,** 555–609.

MILLER, A. (1940) Embryonic membranes, yolk cells and morphogenesis of the stone fly *Pteronarcys proteus. Ann. ent. Soc. Amer.* **33,** 437–47.

MIYA, K. (1965) The embryonic development of a chrysomelid beetle, *Atrachya memtriesi* Faldermann (Coleoptera, Chrysomelidae). I. The stages of development and changes of external form. *J. Fac. Agric. Iwate Univ., Morioka* **7,** 155–66.

MORI, H. (1969) Normal embryogenesis of the water strider *Gerris paludium insularis* Motschulsky, with special reference to midgut formation. *Japanese J. Zool.* **16,** 53–67.

MORI, H. (1970) The distribution of the columnar serosa of eggs among the families of Heteroptera, in relation to phylogeny and systematics. *Japanese J. Zool.* **16,** 89–98.

MOSCONA, A. (1950) Blastokinesis and embryonic development in a phasmid, *Experientia* **6,** 425–6.

MUELLER, K. (1938) Histologische Untersuchungen über die Entwicklungsbeginn bei einem Kleinschmetterling *(Plodia interpunctella). Z. wiss. Zool.* **151,** 192–242.

MÜLLER, M. (1957) Entwicklung und Bedeutung der Vitellophagen im der Embryonalentwicklung in der Honigbiene. Ein Beitrag zur Frage nach der Bedeutung des Dottersystems. *Zool. Jb. Allg. Zool.* **67,** 111–50.

MULNARD, J. (1947) Le dévelopement embryonnaire d'*Acanthoscelides obtectus* Say (Col.). *Archs. Biol., Liége* **58,** 289–314.

NELSEN, O. E. (1931) Life cycle, sex differentiation and testis development in *Melanoplus differentialis* (Acrididae, Orthoptera). *J. Morph.* **51,** 467–526.

NELSEN, O. E. (1934) The segregation of the germ cells in the grasshoppper *Melanoplus differentialis. J. Morph.* **55,** 545–74.

NELSON, J. A. (1915) *The Embryology of the Honey Bee*, University Press, Princeton.

NOACK, W. (1901) Beiträge zur Entwicklungsgeschichte der Musciden. *Z. wiss. Zool.* **70,** 1–57.

NUSBAUM, J. and FULINSKI, B. (1906) Zur Entwicklungsgeschichte des Darmdrüsenblattes bei *Gryllotalpa vulgaris. Z. wiss. Zool.* **93,** 306–48.

OCHIAI, S. (1960) Comparative studies on embryology of the bees—*Apis, Pollistes, Vespula* and *Vespa*, with special reference to the development of the silk gland. *Bull. Fac. Agric. Tamagawa Univ.* **1,** 13–45.

OKADA, M. (1960) Embryonic development of the rice stem borer, *Chilo suppressalis. Sci. Rep. Tokyo Kyoiku Daig.* **B9,** 243–96.

PATERSON, N. F. (1931) A contribution to the embryological development of *Euryope terminalis* Baly (Coleoptera, Phytophaga, Chrysomelidae). Part I. The early embryological development. *S. Afr. J. Sci.* **28,** 344–71.

PATERSON, N. F. (1932) A contribution to the embryological development of *Euryope terminalis* Baly (Coleoptera, Phytophaga, Chrysomelidae), Part II: Organogeny. *S. Afr. J. Sci.* **29,** 414–48.

PATERSON, N. F. (1936) Observations on the embryology of *Corynodes pusis* (Coleoptera, Chrysomelidae). *Q. Jl. microsc. Sci.* **78,** 91–132.

PATTEN, W. (1884) The development of Phryganids, with a preliminary note on the development of *Blatta germanica. Q. Jl. microsc. Sci.* **24,** 549–602.

PAULI, M. E. (1927) Die Entwicklung geschnurter und centrifugierter Eier von *Calliphora erythrocephala* und *Musca domestica. Z. wiss. Zool.* **129,** 483–540.

PIOTROWSKI, F. (1953) The embryological development of the body louse—*Pediculus vestimenti* Nitszch—Part I. *Acta Parasit. pol.* **1,** 61–84.

POULSON, D. F. (1937) *The Embryonic Development of* Drosophila melanogaster, Hermann et Cie, Paris.

POULSON, D. F. (1950) In M. DEMEREC (ed.), *The Biology of* Drosophila, Wiley, New York.

PRESSER, B. D. and RUTCHSKY, C. W. (1957) The embryonic development of the corn earworm *Heliothis zea* (Boddie) (Lepidoptera, Phalaenidae). *Ann. ent. Soc. Amer.* **50,** 133–64.

RABITO, L. (1898) Sull'origine des'intestino medio nella *Mantis religiosa. Naturalista sicil.* **2,** 181–3.

RAKSHPAL, R. (1962) Morphogenesis and embryonic membranes of *Gryllus assimilis* (Fabricius) (Orthoptera, Gryllidea). *Proc. Roy. ent. Soc. Lond.* **37A,** 1–12.

REED, E. M. and DAY, M. R. (1966) Embryonic movements during development of the Light Brown Apple Moth. *Australian J. Zool.* **14,** 253–63.

REIGERT, R. W. (1961) Embryological development of a non-diapause form of *Melanoplus bilituratus* Walker (Orthoptera = Acrididae). *Can. J. Zool.* **39,** 491–4.

REINHARDT, E. (1960) Kernverhaltnisse, Eisystem und Entwicklungsweise von Drohen-und Arbeiterinneneiern der Honigbeine *(Apis mellifera). Zool. Jb. Anat. Ont.* **78,** 167–234.

REMPEL, J. G. (1951) A study of the embryology of *Mamestra configurata* (Walker) (Lepidoptera = Phalaenidae). *Can. Ent.* **83,** 1–19.

REMPEL, J. G. and CHURCH, M. S. (1965) The embryology of *Lytta viridana* Le Conte. 1. Maturation, fertilization and cleavage. *Can. J. Zool.* **43,** 915–25.

REMPEL, J. G. and CHURCH, M. S. (1969) The embryology of *Lytta viridana* Le Conte (Coleoptera = Meloidae). V. The blastoderm, germ layers and body segments. *Can. J. Zool.* **47,** 1157–71.

REMPEL, J. G. and CHURCH, M. S. (1971a) The embryology of *Lytta viridana* Le Conte (Coleoptera: Meloidae). VI. The appendiculate, 72-h embryo. *Can. J. Zool.* **49,** 1563–70.

REMPEL, J. G. and CHURCH, M. S. (1971b) The embryology of *Lytta viridana* Le Conte (Coleoptera: Meloidae). VII. Eighty-eight to 132 h: the appendages, the cephalic apodemes, and head segmentation. *Can. J. Zool.* **49,** 1571–81.

RESSOUCHES, A. P. (1969) Premières observations sur le développement embryonnaire de *Pissodes notatus* F. (Col. Curculionidae). *C.R. Acad. Sci. Paris* **269,** 191–4.

RIES, E. (1931) Die Symbiose der Läuse und Federlinge. *Z. Morph. Ökol. Tiere* **20,** 233–376.

RILEY, W. A. (1904) The embryological development of the skeleton of the head of *Blatta. Amer. Nat.* **38,** 777–810.

ROONWAL, M. L. (1936) Studies on the embryology of the African migratory locust, *Locusta migratoria migratorioides.* I. The early development, with a new theory of multiphase gastrulation among insects. *Phil. Trans. Roy. Soc.* B, **226,** 391–421.

ROONWAL, M. L. (1937) Studies on the embryology of the African migratory locust, *Locusta migratoria migratorioides.* II. Organogeny. *Phil. Trans. Roy. Soc.* B, **227,** 175–244.

RYAN, R. B. (1963) Contribution to the embryology of *Coeloides brumeri* (Hymenoptera; Braconidae). *Ann. ent. Soc. Amer.* **56,** 639–48.

SAITO, S. (1934) A study on the development of the tusser worm, *Antheraea pernyi. J. Coll. Agric. Hokkaido imp. Univ.* **33,** 249–66.

SAITO, S. (1937) On the development of the tusser, *Antheraea pernyi,* with special reference to the comparative embryology of insects. *J. Coll. Agric. Hokkaido imp. Univ.* **40,** 35–109.

SALT, R. W. (1949) A key to the embryological development of *Melanoplus bivittatus* (Say), *M. mexicanus mexicanus* (Sauss.) and *M. packardi* Scudder. *Can. J. Res.* D**27,** 233–5.

SALZEN, E. A. (1960) The growth of the locust embryo. *J. Embryol. exp. Morph.* **8,** 139–62.

SANDER, K. (1956) The early embryology of *Pyrilla perpusilla* Walker (Homoptera), including some observations on later development. *Aligarh Musl. Univ. Publs. (Indian Insect Types)* **4,** 1–61.

SANDER, K. (1959) Analyse der ooplasmatischen Reaktionssystems von *Euscelis plebejus* Fall (Cicadina) durch Isolieren und Kombinieren von Keimteilen. I Mitteilung. Die Differenzierungsleistungen vorderer und Hinterer Eiteile. *Arch. EntwMech. Org.* **151,** 430–97.

SAUER, H. W. (1966) Zeitraffer-Mikro-film-Analyse embryonaler Differezierungsphasen von *Gryllus domesticus. Z. Morph. Ökol. Tiere* **56,** 143-251.

SCHIENERT, W. (1933) Symbiose und Embryonalentwicklung bei Russelkäfern. *Z. Morph. Ökol. Tiere* **27,** 76–128.

SCHNETTER, M. (1934) Morphologische Untersuchungen über das Differenzierungszentrum in der Embryonalentwicklung der Honigbiene. *Z. Morph. Ökol. Tiere* **28,** 114–95.

SCHOELLER, J. (1964) Recherches descriptives et expérimentales sur la cephalogénèse de *Calliphora erythrocephala* (Meigen), au cours des développements embryonnaire et post-embryonnaire. *Archs. Zool. exp. gen.* **103,** 1–216.

SCHOLL, G. (1965) Die Kopfentwicklung von *Carausius* (= *Dixippus*) *morosus. Zool. Anz. Suppl.* **28,** 580–96.

SCHOLL, G. (1969) Die Embryonalentwicklung des Kopfes und Prothorax von *Carausius morosus* Br. (Insecta, Phasmida). *Z. Morph. Ökol. Tiere* **65,** 1–142.

SCHÖLZEL, G. (1937) Die Embryologie der Anopluren und Mallophagen. *Z. Parasitkde.* **9,** 730–70.

SEHL, A. (1931) Furchung und Bildung der Keimanlage bei der Mehlmotte *Ephestia Kühniella,* nebst einer allgemeinen Übersicht über den Verlauf der Embryonalentwicklung. *Z. Morph. Ökol. Tiere* **20,** 535–98.

SEIDEL, F. (1924) Die Geschlechtsorgane in der embryonalen Entwicklung von *Pyrrhocoris apterus. Z. Morph. Ökol. Tiere* **1,** 429–506.

SEIDEL, F. (1929) Untersuchungen über das Bildungsprinzip der Keimanlage im Ei der Libelle *Platycnemis pennipes. Arch. EntwMech. Org.* **119,** 322–440.

SEIDEL, F. (1935) Der Anlageplan im Libellenei, zugleich eine Untersuchung über die allgemeinen Bedingungen für defekte Entwicklung und Regulation bei Dotterreichen Eiern. *Arch. EntwMech. Org.* **132,** 671–751.

SHAFIQ, S. A. (1954) A study of the embryonic development of the gooseberry sawfly *Pteronidea ribesii. Q. Jl. microsc. Sci.* **95,** 93–114.

SHINJI, G. (1919) Embryology of coccids, with special reference to the formation of the ovary, origin and differentiation of the germ cells, germ layers, rudiments of the midgut, and the intracellular symbiotic organisms. *J. Morph.* **33,** 73–167.

SHULOV, A. and PENER, J. P. (1959) A contribution to knowledge of the development of the egg of *Locusta migratoria migratioides* (R. and F.). *Locusta* **6,** 73–88.

SHULOV, A. and PENER, J. P. (1963) Studies on the development of eggs of the desert locust (*Schistocerca gregaria* Førskal) and its interruption under particular conditions of humidity. *Anti.-Locust Bull.* **41,** 1–59.

SINGH, J. P. (1967), Early embryonic development of gonads in *Labidura rifrana* (Pallas) (Labiduridae: Dermaptera). *Agra. Univ. J. Res. Sci.* **16,** 67–76.

SLIFER, E. H. (1932a) Insect development. III. Blastokinesis in the living grasshopper egg. *Biol. Zbl.* **52,** 223–9.

SLIFER, E. H. (1932b) Insect development. IV. External morphology of grasshopper embryos of known age and with a known temperature history. *J. Morph.* **53,** 1–22.

SLIFER, E. H. (1937) The origin and fate of the membranes surrounding the grasshopper egg; together with some experiments on the source of the hatching enzyme. *Q. Jl. microsc. Sci.* **79,** 493–506.

SMART, J. (1963) Explosive evolution and the phylogeny of insects. *Proc. Linn. Soc. Lond.* **174,** 125–6.

SMRECZYŃSKI, S. (1932) Embryologische Untersuchungen über die Zusammensetzung des Kopfes von *Silpha obscura* L. *Zool. Jb. Anat. Ont.* **59,** 1–58.

SMRECZYŃSKI, S. (1934) Beitrag zur Kenntnis der Entwicklungsgeschichte des Russelkäfers *Phyllobius glaucus. Bull. int. Acad. Sci. Lett. Cracovie* B **2,** 287–312.

SMRECZYŃSKI, S. (1938) Entwicklungs-mechanische Untersuchungen am Ei der Käfers *Agelastica alni. Zool. Jb. Anat. Ont.* **59,** 1–58.

SONNENBLICK, B. P. (1950) In DEMEREC, M. (ed.), *The Biology of* Drosophila, Wiley, New York.

SPEICHER, B. R. (1936) Oogenesis, fertilization and early cleavage in *Habrobracon. J. Morph.* **59,** 401–21.

SPRINGER, C. A. (1967) Embryology of the thoracic and abdominal ganglia of the large milkweed bug, *Oncopeltus fasciatus* (Dallas), (Hemiptera: Lygaeidae). *J. Morph.* **122,** 1–18.

SPRINGER, C. A. and RUTSCHKY, C. W. (1969) A comparative study of the embryological development of the median cord in Hemiptera. *J. Morph.* **129,** 375–400.

SRIVASTAVA, U. S. (1966) The development of Malpighian tubules and associated structures in *Philosamia ricini* (Lepidoptera, Saturnidae). *J. Zool.* **150,** 145–63.

STAIRS, G. R. (1960) On the embryology of the spruce budworm *Choristoneura fumiferana* (Clem.) (Lepidoptera, Tortricidae). *Can. Ent.* **92,** 147–54.

STANLEY, M. S. M. and GRUNDMANN, A. W. (1970) The embryonic development of *Tribolium confusum. Ann. ent. Soc. Amer.* **63,** 1248–56.

STEFANI, R. (1961) I fenomoni cariologica nella segmentazione dell'uovo ed i loro rapporti con la partogenesi rudimentale ed accidentale negli Embioptera. *Caryologia* **12,** 1–70.

STEELE, H. V. (1941) Some observations on the embryonic development of *Austroicetes cruciata* Sauss. (Acrididae) in the field. *Trans. Roy. Soc. S. Aust.* **65,** 329–32.

STRIEBEL, H. (1960) Zur Embryonalentwicklung der Termiten. *Acta trop.* **13,** 193–260.

STRINDBERG, H. (1913) Embryologische Studien an Insekten. *Z. wiss. Zool.* **106,** 1–227.

STRINDBERG, H. (1914) Zur Kenntnis der Hymenopteren-Entwicklung. *Vespa vulgaris* nebst einigen Bemerkungen über die Entwicklung von *Trachusa serratulae. Z. wiss. Zool.* **112,** 1–47.

STRINDBERG, H. (1915a) Hauptzüge der Entwicklungsgeschichte von *Sialis lutaria. Zool. Anz.* **46,** 167–85.

STRINDBERG, H. (1915b) Entwicklung der Hymenopteren nebst einigen damit zusammenhangenden Fragen. *Zool. Anz.* **45,** 248–60.

STRINDBERG, H. (1915c) Embryologische Studien über *Forficula auricularia* L. *Zool. Anz.* **45,** 624–31.

STRINDBERG, H. (1915d) Ueber die Bildung und Verwendung der Keimblatter bei *Bombyx mori. Zool. Anz.* **45,** 577–97.

STRINDBERG, H. (1916) Zur Entwicklungsgeschichte und Anatomie der Mallophagen. *Z. wiss. Zool.* **115,** 382–459.

STRINDBERG, H. (1917) Neue Studien über Ameisen Embryologie. *Zool. Anz.* **49,** 177–97.

STRINGER, I. A. N. (1969) Blastokinesis and embryology of the phasmid, *Clitarchus hookeri, Tane* **15,** 141–52.

SUCH, J. (1963) Observations sur l'embryogénèse de la musculature mandibulaire chez le Doryphore. Découverte d'une musculature ventrale abortive de la mandibule. *Soc. Sc. Phys. et Nat. Bordeaux* **1962–3,** pp. 63–65.

SUCH, J. and HAGET, A. (1962) L'apparition des vésicules coelomiques de la tête, chez l'embryon du Doryphore *Leptinotarsa decemlineata* Say. *Soc. Sc. Phys. et Nat. Bordeaux* **1961–62,** pp. 97–100.

SUROWIAK, J. (1958) The early stages of embryonic development of the *Hylobius abietis* L. (Curculionidae, Coleoptera). *Zesz. nauk. Univ. jagiellonsk. Zool.* **3,** 3–29.

THOMAS, A. J. (1936) The embryonic development of the stick-insect, *Carausius morosus. Q. Jl. microsc. Sci.* **78,** 487–512.

TIEGS, O. W. and MURRAY, F. V. (1938) The embryonic development of *Calandra oryzae. Q. Jl. microsc. Sci.* **80,** 159–284.

TOYAMA, K. (1902) Contributions to the study of silkworm. I. On the embryology of the silkworm. *Bull. Coll. Agric. Tokyo imp. Univ.* **5,** 73–118.

Tschuproff, H. (1903) Ueber die Entwicklung der Keimblätter bei den Libellen. *Zool. Anz.* **27**, 29–34.

Ullman, S. L. (1964) The origin and structure of the mesoderm and the formation of the coelomic sacs in *Tenebrio molitor* L. (Insecta, Coleoptera). *Phil. Trans. Roy. Soc.* B, **248**, 245–77.

Ullman, S. L. (1967) The development of the nervous system and other ectodermal derivatives in *Tenebrio molitor* L. (Insecta, Coleoptera). *Phil. Trans. Roy. Soc.* B, **252**, 1–25.

Vaidya, V. G. (1967) Gastrulation in *Popilio polytes* L. *Nature, Lond.* **216**, 936–7.

Viallanes, H. (1891) Sur quelques points de l'histoire due développement embryonnaire de la Mante religieuse. *Annls. Sci. nat. 7th ser. Zool.* **11**, 282–328.

Van Horn, S. N. (1966) Studies on the embryogenesis of *Aulocara elliotti* (Thomas) (Orthoptera, Acrididae). I. External morphogenesis. *J. Morph.* **120**, 83–114.

Vignau-Rogueda, J. (1969) Analyse expérimentale de la genèse de l'épicrâne du *Carausius morosus* Br. par des opérations microchirurgicales effectuées sur les embryons de ce Phasme. *C.R. Acad. Sci., Paris* **268**, 352–5.

Voegele, I. (1967) Embryogenèse de *Aelia cognata* Fieb. (Heteroptera, Pentatomidae). *Al Awamia Rev. Rech. Agron Moracaine* **24**, 41–65.

Wada, S. (1966) Topographie der Anlagenkomplexe der Cephalregion von *Tachycines* (Saltoria) beim Keimstreif. *Z. Naturwiss.* **53**, 414.

Weglarska, B. (1955) The formation of the blastoderm and embryonic membranes in *Polydrosus impar* Gozis (Coleoptera, Curculionidae). *Polskie Pismo ent.* **25**, 193–211.

West, J. A., Cantwell, G. E. and Shortino, T. J. (1968) Embryology of the housefly, *Musca domestica* (Diptera: Muscidae), to the blastoderm stage. *Ann. ent. Soc. Amer.* **61**, 13–17.

Wheeler, W. M. (1889a) The embryology of *Blatta germanica* and *Doryphora decemlineata. J. Morph.* **3**, 291–386.

Wheeler, W. M. (1889b) Ueber drusenartige Gebilde im ersten Abdominalsegment der hemipterenembryonen. *Zool. Anz.* **12**, 500–4.

Wheeler, W. M. (1889c) Homologues in embryo Hemiptera of the appendages to the first abdominal segment of other insect embryos. *Amer. Nat.* **23**, 644–5.

Wheeler, W. M. (1890) New glands in hemipterous embryos. *Amer. Nat.* **24**, 187.

Wheeler, W. M. (1893) A contribution to insect embryology. *J. Morph.* **8**, 1–160.

Wigglesworth, V. B. (1959) Insect blood cells. *Ann. Rev. Entom.* **4**, 1–16.

Williams, F. X. (1916) Photogenic organs and embryology of lampyrids. *J. Morph.* **28**, 145–207.

Wolf, R. (1969) Kinematik und Feinstruktur plasmatischer Faktorebereiche des Eies von *Wachtliella persicariae* L. (Diptera). *Arch. EntwMech. Org.* **162**, 121–60.

Wray, D. L. (1937) The embryology of *Calandra callosa* Olivier, the Southern corn billbug (Coleoptera, Rhynchophoridae). *Ann. ent. Soc. Amer.* **30**, 361–409.

Yajima, H. (1960) Studies on embryonic determination of the Harlequin-fly, *Chironomus dorsalis.* I. Effects of centrifugation and of its combination with constriction and puncturing. *J. Embryol. exp. Morph.* **8**, 198–215.

Young, J. H. (1953) Embryology of the mouthparts of Anoplura. *Microentomology* **18**, 85–133.

Zakhvatkin, Y. A. (1967a) Embryonic development of *Phyllodecta vitellinae* (Coleoptera, Chrysomelidae). *Zool. Zh.* **46**, 88–97.

Zakhvatkin, Y. A. (1967b) Embryonic development of Galerucinae (Coleoptera, Chrysomelidae). *Zool. Zh.* **46**, 1209–18.

Zakhvatkin, Y. A. (1968) Comparative embryology of Chrysomelidae, *Zool. Zh.* **47**, 1333–42.

CHAPTER 8

CRUSTACEANS

IN SPITE of the great diversity in modes of development among modern forms, the terrestrial arthropods discussed in the preceding chapters share a basic mode of development in which a relatively large, yolky egg proceeds through intralecithal cleavage to blastoderm formation and epimorphic later development. In this respect, the Onychophora, Myriapoda and Hexapoda are sharply distinct from the annelids, whose basic mode of development is that of a small egg undergoing total, spiral cleavage and hatching as a planktotrophic trochophore larva.

The development of the Crustacea, at first sight, appears to offer the possibility of an intermediate mode. Basic crustacean development originates from a small egg, passes through a total cleavage which in some species has a spiral basis, and proceeds to hatching as a small planktonic larva, the nauplius. Furthermore, in many crustacean species, this basic sequence is modified to include a larger, yolky egg, intralecithal cleavage, blastoderm formation and epimorphic later development.

Nevertheless, the superficial expectations of intermediacy engendered by the Crustacea are deceptive. As more has become known of small crustacean embryos, it has been increasingly realized that their development is uniquely specialized. For example, although the cleavage of the egg has a spiral basis, it differs fundamentally from the spiral cleavage of annelids. The further development of the embryo is also specialized, commensurate with the emergence of a distinctive naupliar organization at hatching. Then, too, the manner in which development is completed anamorphically during a series of planktonic larval stages with intervening moults has its own underlying peculiarities. We shall see in this chapter that a highly distinctive theme of development obtains in the Crustacea and that they are not intermediate in any way between the annelids and other arthropods.

The elucidation of the basic features of crustacean development from a comparative point of view has been a long and exacting task. Those species in which the basic theme of development is most strongly maintained, i.e. those with small eggs and planktotrophic naupliar development, are also the species which offer the most frustrations to the descriptive embryologist. Although small, the eggs are yolky and the development of the nauplius is a complex sequence of small-scale events obscured by yolk. After hatching, the larval stages combine an elaborate functional morphology with a continued development, interrupted by moults and culminating in a greater or lesser degree of metamorphosis, all within relatively small dimensions. Although a very large literature on crustacean larvae has built up in recent years, following the emergence of new techniques for the laboratory culture of crustacean larvae (e.g. Provenzano, 1967a, 1968a), almost none of this work discusses internal changes during larval development. An adequate level of descriptive analysis of internal events has been attained in recent work on the embryonic and larval development of the Cirripedia (Anderson, 1969; Walley, 1969), but our knowledge of the histological development of other groups with small eggs and a nauplius larva (the Branchiopoda, Copepoda, Euphausiacea, Penaeidea and the recently discovered Cepholocarida) is still poor.

Compounding the difficulty, the many groups of Crustacea in which the egg has a larger yolk volume and a more direct development show evidence of independent evolution of this condition by several different routes. The interpretation of the developmental morphology of heavily yolked crustacean embryos is further complicated by the retention of naupliar development as an embryonic event, in which some of the pathways of formation of naupliar structures persist but others are lost. Finally, although some kinds of larger crustacean embryos complete their development at the embryo stage and hatch as fully formed juveniles (e.g. the Peracarida, Leptostraca and some Astacura), others hatch as complex planktonic larvae (e.g. many Decapoda) and retain an elaborate series of larval stages, often diverse in form and bewildering in variety.

Because of the problems mentioned above, the classical period of crustacean embryology in the fifty years from 1868 to 1918 lacks any central figure who could be said to have established the basis of the subject in the manner of Wheeler and Heymons for the Pterygota (see Chapter 7). Much confusion in terminology and interpretation prevailed before the nineteen-twenties. The subject began tentatively with contributions by several well-known carcinologists and embryologists of the time (e.g. Claus, 1868, 1873, 1875 on the Notostraca, Ostracoda, Branchiura and Hoplocarida; Metschnikoff, 1868 on the Leptostraca; van Beneden and Bessels, 1869, Bessels, 1869 and van Beneden, 1870 on the Copepoda and Amphipoda; Dohrn, 1870 a, b on the Cumacea and Tanaidacea; Bobretsky, 1873, 1874 and Faxon, 1879 on the Isopoda, Caridea and Astacura; Grobben, 1879 on the Cladocera), whose main contribution was to establish the wide range of egg size and developmental type among the Crustacea, the occurrence of total cleavage in some small-egged species and the persistence of a recognizable nauplius stage in the direct development of species with larger eggs.

As classical crustacean embryology grew in volume and diversity, interest continued to centre mainly on common species with conveniently brooded eggs. Apart from the Cladocera, Copepoda and Cirripedia, these were for the most part certain kinds of malacostracans—the mysids, amphipods and isopods among the Peracarida and the carid shrimps in the Eucarida.

Cladoceran development was studied by a series of workers (Lebedinsky, 1891; Samassa, 1893 a, b, c; 1897; Grobben, 1893; Samter, 1900; Agar, 1908; Vollmer, 1912), but the small dimensions, dense yolk and specialized development of the embryos belied their easy accessibility in the brood pouches of quick-breeding females and led to little more than a number of superficial observations. Only Kühn (1908, 1911, 1913) contributed significantly to cladoceran embryology in this period, through a detailed cell-lineage study of cleavage in *Polyphemus*. Although Kühn had some difficulty in comparing his results with other cell-lineage studies, due to a history of unfortunate choices made by other workers in establishing a notation for crustacean cell lineage, his results have recently proved of great value when reinterpreted in a more meaningful way (Anderson, 1969). Occasional contributions on the embryos of other branchiopods were interspersed with these studies (Claus, 1886, Nassonow, 1887 and Snetlage, 1905 on Anostraca; Claus, 1886 and Brauer, 1892 on Notostraca; Sars, 1887, 1896 a, b on Conchostraca), but little insight was gained by them into branchiopod cleavage, gastrulation and organogeny.

Among other non-malacostracan groups, little was done with the Ostracoda (Woltereck, 1898; Muller-Calé, 1913) or Branchiura (Grobben, 1911), but the Copepoda and Cirripedia attracted attention from several embryologists and were eventually the subject of important cell-lineage studies. The small embryos of free-living copepods were studied by Grobben (1881), Urbanowicz (1884, 1886), Claus (1893, 1895), Häcker (1897) and Amma (1911), and the larger embryos of the parasitic copepod *Lernaea branchialis* by Pedaschenko (1893, 1897, 1898). Only Fuchs (1914), however, working on cleavage in *Cyclops*, obtained results of sufficient accuracy and clarity on copepod cell lineage to make a significant contribution to present understanding. Similarly, the development of cirripede eggs was described by Nassonow (1885, 1887), Nusbaum (1887, 1890 a, b), Groom (1894) and Bigelow (1902), but a satisfactory interpretation of cell lineage and the development of the nauplius in the Cirripedia was approached only by Delsman (1917).

For the Malacostraca, the workers on the Peracarida met with a greater success. The brood-pouch development of the peracaridans and their direct development to a juvenile form make this group one of the most accessible among crustaceans for embryological study. Parallel series of investigations were carried out on the Mysidacea (Boutchinsky, 1890; Bergh, 1893; Wagner, 1895, 1898), the Isopoda (Nusbaum, 1886, 1891, 1898, 1904; Reinhard, 1887; Roule, 1889, 1890, 1891, 1892 a, b, 1895, 1896a; Patten, 1890; McMurrich, 1895) and the Amphipoda (Tichomirov, 1883; Pereyaslawzewa, 1888 a, b; Wagner, 1891; Rossiiskaya-Kojewnikowa, 1890, 1893, 1896; Bergh, 1893; Langenbeck, 1898; Heidecke, 1904), all of which revealed the underlying unity in peracaridan development, including teloblastic development of the post-naupliar segments and vitellophage modifications in the early development of the midgut. Interpretations of the origins and subsequent development of the components of the embryo differed, however, and many important points remained unresolved. The lack of proper knowledge of the development of the small crustacean embryos which hatch as nauplii also made the interpretation of the secondarily modified peracaridan embryo difficult at that time. The minor contributions made by Boutchinsky (1897, 1900) and Robinson (1906) on the Leptostraca and by Blanc (1885), Boutchinsky (1893) and Grschebin (1910) on the Cumacea did little more than confuse the situation further.

Much the same tentative picture was built up of decapod development. Various species of Caridea were examined by Ishikawa (1885, 1902), Kingsley (1887, 1889), Brooks and Herrick (1892), Weldon (1892), Gorham (1896), Roule (1896b) and Moroff (1912). A parallel series of studies on astacuran embryos was carried out by Morin (1886), Reichenbach (1877, 1886), Waite (1889), Bumpus (1891), Herrick (1891, 1895) and Fulinski (1908), while occasional papers appeared on the Penaeidea (Brooks, 1880, 1882), Thalassinidea (Boutchinsky, 1894) and Brachyura (Lebedinsky, 1890; Urbanowicz, 1893). The basic pattern of embryonic development common to all decapods can be discerned from these studies, including the retention of a transient nauplius stage in the embryo, the teloblastic development of the post-naupliar segments as a forwardly flexed caudal papilla and the vitellophage function of the early rudiments of the midgut; but as in the case of the Peracarida, precision and reliability in the analysis of crucial questions of origins and fates was always lacking. Taube (1909, 1915) attempted to improve this situation by investigating the small, totally cleaving eggs of euphausiaceans but was unable to provide an adequate interpretation of their embryonic development. This problem is still unrevolved.

Before the First World War, then, crustacean embryology had ranged widely, but not deeply. Intimations of the occurrence of spiral cleavage in the small eggs of cladocerans, copepods, cirripedes and euphausiids were confounded by eccentricities of notation and terminology in the various descriptions, making any comparison with annelids impossible. The basic sequence of embryonic development was reasonably well established for the peracaridans and the eucaridans with larger eggs, but wide variations existed in the interpretation of the origins and fates of the components of the embryo. In the nineteen-twenties and thirties, this situation was to be radically changed through major contributions by only a few workers. An interesting contrast obtains with pterygote embryology, to which numerous important contributions were made during the same years (see Chapter 7). In general, crustacean embryology was severely neglected during this period, but at the same time it became transformed. The first important steps were taken by Cannon (1921, 1924, 1925, 1926) in his detailed studies of segment formation and the elaboration of the heart, musculature, haemocoel and segmental organs of branchiopods. Cannon's work set the stage for two superbly detailed studies by S. M. Manton, which for the first time elucidated all those features of malacostracan development that had so baffled the earlier workers. The first, on the mysid *Hemimysis lamornae*, published in 1928, constitutes the turning point in the history of crustacean embryology. Earlier work was comprehensively reviewed and reinterpreted in accordance with a wealth of new, detailed information, establishing the basis for a new era of progressive discovery. The crucial importance of this paper is emphasized by the fact that it has been quoted in almost every subsequent contribution to the subject. The second paper by Manton, published in 1934, on the leptostracan *Nebalia bipes*, consolidated the position by analysing a malacostracan with a different pattern of direct development, emphasizing the true variability that exists among malacostracan embryos and pointing the way to further studies.

Three such studies were soon forthcoming. Hickman (1937) described in detail the embryology of the syncaridan *Anaspides tasmaniae*, finding many peculiarities in the embryos of this relict species, some as yet unexplained. Shiino (1942) also gave a first detailed account of hoplocaridan development, so that by the end of the Second World War, knowledge of the embryology of the Leptostraca, Syncarida, Hoplocarida and mysid Peracarida among the Malacostraca had been vastly improved. Somewhat more remotely, Cannon's contributions on the Branchiopoda were supplemented by the first detailed studies of cladoceran development by Baldass (1937, 1941). The way was thus clear for a rapid expansion of the subject after 1945.

There were, of course, other studies during the nineteen-twenties and thirties, but mostly they were scattered and incidental observations. Humperdinck (1924), Kühnemund (1929) and Wotzel (1937) presented a few facts on cladoceran embryos, as did Witschi (1934) and Filhol (1934, 1936) on Copepoda, Tokioka (1936) on Branchiura and Krüger (1922) and Kühnert (1934) on Cirripedia. For the Malacostraca, some brief observations on penaeid cleavage were made by Heldt (1931, 1938). Sollaud (1923) published an extensive but partly erroneous account of caridean development. Betances (1921), Fulinski (1922), Piatakov (1925), Terao (1921, 1925, 1929), Zehnder (1934 a, b, 1935) and Bonde (1936) commented on the embryos of astacurans and palinurans. The first investigation of a pagurid embryo was presented by Krainska (1934, 1936), while Nair (1939, 1941) wrote briefly on mysid and hoplocaridan embryos and Goodrich (1939) described the development of the isopod gut accurately for the first time. All of these papers supported and augmented Manton's comprehensive analysis, the only dissenter being Needham (1937, 1942), working on mysids and isopods. Needham's views have not been substantiated by more recent workers.

The advances that have taken place in the last twenty-five years bear striking testimony to the groundwork laid by Cannon, Manton, Hickman, Shiino and Baldass. Major studies have now been carried out on most groups of crustaceans, the only notable exceptions being the Notostraca, Copepoda, Euphausiacea and Penaeidea among species with small eggs (and also, of course, the Cephalocarida and Mystacocarida) and the Thalassinidea and Brachyura among species with larger eggs. The lack of interest in brachyuran embryos for a hundred years is one of the most remarkable cases of obvious neglect in the whole of arthropod embryology.

Among the non-malacostracans, the barnacles have yielded the greatest amount of information on crustacean spiral cleavage and the development of the basic structure of the free-swimming nauplius. Beginning with studies by Batham (1945, 1946) and Vaghin (1947) on larger-egged species and continuing with similar studies by Anderson (1965), Kaufmann (1965) and Turquier (1967 a, b) and investigations on the specialized development of rhizocephalan embryos by Bocquet-Védrine (1960, 1964), Veillet (1961) and Utinomi (1961), this work has culminated in a detailed analysis of cirripede embryonic development by Anderson (1969) and of metamorphosis by Walley (1969) and Turquier (1970, 1971). As will be seen below, the information gained in these studies permits for the first time a fundamental unification of crustacean embryology. For other groups, Weygoldt (1960a) published the first comprehensive study of ostracod embryonic development, Anderson (1967) studied segment formation and metamorphosis in the Conchostraca and Benesch (1969) has recently made a detailed analysis of embryonic and larval development in the anostracan *Artemia.* A special mention should also be made of the work of Sanders (1963) and Sanders and Hessler (1964) on the Cephalocarida. Although these papers deal with larval development and not with embryology, the generalized pattern of development revealed in the Cephalocarida has a fundamental bearing in the comprehension of all patterns of development in the Crustacea. Other papers on the non-malacostracans are briefer and deal mainly with cleavage or with larval development (e.g. Anostraca, Weisz, 1947; Ten and Pai, 1949; Pai, 1958; Gauld, 1959; Anteunis *et al.*, 1961; Nourisson, 1959, 1962; Nakanisha *et al.*, 1962; Baquai, 1963; Fautrez and Fautrez-Firlefyn, 1963, 1964; Broch, 1965; Notostraca, Pai, 1958; Chaigneau, 1960; Conchostraca, Botnaruic, 1948; Pai, 1958; Cladocera, Kaudewitz, 1950; Agar, 1950; Esslova, 1959; Murakami, 1961; Copepoda, Harding, 1955; Davis, 1959; Dudley, 1969; Branchiura, Fryer, 1961). For the Malacostraca, outstanding contributions to the further descriptive analysis of embryonic development have recently been made by several workers. Nair (1949), working on the atyid shrimp *Caridina*, and Shiino (1950), working on *Palinurus*, were the first of these. A series of studies by Weygoldt (1958, 1960b, 1961) then greatly improved our knowledge of the embryos of amphipods, isopods and carid shrimps, to be followed by equally comprehensive accounts by Scholl (1963) on the Tanaidacea, Stromberg (1965, 1967, 1971) on the isopods, Dohle (1970) on the Cumacea, Fioroni (1969, 1970 a, b) on astacurans and pagurids and Lang and Fioroni (1971) on the Brachyura. These works, taken together with the classical papers of Manton (1928, 1934), provide a broad comprehension of malacostracan development and have been further supplemented by many contributions from other workers (Hoplocarida, Manning, 1963; Mysidacea, Jepson, 1965; Almeida, 1966; Petriconi, 1968; Berrill, 1969; Thermosbaenacea, Barker, 1962; Isopoda, Ban, 1950; Kajishima, 1952b; Daum, 1954; Dahl, 1955; Akahira, 1957; Lemercier, 1957; Saudrier and Lemercier, 1960; Kuers, 1961; John, 1968; Chia, 1968; Amphipoda, Rappaport, 1960; Sheader and Chia, 1970; Berrill, 1971; Euphausiacea, Mauchline, 1967; Penaeidea, Dobkin, 1961; Caridea, Nataraj, 1947; Aiyar, 1949; Kajishima, 1950; Dahl, 1957; Ling, 1967; Astacura, Suko, 1969; Fioroni, 1969; Pandian, 1970; Palinura, Alves and Tome, 1967; Paguridea, Oishi, 1959; Brachyura, Oishi, 1960).

In the present chapter, the extensive knowledge of malacostracan development is related to the recent advances in non-malacostracan embryology to establish a basic unity for crustacean embryology as a whole. We shall see that a highly distinctive theme of development obtains throughout the Crustacea, in spite of their immense diversity in detail. The emphasis in the present context is on this theme rather than on its variations. The latter are too numerous and diverse to be encompassed in a short account, but the elucidation of the common ground of crustacean development is essential for comparative purposes. We can begin with the question of interpreting crustacean cleavage as modified spiral cleavage.

Cleavage

Unlike other arthropods, the Crustacea have small eggs and total cleavage as a primary feature. The question of whether their primary total cleavage is a modified form of spiral cleavage remained for many years unresolved. Intimations of spirality in the cleavage of cirripedes, copepods and cladocerans were clear in the results of a number of cell lineage studies (Bigelow, 1902; Kühn, 1913; Fuchs, 1914; Delsman, 1917; Baldass, 1937, 1941), but none of these workers was able to attain a satisfactory interpretation of crustacean total cleavage in relation to the classical spiral cleavage patterns of annelids and molluscs. Furthermore, the results obtained for different groups of crustaceans seemed impossible to reconcile one with another. A recent study by Anderson (1969), however, has served to resolve this crucial problem. The best indications of spirality in crustacean cleavage had previously been given by Bigelow (1902) and Delsman (1917) in cell-lineage studies on lepadomorph and balanomorph cirripedes, but were confounded by eccentricities of notation and lack of critical points of data. It is, perhaps, unexpected that a group of animals with a highly specialized metamorphosis and adult morphology should retain a generalized early embryonic development, and for many years the question of their cell lineage was left in abeyance. The cirripedes also include a variety of genera with larger, more yolky eggs than are produced by the majority of thoracican barnacles, and it was these that attracted the attention of embryologists. These genera, although they retain total cleavage (*Scalpellum*, Krüger, 1922; Kaufmann, 1965; *Alcippe*, Kühnert, 1934; *Pollicipes*, Batham, 1946; *Ibla*, Anderson, 1965; *Trypetesa*, Turquier, 1967b), show a variety of cleavage specializations associated with increased yolk and did not serve to disclose a basic cleavage pattern for Crustacea. Anderson (1969), however, re-examined and carefully analysed the cell lineage of three balanomorph barnacles, *Tetraclita*, *Chthamalus* and *Chamaesipho*, with eggs of the dimensions and developmental type previously studied by Bigelow and Delsman in *Lepas* and *Balanus*, taking into account the new understanding of annelid spiral cleavage that had emerged by that time (see Chapter 2). As a result of this work, it is now possible to discuss crustacean cleavage as a phenomenon of spiral cleavage and to understand some of the modifications of cleavage that have arisen during crustacean evolution.

Among those many thoracican barnacles that hatch as planktotrophic nauplii, the zygote is an elongate ovoid cell, uniformly filled with yolk. The length of the egg varies (e.g. *Chamaesipho columna*, 100 μ; *Tetraclita rosea*, 215 μ; *Balanus balanoides*, 300 μ; Groom, 1894; Anderson, 1969), but the long axis is always the anteroposterior axis of development. As cleavage begins (Fig. 105), the first mitotic division of the zygote nucleus is accompanied by a segregation of the yolk towards the posterior half of the egg, leaving the anterior cytoplasm free of yolk. The first two cleavage nuclei occupy the yolk-free and yolk-filled parts of the cytoplasm respectively, and the first cleavage plane then cuts transversely through the boundary between the two regions to separate an anterior yolk-free cell from a posterior, yolk-filled cell. Adopting the Wilsonian spiral cleavage terminology, the anterior cell is AB, the posterior cell CD. At the second cleavage division, AB divides equally into A, which pushes backwards to a left lateral position, and B, which retains an anterior position. CD, in contrast, cuts off a second, yolk-free cell C, anteriorly on the right, leaving the large, yolk-filled D cell in the posterior position.

It is at this stage that we gain confirmation that the quadrants of the cirripede 4-cell stage correspond to the four quadrants of spiral cleavage. As in all spiral cleavage, the cells B and D retain transverse contact ventrally, while the A and C cells establish sagittal contact dorsally. In typical spiral cleavage, as described in Chapter 2, D is dorsal and B is ventral relative to the axes of future development, but the shift of B to an anterior position and D to a posterior position is a typical modification associated with increased yolk, as seen in various clitellate annelids (Chapter 3). The entire sequence of cirripede cleavage is modified in relation to yolk, though in a manner different from clitellate cleavage.

The identification of the cells of the cirripede 4-cell stage as the quadrants of spiral cleavage is confirmed in the subsequent course of cleavage (Figs. 106 and 107). All four quadrants proceed through three division cycles, the third, fourth and fifth cleavage divisions, in which the division of each cell is perpendicular to the previous division. This is the same sequence that obtains in the alternating dexiotropic, laeotropic and dexiotropic, third to fifth cleavage divisions of typical spiral cleavage (see Chapter 2). During these divisions, other features of spiral cleavage are maintained. The most ventral cells of the B and D quadrants, 3B and 3D by the end of the fifth cleavage division, retain the transverse line of contact established initially at the 4-cell stage. Furthermore, the ectodermal, mesodermal and midgut rudiments

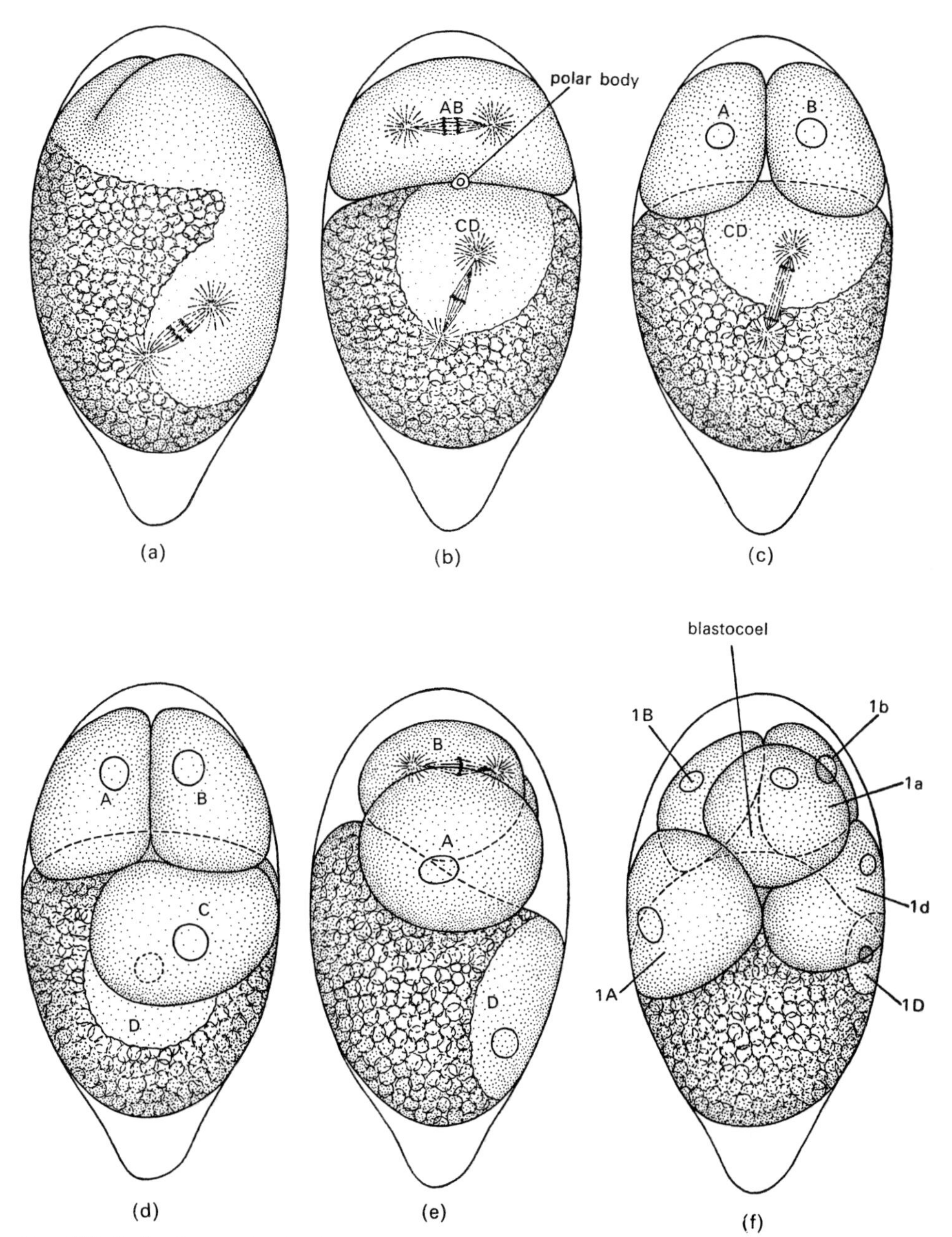

FIG. 105. Early cleavage in the cirripede *Tetraclita rosea*, after Anderson (1969). (a) First cleavage division; (b)–(d) second cleavage division sequence, dorsal view; (e) 4-cell stage, left lateral view; (f) 8-cell stage, left lateral view.

are segregated from one another by the end of the fifth cleavage division, a characteristic spiral cleavage feature.

What we see in cirripede cleavage, in fact, is a functional modification of spiral cleavage appropriate to a small but densely yolky egg. Cleavage proceeds in such a manner that,

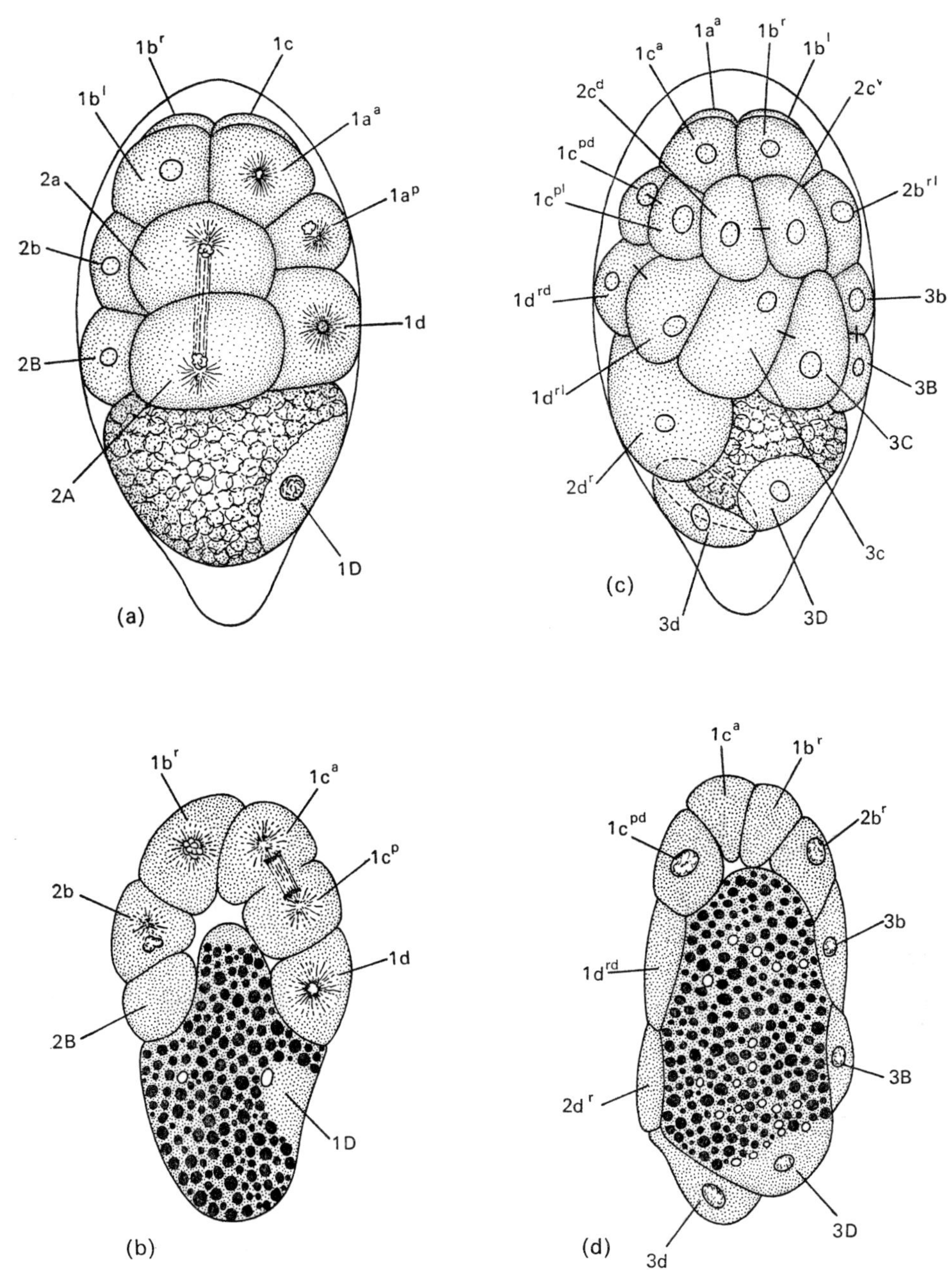

FIG. 106. Late cleavage in the cirripede *Tetraclita rosea*, after Anderson (1969). (a) Fourth cleavage division, left lateral view; (b) sagittal section through the stage shown in (a); (c) 28-cell stage, right lateral view; (d) sagittal section through the stage shown in (a).

during the first five division cycles alluded to above, no cleavage furrows pass through the yolk. During the third to fifth cleavages, the division products of the yolk-free, B-cell spread ventrally backwards over the D-quadrant yolk cell to cover the anteroventral and ventral surfaces of the yolk cell. At the end of the fifth cleavage division the daughter cells

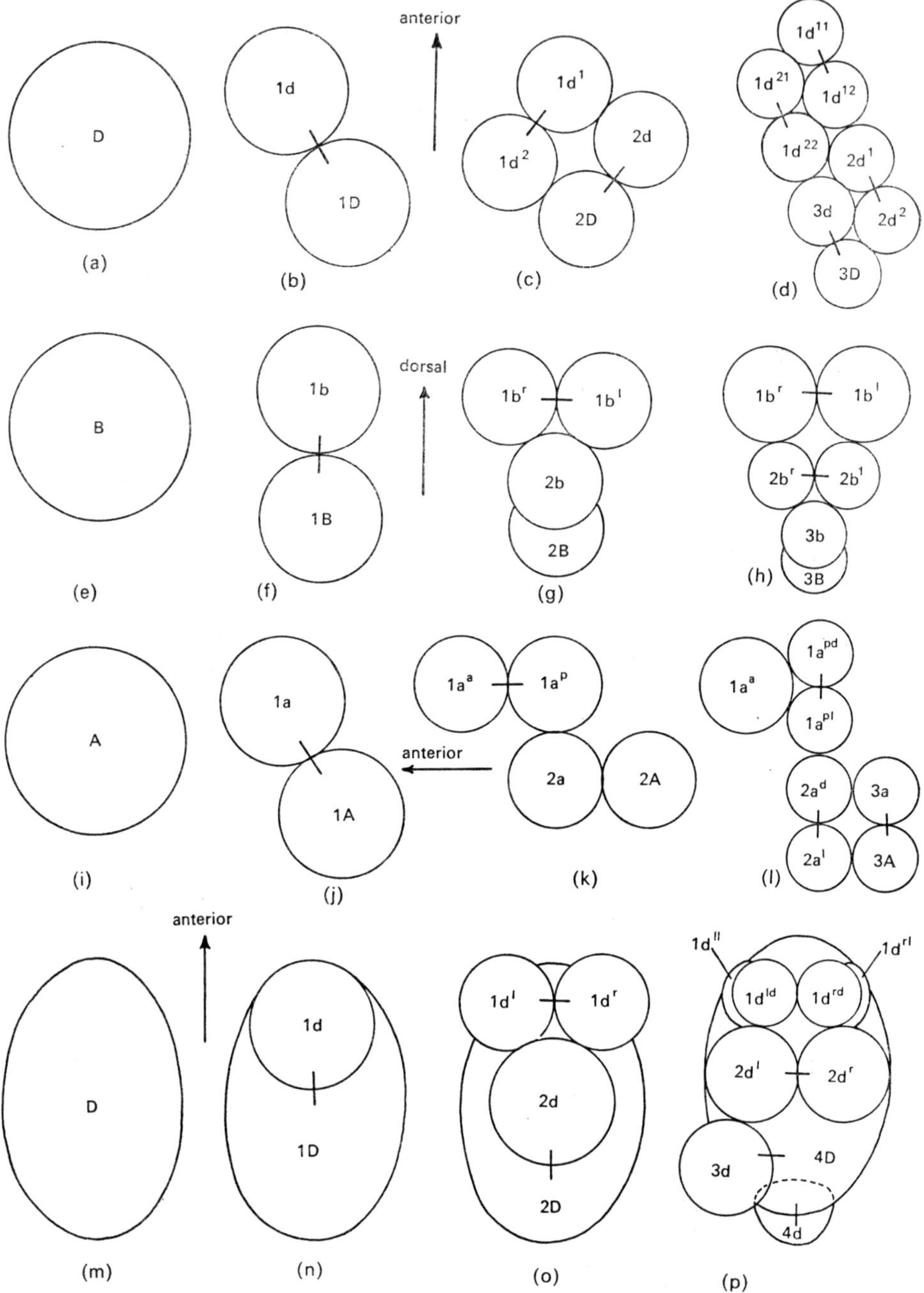

FIG. 107. The modifications of spiral cleavage in the four quadrants of the cirripede egg. (a)–(d) illustrate the third to fifth divisions of basic spiral cleavage. Cleavage is equal, the quadrants are left (A), ventral (B), right (C) and dorsal (D) respectively and the stem cells are posterior. The D quadrant is drawn viewed dorsally, with the anteroposterior axis vertical. The A quadrant viewed from the left would present an identical picture, while the B quadrant viewed ventrally and the C quadrant viewed from the right would present a mirror image. (e)–(p) illustrate the same divisions in the four quadrants of the cirripede egg. In this egg the quadrants are reorientated in relation to the embryonic axes so that B is anterior, D posterior, A and C left and right respectively and the stem cells are ventral. (e)–(h) show the B quadrant viewed from the front. Spindles and cleavage planes are rotated from oblique to bilateral orientations. (i)–(l) show the A quadrant viewed from the left with the anterior end on the left. In association with the rotation of the quadrant from an anteroposterior to a dorsoventral orientation, the third cleavage spindle is rotated through 90°. The C quadrant, viewed from the right with the anterior end on the right, would present a mirror image of the A quadrant. (m)–(p) show the D quadrant viewed from the dorsal surface, with the anteroposterior axis vertical. These diagrams are in the same orientation as (A) to (D) if the latter are taken as D quadrant cells. Once again, all the spindles are rotated into the planes of bilateral symmetry.

of B are, in anteroposterior order, $1b^l$ and $1b^r$, $2b^l$ and $2b^r$, 3b and 3B. The 1b cells do not participate in the fifth cleavage division at this stage. Accompanying the divisions of the B cell, the daughter cells of the yolk-free A and C cells also spread backwards, along the sides of the yolk cell. At the end of the fifth cleavage division, these cells cover the yolk cell anterodorsally and laterally. The anteroposterior order is, on the left, $1a^a$, then $1a^{pd}$, $1a^{pv}$, $2a^d$ and $2a^v$ in a transverse row, and 3a and 3A behind. 3A makes contact with the left side of the midventral cell 3B. The C quadrant cells on the right are in the same order, $1c^a$, then $1c^{pd}$, $1c^{pv}$, $2c^d$ and $2c^v$ in a transverse row, followed by 3c and 3C. The latter is, of course, in contact with the right side of 3B. It will be noticed that, as in the B quadrant, the most anterior cells of the A and C quadrants ($1a^a$ and $1c^a$) do not participate in the fifth cleavage division. We shall see below that this lessening of divisions at the anterior end is compensated by additional cell divisions at the posterior end.

While the yolk-free A, B and C quadrants are dividing and spreading to cover the anterior, lateral and ventral surfaces of the yolk-cell, the latter undergoes the third to fifth divisions of cleavage in such a way as to cover its dorsal and posterior surfaces with its own yolk-free daughter cells. At each of these divisions, the yolk cell cuts off a yolk-free cell dorsally. The daughter cells also divide at the subsequent cleavages, so that by the end of the fifth cleavage division, the anteroposterior order of dorsal, yolk-free, D-quadrant cells is $1d^{ll}$, $1d^{ld}$, $1d^{rd}$ and $1d^{rl}$, followed by $2d^l$ and $2d^r$, then by 3d (slightly displaced to the left at this stage). In contrast to the other quadrants, $1d^l$ and $1d^r$ participate in the fifth cleavage division. In addition, there is a precocious sixth cleavage division in the D quadrant, yielding a yolk-free cell, 4d, at the posterior end.

By the end of the fifth cleavage division, then, the yolk cell is 4D. No cleavage furrows have passed through this cell and it has lost little in size as compared with the initial zygote. At the same time, 4D has become covered, except posteroventrally, by yolk-free blastomeres. The outcome is a blastula in which all cells lie at the surface but one cell, the posteroventral 4D, extends inwards to fill the blastocoel. The nucleus of 4D, surrounded by yolk-free cytoplasm, is just beneath the exposed surface of the cell. Even at this stage, the transverse, ventral contact between the exposed part of 4D and the 3B cell just in front of it on the ventral surface is maintained.

By the end of the fifth cleavage division, allowing for the precocious sixth cleavage division of 3D into 4D and 4d, the segregation of the presumptive areas of the cirripede blastula is complete. We will return to the composition of these areas and the pattern in which they are arranged, in the next section of this chapter. Meanwhile, it is of interest to note that the distinctive modification of spiral cleavage observed in cirripedes perfectly exemplifies a functional intermediate between typical spiral cleavage, yielding a hollow blastula, and intralecithal cleavage, yielding a blastoderm around a yolk mass. As discussed in Chapters 2 and 3, yolky annelid eggs usually retain the blastocoel and exhibit an extensive spreading of small, yolk-free cells to enclose their large, yolky cells after the first five cleavage divisions have occurred. Again, as we have seen in Chapters 4–7, the Onychophora, Myriapoda and Hexapoda basically exhibit direct formation of a blastoderm around a yolk mass and offer no functional intermediates between this condition and an ancestral condition of total, spiral cleavage (see Chapter 10). In the cirripedes, however, we can see how cleavage can still remain total and recognizably spiral, yet yield an enclosing layer of yolk-free cells (essen-

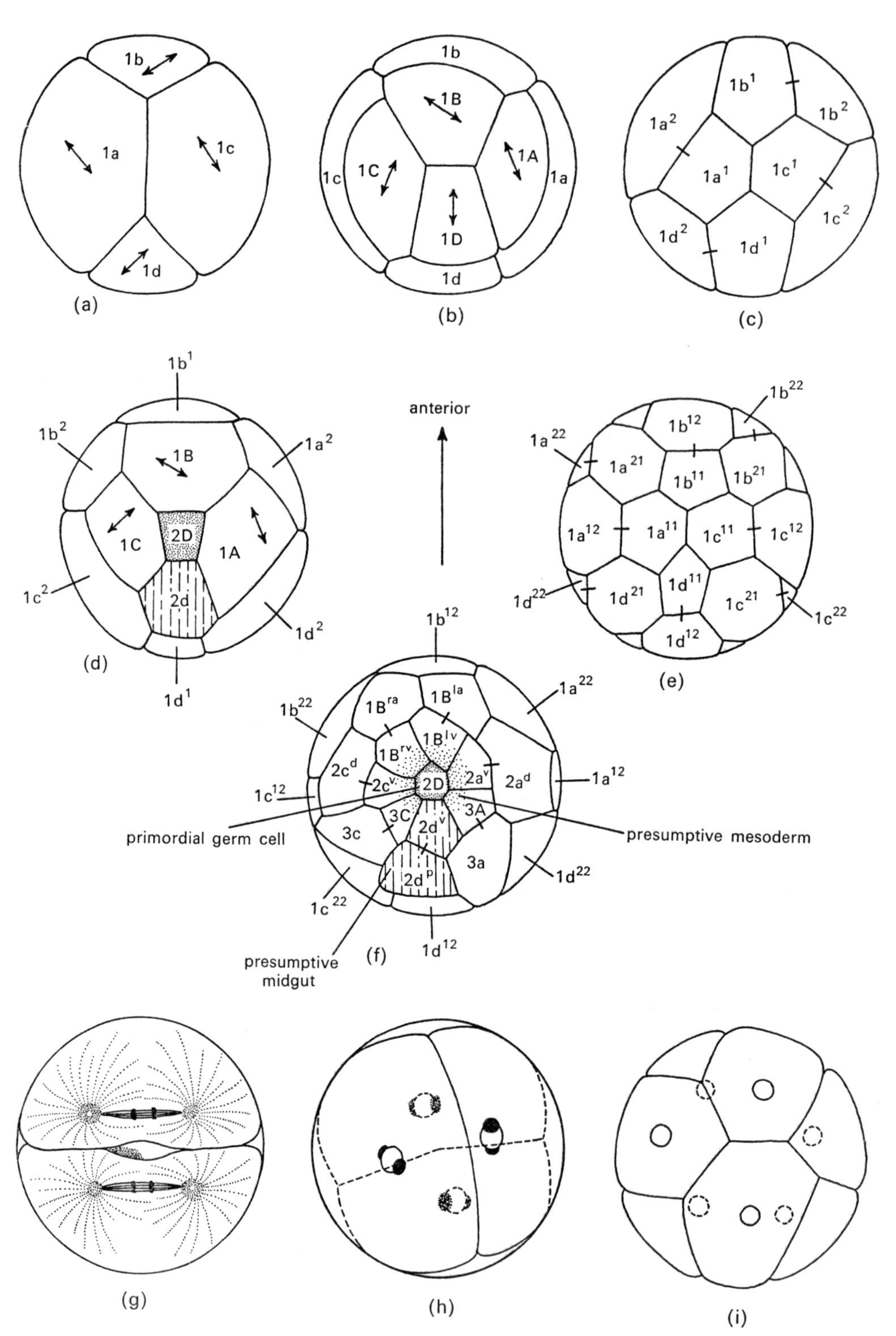

FIG. 108. (a)–(f) The third to fifth cleavage divisions of the cladoceran *Holopedium*, after Baldass (1937) and Anderson (1969). (a) 8-cell stage, dorsal view; (b) 8-cell stage, ventral view; (c) 16-cell stage, dorsal view; (d) 13–16-cell stage, ventral view; (e) 31-cell stage, dorsal view; (f) 31-cell stage, ventral view. All diagrams are drawn with the anteroposterior axis vertical. (g)–(i) Early cleavage in the anostracan *Artemia salina*, after Benesch (1969). (g) Second cleavage division; (h) 4-cell stage; (i) 16-cell stage.

tially, a blastoderm) around a single, large yolk cell (essentially, an internal yolk mass). The basic modification prerequisite to this intermediate condition is that none of the cleavage furrows formed during total cleavage should cut through the yolk. The only further, basic requirement for the transition to blastoderm formation, attained in many Crustacea, is that yolk-free cells should be cut off simultaneously, rather than sequentially, throughout the surface of the yolk cell; in other words, that the cytokinesis associated with the first five or more nuclear divisions of cleavage should be lost. Both are functional corollaries of an increase in the density and volume of yolk in the egg. Both are simple modifications. Clearly, this is not the only route by which the evolution of intralecithal cleavage from spiral cleavage could have occurred. Such evolution has proceeded independently in the onychophoran–myriapod–hexapod assemblage, the Crustacea (probably several times) and the Chelicerata, and a variety of functional transitions can be envisaged. The cirripedes show us one possible route, but, more importantly, they demonstrate that the transition from total, spiral cleavage to intralecithal cleavage in arthropods is not as difficult, functionally, as might appear at first sight.

Interestingly enough, this transition has not been attained in cirripedes with larger eggs. In all of these (e.g. Batham, 1946; Anderson, 1965; Kaufmann, 1965; Turquier, 1967b) the yolk cell is much enlarged relative to the yolk-free cells that are cut off in early cleavage and the latter, as in clitellates, have to undergo much proliferation to enclose the yolk.

Other groups of Crustacea, however, show a variety of further modifications of cleavage in the direction of blastoderm formation. For example, among the Branchiopoda, the cladoceran *Holopedium*, with a spherical egg 120 μ in diameter, retains a sequence of mitotic nuclear divisions whose spiral pattern is even more distinct than that of cirripedes (Baldass, 1937; Anderson, 1969), but has a much more specialized sequence of cytokineses associated with them. Rather than segregation of the yolk into a D-cell, followed by the formation and spread of yolk-free blastomeres at the surface of the yolk-cell, as in cirripedes, we find that no cleavage furrows penetrate the cytoplasm of the *Holopedium* egg until after the third synchronous nuclear division has taken place and the eight resulting nuclei are approaching the egg surface (i.e. the first three cleavage divisions are intralecithal). The eight nuclei (Fig. 108) occupy octants of the egg which subsequent events identify as dorsal, left, anterior, right and posterior (the 1a, 1b, 1c and 1d octants) and ventral left, anterior, right and posterior (the 1A, 1B, 1C and 1D octants). Cleavage furrows then cut in simultaneously from the surface to separate these octants, but the densely yolky, central part of the egg remains uncleaved. A similar radial penetration of cleavage furrows follows the formation of 16 and then 31 nuclei at the fourth and fifth cleavages, but total separation of the cells is attained only at the last of these stages. Thus, in *Holopedium*, the early cleavage planes penetrate the yolky cytoplasm gradually and give rise eventually to a blastula of 31 equal, pyramidal, yolky cells. Anderson (1969) interpreted this condition as secondary to one in which the yolk was confined to a single D-quadrant cell, as in cirripedes, mainly on the grounds that the cleavage divisions of the D-quadrant are still delayed in *Holopedium*, even though the yolk is equally distributed between all cells. A functional explanation of the transition from segregation of yolk into one cell to equal distribution of yolk among all cells is difficult to attain, but there is no doubt that *Holopedium* categorizes the next step in the evolution of blastoderm formation. Following the 31-cell stage, further divisions of the nuclei are accompanied by the segregation of yolk-free cells as an external blastoderm, leaving the yolky inner parts of the blastomeres as anucleate yolk pyramids which fuse to form a yolk mass. The eggs of the cladocerans *Moina* and *Simocephalus* cleave in a similar way. *Daphnia*, however, has evolved one step further in that the nuclear divisions of cleavage follow the same sequence as those of *Holopedium*, but the penetration of cleavage furrows into the larger (250 × 190 μ) and more yolky egg is slight. Blastoderm formation then takes place more or less directly (Baldass, 1941).

With the exception of certain parasitic copepods, all other crustaceans whose cleavage is known show evidence of a present or ancestral condition of equal distribution of yolk among the cleavage blastomeres. At the same time, blastoderm formation is not always the outcome of this condition. In some cases, the reverse trend has manifested itself, yielding a condition of total, radial cleavage. This has been observed in *Polyphemus* (Fig. 109) among the Cladocera, with an egg of approximately the same diameter as that of *Holopedium* but with a lesser density of yolk, and in the free-living copepods, of which *Cyclops* is a well-worked example. The copepod eggs with this cleavage pattern are relatively small (e.g. *Calanus*, 145 μ, Harding, 1955). The radial cleavage of *Polyphemus* was described by Kühn (1913). An identical cleavage sequence was then described for *Cyclops* by Fuchs (1914). When their results were reinterpreted by Anderson (1969), however, in conformity with the spiral cleavage pattern retained in cirripedes, it became obvious that the same basic sequence of spindle orientations, cytokinesis and cell lineages is common to all (compare Figs. 107, 108 and 109). The modifications which yield a hollow spherical blastula of radially disposed cells in *Polyphemus* and *Cyclops* are clearly secondary to an ancestral condition of spiral, unequal cleavage. They must, of course, have evolved convergently in the two examples. By demonstrating that radial cleavage in Crustacea has evolved secondarily from spiral cleavage, this work resolves the paradox that radial cleavage in *Polyphemus* and *Cyclops* leads to a blastula of relatively few cells with a typical "spiral cleavage" mode of early segregation of presumptive areas.

Before passing on to other types of crustacean cleavage in which an equal distribution of yolk is maintained, we can make brief mention of the parasitic copepods. The little that is known of cleavage in the eggs of these animals (e.g. Schimkewitsch, 1896; Pedaschenko, 1898) shows that it is highly variable. It can be total and unequal, as in *Lernaea*, with segregation of the yolk into a single large cell which becomes enclosed by a layer of yolk-free cells. Whether this similarity with cirripede cleavage is primary or secondary requires further study. Alternatively, some parasitic copepods exhibit total equal cleavage (e.g. *Chondracanthus*) while others pass from early total cleavage directly into blastoderm formation. Copepod cleavage, and copepod embryonic development in general, are still poorly understood.

Returning to the branchiopods, little is again known of cleavage in orders other than the Cladocera. Benesch (1969) has recently made a careful study of the anostracan *Artemia*, but the development of conchostracan and notostracan embryos before hatching remains virtually unknown. There is, however, enough fragmentary information to indicate that *Artemia* typifies cleavage in the non-cladoceran branchipods. The egg, 150 μ in diameter, is rich in yolk. Cleavage (Figs. 108 and 109) is total, equal and radial, but has a uniformity which prevents the identification of the axes of development. Consequently, spindle orienta-

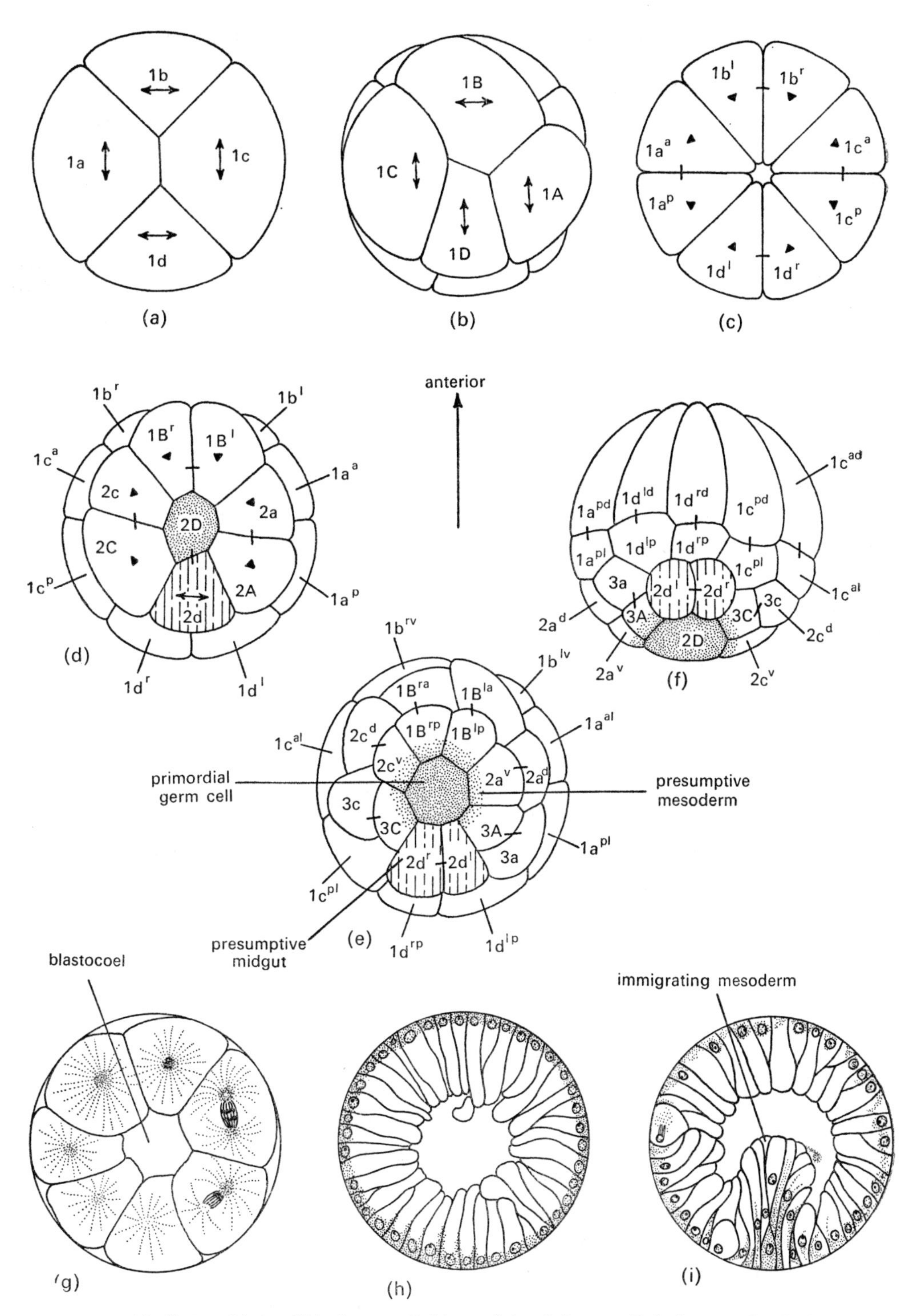

FIG. 109. (a)–(f) The third to fifth cleavage divisions of the cladoceran *Polyphemus*, after Kühn (1913) and Anderson (1969). (a) 8-cell stage, dorsal view; (b) 8-cell stage, ventral view; (c) 16-cell stage, dorsal view; (d) 16-cell stage, ventral view; (e) 31-cell stage, ventral view; (f) 31-cell stage, posterior view. All diagrams except (f) are drawn with the anteroposterior axis vertical. (g)–(i) Late cleavage and onset of gastrulation in *Artemia salina* after Benesch (1969). (g) Onset of fifth cleavage division; (h) 512-cell stage; (i) early gastrulation.

tions and cell lineages cannot be followed. The cleavage divisions continue until at least 512 tall, narrow cells make up the wall of the hollow, spherical blastula. The spiral ancestry of this pattern of cleavage cannot be deduced directly, but can be inferred from comparative considerations which take the Cladocera into account.

The ostracods, in contrast, retain direct evidence of spiral cleavage in their small, but densely yolky eggs. As Weygoldt (1960a) has shown, the early cleavage divisions of the spherical, 100-μ egg of *Cyprideis* follow a course convergently similar to that of *Holopedium*, showing a spiral orientation of the mitotic spindles but only a partial penetration of the cleavage furrows into the egg. While *Holopedium* eventually attains total cleavage at the 31-cell stage, however, the later cleavage divisions of *Cyprideis* are more specialized and result in direct formation of a blastoderm around a yolk mass.

The greatest range of egg size and greatest variety of cleavage patterns in the Crustacea occurs, as might be expected, in the Malacostraca. All of the crustacean examples of intralecithal cleavage and direct formation of a blastoderm are malacostracan. The Leptostraca exhibit this mode (e.g. *Nebalia*, with an egg diameter of 330 μ, Manton, 1934) as do the Hoplocarida (e.g. *Squilla*, with an egg diameter of 490 μ, Shiino, 1942). Since neither the Leptostraca nor the Hoplocarida retain any evidence of ancestral spirality in their cleavage divisions, it is not possible to speculate on the evolution of intralecithal cleavage in these groups of malacostracans. A similar situation is found in the majority of Peracarida. The Mysidacea have intralecithal cleavage (e.g. *Hemimysis*, with a 440-μ egg, Manton, 1928; *Mesopodopsis*, with a 350-μ egg, Nair, 1939) as do the Cumacea (e.g. *Diastylis*, with a 420×290-μ egg, Dohle, 1970), the Tanaidacea (e.g. *Heterotanais*, with a 130×110-μ egg, Scholl, 1963) and the free-living Isopoda (e.g. *Idotea*, with a 450×375-μ egg and *Limnoria*, with a 400×350-μ egg, Stromberg, 1965, 1967). On the other hand, the early cleavage divisions of the amphipods (e.g. *Gammarus*, with a 500×350-μ egg, Rappaport, 1960) and the parasitic epicaridean isopods (e.g. *Bopyroides*, with a 150×170-μ egg, Stromberg, 1971) are total and approximately equal. Since there is no discernible vestige of spirality, and since cleavage subsequently proceeds through pyramid formation to the segregation of a blastoderm around a unitary yolk mass, it seems likely that the early total cleavages of the epicaridean isopods are a secondary consequence of a reduction in egg volume, as in many myriapods and apterygotes. The interpretation of early total cleavage in the Amphipoda is more problematical, since the possibility cannot be ruled out that it is a vestige of the primary total cleavage which must have been a feature of the ancestral Malacostraca.

In other malacostracan groups, much more of the ancestral pattern of total cleavage is retained. Although at present we have no knowledge of cleavage in the egg of the pelagic syncaridan *Paranaspides*, the relatively large yolky egg of *Anaspides* retains a total cleavage in which the early divisions show clear signs of the basic spiral orientations of spindles and cleavage planes. Even though the egg has a diameter of about 1 mm, total equal divisions continue until a hollow blastula of numerous columnar, yolky cells is formed. For its egg volume, therefore, *Anaspides* has a much less modified cleavage pattern than any other malacostracan. The euphausiacean and penaeid Eucarida also retain a total, equal cleavage leading to a hollow, spherical blastula, but their eggs are considerably smaller than that of *Anaspides* (e.g. *Penaeus*, 320 μ in diameter, Dobkin, 1961). Studies on euphausiid and penaeid cleavage have been very superficial (e.g. Brooks, 1882; Taube, 1909; Heldt, 1931;

Dobkin, 1961), but it can be discerned from the work of Taube that the first three cleavage divisions of euphausiid eggs have spindle orientations and cleavage planes similar to those of the cladoceran *Polyphemus* and the copepod *Cyclops*. It seems likely that the secondary radial modification of crustacean spiral cleavage in this way is basic to the eucaridan Malacostraca, since early total cleavage with traces of the same pattern persists in the larger eggs of many decapods (e.g. Weldon, 1892, *Crangon*, 500×400-μ egg; Gorham, 1895, *Virbius*, 360-μ egg; Aiyar, 1949, *Palaemon*, 480×380-μ egg; Nair, 1949, *Caridina*, 950×560-μ egg; Kajishima, 1950, *Leander*, 800×600-μ egg; Shiino, 1950, *Palinurus*, 400-μ egg; Fioroni, 1970a, *Galathea*, 500-μ egg; Lang and Fioroni, 1971, *Macropodia* (a brachyuran), 550-μ egg). In these species, however, early total cleavage is succeeded by the formation of yolk pyramids in the later cleavage divisions (Fig. 110). The nuclei and their associated cytoplasm become peripheral as the number of pyramids increases and are eventually cut off as a yolk-free blastoderm. The yolk pyramids, meanwhile, fuse together, re-establishing a unitary yolk mass within the blastoderm. In a few decapods with large, yolky eggs, cleavage is intralecithal and blastoderm formation takes place directly, in the same functional manner as in the Leptostraca, Hoplocarida and Peracarida (e.g. Zehnder, 1934 a, b, *Astacus*, with a 2·8×2·4-mm egg).

In summary, while cleavage in the Crustacea provides no information on the inter-relationships of crustacean classes, it is possible to discern a basic pattern and a number of adaptive trends in relation to yolk. The basic pattern is the modified spiral pattern of total cleavage, retained in the small eggs of those thoracican cirripedes which hatch a planktotrophic nauplii. Traces of the same pattern can be discerned in the eggs of certain branchiopods, copepods and malacostracans, but here the spindle orientations and cleavage planes have become further modified to yield a secondary radial pattern of total cleavage. A hollow, radial blastula results from cleavage, for instance, in *Polyphemus*, *Artemia*, *Cyclops*, *Euphausia*, *Penaeus* and *Anaspides*. In most groups of crustaceans, however, the outcome of cleavage is a blastoderm around a unitary yolk mass. Some of these species retain traces of the ancestral spiral pattern of total cleavage in the early cleavage divisions (blastodermal Cladocera, e.g. *Holopedium*, *Daphnia*; blastodermal copepods, e.g. some parasitic species; blastodermal ostracods, e.g. *Cyprideis*; blastodermal Eucarida) but others proceed directly through intralecithal cleavage to blastoderm formation (the malacostracan Leptostraca, Hoplocarida and Peracarida). We can therefore state with some confidence that crustacean spiral cleavage has become modified in relation to yolk, firstly by a trend towards total equal cleavage, secondly by the independent evolution of blastoderm formation in the Branchiopoda, Ostracoda, Copepoda and Malacostraca, with the likelihood that a blastoderm has evolved within the Malacostraca at least four times. Curiously enough, those cirripedes which show an increase in egg volume and attendant modifications of cleavage (e.g. *Ibla*, with a 400-μ egg; *Pollicipes*, with a 600-μ egg; Batham, 1946; Anderson, 1965) have followed a different evolutionary route. In these species, cleavage has become highly unequal and the yolk cell is enclosed by micromere proliferation, not by blastoderm formation.

From a comparative viewpoint, the recognition that spiral cleavage is a basic feature in Crustacea, and that blastoderm formation is secondary, is of the greatest interest. Direct comparisons with annelids and other arthropods now become a feasible proposition. As we have already seen in previous chapters, cleavage patterns can only be interpreted satisfactorily in relation to the segregation of presumptive areas in the ensuing blastula or blastoderm. In turning now to this subject, we must bear in mind that the object is to determine, from the variety of fate maps among extant species, a basic fate map for the Crustacea as a whole.

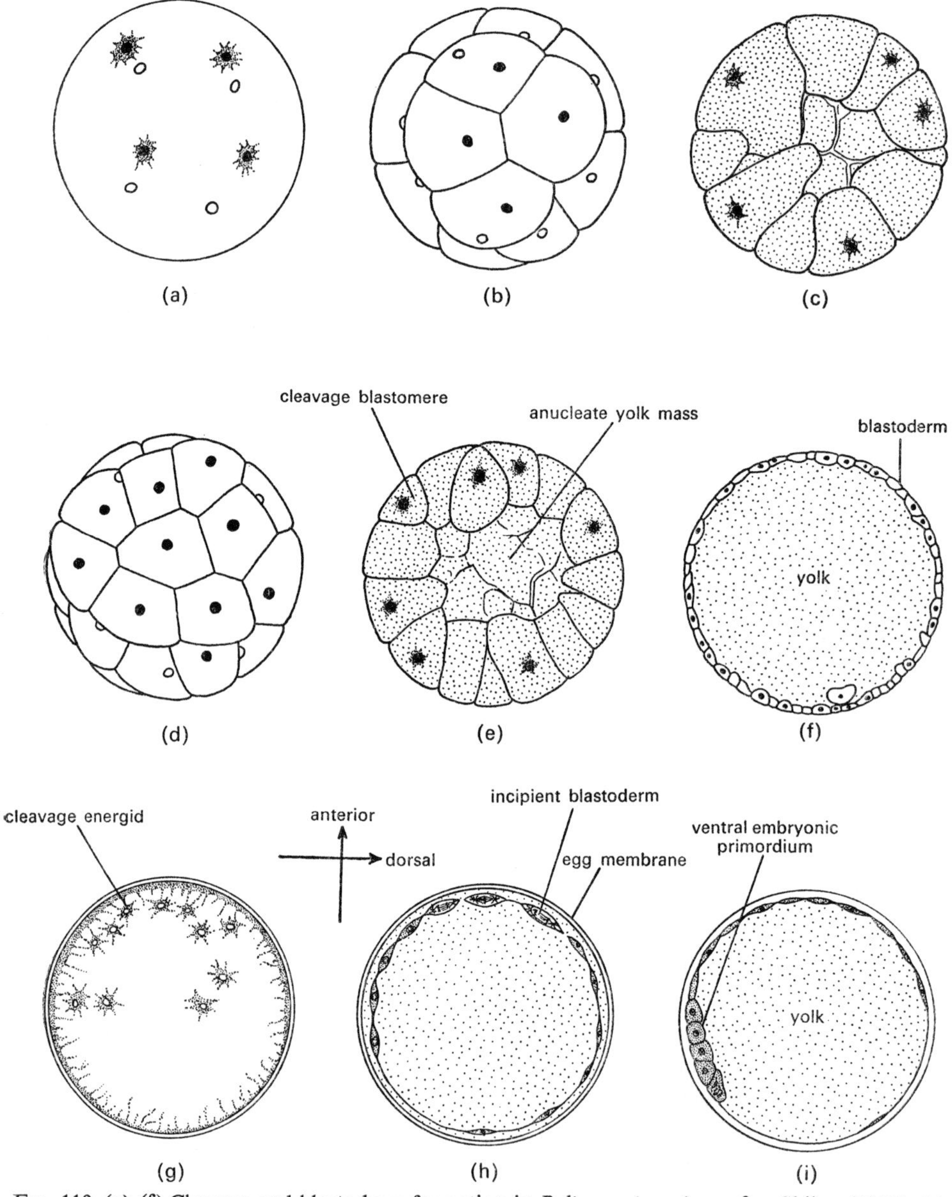

FIG. 110. (a)–(f) Cleavage and blastoderm formation in *Palinurus japonicus*, after Shiino (1950). (a) 8-nucleus stage; (b) 16-cell stage; (c) 32-cell stage in section; (d) 32-cell stage; (e) 64-cell stage in section; (f) blastoderm stage. (g)–(i) Cleavage in *Hemimysis lamornae* after Manton (1928). (g) Twelve-nucleus stage; (h) 64-cell stage, early blastoderm formation; (i) blastoderm with incipient germ disc on ventral surface.

Presumptive Areas of the Blastula or Blastoderm

In spite of a wide range of cleavage types and morphological forms of the blastula or blastoderm, the Crustacea exhibit the same stability in their presumptive areas that we have seen in other unitary groups of annelids and arthropods, such as the polychaetes, clitellates, onychophorans and pterygotes. The major interest of the crustacean fate map from a comparative point of view (Anderson, 1969) is that it does not display the expected pattern

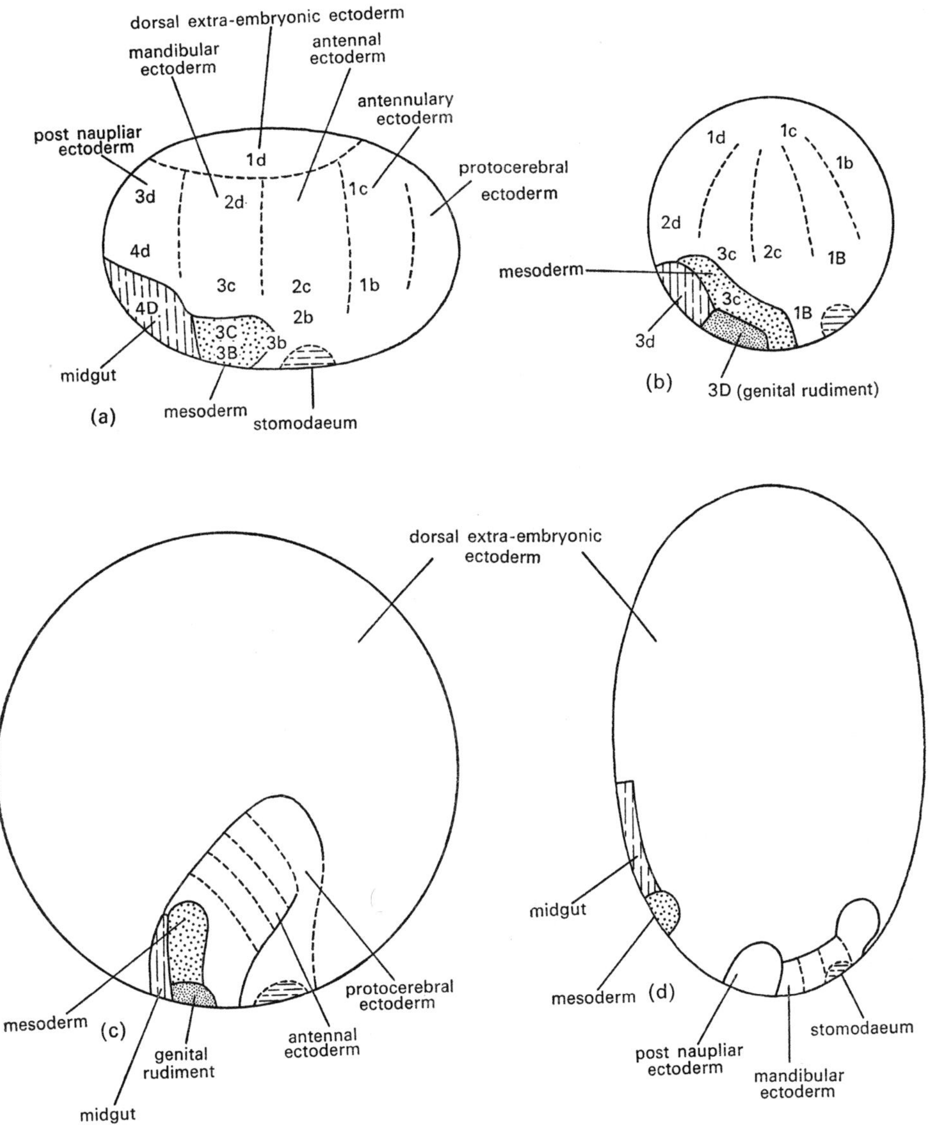

FIG. 111. Fate maps of the crustacean blastula or blastoderm, after Anderson (1969), right lateral view. (a) Cirripedia; (b) Cladocera and Copepoda; (c) the pericaridan *Hemimysis*; (d) the eucaridan *Caridina*. (a) and (b) are blastulae; (c) and (d) are blastoderms. [Based on data of Kühn, 1913; Fuchs, 1914; Manton, 1928; Baldass, 1937; Nair, 1949; and Anderson, 1969.]

for a spiral cleavage group. We shall see below that the presumptive rudiments of crustacean embryos are segregated from one another in a spatial pattern fundamentally different from that of annelids.

As in the annelids and other arthropods, the major variations in the presumptive areas of Crustacea are found in the presumptive midgut and presumptive extra-embryonic ectoderm, the two areas most affected by changes in the volume of yolk in the egg. The mesoderm and the stomodaeum and proctodaeum are conservative in their formation and fates throughout the Crustacea. An early segregation of primordial germ cells as a definite presumptive area takes place in the branchipods, copepods, ostracods and peracaridan Malacostraca, but not in the cirripedes or other Malacostraca. As in other major groups in which the time of first visible differentiation of the primordial germ cells varies (e.g. annelids, pterygotes, chelicerates, molluscs), it is not possible to say whether early or late segregation of germ cells is the more primitive, but the question has no bearing on the integrated development of the embryo as a whole.

Commensurate with the precision and relatively unmodified sequence of their spirally based cleavage, the embryos of cirripedes provide the most basic example of the composition, origins and relative juxtaposition of presumptive areas in the crustacean blastula. We can now examine the fate map of the Crustacea in detail, taking the Cirripedia as a starting point for consideration of each presumptive area in turn (Fig. 111).

The presumptive midgut

In the normal cirripede blastula, excluding those species in which the egg is either enlarged and specialized (e.g. *Pollicipes*, *Ibla*, *Scalpellum*, *Trypetesa*; Batham, 1946; Anderson, 1965; Kaufmann, 1965; Turquier, 1967b) or is reduced and specialized (rhizocephalans, Bocquet-Védrine, 1960, 1964), the presumptive midgut comprises the single large, yolky cell, 4D. This cell fills the interior of the blastula, but is exposed at the surface posteroventrally (Fig. 112). The immediate next step in the division of the presumptive midgut cell (Anderson, 1969) is an equal subdivision into two yolky cells, followed by a further division which segregates two small, yolk-free cells at the posteroventral surface of the blastula from two large, yolky cells which fill the interior. The two large, yolky cells are the anterior midgut rudiment. The two small cells are the posterior midgut rudiment. The further development of the two rudiments will be described in subsequent sections.

Other crustaceans which retain total cleavage and a blastula of 32–64 cells (e.g. *Polyphemus*, *Cyclops*, euphausiids and penaeids) also have a single presumptive midgut cell, which divides into a pair of presumptive midgut cells, located posteroventrally in the blastula wall (Figs. 108 and 109). In these examples, in association with a secondarily radial, equal cleavage and a uniform distribution of yolk among the cleavage blastomeres, the presumptive midgut cell is no longer disproportionately large as compared with the other blastomeres and does not intrude into the blastocoel. Furthermore, the presumptive midgut cell of *Polyphemus* and *Cyclops* has a different division history and a different enumeration from that of cirripedes. As shown in Figs. 108 and 109, this cell is 2d, not 4D (Kühn, 1913; Fuchs, 1914). Whether the equivalent cell in euphausiids and penaeids has the same cell lineage and is also 2d has yet to be determined. As we have seen in previous discussions of cell

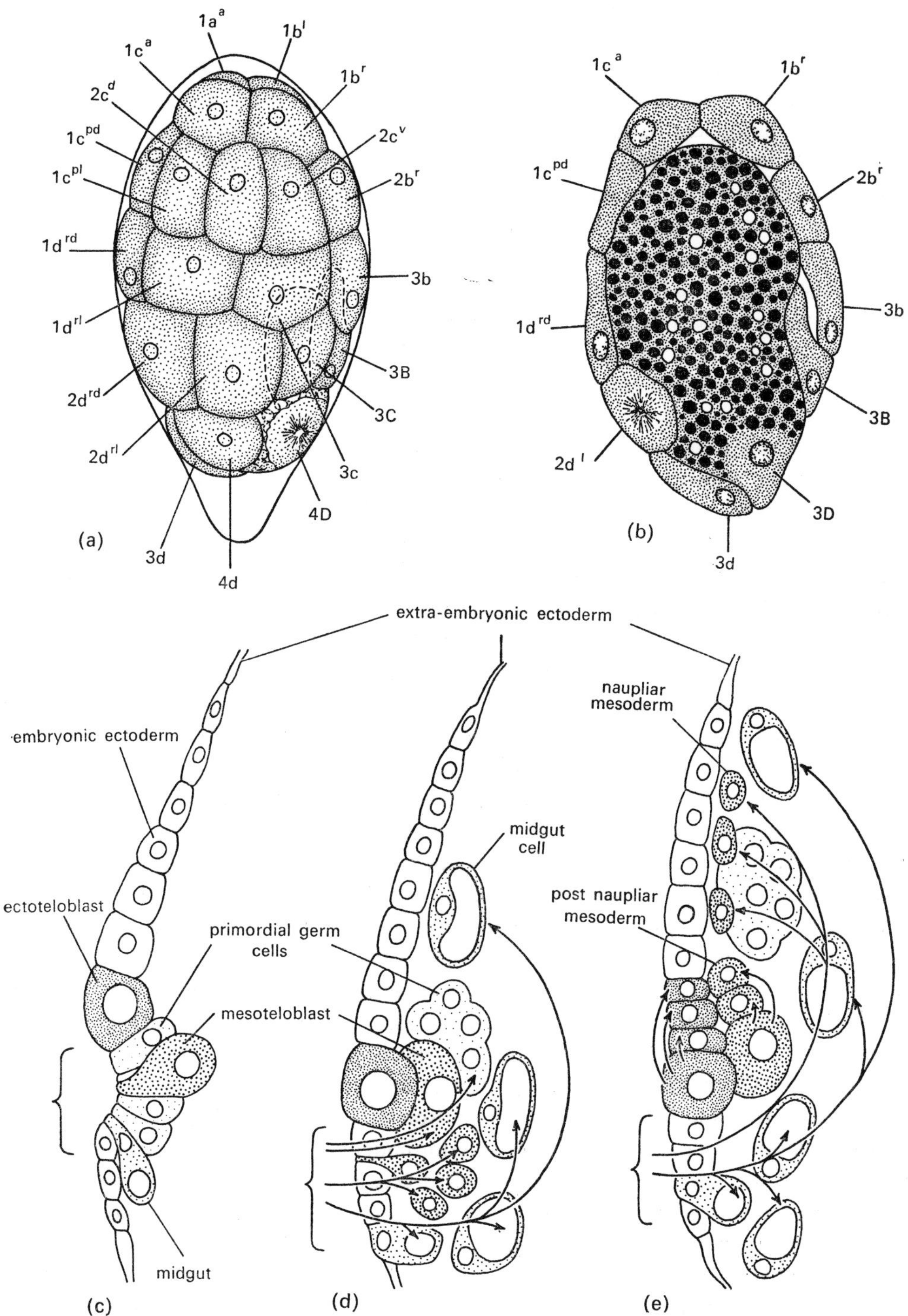

FIG. 112. (a) 31–33-cell stage of *Tetraclita* (Cirripedia), right lateral view showing onset of mesodermal immigration; (b) sagittal section through the stage shown in (a); (c)–(e) diagrammatic sagittal sections showing gastrulation in *Hemimysis* (Malacostraca); (c) onset of gastrulation; (d) gastrulation in progress; (e) gastrulation complete except for continued immigration of midgut cells. [(a) and (b) after Anderson, 1969; (c)–(e) after Manton, 1928.]

lineage in relation to presumptive area formation, however (e.g. in Chapter 3 on the clitellate annelids), homologous presumptive areas are commonly segregated in different ways, and the differences in mode of segregation are less significant than the Wilsonian system of enumeration would appear to suggest. In the present case, the D stem cell of cirripedes (Fig. 107) cuts off four presumptive ectoderm cells, 1d, 2d, 3d and 4d, before assuming its definitive fate as the presumptive midgut cell. In *Polyphemus* and *Cyclops* (Figs. 108 and 109), the D stem cell cuts off one presumptive ectoderm cell, 1d, and then divides into a posteroventral presumptive midgut cell, 2d, and a midventral primordial germ cell, 2D. As far as is known, the latter specialization does not occur in cirripedes, although the primordial germ cells have yet to be identified in the cirripede embryo and may still prove to have a 4D origin. Whether or not this is so, there is no difficulty in recognizing that the 2d presumptive midgut cell of *Polyphemus* and *Cyclops* is homologous with the 4D presumptive midgut cell of cirripedes. It is simply segregated earlier in the cleavage sequence.

After its initial subdivision into two cells, the presumptive midgut of *Polyphemus*, *Cyclops*, euphausiids and penaeids immigrates into the blastocoel and divides repeatedly to form the midgut epithelium. There is no preliminary segregation of a yolky anterior midgut component from a yolk-free posterior midgut component, as there is in the Cirripedia. The difference is a functional corollary of the disproportionately large volume of the presumptive midgut cell in the Cirripedia, but which of the two alternatives should be interpreted as the basic mode of development of the crustacean midgut cannot be decided.

Although the segregation of presumptive midgut as a single, posteroventral, 4D cell, or its equivalent, in the blastula wall after not more than six cleavage cycles is a basic feature for Crustacea, many species have a presumptive midgut consisting of numerous cells. This may take one of two forms. In those species in which cleavage results in a hollow blastula of numerous, narrow, columnar cells, e.g. *Anaspides*, *Artemia*, the presumptive midgut occupies an extensive area of the blastula wall (Hickman, 1937; Benesch, 1969) and entry into the interior is by immigration or invagination of the mass of cells as a whole. Unfortunately, it is not possible on present evidence to establish a precise localization of this area in any species relative to other presumptive areas or to embryonic axes.

Crustaceans in which a blastoderm is formed have a differently modified presumptive midgut. In all species of this type, the presumptive midgut comprises a group of blastoderm cells in the posteroventral midline at the surface of the yolk mass and the internal yolk mass is anucleate. There is no doubt that this modification has evolved independently in the Cladocera, the Ostracoda and several groups of malacostracans (the Leptostraca, Hoplocarida, Peracarida and Decapoda). In comparison with the Cirripedia, it can be seen that early segregation of a pair of yolk-filled, anterior midgut cells in the interior of the embryo and a pair of small, yolk-free, posterior midgut cells at the posteroventral surface of the embryo has been abandoned in favour of an anucleate yolk mass internally and a group of small, posteroventral cells from which the entire midgut develops. The subsequent activity of the midgut rudiment varies. In the Cladocera, the presumptive midgut cells invade the yolk mass, proliferate within the yolk as vitellophages and then assume the definitive, epithelial, midgut form in due course (Baldass, 1937, 1941). In the Ostracoda, in contrast, the presumptive midgut comprises two groups of midventral cells which behave in this way, a posteroventral group from which the posterior part of the midgut arises and a more anterior, ventral group from which the anterior part of the midgut is formed (Weygoldt, 1960a). The cells of both groups act temporarily as vitellophages before giving rise to the definitive midgut epithelium. Among the Malacostraca, the single, posteroventral, presumptive midgut area has two subsequent modes of development. The peracaridan presumptive midgut proliferates cells into the yolk mass, all of which contribute to a yolky, vitellophage epithelium which later becomes the definitive midgut epithelium (Manton, 1928; Goodrich, 1939; Weygoldt, 1958; Scholl, 1963; Stromberg, 1965, 1967, 1971; Dohle, 1970). The Leptostraca, Hoplocarida and Decapoda, in contrast, with a similar blastodermal area of presumptive midgut, first proliferate the cells of an anterior, yolk-digesting epithelium, then the cells of a yolk-free posterior midgut tube which extends behind the yolk mass along the growing trunk (Manton, 1934; Krainska, 1936; Nair, 1939, 1941, 1949; Shiino, 1942, 1950; Aiyar, 1949; Weygoldt, 1961; Fioroni, 1969, 1970 a, b; Lang and Fioroni, 1971). Further details of these processes will be given in the sections in gastrulation and the further development of the midgut, but it is necessary to have a brief acquaintance with them in the present context, in order to appreciate that the common ground of the presumptive midgut throughout the Crustacea resides in the Cirripedia, whose presumptive midgut cell is definitely 4D.

The presumptive stomodaeum and proctodaeum

In all crustacean embryos, the presumptive stomodaeum is a circular area of superficial cells in the ventral midline, anterior to and widely separated from the presumptive midgut. In later development, the presumptive stomodaeum makes an independent invagination into the interior and gives rise to the lining epithelium of the foregut. The first thickening and invagination of the presumptive stomodaeum takes place when the naupliar limb buds are just beginning to form and the stomodaeal rudiment is always seen to lie between the antennal limb buds. In the Cirripedia, the origin of the presumptive stomodaeal cells can be traced to the ventral, superficial cells of the B quadrant, 2b and 3b. The presumptive stomodaeum in all crustaceans is formed by the equivalent cells (i.e. superficial cells in the same relative location) of the blastula or blastoderm.

The presumptive proctodaeum is a more problematical entity in Crustacea, since the proctodaeum always develops late, after the naupliar segments have undergone preliminary development. It does, however, have a definite location at this time, midventrally behind the ectoderm of the pre-telsonic, segmental growth zone. From this knowledge, we can infer the presumptive location of the proctodaeum in the blastula or blastoderm in relation to the presumptive ectoderm, though not always in relation to the presumptive mesoderm or midgut (see below). In all crustaceans, the presumptive proctodaeum is a small group of superficial cells in the posteroventral midline of the presumptive ectoderm, behind the presumptive ectoderm of the post-naupliar growth zone.

The presumptive mesoderm

Unlike the presumptive midgut, the presumptive mesoderm of crustacean embryos varies little with the size and yolk content of the egg. It differs from the presumptive mesoderm of annelids in two fundamental ways which are best displayed in the cirripede blastula. Firstly, the presumptive mesoderm lies in the ventral midline as an arc of superficial cells *in front of* the presumptive midgut. Secondly, the three cells which comprise this arc in the Cirripedia are the cells 3A, 3B and 3C (Anderson, 1969). In annelids, these three stem cells, together with 4D, constitute the basic presumptive midgut, while 4d, which lies behind them, is the presumptive mesoderm cell (Chapters 2 and 3). The 4d cell in Cirripedia has no role in presumptive mesoderm formation and is merely one of the many cells which make up the presumptive ectoderm. The implications of this fundamental difference will be discussed in Chapter 10. Meanwhile, the universality of the cirripede arrangment among crustaceans is seen in the fact that all species retain the same relative juxtaposition of presumptive mesoderm and midgut, even though in most cases it is not possible to establish a definite lineage for the cells. When their lineage can be identified, as in *Holopedium*, *Polyphemus* and *Cyclops* (Kühn, 1913; Fuchs, 1914; Baldass, 1937), the arc of presumptive mesoderm cells is found to have the same ventral A-, B- and C-quadrant lineage as in the Cirripedia, though the number of cells in the arc is great than three. Furthermore, the further development of the presumptive mesoderm arc is the same in all of these cases. The cells migrate into the interior, undergo further divisions and move in a posterior direction to lie behind the midgut rudiment at the posterior end of the embryo. From here, they proliferate a pair of lateral bands of mesoderm cells forwards on either side of the midgut. The cells of the bands aggregate as three pairs of naupliar somites, leaving a residual posterior mass which gives rise to all of the post-naupliar mesoderm.

In larger eggs such as those of most Malacostraca, in which the presumptive mesoderm is a blastodermal area of cells in front of the presumptive midgut at the surface of the yolk mass, this pattern of further development is little changed (Manton, 1928, 1934; Nair, 1949; Shiino, 1950; Weygoldt, 1958, 1960b, 1961; Scholl, 1963; Stromberg, 1965, 1967, 1971; Dohle, 1970). The cells of the presumptive mesoderm move beneath the surface and are overgrown from in front by the presumptive ectoderm. The presumptive mesoderm then proliferates the paired mesodermal bands which break up into the three pairs of naupliar somites, and persists as a residual post-naupliar mass from which all of the post-naupliar mesoderm arises. Only the syncaridan *Anaspides* is said to have an aberrant mode of development of the naupliar mesoderm (Hickman, 1937), to which we will return in a later section. In some malacostracans (e.g. Isopoda, Stromberg, 1965, 1967) the presumptive mesoderm and presumptive midgut cannot be distinguished from one another until after they have become internal as mesendoderm, but the same relative juxtaposition of mesoderm anterior to midgut still persists.

The larger, blastodermal embryos among the Malacostraca also develop a pair of pre-antennulary somites by the independent immigration of two groups of blastoderm cells from the areas of presumptive pre-antennulary ectoderm (see below) in front of the naupliar somites (Manton, 1928; Stromberg, 1965, 1967; Dohle, 1970). Presumptive pre-antennulary mesoderm can therefore be identified at these sites. The significance of the separate development of this mesoderm is not clear, since the small embryos of crustaceans hatching as nauplii do not usually have pre-antennulary somites. Benesch (1969) has identified pre-antennulary somites in the embryo of *Artemia* but was unable to trace their exact origin (see below).

The presumptive ectoderm

The presumptive areas described above (midgut, stomodaeum, proctodaeum and mesoderm) lie along the ventral midline of the crustacean blastula or blastoderm and occupy a relatively small proportion of the surface. The remainder of the surface layer of cells is presumptive ectoderm, divisible between presumptive embryonic ectoderm adjacent to the ventral areas and presumptive extra-embryonic ectoderm more dorsally. The presumptive ectoderm of the Crustacea takes a number of forms, depending on the mode of cleavage. In the Cirripedia, it is a layer of superficial, yolk-free cells laid down around the yolk cell 4D (presumptive midgut) during cleavage. The basic cells making up the presumptive ectoderm are the cells of the first three quartets 1a–1d, 2a–2d and 3a–3d, together with 4d. The only other components deriving from any of these cells are the presumptive stomodaeum, of 2b and 3b origin, and the presumptive proctodaeum, probably of 4d origin. Little ectodermal spread is involved in completing the coverage of the internal rudiments during gastrulation, since only the relatively small mesodermal and posterior midgut rudiments move from the surface to the interior after cleavage has been completed. The presumptive ectoderm then develops as the ectoderm of the acron (pre-antennulary region), labrum, naupliar segments (antennulary, antennal and mandibular) and post-naupliar region (caudal papilla), all of which can be presumptively zoned within the blastula as presumptive embryonic ectoderm. Dorsally, even in small crustacean embryos, the presumptive ectoderm undergoes temporary spread and attenuation around the yolky anterior midgut before attaining its definitive fact as the dorsal epithelium of the naupliar region. The dorsal presumptive ectoderm is, therefore, presumptive extra-embryonic ectoderm.

In other small crustacean eggs, the presumptive ectoderm has the same relative distribution and pattern of sub-areas, though it may be laid down as a wall of yolky cells around a blastocoel (e.g. *Polyphemus*, *Cyclops*, *Artemia*, *Euphausia*, penaeids; Kühn, 1913; Fuchs, 1914; Taube, 1909; Heldt, 1931, 1938; Dobkin, 1961; Benesch, 1969) or as the major part of a blastoderm around a yolk mass (e.g. *Holopedium*, *Daphnia* and other cladocerans, ostracods; Baldass, 1937, 1941; Weygoldt, 1960a). In the larger eggs of certain cirripedes, the presumptive extra-embryonic ectoderm is relatively more extensive than the embryonic ectoderm (e.g. *Pollicipes*, *Ibla*, *Scalpellum*, *Trypetesa*; Batham, 1946; Anderson, 1965; Kaufmann, 1965; Turquier, 1967b), and in the blastoderm of larger malacostracan embryos, more extensive again. In those embryos (Manton, 1928, 1934; Nair, 1939, 1941, 1949; Shiino, 1942, 1950; Aiyar, 1949; Weygoldt, 1958, 1960b, 1961; Scholl, 1963; Stromberg, 1965, 1967; Dohle, 1970), the presumptive post-naupliar embryonic ectoderm lies either around or in front of the presumptive midgut and presumptive mesoderm, as it does in the smaller crustacean embryos. From this focus, the presumptive naupliar embryonic ectoderm extends forwards as a pair of divergent short bands, each comprising presumptive mandibular, antennal and antennulary segmental ectoderm and a terminal area of presumptive pre-antennulary ectoderm. A small, triangular area of presumptive ventral, extra-embryonic ectoderm lies between the bases of the naupliar ectodermal bands. At the front of this area, in the midventral line between the halves of the presumptive antennal ectoderm, lies the presumptive stomodaeum, with labral presumptive ectoderm in front of it and median, pre-antennulary presumptive ectoderm between the paired pre-antennulary areas. The presumptive embryonic ectoderm occupies less than half of the surface of the blastoderm. The remainder, covering the yolk mass dorsally and laterally, is an extensive presumptive extra-embryonic ectoderm. The small area of ventral extra-embryonic ectoderm is soon incorporated into the ventral ectoderm of the naupliar segments as embryonic development proceeds. The dorsal extra-embryonic ectoderm, which has a temporary role as an attenuated, provisional dorsal wall enclosing the large mass of yolky midgut, is later partially transformed into dorsal segmental ectoderm and partially resorbed in various ways (see below). Thus, apart from an exaggerated and precocious attenuation of the dorsal ectoderm in embryos with a large yolk mass, the presumptive ectoderm of all crustacean embryos has the same basic pattern of sub-areas in all species, laid down as a superficial layer of blastula or blastoderm cells which already occupies most of the surface of the embryo. Similarly, the presumptive stomodaeum, mesoderm, midgut and proctodaeum have the same basic location and fate in all Crustacea, allowing for the functional variations manifested by the presumptive midgut in relation to yolk. A basic crustacean fate map is thus not difficult to elucidate. Further discussion of this matter will be deferred to the end of the present chapter, and the significance of the outcome will be considered in the comparative context of Chapter 10, but two points are worth emphasizing at this stage. Firstly, the crustacean fate map is uniquely different from that of annelids, in spite of the common ground of total, spirally based cleavage. Secondly, the future development of the crustacean embryo as a nauplius is already specified at the blastula stage in the pattern of sub-areas of the presumptive ectoderm. We can now consider in more detail how the further development of the presumptive areas of the crustacean blastula or blastoderm proceeds during gastrulation and organogeny.

Gastrulation

As in all yolky arthropod embryos, gastrulation in the Crustacea is a relatively minor aspect of development. Once again, we can take the Cirripedia as a starting point. Gastrulation in *Tetraclita* and other similar thoracican embryos (Anderson, 1969) begins when the fifth cleavage division has been completed. The first event of gastrulation (Fig. 112) is an inward and forward migration of the three presumptive mesoderm cells 3A, 3B and 3C beneath the ventral ectoderm cells in front of them. While this is happening, the yolk-cell 4D divides equally into two cells by an approximately longitudinal division, and then cuts off the two small, posterior midgut cells at the posteroventral surface in the manner described in the previous section. The two presumptive posterior midgut cells, which lie behind the presumptive mesoderm, now perform a gastrulation movement similar to the mesoderm, slipping inwards and forwards to join the mesoderm cells in the interior (compare Fig. 114a). By this time the three mesoderm cells have begun to divide. The presumptive embryonic ectoderm bordering the presumptive mesoderm and posterior midgut spreads to cover the small area of ventral surface vacated by the immigrating mesoderm and midgut cells.

The major events of gastrulation are then complete and the external surface of the embryo retains no surface openings. The gastrulation movement of the presumptive stomodaeum is a later, independent invagination (see Fig. 129), and that of the proctodaeum, independent and later still. To take full account of our previous definition of gastrulation as the movement of presumptive areas into their organ-forming positions, however, we must include as part of the gastrulation movement of the presumptive mesoderm and posterior midgut, the migration of these components to the posterior end of the embryo after they have become internal. Both components undergo cell divisions during this migration, but do not enter into definitive organogeny until the migration is complete. The small group of posterior midgut cells, which are distinctively round and pale staining, settle in a position immediately posterior to the yolky, anterior midgut cells (see Fig. 129). The mesoderm fills the posterior end of the gastrula below and behind them. The presumptive ectoderm, other than performing a minimal spread to complete the ectodermal coverage of the gastrula posteroventrally, undergoes no gastrulation movements. As explained in the preceding discussion on crustacean cleavage, ectodermal covering of the embryo in the Cirripedia is attained largely as a result the segregation of yolk-free blastomeres at the surface of a cleaving yolk cell.

Although the patterns of gastrulation movements in other crustacean embryos are highly variable, commensurate with the variations in the composition and mode of segregation of presumptive areas in different species, all movements can be interpreted in accordance with the basic sequence of gastrulation observed in the Cirripedia. The larger-egged cirripedes, such as *Pollicipes*, *Ibla*, *Scalpellum* and *Trypetesa*, show little modification of this sequence (Batham, 1946; Anderson, 1965; Kaufmann, 1965; Turquier, 1967b). At the end of cleavage in these species, large yolky cells fill the interior of the blastula and are covered by a layer of small yolk-free cells except posteroventrally, where the yolky cells remain exposed at the surface in the usual way. Superficial cells at the margin of the exposed part then migrate into the interior as mesoderm cells and the exposed part is subsequently covered by ectodermal spread. A segregation of yolk-free, presumptive posterior midgut cells from the internal yolk cells also occurs in these species, but whether the posterior midgut cells are first cut off at the surface of the embryo and subsequently migrate into the interior, or are cut off directly from the yolk cells in their posteroventral internal position, is not known.

In giving consideration to gastrulation in other Crustacea, it is advantageous to treat the gastrulation movements of each presumptive area separately. The presumptive midgut behaves during this phase of development in one of two ways. In species which develop a hollow blastula, the presumptive midgut exhibits a clear gastrulation movement which carries it from the blastula wall into the blastocoel. This movement may be an immigration by two cells (e.g. *Polyphemus*, *Cyclops*, euphausiids, penaeids; Brooks, 1882; Taube, 1909; Kühn, 1913; Fuchs, 1914), an immigration by many cells (e.g. *Artemia*, Benesch, 1969), or exceptionally, an invagination of many cells, as described by Hickman (1937) in *Anaspides*. Species which develop a blastoderm, in contrast, and have a presumptive midgut in the form of a group of cells at the surface of the yolk mass, retain no gastrulation movement of this component of the embryo. The presumptive midgut cells enter into proliferative organogeny at the yolk surface (Fig. 112) and their products invade the yolk mass. This process has been described in the Cladocera (Baldass, 1937, 1941), the Ostracoda (Weygoldt, 1960a) and many Malacostraca (Manton, 1928, 1934; Krainska, 1936; Goodrich, 1939; Aiyar, 1949; Nair, 1941, 1949; Shiino, 1942, 1950; Weygoldt, 1958; Scholl, 1963; Stromberg, 1965, 1967; Fioroni, 1970a; Lang and Fioroni, 1971).

Gastrulation movements remain much more distinct throughout the Crustacea in respect of the presumptive mesoderm. The immigration of presumptive mesoderm described above for the Cirripedia has been noted by other workers on small-egged species (Baldass, 1937, 1941; Weygoldt, 1960a; Benesch, 1969) and is little changed in the larger-egged Malacostraca, in which it has been described many times (e.g. Manton, 1938, 1934; Nair, 1949; Weygoldt, 1958, 1960b; Stromberg, 1965, 1967). The cells of the presumptive mesodermal area migrate beneath the surface of the blastoderm and are overgrown by ectoderm from in front (Fig. 112). The independent immigration of the presumptive cells of a pair of pre-antennulary somites associated with the pre-antennulary ectoderm of the Malacostraca has also been noted in a previous section and must be regarded as a component of the gastrulation movement of malacostracan mesoderm.

The presumptive stomodaeum exhibits a uniform gastrulation movement throughout the Crustacea. As exemplified by the Cirripedia, the presumptive stomodaeum invaginates by an independent thickening and infolding (Figs. 127–9) after the gastrulation movements of the mesoderm and midgut have been completed.

Taking into account the very large proportion of the surface already occupied by presumptive ectoderm in the crustacean blastula or blastoderm (Fig. 111), it is not surprising that little ectodermal spread is involved in any species in covering the superficial, posteroventral areas of surface vacated by the immigrating mesoderm and midgut cells. Gastrulation with respect to ectoderm is thus minimal in Crustacea, reflecting the fact that the presumptive ectoderm is laid down in extended form during all types of crustacean cleavage.

As the mesodermal and midgut rudiments complete their entry into the interior and the presumptive ectoderm spreads to replace them at the surface, the embryo begins to show the first signs of those changes in external form which will define the naupliar rudiments at the surface of the body and lead to the further elaboration of external structure. The external development of crustaceans after gastrulation is a highly complex and varied process, since it includes the details of functional larval elaboration and metamorphosis in species with indirect development and the retention of embryonic anamorphosis even in those species with direct development. No attempt will be made in the present context to encompass all of these variations, but a preliminary consideration of the general features of external development in Crustacea is essential to the discussion of the important comparative aspects of their organogeny.

Development of External Form

In those crustaceans which hatch as a nauplius larva, the external development of the embryo before hatching is, in general terms, a relatively simple process. The first notable change is the development of the three pairs of naupliar limb buds, ventrolaterally along the sides of the body. As shown in Figs. 113 and 114, the naupliar limb buds of cirripede and ostracod embryos occupy the major part of the body length, indicating that most of

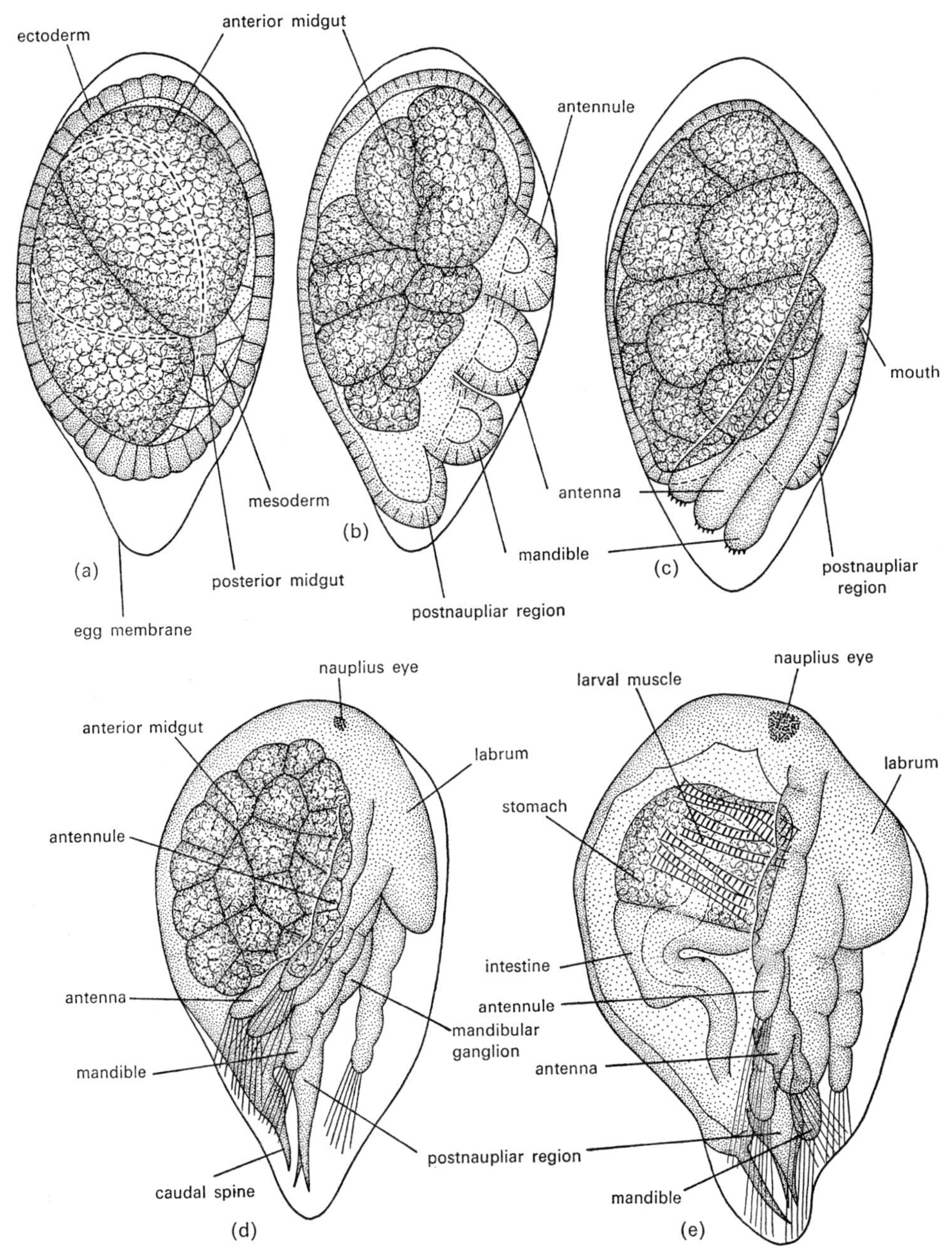

FIG. 113. (a)–(e) External development of the embryo of *Tetraclita* (Cirripedia) after the completion of gastrulation. (After Anderson, 1969.)

the embryo is made up of the rudiments of the three naupliar segments, antennulary, antennal and mandibular. A relatively short, pre-antennulary region completes the anterior end and an even smaller post-naupliar region completes the posterior end. Other than the limb buds, however, the embryo does not manifest any external signs of segmental subdivision. No intersegmental annuli are developed over the convex dorsal surface of the nauplius.

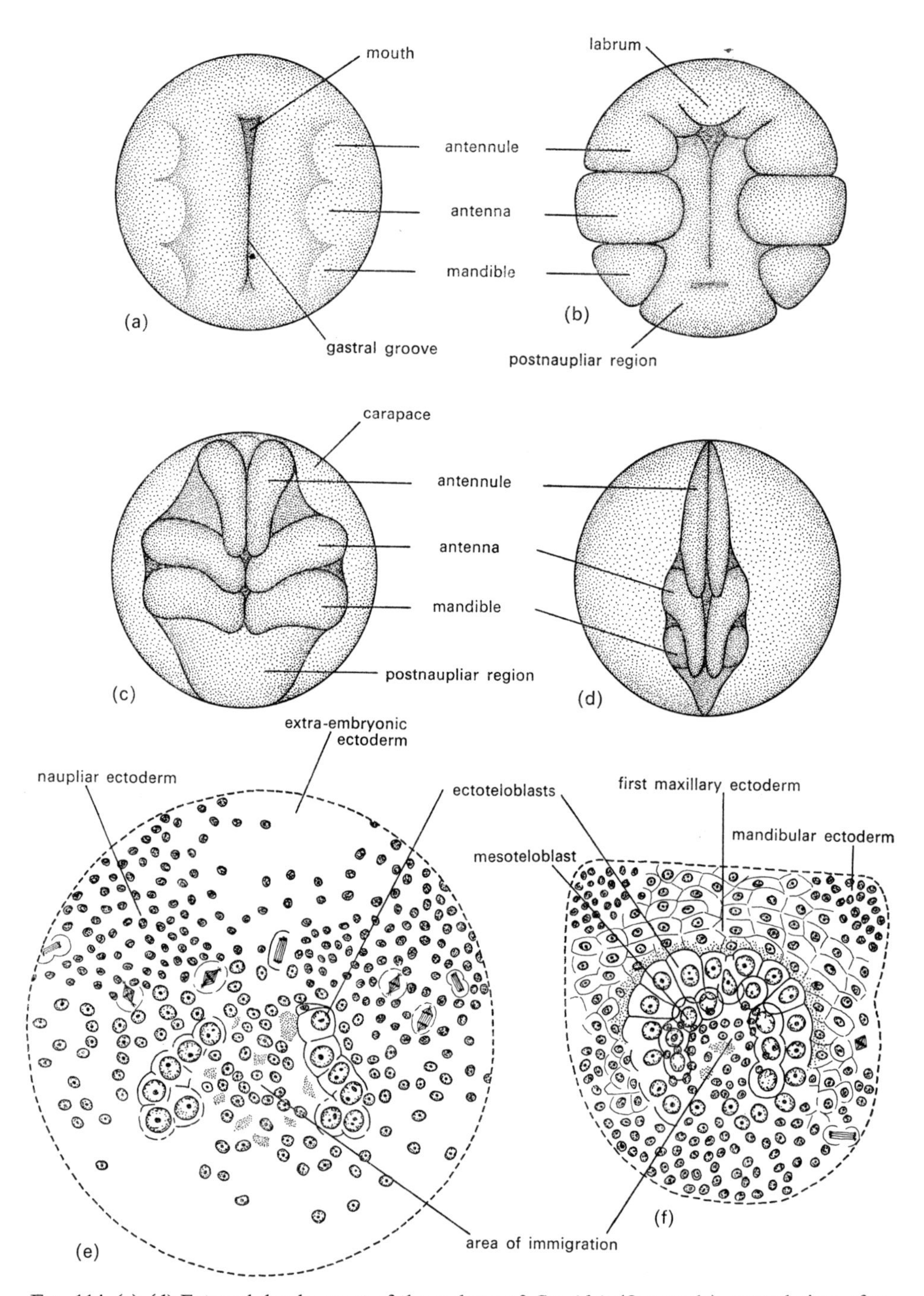

FIG. 114. (a)–(d) External development of the embryo of *Cyprideis* (Ostracoda), ventral view, after Weygoldt (1960). (e) and (f) formation of ectoteloblasts (e) and mesoteloblasts (f) in *Nebalia* (Leptostraca), in surface view, after Manton (1934).

The early development of the naupliar limb buds of copepod embryos proceeds in the same way as in the cirripedes and ostracods, leaving short pre-antennulary and post-naupliar regions at the ends of the embryo. The post-naupliar region is the rudiment from which all post-mandibular segments are developed anamorphically during post-embryonic development. In other groups of Crustacea which hatch as nauplii, the post-naupliar region is proportionately longer than in the cirripedes, copepods and ostracods. This is seen, for example, in the Branchiopoda (Fig. 116) and penaeid Malacostraca (Fig. 119) and also in the Cephalocarida and Mystacocarida. The greater length is due to the precocious development of the rudiments of several of the anterior post-naupliar segments in the embryo. In cirripedes and copepods, the proliferation of these segments is delayed until hatching has occurred, while in ostracods, the post-naupliar region of the body is greatly reduced and undergoes little further development, even after hatching.

As development of the embryo continues, the naupliar limb buds elongate and protrude, usually in a posteroventral direction, though in cirripedes the limb buds rotate to a posterodorsal orientation at an early stage in their elongation (Fig. 113). The bases of the antennal and mandibular limb buds also move forwards and downwards, so that they become crowded around the mouth. At the same time, the labrum arises as a protrusion in front of the mouth and bulges posteriorly along the ventral midline to overhang the mouth. The bases of the antennules lie on either side of the base of the labrum, the bases of the antennae lie on either side of the mouth and the bases of the mandibles lie on either side just behind the mouth.

Associated with the ventral concentration of the limb buds and labrum in the embryo nauplius, the dorsal surface of the embryo becomes more convex. In the majority of nauplii, there is no sign of the development of a carapace fold until hatching has occurred and naupliar development is proceeding (see later). The Ostracoda alone exhibit a precocious development of the carapace fold in the embryo, as a bivalve fold which extends downwards at the side of the body to partially enclose the naupliar limbs (Fig. 114).

As the limb buds continue their elongation, they begin to show the external annulations indicative of subdivision into podomeres, and the antennae and mandibles become biramous. Secretion of the stage I nauplius cuticle begins at about this stage, accompanied by the onset of pigmentation of the median eye (nauplius eye) in the anterior midline of the pre-antennulary region, in front of the labrum. Cuticular secretion includes the formation of a unitary carapace shield over the convex dorsal surface of the embryo and the elaboration of a complex setation on the naupliar limbs. Further discussion of the limbs and their setation is deferred to later paragraphs dealing with the post-embryonic development of the nauplius larva. In most nauplii, the carapace continues smoothly over the anterior end of the body, but the cirripedes are exceptional in developing a pair of projecting frontal horns at the sides of the pre-antennulary region. Similarly, in most nauplii the post-naupliar region of the body remains ovoid and is covered by a smooth cuticle, sometimes bearing a terminal pair of caudal spines, but the cirripedes show a more complex development of this region. The region subdivides into a narrow dorsal protrusion, which forms a median posterior prolongation of the carapace, called the caudal spine, and a larger posteroventral protrusion called the caudal papilla. When the secretion of the nauplius cuticle takes place, the caudal papilla develops a terminal cuticular bifurcation which corresponds to the caudal spines of other nauplii. As will be discussed later, and can be seen in Fig. 118, the posterior infolding which separates the caudal spine from the caudal papilla in the cirripede embryo is an incipient carapace fold.

After hatching, when the limbs of the stage I nauplius are extended and begin to act as propulsive organs, the details of limb setation and of the general external structure of the nauplius can be seen much more clearly. From the numerous descriptions of naupliar development that have been published, I shall select only those which provide the most significant details bearing on the comparative post-embryonic development of the nauplius in different groups of Crustacea. The main points to be considered are the further development of the nauplius region and its metamorphosis to form the anterior part of the adult head, and the formation and further development of the post-naupliar segments during larval development and metamorphosis. The discussion of these matters will be confined, for the sake of brevity, to the Cephalocarida, Branchiopoda, Cirripedia, Copepoda and penaeid Malacostraca. The Ostracoda, Mystacocarida and Euphausiacea offer other variants (e.g. Delmare-Deboutville, 1960; Hessler and Sanders, 1966; Weygoldt, 1960a; Mauchline, 1965, 1970) but have little bearing at the present time on the elucidation of a basic pattern of external and internal development in Crustacea.

A great advance in the understanding of comparative larval development in the Crustacea was achieved only a few years ago with the outstanding work of Sanders (1963 a, b) and Sanders and Hessler (1964) on the larval development of the Cephalocarida. This work, especially that of Sanders on *Hutchinsoniella macracantha*, revealed a hitherto unknown and unsuspected mode of generalized larval development among extant Crustacea. The importance of this discovery from a comparative point of view was further emphasized by Anderson (1967) in a new interpretation of larval development of the Branchiopoda, and will be further developed here to include consideration of the cirripedes, copepods and penaeids.

In the Cephalocarida (Fig. 115), hatching occurs as a nauplius, 0·7 mm long in *H. macracantha*, with a functionally elaborated naupliar region and a conspicuous post-naupliar region. The anterior part of the post-naupliar region carries a pair of maxillules, which have already attained the immediately pre-functional stage of development, and a pair of rudiments of the maxillae in the form of minute lobes. Two or three limbless trunk segments are delineated behind the maxillae. The post-naupliar region then terminates in a pre-telsonic growth zone and telson. The larva still contains yolk and probably does not feed at this stage. The setae of the antennules, antennae, mandibles and maxillules attain their full functional condition at the first larval moult and feeding probably begins at stage II. The larvae are benthic in habit throughout their development, feeding on particulate material stirred up from the surface of the substratum. Development proceeds through a series of moults and metamorphosis to the juvenile adult form takes place during the fourteenth and fifteenth moults, when the larva becomes 1·8 mm long. By this stage, the full complement of nineteen trunk segments has already been developed. The most informative approach to cephalocaridan development from a comparative point of view concerns the changes undergone by each major component of the animal as development proceeds. We can consider first the naupliar limbs, then the development of the post-naupliar region of the animal.

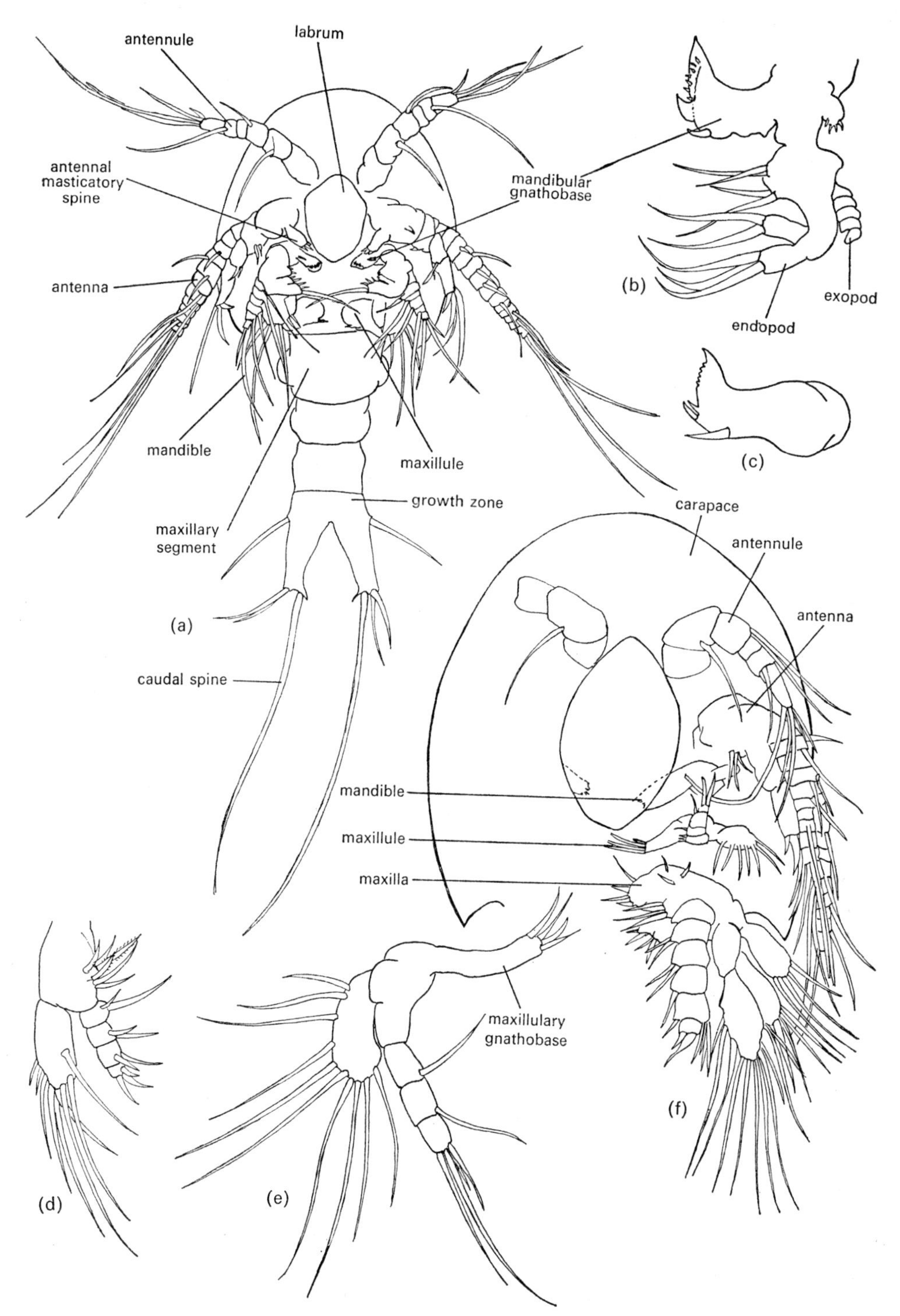

FIG. 115. (a)–(f) Larval development of *Hutchinsoniella* (Cephalocarida), after Sanders (1963a). (a) Stage I nauplius, ventral view; (b) mandible of stage XI nauplius; (c) mandible of juvenile, stage XIV; (d) maxillule of stage VI nauplius; (e) maxillule of juvenile, stage XIV; (f) ventral view of anterior end after metamorphosis.

The antennules show little change other than increase in size as development proceeds. During the naupliar stages, they are elongate, setose, uniramous limbs which function as locomotory organs and also act in stirring up food particles from the substratum. The same form and function then persist after metamorphosis has occurred.

The antennae also function in locomotion and feeding in the nauplius, but in a more complex manner than the antennules. Each antenna is long and biramous, with long setae on the exopod and shorter setae on the endopod. During the oar-like locomotory action of the antennae, the setal tips of the exopods stir up particles from the substratum in the same manner as the antennules. As part of the antennal action, the agitated particles are swept towards the labrum by the endopodal setae. The protopod of each antenna also carries a group of median setae near the distal end and a bifid masticatory spine near the basal end. On the backstroke of each antennal cycle, the protopodal setae assist in pushing the food particles into the sub-labral space. The naupliar form and function of the antennae persist, with increasing size, through fourteen moults. At the fifteenth moult, metamorphosis of the antennae occurs. The distal protopodal setae and masticatory spine, which form part of the naupliar ingestive apparatus, are eliminated, but the remainder of the antennal form and function continue into the adult.

The mandibles of the cephalocaridan nauplius are also long, biramous limbs, with long setae on the exopod, shorter setae on the endopod and a group of median setae on the protopod. The mandibles function in locomotion and feeding in the nauplius in the same manner as the antennae. Late in naupliar development, at the twelfth moult, the base of each mandible enlarges to form a gnathobase which projects medially beneath the labrum towards the mouth. Coincident with this change, the exopod of the mandible becomes vestigial and the endopod is reduced. After two further moults, only the gnathobase of the mandible remains as the mandible of the adult. The naupliar functions of the mandibles in the Cephalocarida are thus lost at metamorphosis and are replaced by a masticatory function.

In sum, then, the naupliar ingestive apparatus of the Cephalocarida (antennal protopodal setae and masticatory spine, mandibular protopodal setae) is lost at metamorphosis, as is the locomotory and food-stirring component of the mandible; but the locomotory and food-stirring structure of the antennules and antennae is retained and a new ingestive apparatus is developed in the form of mandibular gnathobases.

In the Cephalocarida, unlike other Crustacea, a serial functional development of the post-naupliar region accompanies the enlargement of the naupliar region during the moults which precede metamorphosis. The two regions of the body are linked functionally by the maxillules, which become functional with the onset of feeding after the first moult. At this stage, the maxillules are structurally similar to the trunk limbs which subsequently develop behind them. The exopod is flattened and has a natatory function, supplementing the locomotory action of the naupliar limbs. The endopod is cylindrical and has an ambulatory and food-stirring function. The protopod carries a number of median endites, which assist in pushing food particles forwards beneath the labrum. This form and function persists through the first nine stages of naupliar development. Then, at the ninth moult, the basal endite of each maxillule shows its first sign of enlargement as a gnathobase and the remainder of the limb is slightly reduced. This trend continues during several moults until, by the time metamorphosis is complete, the gnathobase is fully elongated and the remaining endites have been lost. The exopods of the maxillules retain their natatory function after metamorphosis but the endopods assume a new orientation and function at this stage. They are no longer ambulatory, but now serve to close off the sublabral space laterally. This change is functionally associated with the transition from naupliar feeding to feeding exclusively in the adult manner by transferring food forwards from the trunk limbs (see below).

Unlike other Crustacea, the development of trunk limbs in the Cephalocarida begins with the first post-maxillulary pair. Since the post-maxillulary segment of the Cephalocarida forms the posterior end of the head in the usual manner, its limbs can be conveniently termed the maxillae; but one of the most important aspects of cephalocaridan structure and function is that the maxillae are structurally and functionally identical with the succeeding trunk limbs. From a condition of minute lobes at hatching, the maxillae become functional after the third naupliar moult, when the naupliar limbs and maxillules are already active in locomotion and feeding. Like the maxillules, the maxillae have a flattened, natatory exopod, a cylindrical ambulatory and food-stirring endopod and a series of median endites which act in the forward transfer of food particles on to the endites of the maxillules. Unlike the maxillules, however, this form and function persists through metamorphosis. The maxillae do not become more jaw-like as development progresses.

The trunk limbs develop slowly in a serial manner behind the maxillae. Eight pairs of trunk limbs are developed in *H. macracantha*, leaving eleven limbless trunk segments at the posterior end of the body. The first pair become functional coincident with the maxillae at the third moult, but the eighth pair does not become functional until the twentieth moult (i.e. five moults after metamorphosis has been completed). Each pair of trunk limbs resembles the maxillae in form and function, and contributes to locomotion, food collection and the forward transfer of food to the maxillules.

At the fifteenth moult, the stage at which the fifth pair of trunk limbs becomes functional in *H. macracantha* (giving six pairs of functional trunk limbs including the maxillae), the naupliar ingestive apparatus is lost and the mandibles and maxillules assume their adult form and function. It is therefore reasonable to conclude that six pairs of "trunk limbs" is the minimum number required in the Cephalocarida for feeding exclusively in the adult manner (Anderson, 1967). As we shall see, this association of metamorphosis with the development of six pairs of trunk limbs is deeply ingrained in crustacean development and has often persisted even when the naupliar and post-naupliar regions of the body have undergone great structural and functional specialization. The development of the Cephalocarida reveals that the association had its origins in the functional requirements of a transition from larval to adult feeding during the anamorphic development of generalized, ancestral Crustacea. In providing a living example of generalized limbs and of the changing method of combined swimming and feeding action of these limbs during development, the Cephalocarida form a basis for the interpretation of the patterns of functional relationship between naupliar development, swimming, feeding and metamorphosis in other Crustacea. The only point to be kept firmly in mind during the ensuing discussion is that all other Crustacea have the maxillae modified as a third pair of jaws and consequently have their first functional trunk limbs one segment further back than the Cephalocarida.

All of the other main groups of Crustacea which hatch as nauplius larvae are developmentally more specialized than the Cephalocarida. Firstly, their nauplius larvae are planktonic and, except in the Penaeidea, planktotrophic. Secondly, the developmental sequences through which they pass after hatching are more specialized than that of the Cephalocarida and differ markedly in each group. Thirdly, the specializations of the nauplius larva differ in the Branchiopoda from those in the Cirripedia and Copepoda, with the eucaridan nauplius being more like the latter than the former. In order to appreciate the nature of these specializations, it is necessary to discuss each group in turn.

Among the Branchiopoda, detailed investigations of the larval development of the anostracan *Artemia* have been carried out by Heath (1924) and Anderson (1967), and of the conchostracan *Limnadia* by Anderson (1967). These studies confirmed and extended the results of other workers on other species (Anostraca, Claus, 1873, 1886; Spandenberg, 1875; Packard, 1883; Sars, 1896a; Oehmichen, 1921; Cannon, 1928; Hsu, 1933; Pai, 1958; Nourisson, 1959; Baquai, 1963; Conchostraca, Packard, 1883; Chambers, 1885; Sars, 1896 a, b; Berry, 1926; Mattox, 1937, 1950; Botnaruic, 1947, 1948; Pai, 1958) and showed that *Artemia* and *Limnadia* can be taken as representative of development in the Anostraca and Conchostraca. Anderson (1967) also showed that the Anostraca and Conchostraca share much in common in their early larval development, though the later development and metamorphosis of the Conchostraca is more specialized than that of the Anostraca. The larval development of the Notostraca is further specialized in the direction of increased yolk, a more advanced hatching stage and a precocious metamorphosis (Claus, 1873; Brauer, 1874; Campan, 1929; Longhurst, 1955; Pai, 1958; Anderson, 1967), but is not well understood at the present time and requires further study. The Cladocera, of course, also have a direct development in most species, which we will examine briefly after discussing the Anostraca and Conchostraca.

Like the nauplius of the Cephalocarida, the newly hatched nauplius of the Branchiopoda has an elaborate naupliar region and a conspicuous post-naupliar region. In comparison with *Hutchinsoniella*, the naupliar region of *Artemia* and *Limnadia* (Fig. 116) is structurally and functionally specialized in relation to active swimming and planktotrophic feeding, while the post-naupliar region is externally simplified. The structural and functional elaboration of the naupliar region is completed during the first three larval stages and feeding begins during stage III. The newly hatched nauplius of *Artemia* is 450 μ long and that of *Limnadia* is 400 μ long. When feeding begins, these lengths have increased to 725 μ and 500 μ respectively. Naupliar development in *Artemia* continues through a further five stages, during which the post-naupliar region increases in length and becomes further elaborated. Metamorphosis to the juvenile adult form then takes place at the ninth moult, the resulting juvenile being about 2 mm long. The development of *Limnadia* is completed through fewer moults (Fig. 116). Metamorphosis takes place at the fifth moult, to yield a juvenile 1 mm long. We shall see below that the modifications which permit the specialized mode of planktonic larval life and early metamorphosis in the Branchiopoda comprise a particular adaptation of the more generalized mode of development exemplified by the Cephalocarida. Discussion of this topic is facilitated by taking the naupliar region first, then the post-naupliar region.

The full structural differentiation of the naupliar region is attained by the time feeding begins. The uniramous antennules are small and project forwards as sensory processes.

The antennae, in contrast, are long, biramous limbs with long, hinged setae on the exopod and endopod. The mandibles are short and uniramous, but also carry terminal, hinged setae. The antennules play no part in locomotion, but the antennae, with their terminal fans of setae, act as the major propulsive organs of the nauplius. Their action is oar-like, with the setal fans spread during the backstroke and partially folded during the forward

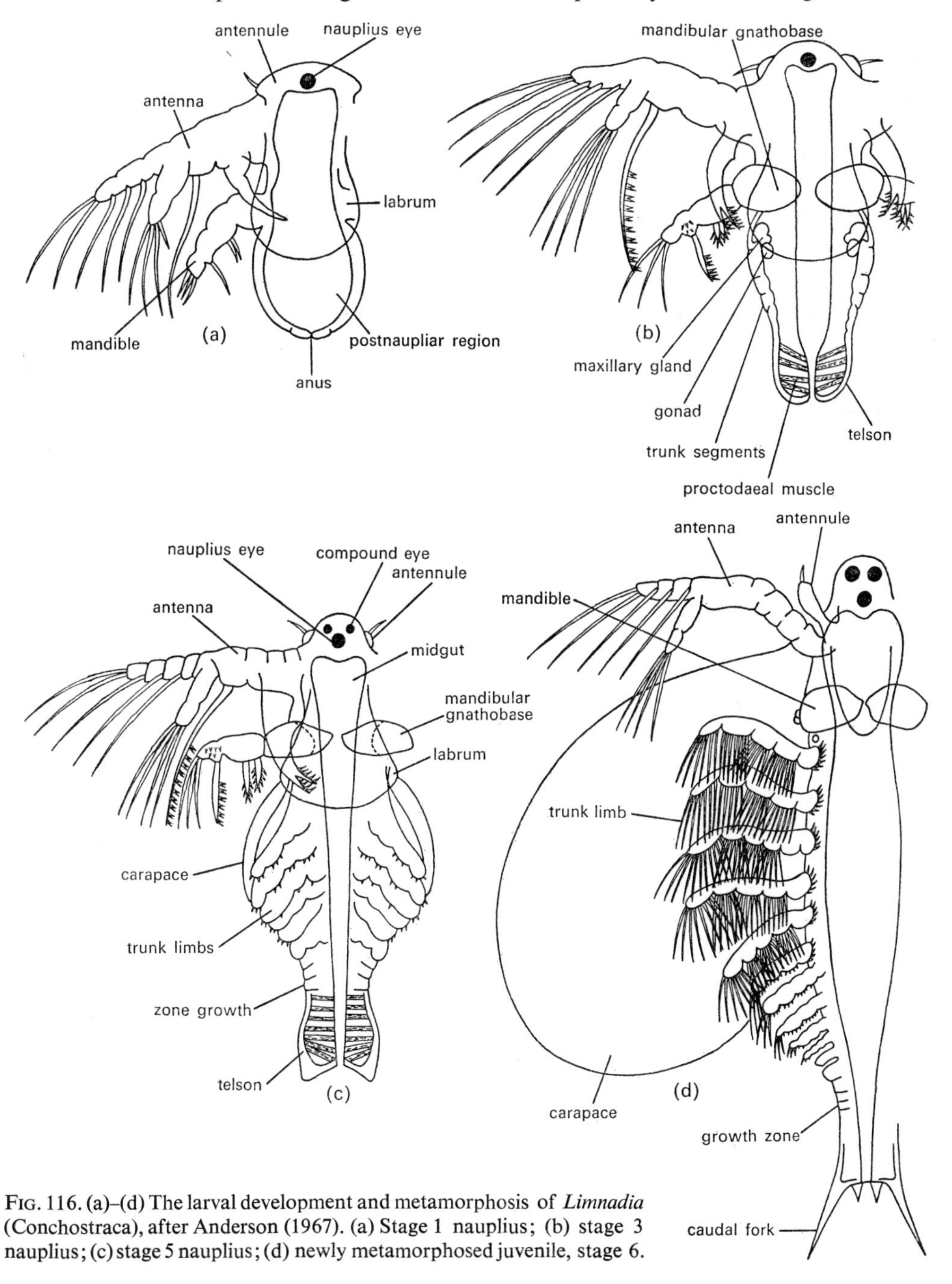

FIG. 116. (a)–(d) The larval development and metamorphosis of *Limnadia* (Conchostraca), after Anderson (1967). (a) Stage 1 nauplius; (b) stage 3 nauplius; (c) stage 5 nauplius; (d) newly metamorphosed juvenile, stage 6.

stroke. An associated oar-like action of the mandibles also assists propulsion through the water.

Swimming in branchiopod nauplii is also accompanied by feeding on particulate matter in suspension in the water. Cannon (1928) and Gauld (1959) showed that the long setae of the antennae act as food-collecting fans, while those of the mandibles play a part in food transfer from the antennae to the naupliar ingestive apparatus. The components of this apparatus are:

1. A masticatory spine at the base of the antennal protopod.
2. A distal, protopodal seta on the median face of the antenna.
3. Proximal and distal setae on the protopod of the mandible. The mandibular gnathobases also develop precociously, becoming functional coincident with the onset of naupliar feeding at stage III.

The broad, flattened labrum forms the floor of a sub-labral space in which the ingestive setae operate. The roof of the sub-labral space is the ventral surface of the maxillulary–maxillary part of the post-naupliar region. The ingestive setae sweep through the sub-labral space on each backstroke of the antennae and mandibles, brushing particulate material towards the mouth. This action sustains the larva until metamorphosis, when the naupliar ingestive apparatus is eliminated and feeding by means of the trunk limbs suddenly commences.

The sharp transition from naupliar feeding to adult feeding at metamorphosis in the Branchiopoda is in marked contrast to events in the Cephalocarida. As we have seen above, naupliar feeding in the Cephalocarida is gradually augmented by the trunk limbs and metamorphosis takes place when the sixth pair of trunk limbs becomes functional at the fifteenth moult. The development of the branchipod postnaupliar region is modified in such a way that the first six pairs of trunk limbs become functional simultaneously, coincident with loss of the specialized naupliar ingestive apparatus, at the ninth moult in Anostraca and the fifth moult in Conchostraca. The metamorphosis of the naupliar region also includes other structural and functional changes in the antennae and mandibles. The mandibles, as in the Cephalocarida, become reduced to gnathobases. The antennae lose their food-collecting function and, in Anostraca, their swimming function; but in the Conchostraca, the swimming function of the biramous antennae persists in conjunction with the enclosure of the trunk at metamorphosis by a bivalve carapace.

The specialization of development of the branchiopod post-naupliar region is evident even at hatching. Although not marked externally by intersegmental annuli or limb buds at this stage, the region already includes the rudiments of the first eight post-naupliar segments (the maxillulary, maxillary and first six trunk segments) together with a pre-telsonic growth zone and a terminal telson which carries the anus. The maxillulary and maxillary segments remain vestigial, but the first six trunk segments develop rapidly and more or less synchronously, exhibiting limb buds at stage IV. The six pairs of limb buds become functional, coincident with metamorphosis, at stage X in Anostraca and stage VI in Conchostraca, but the development of the seventh and succeeding trunk segments and their limb buds is relatively delayed (Fig. 116).

In branchiopods, therefore, development commences with the formation and hatching of a nauplius, as in the Cephalocarida, but is adapted to a more direct attainment of the first functional stage at which feeding can proceed by trunk limbs alone, that is, the stage with six pairs of functional trunk limbs. The adaptation comprises specialization of the naupliar locomotory and feeding apparatus, a synchronous, precocious development of the first six trunk segments and a sudden transition from one to the other at metamorphosis, after fewer naupliar stages than in the Cephalocarida. It is also necessary, of course, to allow for a further factor in this sequence, the specialized vestigiality of the branchiopod maxillules and maxillae. In the Cephalocarida, the maxillules are relatively leg-like until metamorphosis and the maxillae are functionally the first pair of trunk limbs.

Before leaving the Branchiopoda, mention must be made of some facts bearing on the development of the Cladocera, since these point to a conclusion which is relevant to the interpretation of development in the Cirripedia and Copepoda. Among the Conchostraca, there is one species in which the embryo does not hatch as a nauplius larva, but undergoes a more direct development which yields, as a first hatched stage, a juvenile conchostracan with six pairs of functional trunk limbs. The species is *Cyclestheria hislopi* and its unusual mode of development was described by Sars (1896a). After hatching, juvenile development continues in the usual way, with the addition of further trunk segments and limbs. A similar embryonization of naupliar development and hatching as a juvenile with six (or five) pairs of functional trunk limbs is the dominant feature of development in the Cladocera (Grobben, 1879, 1893; Lebedinsky, 1891; Samassa, 1893 a, b, c, 1897; Samter, 1900; Agar, 1908; Vollmer, 1912; Cannon, 1921; Humperdinck, 1924; Kühnemund, 1929; Wotzel, 1937; Kaudewicz, 1950; Agar, 1950; Esslova, 1959; Murakami, 1971), in which only two examples are known of retention of hatching as a nauplius (the winter eggs of *Polyphemus* and *Leptodora*). As in the Conchostraca, the natatory action of the biramous antennae is retained in conjunction with the enclosure of the six pairs of trunk limbs by the bivalve carapace. In the Cladocera, however, the remainder of the trunk remains vestigial and no more trunk limbs are added during subsequent growth and development. It is clear, then, that the Cladocera retain into the adult the body form displayed by the first post-metamorphic stage of their bivalve branchiopod ancestors. It is probable (e.g. Tasch, 1963) that these were closely related to the Conchostraca. The evolution of the Cladocera, therefore, has proceeded through a neoteny which emphasized the significance of the formation of six pairs of trunk limbs as a prerequisite to metamorphosis. The subsequent embryonization of development up to this point must be regarded as a secondary feature.

The Cirripedia and the Copepoda exhibit a parallel neotenic condition of six pairs of trunk limbs as the basic adult number. One of the crucial questions of comparative crustacean development, therefore, is whether the development of the Cirripedia and Copepoda can also be interpreted as a secondary modification of the generalized mode of development retained in the Cephalocarida. In discussing this question, attention will be confined to species which hatch as planktotrophic nauplii.

The numerous descriptions of cirripede development have revealed a constant larval sequence in many species (e.g. Bassindale, 1936; Ishida and Yasugi, 1937; Hudinaga and Kasahara, 1941; Pyefinch, 1948 a, b, 1949; Knight-Jones and Waugh, 1949; Buchholz, 1951; Norris and Crisp, 1953; Jones and Crisp, 1954; Sandison, 1954; Costlow and Bookhout, 1957, 1958; Barnes and Barnes, 1959 a, b; Moyse and Knight-Jones, 1967; Sandison,

1967; Walley, 1969). The swimming and feeding action of cirripede nauplii have been studied by Lochhead (1936), Norris and Crisp (1953), Gauld (1959) and Moyse (1964). The metamorphosis of cirripedes has been elucidated in detail by Walley (1969).

After hatching, the larva (Fig. 117) passes through six naupliar stages with intervening moults. The stage I nauplius, which usually has a length of 200–400 μ, subsists on yolk and moults within 2 hours of hatching. Planktotrophy begins at stage II and persists, with retention of the naupliar swimming and feeding apparatus, until stage VI. The stage VI nauplius usually varies between 500 μ and 1200 μ in length in different species. At the sixth moult, the larva metamorphoses to a cypris (Fig. 118), 500–1100 μ long, in which the naupliar swimming and feeding apparatus are lost, feeding is temporarily suspended, and swimming is performed by six pairs of biramous trunk limbs. After a period of swimming, the cypris settles. Metamorphosis to the adult form is then attained through a further moult.

During the naupliar phase of development, the naupliar region makes up most of the body. The antennules are uniramous but, unlike those of the branchiopod nauplius, are long, setose limbs. Although the antennules are at times pointed forwards and held motionless while the larva swims along, they are also frequently used as swimming organs in conjunction with the antennae and mandibles. The biramous antennae, which carry long (but not hinged) setae on the exopod and endopod, are the main propulsive organs. A masticatory spine near the base of the antennal protopod and a number of distal, median setae on the protopod act in the usual way to sweep food particles beneath the short labrum on the antennal backstroke. The long biramous mandibles also act as propulsive organs, but are, in addition, the main food-collecting limbs of the nauplius. Particles in suspension in the water are gathered mainly by the long setae on the mandibular endopods and swept beneath the ventral surface of the nauplius. The cuticle of this surface bears areas of short, denticulate spines, which appear to scrape the particles from the mandibular setae during the next forward stroke of the mandibles. The latter also carry a median gnathobase, developed precociously at stage II, and a number of distal median setae on the protopod. During the backstroke of the mandibles, these processes assist in the forward transfer of food particles towards the mouth.

The naupliar swimming and feeding action of cirripedes is thus based on a different configuration of functional morphology from that of branchiopods (compare Figs. 116 and 117). It can still, however, be interpreted as a planktotrophic specialization of the cephalocaridan type of nauplius (Sanders, 1963 a, b), in which the locomotory action of the three pairs of limbs is emphasized, food-gathering is restricted to the mandibles, a specialized denticulation of the ventral surface and a precocious development of the mandibular gnathobases assist forward food transfer, and the maxillules are excluded from the mechanism.

In a similar manner, the development of the post-naupliar region in cirripedes can be interpreted as a functional modification of the cephalocaridan pattern, but again differently from that of branchiopods. In the latter, as we have seen, the first eight post-naupliar segments are laid down precociously in the embryo, continue their development more or less simultaneously during the nauplius stages and become functional coincident with loss of the naupliar feeding apparatus (and, in Anostraca, the naupliar swimming apparatus) at a single moult. The cirripedes contrast with this in retaining a sequential development of the

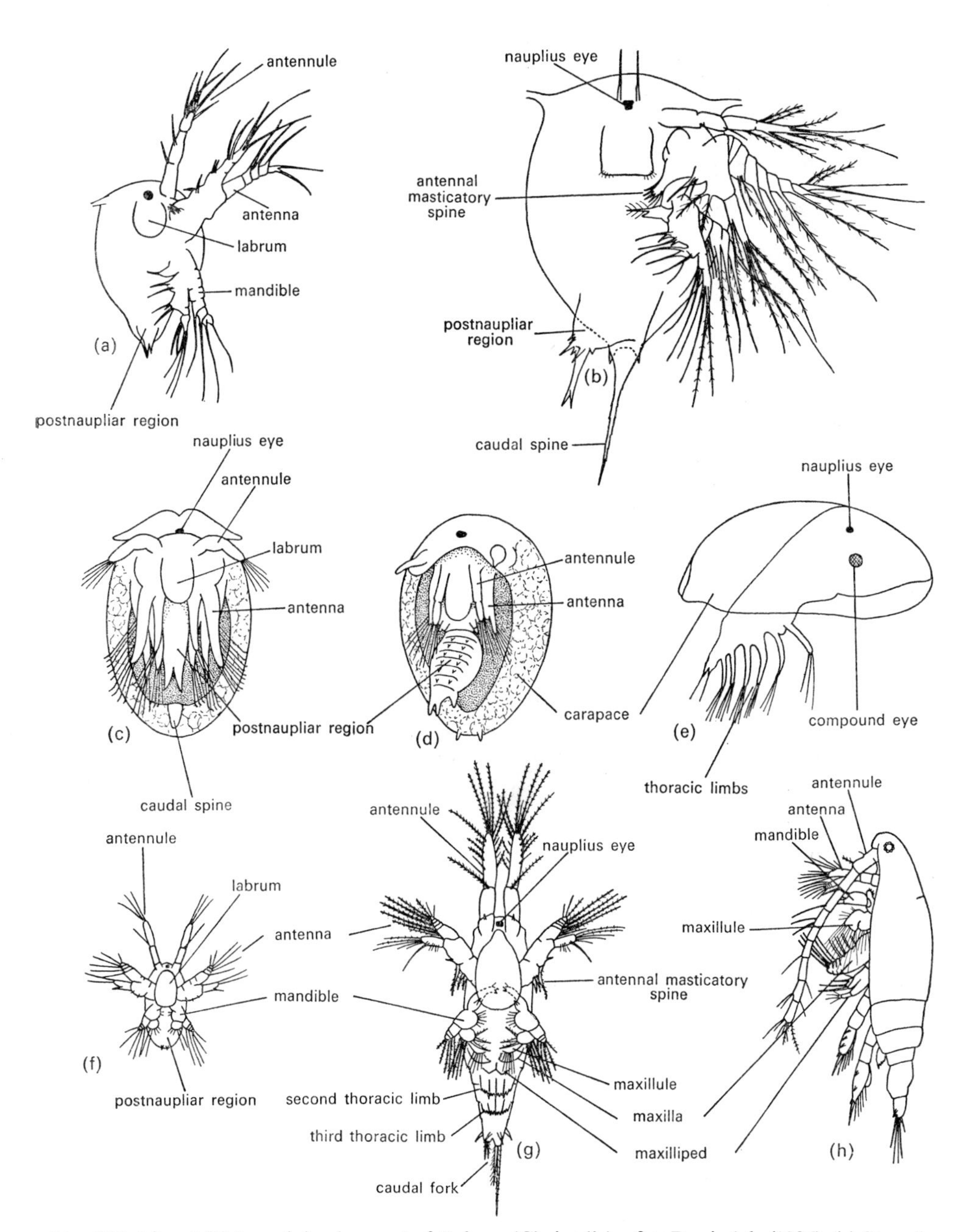

FIG. 117. (a) and (b) Larval development of *Balanus* (Cirripedia), after Bassindale (1936). (a) Stage 1 nauplius; (b) stage 5 nauplius; (c)–(e) larval development and metamorphosis of *Ibla* (Cirripedia), after Anderson (1965). (c) Stage 1 nauplius; (d) stage 5 nauplius; (e) cypris stage. (f)–(h) Larval development and metamorphosis of *Labidocera* (Copepoda), after Johnson (1935). (f) Stage 1 nauplius; (g) stage 6 nauplius; (h) stage 1 copepodid.

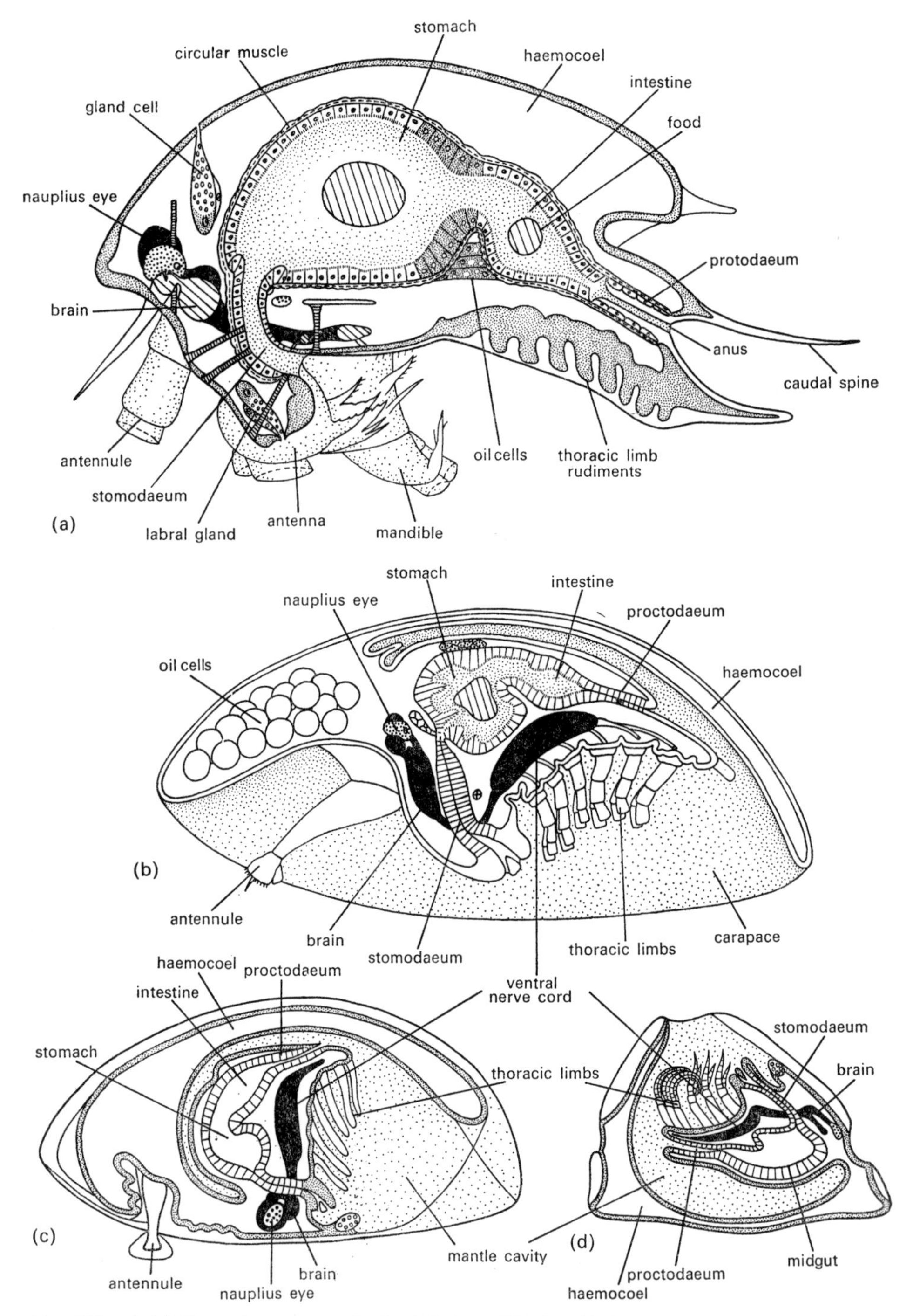

FIG. 118. (a)–(d) Stages in metamorphosis of *Balanus* (Cirripedia) in diagrammatic sagittal section, after Walley (1969). (a) Fully developed stage 6 nauplius; (b) cypris stage; (c) and (d) completion of metamorphosis after settling.

post-naupliar region during the nauplius stages. At hatching, the cirripede post-naupliar region comprises little more than a pre-telsonic growth zone. It is possible that the rudiment of the maxillulary segment is already individuated at this stage (Anderson, 1965; Walley, 1969), but the maxillary and first six trunk segments are budded in succession from the growth zone as the nauplius stages progress. Associated with the retention of this generalized mode of segment formation, however, the Cirripedia have a specialized development of limb buds, functionally convergent with that of branchiopods. The maxillules, maxillae and first six pairs of trunk limbs develop more or less simultaneously as rudiments beneath the cuticle of the stage VI nauplius, and become exposed simultaneously at metamorphosis at a single moult, coincident with the loss of the naupliar swimming and feeding apparatus (Fig. 118). The maxillules and maxillae remain rudimentary in the resulting cypris larva, but the trunk limbs become functional as the six pairs of biramous swimming legs. The shift of locomotory function in this way from the naupliar limbs to six pairs of postnaupliar limbs constitutes the common ground of metamorphosis in the Cephalocarida, Branchiopoda and Cirripedia. The corresponding shift in the feeding function is delayed in the Cirripedia until the second phase of their elaborate metamorphosis has occurred (see below).

The detailed recent study by Walley (1969) of the metamorphosis of *Balanus balanoides* shows that a complex series of external changes accompanies the exposure and onset of function of the six pairs of trunk limbs at the nauplius-cypris moult. While the post-naupliar region elongates and becomes newly functional, the naupliar region shortens and is functionally transformed. Simultaneously, the carapace fold extends laterally and anteriorly, and is greatly extended posteriorly, to form a deep bivalve carapace enclosing the body.

The antennules, which now lie within the anterior extension of the carapace fold, remain large, but take on a new form as flexed, forwardly pointing limbs with terminal adhesive organs. Their functions are now in walking, exploratory sensing and adhesion. The antennae, the main locomotory organs of the nauplius, are wholly resorbed. The labrum is reduced. The mandibles are reduced to gnathobasal rudiments at the sides of the mouth. Behind the mandibles, the maxillules are exposed as ventrolateral rudiments, with the maxillae as short ventral rudiments between them. None of the mouth-parts is functional at this stage. The mouth is closed and feeding is temporarily suspended. Behind the maxillae, as we have seen, the six pairs of trunk limbs are well developed as locomotory organs.

When settlement by antennular attachment has been achieved, the cypris moults again and undergoes a further spectacular change of form (Fig. 118). The anterior extension of the carapace fold is eliminated. The body is rotated through 180°, bringing the postero-ventral opening of the mantle cavity to the apex of the now upper surface and the trunk limbs to a position in which they project upwards rather than downwards. With this change, the synchronous action of the six pairs of trunk limbs becomes a feeding action. The mouth is reopened, the mouth-parts become functional and feeding begins in the adult manner.

The metamorphosis of the Cirripedia is without parallel among the Crustacea. Furthermore, although the process is mainly concentrated into the two stages following the stage VI nauplius, much of the specialized development preliminary to metamorphosis takes place in the last nauplius stage. In spite of this, however, we can easily discern that the extreme specialization of structure and function associated with the evolution of sessility in the Cirripedia is a post-metamorphosis specialization. As pointed out by Newman *et al.* (1969), the ancestors of the Cirripedia were most probably free-swimming bivalve crustaceans with elongate antennules, short biramous antennae, functional mouth-parts, biramous natatory limbs on the first six trunk segments and five limbless segments at the posterior end of the trunk. Apart from the probable addition of the last few trunk segments at successive moults after metamorphosis, this form could easily have been attained at a single mouth following six naupliar stages similar to those retained by extant cirripedes. It is not difficult, therefore, to envisage an ancestral pattern of larval development for the Cirripedia, nor to interpret it as a functional modification of a more generalized cephalocaridan mode of development. The pattern of modification has certain functional parallels with that of the branchiopods (e.g. planktotrophic specialization of the naupliar region, precocious development of the first eight post-naupliar segments, sudden metamorphosis with transition from naupliar swimming and feeding to post-naupliar swimming and feeding after only a few moults), but the details of specialization differ fundamentally in the two groups. When we look at copepod development from the same point of view, it becomes immediately apparent that an equally fundamental similarity exists between copepod and cirripede development.

The development of copepod larvae has been described by many workers on many species (e.g. Lebour, 1916; Gibbons, 1934, 1936, 1938; Campbell, 1934; Gurney, 1934a; Johnson, 1934 a, b, 1937, 1948, 1966; Nicholls, 1934, 1935; Somme, 1934; Davis, 1943; Marshall and Orr, 1955; Humes, 1955; Comita and Thommerdahl, 1960; Koga, 1960; Bernard, 1964; Bjornberg, 1966, 1967; Conover, 1967; Corkett, 1967; Gonzales, 1968; Constanzo, 1969; Grice, 1969, 1971; Elgmark and Langeland, 1970; Evans, 1970; Alvarez and Kewelramani, 1971). On the basis of these studies, it is possible to outline a general course of structural and functional development for copepod species with planktotrophic nauplii (Fig. 117). At hatching, the larva has an elongate ovoid body, lacking the carapace spines of the cirripede nauplius. The length of the stage I nauplius varies, but is typically represented by *Undinula vulgaris* and *Eucalanus pileatus*, with lengths of 160 μ and 245 μ respectively (Bjornberg, 1966, 1967). Development proceeds through six naupliar stages, with feeding commencing at stage III. During these stages, the naupliar region enlarges and the post-naupliar region increases in length and displays the beginnings of external segmentation and limb buds. The body lengths of the stage VI nauplii of *U. vulgaris* and *E. pileatus* are 480 μ and 640 μ.

At the sixth moult, metamorphosis occurs to a first copepodid stage, a juvenile form in which the first five post-naupliar segments become functional. Further development proceeds through a total of five copepodid stages before the moult to the adult takes place. The length of the fifth copepodid stage of *Eucalanus pileatus* is 1·75 mm. The locomotory and feeding action of the copepod nauplius are basically like those of the cirripede nauplius. The uniramous antennules are long, setose limbs, sometimes held in a forward-pointing position and sometimes moving in unison with the antennae and mandibles. The biramous antennae, which have fans of setae on the exopod and endopod and a masticatory spine and other short setae on the median face of the protopod, are the main propulsive organs. The biramous mandibles also carry fans of setae on the exopod and endopod and have short median setae on the protopod. The mandibles assist the antennae in propulsion, but also filter food particles from the water and sweep them beneath the body. As in cirripedes, these particles are scraped from the mandibular setae by a denticulate area on the ventral surface behind the short labrum and are transferred forwards beneath the labrum by the protopodal setae of the mandibles and antennae and the masticatory spines of the antennae. The mandibles also develop functional gnathobases at the third naupliar moult, protruding forwards beneath the labrum as part of the naupliar ingestive apparatus.

It is clear, then, that a common planktotrophic specialization of the naupliar region is shared by copepods and cirripedes before metamorphosis at the sixth moult. After this moult, the form of the naupliar region in the resulting copepodid differs from the specialized form of the same region in the cypris larva, but conforms in general to the postulated form of the region in the ancestors of cirripedes. The antennules become further elongated and are held motionless or used for sudden jumping movements. The antennae become relatively reduced, lose their long swimming setae and ingestive setae and no longer act in ingestion. The mandibular gnathobases become enlarged, but the rami are reduced and are no longer significant in locomotion (though they still contribute to feeding in calanoid copepods). Both the locomotory function and the feeding function, with the exception of continued mandibular involvement in the latter, are shifted to the postnaupliar region.

In branchiopods and cirripedes, the specialized development of the post-naupliar region is a necessary prerequisite to this sudden functional shift. We need to consider, therefore, how the post-naupliar region develops in copepods. Taking the calanoids as an example, the elongation of the post-naupliar region in the nauplius is accompanied by external demarcation of the maxillulary and maxillary segments at stage V and of the first two thoracic segments at stage VI. At metamorphosis to the first copepodid, the region elongates further and the third thoracic segment is delineated. During the succeeding four copepodid stages, the last four thoracic and the five abdominal segments are delineated. In general, segment delineation takes place in a serial manner.

The formation and onset of function of the postnaupliar limbs follows a more complex sequence. The maxillulary rudiments are first exposed in the stage IV nauplius, and the maxillary rudiments in the stage VI nauplius, but neither becomes functional until metamorphosis takes place. Both pairs then assume approximately their adult form and function and become associated with the mandibles as components of the post-naupliar feeding apparatus. In copepods, as in the postulated ancestors of cirripedes, the trunk limbs are excluded from this apparatus, save that the anterior pairs of trunk limbs in copepods is specialized as uniramous maxillules. These, too, first become visible as rudiments in the stage VI nauplius and emerge in their functional maxilliped form at metamorphosis. Modern copepods do not retain a transitional stage in which the first thoracic limbs express their ancestral function as locomotory organs. At the same moult (stage VI nauplius to stage I copepodid), the second and third thoracic segments are exposed as biramous, functional swimming limbs. The fourth, fifth and sixth pairs of thoracic limbs become functional as swimming limbs in the third, fourth and fifth copepodid stages of development respectively.

In the copepods, therefore, the pattern of specialization in the development of the post-naupliar region is different from that of cirripedes, in spite of the similarities in the naupliar region. It is not difficult, however, to envisage the post-naupliar development of copepods as a specialization of a generalized cephalocaridan pattern of development. Associated with the planktotrophic specialization of the naupliar region, the post-naupliar region develops rapidly and metamorphosis occurs at an early moult. Before metamorphosis, the postnaupliar region is not functional in locomotion and feeding. At metamorphosis, it takes over these functions. The takeover involves

1. A simultaneous development of the maxillules, maxillae and first thoracic limbs (maxillipeds) as a postnaupliar feeding apparatus, associated with the metamorphosed mandibles.
2. A coincident, specialized development of the second and third pairs of thoracic limbs as specialized swimming limbs.

The functional specialization of the mouth-parts and the first three pairs of trunk limbs in this way obviates the need for the establishment of six pairs of functional trunk limbs as a prerequisite to moving and feeding in the adult manner. Associatedly, the serial development of the last three pairs of trunk limbs is delayed until after metamorphosis.

Although a cephalocaridan-like ancestry is not difficult to envisage for the developmental patterns of the Branchiopoda, Cirripedia and Copepoda, a more difficult problem is posed by the Malacostraca. In these crustaceans, we do not have the guiding principle of metamorphosis associated with the functional development of six pairs of trunk limbs. The basic sequence of larval development of the Malacostraca, as exemplified by the development of the Penaeidea (e.g. Heldt, 1938; Pearson, 1939; Hudinaga, 1942; Gurney, 1943; Morris, 1948; Anderson *et al.*, 1949; Menon, 1952; Morris and Bennett, 1952; Heegaard, 1953; Johnson and Fielding, 1956; Broad, 1958; Dobkin, 1961; Renfro and Cook, 1963; Ewald, 1965; Rao, 1968; Cook and Murphy, 1971), involves a complex series of larval stages (Figs. 119 and 120) and includes the development of eight thoracic and six abdominal pairs of limbs. Sanders (1963a) regarded this sequence as a likely derivative of a generalized cephalocaridan type of development. In the present context, this possibility will be examined further, to providing an appropriate background against which the more direct development of the larger types of malacostracan embryo can be viewed.

Penaeid naupli at hatching are larger than those of copepods and cirripedes (e.g. *Penaeus duorarum*, 350–400 μ long) and have a relatively long post-naupliar region. The nauplius stages of development, which vary in number (e.g. *P. duorarum*, five naupliar stages), are sustained by yolk carried over from the egg. In association with this, they lack the antennal masticatory spines, antennal and mandibular protopodal setae and mandibular gnathobases of the naupliar ingestive apparatus, but retain naupliar natatory limbs similar to those of copepods and cirripedes. The long, uniramous antennules are usually held motionless in a forward-projecting position but occasionally assist the antennae and mandibles as propulsive organs. The long biramous antennae and shorter, biramous mandibles, both with fans of long setae, are the major propulsive organs of the nauplius. The rudiments of the mandibular gnathobases develop at the last nauplius stage but do not become functional until the next stage of development.

During the naupliar stages, the post-naupliar region increases in length, due to the proliferation of the rudiments of at least ten post-naupliar segments. The nauplius of *P. duorarum*, for example, is about 600 μ long at stage V. The rudiments of the first four pairs of post-naupliar limbs (maxillules, maxillae and first two trunk segments) are also exposed

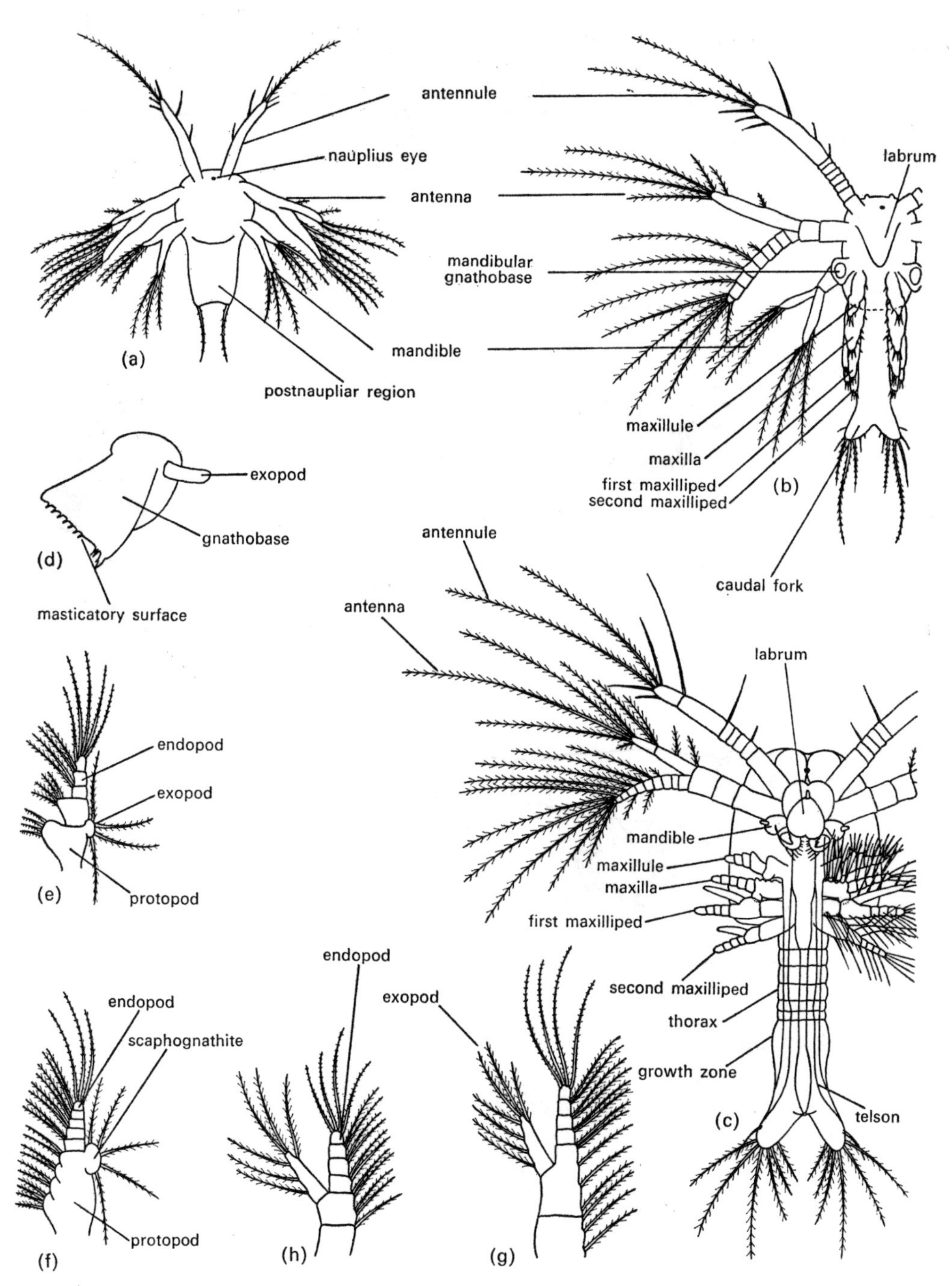

FIG. 119. (a)–(c) Early larval development of *Penaeus* (Malacostraca). (a) Stage 2 nauplius; (b) stage 5 nauplius; (c) first protozoea. (d)–(h) Limbs of first protozoea. (d) Mandible; (e) first maxilla; (f) second maxilla; (g) first maxilliped; (h) second maxilliped. (After Dobkin, 1961.)

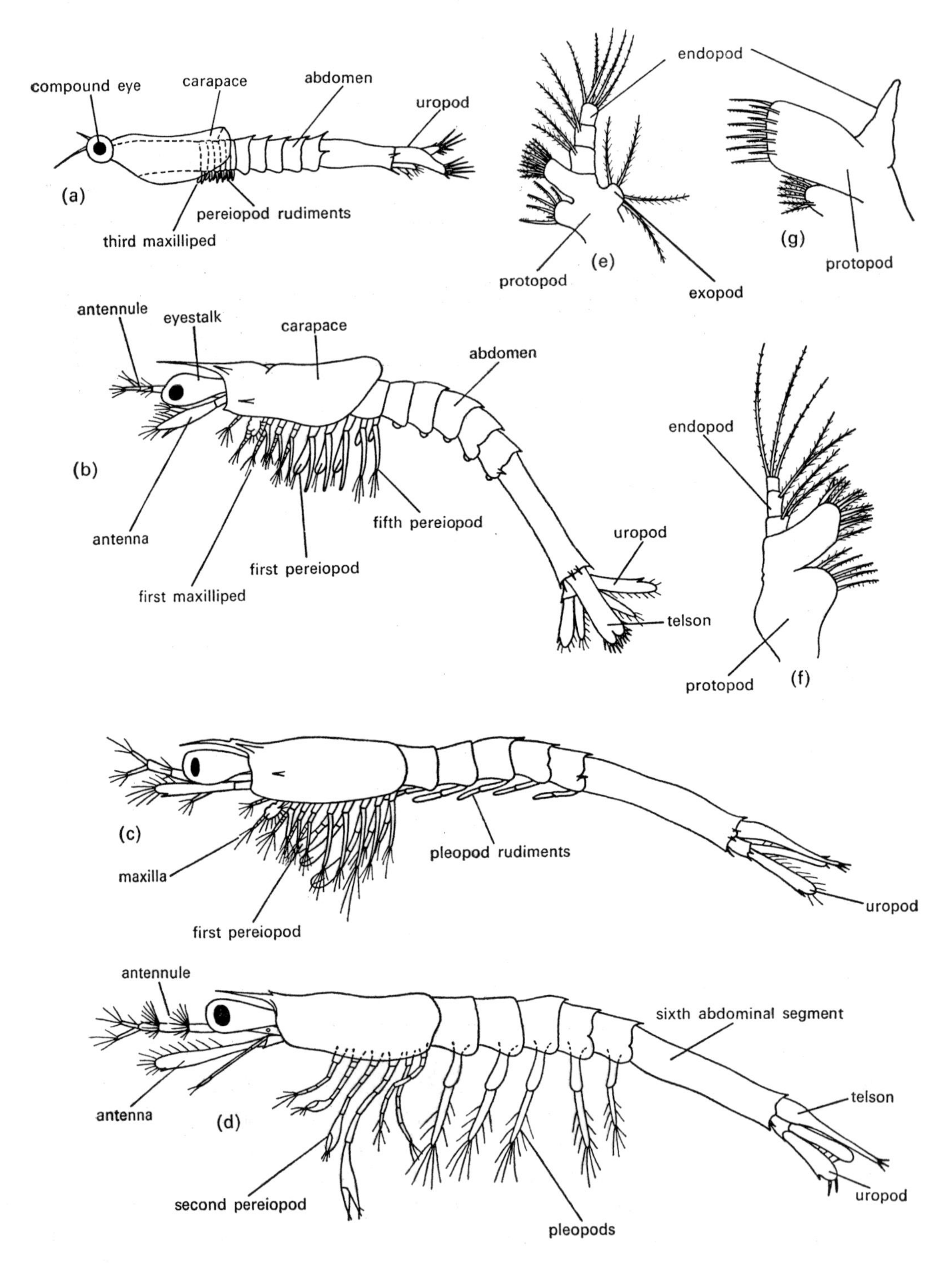

FIG. 120. (a)–(d) Later larval development of *Penaeus* (Malacostraca). (a) Third protozoea; (b) first mysis; (c) third mysis; (d) first post-larva; (e) first maxilla of third protozoea; (f) first maxilla of third mysis; (g) first maxilla of first post-larva.

at the surface in the stage IV nauplius and become enlarged at stage V, but do not become functional until the next moult.

With the next moult (the fifth in *P. duorarum*) the nauplius undergoes the first major step in the development of the post-naupliar region and becomes a protozoea. There is a considerable increase in length (e.g. in *P. duorarum*, to about 1 mm), mostly in the post-naupliar region. The antennules and antennae retain their naupliar form and function, but the mandibles are reduced to functional gnathobases. The carapace fold is also exposed for the first time at this stage of development and extends back over the dorsal surface of the body as far as the second trunk segment.

In the post-naupliar region of the stage I protozoea, the eight thoracic segments are now delineated externally, though only the first two carry limbs. Behind the eighth thoracic segment is the abdominal growth zone, in front of the forked telson. The maxillules and maxillae become elongate and setose, and commence their function as ingestive limbs in association with the mandibles. The first two pairs of trunk limbs develop as setose biramous limbs, the first and second maxillipeds, whose exopods augment the swimming action of the antennae and whose endopods play a part in food capture.

Protozoeal form and function now persists through several moults (e.g. *P. duorarum* has three protozoea stages), during which the body length increases further (e.g. in *P. duorarum*, to about 2·5 mm). During this phase of development, the stalked, compound eyes are exposed on the anterior part of the head and the carapace extends back to cover the posterior part of the thorax. The rudiments of the last six pairs of thoracic limbs are developed, but the thoracic segments on which they lie remain short. In contrast, the abdomen increases greatly in length, with demarcation of the six abdominal segments in front of the telson, and shows a precocious development of the sixth pair of abdominal limbs as functional uropods. The onset of function of the abdomen and tail fan as a locomotory organ for rapid sudden movement through the water is associated with this precocious development, but the first to fifth abdominal limbs are not developed during the protozoea phase.

At its next moult (e.g. the eighth in *P. duorarum*) the protozoea undergoes a further structural and functional transformation, emerging as a first mysis stage. The increase in length at this moult is not excessive (e.g. in *P. duorarum* to about 3·2 mm), but development takes a major step towards the juvenile adult form. Anteriorly, the antennules become relatively reduced and wholly sensory in function, as do the antennae. At this moult, therefore, the naupliar propulsive apparatus is lost. Coincidentally, the last six thoracic segments become enlarged and their limb buds become functional, with long exopodites which join those of the first and second maxillipeds as the swimming organs of the mysis larva. The mandibles, maxillules, maxillae and first two pairs of maxillipeds retain their feeding and ingestive role. Similarly, the long abdomen and tail fan are unchanged, except that rudimentary limb buds are now developed on the first five abdominal segments.

The mysis form, like the preceding protozoea form, is retained through several moults (e.g. there are three mysis stages in *P. duorarum*), during which the length of the larva increases further (e.g. to about 4 mm at the third mysis stage in *P. duorarum*). The first five pairs of abdominal limb buds become elongated during this phase, but do not become functional. The swimming and feeding processes of the larva remain unchanged until, at a single moult (the eleventh moult in *P. duorarum*), the larva undergoes a final transformation into a first post-larva or juvenile stage. At this moult, the mouth-parts and maxillipeds assume their basic adult form, the fourth to eighth thoracic limbs become pereiopods with enlarged endopods and reduced exopods, and the first five pairs of abdominal limbs become functional pleopods. The swimming function thus shifts at a single moult from the thorax to the abdomen.

Basic malacostracan development as exemplified by the Penaeidea therefore exhibits the elimination of naupliar feeding but retains a stict anteroposterior sequence of postnaupliar development, combined with a progressive shift of natatory function to successively more posterior groups of limbs as development proceeds. In spite of the similarities in naupliar locomotion (Gauld, 1959; Sanders, 1963a), the pattern of later larval specialization of the Malacostraca is fundamentally different from those of cirripedes and copepods. On the other hand, a functional derivation from a generalized ancestral development of the cephalocaridan type is easily envisaged. During the period in which naupliar swimming is retained (i.e. the nauplius and protozoea stages), post-naupliar development proceeds sequentially, as in the Cephalocarida, and all of the trunk segments are laid down before metamorphosis occurs. The functional specializations of the Malacostraca before metamorphosis are:

1. The precocious development of the first four post-naupliar limb-pairs as a specialized feeding apparatus (maxillules, maxillae and first two trunk limbs) in association with mandibular gnathobases. This apparatus becomes functional at the moult from nauplius to protozoea.
2. The associated retardation of development of the next six trunk segments and their limbs.
3. The precocious development of the last six trunk segments (but not their limbs, except for the terminal pair) as a specialized "escape mechanism".

At the moult at which the naupliar swimming mechanism is lost, i.e. the moult corresponding to metamorphosis in the Cephalocarida, the malacostracan specializations continue as:

4. A retention and further elaboration of the feeding apparatus already established, including the first two trunk limbs.
5. A sudden onset of function, emphasizing natatory exopods, of the next six pairs of trunk limbs, but no direct involvement of these limbs in feeding.

The later transition from thoracic to abdominal natation at the moult from mysis to post-larva is clearly a secondary specialization.

Even a brief consideration, therefore, suffices to indicate that the ancestors of the penaeid Malacostraca could have had a cephalocaridan-like development, although with functional specialization the association between metamorphosis and the development of six pairs of trunk limbs can no longer be discerned. Since the same developmental ancestry can be proposed on functional grounds for the Cirripedia, Copepoda and Branchiopoda, the larval development of the Cephalocarida must weigh heavily in any attempted elucidation of a basic theme of development for Crustacea. We shall return to this question at the end of the present chapter.

Against a background of information about the development of crustaceans which hatch as nauplii, the external features of development of species with larger eggs and a more advanced hatching stage can be dismissed relatively briefly, since all of them display secondary modifications. To a greater or lesser extent, ancestral larval development is embryonized in these species. The only major groups in which, as far as is known at the present, the embryonization of larval development has not evolved, are the Cephalocarida and Mystacocarida. Embryonization of development is rare in the Ostracoda, but is seen in the large, deep-water, marine ostracod *Gigantocypris*, whose relatively large eggs hatch as juvenile ostracods. Again, embryonization of development has a restricted occurrence in the Branchiopoda, where it occurs only in one conchostracan (*Cyclestheria*) and in the Cladocera. We have already discussed in earlier paragraphs how *Cyclestheria* and the Cladocera hatch as juveniles with six pairs of trunk limbs and how this form persists neotenically in the Cladocera during subsequent development to the adult.

Before going on to discuss the occurrence of embryonized development in the other major groups of crustaceans, another example of what is meant by embryonization will be given. The example chosen is the brachyuran crab *Macropodia longirostris* (Fig. 121), the embryos of which were recently described and figured by Lang and Fioroni (1971). We have seen above that the penaeid decapod *Penaeus duorarum* hatches as a nauplius and passes through several nauplius, protozoea and mysis stages before becoming a juvenile at the eleventh moult. In *Macropodia*, the nauplius and protozoea stages are passed during embryonic development. The stages are clearly recognizable in the embryo, the nauplius stage by its precocious development of the three naupliar segments and their limb buds, the protozoea stage by its precocious development of the maxillules, maxillae, first and second maxillipeds and abdominal segments; but functional differentiation of the embryo proceeds directly to a hatching stage equivalent to the first mysis stage of *Penaeus* (compare Figs. 120b and 121i). The continued suppression of the posterior thoracic region of *Macropodia* at the hatching stage is a specialized brachyuran feature to which we shall return below. The development of *Macropodia* prior to the "first mysis equivalent" has thus been embryonized, and the *functional* differentiation of ancestral larval components prior to this stage has been lost. It has, of course, been replaced by other functions associated with the digestion and distribution of yolk reserves, but these are more pertinent to organogeny and will be discussed in later sections.

Embryonization of development is the rule rather than the exception in the Malacostraca and is also common in the Branchiura, Copepoda and Cirripedia. In the Cirripedia and Copepoda, it is most unusual for embryonization to take in stages beyond the nauplius. Hatching, therefore, still occurs as a nauplius although the first moult may yield a cypris or a copepodid after a very brief, lecithotrophic naupliar phase. The only major exception associated with this generalization is in the development of the Branchiura. In spite of their close relationship with the Copepods, the Branchiura have a fully embryonized development and hatch as juveniles.

The cirripedes in which development is embryonized are those with relatively large eggs, such as the thoracicans *Ibla*, *Scalpellum* and *Pollicipes* and the acrothoracican *Trypetesa* (Batham, 1945, 1946; Anderson, 1965; Kaufmann, 1965; Turquier, 1967 a, b, 1970, 1971). The nauplius is relatively large and yolky when hatching takes place and lacks the feeding setae of a planktotrophic cirripede nauplius. *Ibla quadrivalvis* is a useful example (Fig. 117).

Although the development of the naupliar region has been embryonized and this region undergoes no further growth after hatching, the post-naupliar region grows in the usual way during several moults. Associated with the lecithotrophy of the larva, the metamorphosis to a cypris is anticipated by preliminary changes in the later naupliar stages. The carapace becomes arched over the body, partially enclosing an incipient mantle cavity, and the naup-

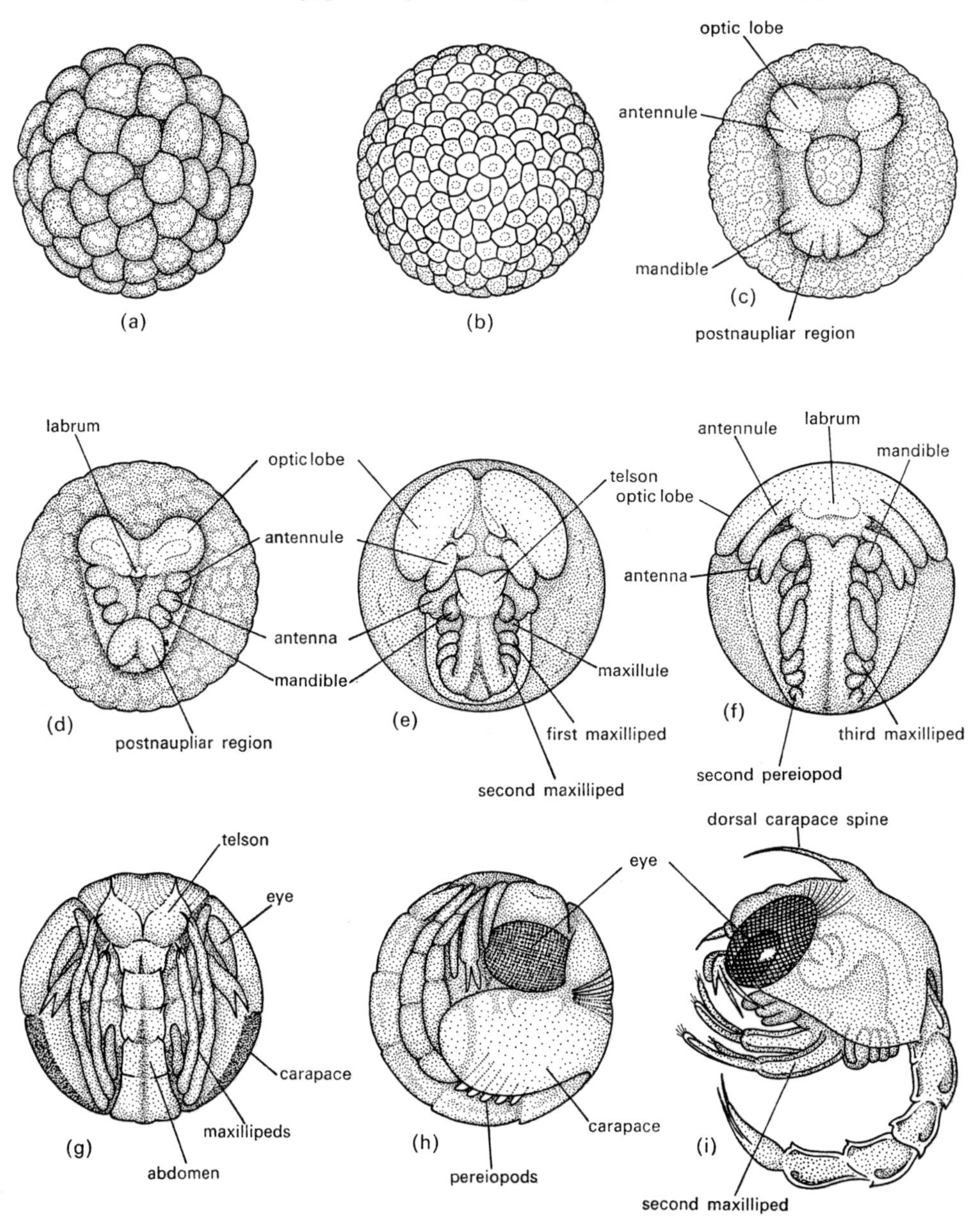

FIG. 121. (a)–(i) External development of the embryo of *Macropodia* (Brachyura), after Lang and Fioroni (1971). (a) Cleavage; (b) blastoderm; (c) nauplius rudiment; (d) formation of caudal papilla; (e) and (f) formation of post-naupliar segments and limbs; (g) and (h) completion of development; (i) newly hatched first zoea.

liar limbs become smaller. Then, at a single moult, metamorphosis to a typical cypris larva takes place and development continues in the normal cirripede manner. Other cirripedes with large, lecithotrophic nauplii show the same lack of growth of the naupliar region during larval development.

Although they are not an example of embryonization of development, we cannot leave the Cirripedia without mention of the Rhizocephala, since this group of parasitic barnacles exemplifies another important phenomenon of crustacean development, the onset of specialization after metamorphosis following a relatively generalized development up to this point. The early development of rhizocephalans proceeds through a series of planktonic nauplius stages, followed by metamorphosis to a cypris in the usual manner. After settlement however, extreme specializations supervene in association with the adoption of parasitic life. These specializations are well illustrated by the work of Yanagimachi (1961 a, b) on *Peltagasterella gracilis* (Boschma), one of the few dioecious rhizocephalans. Sexual dimorphism is displayed even in the cypris larvae of *P. gracilis*, which are of two sizes, 300–340 μ long and 230–260 μ long. The smaller cypris larvae attach to the body surface of their crab host and develop as females. After attachment, the cypris undergoes the second stage of its metamorphosis, emerging as a small, oblong sac enclosed by a thin layer of chitin. This stage is called a kentrogen. The internal tissues dissociate into a mass of cells. A dart-shaped tube differentiated in the anterior region of the cell mass pierces the host integument and the cell mass passes through the tube into the host. Development as a rhizocephalan parasite now proceeds, but sexual maturity is not attained until the rooted parasite is invaded by a male cypris.

The male cypris of *P. gracilis* is the larger of the two cypris forms. In contrast to the females, these larvae attach only to the mantle cavity openings of juvenile, rooted females. The cellular mass of male cypris tissue then migrates through one of the antennules and enters the female mantle cavity to take up residence in a spermatheca. In this location, cell proliferation and sperm formation proceed.

Parasitic cirripedes thus have specialized development after metamorphosis, while the embryonization of naupliar development is a feature of certain free-living cirripedes. In the Copepoda, in contrast, the embryonization of naupliar development is particularly a feature of parasitic species and is combined with specializations of post-metamorphosis development during the invasive copepodid stages. Considerable variation obtains in the way in which these specializations are expressed in different species (e.g. Canu, 1892; Wilson, 1905, 1907 a, b, 1911 a, b; Henderson, 1926; Connolly, 1929; Gray, 1933 a, b; Gurney, 1933, 1934b; Caspers, 1939; Sproston, 1942; Heegaard, 1947; Lang, 1948; Gnadeberg, 1949; Gnanamutha, 1951 a, b; Baer, 1952; Gotto, 1957, 1962; Bresciani and Lutzen, 1961; Lewis, 1963; Dudley, 1966; Gage, 1966; Carton, 1968; Jones and Mathews, 1968; Anderson and Rossiter, 1968 a, b; Mirzoyeva, 1969) but a suitable example is provided by *Dissonus nudiventris* Kabata, a gill parasite of the Port Jackson Shark, *Heterodontus phillipi* Blainville. As described by Anderson and Rossiter (1968b), *D. nudiventris* hatches as a nauplius, but retains none of the ancestral naupliar functions. The naupliar limbs are small and anticipate their form and function in the copepodid. The post-naupliar region is long and well developed. The nauplius remains attached to the egg string of the female parent by a pair of threads secreted from the telson. At a single moult, the first copepodid stage emerges and becomes free-swimming, but is lecithotrophic and already shows the limb specializations of the adult. It seems likely that this stage parasitizes the host and completes its further copepodid stages after attachment. The anchoring strings of the nauplius of *Dissonus* are unique, but the embryonization of naupliar development and the emergence of a specialized copepodid at metamorphosis are typical of parasitic copepods.

The embryonization of development in the Malacostraca is much more diverse and far-reaching than in cirripedes and copepods. It is useful to continue first with the Eucarida, since the embryonized larval stages are easily recognized in this group. The extent of embryonization varies greatly among decapods, especially in the Caridea (see Gurney, 1942; also Cook and Murphy, 1965; Dobkin, 1965, 1967, 1971; Lewis and Ward, 1965; Squires, 1965; Babu, 1967; Herring, 1967; Modin and Cox, 1967; Hubschman and Rose, 1968; Markarov, 1968; Bensam and Kartha, 1969; Heegaard, 1969; Little, 1969; Choudhury, 1970, 1971 for recent accounts). At the least, carid development is embryonized to the mysis stage and the larva hatches with the antennules and antennae in their post-naupliar form, the mouth-parts and maxillipeds functional in feeding and swimming, all thoracic segments defined and the abdominal segments and tail fan well developed. Subsequent development then proceeds through a series of mysis stages with thoracic natation, before the moult to a post-larva with functional abdominal pleopods takes place. In many carids, however, development is embryonized to a much later stage and hatching takes place as a post-larva. The embryonized nauplius, protozoea and mysis stages can be recognized in the embryo. Large variations in hatching stage occur in related genera. For example, within the genera *Alpheus* and *Synalpheus*, every gradation occurs, from nine larval stages in *A. ruber* to a complete embryonization of development in some species of *Synalpheus*. Freshwater carids often retain long larval sequences, e.g. the atyids *Caridina* and *Paratya*, which hatch at an early mysis stage, but some freshwater genera have direct development. From the point of view of comparative development, the carids are easily interpreted as secondary modifications of the larval sequence observed in penaeids. The same conclusion can be reached for the Astacura (see Jorgensen, 1925; W. C. Smith, 1933; Templeman, 1936, 1939; Boas, 1939; E. W. Smith, 1953; Davis, 1964; Pandian, 1970 for marine species; Baumann, 1932; Zehnder, 1934 a, b, 1935; Bieber, 1940; Suko, 1962; Tzukerkis, 1964; Fioroni, 1969 for freshwater species), even though no species of this group hatches at a stage earlier than a mysis in which all thoracic limbs are present. This is the hatching stage, for example, in *Hormarus*. The abdominal appendages develop in anteroposterior succession in a series of larval stages, the uropods being the last to develop. In freshwater astacurans, development is fully embryonized and hatching takes place as a juvenile.

Among the Thalassinidea, Paguridea and Brachyura, it is usual for hatching to take place at an early mysis stage, with only the first two pairs of thoracic limbs developed to the functional condition, but the hatched larva is more modified than a mysis and is better known as a zoea (Figs. 122 and 123). Thalassinid and pagurid zoea larvae differ from the first mysis stage of penaeids in the continued suppression of the last six thoracic segments. Brachyuran zoea larvae differ in this and in other distinctive ways. The cephalothorax in front of the suppressed thoracic segments is relatively enlarged, the carapace carries a rostral and a dorsal spine and the compound eyes are large (Fig. 123). In spite of these differences, however, the further development of the larva proceeds in much the same way

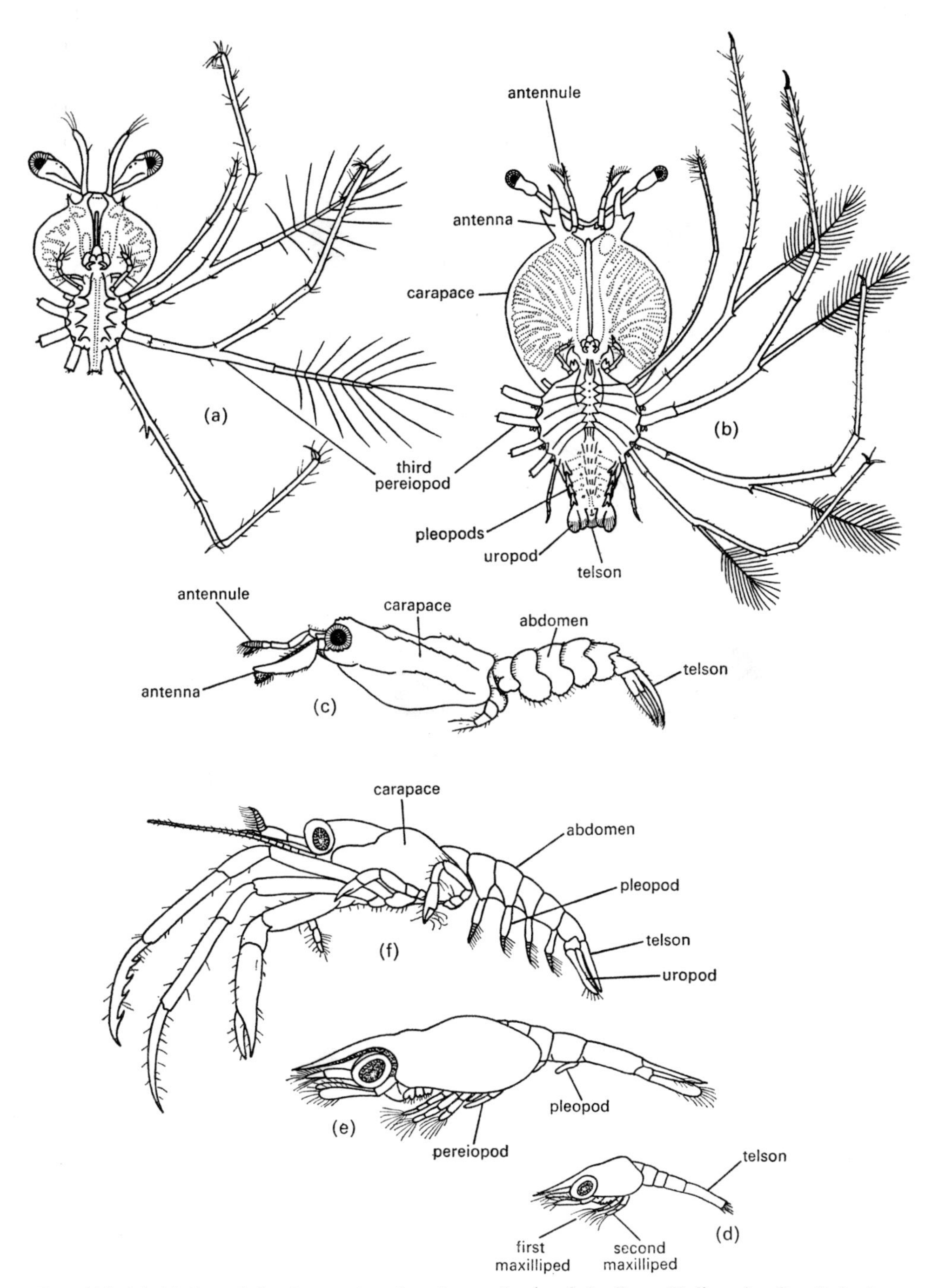

FIG. 122. (a)–(c) Larval development and metamorphosis of *Scyllarus* (Palinura), after Robertson (1968). (a) Stage 1 phyllosoma larva; (b) stage 7 phyllosoma; (c) post-larva. (d)–(f) Larval development and metamorphosis of *Petrochirus* (Paguridea), after Provenzano (1968). (d) Stage 1 zoea; (e) stage 5 zoea; (f) glaucothoë.

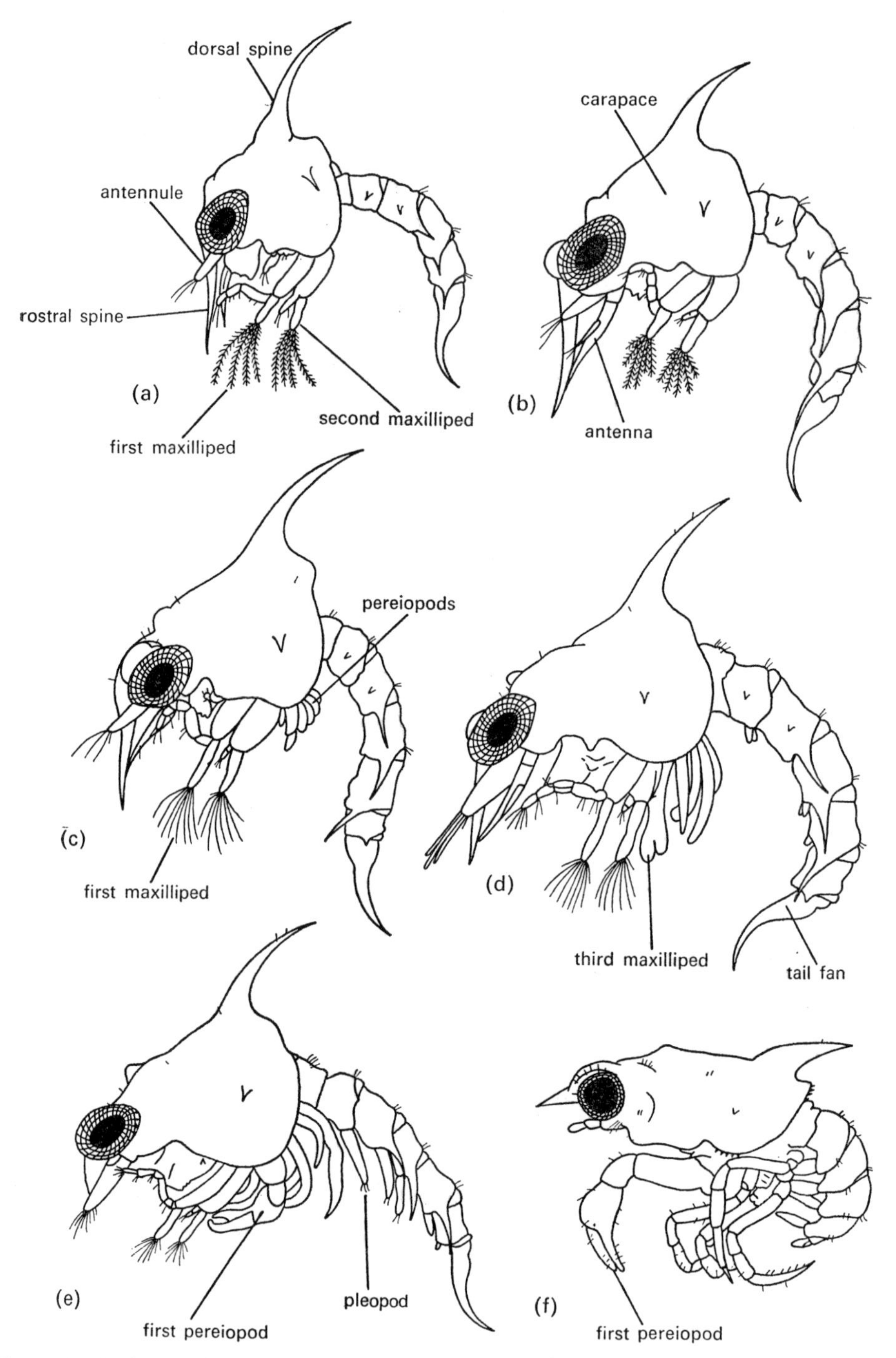

FIG. 123. (a)–(f) Larval development and metamorphosis of *Parthenope* (Brachyura), after Yang (1971). (a) Stage 1 zoea; (b) stage 3 zoea; (c) stage 4 zoea; (d) stage 5 zoea; (e) stage 6 zoea; (f) megalopa.

in the three groups. Several moults are passed as a zoea, during which the last six thoracic segments increase in size and develop limbs, and the abdominal segments develop limbs, but none of the limbs behind the second thoracic pair becomes functional. The swimming and feeding functions of the zoea continue to be carried out by the mouth-parts and the first and second maxillipeds, with assistance from the precociously developed abdomen and tail fan. Then, at a single moult, the zoea transforms into a stage in which the zoeal swimming function is lost, the thoracic limbs become functional in their adult form and the abdominal limbs become functional as pleopods. This larval stage, which is a post-larva in the Thalassinidea but is called a glaucothöe in the Paguridea and a megalopa in the Brachyura, remains free-swimming, but immediately precedes the first juvenile stage attained at the next moult. Informative descriptions of larval development have been given for the Thalassinidea by Lebour (1938, 1941), Dakin and Colefax (1941), and Bahamonde and Lopez (1960). The Paguridea have been studied intensively in recent years, especially by Provenzano (1962 a, b), Provenzano and Rice (1964), Hazlett and Provenzano (1965), Rice and Provenzano (1965), Wear (1965, 1966), Davis (1966), Greenwood (1966), Knight (1966), Provenzano and Rice (1966), Roux (1966), Provenzano (1967a, 1968 b, c, 1971 a, b), Pike and Wear (1969), Shepherd (1969), Nyblade (1970), Roberts (1970), Fagetti and Campodiconico (1971a), Gore (1970) and Macmillan (1972). The larval development of the Brachyura has also been actively studied in the last decade and numerous species have now been reared in the laboratory (e.g. Costlow and Bookhout, 1959, 1960 a, b, 1961 a, b, 1962 a, b, 1966, 1967; Sankolli, 1961, 1962; Hartnoll, 1964; Williamson, 1965, 1967; Rice and Provenzano, 1966, 1970; Boschi *et al.*, 1967; Costlow and Fagetti, 1967; Knight, 1967; Boschi and Scelzo, 1968; Gore, 1968; Herrnkind, 1968; Wear, 1968 a, b, 1970 a, b, c; Yang, 1968, 1971; Fagetti, 1969; Lawinski and Pautsch, 1969; Hashmi, 1970 a, b, c, d; Trask, 1970; Fagetti and Campodiconico, 1971; Ingle and Rice, 1971; Rice *et al.*, 1971; Sandifer and van Engel, 1971).

Only in the Palinura among the decapod Eucarida has a more extreme larval specialization evolved (Fig. 122). Hatching occurs at a relatively advanced stage, in which the antennules and antennae retain no naupliar features, the mouth-parts and maxillipeds approximate to the adult form and the first four pairs of pereiopeds are well developed and functional. Thus, development is embryonized up to a stage which corresponds to a post-larva in respect of the head and most of the thorax. The last thoracic segment and the abdominal segments, however, are still rudimentary at the hatching stage, and the head and thorax are highly specialized for flotation. The antennal–mandibular region is expanded as a flat plate, while the four pairs of pereiopods are greatly elongated and carry long setal plumes on their exopods. The phyllosoma larva, as this stage is called, remains in the plankton for several months and undergoes a number of moults. During this phase of development, the fifth pair of pereiopods develops as long, flotatory limbs and the abdomen enlarges and develops limb buds. Then, at a single moult, the phyllosoma transforms into a juvenile. Important recent studies on palinurid larval development have been carried out by Johnson and Knight (1966), Simms (1966), Batham (1967), Provenzano (1968), Robertson (1968, 1969), Roberts (1968) and Johnson (1971).

All decapods with the exception of the Penaeidea, therefore, have embryonized nauplius and protozoea stages. Many carids and all astacurans exhibit a further embryonization of the mysis phase of development and hatch as a late mysis, a post-larva or a juvenile. The

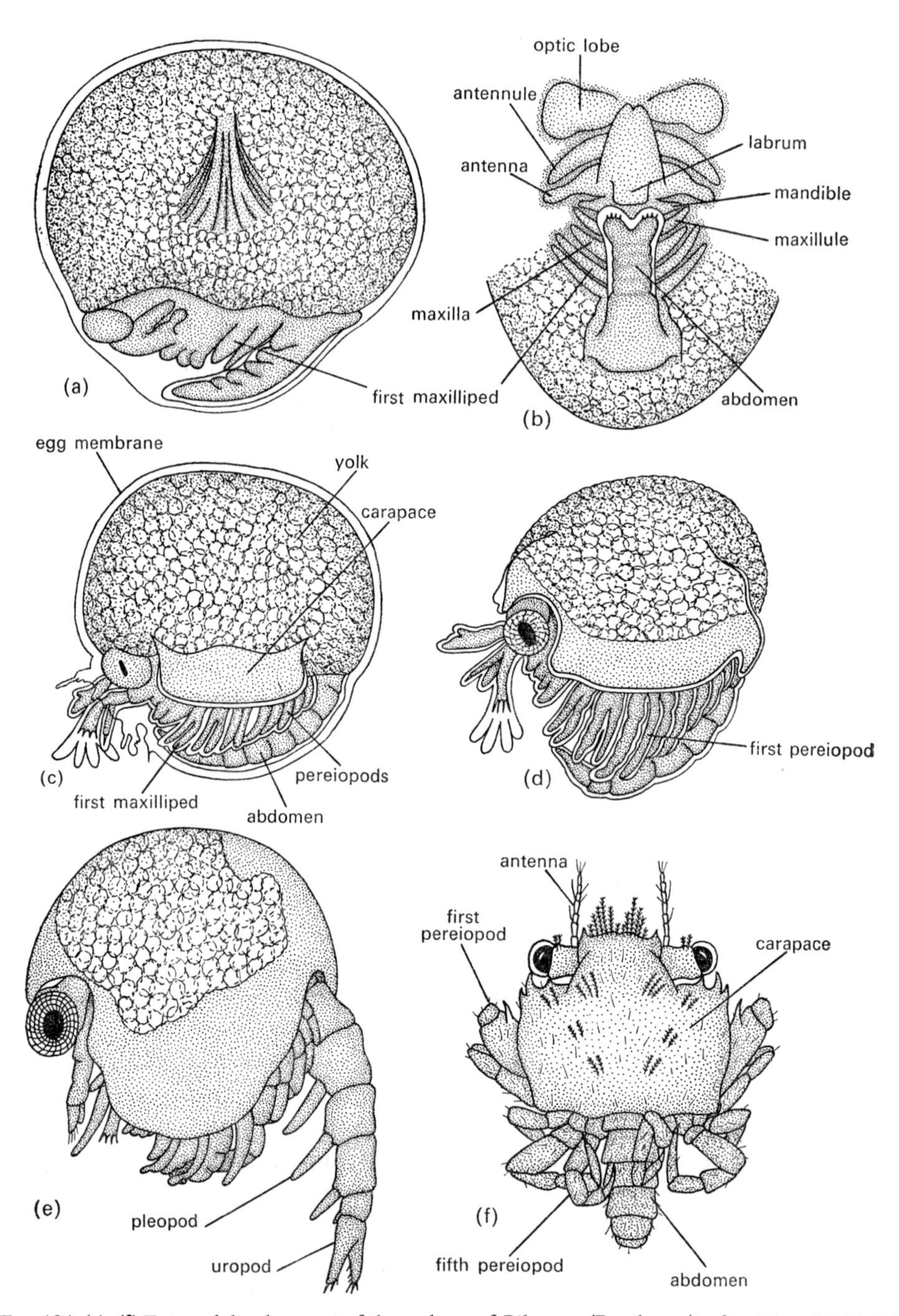

FIG. 124. (a)–(f) External development of the embryo of *Pilumnus* (Brachyura), after Wear (1967). (a) and (b) Metanauplius stage in lateral and ventral view; (c) second embryonic zoeal stage; (d) third embryonic zoeal stage; (e) fourth embryonic zoeal stage; (f) newly hatched megalopa.

Palinura hatch as a modified post-larva, the phyllosoma. The Thalassinidea, Paguridea and Brachyura usually hatch as a modified early mysis stage, the zoea, but a few species hatch at at a later stage. The thalassinid *Upogebia savignyi*, for example, hatches as a juvenile (Gurney, 1937). The Brachyura include several species with late larval hatching (e.g. the xanthid

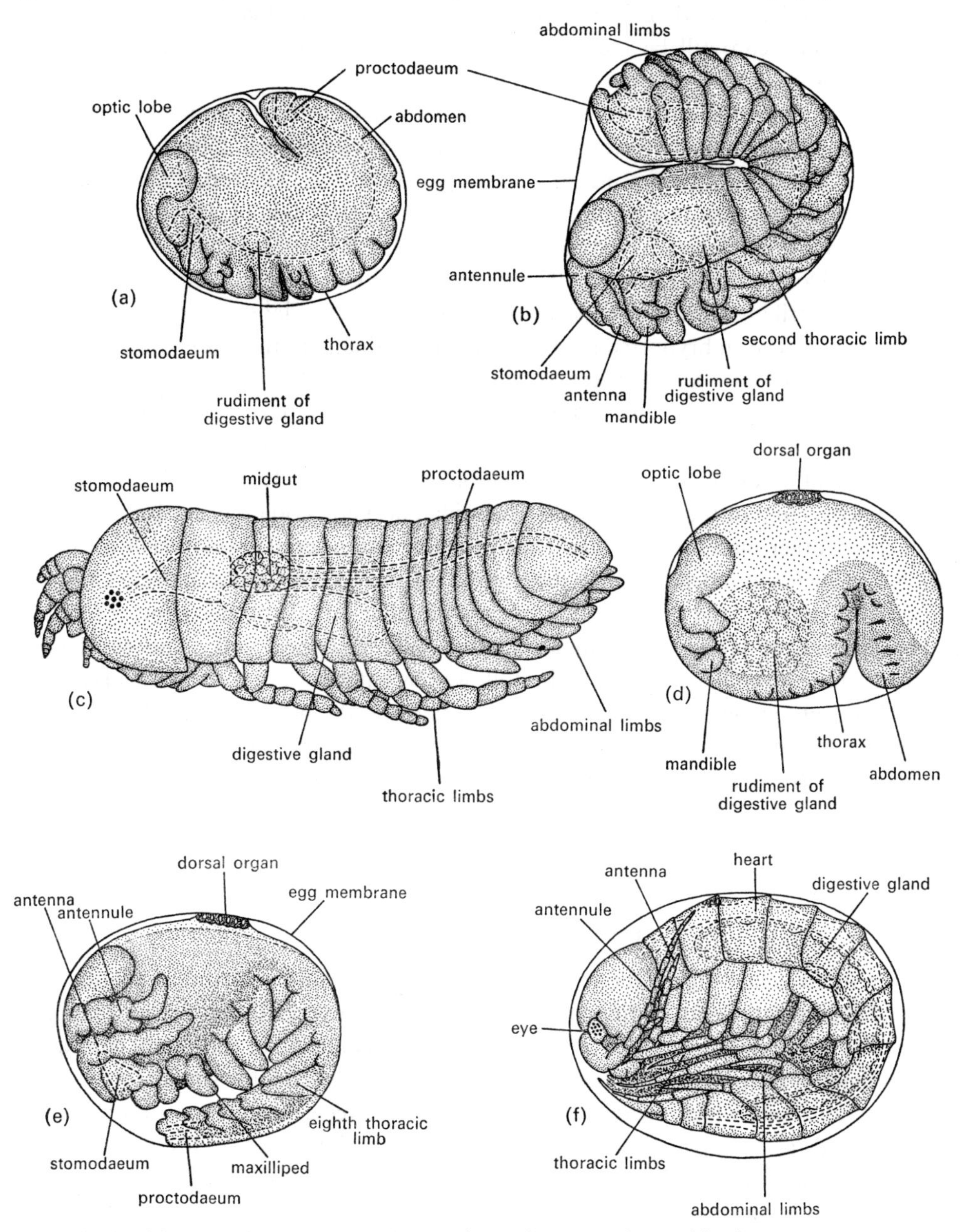

FIG. 125. (a)–(c) External development of the embryo of *Limnoria* (Isopoda), after Stromberg (1967). (a) Segment formation; (b) dorsal closure; (c) newly hatched embryo. (d)–(f) External development of *Gammarus* (Amphipoda), after Weygoldt (1958). (d) Segment formation; (e) dorsal closure; (f) fully developed embryo.

Pilumnus novae-zealandiae; the grapsid *Sesarma perraceae*; Wear, 1967; Soh Cheng Lam, 1969) and one family of freshwater crabs, the Potamidae, which hatches as juveniles (e.g. Bishop, 1967). A comparison between Fig. 123 and 124 shows that the mode of development in these species is indicative of a secondary embryonization of the zoeal phase of brachyuran development. This lends further support to the idea that ancestral larval stages in the Decapoda can still be recognized after their embryonization in association with increased yolk.

When we examine the direct development of other malacostracan groups for evidence of embryonized larval stages, however, a different picture presents itself. Direct development to hatching as a juvenile is a feature of the Leptostraca (e.g. Manton, 1934; Brahm and Geiger, 1966), Syncarida (e.g. Hickman, 1937; Bartok, 1944; Jakobi, 1954) and Peracarida (e.g. Manton, 1928; Jepson, 1965; Stromberg, 1965, 1967, 1971; Almeida, 1966; Berrill, 1969, 1971; Dohle, 1970; Klapow, 1970; Sheader and Chia, 1970), and is exemplified in Fig. 125. The striking feature of these malacostracans is the serial development of the post-naupliar region. The embryonized nauplius remains distinct, but there is no indication of the embryonization of later larval stages in the development of the post-naupliar region. Two interpretations are possible. One is that all traces of ancestral larval stages other than the nauplius have been lost. The alternative interpretation, which is the more likely in view of the clear retention of embryonized larval specializations in the Eucarida, is that the malacostracan ancestors of the Leptostraca, Syncarida and Peracarida did not have specialized larval stages other than the nauplius. They probably had a serial development of the post-naupliar segments in much the same generalized manner as the Cephalocarida. Only in the Eucarida, it appears, did an evolution of post-naupliar larval specializations precede the embryonization of larval stages in association with increased yolk. The Hoplocarida, which also have an embryonized nauplius followed by hatching as a specialized larva, should probably be interpreted in the same manner as the Eucarida, but too little is known of their embryonic development (e.g. Nair, 1941; Shiino, 1942) and larval development (e.g. Manning and Provenzano, 1963; Michel and Manning, 1972) at the present time to warrant further discussion. We can now examine the relationship between embryonization and organogeny in crustacean development.

Further Development of the Gut

In view of the great range of modes of development in Crustacea, including adaptations to yolk and specializations as feeding larvae, it is only to be expected that organogeny in this group of arthropods is highly variable. In the present context, we shall only consider the main outlines of the process. For the gut, these can be expressed quite simply.

The stomodaeum, which invaginates midventrally between the antennae, grows inwards and backwards as a simple epithelial tube and develops directly as the foregut epithelium. During functional differentiation of the foregut, which may take place before hatching of the nauplius larva or may be delayed to a later stage, even to the hatching of a juvenile with the full complement of segments and limbs, the stomodaeal epithelium secretes a chitinous lining to the foregut lumen.

The midgut has a more varied developmental history. When development results in the hatching of a feeding nauplius, the midgut of the nauplius is functionally differentiated as an epithelial tube connecting the foregut to the short hindgut (Fig. 118). The midgut tube may be uniform throughout its length, as in branchiopods, or may be further differentiated into an anterior midgut or stomach and a posterior midgut or intestine, as in cirripedes, copepods and ostracods. The mode of formation of the midgut epithelium in small crustacean embryos varies with the composition of the presumptive midgut and its mode of entry into the interior during gastrulation. In those species in which the presumptive midgut comprises one or two small cells which migrate into the interior during gastrulation (e.g. *Cyclops*, *Polyphemus*, euphausiids and penaeids), the cells divide when internal and give rise directly to the midgut epithelium. When the midgut rudiment is modified, in association with yolk, as a pair of large, yolky anterior midgut cells which become internal during cleavage and a pair of small, yolk-free posterior midgut cells which become internal during gastrulation (Anderson, 1969), the development of the midgut epithelium is little changed (Fig. 126). The yolky anterior midgut cells divide and give rise to the anterior part of the midgut epithelium. The posterior midgut cells proliferate to form a tube which initially extends forwards beneath the mass of yolky, anterior midgut cells, but later becomes connected with the posterior end of the epithelial sac arising from the anterior midgut cells. During its forward growth, the posterior midgut approaches the inner end of the stomodaeum, but the two tubes do not become connected to one another. In cirripedes with larger eggs, however, in which the anterior midgut cells are large and yolk-filled (e.g. *Pollicipes*, *Ibla*; Batham, 1946; Anderson, 1965), the posterior midgut tube establishes connection with the stomodaeum and the yolk-filled cells are excluded from the gut wall. The posterior midgut tube then gives rise to the stomach and intestine, while the yolk cells, which lie in the haemocoel, are gradually resorbed. This specialization is unusual among Crustacea. The typical midgut adaptation to yolk is a temporary vitellophage action of the midgut cells followed by further development of the same cells as epithelial cells of the midgut wall. In ostracods, for example, two groups of midgut cells invade the yolk mass, act as vitellophages and then differentiate as anterior midgut and posterior midgut epithelia (Fig. 127). The cells cluster initially in the centre of the yolk mass and gradually draw in and digest the more peripheral yolk (Weygoldt, 1960a). The majority of cladoceran embryos show a similar invasion of the yolk mass by a single group of midgut cells which act as vitellophages and then differentiate as midgut epithelium (Baldass, 1937, 1941). Of all the modes of development of the midgut in small crustacean embryos, the most generalized is probably that in which a single presumptive midgut cell becomes internal and gives rise directly to the midgut epithelium as a result of further divisions. The digestive gland in small crustacean embryos develops as an outgrowth of the midgut (e.g. Walley, 1969).

The greatest range of specializations in the development of the midgut epithelium and digestive glands is seen in the Malacostraca. Three major patterns have been recognized. The least specialized are the Syncarida, exemplified by *Anaspides* (Hickman, 1937). After the presumptive midgut has invaginated during gastrulation to form a yolky epithelium around a central cavity, the cells become massed together and the cavity is obliterated. At the surface of the mass of cells, a uniform epithelium of yolky cells then differentiates. The central cells enclosed by the epithelium fuse together and are gradually resorbed. The yolky epithelial sac extends along the length of the growing embryo and at the terminal

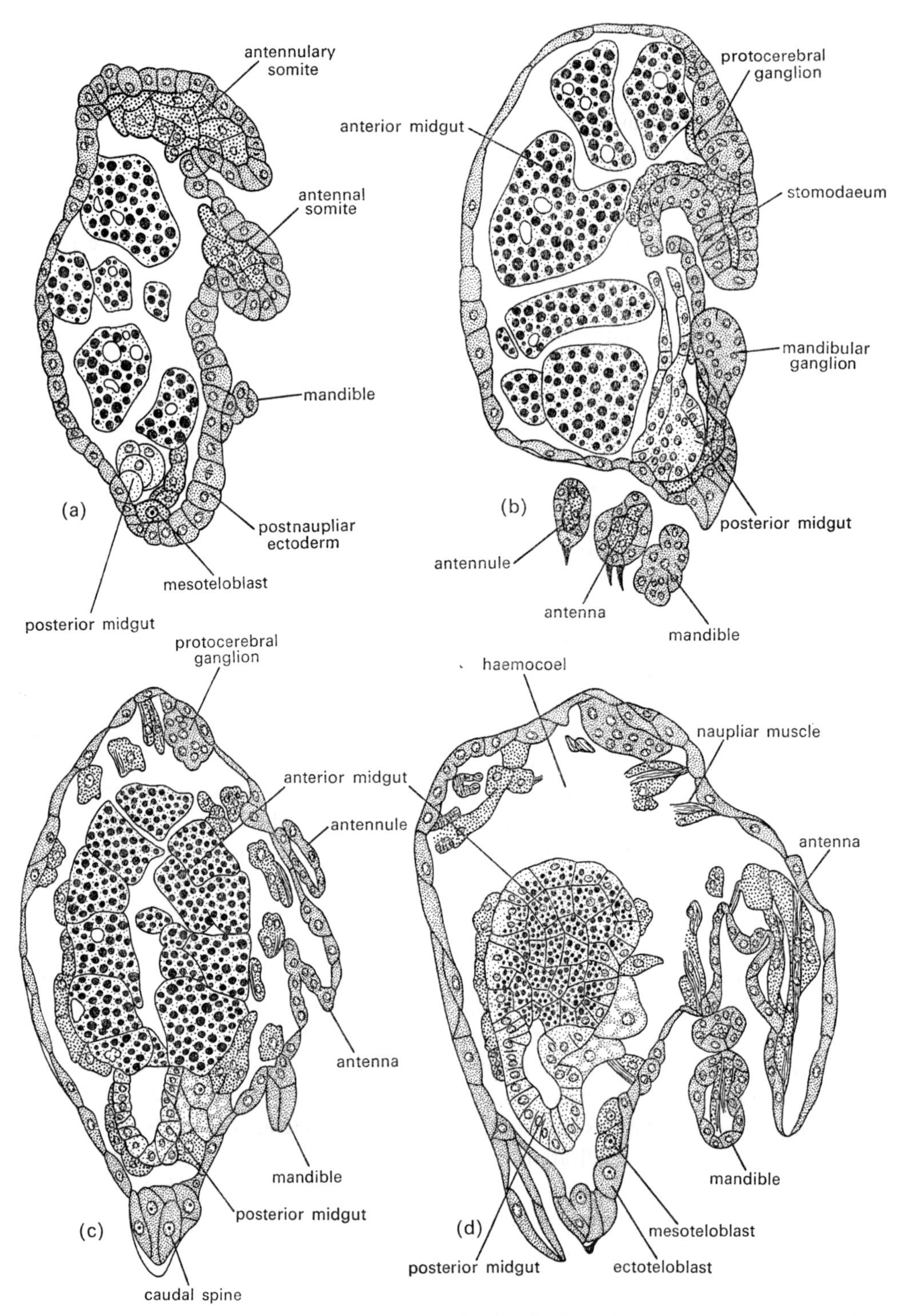

FIG. 126. (a)–(d) Development of midgut in *Tetraclita* (Cirripedia), after Anderson (1969), in longitudinal section. (a) Stage with ventrally directed naupliar limb buds as in Fig. 113b; (b) stage with dorsally directed naupliar limb buds as in Fig. 113c; (c) and (d) later embryo stages as in Figs. 113d and 113e.

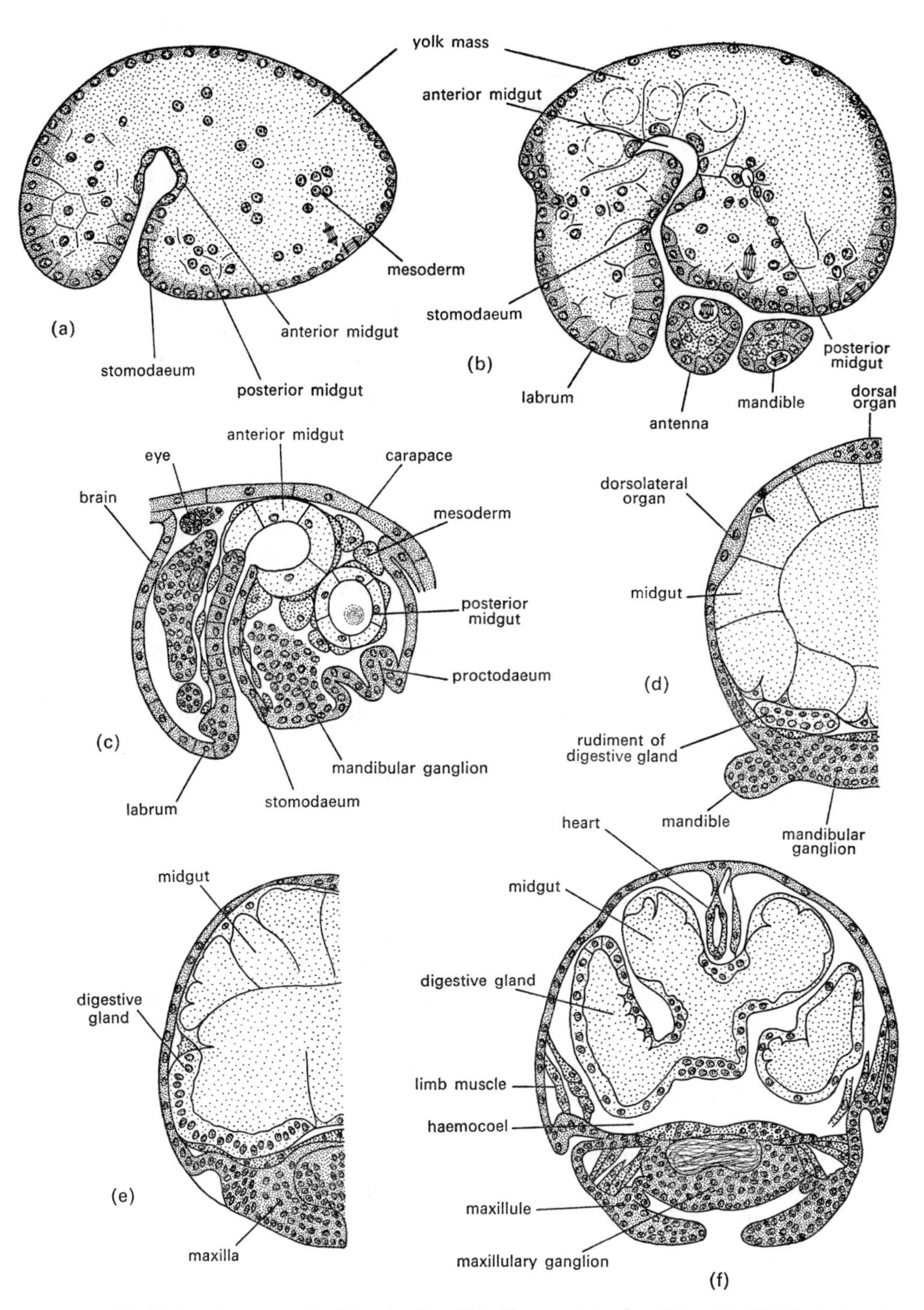

FIG. 127. (a)–(c) Development of midgut in *Cyprideis* (Ostracoda), after Weygoldt (1960), in longitudinal section. (a) Stage shown in Fig. 114b; (b) stage shown in Fig. 114c; (c) newly hatched nauplius stage. (d)–(f) Development of midgut in *Hemimysis* (Malacostraca), after Manton (1928), in transverse section. (d) Section through the mandibular segment shortly after formation of the rudiments of the digestive glands; (e) section through the second maxillary segment as the digestive glands enlarge and begin to become tubular; (f) section through the first maxillary segment of an embryo approaching hatching.

points of contact with the stomodaeum and proctodaeum, begins to differentiate as definitive epithelial cells. This process spreads from each end along the length of the sac, which thus transforms directly into the midgut epithelium. The digestive glands of *Anaspides* arise directly as outgrowths of the midgut wall. The posterior lobes bulge out at the level of the maxillary segment and grow back along the trunk. The anterior lobes arise bilaterally just behind the stomodaeal–midgut junction.

Anaspides provides an example among Malacostraca of a primitive adaptation of the malacostracan midgut to increased yolk, functionally analogous to that of clitellates (Chapter 3). Some cells are digested, while others act temporarily in intracellular yolk digestion and then differentiate as the midgut epithelium. In all other examples of malacostracan lecithotrophy, the presumptive midgut cells are first cut off as small cells at the surface of the yolk mass. Their subsequent development in relation to yolk digestion and midgut function is therefore more complicated and follows one of two routes.

In the Peracarida, all of the cells proliferated from the presumptive midgut invade the yolk mass and assume a temporary vitellophage role. In *Hemimysis*, cell proliferation from the presumptive midgut continues until most of the abdominal segments of the embryo have been formed (see Fig. 133b). The invading midgut cells move over the surface of the yolk mass, dividing as they go and then take up yolk and become enlarged as components of a vitellophage epithelium (Fig. 127). The formation of this epithelium proceeds slowly and is not completed until after the presumptive midgut area has ceased its proliferative activity. The central part of the yolk mass remains acellular. Once the vitellophage epithelium has been completed, the outer nucleated parts of the vitellophage cells in the regions of contact with the stomodaeum and proctodaeum begin to be cut off as a definitive midgut epithelium. This transformation spreads from both ends towards the middle of the midgut. By the time most of the yolk mass has been resorbed, the definitive midgut epithelium is complete.

Apart from the more extreme specialization in the mode of formation of the vitellophage epithelium, the development of the midgut in *Hemimysis* is similar in general to that of *Anaspides*. The formation of the digestive glands, however, proceeds in a different manner. The rudiments of the digestive glands arise precociously, in the form of a pair of ventrolateral cell masses external to the vitellophage epithelium in the mandibular segment (Fig. 127). The masses are segregated from the initial group of cells which proliferates from the presumptive mesodermal and midgut areas when gastrulation begins. It is difficult, therefore, to assign a specific origin to them. Manton (1928) interpreted the digestive gland rudiments of *Hemimysis* as mesodermal, but the results of more recent work on other peracaridan embryos (Stromberg, 1965, 1967) tend more to support their interpretation as early products of the presumptive midgut. The paired cell masses become united by a transverse bar of cells beneath the yolk mass and behind the stomodaeum. Each digestive gland rudiment then grows back along the side of the yolk mass and becomes tubular before uniting with the anterior part of the midgut.

Other peracaridan embryos show further modifications (Weygoldt, 1958, 1961; Scholl, 1963; Stromberg, 1965, 1967, 1971). The vitellophage epithelium of the isopods and tanaids develops in the same manner as in *Hemimysis*, but that of amphipods develops in a different way (Weygoldt, 1958). The vitellophages which invade the yolk mass soon degenerate, and the vitellophage epithelium arises by proliferation of cells from the rudiments of the digestive glands.

In all pericaridans, the rudiments of the digestive glands are set aside early in gastrulation, as a pair of cell plates apposed to the yolk mass. Those of the Tanaidacea and Amphipoda, like those of *Hemimysis*, lie in the mandibular segment of the embryo. Those of the Isopoda and probably the Cumacea lie in the maxillulary segment. At first, there are vitellophages beneath the digestive gland rudiments, but an early local degeneration of the vitellophages soon puts the digestive glands in direct contact with the yolk. The paired rudiments of the digestive glands also become united, as in *Hemimysis*, by a transverse, ventral bridge of cells. The rudiments then begin to extend anteriorly and posteriorly as paired lobes, into which the yolk gradually shifts as the vitellophage midgut sac shrinks. Only the amphipods retain a definitive midgut epithelium, formed by transformation of the vitellophage epithelium, as the central part of the gut. In isopods, tanaids and probably cumaceans, the vitellophage midgut epithelium is completely resorbed as the residual yolk passes into the digestive glands. The proctodaeum in these peracaridans grows longer as the midgut shrinks, eventually uniting with the stomodaeum and digestive glands to make up the definitive gut. In isopods and tanaids, therefore, the precocious segregation of the digestive gland components of the presumptive midgut is accompanied by a complete vitellophage specialization of the central midgut component.

The pattern of development of the midgut in the Leptostraca, Hoplocarida and many decapod Eucarida contrasts markedly with events in the Peracarida (Fig. 128; Manton, 1934; Krainska, 1936; Nair, 1941, 1949; Shiino, 1942, 1950; Aiyar, 1949; Fioroni, 1970 a, b; Lang and Fioroni, 1971). *Nebalia* provides a suitable example of this mode (Manton, 1934). Cell proliferation from the presumptive midgut area begins with the onset of gastrulation but continues only until the sixth post-naupliar segment (the third trunk segment) has been formed. The cells proliferated early in this process, before the naupliar segments have been individuated at the surface of the embryo, initially form a compact mass between the proliferative area and the surface of the yolk mass. As the formation of the post-naupliar segments proceeds, the proliferative area continues to bud off cells which fill the centre of each segment rudiment, resulting in the formation of a midgut strand along the midline of the anterior post-naupliar segments. With cessation of proliferation, the cells of the proliferative area slip into the interior and are added to the posterior end of the midgut strand.

The initial mass of midgut cells, and the strand of cells which forms behind it, behave in different ways during subsequent development. As soon as the initial mass of cells is formed, some of its cells begin to absorb yolk and migrate out in all directions over the surface of the yolk mass. Escape of cells from the initial cell mass continues in this way until the cells form a continuous vitellophage epithelium around the yolk mass. This component, the anterior midgut, has a temporary vitellophage function. The midgut strand proliferated along the centre of the anterior post-naupliar segments does not have this function. The strand develops a lumen and differentiates directly as a posterior midgut epithelium. In later development, the outer parts of the cells of the temporary vitellophage epithelium are cut off as definitive epithelial cells of the anterior part of the midgut. The paired lobes of the digestive gland arise as folds of the anterior midgut wall in the same generalized manner as in *Anaspides*.

Nebalia, therefore, has a less specialized mode of development of its digestive glands than the Peracarida, but a more specialized mode of development of the midgut. The early segregation of the midgut rudiment into anterior and posterior components, of which only the former performs a temporary vitellophage function before developing as definitive midgut epithelium, is a specialization convergently evolved in the Leptostraca, the Cirripedia (this chapter), certain polychaetes (Chapter 2) and yolky onychophorans (Chapter 4).

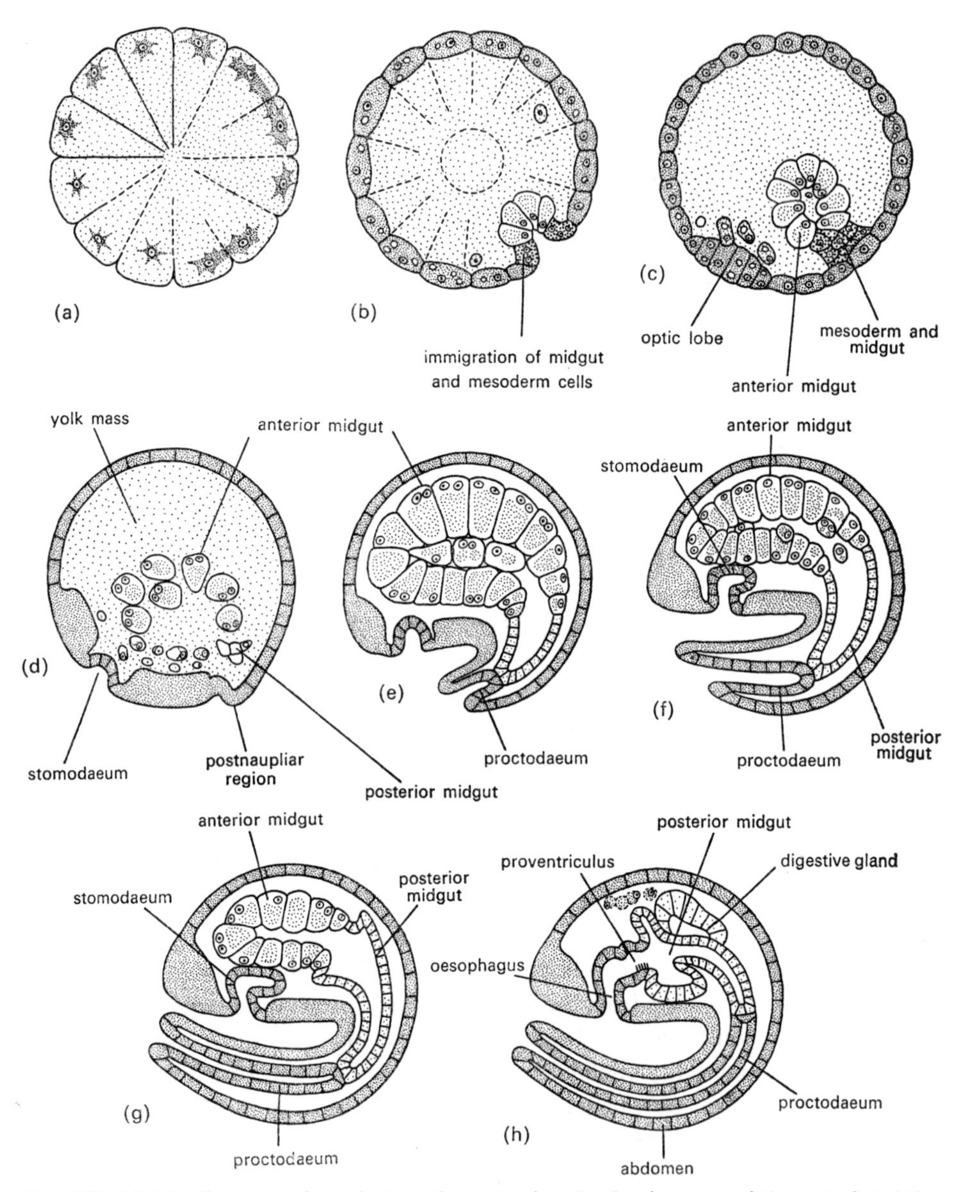

FIG. 128. (a)–(h) Diagrammatic sagittal sections showing the development of the gut of *Galathea* (Paguridea), after Fioroni (1970). (a) Cleavage and blastoderm formation; (b) early gastrulation; (c) later gastrulation; (d)–(h) sequence of formation of the stomodaeum (foregut), midgut and proctodaeum (hindgut).

The same solution to the problem of combining vitellophage function with midgut formation is seen in the Hoplocarida and in the majority of decapod Eucarida. In the latter, the vitellophage specialization of the anterior midgut is greater than in *Nebalia* and leads to a condition analogous to that of isopods. The anterior midgut rudiment of decapods gives rise to the digestive glands but contributes little, if at all (Fioroni, 1969, 1970 a, b; Lang and Fioroni, 1971) to the definitive midgut epithelium. In contrast to the isopods, however, the tubular posterior section of the decapod gut behind the junction with the digestive glands is formed mainly by the posterior midgut, with only a short, terminal proctodaeum. The corresponding part of the isopod gut is formed, as we have seen above, by elongation of the proctodaeum.

Although the development of the decapod midgut is functionally analogous to that of the Leptostraca, the details of the process differ. Proliferation of midgut cells from the midgut presumptive area begins precociously, immediately after the completion of blastoderm formation, and also ends early, when the formation of the post-naupliar segment is just beginning (Fig. 128). Most of the proliferated midgut cells invade the yolk mass, absorb yolk and move out to form a peripheral vitellophage epithelium. The last few cells to be proliferated from the presumptive midgut area aggregate as a small plate of cuboidal cells, contiguous at its edges with the vitellophage epithelium, on the posteroventral surface of the yolk mass. This plate is the posterior midgut rudiment. As post-naupliar segment formation proceeds, the centre of the plate grows out as a posterior midgut tube along the midline of the post-naupliar segments. Numerous cell divisions can be detected in the wall of the tube. Transformation of the vitellophage epithelium into definitive epithelium begins at two paired sites in the decapod embryo, posterolaterally at the junction with the posterior midgut tube and anterolaterally at the junction with the stomodaeum. Both regions bulge out as paired digestive gland rudiments. Epithelial transformation spreads from these sites as the digestive glands grow larger and the yolk mass shrinks, but the central part of the vitellophage epithelium of the anterior midgut is either partly or wholly resorbed.

In general, then, the development of the crustacean gut shows a variety of adaptations associated with yolk, as a result of which the midgut epithelium is either wholly or partly involved in yolk digestion before becoming the definitive midgut. It seems clear, however, that the ancestral crustacean midgut was formed in a simple manner by the repeated division of one or two small, yolky cells. The crustacean stomodaeum and proctodaeum are invaginated as simple epithelial tubes and become connected with the midgut during later embryonic development.

Further Development of the Mesoderm

We have seen in a previous section of this chapter that the crustacean presumptive mesoderm moves into the interior during gastrulation and clumps together as a group of cells internal to the post-naupliar ectoderm. From this mesodermal source, paired bands of mesoderm cells are proliferated forwards bilaterally beneath the naupliar embryonic ectoderm. The cells of the naupliar mesodermal bands then become grouped as three pairs of naupliar somites (Fig. 129), leaving a residual mass of post-naupliar mesoderm at the site of proliferation (e.g. Manton, 1928, 1934; Nair, 1939, 1949; Shiino, 1950; Weygoldt, 1958, 1960a; Scholl, 1963; Stromberg, 1965, 1967, 1971; Anderson, 1969). In front of the antennulary

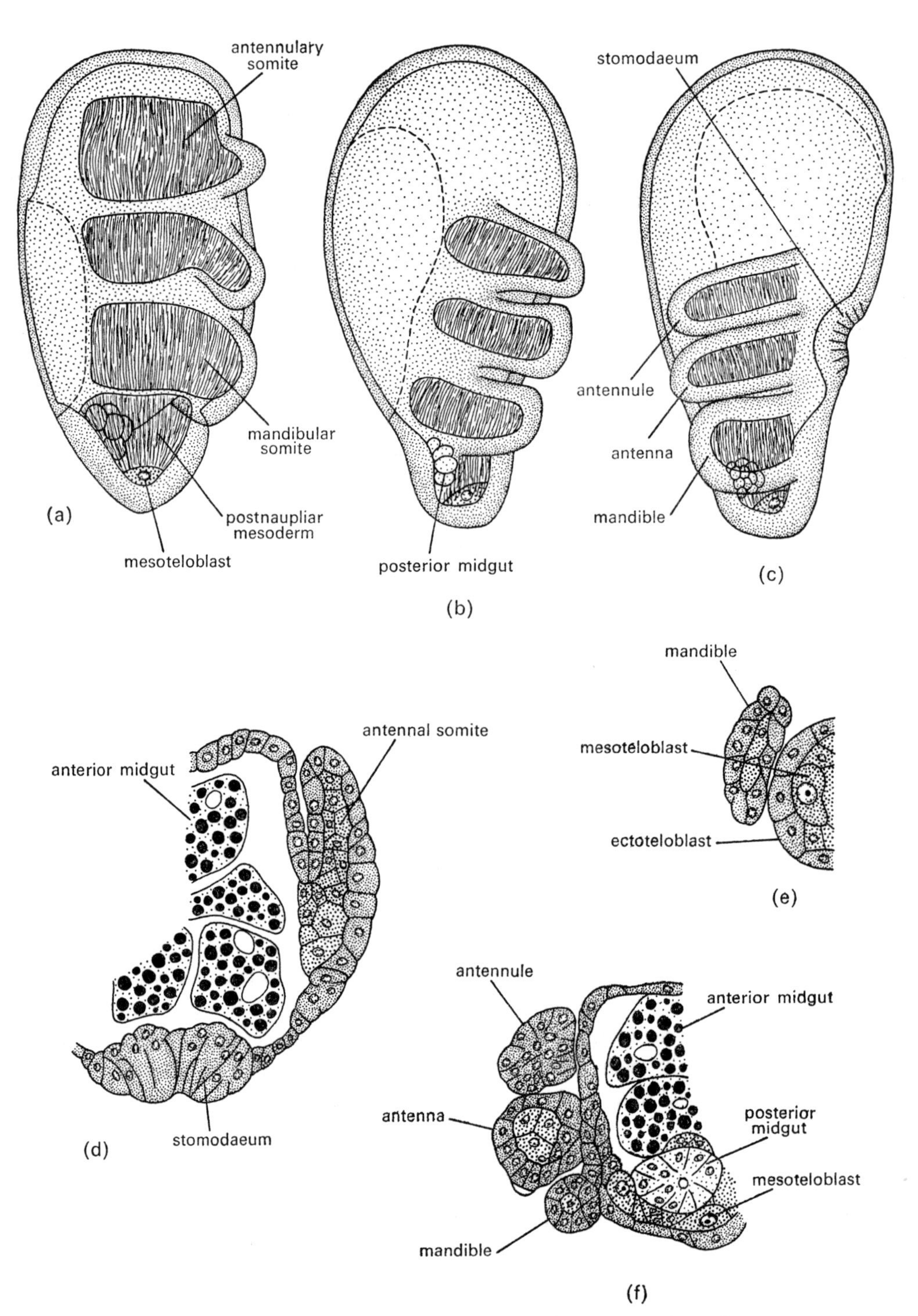

FIG. 129. (a)–(f) development of mesoderm in the embryo of *Tetraclita* (Cirripedia), after Anderson (1969). (a)–(c) Distribution of the naupliar somites and post-naupliar mesoderm during early development of the naupliar limb buds; (d) transverse section through the embryo shown in Fig. 129e, at the level of the stomodaeum; (e) transverse section through the embryo shown in Fig. 129e, at the level of the caudal papilla; (f) transverse section through the embryo shown in Fig. 113c, at the level of the caudal papilla.

somites, pre-antennulary mesoderm is formed. The origin of this mesoderm is variable and will be discussed in later paragraphs. Within the residual post-naupliar mesoderm, a number of cells enlarge to form mesoteloblasts, bilaterally arranged on either side of the ventral midline. The mesoteloblasts are the source from which all of the mesoderm of the post-naupliar segments is proliferated, except for some or perhaps all of the mesoderm cells of the maxillulary and maxillary segments (see Oishi, 1959, 1960; Weygoldt, 1961; Anderson 1965, 1969; Walley, 1969; Dohle, 1970; Stromberg, 1971). The remainder of the residual mesoderm, left after differentiation of the mesoteloblasts, persists at the posterior end of the growing trunk and gives rise to the telson mesoderm and proctodaeal mesoderm.

Two useful generalizations can be made about mesodermal development in the Crustacea. Firstly, the mode of formation of the mesodermal somites observed in species with small eggs and a naupliar larval development persists with little change in species with larger eggs and embryonized later development. Secondly, the external heteronomy of development of the naupliar and post-naupliar segments in Crustacea is also clear in their mesodermal development and persists even when the precocious functional differentiation of the naupliar mesoderm as larval muscles no longer takes place. These factors create some difficulties in attaining a general understanding of somite development and mesodermal organogeny in the Crustacea, since the development of the naupliar somites is specialized even in its basic condition in species which hatch as nauplii and the origin of the post-naupliar somites is specialized in all species. On the other hand, in some species, the specialized origin of the post-naupliar somites is followed by a relatively generalized mode of further development of these somites. On comparative grounds, in the absence of suitable evidence on the internal development of the Cephalocarida, the trunk segments of anostracan and conchostracan branchiopods provide the best example of a generalized somite development basically representative of the Crustacea as a whole (Cannon, 1924, 1926; Anderson, 1967; Benesch, 1969). It is useful to establish a preliminary understanding of this process (Figs. 130 and 132) as a background to the major variations in crustacean mesodermal organogeny.

Although some aspects of mesodermal development in the Branchiopoda are variable and others are still controversial, Anderson (1967) has shown that the trunk somites of the conchostracan *Limnadia* are proliferated from a pair of bilateral arcs of six to seven mesoteloblasts, situated beneath the ectoderm at the junction of the growth zone and telson. As somite formation proceeds, the number of mesoteloblasts increases to nine or ten on each side. The cells proliferated forwards from the mesoteloblasts are grouped at first as a pair of mesodermal bands at the sides of the growing trunk but, with further cell divisions, the newly formed parts of the mesodermal bands extend downwards to meet in the ventral midline. At the anterior border of the resulting U-shaped sheet of mesoderm, transverse bands of cells are segregated successively as U-shaped somite rings. At this stage, each somite ring is attached to the overlying trunk ectoderm and is separated by a wide haemocoelic space from the central, tubular midgut.

Whether the ventral conjunction of the two halves of the somite ring at this stage of somite development is a basic crustacean feature or a peculiarity of the Branchiopoda, is a matter of conjecture. In other crustaceans, as far as is known, the paired somites do not meet temporarily in the ventral midline. The important comparative point is that the somites

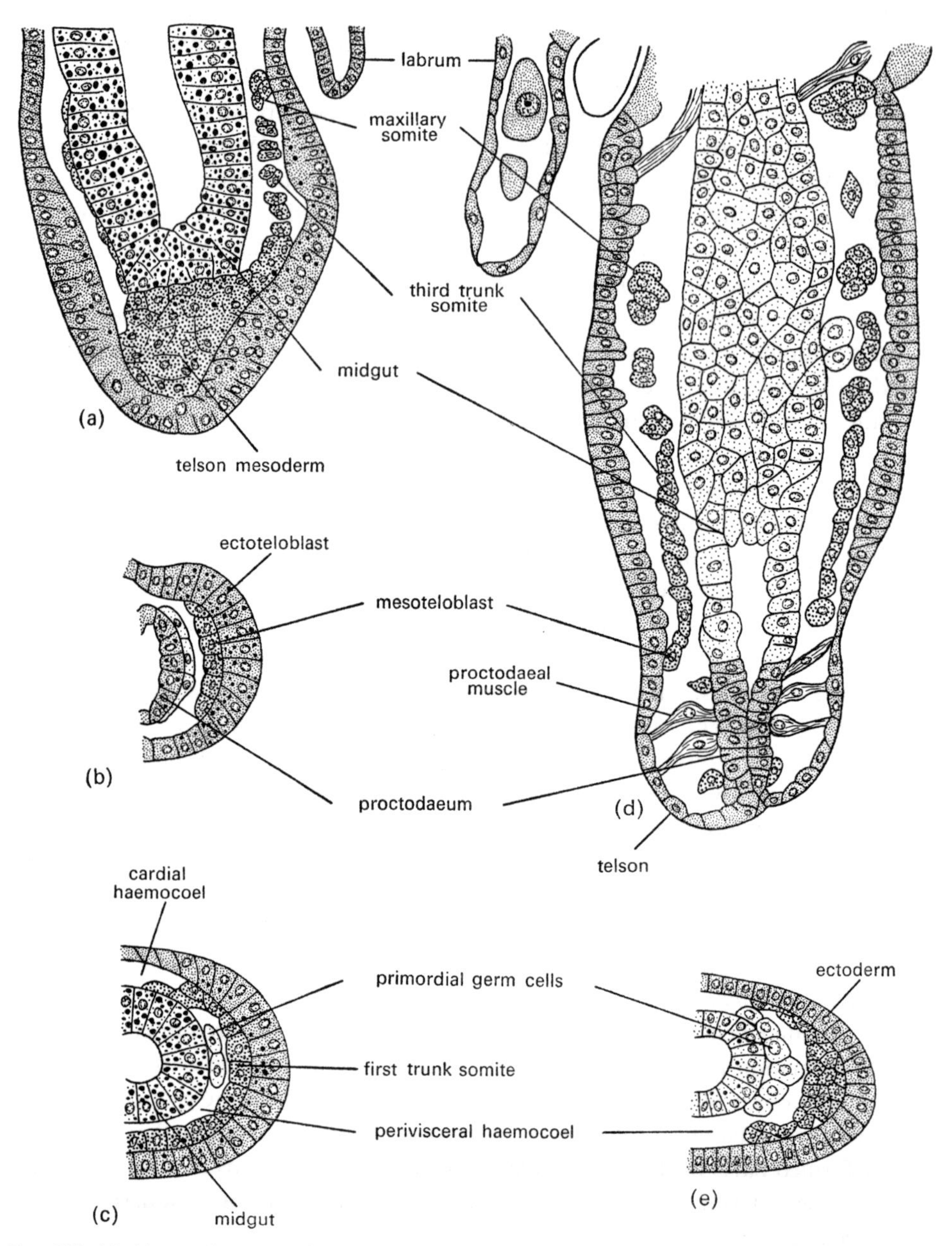

FIG. 130. (a)–(e) Development of post-naupliar mesoderm in the larva of *Limnadia* (Conchostraca). (a) Sagittal section through the postnaupliar region of the stage 1 nauplius; (b) transverse section at the level of the growth zone of the stage 1 nauplius; (c) transverse section at the level of the first trunk segment of the stage 1 nauplius; (d) frontal section through the postnaupliar segment of the stage 3 nauplius shown in Fig. 116b; (e) transverse section through the first trunk segment of the stage 3 nauplius.

in Crustacea are narrow transverse bands of cells and show no resemblance to the paired, hollow somites of other arthropods. The coelomic cavities of crustacean somites are never more than vestigial and are formed only when the somites have entered into further organogeny (see below).

The further development of each somite ring in the Conchostraca and Anostraca begins bilaterally near the dorsolateral ends of the ring. From each side, a transverse sheet of mesoderm cells extends medially across the intervening haemocoelic space to make contact with the dorsolateral surface of the midgut. In this way, a dorsal cardiac haemocoel above the midgut is separated from a wide perivisceral haemocoel around the midgut. A pair of small coelomic cavities is now usually formed by internal splitting within the thickened, dorsolateral region of each somite half.

When this stage is reached, the development of each somite half becomes more complicated. Several events proceed simultaneously. Dorsolaterally, the somite tissue separates from the overlying ectoderm, creating a space which becomes the pericardial haemocoel. The transverse sheet of somite tissue between the ectoderm and midgut persists as the floor of the pericardial haemocoel. With the formation and enlargement of the latter, the coelomic cavity in the underlying segmental component of the pericardial floor is gradually obliterated. At the same time, a mass of cells is proliferated at the lateral, upper edge of the pericardial floor and separates from this edge to lie attached to the ectoderm in the pericardial cavity. This cell mass is a segmental rudiment of the dorsal longitudinal muscle of its side.

Medially above the gut, the upper and lower median edges of the pericardial septum grow towards the midline to meet their counterparts from the other half of the segment. The upper edge grows in contact with the overlying dorsal ectoderm. The lower edge grows in contact with the dorsal surface of the midgut. As a result, when the growing sheets of tissue meet in the midline, they enclose the cardiac haemocoel in a cylinder of mesoderm which forms the wall of the heart. The paired, lateral ostia of the heart are persistent intersegmental spaces between successive segmental components of the heart wall. The mesoderm which forms the ventral part of the heart wall also gives rise to the dorsal part of the splanchnic mesoderm on the surface of the midgut. The remainder of the splanchnic mesoderm is formed by the downward proliferation of a sheet of tissue over the midgut surface from the conjunction of the dorsal splanchnic mesoderm and the median edge of the pericardial floor.

We are left, finally, with the main ventrolateral mass of each somite half, beneath the level of the pericardial floor. This mass is attached to the ectoderm and invades the limb bud of its segment half when the latter is formed. The mass of cells breaks up into subunits which give rise to the intrinsic muscles of the limb, the extrinsic limb muscles, other somatic segmental muscles and, ventromedially, a segmental component of ventral longitudinal muscle. Cannon (1924, 1926) and Manton (1928, 1934) gave evidence of an ectodermal origin of certain of the somatic segmental muscles in Crustacea, but this interpretation has not been further tested in recent studies on the subject.

The distinctive pattern of mesodermal organogeny described in the preceding paragraphs is serially repeated, with minor variations, along the length of the conchostracan and anostracan trunk. As Figs. 131 and 132 show, the same fundamental pattern persists in the trunk segments of the Malacostraca (Manton, 1928, 1934; Nair, 1949; Shiino, 1950; Weygoldt,

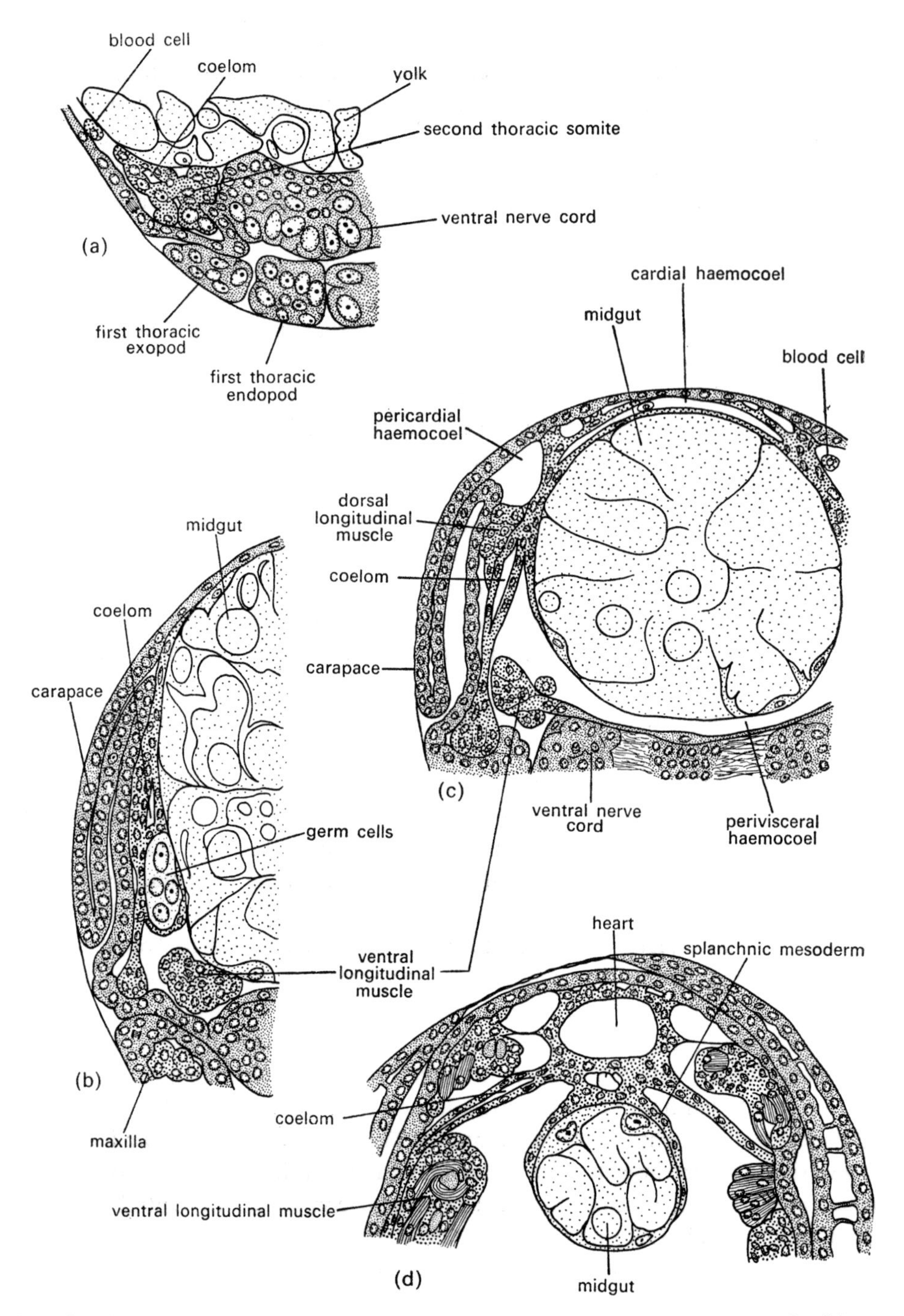

FIG. 131. (a)–(d) Development of post-naupliar mesoderm of *Hemimysis* (Malacostraca), after Manton (1928), in transverse section. (a) Second thoracic segment, showing early development of segmental somite; (b) first thoracic segment, showing dorsal upgrowth and further differentiation of somite; (c) second thoracic segment, showing formation of the heart; (d) sixth thoracic segment, after completion of the heart at this level.

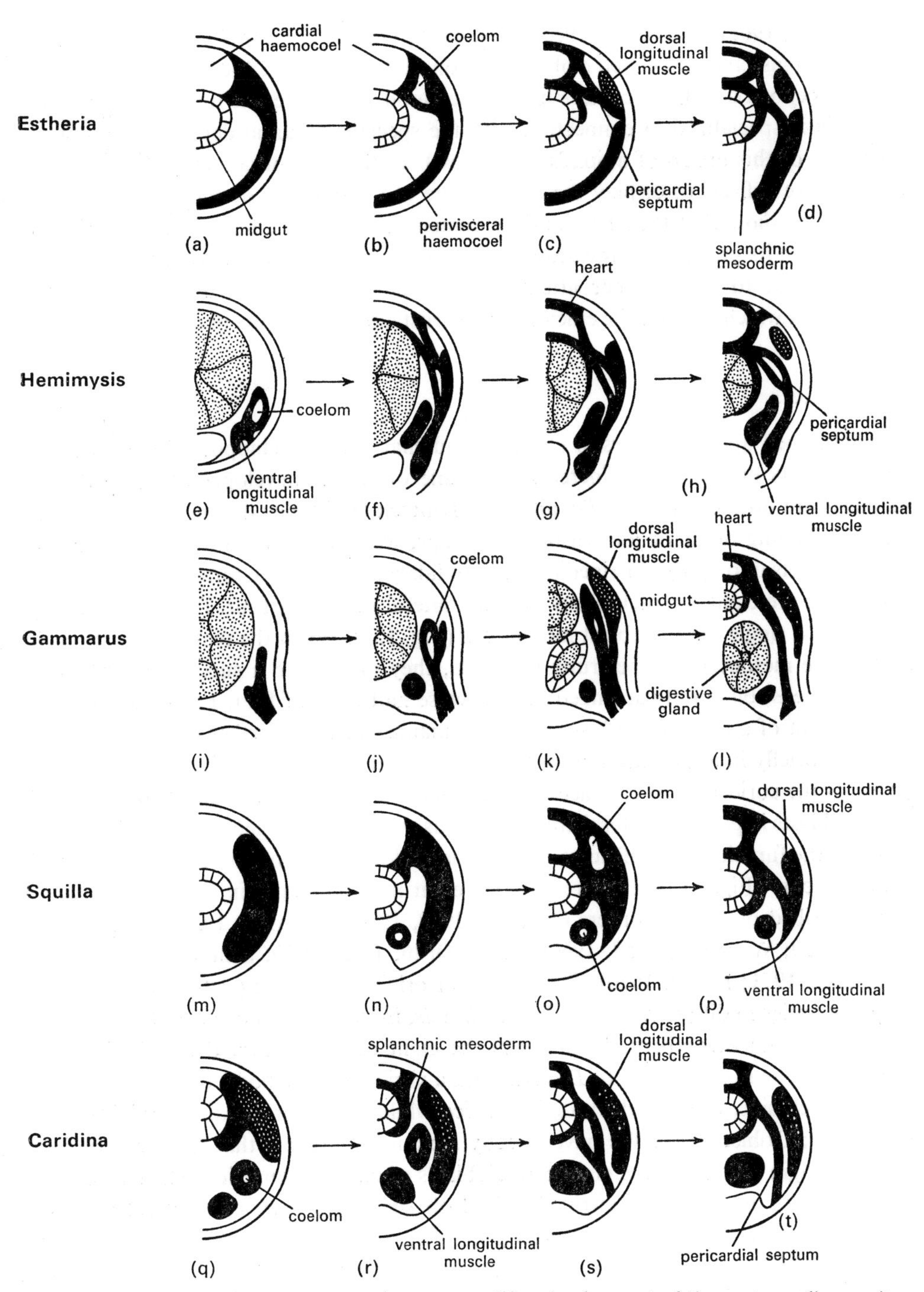

FIG. 132. Diagrammatic transverse sections summarizing development of the post-naupliar somite mesoderm in Crustacea, after Weygoldt (1958). (a)–(d) *Estheria* (Conchostraca); (e)–(h) *Hemimysis* (Peracarida, Mysidacea); (i)–(l) *Gammarus* (Peracarida, Amphipoda); (m)–(p) *Squilla* (Hoplocarida); (q)–(t) *Caridina* (Eucarida, Decapoda).

1958; Scholl, 1963; Stromberg, 1965, 1967, 1971; Dohle, 1970), even in the presence of a larger yolk mass and associated embryonization of development. Mesodermal development in the Copepoda and Ostracoda is little understood at the present time, while that of the Cirripedia is specialized in connection with the specialized larval development and metamorphosis of this group of animals (see below). In the present context, the evidence discussed above suffices to display the basic theme of mesodermal development in the Crustacea and to show that their somite development is uniquely different from that of other arthropods in its progression towards a convergently similar adult morphology. It is necessary, however, to give further attention to certain aspects of crustacean mesodermal development which bear on the elucidation of crustacean development as a whole.

Mesoteloblasts

As described above, Anderson (1967) found that the Branchiopoda have six or seven mesoteloblasts on each side, increasing to nine or ten as development proceeds. These teloblasts are distinct in *Limnadia*, although Benesch (1969) has doubted their presence in *Artemia.* Among other major crustacean groups, we have significant information on the mesoteloblasts of only the Cirripedia and the Malacostraca. In the former (Anderson, 1965, 1969) there are eight mesoteloblasts in two bilateral groups of four. Each group arises by division of a single mesoteloblast mother cell (Fig. 129). When the mesoteloblasts begin to bud off the mesoderm of the trunk segments, they proliferate a succession of transverse rows of 8 cells, bilaterally arranged in transverse half-rows of four. Each row of 8 cells is the rudiment of a pair of trunk somites. The manner of their further development will be discussed briefly in a later section on the mesodermal aspects of cirripede metamorphosis.

One of the curiosities of crustacean development is that the same specialization of eight mesoteloblasts in paired, bilateral arcs of four has evolved convergently in the Malacostraca, although the functional significance of eight mesoteloblasts, as opposed to another number, is still obscure (Fig. 133). The presence of eight mesoteloblasts in the Malacostraca has been described by many workers (Manton, 1928, 1934; Krainska, 1936; Hickman, 1937; Shiino, 1950; Weygoldt, 1958, 1960b, 1961; Oishi, 1959, 1960; Scholl, 1963; Stromberg, 1965, 1967, 1971; Dohle, 1970). Their mode of origin within the post-naupliar mesoderm rudiment varies in different species. In *Hemimysis*, isopods and *Gammarus* (Manton, 1928; Weygoldt, 1958; Stromberg, 1965, 1967) the eight cells are formed by division of two mother cells, as in cirripedes. In *Diastylis* and a variety of decapods, there are two mother cells on each side, one of which divides twice while the other remains undivided (Oishi, 1959, 1960; Dohle, 1970). We have already seen in Chapter 3 that similar variations in mode of origin obtain for the eight ectoteloblasts of clitellate annelids. The malacostracan mesoteloblasts, like those of cirripedes, bud off successive rows of eight cells, each row being the rudiment of a pair of somites (Fig. 133). The number of mesoteloblasts and the manner in which they give rise to the post-naupliar somites in the Cephalocarida, Ostracoda, Copepoda, Branchiura and Mystacocarida has yet to be accurately determined.

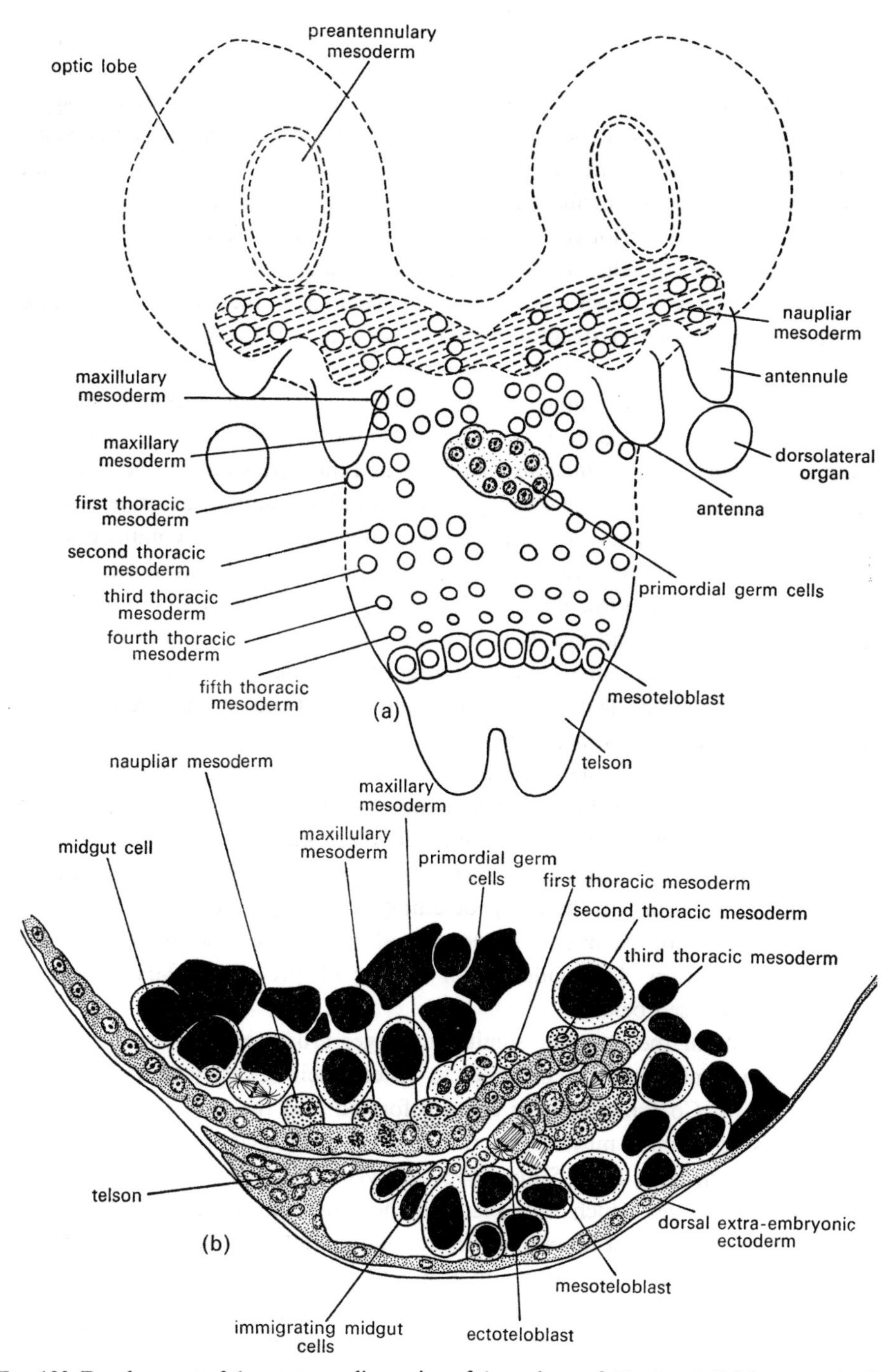

FIG. 133. Development of the post-naupliar region of the embryo of *Hemimysis* (Malacostraca), after Manton (1928). (a) Diagrammatic ventral view of the germ band early in the growth of the caudal papilla, showing the distribution of mesoderm at this stage; (b) sagittal section through the forwardly flexed caudal papilla of the same embryo.

Naupliar somites

The further development of the naupliar somites in embryos which hatch as nauplii has been studied mainly in the Cirripedia and the anostracan Branchiopoda. Anderson (1969) showed that the naupliar somites of the cirripede embryo give rise to the musculature of the three naupliar segments, including the intrinsic and extrinsic muscles of the naupliar limbs (Figs. 126 and 129). Prior to their dispersion and differentiation as musculature, the somites are solid blocks of cells, which extend an appendicular lobe into the corresponding limb bud, but do not develop coelomic cavities or participate in the formation of the heart or pericardial septum. The latter structures are not present in the cirripede nauplius. A thin layer of circular splanchnic muscle covers the external surface of the midgut of the cirripede nauplius, but its origin as splanchnic mesoderm has not yet been established. The elaborate somatic musculature developed from the naupliar somites was described in detail by Walley (1969) for the nauplius larva of *Balanus balanoides* and illustrates well the point made at the beginning of this chapter, that the development of small crustacean embryos is a complex process. The extrinsic muscles of the three pairs of naupliar limbs comprise a series of dorsal muscles attached to the carapace of the larva and a series of ventral muscles attached to ventral, skeletal endosternites. The dorsal muscles consist of four pairs to the antennules, ten pairs to the antennae and five pairs to the mandibles. The ventral muscles consist of one antennulary pair, six antennal pairs and three mandibular pairs. In addition, there are muscles which run between the dorsal carapace and ventral endosternites and others which run between the endosternites themselves.

The development of a complex naupliar musculature from the three pairs of naupliar somites has also been described in detail by Benesch (1969) for the anostracan *Artemia*. In addition, Benesch has identified a set of pre-antennulary muscles which appear to indicate the presence of a pair of vestigial pre-antennulary somites in the Branchiopoda. These somites cannot be identified in the cirripede embryo, although median mesoderm derived from the anterior ends of the naupliar mesodermal bands gives rise to the muscles of the stomodaeum and labrum. The evidence for a pair of pre-antennulary somites in the Malacostraca is discussed below.

The development of the naupliar somites in larger crustacean embryos, in which the naupliar region is embryonized, continues to reflect their ancestry as components of a functional nauplius. Coelomic cavities are not formed and the major product of somite differentiation is the somatic musculature of the three segments, antennulary, antennal and mandibular (Manton, 1928, 1934; Shiino, 1950; Weygoldt, 1958; Scholl, 1963; Stromberg, 1965, 1967). The changes which take place in the musculature of these three segments between the stage of their first functional differentiation and the stage of completion of the adult configuration of the musculature are little understood. Many patterns of change can be expected, commensurate with the variety of functional sequences expressed in the development of the naupliar region in the embryos and larvae of different species. The extent of possible change is elegantly displayed in the work of Walley (1969) on cirripede metamorphosis.

Mesodermal aspects of cirripede metamorphosis

During this process, the naupliar limb muscles and labral muscles are all histolysed and their fragments are digested by phagocytic haemocytes. The antennal and mandibular segments become vestigial and do not redevelop a functional musculature in the cypris stage, but the antennulary segment develops a new set of ten pairs of antennulary extrinsic limb muscles, entirely different in their functional configuration from those of the nauplius. Other new muscles also develop in the naupliar region, in association with the formation of the bivalve carapace. The source of the cells which give rise to the cypris muscles is at present unknown. The functional transition is achieved through the formation of the cypris muscles as rudiments in the stage VI nauplius, while the naupliar musculature is still intact, and by a rapid histolysis of naupliar muscles and simultaneous onset of function of the cypris muscles at the nauplius–cypris moult.

As we have already seen in an earlier section of this chapter, the onset of function of the trunk segments in cirripedes also coincides with metamorphosis from the stage VI nauplius to the cypris. The onset of maxillulary and maxillary function is even further delayed, until the cypris has settled, and need not concern us at the present time. In the trunk segments, the musculature of the six pairs of trunk limbs is formed from the segmental somites in the usual way, although these somites do not give rise to heart, pericardial floor or gonad components. The limb musculature of the trunk differentiates mainly in the stage VI nauplius, at the same time as the antennulary and carapace muscles of the cypris, but does not become functional until the nauplius–cypris moult has taken place. There is no evidence that the trunk somites contribute to the splanchnic mesoderm of the cirripede gut in the usual crustacean manner. The splanchnic musculature formed during embryonic development, before the trunk somites have been proliferated, appears to persist in a modified form after metamorphosis.

Pre-antennulary mesoderm

Whether the mesodermal components anterior to those derived from the antennulary somites should be assigned a segmental origin in the Crustacea is a question of long standing. The evidence for the presence of a pair of pre-antennulary somites in the embryos of species which hatch as nauplii is very limited. Cirripede embryos do not have these somites (Anderson, 1969). Their pre-antennulary mesoderm, derived from the anterior ends of the mesodermal bands, differentiates into labral and stomodaeal musculature and shows no trace of the pattern of differentiation characteristic of the naupliar somites. On the other hand, Benesch (1969) has found that the pre-antennulary musculature of the nauplius of *Artemia* is differentiated in a recognizable segmental pattern, even though the segment has no limbs.

The development of the Malacostraca offers the same kind of conflicting evidence. Pre-antennulary mesoderm in the Malacostraca again gives rise to the labral and stomodaeal musculature, but in addition gives rise to the anterior aorta above the stomodaeum. In *Anaspides* and *Heterotanais*, the pre-antennulary mesoderm is said to arise from the anterior ends of the naupliar mesodermal bands (Hickman, 1937; Scholl, 1963). More usually, it originates as a pair of cell masses (pre-antennulary somites) formed independently of the mesodermal bands. This phenomenon was first described in detail by Manton (1928) in *Hemimysis* and has subsequently been confirmed by many workers on other species (*Nebalia*, Manton, 1934; *Squilla*, Nair, 1941; *Mesopodopsis*, Nair, 1939; *Neomysis*, Petriconi, 1968; *Asellus*, Weygoldt, 1960b; *Idotea*, *Limnoria*, Stromberg, 1965, 1967; *Gammarus*, Weygoldt, 1958; *Caridina*, Nair, 1949; *Palaemon*, Aiyar, 1949; *Palaemonetes*, Weygoldt, 1961; *Palinurus*, Shiino, 1950; *Leptograpsus*, Anderson, unpublished). The two cell masses are formed by the immigration of two groups of cells from the protocerebral ectoderm, which subsequently gives rise to the optic lobes. The pre-antennulary somites, therefore, are developed in these species independently of the presumptive mesoderm from which all other somites arise. In view of the origin of pre-antennulary mesoderm from the anterior ends of the mesodermal bands in some examples, it is tempting to interpret the malacostracan "pre-antennulary somites" as a secondary feature and to doubt their validity as somites of a basic pre-antennulary segment. Yet, as Stromberg (1971) points out, a cautious approach is required. He has found that the vestigial pre-antennulary mesoderm of epicaridean isopods is derived from the anterior ends of the mesodermal bands, but interprets this as a secondary condition. At the present time, therefore, the segmental status of the pre-antennulary mesoderm of Crustacea is unresolved.

The further development of the pre-antennulary mesoderm of the Malacostraca can be exemplified by reference to *Hemimysis*. The account given by Manton (1928) has recently been substantiated in the work of Petriconi (1968) on *Neomysis*. Once internal beneath the protocerebral ectoderm, the two groups of pre-antennulary mesoderm cells proliferate and grow back as solid strands, which become apposed to one another along the midline of the embryo and spread out posteriorly over the surface of the stomodaeum. Posterodorsally, the paired strands of pre-antennulary mesoderm extend upwards to meet the anterior end of the heart. Median separation results in the formation of the anterior aorta. The remainder of the pre-antennulary mesoderm forms musculature of the labrum and stomodaeum.

Two pairs of cavities develop in the pre-antennulary mesoderm of mysids. One pair is formed early, in front of the stomodaeum, and is soon obliterated. The second pair develops later at the sides of the stomodaeum and persists for a longer time. In other genera, only the equivalent of the second pair of cavities develops. They are customarily called the coelomic cavities of the pre-antennulary somites.

Segmental organs

The existence of coelomoduct segmental organs in Crustacea was recognized towards the end of the nineteenth century, but the details of their development have only been confirmed in recent investigations. In addition to the well-known pairs specialized as excretory organs in the antennal and maxillary segments, it now seems possible that vestigial segmental organs develop in a number of other segments (Benesch, 1969). The gonoducts of Crustacea are also of coelomoduct origin, as will be seen in the next section.

A pair of antennal segmental organs develops in the embryo of all crustaceans and becomes functional as larval excretory organs in those species which hatch as nauplii. In the majority of groups, the antennal segmental organs are transitory and are replaced by the maxillary segmental organs as development continues. This is found in the Branchiopoda,

Ostracoda, Cirripedia, Copepoda and Branchiura, and also occurs in the Leptostraca, Syncarida, Hoplocarida and certain kinds of Peracarida (the Cumacea, Tanaidacea and Isopoda). The Mysidacea, Amphipoda and Eucarida, in contrast, retain functional antennal segmental organs in the adult and develop a maxillary pair either transitorily or not at all.

The development of the antennal segmental organs from cells of the antennal somites has been described by workers on several, widely scattered species (Branchiopoda, Benesch, 1969; Malacostraca, Manton, 1928, 1934; Weygoldt, 1958; Stromberg, 1965, 1967). The gland usually develops from two groups of antennal mesoderm cells. One group hollows out and gives rise to the end sac, while the other group of cells forms the duct. Only the short, terminal exit duct is formed by ectoderm. The development of the maxillary segmental organs has been studied in more detail (Branchiopoda, Lebedinsky, 1891; Samassa, 1893; Cannon, 1924, 1926; Benesch, 1969; Malacostraca, Manton, 1934; Hickman, 1937; Scholl, 1963; Stromberg, 1965). Once again, the gland usually develops from two components derived from the mesoderm of the maxillary somite, a dorsal component which forms the end sac and a lateral component which forms the coiled duct. The duct opens on the maxilla by a short, ectodermal exit duct.

The fact that the antennal and maxillary glands of crustaceans are coelomoduct segmental organs is thus indisputable. The tentative identifications of vestigial segmental organs in other segments in several crustacean embryos are less well founded. The question is discussed by Benesch (1969), who has identified segmental organs in the mandibular and thoracic segments of *Artemia*, in the form of mesodermal cell masses derived from the segmental somites. The cell masses are sites of proliferation of haemocytes, but whether they should be interpreted as vestigial segmental organs is open to doubt.

Gonads and gonoducts

The development of the gonads in Crustacea always begins with the differentiation of primordial germ cells, but the site of origin of these cells and the timing of their first visible differentiation is highly variable among species. A segmental distribution of primordial germ cells is found in *Nebalia* and *Anaspides* (Manton, 1934; Hickman, 1937). In these malacostracans, a pair of germ cells differentiates in the ventral walls of each pair of segmental coelomic cavities in the pericardial septum. The germ cells, which have characteristic pale-staining cytoplasm and nuclei with dense chromatin granules, proliferate to form small groups of cells, which link up as paired strands suspended from the under surface of the pericardial floor. The segmental coelomic cavities in the pericardial floor are obliterated during this process. Each strand of germ cells is invested by mesodermal cells derived from the pericardial floor and hollows out by internal splitting. The cavity of the gonad is thus formed independently of the vestigial coelomic cavities.

In other Malacostraca, the gonad rudiments are more compact and temporarily occupy the median faces of the maxillary or first trunk somites before they extend back beneath the pericardial floor. The first visible differentiation of the germ cells may be at this site, as in *Squilla* and the decapods (Nair, 1941, 1949; Shiino, 1950; Weygoldt, 1961), but may occur earlier in development, as in the Peracarida. In the Isopoda, Amphipoda and Tanaidacea (Weygoldt, 1958, 1960b; Scholl, 1963; Stromberg, 1965, 1967) the primordial germ cells first become differentiated during gastrulation, in the form of a group of internal cells in the ventral midline between the presumptive mesoderm and presumptive midgut. During the proliferation of the post-naupliar segment rudiments, the germ cells remain associated with the mesoteloblasts until the mesoderm of the first pair of thoracic somites is formed. They then attach in two bilateral groups to the median faces of the first thoracic somites and are carried upwards to a dorsolateral position beneath the pericardial flood as the somites continue their development. The same mode of development of the genital rudiments obtains in *Hemimysis* (Manton, 1928; Fig. 133), but here the first visible differentiation of the primordial germ cells is attained precociously, as a midventral presumptive area in the blastula, between the presumptive mesoderm and presumptive midgut (Fig. 111). The cells become internal at an early stage of gastrulation (Fig. 112).

The precocious differentiation of a single primordial germ cell in the wall of the blastula has been observed in several small-egged crustaceans (Fig. 111). In the cladocerans *Holopedium* and *Polyphemus* and the copepod *Cyclops*, the primordial germ cell can be identified as blastomere 2D, lying between the presumptive mesoderm and presumptive midgut in the ventral midline (Fuchs, 1914; Kühn, 1914; Baldass, 1937). A similar early segregation of a primordial germ cell may occur in euphausiids (Taube, 1909, 1915). Little attempt has yet been made to trace the development of the gonads from this cell, but there is evidence (Grobben, 1881; Amma, 1911) that it moves into the interior during gastrulation and divides into two cells which become associated with the post-naupliar growth zone. In copepods, the two cells then come to lie dorsolaterally in the last thoracic segment. Here, each primordial germ cell divides into a large cell and several small cells. The large cells, with further divisions, form the gonad, a median dorsal structure above the gut. The small cells on each side in the last thoracic segment give rise to the paired gonoducts. The latter is a specialization, since the gonoducts are usually developed as mesodermal coelomoducts of the genital segment, growing out from the pericardial floor.

In the anostracan and conchostracan branchiopods, as in the decapod Malacostraca, the primordial germ cells first become differentiated in the form of two groups of cells on the median faces of the first trunk somites (Fig. 130; Cannon, 1924; Anderson, 1967; Benesch, 1969). After becoming associated with the pericardial floor in this segment, each group of germ cells then proliferates as a solid rod along the pericardial floor and hollows out to form a gonad. The development of the gonads in the Branchiopoda thus has much in common with that of the peracaridan Malacostraca. The origin of the primordial germ cells and development of the gonads in the Cirripedia is still unresolved, but must occur late in development. No primordial germ cells are present in the blastula and none can be detected in the embryo, nauplius larval stages or cypris larva (Anderson, 1969; Walley, 1969).

The variations in the mode of development of the gonads in the Crustacea preclude the formation of a generalized statement on this topic. It is clear that the primordial germ cells are closely associated with the mesoderm, but vary in their time of first visible differentiation. The cavities of the gonads develop independently of the segmental coelomic cavities, but the gonoducts are coelomoducts of a variable genital segment. It is not clear, therefore, whether the generalized, cephalocaridan-like ancestors of modern Crustacea had an early or late segregation of primordial germ cells, a derivation of gonads from segmental, hollow somites, or more than one pair of coelomoduct gonoducts.

Further Development of the Ectoderm

One of the most useful results of a comparison of fate maps in the Crustacea (Fig. 111) is the recognition of a stable pattern of sub-areas within the presumptive ectoderm. This pattern is of interest in two ways. Firstly, it shows that the basic ectodermal components of the nauplius are already established at the blastula or blastoderm stage. There is a median protocerebral area anteriorly, paired antennulary, antennal and mandibular areas laterally and a median, postnaupliar area posteriorly. The further development of the ectoderm can therefore be discussed in terms of these areas, whether the outcome at the hatching stage is a nauplius larva or a fully formed juvenile. Secondly, the pattern displays a dorsal area of presumptive ectoderm which, in small embryos, is presumptive carapace ectoderm, but in larger embryos is modified as presumptive extra-embryonic ectoderm enclosing the large yolk mass. In later development, the extra-embryonic ectoderm partially reverts to a definitive fate as dorsal ectoderm and partially undergoes resorption in various ways.

In general, the further development of the naupliar ectoderm proceeds in a relatively simple manner, yielding the epidermis and epidermal glands, the ganglia of the brain and ventral nerve cord, the ventral endoskeletal invaginations to which many of the somatic muscles are attached, and the short, ectodermal exit ducts of the antennal and maxillary segmental organs and gonoducts. The major event which must preceed much of this differentiation is the proliferation of the ectoderm of the postnaupliar segments from the cells of the post-naupliar ectoderm. As we have seen, this proliferation may occur during larval development, after hatching as a nauplius, or may be completed to differing extents before hatching takes place. Present evidence shows that, whatever the degree of post-naupliar segment formation before or after hatching, the ectoderm of these segments is proliferated by ectoteloblasts.

Ectoteloblasts

Ectoteloblasts which bud off the segmental, post-naupliar ectoderm have been described in branchiopods (Anderson, 1967), cirripedes (Anderson, 1965, 1969; Walley, 1969) and many malacostracans (see below). The ectoteloblasts cells originate by cell enlargement in a transverse band, sometimes curved posteriorly at its ends (Fig. 114), within the post-naupliar presumptive ectoderm. Anterior to the ectoteloblasts, this ectoderm contributes to a greater or lesser extent to the ectoderm of the maxillulary and maxillary segments, but there seems no doubt that all of the ectoderm of the trunk segments is of ectoteloblast origin. The post-naupliar presumptive ectoderm posterior to the ectoteloblasts gives rise to the ectoderm of the telson.

The ectoteloblasts of branchiopods are indefinite in number and form a transverse ring of cells at the posterior end of the growing trunk (Fig. 130). Those of cirripedes form a U-shaped band in the same relative position (Fig. 129). In ***Ibla** quadrivalvis* there are seven ectoteloblasts (Anderson, 1965). Ectoderm cells are budded forwards as a generalized layer at the surface of the growing trunk and subsequently become segmentally delineated in association with the paired mesodermal somites. The proliferation of the trunk ectoderm in the copepods, the cephalocaridans and the euphausiid and penaeid malacostracons has not been satisfactorily described, but that of the larger embryos in the Malacostraca has been studied in detail. The number of ectoteloblasts is variable in these embryos, even within a species, but always includes a midventral cell and at least seven pairs of more lateral cells. Representative numbers include *Nebalia*, 19, *Anaspides*, 19, *Hemimysis*, 15, *Idotea*, 21 to 25, *Gammarus*, 21 to 25, *Caridina*, 21, *Pagurus*, 19 and *Hemigrapsus*, 19 (Manton, 1928, 1934; Hickman, 1937; Nair, 1949; Weygoldt, 1958; Oishi, 1959, 1960; Stromberg, 1965). In *Nebalia*, the peracaridans and the decapods, each ectoteloblast is formed by the enlargement of a single ectoderm cell. A few cells enlarge on each side of the midline, and are then augmented by the enlargement of a median cell and further lateral cells (Manton, 1928, 1934; Nair, 1939; Weygoldt, 1958, 1960b; Scholl, 1963; Stromberg, 1965, 1967, 1971; Petriconi, 1968; Dohle, 1970). In the decapods, in contrast, a median and five pairs of lateral cells undergo initial enlargement and the full number of ectoteloblasts is then attained through further divisions of the second, third and fourth enlarged cells on each side (Oishi, 1959, 1960). The usual site of differentiation of the ectoteloblasts in the Malacostraca is near the anterolateral margin of the postnaupliar presumptive ectoderm, in front of the immigrating presumptive mesoderm and midgut, but in a few examples they have been found to lie behind the latter (*Gammarus*, *Diastylis*; Weygoldt, 1958; Dohle, 1970).

The ectoteloblasts of malacostracan embryos bud off successive transverse rows of ectoderm cells in a forward direction (Figs. 112 and 133). With further divisions of the cells, the ectoderm rows become a generalized trunk ectoderm, which subsequently becomes segmentally delineated in association with the paired mesodermal teloblasts. In the majority of peracaridans, the ectoteloblasts remain in a transverse row throughout this process and the elongating band of ectoderm lies flat against the yolk mass (Fig. 125). Customarily among the Malacostraca, however, the ectoteloblast row curves towards the midline at its free ends during early budding, so that it becomes a ring of cells. The further products of this ring then form a tubular continuation of the trunk, turned forwards beneath the ventral surface of the yolk-filled, anterior part of the embryo (Figs. 121, 124 and 133). When the ectoderm of the last abdominal segment is formed, the ectoteloblasts can no longer be distinguished.

Nervous system

The ganglia of the central nervous system in the Crustacea arise in the usual way, by the inwards proliferation of numerous small cells from the ectoderm. The basic components of the nervous system are a pair of relatively large protocerebral ganglia developed from the protocerebral ectoderm, paired antennulary and antennal ganglia developed from the ventrolateral presumptive ectoderm of these segments, and a paired, midventral ganglion formed from the ventral ectoderm of each post-naupliar segment (Figs. 127 and 131). All of the ganglia separate into the interior, leaving epidermis at the surface. The antennulary or deutocerebral ganglia and the antennal or tritocerebral ganglia make up paired areas of nervous tissue which pass on either side of the stomodaeum and link the protocerebral ganglia with the mandibular ganglion (Fig. 118). The tritocerebral commissure is postoral (Walley, 1969). In embryos which hatch as nauplii, the anterior part of the nervous system, from the protocerebral ganglia to the mandibular ganglion, is fully developed and functional (Figs. 118, 126 and 127), though it undergoes further changes as larval development proceeds. The ganglia of the maxillulary, maxillary and trunk segments are then developed as part of the elaboration of their respective segments. When naupliar development is embryonized and hatching occurs at a later stage, the distinction between the development of the naupliar and post-naupliar parts of the nervous system is less marked, but still persists as a residual aspect of heteronomous development. The nauplius eye and compound eyes are developed from the protocerebral ectoderm in conjunction with the protocerebral ganglia.

The only question which is not fully resolved on the basic composition of the crustacean nervous system is the existence of a pair of pre-antennulary segmental ganglia distinct from the main protocerebral ganglia. A pair of small ganglia has been identified during the development of the brain of several species, lying between the protocerebral ganglia and antennulary ganglia. Weygoldt (1958, 1961) described them clearly in *Gammarus* and *Palaemonetes*, and they have subsequently been identified in several isopods (Scholl, 1963; Stromberg, 1965, 1967) and, most recently, in *Artemia* (Benesch, 1969). Anderson (1969) was not able to distinguish a pair of pre-antennulary ganglia in cirripede embryos, but the weight of evidence seems to favour their existence as a basic component of the developing crustacean head. In view of the absence of corresponding limb buds and the difficulties, discussed in the previous section, of interpreting the pre-antennulary mesoderm as a pair of segmental somites, the segmental significance of the pre-antennulary ganglia remains undecided. It is possible that a vestigial pre-antennulary segment forms part of the naupliar structure, in addition to the clearly recognizable antennulary, antennal and mandibular segments, but the present evidence on this point is not conclusive. As will be seen in Chapter 10, however, the segmental composition of the crustacean head is no longer a significant factor in considerations of comparative arthropod embryology.

Dorsal organs

Like other yolky arthropod embryos with an extensive extra-embryonic ectoderm, most malacostracan embryos develop temporary dorsal organs during their later embryonic development. The functions of these temporary structures have not been investigated in detail. The embryos of *Nebalia* and *Anaspides* (Manton, 1934; Hickman, 1937) and those of many decapods (*Palaemonetes*, *Leander*, *Crangon*, *Alpheus*, *Homarus*, *Astacus*, *Palinurus*, *Leptograpsus*; Kingsley, 1887; Herrick, 1891, 1895; Sollaud, 1923; Piatakov, 1925; Terao, 1929; Shiino, 1950; Weygoldt, 1961) develop a median dorsal organ as a cellular thickening in the dorsal midline of the extra-embryonic ectoderm. The thickening becomes a site of cellular degeneration and eventually sinks into the yolk mass to be resorbed during completion of the definitive dorsal body wall. One function of the median dorsal organ in these examples therefore seems to be the concentration and histolysis of cells of the temporary extra-embryonic ectoderm.

The embryos of peracaridan Malacostraca also develop a median dorsal organ (mysids, Manton, 1928; cumaceans, Boutchinsky, 1894; amphipods, Weygoldt, 1958; isopods, Stromberg, 1965, 1967), but it has a more organized structure than those of other malacostracans. The organ is formed by the concentration and elongation of extra-embryonic cells in a circular patch, which becomes invaginated at the centre. The cells fuse, but the nuclei retain the invaginated rosette pattern set up during formation of the structure. In later development the median dorsal organ of the Peracarida invaginates more deeply into the interior of the embryo and is resorbed after hatching has occurred. The function ascribed to this organ is the same as that usually attributed to the median dorsal organ of other Malacostraca, namely, the partial concentration and histolysis of the extra-embryonic ectoderm. In addition, however, peracaridan embryos develop a pair of dorsolateral organs in the extra-embryonic ectoderm (e.g. Manton, 1928; Weygoldt, 1960b; Scholl, 1963; Stromberg, 1965). These are also rosette-shaped invaginations composed of elongated cells which later become syncitial. The narrow apical regions of the cells around the invagination cavity become transversely striated, possibly indicating a secretory function. The composition and function of the secretion are not known. Like the median dorsal organ, the dorsolateral organs eventually sink beneath the surface of the embryo and are resorbed. In *Asellus*, the dorsolateral organs undergo enlargement and protrude temporarily through the egg membranes as "embryonic gills" (Weygoldt, 1960b).

The Basic Pattern of Development in Crustacea

Comparative studies leave no doubt that the basic pattern of development in the Crustacea includes a small yolky egg, a modified spiral cleavage, hatching as a nauplius larva and anamorphic development of the post-naupliar segments during a series of larval stages with intervening moults. Although extant species with this mode of development are developmentally specialized in various ways, it is possible to set out in some detail the general theme underlying these specializations.

The basic crustacean egg can probably be envisaged as having a diameter of 100 to 200 μ. Cleavage is spiral, but is modified in such a way that the yolk is segregated into a single, posteroventral cell, 4D, partially covered at the blastula stage by the remaining, yolk-free blastomeres. The yolk cell 4D is the presumptive midgut. The three ventral stem cells anterior to the yolk cell, 3A, 3B and 3C, are the presumptive mesoderm cells. Some of the descendants of blastomeres 2b and 3b in the ventral midline in front of the presumptive mesoderm constitute the presumptive stomodaeum. The remaining blastomeres make up the presumptive ectoderm, already zoned as presumptive protocerebral ectoderm anteriorly, antennulary, antennal and mandibular ectoderm laterally, post-naupliar ectoderm posteriorly and carapace ectoderm dorsally.

At gastrulation, the presumptive midgut and mesoderm cells move into the interior and are replaced at the surface by a ventral spreading of the mandibular and post-naupliar ectoderm. Once internal, the yolky presumptive midgut cell divides several times and forms the midgut epithelium, functionally differentiated as the midgut of the nauplius larva by the time hatching occurs. The presumptive mesoderm cells migrate to the posterior end of the embryo to lie internal to the post-naupliar ectoderm, and proliferate a pair of lateral mesodermal bands. These grow forwards on either side of the developing midgut as far as the protocerebral ectoderm, and become subdivided into pre-antennulary mesoderm anteriorly, three pairs of naupliar somites laterally and a residual mass of post-naupliar mesoderm at the posterior end. Meanwhile, the presumptive stomodaeum invaginates midventrally at the level of the antennal segment and becomes connected with the anterior end of the midgut. The presumptive proctodaeum also invaginates at the posterior midpoint of the post-naupliar ectoderm and becomes connected with the posterior end of the midgut, thus completing the functional gut of the nauplius. As it grows inwards, the proctodaeum pushes through the residual, post-naupliar mesoderm.

In the further development of the mesoderm, the pre-antennulary mesoderm gives rise to the labral and other musculature at the anterior end of the nauplius, and to the stomodaeal musculature. It is possible that the pre-antennulary mesoderm includes the vestiges of a pair of pre-antennulary somites. The naupliar somites give rise to the complex musculature of the three pairs of naupliar limbs and perhaps contribute a thin layer of splanchnic mesoderm on the external surface of the midgut. A pair of antennal glands (coelomoduct segmental organs) is also formed from cells of the antennal somites. The post-naupliar mesoderm develops into four main components, the maxillulary and maxillary somites, a bilateral ring of mesoteloblasts behind these somites, and the mesoderm of the telson and proctodaeum. The mesoteloblasts may bud off the somite rudiments of one or more post-maxillary segments before hatching takes place, but none of the post-naupliar segments is functional in the newly hatched nauplius.

In the further development of the ectoderm, the general product is the epidermis of the nauplius. The protocerebral ectoderm also gives rise to the protocerebral ganglia, possibly including a pair of small ganglia of a pre-antennulary segment, and to the median nauplius eye. The ventral protocerebral ectoderm pouches out as the labrum, with ectodermal labral glands. The naupliar segmental ectoderm pouches out ventrolaterally as a pair of limb buds in each segment, the antennules being pre-oral, the antennae para-oral and the mandibles post-oral. The antennules remain uniramous, but the antennae and mandible become biramous. Each pair of naupliar limbs develops to a functional condition, with long rami and, on the antennae and mandibles, basal ingestive setae. The ventral ectoderm of the naupliar segments gives rise to paired segmental ganglia. A pre-oral, antennulary pair are attached to the protocerebral ganglia. The antennal ganglia are para-oral, with a post-oral comissure, and the mandibular ganglia occupy a median, ventral position behind the mouth. The post-naupliar ectoderm develops as four components which correspond to those of the mesoderm. These comprise a ring of maxillulary ectoderm, a ring of maxillary ectoderm, a ring of ectoteloblasts and a posterior cap of telson ectoderm with the anus at its midpoint. The ectoderm of one or more post-maxillary segments may be budded from the ectoteloblasts before hatching takes place, but none of the post-naupliar segments has functional limbs in the newly hatched nauplius.

In the generalized condition, there is little change in the fundamental structure of the naupliar region during larval development and metamorphosis, except for some change in the mandibular musculature associated with the loss of the naupliar ingestive setae and onset of function of the mandibular gnathobases at metamorphosis, and for the development of the compound eyes as outgrowths of the protocerebrum. The development of the post-naupliar region, in contrast, proceeds in anteroposterior succession, with metamorphosis from the nauplius to the juvenile coinciding with the completion of six pairs of functional trunk limbs. The first pair of post-naupliar limbs, the maxillules, is cephalized before metamorphosis takes place, indicating that the maxillary segment is part of the generalized crustacean head. The second pair of post-naupliar limbs, the maxillae, is functionally the first pair of trunk limbs in its most generalized condition, indicating that the maxillary segment is a later addition to the crustacean head. The question of the number of segments in the generalized crustacean head, however, is irrelevant to considerations of comparative arthropod embryology and phylogeny. Even in the simplest condition that can be envisaged, the segmental and other components of the crustacean anterior end are fundamentally specialized as crustacean components and give no indication of what the ancestors of the Crustacea might have been.

This same uniqueness characterizes the further development of the post-naupliar segments during larval development. The segments are proliferated serially from the teloblastic growth zone in front of the telson and are built around the elongating midgut. The mesoteloblasts proliferate continuations of the lateral mesodermal bands, which divide anteroposteriorly into bilateral arcs of cells, the somites, lying against the ectoderm. The ectoteloblasts proliferate a generalized cylindrical ectoderm which becomes segmentally delineated in association with the paired somites.

When first formed, the somites are separated from the midgut by a wide haemocoel. Each segment then develops directly to the juvenile adult condition, with no temporary larval specializations.

The segmental ectoderm pouches out ventrolaterally as limb buds and gives rise midventrally to the paired ganglia of the segment. Invaginations of ectoderm form endophragmal skeletal bars and also form the short exit ducts of the maxillary segmental organs and the gonoducts. The gonopores lie in the middle region of the trunk.

The paired somites first produce median extensions across the haemocoel to meet the midgut dorsolaterally, forming the pericardial floor. A pair of vestigial coelomic cavities then develops in the pericardial floor and the somites become elaborated as heart middorsally, dorsal longitudinal muscle dorsolaterally in the pericardial haemocoel, and limb musculature and ventral longitudinal muscle ventrolaterally. The splanchnic mesoderm at each segmental level is formed by a downward growth of mesoderm over the midgut from the median edge of the pericardial floor. It is notable that the three pairs of naupliar somites are specialized and retain only certain aspects of this generalized pattern (limb musculature and perhaps splanchnic mesoderm). They do not contribute to the heart, pericardial floor or longitudinal muscles.

Among the post-naupliar somites, only the maxillary pair give rise to a pair of coelomoduct segmental organs, the maxillary glands, but there is evidence to suggest that segmental organs were a feature of all trunk somites in ancestral crustaceans. The gonoducts are coelomoducts, but the development of the gonads is not closely associated with the vestigial coelomic cavities. The primordial germ cells temporarily occupy the ventral walls of these cavities, at least in the anterior part of the trunk, but as the germ cells proliferate and join up to form paired strands, the strands move down to become depended from the ventral surface of the pericardial floor. Each strand, which is invested by mesoderm cells from the pericardial floor, then hollows out to form a cylindrical gonad. The gonocoels are thus developed independently of the segmental coelomic cavities, though they later gain communication posteriorly with the coelomoduct gonoducts. It is not certain whether, in the generalized condition, the primordial germ cells of Crustacea show their first visible differentiation in the somites or are differentiated precociously and migrate into the somites secondarily.

The Crustacea thus exhibit a distinctive theme of generalized development, quite different from those of other arthropods. The theme persists in all extant species, in spite of manifold and extreme variations, and provides strong evidence of the monophyletic nature of the Crustacea as a group. It is also unique in almost all respects, save for the inclusion of spiral cleavage, metameric segmentation and paired coelomic cavities. Furthermore, the uniqueness of crustacean development is effectively epitomized in the fate map of the crustacean blastula, which contains all of the information necessary for comparative purposes. In Chapter 10 the relationships between the Crustacea and the annelids and other arthropods will be discussed from this point of view.

References

AGAR, W. E. (1908) Notes on the early development of a cladoceran *(Holopedium gibberum)*. *Zool. Anz.* **33,** 420–7.

AGAR, W. E. (1950) The swimming setae of *Daphnia carinata. Q. Jl. microsc. Sci.* **91,** 353–68.

AIYAR, R. P. (1949) On the embryology of *Palaemon idae* Heller. *Proc. zool. Soc. Bengal* **2,** 101–31.

AKAHIRA, Y. (1957) Effects of certain physiological solutions on the embryo of a terrestrial isopod. *Porcellio scaber. J. Fac. Hokkaido Imp. Univ. (Zool.)* **13,** 359–63.

ALMEIDA, P. M. S. DE (1966) Notes on the development stages of *Schistomysis spiritus* (Norman 1860). *Acad. Brasil Cienc. An.* **38,** 349–53.

ALVAREZ, V. and KEWELRAMANI, H. G. (1971) Nauplius development of *Pseudodiaptomus ardjuna* Brehm (Copepoda). *Crustaceana* **18,** 269–76.

ALVES, T. and TOME, G. DE S. (1967) Contribuicas ao conhecimento do ultimo stadio embrionaire da lagosta *Palinurus argus* (Latr.). *Estac. Biol. Mar. Univ. Fed. Ceara* **7,** 63–66.

AMMA, K. (1911) Über die Differenzierung der Keimbahnzellen bei Copepoden. *Arch. Zellforsch.* **6,** 497–576.

ANDERSON, D. T. (1965) Embryonic and larval development and segment formation in *Ibla quadrivalvis* (Cuv.) (Cirripedia). *Aust. J. Zool.* **13,** 1–15.

ANDERSON, D. T. (1967) Larval development and segment formation in the branchiopod crustaceans *Limnadia stanleyana* King (Conchostraca) and *Artemia salina* (L.) (Anostraca). *Aust. J. Zool.* **15,** 47–91.

ANDERSON, D. T. (1969) On the embryology of the cirripede crustaceans *Tetraclita rosea* (Krauss), *T. purpurascens* (Wood), *Chthamalus antennatus* (Darwin) and *Chamaesipho columna* (Spengler) and some considerations of crustacean phylogenetic relationships. *Phil. Trans. R. Soc.* B, **256,** 183–235.

ANDERSON, D. T. and ROSSITER, G. T. (1968a) Hatching and larval development of *Haplostomella australiensis* Gotto (Copepoda: Fam. Ascidicolidae), a parasite of the ascidian *Styela etheridgii* Herdman. *Proc. Linn. Soc. N.S.W.* **93,** 464–75.

ANDERSON, D. T. and ROSSITER, G. T. (1968b) Hatching and larval development of *Dissonus nudiventris* Kabata (Copepoda: Fam. Dissonidae) a gill parasite of the Port Jackson Shark. *Proc. Linn. Soc. N.S.W.* **93,** 476–81.

ANDERSON, W. W., KING, J. E. and LINDNER, M. J. (1949) Early stages in the life history of the common marine shrimp, *Penaeus setiferus* (Linnaeus). *Biol. Bull., Woods Hole* **86,** 168–72.

ANTEUNIS, A. N., FAUTREZ-FIRLEFYN, N. and FAUTREZ, J. (1961) Early blastocoel formation in the egg of *Artemia salina. Exptl. Cell Res.* **25,** 463–5.

BABU, N. (1967) Observations on the biology of *Caridina propinqua* De Man. *Indian J. Fish.* **10A,** 107–17.

BAER, J. G. (1952) *Ecology of Animal Parasites*, Univ. of Illinois Press, Urbana.

BAHAMONDE, N. and LOPEZ, J. T. (1960) Estudios biologicos en la población de *Aegla laevis laevis* (Latreille) de El Monte (Crustacea, Decapoda, Anomura). *Invest. Zool. Chilenco* **7,** 19–58.

BALDASS, F. VON (1937) Entwicklung von *Holopedium gibberum. Zool. Jb. Anat. Ont.* **63,** 399–454.

BALDASS, F. VON (1941) Die Entwicklung von *Daphnia pulex. Zool. Jb. Anat. Ont.* **67,** 1–60.

BAN, H. (1950) A culture method of the eggs of a terrestrial isopod, *Armadillidium vulgare* (Latreille). *Sci. Rep. Tohoku Univ.*, ser. 4, **18,** 276–8.

BAQUAI, I. U. (1963) Studies on the post-embryonic development of the fairy shrimp *Streptocephalus seali* Ryder. *Tulane Stud. Zool.* **10,** 92–120.

BARKER, D. (1962) A study of *Thermosbaena mirabilis* (Malacostraca, Peracarida) and its reproduction. *Q. Jl. microsc. Sci.* **103,** 261–86.

BARNES, H. and BARNES, M. (1959a) Note on stimulation of cirripede nauplii. *Oikos* **10**, 19–23.

BARNES, H. and BARNES, M. (1959b) The naupliar stages of *Balanus hesperius* Pilsbury. *Can. J. Zool.* **37**, 237–44.

BARTOK, P. (1944) Die morphologische Entwicklung von *Bathynella chappiusi. Acta Sci. Math. Nat. Kolozsvár* **21**, 1–46.

BASSINDALE, R. (1936) The developmental stages of three English barnacles, *Balanus balanoides* (Linn.), *Chthamalus stellatus* (Poli) and *Verruca stroemia* (O. F. Müller). *Proc. zool. Soc. Lond.* **106**, 57–74.

BATHAM, E. J. (1945) Description of female, male and larval forms of a tiny stalked barnacle, *Ibla idiotica* n.sp. *Trans. R. Soc. N.Z.* **75**, 347–56.

BATHAM, E. J. (1946) *Pollicipes spinosus* Quoy and Gaimard. II. Embryonic and larval development. *Trans. R. Soc. N.Z.* **75**, 405–18.

BATHAM, E. J. (1967) The first three larval stages and feeding behaviour of phyllosoma of the New Zealand palinurid crayfish *Jasus edwardsii* (Hutton, 1875). *Trans. R. Soc. N.Z. (Zool.)* **9**, 53–64.

BAUMANN, H. (1932) Die postembryonale Entwicklung von *Potamobius astacus* L. bis zur zweiten Hautung. *Z. wiss. Zool.* **141**, 36–51.

BAZIN, F. (1971) Le développement embryonnaire des organes deutocerebraux chez *Astacus leptodactylus* Esch. (Crustacé decapode Reptantia). *Ann. Embryol. Morphog.* **4**, 137–44.

BENESCH, R. (1969) Zur Ontogenie und Morphologie von *Artemia salina* L. *Zool. Jb. Anat. Ont.* **86**, 307–458.

BENSAM, P. and KARTHA, K. N. R. (1969) Notes on the eggs and larval stages of *Hippolysmate ensirostris* Kemp. *Mar. Biol. Assoc. India, Symp. Ser.* **2**, 736–43.

BERGH, R. S. (1893) Beiträge zur Embryologie der Crustacean. 1. Zur Bildungsgeschichte des Keimstreifens von *Mysis. Zool. Jb. Anat. Ont.* **6**, 491–526.

BERNARD, M. (1964) Le développement nauplien de deux copépodes carnivores: *Euchaeta marina* (Prestandr.) et *Candacia armata* (Boeck). *Pelagos Bull. Inst. Oceanogr. Alger.* **2**, 51–70.

BERRILL, M. (1969) The embryonic behaviour of the mysid shrimp, *Mysis relicta. Can. J. Zool.* **47**, 1212–21.

BERRILL, M. (1971) The embryonic behaviour of *Caprella unica* (Crustacea Amphipoda). *Can. J. Zool.* **49**, 499–504.

BERRY, E. W. (1926) Description and notes on the life history of a new species of *Eulimnadia. Amer. J. Sci.* (5) **11**, 429–33.

BESSELS, E. (1869) Einige Worte über die Entwicklungsgeschichte und den morphologischen Wert des kugelförmigen Organes der Amphipoden. *Jena Zeitschr.* **5**, 91–101.

BETANCES, L. M. (1921) Les cellules du sang de l'*Astacus fluviatilis. Arch. anat. microsc.* **18**, 1–45.

BIEBER, A. (1940), Beiträge zur Kenntnis der Jugendphasen des Flusskrebes *Astacus fluviatilis* (Rond.) L. mit besonderer Berücksicktigung der Metamorphose. *Rev. suisse Zool.* **47**, 389–470.

BIGELOW, M. A. (1902) The early development of *Lepas*. A study of cell lineage and germ layers. *Bull. Mus. comp. Zool. Harvard* **40**, 61–144.

BISHOP, J. A. (1967) The zoogeography of the Australian freshwater decapod Crustacea. In WEATHERLY, A. H. (ed.), *Australian Inland Waters and their Fauna*, Aust. Nat. Univ., Canberra.

BJORNBERG, T. K. S. (1966) The developmental stages of *Undinula vulgaris* (Dana) (Copepoda). *Crustaceana* **11**, 65–76.

BJORNBERG, T. K. S. (1967) The larvae and young forms of *Eucalanus* Dana (Copepoda) from tropical Atlantic waters. *Crustaceana* **12**, 59–73.

BLANC, H. (1885) Développement de l'œuf et formation des feuillets primitifs chez la *Cuma Rathkii* Krøyer. *Rec. Zool. Suisse* **2**, 253–74.

BOAS, J. E. V. (1939) Die Gattung *Polycheles*, ihre verwandschaftliche Stellung und ihre postembryonal Entwicklung. *Biol. Meddel. København.* **14**, 1–32.

BOBRETSKY, N. (1873) Sur l'embryologie des arthropodes (écrevisse et *Palaemon*). *Jb. Anat. Physiol.* **2**, 312–18.

BOBRETSKY, N. (1874) Zur Embryologie des *Oniscus murarius. Z. wiss. Zool.* **24**, 179–303.

BOCQUET-VÉDRINE, J. (1960) Premiers stades de segmentation de l'œuf de *Chthamalophilus delagei* J. Bocquet-Védrine (Crustacé-Rhizocephale). *C.R. Acad. Sci., Paris* **250**, 1557–9.

BOCQUET-VÉDRINE, J. (1964) Embryologie précoce de *Sacculina carcini* Thompson. *Zool. Meded.* **39**, 1–11.

BONDE, C. VON (1936) The reproduction, embryology and metamorphosis of the Cape crawfish *Jasus lalandii* (Milne Edwards) (Ortmann). *Invest. Rep. Fish Survey S. Africa* **6**, 5–25.

BOSCHI, E. E. and SCELZO, M. A. (1968) Larval development of the spider crab *Libinia spinosa* H. Milne Edwards, reared in the laboratory (Brachyura, Majidae). *Crustaceana suppl.* **2**, 170–80.

BOSCHI, E. E., SCELZO, M. A. and GOLDSTEIN, B. (1967) Desarollo larval de dos espécies de Crustáceos

Décapodos en la laboratorio *Pachycheles haigae* Rodrigues Da Costa (Porcellanidae) y *Chasmagnathus granulata* Dana (Grapsidae). *Inst. biol. mar. Mar des Plata* **12,** 1–46.

BOTNARUIC, N. (1947) Contributions à la connaissance des phyllopodes conchostracés de Roumanie. *Notationes biol.* **5,** 1–3.

BOTNARUIC, N. (1948) Contribution à la connaissance du développement des Phyllopodes conchostracés. *Bull. biol. Fr.-Belg.* **82,** 31–36.

BOUTCHINSKY, P. (1890) Observations on the development of *Parapodopsis cornuta* Czern. *Zapiski Novoross. Obsheh. Odessa* **14,** 74–170.

BOUTCHINSKY, P. (1893) Zur Embryologie der Cumaceen. *Zool. Anz.* **16,** 386–7.

BOUTCHINSKY, P. (1894) Zur Entwicklungsgeschichte un *Gebia litoralis. Zool. Anz.* **17,** 253–6.

BOUTCHINSKY, P. (1897) Die Furchung des Eies und die Blastodermbildung der *Nebalia. Zool. Anz.* **20,** 219–20.

BOUTCHINSKY, P. (1900) Zur Entwicklungsgeschichte der *Nebalia geoffroyi. Zool. Anz.* **23,** 493–5.

BRAHM, C. and GEIGER, S. R. (1966) On the biology of the pelagic crustacean *Nebaliopsis typica* G. O. Sars. *Bull. Southern Calif. Acad. Sci.* **65,** 41–46.

BRAUER, A. (1874) Vorläufige Mittheilungen über die Entwicklung und Lebenweise des *Lepidurus productus* Bosc. *Sitzb. Akad. Wiss. Wien* **69,** 130–41.

BRAUER, A. (1892) Über das Ei von *Branchipus Grubii* v. Dyb. von der Bildung bis zur Ablage. *Deut. Akad. Wiss. Berlin* **1892,** 1–66.

BRESCIANI, J. and LUTZEN, J. (1961) *Gonophysema gullmarensis* (Copepoda parasitica). An anatomical and biological study of an endoparasite living in the ascidian *Ascidiella aspersa*. II. Biology and development. *Cah. Biol. mar.* **2,** 347–72.

BROAD, A. C. (1958) Larval development of the crustacean *Thor floridanus* Kingsley. *J. Elisha Mitchell Sci. Soc.* **73,** 317–28.

BROCH, E. S. (1965) Mechanisms of adaptation of the fairy shrimp *Chirocephalopsis bundyi* Forbes to the temporary pond. *New York St. Coll. Ag. Mem.* **392,** 1–48.

BROOKS, W. K. (1880) The embryology and metamorphosis of the Sergestidae. *Zool. Anz.* **3,** 563–7.

BROOKS, W. K. (1882) *Leucifer*, a study in morphology. *Phil. Trans. R. Soc.* B, **173,** 57–137.

BROOKS, W. K. and HERRICK, F. H. (1892) The embryology and metamorphosis of the Macrura. *Mem. Nat. Acad. Sci. Washington* **5,** 325–576.

BUCHHOLZ, H. (1951) Die Larvenformen von *Balanus improvisus*. Beiträge zur Kenntnis des Larvenplankton. *Kieler Meeresforsch* **8,** 49–57.

BUMPUS, H. C. (1891) The embryology of the American lobster. *J. Morph.* **5,** 215–62.

CAMPAN, F. (1929) Contributions à la connaissance des Phyllopodes notostracés. *Bull. Soc. zool. Fr.* **54,** 95–118.

CAMPBELL, M. H. (1934) The life history and post-embryonic development of the copepods *Calanus torsus* Brady and *Euchaeta japonica* Murakavoa. *J. Biol. Bd. Toronto* **1,** 1–65.

CANNON, H. G. (1921) The early development of the summer eggs of the cladoceran *(Simocephalus vetulus)*. *Q. Jl. microsc. Sci.* **65,** 627–42.

CANNON, H. G. (1924) On the development of an estheriid crustacean. *Phil. Trans. R. Soc.* B, **212,** 395–430.

CANNON, H. G. (1925) Ectodermal muscles in a crustacean. *Nature, Lond.* **115,** 458–9.

CANNON, H. G. (1926) On the post-embryonic development of the fairy shrimp *(Chirocephalus diaphanus)*. *J. Linn. Soc. London* **36,** 401–16.

CANNON, H. G. (1928) On the feeding mechanism of the fairy shrimp *Chirocephalus diaphanus* Prevost. *Trans. R. Soc. Edinb.* **55,** 807–22.

CANU, E. (1892) Les copepodes du Boulonnais. *Trav. Sta. zool. Wimereux* **6,** 1–345.

CARTON, Y. (1968) Développement de *Cancerilla tubulata* Dalyell parasite de l'ophiure *Amphipholis squamata* Della Chiaje. *Crustaceana*, suppl. **1,** 11–28.

CASPERS, H. (1939) Über Vorkommen und metamorphose von *Mytilicola intestinalis* Stauer (Copepoda para) in südlich Nordsee. *Zool. Anz.* **126,** 161–71.

CHAIGNEAU, J. (1960) Action de la dessication et de la temperature sur l'eclosion de l'œuf de *Lepidurus apus* (Crustacé, Phyllopode). *Bull. Soc. zool. Fr.* **84,** 398–407.

CHAMBERS, V. T. (1885) The larva of *Estheria mexicana. Amer. Nat.* **19,** 190–1.

CHIA, K. T. (1968) Oogenesis and embryonic development in *Porcellio scaber* (Crustacea, Isopoda). Thesis, University of Sydney.

CHOUDHURY, P. C. (1970) Complete larval development of the palaemonid shrimp *Macrobrachium acanthurus* (Wiegmann, 1836) reared in the laboratory. *Crustaceana* **18,** 113–32.

CHOUDHURY, P. C. (1971) Complete larval development of the palaemonid shrimp *Macrobrachium carcinus* (L.) reared in the laboratory (Decapoda, Palaemonidae). *Crustaceana* **20,** 51–69.

CLAUS, C. (1868) Beiträge zur Kenntnis der Ostrakoden. I. Entwicklungsgeschichte von *Cypris* Schrift. *Ges. Bef. ges. Naturwiss. Marburg* **9,** 151–64.

CLAUS, C. (1873) Zur Kenntnis des Baues und der Entwicklung von *Branchipus stagnalis* und *Apus cancriformis. Abh. Kg. Ges. Wiss. Göttingen* **18,** 93–136.

CLAUS, C. (1875) Ueber die Entwicklung, Organisation und systematische Stellung der Argulidae. *Z. wiss. Zool.* **25,** 217–24.

CLAUS, C. (1886) Untersuchungen ueber die Organisations und Entwicklung von *Branchipus* und *Artemia. Arb. zool. Inst. Wien* **6,** 267–370.

CLAUS, C. (1893) Neue Beobachtungen über die Organisation und Entwicklung von *Cyclops. Arb. zool. Inst. Wien* **10,** 283–356.

CLAUS, C. (1895) Beiträge zur Kentniss den Süsswasser-Ostracoden. *Arb. zool. Inst. Wien* **11,** 17–48.

COMITA, G. W. and THOMMERDAHL, D. M. (1960) The postembryonic developmental instars of *Diaptomus siciloides* Lilljeborg. *J. Morph.* **107,** 297–320.

CONNOLLY, C. J. (1929) A new copepod parasite *Choniosphaera cancrorum* gen. et. sp. n., representing a new genus and its larval development. *Proc. zool. Soc. Lond.* **99,** 415–27.

CONOVER, R. J. (1967) Reproductive cycle, early development, and fecundity in laboratory populations of the copepod *Calanus hyperboreus. Crustaceana* **13,** 61–72.

CONSTANZO, G. (1969) Stadi naupliari e primo copepodite di *Lichomolgus canni* G. O. Sars (Copepoda, Cyclopoida) del lago di fars (Messina) allevata sperimentalmente. *Bull. Zool.* **36,** 148–53.

COOK, H. L. and MURPHY, A. M. (1965) Early developmental stages of the rock shrimp, *Sicyonia brevirostris* Stimpson, reared in the laboratory. *Tulane Stud. Zool.* **12,** 109–28.

COOK, H. L. and MURPHY, M. A. (1971) Early developmental stages of the brown shrimp, *Penaeus aztecus* Ives, reared in the laboratory. *Fish. Bull. U.S. Dep. Commer.* **69,** 223–39.

CORKETT, C. J. (1967) The copepodid stages of *Temora longicornis* (O. F. Müller, 1792) (Copepoda). *Crustaceana* **12,** 261–73.

COSTLOW, J. D. JR. and BOOKHOUT, C. G. (1957) Larval development of *Balanus eburneus* in the laboratory. *Biol. Bull., Woods Hole* **112,** 313–24.

COSTLOW, J. D. JR. and BOOKHOUT, C. G. (1958) Larval development of *Balanus amphitrite* var. *denticulata* Broch reared in the laboratory. *Biol. Bull., Woods Hole* **114,** 284–95.

COSTLOW, J. D. JR. and BOOKHOUT, C. G. (1959) The larval development of *Callinectes sapidus* Rathbun reared in the laboratory. *Biol. Bull., Woods Hole* **115,** 373–96.

COSTLOW, J. D. JR. and BOOKHOUT, C. G. (1960a) The complete larval development of *Sesarma cinereum* (Bosc.) reared in the laboratory. *Biol. Bull., Woods Hole* **118,** 203–14.

COSTLOW, J. D. JR. and BOOKHOUT, C. G. (1960b) A method for developing brachyuran eggs *in vitro. Limnol. Oceanogr.* **5,** 212–15.

COSTLOW, J. D. JR. and BOOKHOUT, C. G. (1961a) The larval development of *Eurypanopeus depressus* (Smith) under laboratory conditions. *Crustaceana* **2,** 6–15.

COSTLOW, J. D. JR. and BOOKHOUT, C. G. (1961b) The larval stages of *Panopeus herbstii* Milne-Edwards reared in the laboratory. *J. Elishah Mitchell Sci. Soc.* **77,** 33–42.

COSTLOW, J. D. JR. and BOOKHOUT, C. G. (1962a) The larval development of *Sesarma reticulatum* Say reared in the laboratory. *Crustaceana* **4,** 281–94.

COSTLOW, J. D. JR. and BOOKHOUT, C. G. (1962b) The larval development of *Hepatus epheliticus* (L.) under laboratory conditions. *J. Elishah Mitchell Sci. Soc.* **78,** 113–25.

COSTLOW, J. D. JR. and BOOKHOUT, C. G. (1966) The larval development of *Ovalipes ocellatus* (Herbst) under laboratory conditions. *J. Elishah Mitchell Sci. Soc.* **82,** 160–71.

COSTLOW, J. D. JR. and BOOKHOUT, C. G. (1967) The larval stages of the crab *Neopanope packardii* (Kingsley) in the laboratory. *Bull. mar. Sci.* **17,** 52–63.

COSTLOW, J. D. JR. and FAGETTI, E. (1967) The larval development of the crab *Cyclograpsus cinereus* Dana, under laboratory conditions. *Pacific Sci.* **21,** 166–77.

DAHL, E. (1955) Some aspects of the ontogeny of *Mesamphisepus capensis* (Barnard) and the affinities of the Isopoda Phreatocoidea. *Kungl. Fyziogr. Salisk z. Lund Forhandl.* **24,** 83–88.

DAHL, E. (1957) Embryology of X-organs in *Crangon allmanni. Nature, Lond.* **179,** 482.

DAKIN, W. J. and COLEFAX, A. N. (1941) The plankton of the Australian coastal waters off New South Wales. *Publ. Univ. Sydney Dept. Zool.* **1,** 1–215.

DAUM, J. (1954) Zur Biologie einer Isopodenart unterirdischer Gewässer: *Caecosphaeroma (Vireia) burgundun* Dollfus. *Ann. Univ. Saraviensis* **3,** 140–60.

DAVIS, C. C. (1943) The larval stages of the calanoid copepod *Eurytemora hirundoides* (Nordquist). *Chesapeake biol. Lab. Publ.* **58**, 1–52.

DAVIS, C. C. (1959) Osmotic hatching in the eggs of some freshwater copepods. *Biol. Bull., Woods Hole* **116**, 15–29.

DAVIS, C. C. (1964) A study of the hatching process in aquatic invertebrates. XIII. Events of eclosion in the American lobster, *Homarus americanus* Milne Edwards (Astacura, Hormidae). *Amer. Midland Nat.* **72**, 203–10.

DAVIS, C. C. (1966) A study of the hatching process in aquatic invertebrates. XXIII. Eclosion in *Petrolisthes armatus* (Gibbes) (Anomura, Porcillanidae). *Int. Rev. Gesamten Hydrobiol.* **51**, 791–6.

DELAMARE-DEBOUTVILLE, C. (1960) Biologie des eaux souterraines littorales et continentales. *Vie et Milieu suppl.* **9**, 1–740.

DELSMAN, H. C. (1917) Die Embryonalentwicklung von *Balanus balanoides* Linn. Helder. *Tijdschr. Ned. Dierk. Ver.* (2) **15**, 419–520.

DOBKIN, S. (1961) Early developmental stages of pink shrimp *Penaeus duorarum*, from Florida waters. *U.S. Fish and Wildlife Serv. Fish. Bull.* **61**, 321–49.

DOBKIN, S. (1965) The first post-embryonic stage of *Synalpheus brooksi* Contiere. *Bull. mar. Sci. Gulf and Carib.* **15**, 450–62.

DOBKIN, S. (1967) Abbreviated larval development in caridean shrimps and its significance in the artificial culture of these animals. *F.A.O. World Conf. Mexico.* F.A.O. Experience Paper No. 53, 1–2.

DOBKIN, S. (1971) The larval development of *Palaemonetes cummingi* Chace, 1954 (Decapoda, Palaemonidae), reared in the laboratory. *Crustaceana* **20**, 285–97.

DOHLE, W. (1970) Die Bildung und Differenzierung des post-nauplialen Keimstreifs von *Diastylis rathkei* (Crustacea, Cumacea). 1. Die Bildung der Telobasten und ihrer Derivate. *Z. Morph. Tiere* **67**, 367–92.

DOHRN, A. (1870a) Über den Bau und die Entwicklung der Cumacean. *Jena. Zeitschr.* **5**, 54–80.

DOHRN, A. (1870b) Zur Kenntniss von Bau und der Entwicklung von *Tanais*. *Jena Zeitschr.* **5**, 293–306.

DUDLEY, P. L. (1966) *Development and Systematics of some Pacific Marine Symbiotic Copepods*, University of Washington Press, Seattle.

DUDLEY, P. L. (1969) The fine structure and development of the nauplius eye of the copepod *Doropygus serluses* Illg. *Cellule* **68**, 7–42.

ELGMARK, K. and LANGELAND, A. L. (1970) The number of naupliar instars in Cyclopoida (Copepoda). *Crustaceana* **18**, 277–82.

ESSLOVA, M. (1959) Embryonic development of parthenogenetic eggs of *Daphnia pulex*. *Vest. csz. zool. Spol.* **23**, 80–88.

EVANS, A. J. (1970) Some aspects of the ecology of a calanoid copepod, *Pseudoboeckella brevicandata* Brady 1875, on a subantarctic island, *A.N.A.R.E. Sci. Rep.* Series B (11), *Zoology* No. 110, pp. 1–100.

EWALD, J. J. (1965) The laboratory rearing of pink shrimp, *Penaeus duorarum* Burkenroad. *Bull. mar. Sci. Gulf and Carib.* **15**, 436–49.

FAGETTI, E. (1969) Larval development of the spider crab *Pisoides edwardsi* (Decapoda, Brachyura). *Mar. Biol. Berlin* **4**, 160–5.

FAGETTI, E. and CAMPODICONICO, I. (1971a) Larval development of the red crab *Pleuroncodes monodon* (Decapoda Anomura: Galatheidae) under laboratory conditions. *Mar. Biol., Berlin* **8**, 70–81.

FAGETTI, E. and CAMPODICONICO, I. (1971b) The larval development of the crab *Cyclograpsus punctatus* H. Milne Edwards, under laboratory conditions (Decapoda, Brachyura, Grapsidae, Sesarminae). *Crustaceana* **21**, 183–95.

FAUTREZ, J. and FAUTREZ-FIRLEFYN, N. (1963) A propos de la localisation du noyau dans les blastomères de l'œuf d'*Artemia salina*. L'influence du mercapto-ethanol et du dithiodiglycol sur cette localisation. *Dev. Biol.* **6**, 250–61.

FAUTREZ, J. and FAUTREZ-FIRLEFYN, N. (1964) Sur la presence et la persistance d'un noyau vitellin atypique dans l'œuf d'*Artemia salina*. *Dev. Biol.* **9**, 81–91.

FAXON, W. (1879) On the development of *Palaemonetes vulgaris*. *Bull. Mus. comp. Zool. Harvard* **5**, 303–30.

FILHOL, F. (1934) Embryologie et développement de *Lamproglena pulchella* Nordmann, description du mâle. *Bull. int. Acad. Cracovie* (B) **1934**, 225–32.

FILHOL, F. (1936) Nouvelles observations sur le développement et le biologie du *Lamproglena pulchella* Loidman. *Ann. Parasitol., Paris* **14**, 246–55.

FIORONI, P. (1969) Zum embryonalen und postembryonalen Dotterabbau des Flusskrebses (*Astacus*, Crustacés Malacostraca, Decapoda). *Rev. suisse Zool.* **76**, 919–46.

FIORONI, P. (1970a) Die organogenetische Rolle der vitellophagen in der Darmentwicklung von *Galathea* (Crustacea, Decapoda, Anomura). *Z. Morph. Tiere* **67**, 263–306.

FIORONI, P. (1970b) Am Dotteraufschluss beteiligte Organ und Zelltypen bei höheren Krebsen; der Versuch zu einer einheitlichen Terminologie. *Zool. Jb. Anat. Ont.* **87**, 481–522.

FRYER, G. (1961) Larval development in the genus *Chonopeltus* (Crustacea, Branchiura). *Proc. zool. Soc., Lond.* **137**, 61–69.

FUCHS, F. (1914) Die Keimblätterentwicklung von *Cyclops viridis* Jurine. *Zool. Jahb. Anat. Ont.* **38**, 103–56.

FULINSKY, B. (1908) Zur Embryonalentwicklung des Flusskrebses. *Zool. Anz.* **33**, 20–28.

FULINSKY, B. (1922) Ueber die Entwicklung der Keimdrusen bei *Astacus fluviatilis* Rond. *Prace. Kom. mat. przyr. poznan Tow. przyj hank, (B)* **1**, 244–59.

GAGE, J. (1966) Seasonal cycles of *Notodelphys* and *Ascidicola*, copepod associates with *Ascidiella* (Ascidiacea). *J. Zool.* **150**, 223–33.

GAULD, D. T. (1959) Swimming and feeding in crustacean larvae; the nauplius larva. *Proc. zool. Soc., Lond.* **132**, 31–50.

GIBBONS, S. G. (1934) A study of the biology of *Calanus finmarchicus* in the North-western North Sea. *Sci. Invest. Fish., Scotl.* **1**, 1–24.

GIBBONS, S. G. (1936) Early developmental stages of Copepoda. I. *Rhinocalanus nasutus* and *Eucalanus elongatus. Ann. Mag. nat. Hist.* (10) **18**, 384–92.

GIBBONS, S. G. (1938) Early development stages of Copepoda. II. *Metridia lucens* Boech. *Ann. Mag. nat. Hist. (11)* **2**, 493–7.

GNADEBERG, W. (1949) Beiträge zur Biologie und Entwicklung des *Ergasilis sieboldii* v. Nordmann (Copepoda parasitica). *Z. Parasitenk.* **14**, 1–2.

GNANAMUTHU, C. P. (1951a) Notes on the life history of a parasitic copepod *Lernea chackoensis. Parisitology* **41**, 148–55.

GNANAMUTHU, C. P. (1951b) Studies on a lernaeid copepod *Cardiodectes anchorellae* Brain and Grey. *Proc. zool. Soc., Lond.* **121**, 237–52.

GONZALES, S. A. (1968) Desarrollo larvaris de *Diaptomus proximus* Kiefer (Copepoda, Calanoida). *Hydrobiologia* **32**, 523–44.

GOODRICH, A. I. (1939) The origin and fate of the endoderm elements in the embryogeny of *Porcellio laevis* Latr. and *Armadillidium nasutum* B. L. (Isopoda). *J. Morph.* **64**, 401–29.

GORE, R. H. (1968) The larval development of the commensal crab *Polyonyx gibbesi* Haig, 1956 (Crustacea: Decapoda). *Biol. Bull., Woods Hole* **135**, 111–29.

GORE, R. H. (1970) The complete larval development of *Porcellana sigsbeiana* (Crustacea; Decapoda) under laboratory conditions. *Mar. Biol., Berlin* **11**, 344–55.

GORHAM, F. P. (1895) The cleavage of the egg of *Virbius zostericola*, a contribution to crustacean cytogeny. *J. Morph.* **11**, 741–6.

GOTTO, R. V. (1957) The biology of a commensal copepod, *Ascidicola rosea* Thorell, in the ascidian *Corella parallelogramma* (Muller). *J. mar. biol. Ass. U.K.* **36**, 281–90.

GRAY, P. (1933a) The nauplii of *Notodelphys agilis* Thorell and *Doropygus porcicarida* Brady. *J. mar. biol. Ass. U.K.* **18**, 519–22.

GRAY, P. (1933b) *Mycophilus rosovula* n.sp.a notodelphyoid copepod parasitic within *B.(Botrylloides) leachii* Sav., with a description of the nauplius and notes on its habits. *J. mar. biol. Ass. U.K.* **18**, 523–7.

GREENWOOD, J. V. (1966) Some larval stages of *Pagurus novae-zealandiae* (Dana), 1852 (Decapoda, Anomura). *New Zealand J. Sci.* **9**, 545–58.

GRICE, G. D. (1969) The developmental stages of *Pseudodiaptomus coronatus* Williams (Copepoda, Calanoida). *Crustaceana* **16**, 291–301.

GRICE, G. D. (1971) The developmental stages of *Eurytemora americana* Williams 1906, and *Eurytemora herdmani* Thompson and Scott, 1897 (Copepoda, Calanoida). *Crustaceana* **20**, 145–58.

GROBBEN, C. (1879) Die Entwicklungsgeschichte der *Moina rectirostris. Arb. zool. Inst. Wien.* **2**, 203–68.

GROBBEN, C. (1881) Die Entwicklungsgeschichte von *Cetochilus septentrionalis* Goodsir. *Arb. zool. Inst. Wien.* **3**, 243–82.

GROBBEN, C. (1893) Einige Bemerkungen zu Dr. P. Samassa's Publikationen über die Entwicklung von *Moina rectirostris. Arch. mikrosk. Anat.* **42**, 213–17.

GROBBEN, C. (1911) Die Bindesubstanzen von *Argulus*. Ein Beitrag zur Kenntnis der Bindesubstanz der Arthropoden. *Arb. zool. Inst. Wien.* **19**, 75–98.

GROOM, T. T. (1894) On the early development of Cirripedia. *Phil. Trans. R. Soc.* B, **185**, 119–232.

GRSCHEBIEN, S. (1910) Embryologie von *Pseudocumo pectinata* Sowinsky. *Zool. Anz.* **35**, 808–13.

GURNEY, R. (1933) Notes on some Copepoda from Plymouth. *J. mar. biol. Ass. U.K.* **19**, 299–304.

GURNEY, R. (1934a) The development of *Rhinocalanus. Discovery Reps.* **9**, 107–14.

GURNEY, R. (1934b) The development of certain parasitic Copepoda of the families Caligidae and Clavellidae. *Proc. zool. Soc., Lond.* **103**, 177–217.

GURNEY, R. (1937) Notes on some decapod and stomatopod Crustacea from the Red Sea. *Proc. zool. Soc., Lond.* **107**, 319–36.

GURNEY, R. (1942) *Larvae of Decapod Crustacea*, Royal Society, London.

GURNEY, R. (1943) The larval development of two penaeid prawns from Bermuda of the genera *Sicyonia* and *Penaeopsis*. *Proc. zool. Soc., Lond.* **113**, 1–16.

HÄCKER, V. (1897) Die Keimbahn von *Cyclops*. *Arch. mikrosk. Anat.* **49**, 35–91.

HARDING, J. P. (1955) Development. In MARSHALL, S. M. and ORR, A. P., *Biology of a Marine Copepod*, Oliver & Boyd, Edinburgh.

HARTNOLL, R. G. (1964) The zoeal stage of the spider crab *Microphrys bicornutus* (Latr.). *Ann. Mag. Nat. Hist.* **(13) 7**, 241–6.

HASHMI, S. S. (1970a) The brachyuran larvae of W. Pakistan hatched in the laboratory. II. Portunidae: *Charybdis* (Decapoda: Crustacea). *Pak. J. Sci. Ind. Res.* **12**, 272–8.

HASHMI, S. S. (1970b) The larva of *Elamena* (Hymenosomidae) and *Pinnotheres* (Pinnotheridae) hatched in the laboratory (Decapoda: Crustacea). *Pak. J. Sci. Ind. Res.* **12**, 279–85.

HASHMI, S. S. (1970c) Study on larvae of the family of Xanthidae *(Pilumnus)* hatched in the laboratory (Decapoda: Brachyura). *Pak. J. Sci. Ind. Res.* **13**, 420–6.

HASHMI, S. S. (1970d) The larval development of *Philyra corallicola* (Alcock) under laboratory conditions (Brachyura, Decapoda). *Pak. J. Zool.* **2**, 219–33.

HAZLETT, B. A. and PROVENZANO, A. J. (1965) Development of behaviour in laboratory-reared hermit crabs. *Bull. mar. Sci. Gulf and Carib.* **15**, 616–33.

HEATH, H. (1924) The external morphology of certain phyllopods. *J. Morph.* **38**, 453–83.

HEEGAARD, P. (1947) Contribution to the phylogeny of the arthropods. *Skrif. Univ. Zool. Mus. København* **8**, 1–236.

HEEGAARD, P. (1953) Observations on spawning and larval history of the shrimp *Penaeus setiferus* (L.). *Publ. Inst. mar. Sci. Univ. Texas* **3**, 73–105.

HEEGAARD, P. (1969a) The first larval stage of *Chlorotocus crassicornis* (Decapoda, Pandalidae). *Crustaceana* **17**, 151–8.

HEEGAARD, P. (1969b) Larvae of decapod Crustacea: the Amphionidae. *Dana Rep. Carlsberg Found.* **17**, 5–82.

HEIDECKE, P. (1904) Untersuchungen über die ersten Embryonalstadien von *Gammarus locusta*. *Jena Zeitschr.* **38**, 505–52.

HELDT, J. H. (1931) Observations sur la ponte, la fécondation et les premiers stades du développement de l'œuf chez *Penaeus caramote* Risso. *C.R. Acad. Sci., Paris* **193**, 1039–41.

HELDT, J. H. (1938) La reproduction chez les Crustacés Decapodes de la famille des Penéides. *Ann. Inst. oceanogr.* **18**, 31–206.

HENDERSON, J. T. (1926) Description of a copepod gill parasite of pike-perches in lakes of northern Quebec, including an account of the free-swimming male and of some developmental stages. *Contrib. Canad. Biol.* (N.S.) **3**, 235–45.

HERRICK, F. H. (1891) The development of the American lobster. *Zool. Anz.* **14**, 133–7.

HERRICK, F. H. (1895) The American lobster, a study of its habits and development. *Bull. U.S. Fish. Comm.* **13**, 1–252.

HERRING, P. J. (1967) Observations on the early larvae of three species of *Acanthephyra* (Crustacea, Decapoda, Caridea). *Deep Sea Research* **14**, 325–9.

HERRNKIND, W. F. (1968) The breeding of *Uca pugilator* (Bosc) and mass rearing of the larvae with comments on the behaviour of the larval and early crab stages. *Crustaceana*, suppl. 2, pp. 214–24.

HESSLER, R. R. and SANDERS, H. L. (1966) *Derocheilocaris typicus* Pennak and Zinn (Mystacocarida) revisited. *Crustaceana* **11**, 141–55.

HICKMAN, V. V. (1937) The embryology of the syncarid crustacean *Anaspides tasmaniae*. *Pap. Proc. R. Soc. Tasm.* **1936**, 1–36.

HSU, F. (1933) Studies on the anatomy and development of a freshwater phyllopod, *Chirocephalus nankingensis* (Shen). *Contr. Biol. Lab. Sci. Soc. China Nanking Zool.* **9**, 119–63.

HUBSCHMANN, J. H. and ROSE, J. A. (1968) *Palaemonetes kodiakensis* Rathbun, post-embryonic growth in the laboratory (Decapoda, Palaemonidae). *Crustaceana* **16**, 81–87.

HUDINAGA, M. (1942) Reproduction, development and rearing of *Penaeus japonicus* Bate. *Jap. Jl. Zool.* **10**, 305–93.

HUDINAGA, M. and KASAHARA, H. (1941) Larval development of *Balanus amphitrite hawaiensis*. *Zool. Mag., Tokyo* **54,** 108–18.

HUMES, A. G. (1955) Post-embryonic development of the calanoid copepod *Epischura*. *J. Morph.* **96,** 441–72.

HUMPERDINCK, I. (1942) Über Muskulatur und Endoskelett von *Polyphemus pediculus* de Geer. Topographisches und Embryologisches. *Z. wiss. Zool.* **121,** 621–55.

INGLE, R. W. and RICE, A. L. (1971) The larval development of the masked crab, *Corytes cassivelaunus* (Pennant) (Brachyura, Corystidae) reared in the laboratory. *Crustaceana* **20,** 271–84.

ISHIDA, S. and YASUGI, R. (1937) Free swimming stages of *Balanus amphitrite albicostatus*. *Botany and Zool., Tokyo* **5,** 1659–66.

ISHIKAWA, C. (1885) On the development of the freshwater macrurous crustacean *Atyephira compressa* de Haan. *Q. Jl. microsc. Sci.* **25,** 391–428.

ISHIKAWA, C. (1902) Über rhythmische Auftreten der Furchungslinie bei *Atyephyra compressa* de Haan. *Arch. EntwMech. Org.* **15,** 535–42.

JAKOBI, H. (1954) Biologie, Entwicklungsgeschichte und Systematik von *Bathynella natans* Vedj. *Zool. Jahrb. Syst.* **83,** 1–62.

JEPSON, J. (1965) Marsupial development of *Boreomysis arctica* (Kroyer, 1861). *Sarsia* **20,** 1–8.

JOHN, P. A. (1968) *Habits, Structure and Development of* Sphaeroma terebrans *(a wood-boring isopod)*, University of Kerala, Trivandrum.

JOHNSON, M. C. and FIELDING, J. R. (1956) Propagation of the white shrimp, *Penaeus setiferus* (Linn.), in captivity. *Tulane Stud. Zool.* **4,** 173–190.

JOHNSON, M. W. (1934a) The life history of the copepod *Tortanus descandatus* (Thompson and Scott). *Biol. Bull., Woods Hole* **67,** 182–200.

JOHNSON, M. W. (1934b) The developmental stages of the copepod *Epilabidocera amphitrites* (McMurrich). *Biol. Bull., Woods Hole* **67,** 466–83.

JOHNSON, M. W. (1937), The developmental stage of the copepod *Eucalanus elongatus* Dana var. *bungii* Giesbrecht. *Trans. Amer. micr. Soc.* **56,** 79–98.

JOHNSON, M. W. (1948) The post-embryonic development of the copepod *Pseudodiaptomus euryhalimus* Johnson and its phylogenetic significance. *Trans. Amer. micr. Soc.* **67,** 319–30.

JOHNSON, M. W. (1966) The nauplius larva of *Eurytemora herdmani* Thompson and Scott 1897 (Copepoda, Calanoida). *Crustaceana* **11,** 307–13.

JOHNSON, M. W. (1971) The phyllosoma larva of *Scyllarus delfini* (Bouvier) (Decapoda, Palurinidea). *Crustaceana* **21,** 161–4.

JOHNSON, M. W. and KNIGHT, M. (1966) The phyllosoma larvae of the spiny lobster *Palinurus inflatus* (Bouvier). *Crustaceana* **10,** 31–47.

JONES, D. H. and MATHEWS, J. B. L. (1968) On the development of *Sphyrion lumpi* (Kroyer). *Crustaceana*, suppl. I, pp. 177–85.

JONES, L. W. G. and CRISP, D. J. (1954) The larval stages of the barnacle *Balanus improvisus* Darwin. *Proc. zool. Soc., Lond.* **123,** 765–80.

JORGENSEN, O. M. (1925) The early stages of *Nephrops norvegicus* from the Northumberland plankton together with a note on the post-larval development of *Homarus vulgaris*. *J. mar. biol. Ass. U.K.* **13,** 870–6.

KAJISHIMA, T. (1950) Studies on the embryonic development of *Leander pacificus* Stimpson. II Early development. III Later embryonic development. *Zool. Mag., Tokyo* **59,** 82–86 and 108–11.

KAJISHIMA, T. (1952a) The development of half blastomeres of *Balanus tintinnabulum*. *Zool. Mag., Tokyo* **61,** 18-21.

KAJISHIMA, T. (1952b) Experimental studies on the embryonic development of the isopod crustacean *Megaligia exotica* Raux. *Annot. zool. Jap.* **25,** 172–81.

KAUDEWITZ, F. (1950) Zur Entwicklungsphysiologie von *Daphnia pulex*. *Arch. EntwMech. Org.* **144,** 410–77.

KAUFMANN, R. (1965) Zur embryonal und Larvalentwicklung von *Scalpellum scalpellum* L. (Crust. Cirr.) mit einen Beitrag zur Autokologie dieser Art. *Z. Morph. Ökol. Tiere* **55,** 161–232.

KINGSLEY, J. S. (1887) The development of *Crangon vulgaris*. *Bull. Essex Inst.* **18,** 99–153.

KINGSLEY, J. S. (1889) The development of *Crangon vulgaris*, Pt. 3. *Bull. Essex Inst.* **21,** 1–41.

KLAPOW, L. A. (1970) Ovoviviparity in the genus *Exocirolana* (Crustacea, Isopoda). *J. Zool.* **162,** 359–69.

KNIGHT, M. D. (1966) The larval development of *Polyonyx quadriungulatus* Glassell and *Pachycheles rudis* Stimpson (Decapoda, Porcellanidae) cultured in the laboratory. *Crustaceana* **10,** 75–97.

KNIGHT, M. D. (1967) The larval development of the sand crab *Emerita rathbunae* Schmidt (Decapoda, Hippidae). *Pacific Sci.* **21,** 58–76.

KNIGHT-JONES, E. W. and WAUGH, G. D. (1949) On the larval development of *Elminius modestus* Darwin. *J. mar. biol. Ass. U.K.* **28**, 413–28.

KOGA, F. (1960) The developmental stages of nauplius larvae of *Parenchauta russelli* (Farran). The nauplius larvae of *Centropages abdominalis* Sato. *Bull. jap. Soc. Sci. Fish.* **26**, 792–6 and 877–81.

KRAINSKA, M. K. (1934) Recherches sur le développement d'*Eupagurus prideauxi* Leach. I. Segmentation et gastrulation. *Bull. int. Acad. Cracovie* B.**2**, 141–65.

KRAINSKA, M. K. (1936) On the development of *Eupagurus prideauxi* Leach. *C.R. Int. Congr. Zool.* **12**, 554–65.

KRÜGER, P. (1922) Die Embryonalentwicklung von *Scalpellum scalpellum* L. *Arch. mikrosk. Anat.* **96**, 355–86.

KUERS, L. M. (1961) Zur Augenentwicklung von *Porcellio scaber*. *Crustaceana* **3**, 24–30.

KÜHN, A. (1908) Die Entwicklung der Keimzellen in den parthenogenetischen Generationen der Cladoceren *Daphnia pulex* (de Geer). *Arch. Zellforsch.* **I**, 538–86.

KÜHN, A. (1911) Über determinative Entwicklung der Cladoceren. *Zool. Anz.* **38**, 345–57.

KÜHN, A. (1913) Die Sonderung der Keimesbezirke in der Entwicklung der Sommereier von *Polyphemus pediculus* de Geer. *Zool. Jb. Anat. Ont.* **35**, 243–340.

KÜHNEMUND, E. (1929) Die Entwicklung der Scheitel-platte von *Polyphemus pediculus* de Geer von der Gastrula bis zu Differenzierung der aus ihr hervogendes Organe. *Zool. Jb. Anat. Ont.* **50**, 385–432.

KUHNERT, L. (1934) Beiträge zur Entwicklungsgeschichte von *Alcippe lampas* Hancock. *Z. Morph. Ökol. Tiere* **29**, 45–78.

LANG, K. (1948) Copepoda "Notodelphyoida" from the Swedish west coast, with an outline of the systematics of the copepods. *Ark. Zool.* **40**, 1–36.

LANG, R. and FIORONI, P. (1971) Darmentwicklung und Dotteraufschluss bei *Macropodia* (Crustacea, Malacostraca, Decapoda, Brachyura). *Zool. Jb. Anat. Ont.* **88**, 84–137.

LANGENBECK, C. (1898) Formation of the germ layers in the amphipod *Microdeutopus gryllotalpa* Costa. *J. Morph.* **14**, 301–36.

LAWINSKI, L. and PAUTSCH, F. (1969) A successful trial to rear larvae of the crab *Rithropanopeus harrisi* (Gould) subsp. *tridentatus* (Maitland) under laboratory conditions. *Zool. Pol.* **19**, 495–504.

LEBEDINSKY, J. (1890) Einige Untersuchungen über die Entwicklungsgeschichte der Seekrabben. *Biol. Zbl.* **10**, 178–85.

LEBEDINSKY, J. (1891) Die Entwicklung de *Daphnia* aus dem Sommereie. *Zool. Anz.* **14**, 149–52.

LEBOUR, M. V. (1916) Stages in the life history of *Calanus finmarchicus* (Gennenes), experimentally reared by Mr. L. R. Crawshay in the Plymouth laboratory. *J. mar. biol. Ass. U.K.* **11**, 1–17.

LEBOUR, M. V. (1938) The newly-hatched larva of *Callianassa affinis* Holmes. *Proc. zool. Soc. Lond.* **108**, 47–48.

LEBOUR, M. V. (1941) Notes on thalasiniid and processid larvae (Crustacea Decapoda) from Bermuda. *Ann. Mag. Nat. Hist.* (11), **7**, 402–20.

LEMERCIER, A. (1957) Sur le développement in vitro des embryons d'un Crustacé isopode asillate, *Jaera marina*, Fabr. *C.R. Acad. Sci., Paris* **244**, 1280–3.

LEWIS, A. G. (1963) Life history of the caligid copepod *Lepeophtheirus dissimulatus* Wilson 1905 (Crustacea Caligoida). *Pacific Sci.* **17**, 195–242.

LEWIS, J. B. and WARD, J. (1965) Developmental stages of the palaemonid shrimp *Macrobrachium carcinus* (Linnaeus 1758). *Crustaceana* **9**, 137–48.

LING, S. W. (1967) The general biology and development of *Macrobrachium rosenbergii* (de Man). *F.A.O. World Sci. Conference on Biology and Culture of Shrimps and Prawns.* Mexico F.A.O. Experience Paper No. 30, pp. 1–18.

LITTLE, G. (1969) The larval development of the shrimp, *Palaemon macrodactylus* Rathbun, reared in the laboratory, and the effect of eyestalk extirpation on development. *Crustaceana* **17**, 69–87.

LOCHHEAD, J. H. (1936) On the feeding mechanism of the nauplius of *Balanus perforatus* Brugière. *Proc. Linn. Soc. Lond. (Zool.)* **39**, 429–42.

LONGHURST, A. (1955) A review of the Notostraca. *Bull. Brit. Mus. (Nat. Hist.) Zoology* **3**, 1–15.

MACMILLAN, F. E. (1972) The larval development of Northern California Porcellanidae (Decapoda, Anomura). I. *Pachycheles pubescens* Hohnes in comparison to *Pachycheles rudis* Stimpson. *Biol. Bull., Woods Hole* **142**, 57–70.

MCMURRICH, J. P. (1895) Embryology of the isopod Crustacea. *J. Morph.* **11**, 63–154.

MANNING, R. B. (1963) Notes on the embryology of the stomatopod crustacean *Goniodactylus oestadii* Hansen. *Bull. mar. Sci. Gulf and Carib.* **13**, 422–32.

MANNING, R. B. and PROVENZANO, A. J. (1963) Studies on the development of stomatopod Crustacea. I. Early larval stages of *Gonodactylus oerstedii* Hansen. *Bull. mar. Sci. Gulf and Carib.* **13,** 467–87.
MANTON, S. M. (1928) On the embryology of a mysid crustacean *Hemimysis lamornae. Phil. Trans. R. Soc.* B, **216,** 363–463.
MANTON, S. M. (1934) On the embryology of *Nebalia bipes. Phil. Trans. R. Soc.* B, **223,** 168–238.
MARKAROV, R. R. (1968) Abbreviated larval development in decapods (Crustacea: Decapods). *Zool. Zhur.* **47,** 348–59.
MARSHALL, S. M. and ORR, A. P. (1955) *Biology of a Marine Copepod,* Oliver & Boyd, Edinburgh.
MATTOX, N. T. (1937) Studies on the life history of a new species of fairy shrimp, *Eulimnadia diversa. Trans. Amer. microsc. Sci.* **56,** 249–55.
MATTOX, N. T .(1950) Notes on the life history and description of a new species of conchostracan phyllopod, *Caenestheriella gynoecia. Trans. Amer. microsc. Sci.* **69,** 50–53.
MAUCHLINE, J. (1965) The larval development of the euphausiid *Thysanoessa rashii* (M. Sars). *Crustaceana* **9,** 31–40.
MAUCHLINE, J. (1967) Feeding appendages of the Euphasiacea (Crustacea). *J. Zool.* **153,** 1–43.
MAUCHLINE, J. (1970) The biology of *Thysanoessa rashii* (M. Sars) with a comparison of its diet with that of *Meganyctiphanes norvegica* (M. Sars). In BARNES, H. (ed.), *Some Contemporary Studies in Marine Science,* Allen & Unwin, London.
MENON, M. K. (1952) The life history and bionomics of an Indian penaeid prawn *Metapenaeus dobsoni* Miers. *Proc. Indo. Pacif. Fish Conc.* 3, sect. 2, 80–93.
METCHNIKOFF, E. (1868) On the development of *Nebalia. Mem. Acad. Sci. Petersb.* **13,** 1–48.
MICHEL, A. and MANNING, R. B. (1972) The pelagic larvae of *Cherisquilla tuberculata* (Borradaile, 1970) (Stomatopoda). *Crustaceana* **22,** 113–26.
MIRZOYEVA, L. M. (1969) Nauplius stages of *Sinergasilus lieni* Yin, 1949 (Crustacea, Copepoda parasitica). *Dopov. Akad. Nauk URSR* **31,** 753–5.
MODIN, J. C. and COX, K. W. (1967) Post-embryonic development of laboratory-reared Ocean shrimp, *Pandalus jordaoni* Rathbun. *Crustaceana* **13,** 197–219.
MORIN, J. (1886) Zur Entwicklungsgeschichte des Flusskrebses. *Zap. Novoross. Obsch.* **10,** 1–22.
MOROFF, T. (1912) Entwicklung und physiologische Bedeutung des Medianauges bei Crustacean. *Zool. Jb. Anat. Ont.* **34,** 473–620.
MORRIS, M. C. (1948) Life history of an Australian crustacean *Acetes australis* (Decapoda, Tribe Penaeidae). *Proc. Linn. Soc. N.S.W.* **73,** 1–15.
MORRIS, M. C. and BENNETT, I. (1952) The life history of a penaeid prawn *(Metapenaeus)* breeding in a coastal lake (Tuggerah, New South Wales). *Proc. Linn. Soc. N.S.W.* **77,** 164–82.
MOYSE, J. (1964) Feeding in barnacle nauplii. *Rep. Challenger Soc.* **3** (16), 23.
MOYSE, J. and KNIGHT-JONES, E. W. (1967) Biology of cirripede larvae. *Mar. Biol. Assoc. India Symp.,* Ser. 2, 595–611.
MULLER CALÉ, C. (1913) Über die Entwicklung von *Cypris incongruens. Zool. Jb. Anat. Ont.* **36,** 113–70.
MURAKAMI, Y. (1961) Studies on the winter eggs of the water flea *Moina macropa* Straus. *J. Fac. Fish. Hiroshima Univ.* **3,** 323–49.
NAIR, K. B. (1939) The reproduction, oogenesis and development of *Mesopodopsis orientalis* Tatt. *Proc. Ind. Acad. Sci.* B, **9,** 175–223.
NAIR, K. B. (1941) On the embryology of *Squilla. Proc. Ind. Acad. Sci.* B, **14,** 543–76.
NAIR, K. B. (1949) On the embryology of *Caridina laevis. Proc. Ind. Acad. Sci.* B, **29,** 211–88.
NAKANISHA, Y. H., IWASAKI, T., OKIGAKI, T. and KATO, H. (1962) Cytological studies of *Artemia salina.* 1. Embryonic development without cell multiplication after the blastula stage in encysted dry eggs. *Annot. Zool. Japan* **35,** 223–8.
NASSANOW, W. N. (1885) Zur Embryonalentwicklung von Balanus. *Zool. Anz.* **8,** 44–47.
NASSANOW, W. N. (1887) On the ontogeny of the crustaceans *Balanus* and *Artemia. Izsvest. imp. Obshch. Ligubit Estestv. Antrop. i Ethnog. Moscow* **52,** 1–14.
NATARAJ, S. (1947) Preliminary observations on the bionomics, reproduction and embryonic stages of *Palaemon idae* Hiller (Crustacea Decapoda). *Rec. Ind. Mus. Calcutta* **45,** 89–96.
NEEDHAM, A. E. (1937) The development of *Neomysis. Q. Jl. microsc. Sci.* **79.**
NEEDHAM, A. E. (1942) The structure and development of the segmental excretory organs of *Asellus aquaticus. Q. Jl. microsc. Sci.* **83,** 205–43.
NEWMAN, W. A., ZULLO, V. A. and WITHERS, T. H. (1969) In MOORE, R. C. (ed.), *Treatise on Invertebrate Palaeontology,* Part R, Arthropoda 4, University of Kansas and Geological Society of America.

NICHOLLS, A. G. (1934) The developmental stages of *Euchaeta norvegica* Broch. *Proc. R. Soc. Edin.* **54,** 31–50.

NICHOLLS, A. G. (1935) The larval stages of *Longipedia coronata* Claus, *L. scotti* G. O. Sars, and *L. minor* T. and A. Scott, with a description of the male of *L. scotti*. *J. mar. biol. Ass. U.K.* **20,** 29–45.

NORRIS, E. and CRISP, D. J. (1953) The distribution and planktonic stages of the cirripede *Balanus perforatus* Brugière. *Proc. zool. Soc. Lond.* **123,** 393–409.

NOURISSON, M. (1959) Quelques données relatives au développement post-embryonnaire de *Chirocephalus stagnalis* Shaw. *Terre et la Vie* **106,** 174–82.

NOURISSON, M. (1962) Maturation, fecondation et segmentation de l'œuf de *Chirocephalus stagnalis* Shaw (Crustacé Phyllopode). *C.R. Acad. Sci., Paris* **254,** 2567–3569.

NUSBAUM, J. (1886) L'embryologie d'*Oniscus murarina. Zool. Anz.* **9,** 454–8.

NUSBAUM, J. (1887) L'embryologie de *Mysis chameleo* (Thompson). *Arch. Zool. exp. gen.* (2), **5,** 123–202.

NUSBAUM, J. (1890a) Bildung und Anzahl der Richtungskörper bei Cirripedien. *Zool. Anz.* **12,** 122.

NUSBAUM, J. (1890b) *Anatomische Studien an Californischen Cirripedien*, Bonn.

NUSBAUM, J. (1891) Beiträge zur Embryologie der Isopoden. *Biol. Centrabl.* **11,** 42–49.

NUSBAUM, J. (1898) Zur Entwicklungsgeschichte des Mesoderms bei den Parasitischen Isopoden. *Biol. Centrabl.* **18,** 557–69.

NUSBAUM, J. (1904) Nouveaux recherches sur l'embryologie des Isopodes *(Cymothoa)*. *Zool. Centrabl.* **12,** 130–3.

NYBLADE, C. F. (1970) Larval development of *Pagurus annulipes* (Stimpson, 1862) and *Pagurus pollicaris* Say, 1817, reared in the laboratory. *Biol. Bull., Woods Hole* **139,** 557–73.

OEHMICHEN, A. (1921) Die Entwicklung der äusseren Form des *Branchipus grubei* Dyb., *Zool. Anz.* **53,** 241–53.

OISHI, S. (1959) Studies on the teloblasts in the decapod embryo. I. Origin of teloblasts in *Heptacarpus rectirostris* Stimpson. *Embryologia* **4,** 283–309.

OISHI, S. (1960) Studies on the teloblasts in the decapod embryo. II. Origin of teloblasts in *Pagurus samuelis* (Stimpson) and *Hemigrapsus sanguineus* (de Haan). *Embryologia* **5,** 270–82.

PACKARD, A. S. (1883) A monograph of the phyllopod Crustacea of North America with remarks on the order Phyllocarida. *Ann. Rep. U.S. Geol. Geogr. Surv. Terr.* **12,** sect. 2, 295–592.

PAI, P. G. (1958) On post-embryonic stages of phyllopod crustaceans, *Triops (Apus)*, *Streptocephalus* and *Estheria*. *Proc. Ind. Acad. Sci.* B, **48,** 229–50.

PANDIAN, T. J. (1970) Yolk utilization and hatching time in the Canadian lobster *Homarus americanus*. *Mar. Biol. Berlin* **7,** 249–54.

PATTEN, W. (1890) On the origin of vertebrates from arachnids. *Q. Jl. microsc. Sci.* **31,** 317–78.

PEARSON, J. C. (1939) The early life histories of some American Penaeidae, chiefly the commercial shrimp. *Penaeus setiferus* (Linn.). *Bull. U.S. Bureau Fish.* **49,** 1–73.

PEDASCHENKO, D. (1893) Sur la segmentation de l'œuf et la formation des feuillets embryonnaires chez la *Lernaea branchialis. Rev. Sci. Nat. Petersb.* **4,** 186–99.

PEDASCHENKO, D. (1897) Über die Entwicklung des Nerven-systems und des Genitalzellen und die Dorsalorgane von *Lernaea branchialis. Trav. Soc. Nat. St. Petersb.* **27,** 187–94 and 208–10.

PEDASCHENKO, D. (1898) Embryonalentwicklung und Metamorphose von *Lernaea branchialis. Trav. Soc. Nat. St. Petersb.* **26,** 247–307.

PEREYASLAWZEWA, S. (1888a) Le développement de *Gammarus poecilurus* Rathké. *Bull. Soc. Imp. Nat. Mosc.* **2,** 187–219.

PEREYASLAWZEWA, S. (1888b) Le développement de *Caprella ferox* Chun. *Bull. Soc. Imp. Nat. Mosc.* **2,** 583–97.

PETRICONI, V. (1968) Zur Bildung des präantennalen Mesoderms bei *Neomysis integer* un Honblick auf die Kopfsegmentierung. *Zool. Jb. Anat. Ont.* **85,** 579–96.

PIATAKOV, M. L. (1925) Zur Embryonalentwicklung von *Lepidurus apus* und *Triops cancriformis. Zool. Anz.* **62,** 234–6.

PIKE, R. B. and WEAR, R. G. (1969) Newly hatched larvae in the genera *Gastroptychus* and *Uroptychus* (Crustacea, Decapoda, Galatheidae) from New Zealand waters. *Trans. R.S. New Zealand Biol. Sci.* **11,** 189–95.

PROVENZANO, A. J. (1962a) The larval development of the tropical land hermit *Coenobita clypeatus* (Herbst) in the laboratory. *Crustaceana* **4,** 207–28.

PROVENZANO, A. J. (1962b) The larval development of *Calcinus tibicens* (Herbst) (Crustacea Anomura) in the laboratory. *Biol. Bull., Woods Hole* **123,** 179–202.

PROVENZANO, A. J. (1967a) Recent advances in the laboratory culture of decapod larvae. *Proc. Symp. Crustacea*, Part II, 940–5.

PROVENZANO, A. J. (1967b) The zoeal stages and glaucothöe of the tropical eastern Pacific hermit crab *Trizopagurus magnificus* (Bouvier 1898) (Decapoda: Diogenidae) reared in the laboratory. *Pacific Sci.* **21,** 457–73.

PROVENZANO, A. J. (1968a) Recent experiments on the laboratory rearing of tropical lobster larvae. *Proc. Gulf Caribb. Fish. Inst.* **21,** 152–7.

PROVENZANO, A. J. (1968b) The complete larval development of the West Indian hermit crab *Petrochirus diogenes* (L.) (Decapoda, Diogenidae) reared in the laboratory. *Bull. mar. Sci. Gulf and Carib.* **18,** 143–81.

PROVENZANO, A. J. (1968c) Biological investigations of the deep sea. 37. *Lithopagurus yucatanicus*, a new genus and species of hermit crab with a distinctive larva. *Bull. mar. Sci. Gulf and Carib.* **18,** 627–44.

PROVENZANO, A. J. (1971a) Zoeal development of *Pylopaguropsis atlantica* Wass, 1963, and evidence from larval characters of some generic relationships within the Paguridae. *Bull. mar. Sci. Gulf and Carib.* **21,** 237–55.

PROVENZANO, A. J. (1971b) Rediscovery of *Munidopagurus macrocheles* (A. Milne Edwards, 1880) (Crustacea, Decapoda, Paguridae) with a description of the first zoeal stage. *Bull. mar. Sci. Gulf and Carib.* **21,** 256–66.

PROVENZANO, A. J. and RICE, A. L. (1964) The larval stages of *Pagurus marshi* Benedict (Decapoda, Anomura) reared in the laboratory. *Crustaceana* **7,** 217–35.

PROVENZANO, A. J. and RICE, A. L. (1966) Juvenile morphology and the development of taxonomic characters in *Paguristes sericeus* A. Milne Edwards (Decapoda, Diogenidae). *Crustaceana* **10,** 53–69.

PYEFINCH, K. A. (1948a) Methods of identification of the larvae of *Balanus balanoides* (L.), *B. crenatus* Brig. and *Verruca stroemia* O. F. Müller. *J. mar. biol. Ass. U.K.* **27,** 451–63.

PYEFINCH, K. A. (1948b) Notes on the biology of cirripedes. *J. mar. biol. Ass. U.K.* **27,** 464–503.

PYEFINCH, K. A. (1949) The larval stages of *Balanus crenatus* Brugière. *Proc. zool. Soc., Lond.* **118,** 916–23.

RAO, P. V. (1968) A new species of shrimp, *Acetes cochinensis* (Crustacea Decapoda, Sergestidae) from south-west coast of India with an account of its larval development. *J. mar. biol. Assoc., India* **10,** 298–320.

RAPPAPORT, R. JR. (1960) The origin and formation of blastoderm cells of gammarid Crustacea. *J. exp. Zool.* **144,** 43–60.

REICHENBACH, H. (1877) Die Embryonalanlage und erste Entwicklung des Flusskrebses. *Z. wiss. Zool.* **29,** 123–96.

REICHENBACH, H. (1886) Studien zur Entwicklungsgeschichte de Flusskrebses. *Abh. Senckenb. Ges., Frankfurt* **14,** 1–137.

REINHARD, W. (1887) Zur Ontogenie des *Porcellio scaber*. *Zool. Anz.* **10,** 9–13.

RENFRO, W. C. and COOK, H. L. (1963) Early larval stages of the seabob *Xiphopenaeus kroyeri* (Heller). *U.S. Fish Wildlife Service, Fishery Bull.* **63,** 165–77.

RICE, A. L., INGLE, R. W. and ALLEN, E. (1971) The larval development of the sponge crab *Dromia personata* (L.) (Crustacea, Decapoda, Dromiidea) reared in the laboratory. *Vie Milieu ser. A. Biol. Mar.* **21,** 223–40.

RICE, A. L. and PROVENZANO, A. J. (1965) The zoeal stages and the glaucothöe of *Paguristes sericens* A. Milne Edwards (Anomura, Diogenidae). *Crustaceana* **8,** 239–54.

RICE, A. L. and PROVENZANO, A. J. (1966) The larval development of the West Indian sponge crab *Dromidia antillensis* (Decapoda, Dromiidae). *J. Zool.* **149,** 297–319.

RICE, A. L. and PROVENZANO, A. J. (1970) The larval stages of *Homola barbata* (Fabricius) (Crustacea, Decapoda, Homolidea) reared in the laboratory. *Bull. mar. Sci. Gulf and Carib.* **20,** 446–71.

ROBERTS, M. H. (1968) Larval development of *Bathynectes superba* (Costa) reared in the laboratory. *Biol. Bull., Woods Hole* **137,** 338–51.

ROBERTS, M. H. (1970) Larval development of *Pagurus longicarpus* Say reared in the laboratory. I. Description of larval instars. *Biol. Bull., Woods Hole* **139,** 188–202.

ROBERTSON, R. B. (1968) The complete larval development of the sand lobster *Scyllarus americanus* (Smith) (Decapoda Scyllaridae) in the laboratory, with notes on larvae from the plankton. *Bull. mar. Sci. Gulf and Carib.* **18,** 294–342.

ROBERTSON, R. B. (1969) The early larval development of the scyllarid lobster *Scyllarides aequinoctialis* (Lund) in the laboratory, with a revision of the larval characters of the genus. *Deep-sea Res. Oceanogr. Abst.* **16,** 557–86.

ROBINSON, M. (1906) On the development of *Nebalia*. *Q. Jl. microsc. Sci.* **50,** 383–433.
ROSSIISKAYA-KOJEWNIKOWA, M. (1890) Le développement de la *Sunamphitoe valida* Czerniquski et de l'*Amphitoe picta* Rathke. *Bull. Soc. Imp. Nat. Mosc.* **4,** 82–103.
ROSSIISKAYA-KOJEWNIKOWA, M. (1893) Sur la formation des organes genitaux chez les amphipodes. *Zool. Anz.* **16,** 33–35.
ROSSIISKAYA-KOJEWNIKOWA, M. (1896) Le développement embryonnaire de *Gammarus*. *Bull. Soc. Imp. Nat. Mosc.* **10,** 53–62.
ROULE, L. (1889) Sur l'évolution initiale des feuillets blastodermiques chez les Crustacés isopodes (*Asellus aquaticus* L. et *Porcellio scaber* Latr.). *C.R. Acad. Sci., Paris* **109,** 78–79.
ROULE, L. (1890) Sur le développement du blastoderme chez les Crustacés isopodes *(Porcellio scaber)*. *C.R. Acad. Sci., Paris* **110,** 1373–4.
ROULE, L. (1891) Sur le développement des feuillets blastodermiques chez les Crustacés isopodes *(Porcellio scaber)*. *C.R. Acad. Sci., Paris* **112,** 1460–2.
ROULE, L. (1892a) Sur le développement du mesoderme des Crustacés et sur celui de ces organes derivées. *C.R. Acad. Sci., Paris* **113,** 153–5.
ROULE, L. (1892b) Sur les premieres phases de développement des Crustacés edriophthalmes. *C.R. Acad. Sci., Paris* **113,** 868–70.
ROULE, L. (1895) Études sur le développement des Crustacés. I. Le développement du *Porcellio scaber*. *Ann. Sci. Nat. Zool.* (7) **18,** 1–156.
ROULE, L. (1896a) Études sur le développement des Crustacés. La segmentation ovulaire et le façonnement du corps chez l'*Asellus aquaticus*. *Ann. Sci. Nat. Zool.* (8) **2,** 163–96.
ROULE, L. (1896b) Études sur le développement embryonnaire des Crustacés. 2. Développement du *Palaemon serratus* Lab. *Ann. Sci. Nat. Zool.* **8,** 1–116.
ROUX, A. LE (1966) Le développement larvaire de *Porcellana longicornis* Pennant (Crustacea, Decapoda, Anomura, Galatheidae). *Cah. Biol. Mar.* **7,** 69–78.
SAMASSA, P. (1893a) Die Keimblätterbildung bei den Cladoceren. I. *Moina rectirostris*, Baird. *Arch. mikrosk. Anat.* **41,** 339–66.
SAMASSA, P. (1893b) Die Keimblätterbildung bei *Moina*. *Zool. Anz.* **16,** 434.
SAMASSA, P. (1893c) Die Keimblätterbildung bei den Cladoceren. II. *Daphnella* und *Daphnia*. *Arch. mikrosk. Anat.* **41,** 650–88.
SAMASSA, P. (1897) Die Furchung der Wintereiern der Cladoceren. *Zool. Anz.* **20,** 51–55.
SAMTER, M. (1900) Studien zur Entwicklungsgeschichte der *Leptodora hyalina* Lillij. *Z. wiss. Zool.* **68,** 169–260.
SANDERS, H. L. (1963a) The Cephalocarida. Functional morphology, larval development, comparative external anatomy. *Memoirs Connecticut Acad. Arts Sci.* **150,** 1–80.
SANDERS, H. L. (1963b) Significance of the Cephalocarida. In WHITTINGTON, H. B. and ROLFE, W. D. I. (eds.), *Phylogeny and Evolution of the Crustacea*, Museum Comp. Zool., Cambridge, Massachusetts.
SANDERS, H. L. and HESSLER, R. R. (1964) The larval development of *Lightiella incisa* Gooding (Cephalocarida). *Crustaceana* **7,** 81–97.
SANDIFER, P. A. and VAN ENGEL, W. A. (1971) Larval development of the spider crab *Libinia dubia* H. Milne Edwards (Brachyura, Majidae, Pisinae) reared in laboratory culture. *Chesapeake Sci.* **12,** 18–25.
SANDISON, E. E. (1954) The identification of the nauplii of some South African barnacles with notes on their life histories. *Trans. R. Soc. S. Africa* **34,** 69–101.
SANDISON, E. E. (1967) The naupliar stages of *Balanus pallidus stutsburi* Darwin and *Chthamalus aestuarii* Stubbings (Cirripedia Thoracica). *Crustaceana* **13,** 161–74.
SANKOLLI, K. N. (1961) On the early larval stages of two leucosiid crabs, *Philyra corallicola* Alcock and *Arcania septemspinosa* (Fabricius). *J. mar. biol. Ass. India* **3,** 87–91.
SARS, G. O. (1887) On *Cyclestheria hislopi* (Baird), a new generic type of bivalve Phyllopoda, raised from dried Australian mud. *Forhandl. Vidensk. selsk. Christiana* **1,** 1–65.
SARS, G. O. (1896a) Phyllocarida og Phyllopoda. *Fauna Norvegiae* **1,** 1–140.
SARS, G. O. (1896b) Development of *Estheria packardi* Brady, as shown by artificial hatching from dried mud. *Arch. Math. Naturvidensk.* **18,** 3–27.
SAUDRIER, Y. and LEMERCIER, A. (1960) Observations sur le développement des œufs de *Ligia oceanica* Fabr. Crustacé Isopode oniscoide. *Bull. Inst. oceanogr. Monaco*, No. 1162, pp. 1–11.
SCHIMKEWITSCH, W. (1896) Studien über parasitische Copepoden. *Z. wiss. Zool.* **61,** 339–62.
SCHOLL, G. (1963) Embryologische Untersuchungen an Tanaidaceen. *Zool. Jb. Anat. Ont.* **80,** 500–54.
SHEADER, M. and CHIA, F. S. (1970) Development, fecundity and brooding behaviour of the amphipod *Marinogammarus obtusatus*. *J. mar. biol. Ass. U.K.* **50,** 1079–99.

SHEPHERD, M. C. (1969) The larval morphology of *Myonyx transversus* (Haswell), *Pisidia dispar* (Stimpson) and *Pisidia streptochirhoides* (De Man) (Decapoda: Porcellanidae). *Proc. R. Soc. Queensland* **80**, 97–123.

SHIINO, S. M. (1942) Studies on the embryology of *Squilla oratoria* de Haan. *Mem. Coll. Sci. Kyoto Univ.* B, **28**, 77–174.

SHIINO, S. M. (1950) Studies on the embryonic development of *Palinurus japonicus* (von Siebold). *J. Fac. Fisheries, Prefecterate Univ. of Mie, Otanimachi* **1**, 1–168.

SIMMS, H. W. JR. (1966) Notes on the newly hatched phyllosoma of the sand lobster *Scyllarus americanus* (Smith). *Crustaceana* **11**, 288–90.

SMITH, E. W. (1953) The life history of the crawfish *Orconectes (Faxonella) clypeatus* (Hay). *Tulane Stud. Zool.* **1**, 79–96.

SMITH, W. C. (1933) A lobster rearing experiment contributing some addition to knowledge of the early life history of *Homarus vulgaris*. *Proc. Lpool. Biol. Soc.* **47**, 5–16.

SNETLAGE, E. (1905) Über die Frage vom Muskelansatz und der Herkunft der Muskulatur bei den Arthropoden. *Zool. Jb. Anat. Ont.* **21**, 495–515.

SOH CHENG LAM (1969) Abbreviated development of a non-marine crab, *Sesarma (Geosesarma) perracae* (Brachyura, Grapsidae) from Singapore. *J. Zool.* **158**, 357–70.

SOLLAUD, E. (1923) Recherches sur l'embryogenie des Crustacés decapodes de la sous-famille des Palaemoninae. *Bull. Biol. Fr.-Belg.* **5** suppl., 1–234.

SOMME, J. D. (1934) Animal plankton of the Norwegian coast waters and the open seas. I. Production of *Calanus finmarchicus* (Gunner) and *Calanus hyperboreus* (Kroyer) in the Lofoten Area. *Fiskeridir. Skr. Havundersog* **4**, 1–163.

SPANDENBERG, F. (1875) Zur Kenntnis von *Branchipus stagnalis*. *Z. wiss. Zool.* **25**, 1–64.

SPROSTON, N. G. (1942) The developmental stages of *Lernaeocera branchialis* (Linn.). *J. mar. biol. Ass. UK.* **25**, 441–66.

SQUIRES, H. J. (1965) Larvae and megalopa of *Argis dentata* (Crustacea, Decapoda) from Ungava Bay. *J. Fish. Res. Bd. Canada* **22**, 69–82.

STROMBERG, J.-O. (1965) On the embryology of the iosopod *Idotea*. *Ark. Zool.* **17**, 421–73.

STROMBERG, J.-O. (1967) Segmentation and organogenesis in *Limnoria lignorum* (Rathke) (Isopoda). *Ark. Zool.* **20**, 91–139.

STROMBERG, J.-O. (1971) Contribution to the embryology of bopyrid isopods with special reference to *Bopyroides*, *Hemiarthrus* and *Pseudione* (Isopoda, Epicaridea). *Sarsia* **47**, 1–46.

SUKO, T. (1962) Studies on the development of the crayfish. VII. The hatching and hatched young. *Sci. Rep. Saitama Univ.*, Ser. B, *Biol. Earth Sci.* **4**, 37–42.

TASCH, P. (1963) Evolution of the Branchiopoda. In WHITTINGTON, H. B. and ROLFE, W. D. I. (eds.), *Phylogeny and Evolution of the Crustacea*, Museum Comp. Zool., Cambridge, Massachusetts.

TAUBE, E. (1909) Beiträge zur Entwicklungsgeschichte der Euphausiden. I. Die Furchung der Eies bis zur Gastrulation. *Z. wiss. Zool.* **92**, 427–64.

TAUBE, E. (1915) Beiträge zur Entwicklungsgeschichte der Euphausiden. II. Von der Gastrula bis zum Furciliastadium, *Z. wiss. Zool.* **114**, 577–656.

TEMPLEMAN, W. (1936) The influence of temperature, salinity, light and food conditions on survival and growth of the larva of the lobster *(Homarus americanus)*. *J. Biol. Bd. Canada* **2**, 485–97.

TEMPLEMAN, W. (1939) Investigations into the life history of the lobster *(Homarus americanus)* on the west coast of Newfoundland, 1938. *J. Fish. Res. Bd. Canada* **5**, 71–83.

TEN, P. Y. and PAI, S. (1949) Über die totale und superfizielle Furchung bei *Chirocephalus nankinensis* und *Branchinella kugenumansis*. *Exptl. Cell. Res. suppl.* **1**, 537–9.

TERAO, A. (1921) On the development of *Palinurus japonicus* (v. Siebold). *Rep. Imp. Fish. Inst.*, Vol. 14, No. 5, 1–79.

TERAO, A. (1925) Zur Piatakov's Entdeckung eines Dorsalorgane bei *Potamobius*. *Zool. Anz.* **65**, 1–2.

TERAO, A. (1929) The embryonic development of the spiny lobster, *Palinurus japonicus*. *Jap. Jl. Zool.* **2**, 387–449.

TICHOMIROV, A. (1883) Sur l'embryologie des Amphipodes à l'eau douce. *Protoc. de la VIIe Ass. des net. Russ. a Odessa.*

TOKIOKA, T. (1936) Larval development and metamorphosis in *Argulus japonicus*. *Mem. Coll. Sci. Kyoto* **12**B, 93–114.

TRASK, T. (1970) A description of laboratory reared larvae of *Cancer productus* Randall (Decapoda–Brachyura) and a comparison to larvae of *Cancer magister* Dana. *Crustaceana* **18**, 133–46.

TURQUIER, Y. (1967a) Le développement larvaire de *Trypetesa nassarioides* Turquier, cirripède acrothoracique. *Arch. Zool. exp. gén.* **108**, 33–47.

TURQUIER, Y. (1967b) L'embryogénèse de *Trypetesa nassarioides* Turquier (cirripède acrothoracique). Ses rapports avec celle des autres cirripèdes. *Arch. Zool. éxp. gén.* **108**, 111–37.

TURQUIER, Y. (1970) Recherches sur le biologie des cirripèdes acrothoraciques. III. La métamorphose des cypris femelles de *Trypetesa nassarioides*. Turquier det de *T. lampas* (Hancock). *Arch. Zool. éxp. gén.* **111**, 573–628.

TURQUIER, Y. (1971) Recherches sur la biologie des cirripèdes acrothoraciques. IV. La metamorphose des cypris mâles de *Trypetesa nassarioides* Turquier et de *Trypetesa lampas* (Hancock). *Arch. Zool. éxp. gén.* **112**, 301–48.

TZUKERKIS, Y. M. (1964) On interspecific relations of *Astacus astacus* L. and *A. leptodactylus* Esch. in the lakes of east Lithuania. *Zool. Zh.* **43**, 172–7.

URBANOWICZ, F. (1884) Zur Entwicklungsgeschichte der Cyclopiden. *Zool. Anz.* **7**, 615–19.

URBANOWICZ, F. (1886) Contributions à l'embryologie des Copépodes. *Arch. Slav. Biol.* **1**, 663–7.

URBANOWICZ, F. (1893) Note préliminaire sur le développement embryonnaire de *Maja squinado*. *Biol. Zbl.* **13**, 348–53.

UTINOMI, H. (1961) Studies on the Cirripedia Acrothoracica. III. Development of the female and male of *Berndtia purpurea* Utinomi. *Publ. Seto mar. biol. Lab.* **9**, 413–46.

VAGHIN, V. L. (1947) On cleavage in the Acrothoracica and its relationship to cleavage in Arthropoda. *C.R. Acad. Sci., Moscow*, N.S., **55**, 367–70.

VAN BENEDEN, E. and BESSELS, E. (1869) Mémoire sur la formation du blastoderms chez les Amphipodes, les Lerneens et les Copepodes. *Mémoires couronnes d l'Acad. roy. Belg.* **24**,

VAN BENEDEN, E. (1870) Recherches sur l'embryongenie des Crustacés. IV. *Anchorella, Lerneapoda, Branchinella, Hessia*. *Bull. Acad. R. Belgique* **29**, 28.

VEILLET, A. (1961) Sur la métamorphose et la determinisme du sexe du cirripède *Scalpellum scalpellum* Leach. *C.R. Acad. Sci., Paris* **253**, 3087–8.

VOLLMER, C. (1912) Über die Entwicklung der Dauereier der Cladoceren. *Z. wiss. Zool.* **102**, 646–700.

WAGNER, J. (1891) Développement de la *Melita palmata*. *Bull. Soc. Nat. Mosc.* **5**, 401–9.

WAGNER, J. (1895) Zur Entwicklungsgeschichte der Schizopoden. Über Bildung der Mitteldarmepithel und die Entstehung der Sexualzellen bei *Neomysis vulgaris*. *Zool. Anz.* **17**, 437–40.

WAGNER, J. (1898) Einige Beobachtungen über die embryonalen Entwicklung von *Neomysis vulgaris* var. *Baltica* Czern. *Trudii St. Petersb. Obsch.* **26**, 1–221.

WAITE, F. C. (1889) The structure and development of the antennal glands in *Homarus americanus* Milne Edwards. *Bull. Mus. comp. Zool., Harvard*, **35**, 151–210.

WALLEY, L. J. (1969) Studies on the larval structure and metamorphosis of *Balanus balanoides* (L.). *Phil. Trans. R. Soc.* B, **256**, 237–80.

WEAR, R. G. (1965a) Larvae of *Petrocheles spinosus* Miers, 1876 (Crustacea, Decapoda, Anomura) with keys to New Zealand Porcellanid larvae. *Trans. R. Soc. N.Z. Zool.* **5**, 147–68.

WEAR, R. G. (1965b) Breeding cycles and pre-zoea larvae of *Petrolisthes elongatus* (Milne Edwards, 1837) (Crustacea, Decapoda). *Trans. Roy. Soc. N.Z. Zool.* **5**, 169–75.

WEAR, R. G. (1966) Pre-zoea larva of *Petrocheles spinosus* Miers, 1876 (Crustacea, Decapoda, Anomura). *Trans. R. Soc. N.Z. Zool.* **8**, 119–24.

WEAR, R. G. (1967) Life history studies on New Zealand Brachyura. I. Embryonic and postembryonic development of *Pilumnus novae-zealandiae* Filhol 1886, and of *P. lumpinus* Bennett, 1964 (Xanthidae, Pilumnae). *N.Z. Jl. mar. fw. Res.* **1**, 482–535.

WEAR, R. G. (1968a) Life history studies on New Zealand Brachyura. 2. Family Xanthidae Larvae of *Heterozius rotundifrons* A. Milne Edwards 1867, *Ozius truncatus* H. Milne Edwards 1834 and *Heteropanope (Pilumonopeus) serratifrons* (Kinahan, 1856). *N.Z. Jl. mar. fw. Res.* **2**, 293–332.

WEAR, R. G. (1968b) Life history studies on New Zealand Brachyura. 3. Family Ocypodidae. First stage zoea larva of *Hemiplax pistipes* (Jacquinot, 1853). *N.Z. Jl. mar. fw. Res.* **2**, 698–707.

WEAR, R. G. (1970a) Some larval stages of *Petalomera wilsoni* (Fulton and Grant, 1902) (Decapoda, Dromiidae). *Crustaceana* **18**, 1–12.

WEAR, R. G. (1970b) Notes and bibilography on the larvae of xanthid crabs. *Pacific Sci.* **24**, 84–89.

WEAR, R. G. (1970c) Life history studies on New Zealand Brachyura. 4. Zoea larvae hatched from crabs of the family Grapsidae. *N.Z. Jl. mar. fw. Res.* **4**, 3–35.

WEISZ, P. B. (1947) The histological pattern of metameric development in *Artemia salina*. *J. Morph.* **81**, 45–96.

WELDON, W. F. R. (1892) The formation of the germ layers in *Crangon vulgaris*. *Q. Jl. microsc. Sci.* **33**, 343–63.

WEYGOLDT, P. (1958) Die Embryonalentwicklung des Amphipoden *Gammarus pulex pulex* (L.). *Zool. Jb. Anat. Ont.* **77**, 51–110.

WEYGOLDT, P. (1960a) Embryologische Untersuchungen an Ostrakoden: Die Entwicklung von *Cyprideis litoralis. Zool. Jb. Anat. Ont.* **78**, 369–426.

WEYGOLDT, P. (1960b) Beitrag zur Kenntnis der Malakostrakenentwicklung. Die Keimblätterbildung bei *Asellus aquaticus* (L.). *Z. wiss. Zool.* **163**, 340–54.

WEYGOLDT, P. (1961) Beitrag zur Kenntnis der Ontogenie der Dekapoden: Embryologische Untersuchungen an *Palaemonetes varians* Leach. *Zool. Jb. Anat. Ont.* **79**, 223–70.

WILLIAMSON, D. I. (1965) Some larval stages of three Australian crabs belonging to the families Homolidae and Raninidae, and observations on the affinities of these families (Crustacea, Decapodea). *Aust. Jl. mar. F.W. Res.* **16**, 369–98.

WILLIAMSON, D. I. (1967) The megalopa stage of the homoloid crab *Latreilla australiensis* Henderson and comments on other homoloid megalopas. *Aust. Zool.* **14**, 206–11.

WILSON, C. B. (1905) North American parasitic copepods belonging to the family Caligidae. Part 1. The Caliginae. *Proc. U.S. Nat. Mus.* **28**, 479–672.

WILSON, C. B. (1907a) North American copepods belonging to the family Caligidae. Part 2. The Trebinae and Euryphorinae. *Proc. U.S. Nat. Mus.* **31**, 669–720.

WILSON, C. B. (1907b) North American copepods belonging to the family Caligidae. Parts 3 and 4. A revision of the Panderinae and Cecropinae. *Proc. U.S. Nat. Mus.* **33**, 323–490.

WILSON, C. B. (1911a) North American parasitic copepods. Part 9. The Lernaeopodidae. *Proc. U.S. Nat. Mus.* **39**, 189–226.

WILSON, C. B. (1911b) North American parasitic copepods belonging to the family Ergasilidae. *Proc. U.S. Nat. Mus.* **39**, 263–400.

WITSCHI, E. (1934) On determinative cleavage and yolk formation in the harpacticoid copepod *Tisbe furcata* Baird. *Biol. Bull., Woods Hole* **67**, 335–40.

WOLTERECK, R. (1898) Zur Bildung und Entwicklung des Ostracodeneies. *Z. wiss. Zool.* **69**, 596–623.

WOTZEL, R. (1937) Zur Entwicklung des Sommereies von *Daphnia pulex. Zool. Jb. Anat. Ont.* **63**, 455–70.

YANAGIMACHI, R. (1961a) Studies on the sexual organization of the Rhizocephala. III. The mode of sex determination in *Peltogastarella. Biol. Bull., Woods Hole* **120**, 272–83.

YANAGIMACHI, R. (1961b) The life cycle of *Peltogasterella* (Cirripedia Rhizocephala). *Crustaceana* **2**, 183–6.

YANG, W. T. (1968) The zoeae, megalopa and first crab of *Epialtus dilatatus* (Brachyura, Majidae) reared in the laboratory, *Crustaceana*, suppl. 2, 181–202.

YANG, W. T. (1970) The larval and post-larval development of *Parthenope serrata* reared in the laboratory and the systemic position of the Parthenopinae (Crustacea, Brachyura). *Biol. Bull., Woods Hole* **140**, 166–89.

ZEHNDER, H. (1934a) Ueber die Embryonalentwicklung des Flusskrebses. *Acta Zool. Stockh.* **15**, 261–448.

ZEHNDER, H. (1934b) Zur Embryologie des Flusskrebses. *Verh. schweiz. naturf. Ges.* **115**, 357–8.

ZEHNDER, H. (1935) Ueber die Embryonalentwicklung des Flusskrebses. *Vjschr. naturf. Ges. Zürich* **80**, 1–16.

CHAPTER 9

CHELICERATES

THE Chelicerata are important in the terrestrial fauna both for the conspicuousness and diversity of their larger members and for the vast infiltration of minute acarines into almost every terrestrial environment. Embryologically, however, these fascinating and successful creatures have not yet received the attention that they merit. In spite of a large literature, the embryonic development of most of the major subdivisions of the Chelicerata is still little known and even the spiders require a great deal more investigation before a critical understanding of their embryos can be achieved. Almost all of the work on chelicerate embryology has been carried out by investigators different from those who have contributed so largely to other aspects of arthropod embryology, with the result that chelicerate embryology has grown up in marked isolation, instigating and elaborating its own terminology. In conjunction with a lack of attention to such crucial aspects as the origin and further development of the mesoderm, the historical isolation of chelicerate embryology has progressively increased the problem of reconciliation with other branches of arthropod embryology and hence the difficulty of employing embryological data in examining the phylogeny of chelicerates. In the present chapter, an attempt is made to reverse this position by re-examining chelicerate embryos in accordance with the principles and terminology applied to the other arthropods in previous chapters. As will be seen, some interesting tentative conclusions emerge, although certain points need further clarification.

All arachnids, if one accepts the suppressed fourth leg segment of acarine larvae as a secondary feature, emerge from the egg (or are born, in the case of scorpions, pseudoscorpions and a few viviparous mites) as juveniles, with the adult body form and full complement of segments already developed. Development is thus direct and epimorphic. Furthermore, the same epimorphic development is evident in the Xiphosura. Even though the juvenile is a so-called trilobite larva, it has the full complement of segments and the basic xiphosuran form when first hatched. In or following the transition to land, the Scorpionida have adopted viviparity (at what stage of their evolution is not known) but all other groups of arachnids have remained basically oviparous.

The pioneer of arachnid embryology, as of so many other groups of invertebrates, was Edouarde Claparède, who published accounts of the embryos of spiders and mites in 1862 and 1868. Other workers, including Salensky (1871), Metschnikoff (1871), and Balbiani (1873) soon supplemented this work, but studies on arachnid embryos progressed only slowly during the next fifteen years (Balfour, 1880; Sabatier, 1881; Henking, 1882; Locy,

1886; Morin, 1887, 1888; Schimkewitsch, 1887, on spiders; Kowalevsky and Shulgin, 1886, on scorpions). As histological techniques improved and the significance of the Xiphosura as primitive chelicerates began to be better appreciated, the twenty years between 1890 and 1910 saw an increase in interest which led to the establishment of much of the basic knowledge of chelicerate internal embryology available to us today. The Xiphosura were studied by Kishinouye (1891a), Kingsley (1892, 1893) and Patten and Hazen (1900). Patten also essayed other studies on *Limulus* embryos, but with generally unacceptable results. Laurie (1890, 1891) and especially Brauer (1894, 1895) made important contributions to the embryology of scorpions. The spiders attracted many workers, though the species which they investigated were almost all entelegynid spiders with rather specialized embryos (Faussek, 1891; Simmons, 1894; Purcell, 1895; Schimkewitsch, 1898, 1906, 1911; Pokrowsky, 1899; Pappenheim, 1903; Wallstabe, 1908; Janeck, 1909; Kautzsch, 1909, 1910; Lambert, 1909; Montgomery, 1909 a, b; Hamburger, 1910; Fulinski, 1912; Ivanic, 1912). A few significant papers on acarine embryos were also published during this period (Wagner, 1894 a, b; Brucker, 1900; Bonnet, 1907; Reuter, 1909) and a little on other arachnid groups (Pereyaslawzewa, 1901; Gough, 1902 on Amblypygi; Heymons, 1904 on Solifugae; Schimkewitsch, 1898 on Opiliones; Schimkewitsch, 1906 on Uropygi). The scattered distribution of these studies among species with many individual specializations, together with the lack of knowledge on primitive spiders, uropygids and other developmentally generalized arachnids, created great difficulties in comprehending the generalities of chelicerate embryology and in comparing chelicerate embryology with that of other arthropods. It was during this period that the terminological peculiarities of chelicerate embryology become firmly rooted. The outcome was a dearth of interest in the subject which lasted for thirty years. Between 1910 and 1940, while pterygote and crustacean embryology made great progress, only a handful of papers was published on chelicerate embryos. One was a long account by Iwanoff (1933) of the development of the xiphosuran *Tachypleus gigas*, useful in certain ways but biased by Iwanoff's preoccupation with the concept of primary and secondary segmentation. Another was a paper by Pflugfelder (1930) on the specialized development of the yolkless viviparous scorpion *Hormurus*, which did little to assist the generalities of chelicerate development and has been subsequently shown to be erroneous in many respects.

The most significant contribution to chelicerate embryology in the years between the two World Wars was the account given by Holm (1941) of the development of haplogynid spiders. Although concerned entirely with the external features of development, this work brought out the fact that the embryos of primitive spiders had been much neglected hitherto and provided a focal point for new comparative studies during the succeeding years. It is true that many of these investigations have treated only the external features of development, leaving fundamental problems unresolved. We are also still awaiting definitive studies on xiphosuran embryology and on important arachnid groups such as the Solifugae and Amblypygi. At the same time, the recent investigations on araneids, including several primitive species (Holm, 1952, 1954; Yoshikura, 1954, 1955, 1958; Rempel, 1957; Ehn, 1962, 1963; Crome, 1963, 1964; Pross, 1966; Seitz, 1966) and on uropygids (Kästner, 1948, 1949, 1950; Yoshikura, 1961), opilionids (Holm, 1947; Moritz, 1957, 1959; Winkler, 1957; Juberthie, 1961, 1964), acarines (Aeschlimann, 1958, 1961; Anderson, 1972), pseudoscorpions (Weygoldt, 1964, 1965, 1968) and scorpions (Abd-el Wahab, 1952, 1954, 1956;

Mathew, 1956) have provided a new basis on which the achievements and outstanding problems in chelicerate embryology can be reassessed. We shall see that the distinctive theme evident throughout chelicerate embryonic development is so far removed from any basic mode of spiral cleavage development that it offers no indication of a relationship between the Chelicerata and the other major groups of arthropods.

Cleavage

Rounded or ovoid eggs rich in yolk are characteristic of the majority of chelicerates. Only the trombidiform and sarcoptiform mites, the pseudoscorpions and the viviparous scorpions produce small eggs showing evidence of a secondary reduction in yolk. The embryos of mites with small eggs hatch as minute hexapod larvae. Those of pseudoscorpions and viviparous scorpions develop to a size similar to that of many yolky arachnid embryos, but are supplied throughout embryonic development with nutrients secreted by the maternal tissues.

Among the Xiphosura, the egg of *Limulus polyphemus* is ovoid, with an average diameter of 2 mm (Kingsley, 1892), while that of *Tachypleus tridentatus* averages $2\frac{1}{2}$ mm diameter (Kishinouye, 1891a) and the spherical egg of *T. gigas* is $3\frac{1}{2}$ mm across (Iwanoff, 1933). The cytoplasm of the egg is packed with uniformly distributed, largish yolk globules and the nucleus is excentrically placed. A single, tough, chitinous membrane, the chorion, encloses the egg.

Among the arachnids, large, spherical eggs are also produced by species of the Solifugae, Amblypygi and Uropygi, though almost nothing is known of the development of the first two of these groups. Exemplifying the Uropygi, *Thelyphonus caudatus* has a spherical egg 3 mm in diameter, covered by a thin chorion (Schimkewitsch, 1904; Kästner, 1948) and *Typopeltis stimpsonii* has a similar egg (Yoshikura, 1961).

Spiders also exhibit relatively large eggs among primitive species, though none has an egg exceeding 2 mm diameter and the majority of spider eggs are less than 1 mm. We can display the range of egg sizes among spiders most easily by means of a table (Table 1).

As Holm (1941) pointed out, the smallest eggs in spiders occur in the Erigonidae and small Theriidae and have a diameter of 0·4 mm, while the largest, with a diameter up to 1·9 mm, are produced by species of *Dolomedes* and *Agelena*.

Spiders eggs of all sizes retain a similar basic structure, described in the liphistiid *Heptathela* by Yoshikura (1955). The nucleus is central, surrounded by a small cytoplasmic halo. A cytoplasmic reticulum, enmeshing the dense and uniformly packed yolk, extends outwards from the central halo to reach a thin periplasm at the surface of the egg. The egg is enclosed by a vitelline membrane and chorion, which in *Heptathela* are both thin and transparent. The same structure has been described for mygalomorph eggs by L. and W. Schimkewitsch (1911) and Holm (1954) and for araneomorph eggs by Locy (1886), Morin (1887), Kishinouye (1891b), Montgomery (1909a), Kautzsch (1910), Holm (1941, 1952) and Ehn (1963). There is some suggestion in recent work (Holm, 1952, 1954; Ehn, 1963) that the yolk globules are arranged as radial columns and that the intervening cytoplasm forms radial threads connecting the periplasm with the halo around the nucleus. The chorion and vitelline membrane are present in all species, but while the vitelline membrane is

TABLE 1. EGG SIZES IN SPIDERS (ARANEAE)

	Species	Egg diameter, mm	Source
Liphistiomorpha			
LIPHISTIIDAE	*Heptathela kimurai*	1·4	Yoshikura, 1955
Mygalomorpha			
ATYPIDAE	*Atypus karschi*	1·0	Yoshikura, 1958
DIPLURIDAE	*Ischnothele karschi*	1·2–1·3	Holm, 1954
Araneomorpha			
Haplogynae			
DYSERIDAE	*Segestria bavarica*	0·8	Holm, 1941
Entelegynae			
DICTYNIDAE	*Amaurobius fenestralis*	1·0	Holm, 1941
SPARASSIDAE	*Torania variata*	1·55	Ehn, 1963
PISAURIDAE	*Dolomedes fimbriatus*	1·8	Pappenheim, 1903
AGELENIDAE	*Agelena labyrinthica*	1·2–1·3	Holm, 1952
THERIDIIDAE	*Theridion tepidariorum*	0·5	Montgomery, 1909
THERIDIIDAE	*Latrodectus mactans*	0·8	Rempel, 1957
OTENIDAE	*Cupiennius salei*	1·2–1·4	Seitz, 1966

thin and transparent, the chorion in mygalomorphs and araneomorphs is thick and opaque. Holm (1941) found that normal development of the embryo proceeds when the egg is immersed in a vegetable oil which renders the chorion transparent. This simple technique has greatly facilitated the study of arachnid embryos in recent years (e.g. Ehn, 1963; Aeschlimann, 1958, 1961; Seitz, 1966).

A few hours after laying, the periplasm of the egg of mygalomorphs and aranaeomorphs becomes subdivided into numerous polygonal fields separated by fine hyaline lines (Balbiani, 1873; Locy, 1886; Kishinouye, 1891b; Kautzsch, 1910; Holm, 1941, 1954; Ehn, 1963; Seitz, 1966). The relationship between the polygonal, periplasmic fields and the events of cleavage will be discussed below.

The Opiliones, as far as is known, also lay ovoid or spherical yolky eggs of approximately the same size range as those of spiders. The structure of the egg has been examined for *Opilio parietinus* by Holm (1947) and for *Platybunus triangularis* by Moritz (1957). Both report a typical arachnid egg structure, with a central nucleus surrounded by a small cytoplasmic halo, a yolk-filled cytoplasm and a thin periplasm. A thin, transparent chorion and vitelline membrane surround the egg. Some typical dimensions for opilionid eggs are as follows:

Opilio parietinus	0·75–0·80 mm	(Holm, 1947)
Phalangium opilio	0·65–0·68 mm	(Winkler, 1957)
Platybunus triangularis	0·56–0·62 mm	(Moritz, 1957)

Juberthie (1964) gives the egg sizes of several species, indicating a range of $\frac{1}{2}$–$1\frac{1}{2}$ mm. This compares with the range of 0·4–1·9 mm given for spiders by Holm (1941). It is to be noted, however, that all of the Opiliones examined embryologically belong to the advanced group Palpatores.

As might be expected, the eggs of acarines are generally smaller than those of spiders, though the two ranges of size overlap. The ticks retain quite large eggs, ovoid in form. Aeschlimann (1958) has made a careful study of the egg of *Ornithodorus moubata*, which is 0·9–1·1 mm long and 0·7–0·9 mm broad, and finds that it has the usual centrolecithal structure. As in spiders, the nucleus lies in the centre, surrounded by a typical cytoplasmic halo, and the cytoplasm is filled with yolk spheres of various sizes, held in a cytoplasmic reticulum which unites with a thin periplasm at the surface of the egg. A thin, transparent vitelline membrane and a thicker, two-layered chorion surround the egg. According to Aeschlimann, both of these membranes are secreted by the egg surface. Ixodid ticks lay similar, but slightly smaller eggs (e.g. *Boophilus calcaratus*, 580×410 μ, *Hyalomma aegyptium*, 645×425 μ; Wagner, 1894a; Bonnet, 1907) and the little that is known of the eggs of mesostigmatid mites indicates that these are again similar (e.g. *Pergamasus brevicornis*, 465×330 μ, *Ophyionyssus natricis*, 340×230 μ; Zukowski, 1964; Camin, 1953). In contrast, the eggs of trombidiform and sarcoptiform mites are generally less than 200 μ long and are usually between 100 and 150 μ long (Claparède, 1868; Henking, 1882; Wagner, 1894a; Brucker, 1900; Reuter, 1909; Thor, 1925; Munchberg, 1935; Hafiz, 1935; Cooper, 1940; Hughes, 1950; Sokolov, 1952; Klumpp, 1954; Perron, 1954; Langenscheidt, 1958; Prasse, 1968; Heinemann and Hughes, 1970). A central nucleus is retained, together with evenly distributed yolk in the cytoplasm, but the presence or absence of a periplasm is not clearly established. Recent work (Klumpp, 1954; Langenscheidt, 1958) suggests that there is no periplasm. The egg membranes of trombidiform and sarcoptiform acarines also need further study. Present evidence indicates that both a chorion and a vitelline membrane are present, as in ticks.

Reduction in the size of the egg is also displayed by scorpions, but is related in this group of arachnids to viviparity. As an example of the ovoviviparous species (Buthidae, Chactidae, Bothriuridae) we can take *Euscorpius italicus*. The egg of this species is elongate ovoid and is about 400 μ long when released into the oviduct, although the embryo increases in size during later development. The nucleus is central in the usual way. The cytoplasm contains numerous yolk globules and a vitelline membrane surrounds the egg (Laurie, 1890). According to Brauer (1894), the similar egg of *Buthus carpathicus* has a polar cap of yolk-free cytoplasm at one pole. In the viviparous scorpions (Scorpionidae, Diplocentridae, Vejovidae), in contrast, the egg is small, spherical and yolk-free (Laurie, 1891; Poliansky, 1903; Pavlovsky, 1926; Pflugfelder, 1930). Mathew (1956) gives the egg diameter of the scorpionid *Heterometrus scaber* as 40 μ. The nucleus of the egg retains a central position and is relatively large, while the cytoplasm is granular but not yolky. A single thin membrane encloses the egg. Development of the embryo proceeds within the lateral ovarian follicle in which the egg is formed. The complex relationship established between the developing embryo and the surrounding follicle is discussed in subsequent sections.

The pseudoscorpions also have secondarily small eggs with little yolk, though in this group of arachnids the eggs are passed into an external brood pouch into which a supply of nutrients is later passed from the maternal body. Weygoldt (1964, 1965, 1968) has described the eggs of several species as spherical, with a central nucleus and little yolk in the cytoplasm. He gives the egg diameters of *Pselaphochernes* as 120–130 μ, *Neobisium* as 80–100 μ and *Chthonius* as 60 μ and 70 μ, respectively, in two different species.

As might be expected, cleavage in chelicerates is considerably modified in relation to yolk and further modified where secondary loss of yolk has occurred (Figs. 134–6). In spite of having relatively large eggs, the Xiphosura still retain a cleavage pattern more primitive than that of any arachnid. Kishinouye (1891a), Kingsley (1892) and Iwanoff (1933) showed that the first few nuclear divisions are intralecithal, but that surface furrowing then begins to develop at the pole adjacent to the dividing nuclei, indicative of the onset of total cleavage. Spreading from this pole, the egg becomes first divided into numerous large, equal, yolky cells with central nuclei. Divisions then become concentrated in the peripheral cells, which give rise to a columnar blastoderm. The central cells remain undivided, filling the space within the forming blastoderm. There seems to be no doubt, as we shall see below, that the central cells give rise to the midgut epithelium.

The blastoderm of the Xiphosura secretes a blastodermal cuticle. According to Sekiguchi (1960), the chorion breaks down some days after the blastodermal cuticle has been formed and the cuticle persists as a protective membrane until hatching.

At the present time, the cleavage processes of arachnids can be most usefully compared with that of Xiphosura through a preliminary consideration of spiders, which have received more attention than other groups. Araneid cleavage, although proceeding in eggs smaller than those of Xiphosura, is more specialized than xiphosuran cleavage, but it retains marked reminiscences of the latter (Locy, 1886; Morin, 1887; Schimkewitsch, 1887; Kishinouye, 1891b; Montgomery, 1909a; Kautzsch, 1910; L. and W. Schimkewitsch, 1911; Holm, 1941, 1952, 1954; Yoshikura, 1954, 1955, 1958; Rempel, 1957; Ehn, 1962, 1963; Seitz, 1966). The details of araneid cleavage can be exemplified from the work of Yoshikura (1954, 1955) on the primitive liphistiid *Heptathela kimurai* and of Holm (1952) on the agelenid *Agelena labyrinthica*, typifying advanced spiders. The eggs of the two species are little different in size (see Table 1). In both, the first three nuclear divisions proceed without cytoplasmic division, giving eight nuclei with cytoplasmic haloes, equally distributed in the octants of the egg. The cytoplasm then divides into yolk pyramids, more than eight in number, around a small central blastocoel. Further nuclear divisions, accompanied by migration of the nuclei towards the periphery of the egg, are linked with further radial subdivisions of the yolk pyramids, though the precise relationship between nuclear division and pyramid division is not clear. When the nuclei and their associated cytoplasmic halos merge with the periplasm at the peripheral surface of the pyramids, the nucleated, yolk-free cytoplasm is cut off as an incipient blastoderm and the anucleate yolk pyramids fuse together once more, with obliteration of the blastocoel, to form a unitary yolk mass within the forming blastoderm. Further cell divisions at the surface complete the formation of a uniform, low, cuboidal blastoderm around the yolk mass. The yolk mass is subsequently reinvaded by scattered blastoderm cells which differentiate as vitellophages and later give rise to the epithelium of the major part of the midgut (see below). Both the formation of yolk pyramids, reminiscent of total cleavage, and the development of the central yolk mass as the midgut epithelium recall events in the Xiphosura.

As a final stage of blastoderm formation in spiders, preceding the onset of gastrulation and formation of the germ band, the blastoderm undergoes a contraction which flattens its ventral surface. Contraction proceeds differently in primitive and advanced spiders. Yoshikura (1955, 1958) has shown that in *Heptathela* and *Atypus*, blastoderm formation is followed immediately by a contraction of the yolk mass away from the blastoderm on

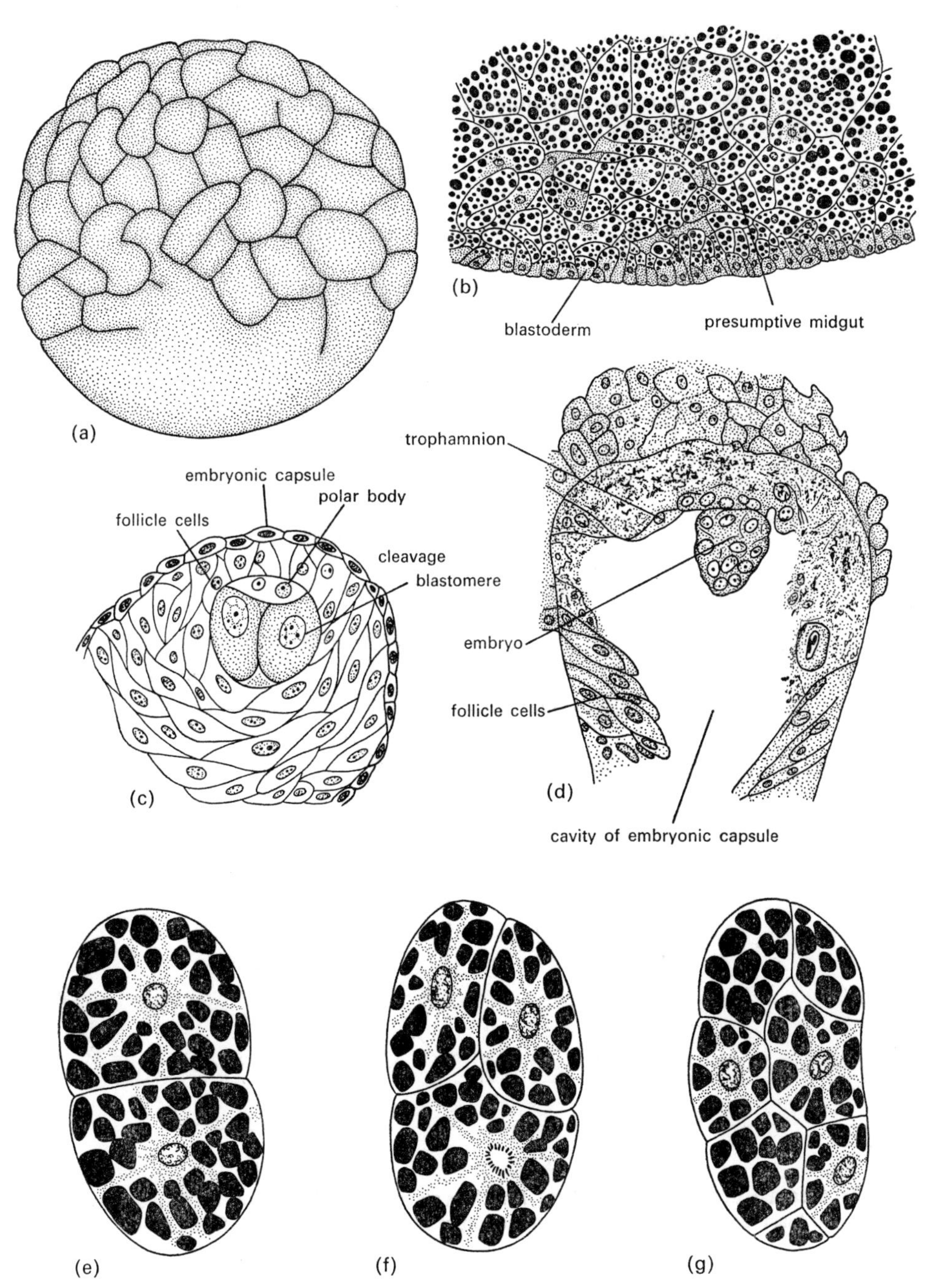

FIG. 134. (a) Polarized total cleavage in *Tachypleus*; (b) transverse section through the blastoderm of *Tachypleus*; (c) *Heterometrus*, 2-cell stage; (d) *Heterometrus*, morula stage; (e)–(g) 2-cell, 4-cell and 8-cell stages of *Caloglyphus*, diagrammatic longitudinal sections. [(a) and (b) after Iwanoff, 1933; (c) and (d) after Mathew, 1956; (e)–(g) after Prasse, 1968].

one side of the egg and an accumulation of fluid in the intervening space. The blastoderm then contracts down onto the flattened yolk surface, with extrusion of the fluid into the space between the blastoderm and vitelline membrane. The flattened surface is the ventral surface of the embryo. In other mygalomorph spiders and in araneomorph species, contraction and flattening of the yolk mass and blastoderm proceed simultaneously, with extru-

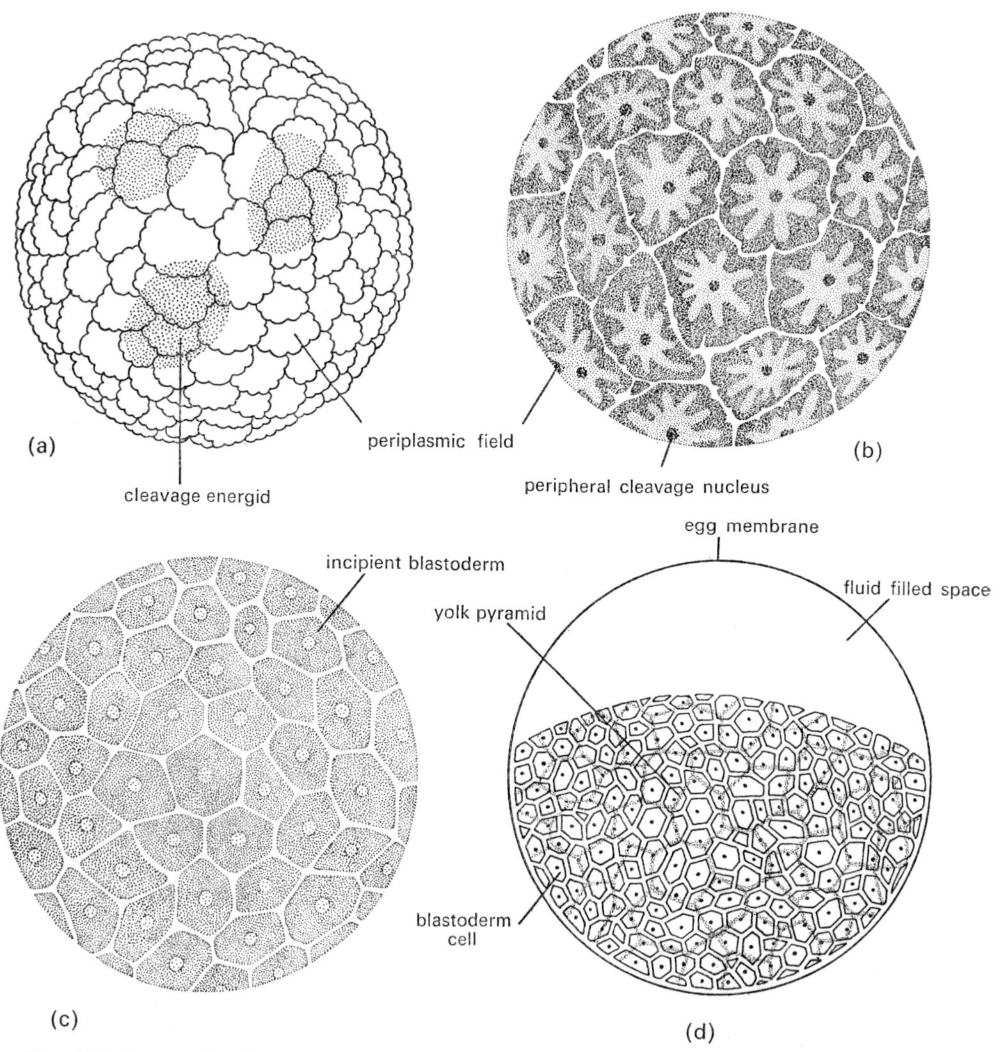

FIG. 135. Cleavage in *Cupiennius salei*. (a) 4-nucleus stage; (b) early blastoderm formation; (c) later blastoderm formation; (d) contraction of the blastoderm. (After Seitz, 1966.)

sion of fluid directly into the resulting space beneath the vitelline membrane (Salensky, 1871; Balbiani, 1873; Schimkewitsch 1887; Holm, 1941, 1952, 1954; Seitz, 1966). The outcome is the same, however, a flattened surface which is the ventral surface of the embryo.

As mentioned above in describing spiders eggs, the periplasm of mygalomorph and araneomorph eggs becomes subdivided, during early intralecithal cleavage, into a number of polygonal fields (e.g. Kautzsch, 1910; Holm, 1941, 1954; Ehn, 1963; Seitz, 1966). When the cleavage nuclei invade the periplasm, these fields are lost, prior to the incorporation of the periplasm into nucleated blastoderm cells. The relationship of the periplasmic fields to other aspects of cleavage and the significance of the fields are still obscure. Similar periph-

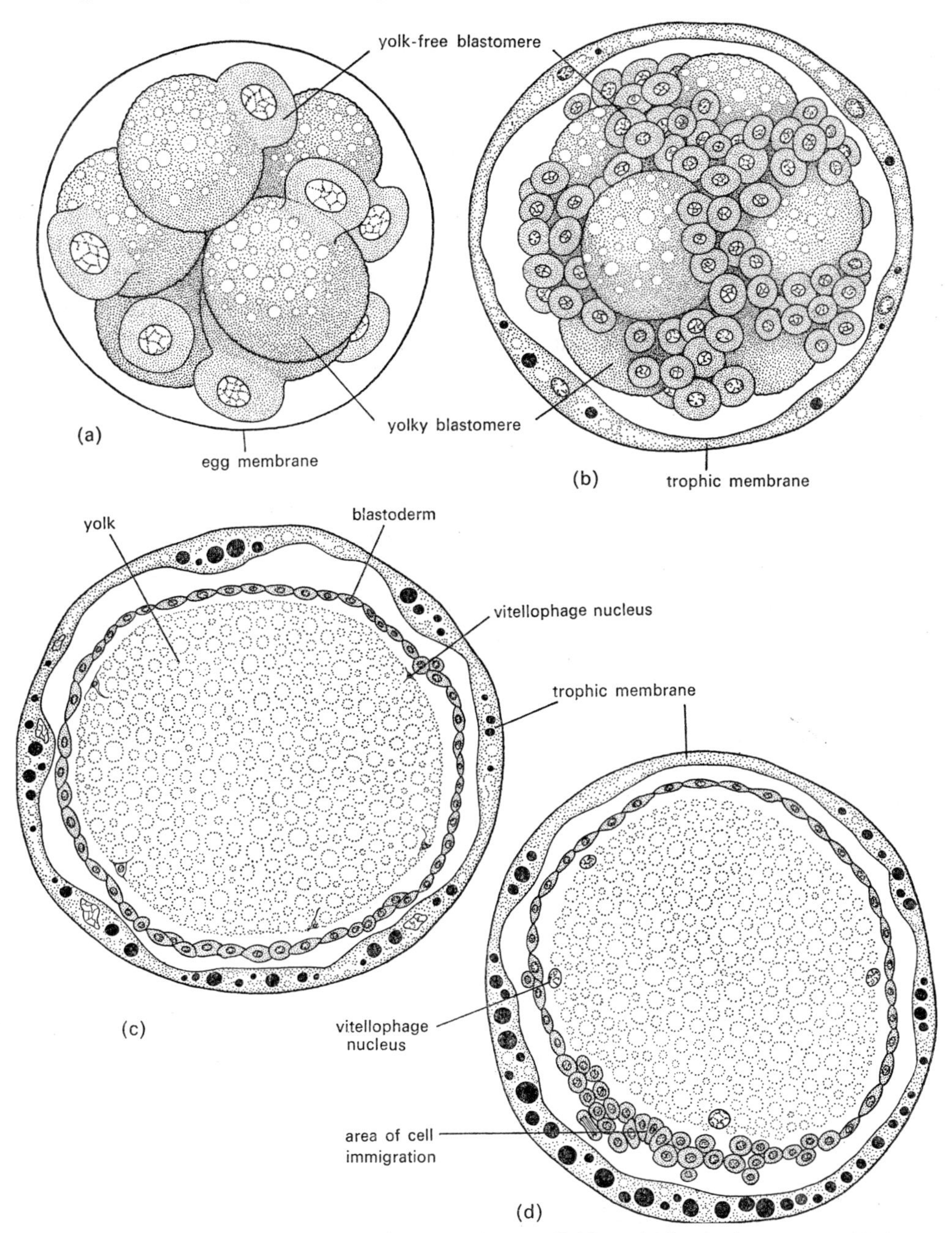

FIG. 136. Cleavage in *Pselaphochernes*. (a) Fourth cleavage division, yielding 8 micromeres and 8 macromeres; (b) later cleavage stage, with 8 macromeres and numerous micromeres; (c) section through blastoderm stage; (d) sagittal section showing onset of gastrulation. (After Weygoldt, 1964.)

eral fields are reported for the periplasm of other arachnid eggs undergoing early cleavage (ticks, Aeschlimann, 1958, 1961; opilionids, Juberthie, 1961, 1964).

Cleavage is less well documented in other arachnids than in the spiders. Schimkewitsch (1906) has indicated unequal pyramid formation, yielding an excentric blastocoel, followed by blastoderm formation, in the egg of the uropygid *Thelyphonus caudatus*. In contrast, in the Opiliones, Schimkewitsch (1898) described a typical intralecithal cleavage preceding blastoderm formation and this has subsequently been confirmed by Holm (1947), Moritz (1957) and Juberthie (1961, 1964). It seems likely that some of the cleavage nuclei of opilionids persist within the yolk mass as vitellophages, a specialization unusual among arachnids.

Intralecithal cleavage leading directly to blastoderm formation without the intervention of pyramid formation is also a feature of developmentally generalized Acarina (e.g. the ticks, see Aeschlimann, 1958, 1961). In *Ornithodorus moubata*, with a relatively large egg, eight synchronous cleavage divisions precede the arrival of 256 nuclei at the egg surface, and the formation of a uniform, low cuboidal blastoderm then follows. A similar cleavage has been more sketchily reported in other ticks (Wagner, 1894 a, b; Bonnet, 1907) and in certain mites (Claparède, 1868; Henking, 1882; Klumpp, 1954; Langenscheidt, 1958). A number of trombidiform and sacoptiform mites with small eggs, however, are known to exhibit a preliminary phase of total cleavage preceding blastoderm formation (Brucker, 1900; Reuter, 1909; Cooper, 1940; Sokolov, 1952; Edwards, 1958; Prasse, 1968; Heinemann and Hughes, 1970). This process is typified by cleavage in *Caloglyphus*. The ovoid egg first divides transversely and equally. Each daughter cell then divides along the anteroposterior axis of the egg, but the planes of division of the two cells are perpendicular to one another. A similar division in the alternate plane separates eight equal blastomeres with more or less peripheral nuclei. Further divisions proceed with increasing asynchrony, first cutting the eight blastomeres into smaller cells with peripheral nuclei, then cutting off a blastodermal layer of yolk-free cells from internal, anucleate yolk masses which fuse together once more. The final result, a uniform low, cuboidal blastoderm around a unitary yolk mass, is common to all acarines. The specialized sequence of acarine total cleavage is a clear indication of the secondary nature of the phenomenon in this group of arachnids and we can reasonably presume that intralecithal cleavage, with associated pyramid formation, leading to formation of a uniform low blastoderm around a unitary yolk mass, is a basic process common to several arachnid groups (Araneae, Uropygi, probably Solifugae and Amblypygi, Opiliones and Acarina). We shall see below that these groups exhibit a common theme in embryonic development in other ways.

In the scorpions and pseudoscorpions, in contrast, in association with their specialized modes of development involving secondary yolk reduction and a supply of nutrients from the maternal body, peculiarities of cleavage are only to be expected. The specializations displayed within the two groups are different. Furthermore, the scorpions, as we shall see, are developmentally more different from the other arachnids than are the pseudoscorpions.

We can start with a consideration of cleavage in the ovoid yolky eggs of ovoviviparous buthid scorpions. This process has been described for *Euscorpius italicus*, *Buthus carpathicus* and *B. quinquestriatus* by Laurie (1890), Brauer (1894) and Abd-el Wahab (1954). The egg has a polar cap of yolk-free cytoplasm at one end, and the zygote nucleus lies close to this pole. At the first cleavage division, the polar cap furrows across the middle and becomes two flattened, equal cells. The second cleavage division proceeds in the same way, perpendicular to the first. As divisions continue, the cells become smaller and the cap of cells spreads more extensively over the surface of the yolk mass. There is no evidence to suggest that any cleavage furrows pass into or through the yolk mass. Eventually, the spreading cells extend to cover the entire surface of the yolk mass, but the onset of gastrulation and further development is observed when the polar cap is still restricted to about 64 cells. We shall return later to the further development of the buthid embryo, noting at this stage only that polar discoidal cleavage is confined to scorpions among the arachnids and can be interpreted as a specialization of the polarized total cleavage observed in Xiphosura, different from the specialization seen in araneids and other groups.

The small yolkless eggs of viviparous scorpions have, of course, lost all traces of discoidal cleavage and display a total cleavage convergently similar to that of the viviparous Onychophora. The process was first described by Pflugfelder (1930) in *Hormurus australiensis* and subsequently in more detail by Mathew (1956) in *Heterometrus scaber*. Cleavage is preceded by the release of two polar bodies which mark the anterior end of the egg. The cleavage divisions then proceed as regular, total, equal divisions, passing through 2-, 4-, 8-, 16- and 28-cell stages before becoming more irregular. The result is a spherical mass of small cells surrounding a small central blastocoel. When the egg first begins to cleave, it is surrounded by a membrane and lies within a lateral ovarian follicle as described previously. As cleavage continues, the follicle wall shrinks away from the egg, leaving a space, except anteriorly, where the cleavage morula remains attached to the distal follicle cells. The latter proliferate as an enveloping layer around the embryo, the embryonic capsule. At the same time, other distal follicle cells produce a terminal plug of large clear cells. These cells subsequently develop into a feeding structure called the appendix, through which nutriment is passed from the maternal tissues to the growing embryo.

The recent work of Weygoldt (1964, 1965, 1968) has emphasized the uniqueness of the embryonic specializations in pseudoscorpions, beginning with a mode of cleavage unique among arachnids (Fig. 136). In *Pselaphochernes*, with a 120–130-μ egg, cleavage is total and equal up to the 8-cell stage, so that each cell or macromere contains a similar amount of yolk. Each macromere then cuts off a small, yolk-free micromere externally. The macromeres undergo no further divisions and subsequently fuse to form a nucleated yolk mass. The micromeres undergo further divisions and spread around the macromeres as an external cell layer. Most of the products of micromere division contribute to a blastoderm around the yolk mass, but some of the cells take up an external position between the forming blastoderm and the thin egg membrane. Here, these external cells spread and unite to form a syncitial embryonic membrane beneath the egg membrane. The embryonic membrane is a temporary trophic membrane acting in the uptake and transfer of nutrients from the brood pouch to the embryo during early development. It recalls the provisional envelope of naidid oligochaete embryos and the trophamnion of the embryos of parasitic Hymenoptera, but its evolutionary origin in pseudoscorpions is obscure. *Neobisium*, with an 80–100-μ egg, proceeds through cleavage in a similar way except that the micromeres are cut off from the macromeres at the 4-cell stage. Alternatively, in *Chthonius*, with a 60–70 μ egg, the micromeres are cut off after equal cleavage has produced 8 cells, as in *Pselaphochernes*, but only one of the resulting eight micromeres gives rise to the embryonic envelope.

In spite of the specializations of cleavage in pseudoscorpions, we can discern two points of interest in comparison with other arachnids. Firstly, there is no indication of a discoidal mode of cleavage such as occurs in scorpions. Secondly, apart from the specialized segregation of an embryonic envelope from the remainder of the blastoderm cells, the cleavage of pseudoscorpions proceeds in the manner to be expected of the normal mode of cleavage and blastoderm formation in arachnids other than scorpions, given a reduction in the size and yolk content of the egg. Total equal cleavage supplants yolk-pyramid formation and the superficial yolk-free blastomeres are cut off at relatively early stage, but the radial symmetry and uniformity of the process is retained and the outcome, a uniform blastoderm around a unitary yolk mass, is the same. As we shall see below, this difference from scorpions and similarity with other arachnids is evident in all phases of pseudoscorpionid embryonic development.

Development of External Form

Although, in previous chapters, I have discussed the fate map of the blastula or blastoderm and the events of gastrulation before describing the development of the external form of the embryo, the present position in chelicerate embryology necessitates a reversal of this procedure. The origin and early development of the midgut and mesoderm of chelicerate embryos are far from being resolved and the available evidence is most easily viewed against a background of knowledge of the general course of development in different groups. The most generalized pattern of external development, as might be expected, is found in the Xiphosura.

The external features of development of the Xiphosura still require further study, but the accounts by Kishinouye (1891a), Kingsley (1892) and Iwanoff (1933) provide a reasonably uniform interpretation for three species. It seems likely that the first external indication of development beyond the blastoderm stage in *Limulus* and *Tachypleus* is the formation of a small ventral area of concentrated blastoderm cells, quickly followed by the formation of a second similar area immediately behind the first. The two areas soon merge by thickening of the intervening region, to form an oval embryonic primordium which is broader posteriorly than anteriorly. While merging is taking place, a median gastral groove arises anteriorly and extends along the midline of the embryonic primordium. The remainder of the blastoderm attenuates to form a dorsal extra-embryonic ectoderm over the large mass of yolky cells.

The embryonic primordium now enters into elongation and segmentation as a germ band (Fig. 137). A transverse line, situated behind the anterior end of the gastral groove, demarcates a small cephalic lobe from a larger growth zone. Behind this line, the pedipalpal and four ambulatory segments of the prosoma are successively delineated as the growth zone augments the length of the germ band. The gastral groove progressively closes anteriorly and extends posteriorly along the midline of the forming segments, but remains open at the extreme anterior end as the stomodaeal invagination.

The prosomal segments broaden as they are formed, so that the prosomal part of the xiphosuran germ band takes up most of the ventral surface of the egg. When the opisthoso-

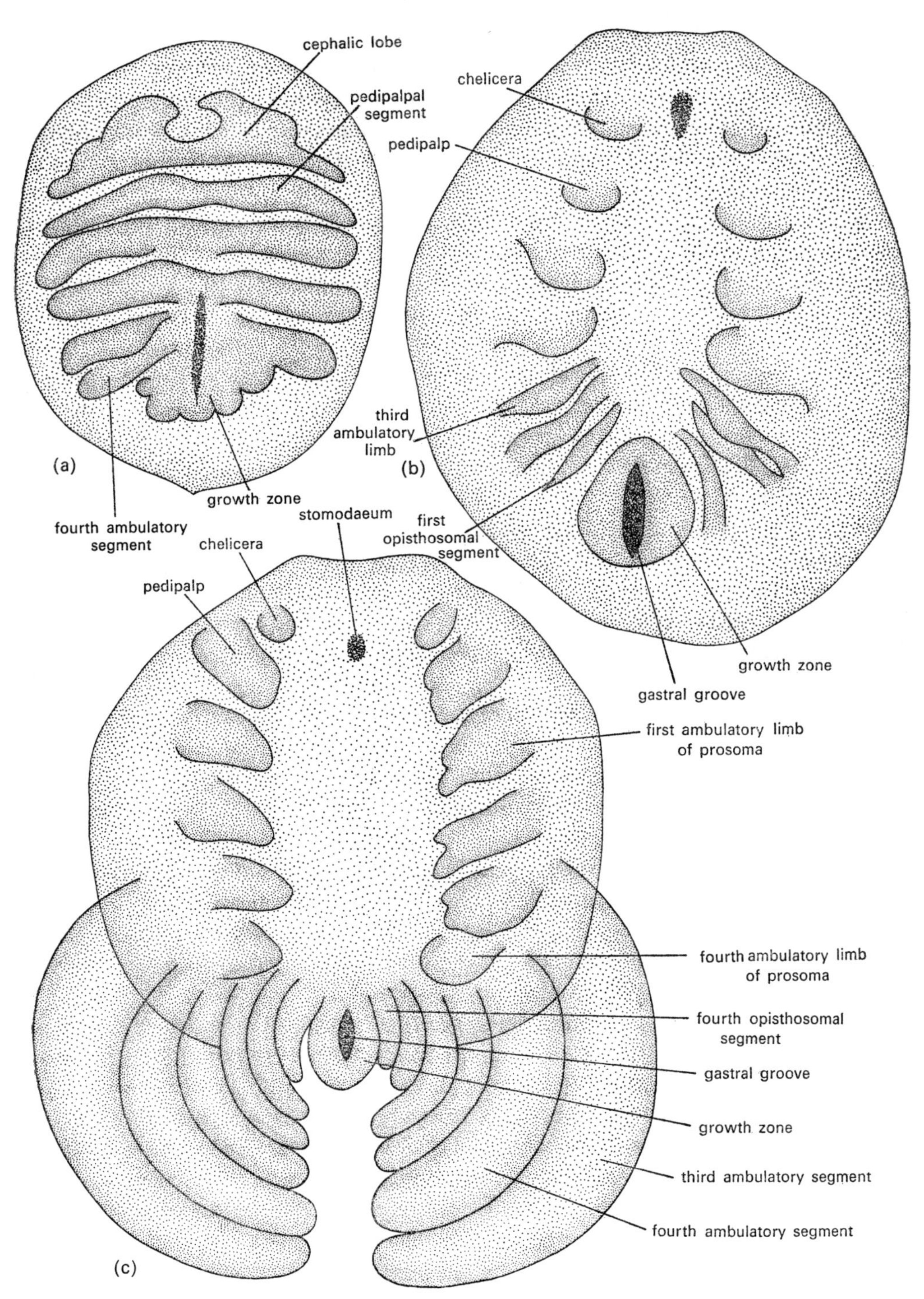

FIG. 137. The segmenting germ band of *Tachypleus*, ventral view. (a) during formation of the prosomal segment rudiments; (b) at the onset of formation of the opisthosomal segment rudiments; (c) late in formation of the opisthosomal segment rudiments. (After Iwanoff, 1933.)

mal segments begin to be proliferated by the growth zone, the segment rudiments are situated at the curved posterior pole of the egg. Furthermore, each segment rudiment is short and broad when first formed. The rudiments are therefore laid down as a concentric series on the posterior pole of the yolk mass. The gastral groove continues to extend in a posterior direction along the midline during formation of the opisthosomal segments, but is then obliterated in an anteroposterior wave.

While the rudiments of the opisthosomal segments are being formed, the prosomal segments develop paired, ventrolateral limb buds on the pedipalpal and four ambulatory segments. A pair of cheliceral limb buds also arises posterolaterally on the cephalic lobe, behind the level of the stomodaeum. No trace of pre-cheliceral limb buds has ever been observed.

In the subsequent development of the opisthosomal segments, each segment rudiment increases in length anteroposteriorly (Fig. 138) and also undergoes dorsal closure, so that the concentric arrangement of the early segment rudiments is supplanted by an anteroposterior succession of gradually diminishing, tubular segments. The large mass of yolky cells remains mainly within the prosoma. Limb buds also develop on the opisthosoma during this phase of development. The first pair, on the first opisthosomal segment, gives rise to the chilaria. The second to fourth pairs, which arise as broad, transverse lobes and become biramous, develop into the genital operculum and first two pairs of book gills. The book gills of the fifth to seventh opisthosomal segments are formed later, during larval development. The eighth and succeeding opisthosomal segments remain limbless and take part in the formation of the caudal spine.

During the outgrowth and development of the anterior opisthosomal limbs, the prosomal limb buds elongate and assume their definitive features. The prosomal segments also complete their dorsal closure around the large yolk mass and the round embryo subsequently becomes elongated and flattened, forming the trilobite larva. The fate of the dorsal extra-embryonic ectoderm during dorsal closure of the prosoma is not clear. At the anterior end of the prosoma, temporary ventrolateral invaginations on the cephalic lobe mark the formation of the precheliceral ganglia. As these are closed off, the chelicerae appear to move forwards to their pre-oral position. In fact, the stomodaeal aperture shifts in a posterior direction by extension of its posterior margin and progressive closure of its anterior margin, until the aperture lies behind the bases of the chelicerae.

The basic process in the development of the Xiphosura is thus the formation of a sequence of short, broad segment rudiments from a growth zone which first originates as the posterior part of an embryonic primordium, the anterior part of which constitutes the rudiment of the cephalic lobe. The latter includes the pre-cheliceral and cheliceral components of the body, of which the cheliceral component is a segment, like the following segments, but the pre-cheliceral component is a region whose morphological composition is more complex and will be discussed in a later section. A midventral gastral groove extends temporarily through all segments. The relatively large prosoma develops by dorsal closure around the large yolk mass. The smaller, conical opisthosoma develops mainly behind the yolk mass.

The development of the external features of the embryo of *arachnids* after the blastoderm stage displays two modifications of this basic pattern. In scorpions the small embryonic primordium, consisting partly of a cephalic lobe and partly of a growth zone from which all

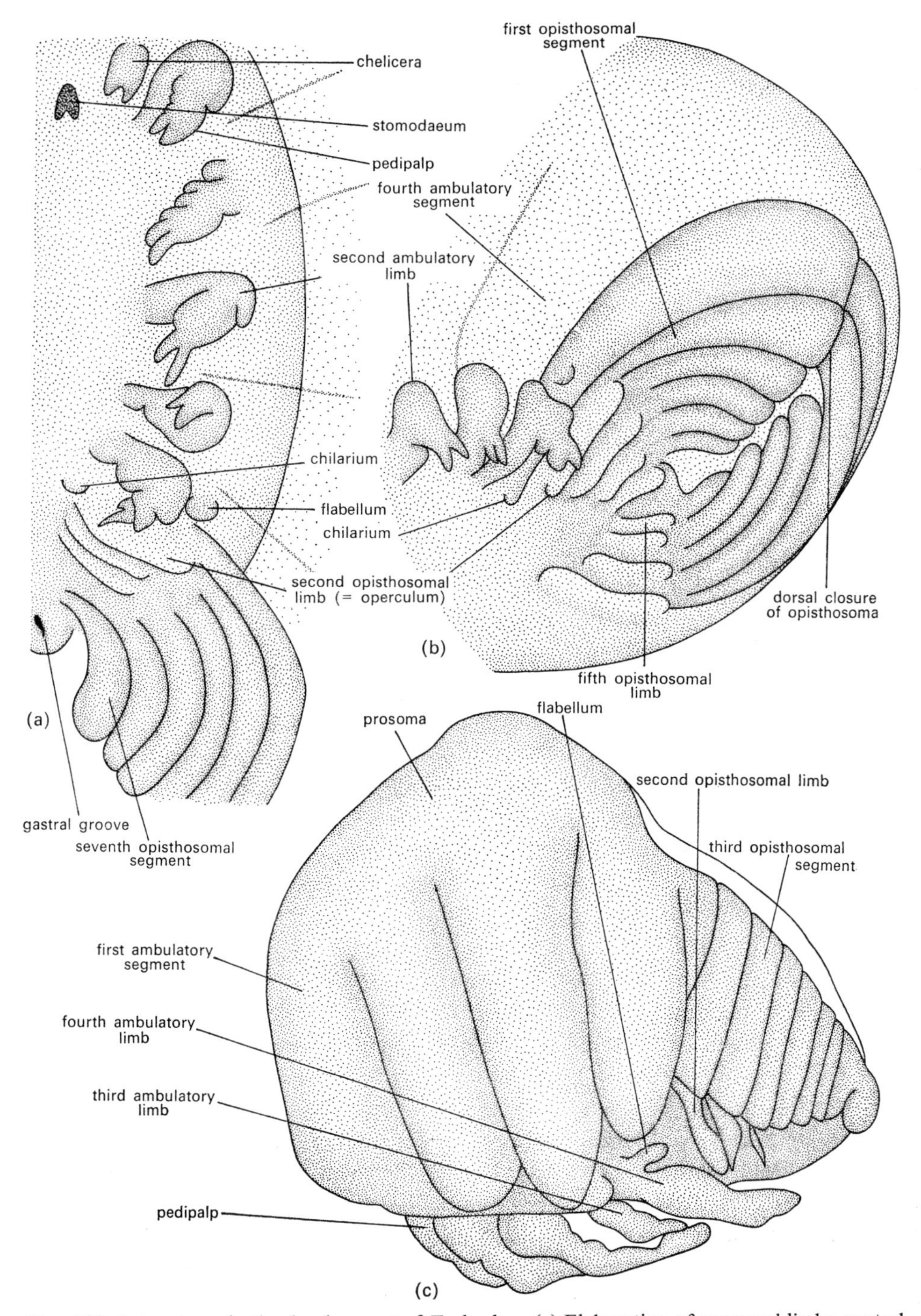

FIG. 138. Later stages in the development of *Tachypleus*. (a) Elaboration of prosomal limbs, ventral view; (b) formation of opisthosomal limbs, posterolateral view; (c) dorsal closure and extension of the opisthosoma, lateral view. (After Iwanoff, 1933.)

post-cheliceral segments are proliferated in succession, is still retained. In all other arachnids, the prosomal part of the germ band, with its cephalic lobes, pedipalpal segment and four ambulatory segments and its delayed delineation of the cheliceral segment, is primitively formed directly from the blastoderm. The growth zone develops behind the fourth ambulatory segment and proliferates only the segments of the opisthosoma. Since the scorpions, in spite of their retention of a primitive pattern of segment formation, display a number of specializations associated with ovoviviparity and viviparity, there are advantages in discussing their development after that of other arachnids. The recent advances that have been made in the investigation of the embryos of primitive spiders make the araneids once again a useful group with which other orders of arachnids can be compared. At the time when Dawydoff (1949) gave the last recent account of chelicerate embryology, almost nothing was known of the embryonic development of orthognath spiders, but the study by Yoshikura (1954, 1955) of the Japanese liphistiid species *Heptathela kimurai* and the accounts of mygalomorph embryos given by Holm (1954), Yoshikura (1958) and Crome (1963, 1964) have altered this situation in a most revealing way.

Heptathela, as might be expected from the primitive nature of the liphistiomorph spiders, retains a basic pattern for araneid embryonic development (Fig. 139). After contraction of the blastoderm, the majority of the blastoderm cells are concentrated onto the flattened ventral surface of the yolk mass, leaving only a thin covering of extra-embryonic cells on the convex dorsal surface. The concentrated cells form an elongate embryonic primordium with a small posterior protuberance. A gastral groove makes a temporary appearance along the ventral midline of the posterior half of the primordium, in front of the posterior protuberance. As we shall see, this is the region of the presumptive ambulatory segments. The stomodaeal invagination arises separately from and anterior to the gastral groove in *Heptathela*.

The embryonic primordium now becomes demarcated by transverse grooves into four ambulatory segments and an anterior lobe. The posterior part of the latter is then marked off as the pedipalpal segment, leaving a cephalic lobe which becomes demarcated into the cheliceral segment and pre-cheliceral region. At the same time, limb buds begin to protrude ventrolaterally on the prosomal segments. The sequence in which they develop is unusual among spiders. The first to appear is the first ambulatory pair, followed by the pedipalpal, the cheliceral and then the remaining ambulatory pairs. Usually the pedipalpal and ambulatory limbs form together or in slight sequence and the cheliceral pair is delayed.

Coincident with the onset of formation of the prosomal limb buds, the protuberant growth zone begins to grow out as a tubular, posterior extension of the body, the opisthosomal rudiment. The ventral and lateral walls of the rudiment are formed by the growth zone, while the dorsal wall is a narrow posterior extension of the dorsal extraembryonic ectoderm covering the yolk mass. In association with the enclosure of the embryo in a confining, double egg membrane, the tubular opisthosomal rudiment soon flexes forward in a ventral direction beneath the yolk-filled prosoma. As segment delineation proceeds in anteroposterior succession along the opisthosoma, it can be seen that the point of forward flexure lies at the fourth opisthosomal segment. Twelve units are delineated along the opisthosoma. The first is the genital segment, followed by ten further segments and a telson. A pre-genital segment can be recognized internally in the embryo of *Heptathela* (see later) but has no external expression.

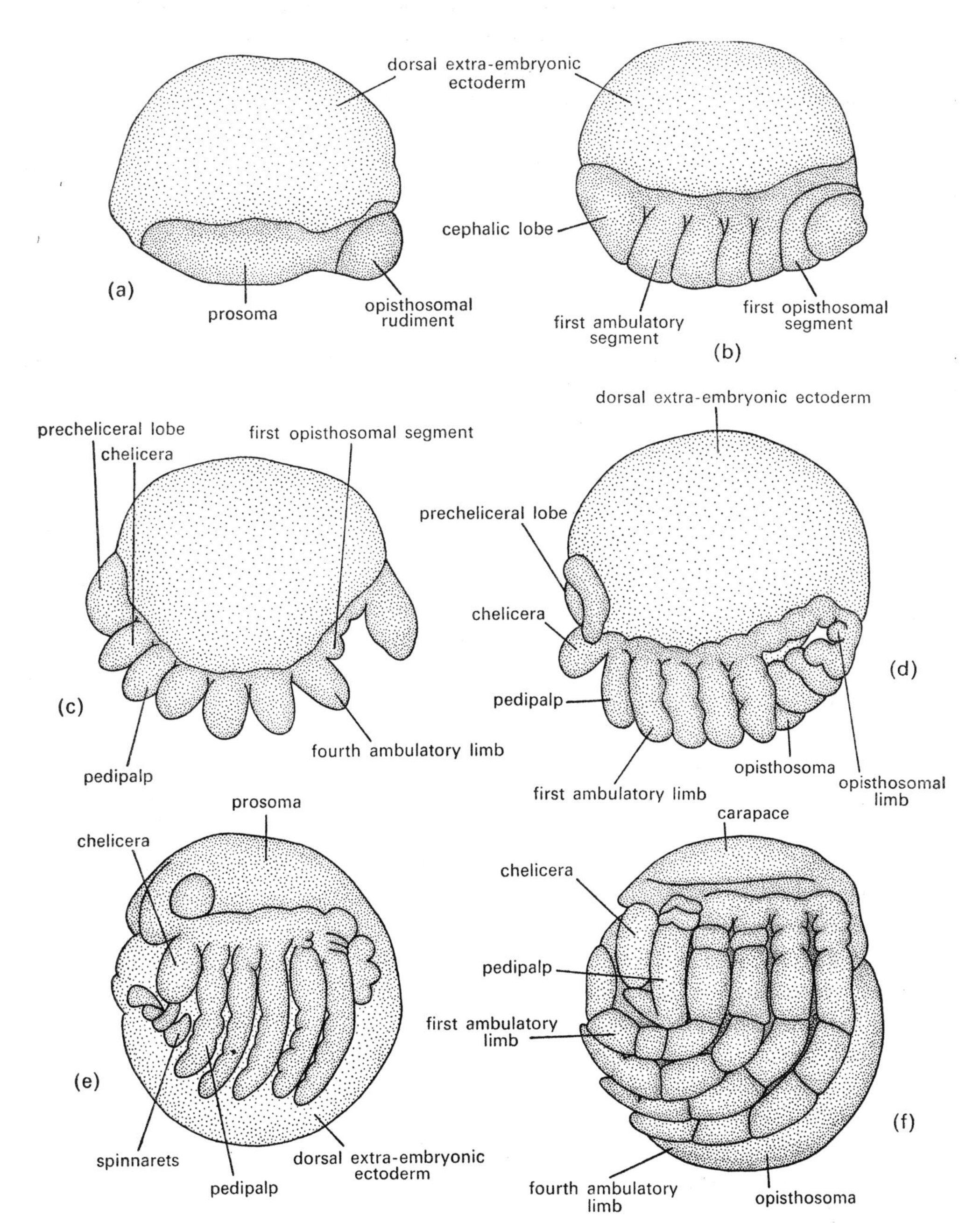

FIG. 139. *Heptathela kumurai*, lateral view. (a) Embryonic primordium; (b) segmentation of the prosoma; (c) early segmentation and growth of the opisthosoma; (d) completion of opisthosomal growth; (e) dorsal closure of the prosoma and translocation of yolk to the opisthosoma; (f) fully developed embryo. (After Yoshikura, 1954.)

When the full complement of segments has developed, the forwardly flexed opisthosoma is a narrow tube but the prosomal region is still swollen with yolk. Now, however, the remaining yolk is gradually moved from the prosoma into the opisthosoma. The area of the dorsal extra-embryonic ectoderm in the prosoma is drastically reduced during this translocation, and dorsal closure quickly follows, accompanied by an invasion of the curvature of the region from ventrally convex to ventrally concave. Simultaneously, the forwardly flexed opisthosoma becomes greatly swollen, the increase in volume being accommodated by a large expansion of the dorsal, extra-embryonic ectoderm of this part of the body. The opisthosomal segment rudiments remain in the ventral midline as the yolk moves into this region of the embryo, and dorsal closure of the opisthosoma is much slower than that in the prosoma.

The posterior translocation of the yolk mass in *Heptathela* is so timed in relation to the other events of development that it plays an essential role in the development of the swollen, spheroidal form of the opisthosoma, by establishing a surface over which the opisthosomal segments can extend dorsally as short, wide rings during dorsal closure. An interesting comparison can be made with the embryos of the decapod Crustacea, whose forwardly flexed caudal papilla is not swollen by translocation of yolk (see Chapter 8). In these animals the posterior part of the trunk remains of the same general diameter as the anterior part. It seems likely that all arachnids which have a disproportionately large opisthosoma use the yolk of the embryo as a means of achieving the initial expansion. We shall look at this question further as we proceed.

Accompanying the translocation of yolk from the prosoma to the opisthosoma in *Heptathela*, the prosomal limbs increase further in length and the cheliceral segment shifts forward bilaterally relative to the mouth, carrying the chelicerae to a preoral position. In the opisthosoma, the formation of the last few segments is also accompanied by the outgrowth of limb buds on the first six visible segments. The first pair, on the genital segment, and the sixth pair are transient and are resorbed by the time translocation of the yolk has begun. The second and third pairs gradually flatten and disappear as book-lungs invaginate and develop behind them (see below). The fourth and fifth pairs become specialized as spinnarets. Each pair becomes bifid, giving four pairs of spinnarets, an anterolateral and an anteromedian pair on segment 4, and a posterolateral and posteromedian pair on segment 5. During this stage, the proctodaeum invaginates on the telson.

A full understanding of the external development of the *Heptathela* embryo has proved of great value as a basis for interpreting the development of other araneid embryos. It is now possible to show that several mygalomorph spiders retain much of this basic pattern (*Atypus*, Yoshikura, 1958; *Evagrus*, Montgomery, 1909b; *Ischnocolus*, L. and W. Schimkewitsch, 1911), as do the haplogynid araneomorphs (*Segestria*, *Harpactes*, Holm, 1941; *Loxosceles*, Montgomery, 1909b; *Pholcus*, Claparède, 1862; Emerton, 1872; Sabatier, 1881; Schimkewitsch, 1887, 1898; Morin, 1887, 1888; Kishinouye, 1891b; Pokrowsky, 1899). Other mygalomorphs (*Ischnothele*, Holm, 1954) and the cribellate araneomorphs (*Dictyna*, *Amaurobius*, Holm, 1941), in contrast, exhibit more specialized patterns of development, while the entelygynous araneomorphs show an extreme specialization in both the mode of growth of the germ band and the relationship between the germ band and the yolk mass (Table 2). Of these accounts, the most informative are those of Wallstabe (1908), Kautzsch (1909,

TABLE 2. DESCRIPTIONS OF THE EXTERNAL FEATURES OF THE EMBRYOS OF ENTELEGYNOUS SPIDERS

Clubionoidea	
Gnaphosidae	*Drassodes*, Morin, 1887, 1888
Clubionidae	*Clubiona*, Salensky, 1871
Sparassidae	*Torania*, Ehn, 1963
Lycosoidea	
Lycosidae	*Lycosa* (= *Trochosa*), Claparéde, 1862; Barrois, 1877; Schimkewitsch, 1887, 1898; Morin, 1887, 1888; Kishinouye, 1891b, 1894; Janeck, 1909
	Pardosa, Pross, 1966
Ctenidae	*Cupiennius*, Seitz, 1966
Pisauridae	*Dolomedes*, Kishinouye, 1891b, 1894; Pappenheim, 1903; Janeck, 1909
Agelenidae	*Agelena*, Balbiani, 1873; Balfour, 1880; Locy, 1886; Kishinouye, 1891b, 1894; Wallstabe, 1908; Kautzsch, 1909, 1910; Holm, 1952
	Tegenaria, Balbiani, 1873; Barrois, 1877; Schimkewitsch, 1887
	Argyroneta, Hamburger, 1910
Araneoidea	
Argiopidae	*Aranea*, Claparède, 1862; Barrois, 1877; Sabatier, 1881; Schimkewitsch, 1887; Kishinouye, 1891b, 1894; Lambert, 1909
	Meta, Holm, 1941
	Singa, Salensky, 1871
Theridiidae	*Theridion*, Salensky, 1871; Morin, 1887, 1888; Kishinouye, 1891b, 1894; Simmons, 1894; Montgomery, 1909a
	Latrodectus, Rempel, 1957.

1910) and Holm (1952) on *Agelena*, Montgomery (1909a) on *Theridion*, Rempel (1957) on *Latrodectus*, Ehn (1963) on *Torania*, Seitz (1966) on *Cupiennius* and Pross (1966) on *Pardosa*.

Unlike *Heptathela*, which develops a long, broad embryonic primordium including the prosoma and an opisthosomal rudiment, all other spiders develop a small embryonic primordium in the centre of the flattened, ventral surface of the blastoderm (Fig. 140). A short gastral groove then arises in the centre of the primordium and a small, rounded swelling pushes out behind the posterior end of the groove. This protuberance results from the formation of a compact group of large cells beneath the surface of the embryonic primordium and is usually called the posterior cumulus. As the gastral groove is obliterated, the posterior cumulus migrates in a posterior direction between the yolk mass and the surface layer of cells. The migrating cumulus continues around the posterior pole of the yolk mass on to the dorsal surface, where the cumulus cells disperse and can no longer be distinguished in external view.

The developmental significance of the posterior cumulus has not been investigated in mygalomorph, haplogynid or cribellate spiders, but experimental investigations on *Agelena* embryos by Holm (1952) and Ehn (1962) have shown that the cumulus of entelegynids has a causal role as an organizer. The direction of migration fixes the anteroposterior axis of the embryo and the migrating cumulus influences the blastoderm to develop a segmenting germ band.

In certain mygalomorph and haplogynid embryos, before migration of the posterior cumulus begins, the embryonic primordium enlarges to form an oval plate occupying most of the ventral surface of the yolk mass (*Atypus*, Yoshikura, 1958; *Ischnocolus*, L. and W. Schimkewitsch, 1911; *Segestria*, Holm, 1941; *Conothele*, Crome, 1963, 1964). After the posterior cumulus has migrated away from the enlarged primordium, the latter then develops a second posterior protuberance. Subsequent events reveal that the embryonic primordium at this stage is similar to that formed directly from the blastoderm of *Heptathela* (compare

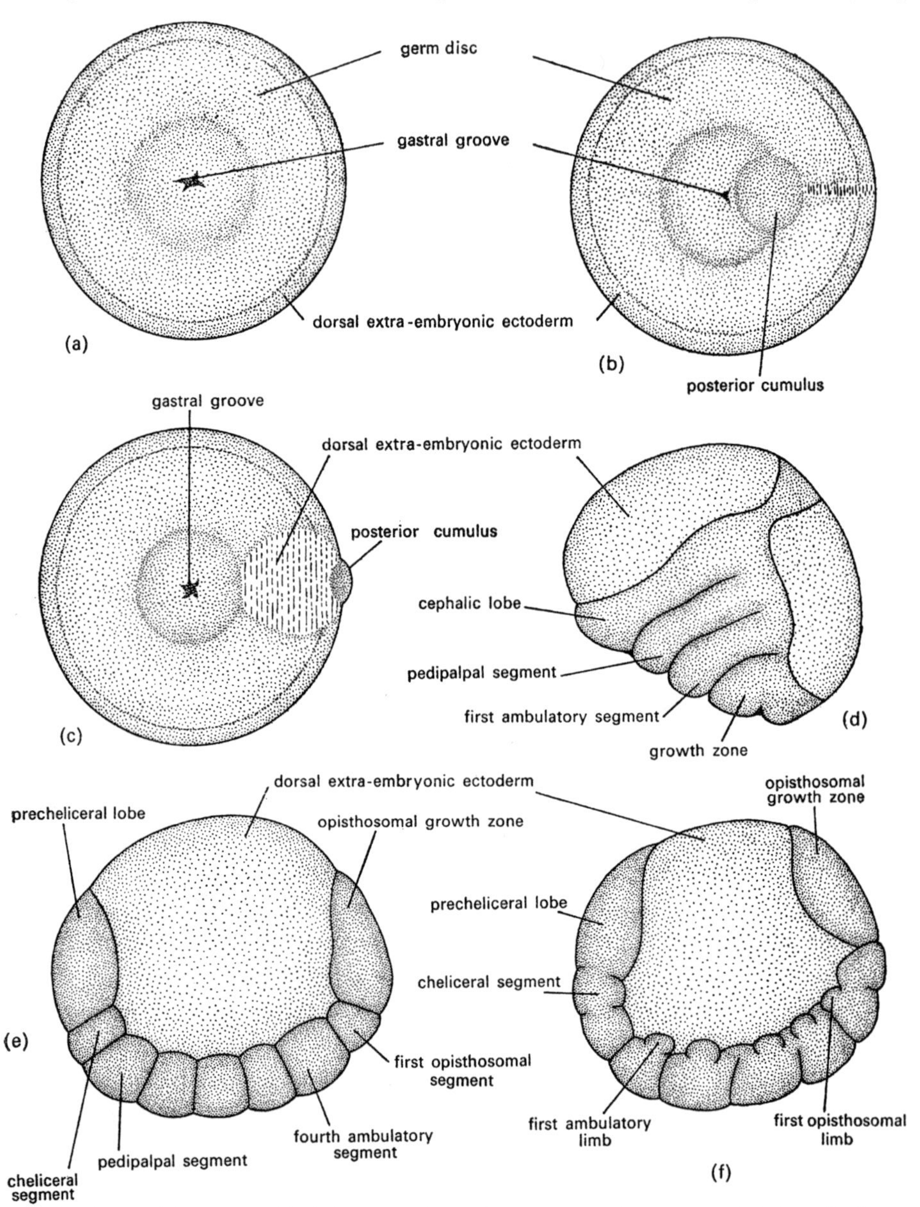

FIG. 140. *Agelena labyrinthica*. (a) Embryonic primordium and gastral groove, ventral view; (b) formation of the posterior cumulus, ventral view; (c) migration of the posterior cumulus, ventral view; (d) early segmentation of the prosoma, lateral view; (e) completion of ventral concentration, antero-posterior elongation and prosomal segmentation of the germ band; (f) formation of prosomal limb buds. (After Holm, 1952.)

Fig. 139). The oval plate is the prosomal rudiment and the posterior protuberance is the opisthosomal rudiment. It is of interest that the opisthosomal rudiment develops at the same site as the posterior cumulus. In *Heptathela*, the opisthosomal rudiment contains the posterior midgut rudiment, with which the posterior cumulus of other spider embryos is thought to be homologous (see below). Furthermore, the gastral groove of *Atypus*, *Ischnocolus* and *Segestria* lies, mainly in front of this site, in that part of the germ disc which gives rise to the posterior ambulatory segments of the prosoma, as it does in *Heptathela*.

Once the embryonic primordium of *Atypus*, *Ischnocolus* or *Segestria* has reached this stage, the further development of the germ band proceeds in the same manner as in *Heptathela* (compare Figs. 139 and 141). The prosomal rudiment increases in length and breadth and becomes demarcated by transverse grooves into the four ambulatory segments, the pedipalpal segment and a cephalic lobe, which subsequently becomes divided into the cheliceral segment and pre-cheliceral lobe. At the same time, the opisthosomal rudiment proliferates the first two opisthosomal segments as broad bands on the surface of the yolk behind the fourth ambulatory segment, and then protrudes from the surface at the posterior pole of the yolk mass.

Other mygalomorphs (*Ischnothele*, Holm, 1954) and haplogynids (*Pholcus*, Emerton, 1872; Morin, 1887, 1888) and all entelegynids (e.g. Claparède, 1862; Salensky, 1871; Balbiani, 1873; Balfour, 1880; Locy, 1886; Morin, 1887, 1888; Schimkewitsch, 1887, 1898; Kishinouye, 1891b; Kautzsch, 1909, 1910; Montgomery, 1909a; Lambert, 1909; Fulinski, 1912; Holm, 1952; Rempel, 1957; Ehn, 1963; Seitz, 1966; Pross, 1966: the species examined are listed in Table 2) exhibit a more specialized mode of development of the germ band (Fig. 140).

While the embryonic primordium still occupies only a small area in the centre of the flattened ventral surface of the blastoderm and retains the gastral groove and primitive cumulus, the primordium extends equally in all directions to form a circular plate on the ventral surface of the yolk mass with the posterior cumulus situated just behind the centre of the plate.

The development of the embryonic primordium as a circular plate with an almost central cumulus appears to contrast strongly with the development of the primordium of more primitive spiders as an oval plate with the cumulus at its posterior end. When the posterior cumulus begins its posterior migration, however, this apparently large difference is revealed as a simple modification. Firstly, the superficial cells along the path of migration become attenuated. After the posterior cumulus has passed the margin of the circular plate and has continued along its path over the posterior pole of the yolk mass, the attenuated zone expands to form a triangular sector, whose apex is near the centre of the germ disc, where the posterior cumulus was first formed, and whose base is posterior. The attenuated sector, in fact, is a posteroventral extension of the dorsal extra-embryonic ectoderm. The sector widens until the apical angle is more than 180° and the margins are directed slightly anterolaterally. In association with this change, the circular embryonic primordium contracts forwards to form a broadly triangular germ band lying in front of the original site of the posterior cumulus. The band now increases in length and width and becomes demarcated by transverse grooves into a cephalic lobe, a pedipalpal segment, a first ambulatory segment and a posterior, growth zone. By comparison with the same stage of the germ band of

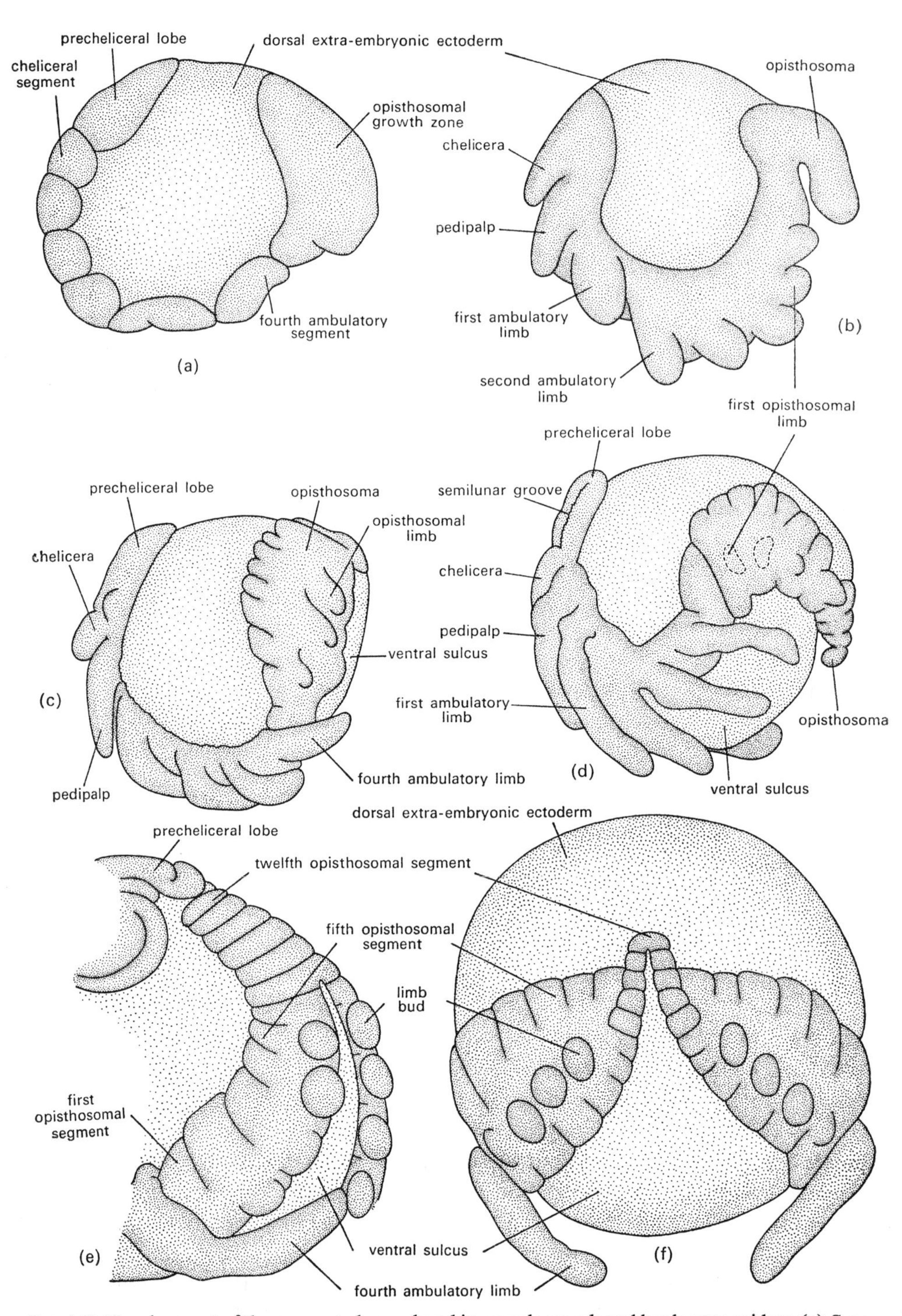

FIG. 141. Development of the segmented germ band in mygalomorph and haplogyne spiders. (a) *Segestria*, germ band at completion of prosomal segmentation; (b) *Segestria*, early segmentation and growth of the opisthosoma; (c) *Ischnothele*, early segmentation and growth of the opisthosoma; (d) *Ischnothele*, completion of segmentation of the opisthosoma, onset of inversion; (e), (f) completion of segmentation of the opisthosoma of *Amaurobius* and *Meta*. [(a), (b), (e) and (f) after Holm, 1941; (c) and (d) after Holm, 1954.]

primitive spiders, the only difference is that the second to fourth ambulatory segments, as well as the opisthosomal rudiment, are concentrated into the growth zone. We can now retrace the sequence of events to see that, with the shortening of the posterior half of the embryonic primordium, the anterior half is initially developed as an arc bent posteriorly around the sides of the growth zone, giving a circular plate with a gastral groove, posterior cumulus and growth zone at its centre, rather than an oval plate with a gastral groove, posterior cumulus and growth zone at its posterior end. Accompanying and following the migration of the posterior cumulus, the two halves of the arc contract forwards towards the ventral midline in front of the growth zone, restoring the normal orientation of the anterior part of the germ band. The growth zone now becomes active, proliferating the second to fourth ambulatory segments and the first two opisthosomal segments. During this time, the segmenting germ band becomes longer and narrower, extending over the poles of the yolk mass to attain the form common to all spider embryos at this stage (Fig. 140). The posterior opisthosomal rudiment remains large but, unlike the protuberant opisthosomal rudiment of more primitive spider embryos, lies flat against the yolk. The consequences of this difference will emerge in due course.

In the mygalomorphs *Atypus*, *Ischnocolus*, *Evagrus* and *Conothele* and the haplogynid *Segestria* (Figs. 141 and 142), the protuberant opisthosomal rudiment continues its growth, as in *Heptathela* (compare Fig. 139), in the form of a tube flexed forward at the fourth opisthosomal segment (Yoshikura, 1958; L. and W. Schimkewitsch, 1911; Montgomery, 1909b; Holm, 1941; Crome, 1963, 1964). Segment proliferation continues until eleven opisthosomal segments and a terminal telson have been formed. Meanwhile, the two halves of the germ band anterior to the fourth opisthosomal segment undergo further elongation and separate from one another along the ventral midline, arching slightly upwards over the sides of the yolk mass to create a median ventral sulcus. The ventral ectoderm is temporarily attenuated as ventral extra-embryonic ectoderm at the surface of the ventral sulcus. In more specialized spider embryos, the protrusion of the opisthosomal rudiment is much delayed and the growth zone continues at first to proliferate opisthosomal segment rudiments flat against the yolk. Associated with this modification, the two halves of the germ band flex widely apart on the sides of the yolk mass, creating a broad ventral sulcus. In the mygalomorph *Ischnothele* (Fig. 141) and the haplogynes *Pholcus* and *Loxosceles* the growth zone eventually becomes protuberant after the fifth opisthosomal segment has been proliferated and gives rise to a short caudal papilla flexed in a ventral direction at the eight segment. The usual total of eleven opisthosomal segments and a telson develops in these species (Holm, 1954; Claparède, 1862; Morin, 1887, 1888; Schimkewitsch, 1887; Montgomery, 1909b). When segment formation is completed, the two halves of the germ band curve ventrally from the anterodorsal head, then flex dorsally in the middle region where the prosoma joins the opisthosoma, and ventrally again as the caudal papilla. Each half therefore described an ∽ -shape around its side of the yolk mass.

The cribellate *Amaurobius* and the entelegynid *Meta* are even more specialized (Fig. 141). In these embryos, the opisthosomal segments are proliferated flat against the yolk until eight and nine segments respectively have been formed. As the length of the two halves of the germ band increases, the halves move apart in the usual way to establish a broad ventral sulcus. The growth zone then protrudes and proliferates the remaining segments (9–12 in

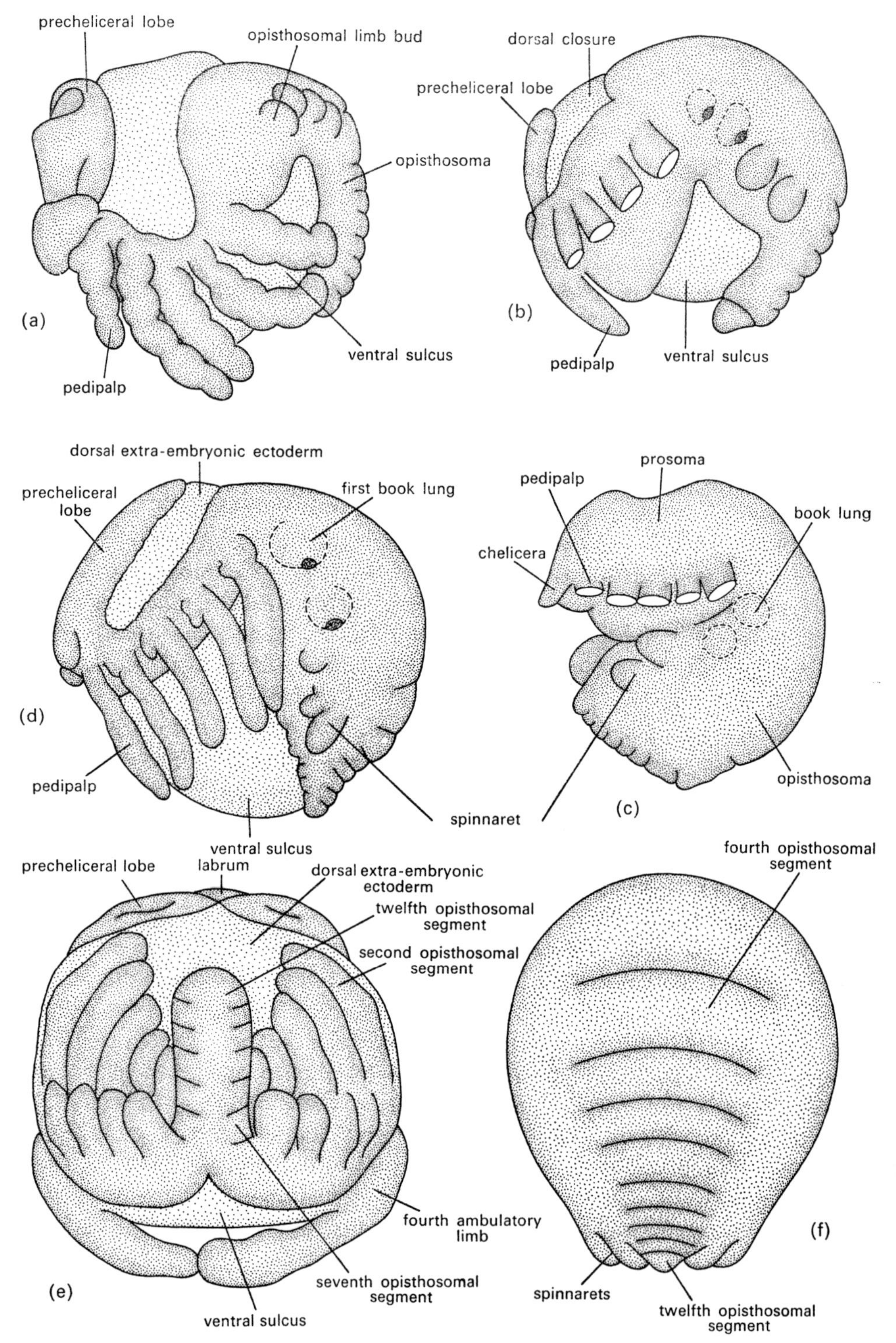

FIG. 142. Inversion and dorsal closure in mygalomorph and haplogyne spiders. (a) *Segestria*, early inversion; (b) *Segestria*, late inversion; (c) *Segestria*, near hatching; (d) *Ischnothele*, late inversion; (e) *Amaurobius*, early inversion, posterior view; (f) *Amaurobius*, opisthosoma after completion of inversion, posterior view. [(a)–(c), (e), (f) after Holm, 1941; (d) after Holm, 1954.]

Amaurobius, 10–12 in *Meta)* as a tubular caudal papilla, but without an accompanying ventral flexure. The caudal papilla thus protrudes in a dorsal direction and the fully segmented germ band has a simplified U-shape in which the anterior and posterior ends approach one another on the dorsal surface of the yolk mass. Finally, in most entelegynids (Fig. 143), the full complement of opisthosomal segments is proliferated flat against the yolk, without

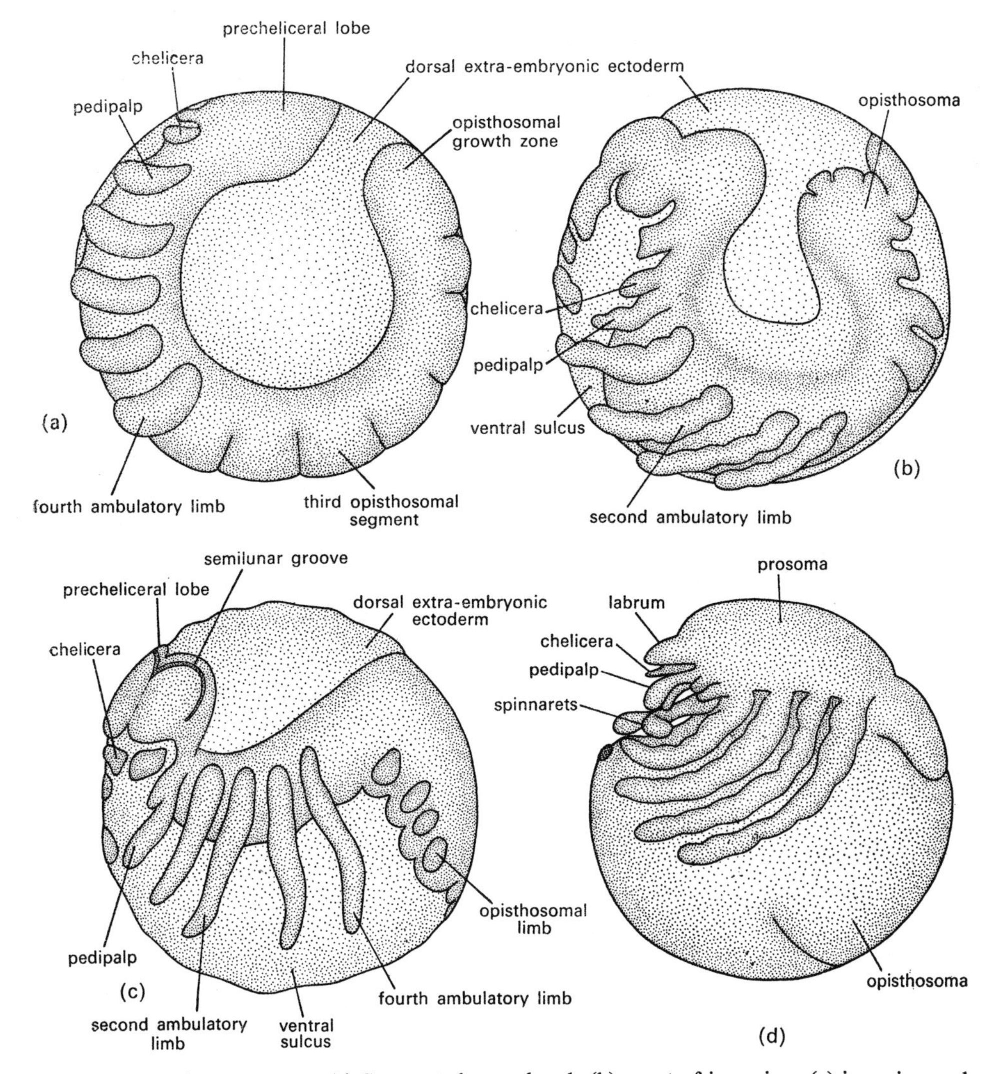

FIG. 143. *Latrodectus mactans.* (a) Segmented germ band; (b) onset of inversion; (c) inversion and dorsal closure nearing completion; (d) ventral closure completed, embryo approaching hatching. (After Rempel, 1957.)

protrusion of a caudal papilla (e.g. *Agelena*, Wallstabe, 1908; Kautzsch, 1909; Lambert, 1909; *Theridion*, Montgomery, 1909a; *Latrodectus*, Rempel, 1957; *Torania*, Ehn, 1963; *Cupiennius*, Seitz, 1966). The segmented germ band remains U-shaped and the two halves curve around the surface of yolk mass, separated by a ventral sulcus, throughout their entire length. Some entelegynid genera develop 11 opisthosomal segments and a telson in the embryo (e.g. *Lycosa*, *Pisaura*, *Aranea*, Holm, 1941), but other genera have a smaller number of opisthosomal segments (e.g. *Agelena*, 9, Wallstabe, 1908; *Cupiennus*, 9, Seitz, 1966; *Theridion*, *Latrodectus* and *Steatoda*, 8, Montgomery, 1909a; Holm, 1941; Rempel, 1957; *Dolomedes*, 8, Pappenheim, 1903).

While the opisthosomal segments are being formed, the embryos of all spiders develop limb buds on the prosomal segments (Figs. 140–3). Development of the pedipalpal and four pairs of ambulatory limb buds is usually simultaneous, but formation of the cheliceral limb buds is slightly delayed. The stomodaeum is also formed as a median invagination between the precheliceral lobes during this period, confirming that the entire cheliceral segment and its associated limb buds are post-oral in spiders when first formed in the embryo. In primitive spiders (e.g. the liphistiomorph *Heptathela*, the haplogynid *Ischnothele*) the four pairs of ambulatory limbs, as well as the pedipalps, develop median protrusions of the coxae, but only the pedipalpal pair persist and become functional gnathobases (Holm, 1954; Yoshikura, 1954, 1955). Entelegynids lack the transient coxal protrusions on the ambulatory limbs.

During the outgrowth of the prosomal limbs, paired limb buds also develop on the anterior opisthosomal segments. It will be recalled that *Heptathela* develops limb buds on the first to sixth opisthosomal segments, the first and sixth pairs being transitory. All other spider embryos develop limb buds on second to fifth opisthosomal segments, but lack the transient pair on the first (genital) segment. The haplogynid *Loxosceles* is said to develop transient limb buds on the sixth and seventh opisthosomal segments (Montgomery, 1909b) and *Segestria* develops a transient pair on segment 6 (Holm, 1941), but other species lack opisthosomal limb buds posterior to segment 5 (*Atypus*, Yoshikura, 1958; *Evagrus*, Montgomery, 1909b; *Conothele*, Crome, 1963, 1964; *Ischnothele*, Holm, 1954; entelegynids, Salensky, 1871; Schimkewitsch, 1887; Kishinouye, 1891b; Wallstabe, 1908; Rempel, 1957; Seitz, 1966).

The further development of the opisthosomal limb buds is completed after dorsal closure has taken place (see below). In mygalomorphs, as in *Heptathela*, book-lungs invaginate behind the first two pairs of these limb buds, on segments 2 and 3, and the limb buds themselves are simultaneously withdrawn into the general surface epithelium of their respective segments (Montgomery, 1909b; Holm, 1954; Yoshikura, 1958). In araneomorph spiders, a single pair of book-lungs develops in the same way behind the limb buds on segment 2, but invaginations behind the limb buds on segment 3 give rise to tracheae. The paired limb buds on segments 4 and 5 develop, as in *Heptathela*, into spinnarets. In mygalomorphs, the pair on segment 4 is resorbed while the pair on segment 5 become bifid and form two pairs of spinnarets corresponding to the intermediate and posterior spinnarets of *Heptathela*. Araneomorph spiders, in contrast, retain both pairs of limb buds. The anterior pair become bifid, with the lateral rami developing as the anterior spinnarets and the median rami merging to form the median colulus. The posterior pair also become bifid in the usual way, forming the intermediate and posterior spinnarets (Salensky, 1871; Locy, 1886; Morin, 1887; Kishinouye, 1891b; Montgomery, 1909 a, b; Rempel, 1957).

Inversion in spider embryos

Dorsal closure in the embryo of *Heptathela*, as we have seen, begins in the prosoma and is accompanied by a translocation of the yolk mass from the prosoma to the forwardly flexed opisthosoma. This shift provides the basis for the subsequent development of the characteristic globular shape of the opisthosoma, by dorsal closure around the translocated yolk mass.

In spiders in which the two halves of the segmented germ band are separated by a ventral sulcus and the flexure of the posterior part of the opisthosoma is reduced or eliminated, the process of dorsal closure and formation of a globular opisthosoma is variously modified (Figs. 141–3). Some of the mygalomorph species which retain a ventrally flexed, tubular opisthosoma (e.g. *Atypus*, *Conothele*, Yoshikura, 1958; Crome, 1963, 1964) show little modification. As dorsal closure proceeds in the prosoma, the yolk mass is translocated in a posterior direction and the two halves of the germ band come together once more in the ventral midline, with elimination of the ventral sulcus. Expansion of the dorsal extra-embryonic ectoderm of the opisthosoma accommodates the shift of yolk, in the same manner as in *Heptathela*, and dorsal closure of the opisthosoma then follows.

In the haplogynid *Segestria*, however, even though the opisthosoma is tubular and flexes forwards at the level of the fourth opisthosomal segment a modified sequence of dorsal closure is in evidence (Fig. 142). Instead of the prosoma closing dorsally and the opisthosoma expanding dorsally in the manner described above, closure proceeds forwards along the dorsal midline from the fourth opisthosomal segment and is achieved mainly by an upward movement of the two halves of the germ band towards one another over the yolk mass. This movement inverts the curvature of the anterior part of each half of the germ band from ventrally convex to dorsally convex and is accommodated by a stretching of the ventral extra-embryonic ectoderm of the ventral sulcus. Ventral closure of the prosoma now takes place, shifting the remaining yolk into the opisthosoma. The translocated yolk is accommodated by further expansion of the ventral sulcus (i.e. the *ventral* extra-embryonic ectoderm) of the opisthosoma, partly through broadening in the region of the first four segments, partly by further extension of the sulcus along the ventral midline of the fifth and succeeding segments. Ventral closure of the opisthosoma then takes place around the yolk mass, establishing the definitive globular form of the opisthosoma.

The mygalomorph *Ischnothele* and the haplogynid *Pholcus*, which have only a short caudal papilla flexed ventrally at the eighth opisthosomal segment (Fig. 141), are more modified again (Holm, 1954; Schimkewitsch, 1887; Morin, 1887, 1888; Pokrowsky, 1899). Dorsal closure in these species proceeds forwards from the eighth opisthosomal segment and is again effected mainly by dorsal migration of the two halves of the germ band, with inversion of flexure. The ventral sulcus widens in the usual way during this process, but in addition, extends back as far as the telson and also shortens in length, so that the entire opisthosoma is flexed ventrally relative to the prosoma. Ventral closure of the prosoma, and then of the swollen opisthosoma, now follows.

The most modified patterns of inversion and dorsal closure, as might be expected, occur in those spider embryos which have either a short, unflexed caudal papilla (e.g. the cribellate *Amaurobius*, Holm, 1941; the entelegynid *Meta*, Holm, 1941) or have lost the caudal papilla entirely, as in most entelegynids. In *Amaurobius*, when the two halves of the germ band begin to invert their curvature and flex dorsally up the sides of the yolk mass, the opisthosomal segments in front of the caudal papilla grow rapidly upwards to meet in the dorsal midline (Fig. 142). During this phase, the caudal papilla retains its original orientation, projecting in a dorsal direction, but the base of the papilla (at about segment 7 of the opisthosoma) is shifted downwards over the posterior pole of the yolk mass. As inversion continues, the two halves of the prosoma come together dorsally, leaving the usual wide, ventral sulcus. Then, as the prosoma undergoes ventral closure and the remaining yolk is shifted into the opisthosoma, the caudal papilla turns downwards and becomes a normal continuation of the swollen opisthosoma. Ventral closure of the opisthosoma now completes the body wall in the usual way.

Inversion in entelegynid embryos (Fig. 143) has been described by many workers (e.g. Claparède, 1862; Salensky, 1871; Morin, 1887, 1888; Kishinouye, 1891b; Wallstabe, 1908; Montgomery, 1909a; Rempel, 1957; Ehn, 1963; Seitz, 1966), but can now be given a more satisfactory interpretation against a background of knowledge of inversion and dorsal closure in the embryos of more primitive spiders. As has been described above, none of the opisthosomal segments in the majority of entelegynids (e.g. *Agelena*, *Theridion*, *Latrodectus*) projects as a caudal papilla, so that the two halves of the U-shaped segmented germ band are separated throughout their length by the ventral sulcus. During the first stage of inversion, the two halves of the opisthosoma move rapidly upwards to meet in the dorsal midline, effecting dorsal closure of this part of the body. At the same time, the posterior end of the opisthosoma slides in a ventral direction over the posterior pole of the yolk mass, so that the opisthosoma quickly acquires a ventrally flexed orientation relative to the prosoma. Inversion of flexure now continues by an upward movement of the prosomal segments until they, too, meet in the dorsal midline. Thus, when inversion is completed, the major part of the yolk mass is accommodated within the thin wall of a much expanded ventral sulcus. Entelegynid inversion takes place rapidly (e.g. *Latrodectus*, 70 hours, Rempel, 1957; *Dolomedes*, 2 hours, Legendre, 1958), presumably because it is accomplished entirely by movement of the two halves of the germ band over the surface of the yolk. Once inversion is complete, the ventral closure of the prosoma shifts the remaining yolk into the opisthosoma in the usual way. Ventral closure of the opisthosoma around the remaining yolk mass then completes the globular body wall of this part of the body.

The remarkable similarity of the outcome of inversion and ventral closure in entelegynids and of translocation of yolk and dorsal closure in primitive liphistiomorph spiders can be seen in a comparison of the late embryos of *Heptathela* and *Latrodectus* (Figs. 139 and 143). Both arrive at the same body form, with a swollen opisthosoma tucked forwards ventrally between the paired appendages hanging down from a flattened prosoma. In fact, all spider embryos terminate their development in this form and orientation within the spherical egg membrane. Again, as has been mentioned earlier, all spider embryos pass through an earlier stage in which a U-shaped germ band extends from the anterior to the posterior pole along the ventral surface of the yolk mass and comprises a cephalic lobe (future pre-cheliceral lobe and cheliceral segment), pedipalpal segment and four ambulatory segments of the prosoma, the first two segments of the opisthosoma and a terminal opisthosomal growth zone. The events which intervene between these two stages vary from a primitive mode in *Heptathela* through various degrees of modification in *Atypus*, *Ischnothele*, *Segestria* and *Amaurobius* to an extreme of specialization in entelegynids such as *Agelena* and *Latrodectus*. We can therefore consider what are the functional advantages of this trend of modification in spider embryos. Two features stand out. One is that, with the replacement of a long, forwardly flexed caudal papilla by a U-shaped germ band, the need for massive translocation of the yolk mass prerequisite to the formation of a swollen opisthosoma is eliminated. The second feature, correlated with the first, is that the modified mode of development effects dorsal closure of the opisthosoma at a relatively early stage of development. Since dorsal closure of this region must precede development of the heart and the onset of circulation of the blood, it seems likely that a relatively precocious development of the heart and circulation results from the modified development, facilitating the digestion and distribution of the remaining yolk reserves. It is significant, perhaps, that the posterior shift of yolk in *Heptathela* takes 7 days, while inversion in *Latrodectus* takes 3 days; furthermore, that development to hatching in *Heptathela* takes about 2 months, in *Latrodectus*, 17 days. We shall see below that the embryos of all arachnids are variously modified in ways which facilitate the solution of problems presented by a large yolk mass, especially in its role in the formation of a swollen opisthosoma in certain groups and its interference with dorsal closure. We have already seen that the xiphosuran solution to these problems is the development of a very short, broad opisthosomal rudiment on the posterior surface of the yolk mass, followed by longitudinal expansion of this rudiment, with retention of the yolk mainly in the prosoma.

The development of the head in spiders proceeds in a generally similar manner in all species. Useful descriptions have been given by Yoshikura (1954, 1955, 1958) for *Heptathela* and *Atypus* (Fig. 144) and by a number of workers for entelegynids (Kishinouye, 1891b; Pappenheim, 1903; Wallstabe, 1908; Montgomery, 1909a; Rempel, 1957; Pross, 1966). While the cheliceral limb buds are developing, the precheliceral lobe enlarges and becomes bilobed. The stomodaeum invaginates in the midline between the pre-cheliceral lobes, while the labrum, usually in the form of paired rudiments (Schimkewitsch, 1887; Lambert, 1909; Montgomery, 1909a; Police, 1933; Yoshikura, 1954, 1955, 1958; Legendre, 1959) but occasionally reported to be a single median lobe (e.g. Rempel, 1957), pouches out in front of the mouth. At the same time, the pre-cheliceral lobes develop a pair of semilunar grooves which extend around the anterior margins of the lobes and curve back on each side of the ventral midline to reach the labrum. On the lateral margins of the precheliceral lobes, a pair of smaller, separate invaginations also develops, forming the lateral vesicles. These invaginations are all components of the development of the pre-cheliceral ganglia, which will be discussed in more detail in a later section. The anterior border of each semilunar groove spreads back as a superficial fold over the surface of the pre-cheliceral lobe and merges with the labrum to form the external wall of the pre-cheliceral part of the head. Accompanying this process, the chelicerae shift forwards into a pre-oral position and the bases of the pedipalps approach the level of the mouth, bringing the median gnathobasal lobes on their coxae into a para-oral position.

Thus, a comparative survey of the external features of development of the embryos of spiders reveals that the liphistiid spiders exhibit a primitive mode of development and that the peculiarities of other spider embryos can all be interpreted as functional modifications of this mode. We can now examine the external development of the embryos of other groups

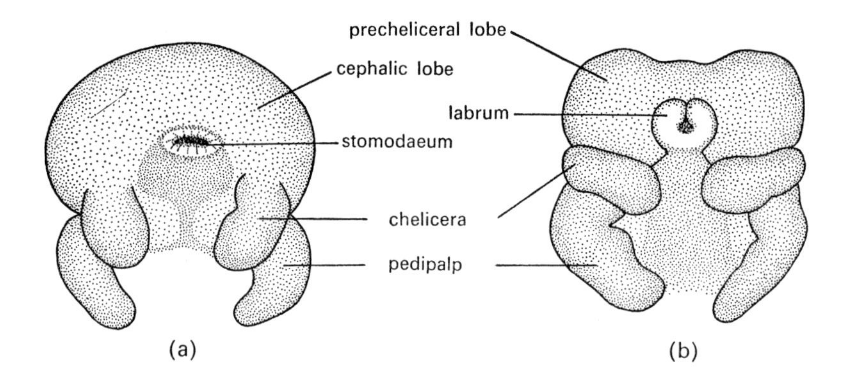

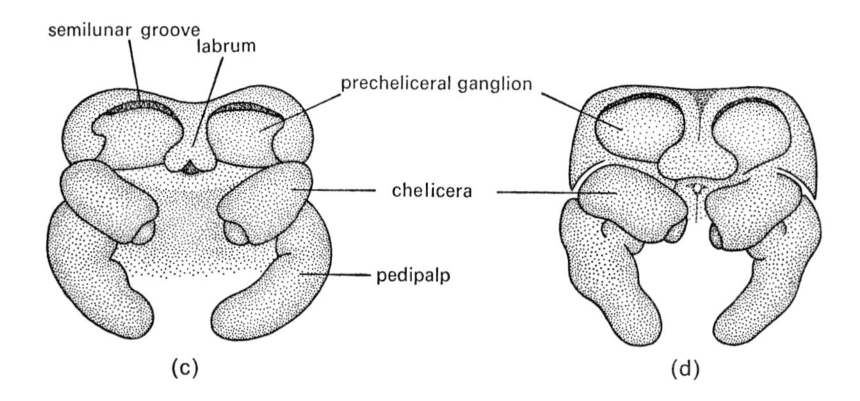

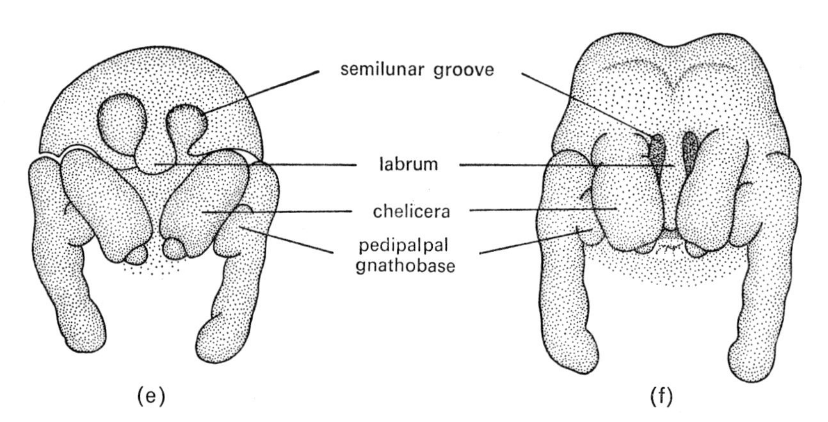

FIG. 144. Development of the anterior end of the prosoma of *Atypus karschi*, seen in anterior view. (a) Cephalic lobe, with postoral chelicerae and pedipalps; (b) formation of the labrum; (c) formation of the semilunar grooves; (d)–(f), overgrowth and closure of the semilunar grooves and pre-oral migration of the chelicerae. (After Yoshikura, 1958.)

of arachnids, bearing in mind that less is known about these (the Amblypygi and Solifugae are virtually unknown, the Palpigradi and Ricinulei wholly unknown) and that the development of pseudoscorpions and scorpions is highly modified. The first question to be considered is, how does the development of opilionid, uropygid and acarine embryos compare with that of liphistiid spiders?

Opiliones

Little attention was paid to opilionid embryos by classical embryologists (Faussek, 1891; Schimkewitsch, 1898), but a number of recent papers (Holm, 1947; Moritz, 1957; Winkler, 1957; Juberthie, 1961, 1964) have shown that development follows the same course in *Opilio parietinus*, *Phalangium opilio*, *Platybunus triangularis*, *Lacinius ephippiatus*, *Odiellus gallicus* and other Palpatores (Fig. 145) and is more generalized than that of araneids in respect of development of the opisthosoma. The first indication of development beyond the blastoderm stage is the formation of a small germ disc by localized cell division at the posteroventral pole of the blastoderm. Immediately behind the germ disc, a posterior cumulus arises (Holm, 1947) and migrates over the posterior pole of the yolk mass onto the dorsal surface, before disappearing from external view. The significance of the posterior cumulus in opilionid development is unknown. During the formation and early migration of the posterior cumulus, the germ disc becomes broader, with its lateral arms extending posteriorly so that the disc is horseshoe-shaped. In the midline of the disc, a second thickening is now developed, which has been interpreted as the genital rudiment (i.e. primordial germ cells), although there is no satisfactory evidence of the subsequent entry of these cells into the gonads.

As external development continues, it becomes clear that the opilionid germ disc is not the embryonic primordium but is merely the opisthosomal rudiment of the embryo. The prosomal segments develop directly from the blastoderm in front of the disc, as narrow thickenings which extend across the ventral surface and up the sides of the yolk mass. The rudiments of the four ambulatory segments occupy most of the ventral surface of the yolk mass. The pedipalpal and cheliceral rudiments, smaller than those of the ambulatory segments, extend around the anteroventral surface of the yolk mass, while the pre-cheliceral lobe lies at the anterior pole. As usual, the cheliceral segment becomes distinct slightly later than the other prosomal segments.

Once the prosomal segment rudiments have formed, they concentrate towards the ventral midline and link up to form a continuous prosomal germ band. At the same time, the two halves of each prosomal segment separate slightly from each other, producing a narrow ventral sulcus, and develop limb buds. The precheliceral lobe becomes bilobed and the stomodaeum invaginates in the midline between the two lobes. The opisthosomal rudiment becomes active as a growth zone, proliferating an unsegmented opisthosomal band over the posterior pole onto the dorsal surface. As the growth zone is carried backwards during this activity, the genital rudiment remains beneath the growth zone and moves with it. At this stage, therefore, the segmenting germ band is U-shaped, with a median ventral sulcus, as it is in entelegynid spiders. In contrast to the latter, however, opilionid embryos do not show either inversion of the germ band or development of a swollen opisthosoma.

Instead, the prosoma shortens as its limb buds grow longer and the opisthosoma is drawn forwards along the ventral surface of the yolk mass into a posteroventral position. During this change of position, the opisthosoma subdivides, in anteroposterior succession, into a series of short broad segments. The first of these in the pregenital segment, followed by the genital segment, then six post-genital segments and a telson. None of the opisthosomal

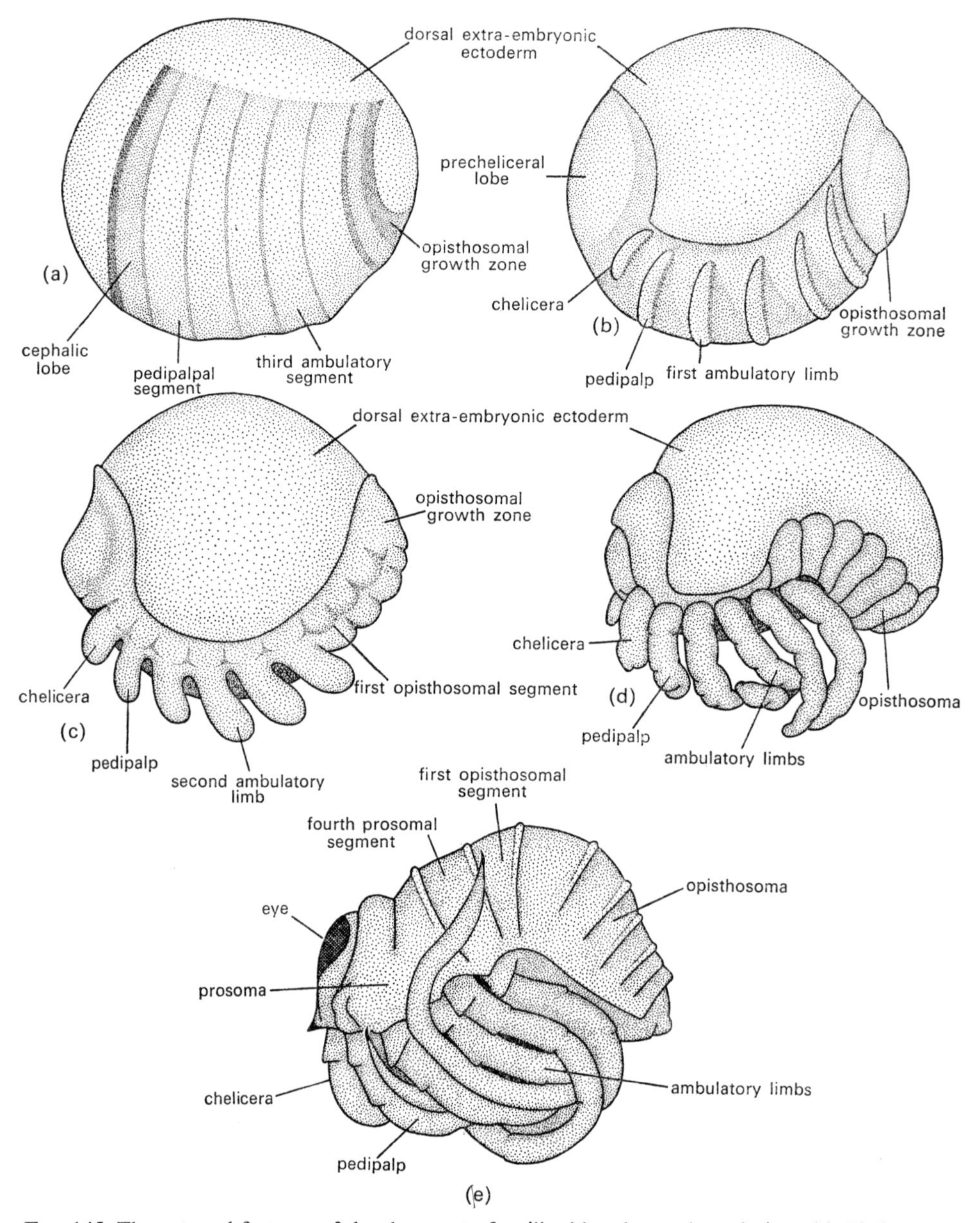

FIG. 145. The external features of development of opilionid embryos, lateral view. (a) *Phalangium*, formation of prosomal segment rudiments; (b) *Platybunus*, germ band with prosomal limb buds; (c)–(e) *Phalangium*, growth and segmentation of the opisthosoma, onset of dorsal closure, embryo approaching hatching. (After Moritz, 1957.)

segments develop limb buds. By the time opisthosomal segmentation is complete, the telson lies at the posterior end of the yolk mass and the segmented germ band extends from pole to pole along the ventral surface of the yolk. The germ band still has a convex ventral surface at this stage, but as the prosomal limbs continue to grow longer and the opisthosomal segments begin to extend up the sides of the yolk mass, initiating dorsal closure, the ventral surface of the embryo gradually becomes slightly concave.

Dorsal closure in opilionid embryos is completed in a simple way by upgrowth of the lateral walls of the segments around the yolk mass. The opisthosomal segments attain dorsal closure more rapidly than the prosomal segments, as they do in the Xiphosura. Neither of these groups exhibits the shifting of the yolk mass associated with precocious dorsal closure of the prosoma and swelling of the opisthosoma in primitive spiders.

In contrast to the conspicuous difference between opilionids and primitive spiders in the mode of development of the opisthosoma, the development of the head continues in the same way in both (compare Fig. 144). When the pre-cheliceral lobe has become bilobed, a pair of semilunar grooves develops around the anterolateral margin of the lobes and a median labral lobe forms in front of the mouth. The semilunar grooves become deeper and are overgrown by their anterior margins, which gradually merge into the labrum. The broad areas covered by overgrowth in this way thicken to form the paired cerebral ganglia. The labrum pushes back over the mouth and the bases of the chelicerae move forwards in the usual way to a pre-oral position. The pedipalpal coxae, which by now have developed median gnathobases, simultaneously come to lie on either side of the mouth. Moritz (1957) has demonstrated that the lower lip is formed in opilionid embryos as a sternal thickening between the coxal bases of the pedipalps.

Uropygi

The embryos of uropygids are larger than those of opilionids, with about nine times the volume of yolk. Together with the amblypygids and Solifugae, they are the only embryos among arachnids which compare in size with those of the Xiphosura. Since these three groups, like the araneids, develop a swollen opisthosoma, their embryonic development might be expected to be specialized in ways which facilitate both dorsal closure and the shaping of the opisthosoma around the yolk. Neither the Amblypygi nor the Solifugae have been studied embryologically during the last sixty-five years, and the early observations on these animals by Pereyaslawzewa (1901), Gough (1902) and Heymons (1904) are inadequate; but the Uropygi have been better served. Studies by Kästner (1948) on *Thelyphonus caudatus* and Yoshikura (1961) on *Typopeltis stimpsonii* have shown that uropygid embryos display an inversion of the germ band analogous to that of entelegynid spiders (Fig. 146). The process is not identical, since the uropygid germ band does not extend over the posterior pole onto the dorsal surface of the yolk mass. Even so, uropygid and araneid embryos have much in common. To begin with, a large, oval, embryonic primordium develops on the ventral surface of the uropygid blastoderm and acquires a protuberant opisthosomal rudiment at the posterior end, as in the liphistiid spider *Heptathela*. The embryonic primordium elongates, with shift of the opisthosomal rudiment towards the posterior pole of the yolk mass, and subdivides into the usual series of short, broad, prosomal segments (pre-cheliceral lobe, cheliceral, pedipalpal and four ambulatory segments), the delineation of the cheliceral segment being slightly delayed. Simultaneously, the opisthosomal rudiment proliferates the first two or three opisthosomal segments as short, broad rudiments on the posteroventral face of the yolk mass.

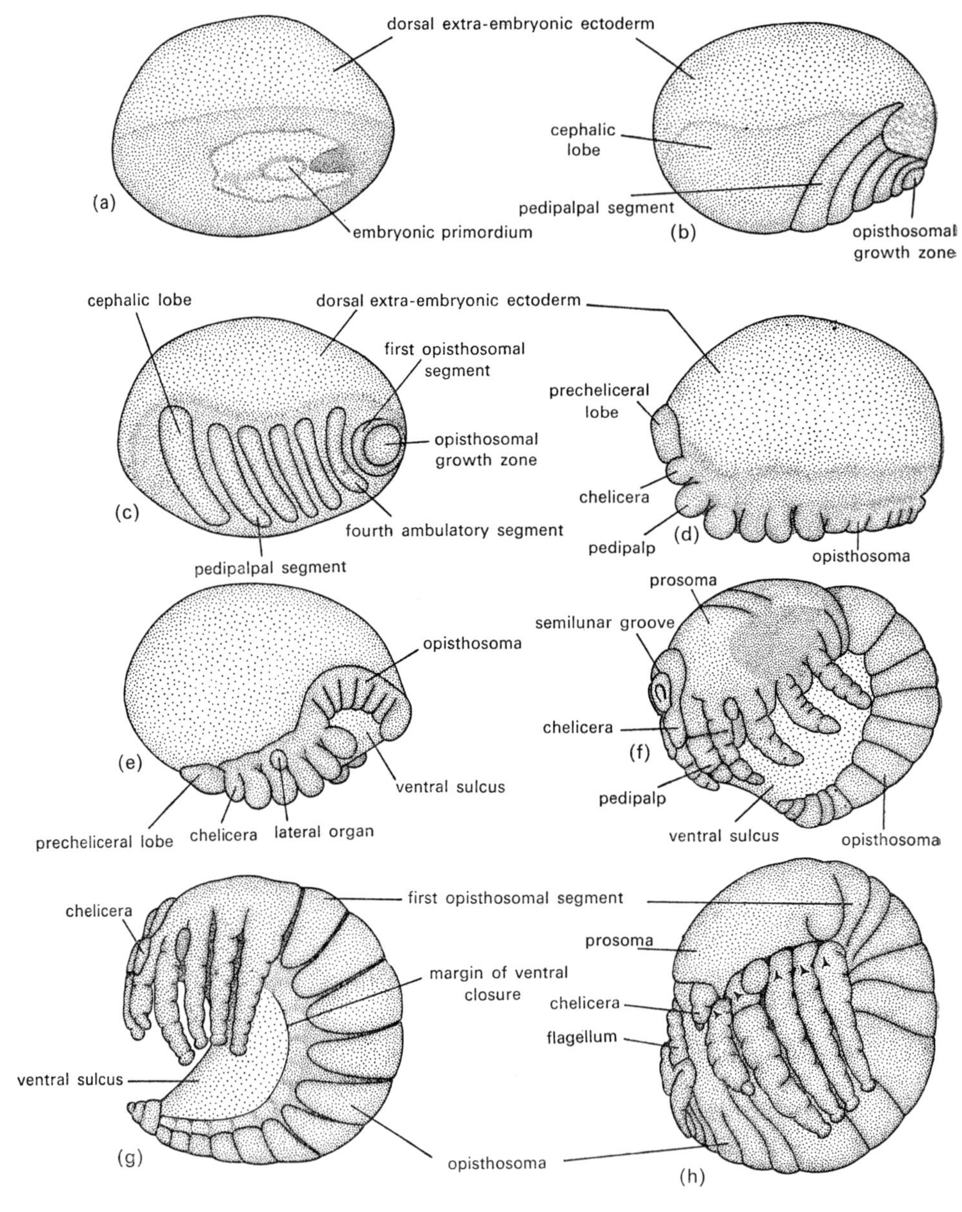

FIG. 146. The external features of development of *Typopeltis stimpsonii*. (a) Formation of the embryonic primordium; (b) early segmentation of the prosoma; (c) completion of prosomal segmentation; (d) completion of opisthosomal segmentation; (e) early inversion; (f) inversion and dorsal closure completed; (g) ventral closure of the prosoma; (h) ventral closure of the opisthosoma, embryo approaching hatching. (After Yoshikura, 1961.)

As limb buds develop on the prosomal segments, the segment rudiments themselves become longer and narrower. The two halves of the prosoma separate along the ventral midline, creating a ventral sulcus. Meanwhile, further segments continue to be proliferated at the posterior end until twelve opisthosomal segments (including an externally visible pre-genital segment) and a telson have been formed. None of these segments develops limb buds. Only during this phase of development does a conspicuous difference between the embryos of uropygids and *Heptathela* become apparent. Instead of forming a forwardly flexed caudal papilla, the opisthosomal segments of uropygids continue to push forwards from a growth zone which retains its position at the posteroventral pole of the yolk mass. Simultaneously, the two halves of the germ band flex upwards around the sides of the yolk mass, widening the ventral sulcus. The inversion of flexure is centred on the anterior opisthosomal segments, and carries the two halves of the germ band towards the dorsal midline. Dorsal closure begins simultaneously at the posterior end of the opisthosoma and the anterior end of the prosoma, and proceeds towards the middle of the embryo. At the same time the ventral sulcus widens to accommodate the large mass of remaining yolk.

During inversion and dorsal closure, the prosomal limbs increase in length. Ventral closure of the prosoma then takes place, accompanied by a shift on the remaining yolk into the opisthosoma and a forward flexure of the swollen opisthosoma between the prosomal limbs just as in spiders. Finally, ventral closure of the opisthosoma is completed around the yolk mass, establishing the globular opisthosoma wall. The final position of the embryo within the egg membrane is similar to that of spiders.

The same similarity to araneids is evident in the development of the anterior end of uropygids (Kästner, 1949). After the pre-cheliceral lobe has become bilobed, with a median stomodaeum, paired labral lobes arise in front of the stomodaeal opening and a pair of semilunar grooves develops on the pre-cheliceral lobes. The margins of the semilunar grooves overgrow the surface of the pre-cheliceral lobes in the usual way, merging with the labrum, which fuses to form a median lobe. The chelicerae become pre-oral and the pedipalpal gnathobases become para-oral. The sternal lower lip of the mouth is separate from the gnathobases, as it is in opilionids and araneids.

In view of the obvious indication of a common basis of development in araneids and uropygids, renewed studies on amblypygids and solifugids would be of great interest. A hint of the occurrence of inversion in the embryos of these groups can be detected in the older accounts, but more facts are needed. Heymons (1904) made several useful observations on *Galeodes*. For example, median coxal processes develop on the four pairs of ambulatory limbs as well as on the pedipalps, though only the pedipalpal pair persist and develop as gnathobases. The same transient vestiges occur in primitive spiders. Furthermore, *Galeodes* has several primitive features on the opisthosoma. The pregenital segment is visible externally in the embryo, as the first of ten opisthosomal segments and a telson. A small pair of limb buds develops on the pregenital segment, followed by a larger pair on the genital segment and small pairs on each of segments 3–10. All pairs are transient, but paired stigmata develop behind the limb buds on the third, fourth and fifth segment.

Thus the primitive Araneae, the Uropygi, and perhaps the Amblypygi and Solifugae, share a number of embryonic features in common in which they differ from the Opiliones. The focal point of these features is the forward flexure of the opisthosoma and the progressive shift of yolk entirely into the opisthosoma, specializations which are lacking in the Opiliones. We can now examine the position of the Acarina with respect to external development. The opisthosoma is, of course, much shortened in the acarines, obliterating the distinction between the two basic parts of the body, but during embryonic development, the prosoma and opisthosoma can be easily distinguished and their relationship to each other and to the yolk mass can be examined.

Acarina

The external features of acarine development have been studied in most detail in the ticks (Wagner, 1894 a, b; Bonnet, 1907; Aeschlimann, 1958, 1961; Anderson, 1972), the observations on mites being more fragmentary and superficial (Figs. 147–50). In general, development follows a similar course in all species, irrespective of the secondary reduction of the size of the egg in tarsonemid, trombidiform and sarcoptiform mites. The first external sign of development is the formation of a small, midventral germ disc with a temporary gastral groove at its centre. The germ disc has been clearly identified in the ticks *Ornithodorus moubata* and *Hyalomma dromadarii* (Aeschlimann, 1958; Anderson, 1972) and was also indicated in mites by Reuter (1909), Hafiz (1935) and Klumpp (1954).

As far as is known, acarine embryos do not develop a posterior cumulus. Instead, the germ disc as a whole migrates in a posterior direction over the surface of the yolk mass and a broad germ band is formed by the aggregation and proliferation of blastoderm cells in front of the germ disc. In spite of the absence of a posterior cumulus, the embryos of *Ornithodorus* and *Hyalomma* resemble that of the spider *Agelena* in an interesting way. Both Aeschlimann (1958) and Anderson (1972) have observed that the axis of bilateral symmetry of the metastigmatid germ band is established by the direction of migration of the germ disc, in the same way that the axis of bilateral symmetry of *Agelena* is established by the migrating posterior cumulus. Normally, the germ disc of *Ornithodorus* and *Hyalomma* migrates along the long axis of the yolk mass, but in some embryos, the direction of migration is oblique or transverse to this axis and the germ band itself is orientated accordingly. Whether similar phenomena exist in other arachnids, for example opilionids, which have a posterior cumulus has yet to be investigated. A time-lapse study of the *Ornithodorus* embryo by Aeschlimann (1961) showed that the entire blastoderm and underlying peripheral yolk participate in the movement leading to the formation of the germ band. The movement begins mid-dorsally and propagates towards the ventral surface in a wave which carries the germ disc in a posterior direction and concentrates the blastoderm cells in front of the migrating disc.

The germ band of *Ornithodorus* increases in length, without exhibiting segmentation, until it is U-shaped, with both ends on the dorsal surface of the yolk mass. How the increase in length takes place is not known, but it seems likely to be due to generalized growth combined with addition at the posterior end by the germ disc acting as a growth zone. When segment delineation sets in, the two halves of the germ band simultaneously separate, creating a broad ventral sulcus. The first three ambulatory segments are delineated, followed by the pedipalpal and fourth ambulatory segment, then five opisthosomal segments and a telson. Delineation of the cheliceral segment from the pre-cheliceral lobes is, as usual, somewhat delayed.

The germ band of *Hyalomma* develops in a more specialized manner (Fig. 147). While the germ band still occupies only the posteroventral surface of the yolk mass, the band becomes precociously separated into two divergent halves which give rise to the rudiments of the third and fourth ambulatory segments and, with further proliferation by the growth zone, to the five opisthosomal segments and terminal telson. The anterior part of the germ band, in contrast, arises *in situ* as bilateral aggregations of cells which form, in succession, the second ambulatory, first ambulatory and pedipalpal segments and the cephalic rudiments. Demarcation of the two halves of the cheliceral segment takes place at the posterior ends of the cephalic lobes. The resulting U-shaped, segmented germ band is similar to that of *Ornithodorus*.

Lack of information prevents any comparison of germ band formation in other acarines with that of *Ornithodorus* and *Hyalomma*, but the majority of species develop a similar, bilateral germ band whose ends lie on the dorsal surface of the yolk mass (Fig. 150). (*Hyalomma aegyptium*, *Boophilus calcaratus*, Wagner, 1894 a, b; Bonnet, 1907; *Myobia*, *Cheyletus*, *Tetranychus*, *Atax*, *Pyemotes*, *Pediculopsis*, *Acarapis*, *Tyrolyphus*, Claparède, 1868; Brucker, 1900; Reuter, 1909; Hafiz, 1935; Hughes, 1950; Klumpp, 1954). The ventral sulcus is usually lacking in small acarine embryos. A modified orientation of the germ band is known only for mesostigmatid mites (*Ophionyssus*, *Pergamasus*, Camin, 1953; Zukowski, 1964). Here, the posterior end of the germ band extends onto the dorsal surface of the yolk mass but the anterior end reaching no further than the anterior pole.

The clear segmentation of the opisthosoma observed in *Ornithodorus* and *Hyalomma* has not been reported in other species, but whether the opisthosomal rudiment has entirely lost an external manifestation of segments in other acarines is still not clear. We also need to know something of the development of the primitive acarines belonging to the Notostigmata and Holothyroida, before being able to decide whether the orientation of the germ band in mesostigmatids is more primitive than that of other acarines. An extension of the anterior end of the germ band onto the dorsal surface of the yolk mass, giving a U-shaped band, is otherwise found only as a secondary condition within the araneids. The anterior end of the germ band usually lies at the anterior pole of the yolk mass, as in liphistiid spiders, uropygids, amblypygids and opilionids. However, the mesostigmatid germ band, as we shall see, undergoes later contortions which are far from primitive.

As soon as the bilateral, segmented germ band has been formed, limb buds develop on the prosomal segments (Fig. 148). The ambulatory limb buds are formed before the pedipalpal pair, and the cheliceral pair is somewhat delayed in the usual chelicerate manner. No limb buds are formed on the opisthosoma. The embryo in acarines is also the stage of temporary regression of the fourth pair of ambulatory limbs, anticipating hatching as a hexapod larval stage. Of the species so far investigated, all display a reduced fourth ambulatory segment in the embryo and most still develop a pair of limb buds on this segment. The limb buds are seen as small external rudiments in the Mesostigmata, Metastigmata and Tarsonemini (e.g. *Pergamasus*, *Ophyionyssus*, *Hyalomma*, *Ornithodorus*, *Acarapis*, *Pyemotes*, *Pediculopsis*) and in some sarcoptiforms (e.g. *Tyroglyphus*, *Histiosoma*). During later embryonic development, the limb buds are retracted within the body (Fig. 149). In the Prostigmata (e.g. *Cheyletus*, *Myobia*, *Tetranychus*, *Atax*, *Trombidium*) the fourth pair of ambulatory limbs is not developed externally at any stage during embryonic development.

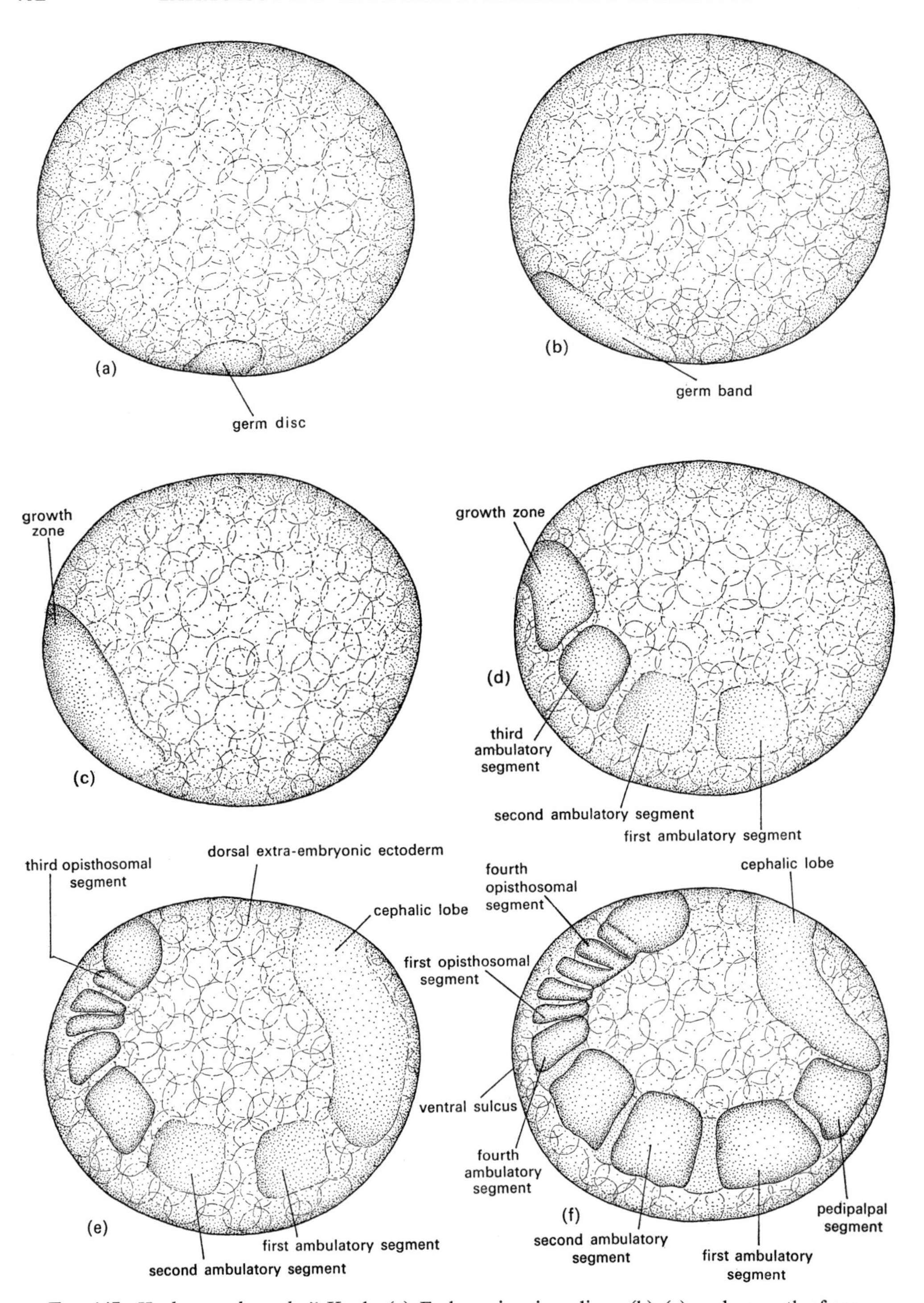

FIG. 147. *Hyalomma dromedarii* Koch. (a) Embryonic primordium; (b), (c), early growth of germ band; (d) onset of formation of anterior prosomal segments; (e) further formation of prosoma and segmentation of opisthosoma; (f) segmented germ band.

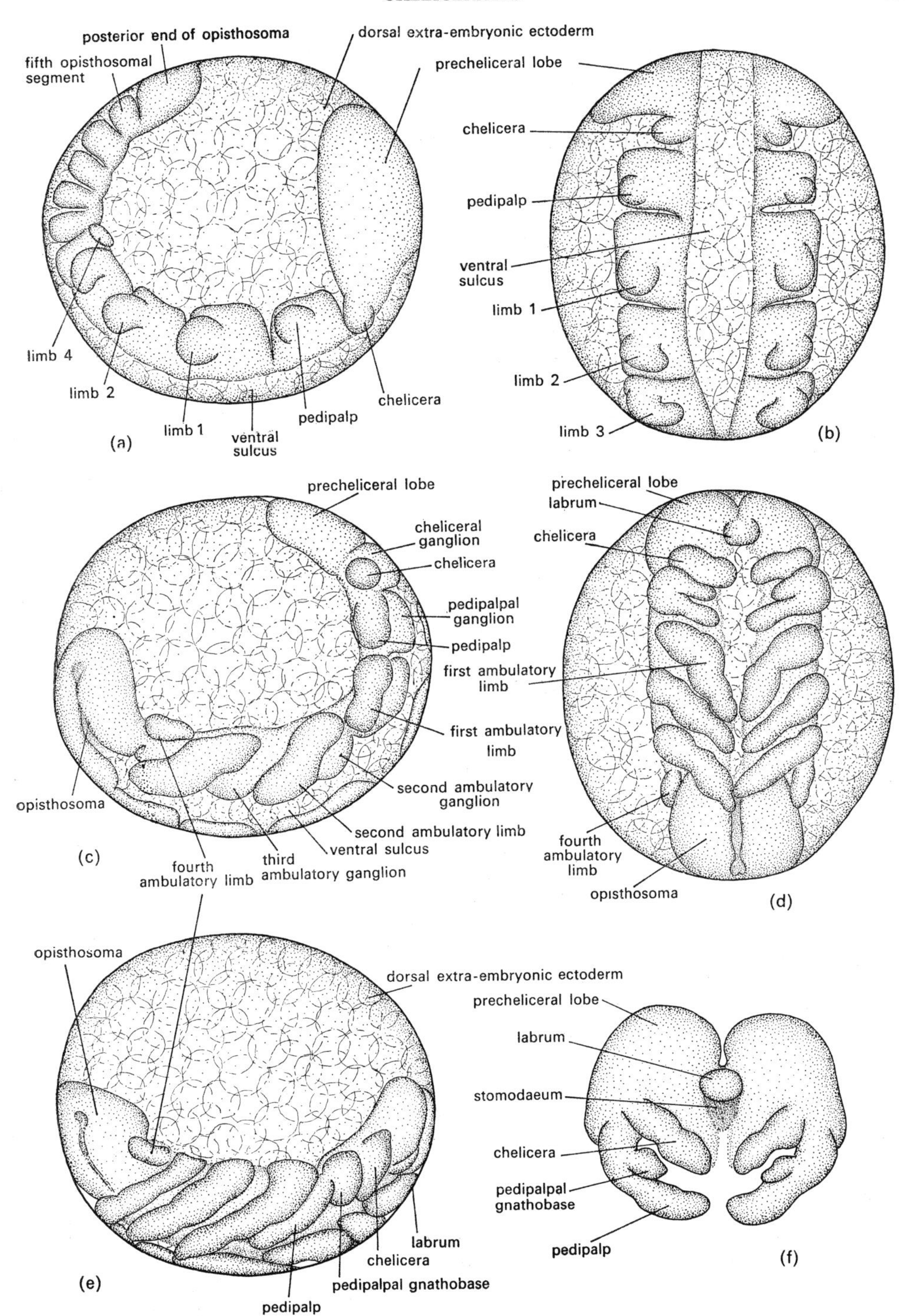

FIG. 148. *Hyalomma dromedarii* Koch. (a), (b) Formation of limb buds, lateral and ventral view; (c)–(e) ventral concentration of germ band, closure of ventral sulcus, elongation of prosomal limbs; (f) anterior end of prosoma in anterior view, chelicerae becoming preoral.

As limb buds begin to develop on the acarine prosoma, the entire segmented germ band undergoes longitudinal contraction. The two halves of the germ band come together in the ventral midline, with elimination of the ventral sulcus, and the germ band becomes short and broad. At the same time, the limb buds grow longer and the external, intersegmental annuli disappear. In most acarines, the anterior end of the contracted germ band

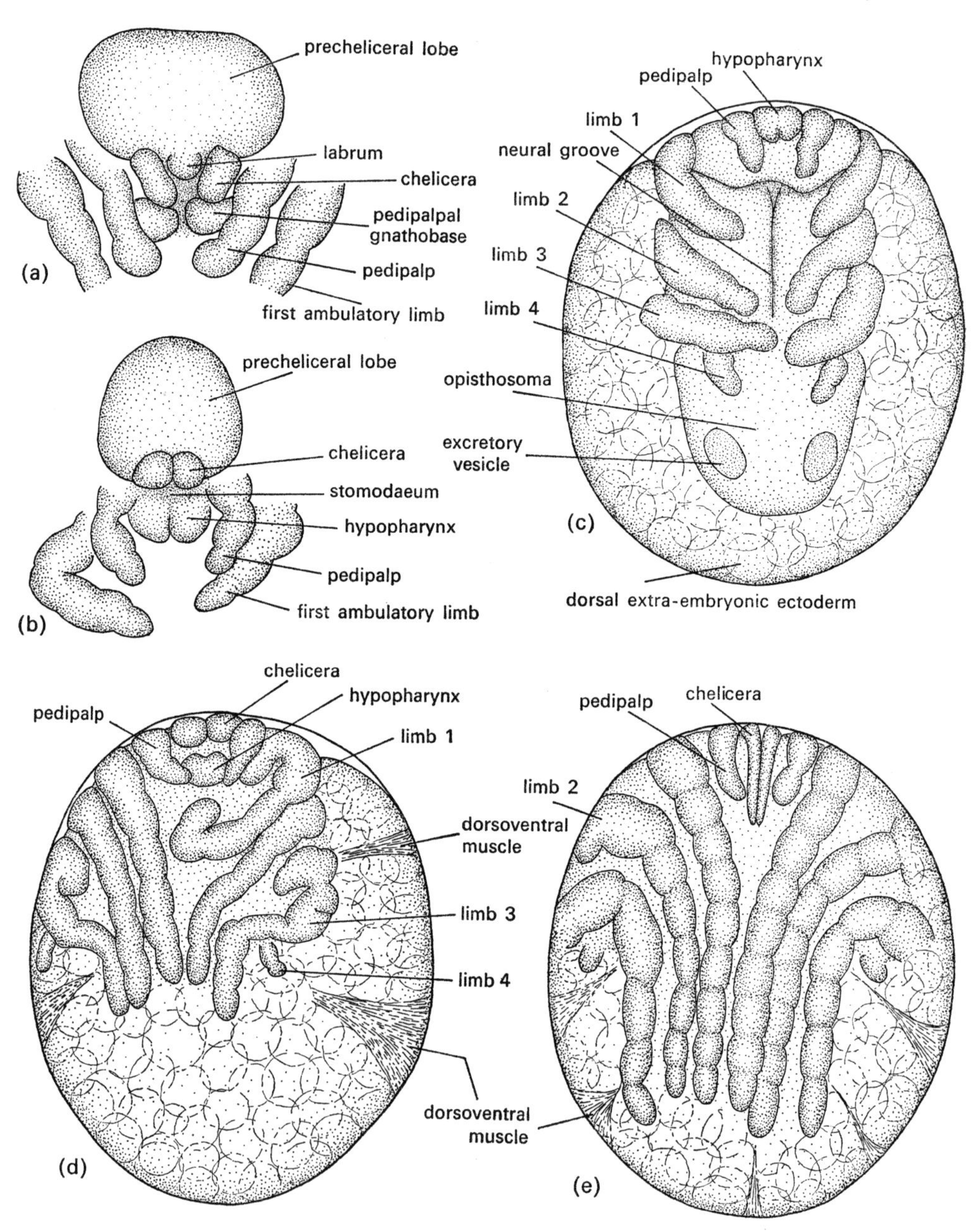

FIG. 149. *Hyalomma dromedarii* Koch. (a), (b) Anterior end of prosoma in anterior view, showing preoral migration of chelicerae and formation of hypostome; (c) shortened germ band, ventral view; (d) embryo during dorsal closure ventral view; (e) embryo approaching hatching.

remains at the anterior end of the yolk mass, while the posterior end withdraws onto the ventral surface of the yolk mass (Claparède, 1868; Wagner, 1894; Brucker, 1900; Reuter, 1909; Hughes, 1950; Klumpp, 1954; Aeschlimann, 1958; Anderson, 1972). A different mode of contraction of the germ band has been observed in the mesostigmatids *Ophionyssus* and *Pergamasus* by Camin (1953) and Zukowski (1964). Here, the posterior end of the germ band remains on the dorsal surface of the yolk mass during contraction and the anterior end withdraws along the ventral surface of the yolk until the contracted germ band is curved around the posterior pole of the yolk mass. Pulsations then occur which shift the germ band forwards to the normal position on the anteroventral surface of the yolk mass.

Dorsal closure now proceeds directly around the yolk mass, being completed more rapidly by the opisthosoma than by the prosoma, as it is in the Opiliones. At the same time, the distinction between these two parts of the body is obliterated and a secondary distinction is established between the gnathostoma and the histiosoma. As the ambulatory limbs grow longer, extending back beneath the ventral surface of the embryo, the opisthosomal wall spreads dorsally to embrace the posterior part of the yolk mass, and the prosomal wall spreads in similar fashion, carrying the bases of the ambulatory limbs to a more lateral position. The completion of the body form is thus attained without any movement of the yolk mass relative to the body wall. When the yolk mass is relatively large, as in *Ornithodorus*, *Boophilus*, *Hyalomma*, *Pergamasus* and *Ophionyssus*, the ventral surface of the embryo remains straight during the final phase of development, but in the small embryos of trombidiform and sarcoptiform mites, the ventral surface shows a slight inversion of curvature during dorsal closure, the anterior and posterior ends being both flexed ventrally (Claparède, 1868; Brucker, 1900; Reuter, 1909; Hughes, 1950; Klumpp, 1954; Perron, 1954).

The development of the anterior end of the prosoma has been examined more carefully in ticks than in other acarine embryos (Aeschlimann, 1958; Anderson, 1972). During contraction of the germ band (Figs. 148 and 149), a median labral lobe arises in the midline between the pre-cheliceral lobes. Simultaneously, the chelicerae and pedipalps increase in length and a median lobe develops on the coxa of each pedipalp. The chelicerae then migrate forwards to a pre-oral position (Claparède, 1868; Wagner, 1894a; Brucker, 1900; Bonnet, 1907; Reuter, 1909; Thor, 1925; Aeschlimann, 1958; Anderson, 1972). The bases of the chelicerae come together in front of the mouth, with elimination of the labrum. Behind the mouth, the median coxal lobes of the pedipalps come together to form the hypostome, while the remainder of each pedipalp forms a functional palp. Semilunar grooves are lacking on the precheliceral lobes of tick embryos.

In the absence of knowledge of the embryos of primitive acarines, it is difficult to compare the known specialized development of acarines with the embryonic development of other arachnids. We can reasonably assume, however, that the development of *Ornithodorus* and *Hyalomma* is not far removed from a generalized condition for acarines and permits certain conclusions to be drawn. For example, as in opilionids, the shift of yolk and inversion of flexure characteristic of spiders do not occur. Again, the mode of formation and bilateral separation of the prosomal part of the germ band are more like those of opilionids than of spiders, and the germ disc is essentially the opisthosomal rudiment, not the full embryonic primordium. Thus, at the present time, the acarine mode of embryonic development appears

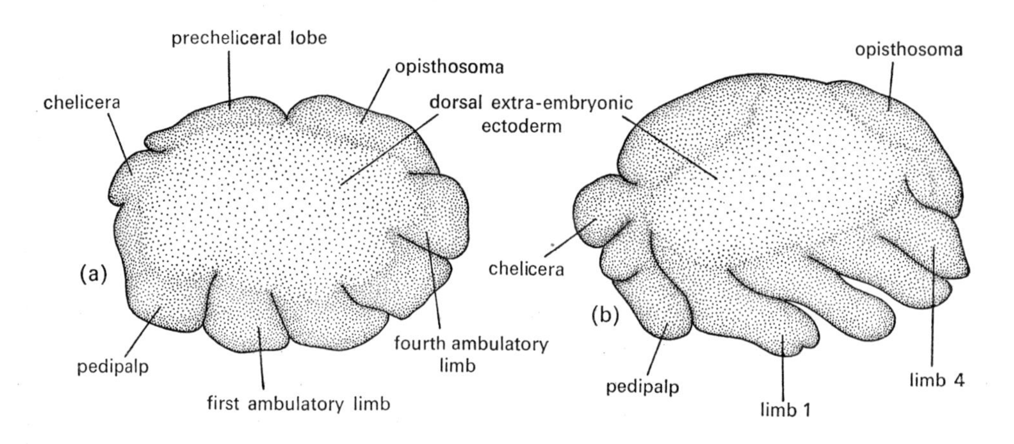

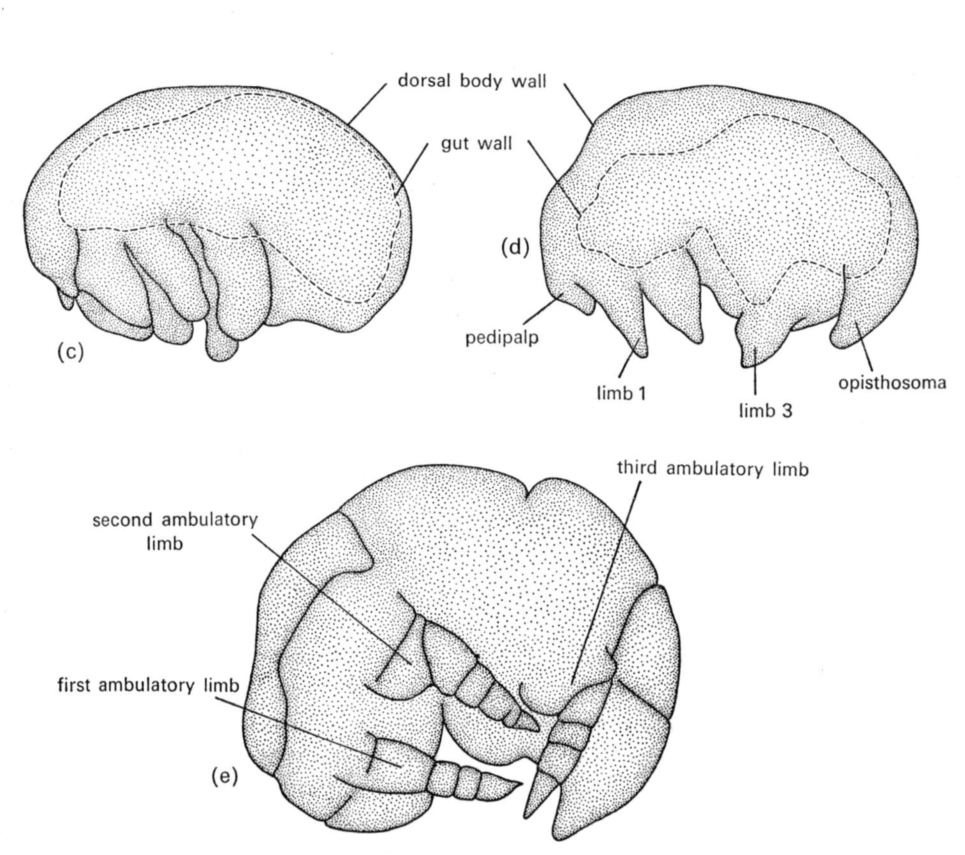

FIG. 150. The external features of development of *Pediculopsis graminum*. (a) segmented germ band; (b) formation of prosomal limb buds; (c) dorsal closure; (d), (e) completion of external development. (After Reuter, 1909.)

to be closer to the basic pattern underlying the development of opilionids than to that shared by spiders, uropygids, amblypygids and perhaps solifugids.

Pseudoscorpiones

We now come to the question of the development of arachnid embryos in which specializations are manifested in relation to viviparity, namely, the scorpions and pseudoscorpions. Of the two, the development of pseudoscorpions is the more like that of the arachnids already discussed, and thus the more suitable for first analysis. Knowledge of the embryos of pseudoscorpions has been greatly enhanced by the outstanding work of Weygoldt (1964, 1965, 1968, 1971), principally on *Pselaphochernes*, *Neobisium* and *Chthonius*. We have already seen that the small eggs of these animals are held within a brood pouch beneath the opisthosoma of the female and that they develop initially as a blastoderm around a small yolk mass, but with an additional, specialized embryonic membrane external to the blastoderm. Development then continues with the brood pouch, sustained by a nutrient secretion exuded into the brood pouch from the oviducts of the female (Fig. 151).

In the first phase of development beyond the blastoderm, a posteroventral embryonic primordium arises and develops a small gastral groove through which cells pass into the interior. Some of these cells move forwards and outwards between the blastoderm and the yolk mass, while others accumulate beneath the posterior part of the embryonic primordium to form a superficial protrusion which becomes the opisthosomal rudiment.

In front of the opisthosomal rudiment, paired limb buds protrude on either side of the ventral midline, accompanied by a precocious labral rudiment at the anterior end of the embryo. The limb buds developed at this stage are the pedipalpal and four ambulatory pairs. The cheliceral pair, which arises on either side of the base of the labrum, develops later. Except for the precocious outpouching of the labrum, the functional significance of which will emerge below, the development of the pseudoscorpionid embryo up to this stage follows the same basic sequence as that of primitive spiders. The similarity is later enhanced by the outgrowth of the opisthosomal rudiment as a tabular, caudal papilla flexed forwards between the prosomal limbs (Figs. 156). The major specializations observed in the embryo are all related to the replacement of yolk by imbibed nutrients.

Accompanying the precocious outpouching of the labrum, the stomodaeum invaginates precociously behind the labrum. With continued enlargement, the labrum dips ventrally between the bases of the pedipalps. In *Neobisium* and *Chthonius*, the labrum and pedipalpal coxae now combine to form a muscular pumping organ which passes nutrients from the exterior into the stomodaeum of the embryo. In *Pselaphochernes*, the labrum shifts further back and the pumping organ is formed by the labrum and the coxae of the first ambulatory limbs. Pumping begins while the embryo is still enclosed within the embryonic membrane and egg membrane. The embryonic membrane is ingested first, followed by the first intake of nutrient fluid. The embryo swells and hatches from the egg membrane into the brood pouch as the second embryonic stage.

As pumping continues, increase in size continues, including growth in length of the caudal papilla and external delineation of the opisthosomal segments. Increase in size takes place differently in different species. *Pselaphochernes* shows emphasis, for instance, on increase in

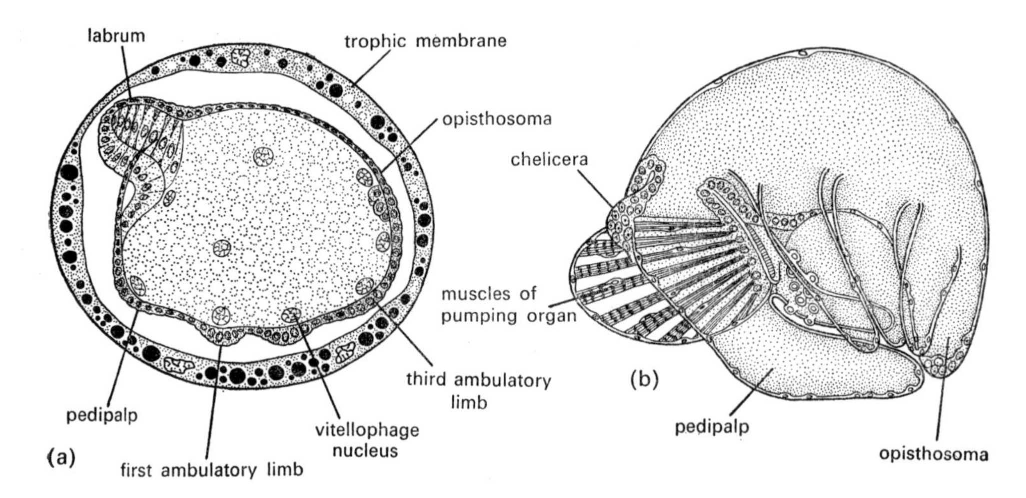

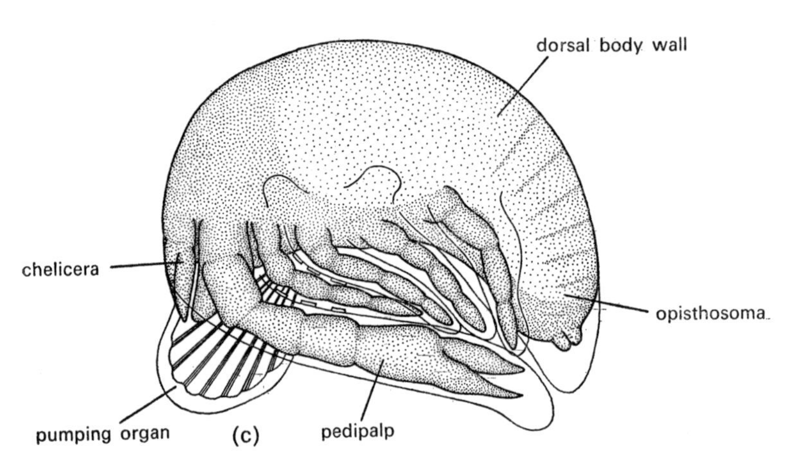

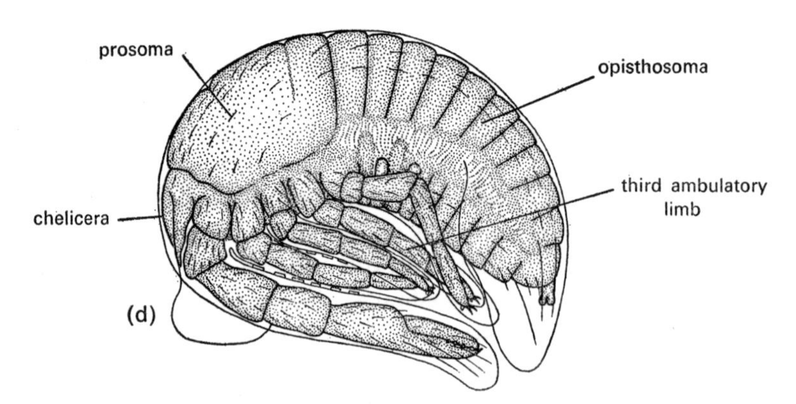

FIG. 151. The external features of development of *Pselaphochernes scorpioides*. (a) Early development of limb buds and labrum; (b) early second embryonic stage with functional pumping organ; (c) late second embryonic stage; (d) embryo almost fully developed. (After Weygoldt, 1964.)

leg length. *Chelifer*, conversely, retains short legs and shows great expansion of the dorsal body wall. During this phase of growth, the embryo undergoes considerable external development, including secretion of a cuticle. A moult terminates this phase of development. At the moult, the embryonic pumping organ is rebuilt into a more definitive suctorial apparatus. A third and final phase of development now ensues, during which suctorial feeding continues and organogenesis proceeds, but no further increase in the size of the embryo is observed. This phase terminates with escape from the brood pouch.

The basic similarity of pseudoscorpionid embryos to araneid embryos, underlying the specializations associated with a changed mode of embryonic nutrition, is concomitant with the comparative cleavage of the two orders and is further borne out by comparative internal features of development which will be discussed below. Scorpion embryos provide a complete contrast.

Scorpiones

Even though the embryos of all scorpions are specialized in relation to development within the maternal body, the ovoviviparous buthids retain sufficient generalized features in their embryos to provide a basis for comparison with other arachnids. The account given by Brauer (1894, 1895) of the development of *Buthus carpathicus* remains the major source of information, though Abd-el Wahab (1952, 1954) confirmed a number of Brauer's observations in a study of *Buthus quinquestriatus*. The work of Mathew (1956) on the viviparous *Heterometrus scaber* further demonstrates the additional modifications manifested in scorpion embryos in association with loss of yolk.

We have already seen that the blastoderm of *Buthus* develops in a polarized fashion, centred on a dense region of cells at the posterior pole of the yolk mass, with a gradual spread of more attenuated blastoderm from this focus over the remainder of the yolk surface. Continued development reveals that the polar mass is a precociously formed embryonic primordium. Before discussing the further development of the primordium, however, we need to take account of the fact that ovoviviparous scorpions develop extra-embryonic membranes analogous with those of primitive pterygotes. As the embryonic primordium begins to increase in length and gives rise to a germ band, the attenuated blastoderm separates from the margin of the primordium and extends to merge over the outer surface of the primordium. As a result, the yolk mass and embryonic primordium lie within a thin external envelope of cells, customarily called the serosa. The serosa plays no further structural role in the development of the embryo. It persists until birth and is sloughed off after birth. Presumably the serosa has a functional role in relation to the development of the embryo within the maternal oviduct, but nothing is known of this function at the present time.

In addition to the serosa, buthid embryos also form a second embryonic membrane, the amnion. This membrane arises during the early elongation of the germ band. The margin of the germ band turns downwards and spreads inwards from all sides to merge along the ventral midline. The amnion thus encloses an amniotic cavity beneath the external surface of the germ band, but the functional significance of the membrane and cavity are again unknown. During elongation of the germ band and subsequently during dorsal closure, the

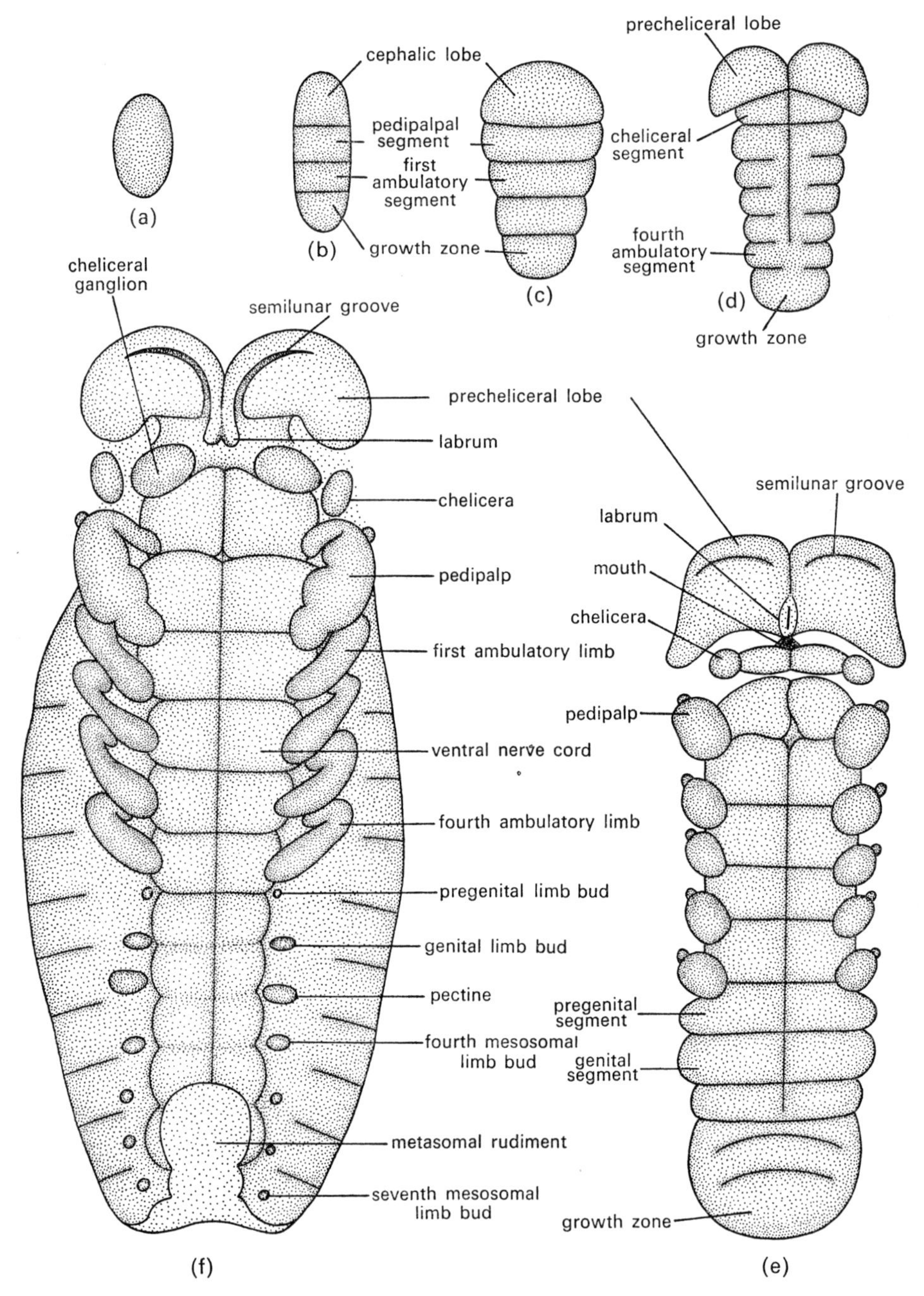

FIG. 152. Development of the germ band of *Buthus carpathicus*, ventral view. (a) Embryonic primordium; (b)–(d) formation of prosomal segments; (e) further development of the prosoma, formation of opisthosomal segments; (f) fully segmented germ band. (After Brauer, 1895.)

amnion continues to cover the external surface of the embryo, so that it eventually forms a second complete envelope beneath the serosa. Like the serosa, the amnion is shed only after birth.

We can now follow the development of the buthid germ band itself (Fig. 152). There is no evidence at present to indicate that a gastral groove is formed in the embryonic primordium. It seems likely that the internal rudiments of the mesoderm and midgut separate from the ectoderm by delamination, though further investigation of this point is required. The early elongation of the embryonic primordium is accompanied by a transverse demarcation into three equal parts, constituting a precociously delineated cephalic lobe, a pedipalpal segment and a posterior growth zone. At this stage, the germ band still lies at the posteroventral pole of the yolk mass. The growth zone retains this position and proliferates further segments in a forward direction. As the length of the germ band increases, the segments which are already formed become more fully elaborated and the anterior end of the germ band pushes forwards until it reaches the anterior pole of the yolk mass.

During formation of the four ambulatory segments of the prosoma, the anterior part of the cephalic lobe expands bilaterally to form a pair of pre-cheliceral lobes and the posterior part is demarcated as a cheliceral segment. The stomodaeum invaginates between the pre-cheliceral lobes. The development of prosomal limb buds does not begin until the first three opisthosomal segments have been formed by the growth zone, but then takes place more or less simultaneously on all prosomal segments, producing small cheliceral buds, large pedipalpal and first ambulatory limb buds and successively smaller pairs on the last three ambulatory segments. Accompanying the ventrolateral outpouching of the limb buds, the ventral parts of each segment rudiment thicken to form a pair of ganglion rudiments on either side of a midventral neural groove, but there is no bilateral separation of the germ band of formation of a ventral sulcus at any stage of development.

The proliferation of opisthosomal segments forwards along the ventral surface of the yolk mass continues until the seven mesosomal segments have been formed. By this stage, the enlarged precheliceral lobes have reached the anterior pole of the yolk mass. Seven pairs of mesosomal limb buds develop, accompanied by ventral ganglia and an intervening neural groove, as the last two mesosomal segments are formed.

After proliferating the seventh mesosomal segment flat against the yolk, the growth zone protrudes and proliferates the six metasomal segments as a tubular, fowardly flexed, caudal papilla. The terminal unit formed by the growth zone is the telson, which subsequently develops as the sting. No limb buds are formed on the metasomal segments. Meanwhile, the prosomal limbs grow longer and the mesosomal limb buds undergo various changes. The first or pre-genital pair are soon resorbed. The second pair develop vestigially and merges to form the genital operculum. The third pair develop into the pectines. Behind the fourth to seventh pairs, invaginations initiate the formation of the four pairs of book-lungs and the limb buds gradually merge into the body surface as lung development continues.

Dorsal closure of the prosoma and mesosoma takes place by direct upgrowth of the lateral margins of the segments towards the dorsal midline, passing over the surface of the yolk mass beneath the external serosa, and carrying the edge of the amnion with them. At the anterior end, a pair of long, narrow labral lobes develops at the median edges of the pre-cheliceral lobes. The swollen posterior ends of the labral rudiments merge and give rise to the labrum, overhanging the mouth. Simultaneously, a pair of semilunar grooves develops around the anterolateral margins of the pre-cheliceral lobes. The outer margins of the semilunar grooves then overgrow the broad surfaces of the pre-cheliceral lobes in a posteromedian direction and merge into the labrum. The chelicerae move forwards to a pre-oral position on either side of the labrum, while the pedipalps become para-oral, with medially directed gnathobases. In spite of the peculiarities of development in the embryos of ovoviviparous scorpions, the development of the anterior end retains the features common to all arachnids except the specialized Acari.

The development of yolkless viviparous scorpions such as *Heterometrus* is, of course, more specialized even than that of buthids, but is recognizable as a further modification of the same basic pattern. As we have already seen, the small spherical embryo is attached by its anterior end to the distal wall of a cavity within a lateral ovarian follicle and is enclosed within an embryonal capsule formed from follicle cells. Before the embryo begins to elongate, the embryonal capsule breaks down and is replaced by a cellular envelope developed by proliferation of the polar body cells which attach the embryo to the follicle wall. Mathew (1956) called this membrane the trophamnion. When the embryo begins to elongate, the trophamnion breaks down and is replaced by a second cellular envelope derived from the follicle cells. According to this interpretation, the serosa and amnion of buthid embryos are not represented in yolkless scorpions.

The embryo elongates along the long axis of the ovarian follicle, the cavity of which enlarges as the embryo develops. The anterior end of the embryo is directed towards the free end of the follicle. When elongation first begins, increasing the length of the embryo to 200 μ, a large mouth develops by invagination of the anterior wall of the embryo. As the length of the embryo continues to increase, obviously due mainly to the activity of a posterior growth zone, the rudiments of the chelicerae develop precociously as a pair of forward protrusions flanking the mouth. By the time the embryo is 750 μ long, the chelicerae have gripped the appendix at the end of the ovarian follicle, pressing the appendix against the mouth. Channels leading through the appendix provide an access for nutrients synthesized in the maternal digestive glands (Vachon, 1953). The precocious feeding apparatus of viviparous scorpions is thus different from that of pseudoscorpions, which do not employ the chelicerae in this role.

During the first elongation of the embryo to 750 μ, the body becomes subdivided into an anterior prosomal rudiment, a median mesosomal rudiment with three pairs of dorsolateral protrusions, and a growth zone. This subdivision gives the first indication of the second important specialization of the embryo in viviparous scorpions, precocious development of the mesosoma. As the length now increases to 2 mm, the prosoma remains short and devoid of pedipalpal and ambulatory limb rudiments, but the mesosoma acquires the full complement of seven segments, each with a pair of dorsolateral diverticula (Fig. 153). The midgut does not extend into these protrusions but the protrusions themselves persist throughout development, presumably as exchange surfaces between the embryo and maternal tissue.

The embryo continues to grow longer as the metasomal segments are added one by one, and the large mesosoma takes on a dorsoventral flexure, with the dorsal surface convex and the dorsolateral diverticula protruding from it. Small pedipalpal and ambulatory limb buds develop on the prosoma and a small labral lobe pushes out between the chelicerae and the anterior end of the body. The metasomal segments develop dorsolateral diverticula smaller than those on the mesosoma.

By the time the embryo is 6–6·5 mm long, the dorsoventral flexure of the large mesosoma is increased. The metasoma shows a reciprocal upward flexure. The prosomal limb buds are now longer, and the mesosoma has developed limb buds on the second to sixth segments. Those of the second and third segments are the rudiments of the genital operculum and

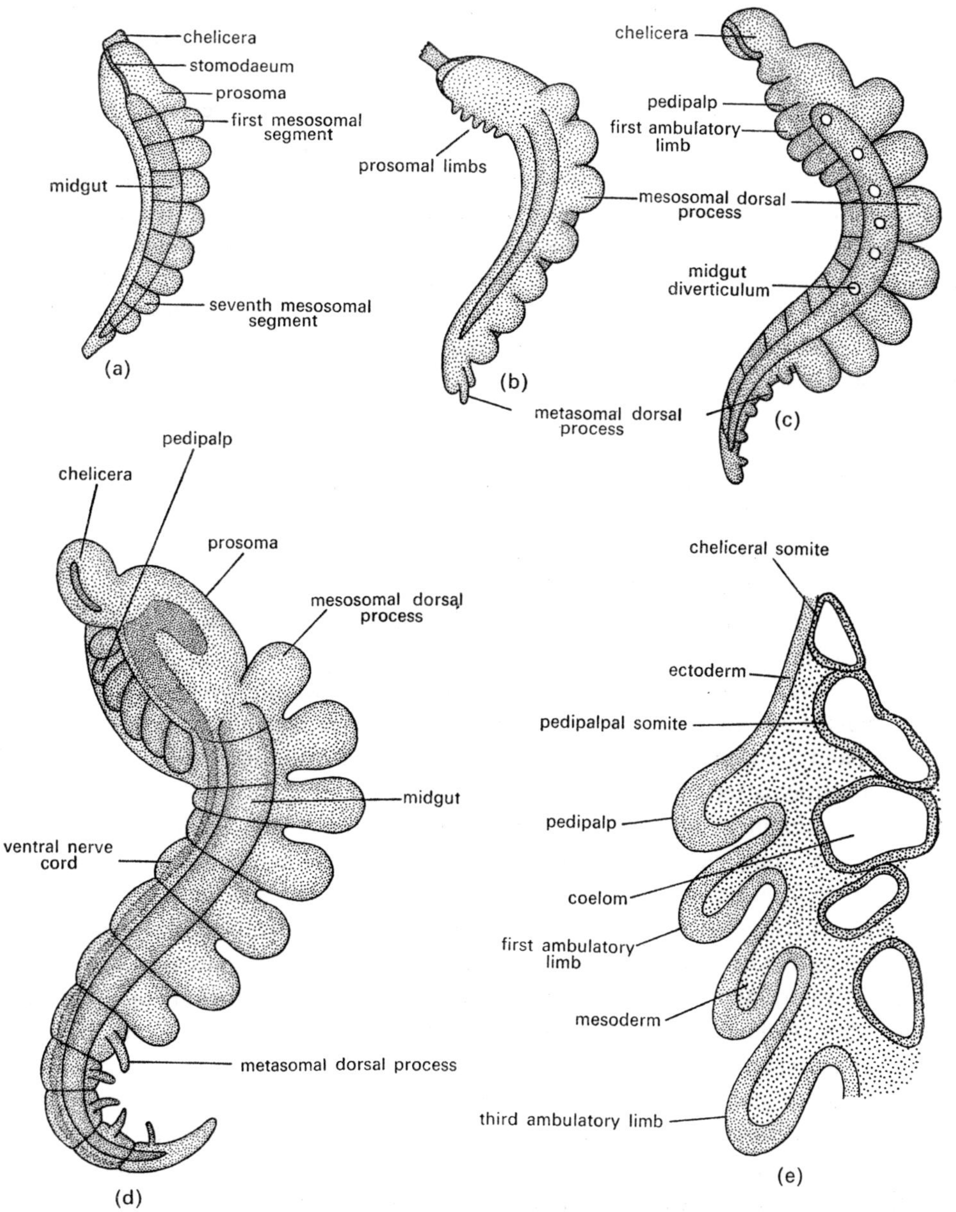

FIG. 153. (a)–(d) External development of the embryo of *Heterometrus*. (a) 3-mm embryo; (b) 4-mm embryo; (c) 6-mm embryo; (d) 9-mm embryo. (e) Longitudinal section through prosomal limb buds and associated coelomic sacs of a 4-mm embryo. (After Mathew, 1956.)

pectines. The limb buds on the third to sixth segments are resorbed while book-lungs invaginate on these and the seventh mesosomal segment in the usual way. No limb vestiges are developed on the pre-genital or seventh mesosomal segment.

Since the embryo is tubular from the beginning, the need for dorsal closure is obviated. The full external appearance of the scorpion is attained during the later growth of the embryo to its maximum length of 19 mm. The dorsolateral protrusions gradually retract, but are not finally resorbed until after birth. Similarly, the chelicerae retain their specialization until after birth, acting as part of the feeding apparatus of the embryo throughout development.

In spite of the specializations of development directly related to feeding, the embryo of *Heterometrus* follows the same sequence of anteroposterior development as that of *Buthus*. Any similarities with pseudoscorpion development, such as precocious development of the stomodaeum and midgut sac are clearly convergent. The scorpions are unique among arachnids in the anteroposterior proliferation of the germ band from an embryonic primordium which comprises the cephalic lobe, pedipalpal segment and growth zone only, sharing this pattern of growth with the Xiphosura alone among modern chelicerates. All other arachnids, as we have seen, develop the prosoma directly from the blastoderm and proliferate only the opisthosoma from a growth zone. A comparison of the external features of development therefore indicates that:

(a) The Xiphosura retain the most generalized pattern of embryonic development among modern chelicerates.

(b) The Scorpionida, in spite of specializations, retain the nearest approximation to this pattern among modern arachnids.

(c) All other arachnids share a derived pattern of development whose basis could lie in that of Xiphosura but not that of scorpions.

(d) The derived pattern of development common to arachnids other than scorpions is expressed as two variants, in the opilionids and Acari, the more generalized of the two, and the araneids, uropygids, amblypygids, solifugids and pseudoscorpions, with a forwardly flexed opisthosoma.

(e) The development of the Palpigrades and Ricinulei is still unknown. In terms of chelicerate phylogeny, these conclusions suggest that the Xiphosura and arachnids diverged from a common ancestral Chelicerate stock but that the scorpions diverged early from the other arachnids or perhaps evolved independently of the other arachnids. It can further be suggested that the arachnids other than scorpions are a unitary group with two extant divergent lines whose relationships to each other are not clear, but of which we can perhaps say that the first includes the Opiliones and Acarina, while the second has radiated to produce the Solifugae, the Uropygi and Amblypygi, the Pseudoscorpionida and the Araneae. How the Palpigrades and Ricinulei fit into this picture remains to be determined. It is also essential, if the phylogenetic implications suggested above are to be upheld, that the facts of internal development should not contradict the conclusions drawn from external development. As we shall see, these facts, in so far as they are available at the present time, do provide the necessary support.

Gastrulation

Gastrulation in chelicerate embryos has not been fully investigated in any species, so that a comprehensive interpretation is not possible at the present time. When the scattered available observations are brought together, however, they suffice to illustrate the broad outlines of the process (Fig. 154). As in other yolky, arthropod embryos, chelicerate gastrulation tends to be prolonged and to be broken up into a number of discrete, though interrelated processes woven around the presence of a large mass of yolk and the requirement of making this yolk available to the developing tissues.

None of the workers who has studied gastrulation in the Xiphosura (Kishinouye, 1891a; Kingsley, 1893; Iwanoff, 1933) has given a satisfactory account of this stage of development. The most we can glean from their accounts is as follows:

The mass of large, polygonal, yolky cells which lies within the blastoderm is the rudiment of the midgut. The presumptive midgut thus comes to occupy its internal position as a result of total cleavage and subsequent delamination of the blastoderm, and exhibits no gastrulation movement. The mesoderm, on the other hand, is proliferated into the interior along the length of the gastral groove, spreading outwards between the superficial ectoderm of the germ band and the large, internal midgut rudiment. We can infer from this that the presumptive mesoderm occupies the midventral line of the embryonic primordium and retains a slight gastrulation movement, in the form of a shallow insinking, before entering into proliferative organogeny. According to Iwanoff (1933), the mesoderm of the first four segments of *Tachypleus gigas* takes origin from a single yolky blastomere which is formed in an internal position during cleavage, only that of the fifth and succeeding somites being derived from the walls of the gastral groove. The evidence in support of this contention is unsatisfactory and requires reassessment. In general, Iwanoff's intimations of primary and secondary segmentation in arthropods have proved, on further investigation, to be mistaken. At the anterior end of the gastral groove, the invaginated cells from the walls of the stomodaeum and the external aperture of this part of the groove persists to form the mouth when the remainder of the groove undergoes anteroposterior closure. The same gastrulation movement which carried the limulid presumptive mesoderm into the interior thus carries the stomodaeal rudiment inwards anteriorly and forms the mouth. The proctodaeum, in contrast, develops very late in xiphosuran embryos, as an independent invagination on the telson (Kingsley, 1893).

Among arachnids, the pattern of gastrulation movements displayed by the Xiphosura is variously modified. Once again we can begin with primitive spiders. According to Yoshikura (1955), the yolk is devoid of nuclei until it is reinvaded by cells released in a generalized fashion throughout the embryonic primordium, principally from the region of the opisthosomal rudiment at the posterior end of the primordium. The invading cells act temporarily as vitellophages but subsequently develop into the definitive rudiment of the anterior part of the midgut. Some of the cells passing inwards at the posterior end of the embryonic primordium accumulate as a group beneath the opisthosomal rudiment, forming the rudiment of the posterior part of the midgut. The precise origin of these cells is uncertain, but there is some indication that they are associated with the posterior end of the gastral groove which now forms. The latter results from an invagination of cells which then proliferate and

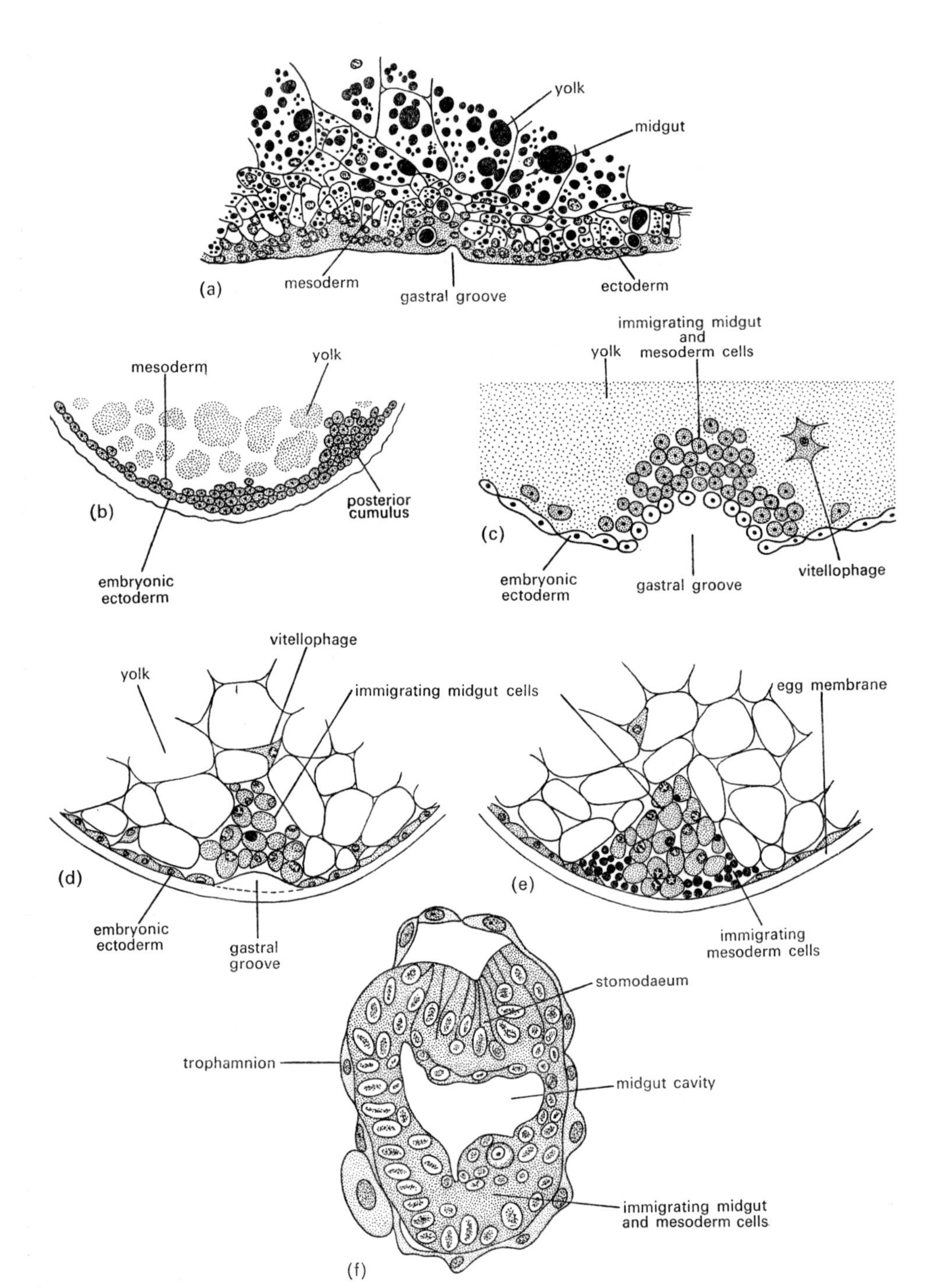

FIG. 154. Gastrulation in chelicerate embryos. (a) Transverse section through early gastral groove of *Tachypleus*, after Iwanoff (1933); (b) sagittal section through the embryonic primordium of *Latrodectus* following onset of gastrulation, after Rempel (1957); (c) transverse section through the gastral groove of *Ischnocolus*, after Davydoff (1949); (d), (e) transverse sections through the gastral groove of *Ornithodorus*, showing median vitellophage proliferation and bilateral mesoderm proliferation, after Aeschlimann (1958); (f) frontal section through gastrulating embryo of *Heterometrus*, after Mathew (1956).

spread as mesoderm, essentially as in Xiphosura; but the invagination is confined to the midventral region of the posterior prosomal segments and is soon obliterated. The stomodaeal invagination takes place at a later stage of development, independently of the gastral groove, as it does in all arachnids. The invagination of the proctodaeum on the telson is also independent and late in all arachnids, as it is in the Xiphosura. In terms of gastrulation movements, therefore, *Heptathela* differs from the Xiphosura in several ways:

1. The spatial and temporal separation of the stomodaeal invagination from the gastral groove, a difference universal among arachnids.
2. The shorter length of the gastral groove and more rapid invagination of the mesoderm.
3. The entry of a group of posterior midgut cells via the posterior end of the gastral groove.

This trend in specialization is more fully expressed in other spiders. It seems probable (see below) that the vitellophages within the yolk mass give rise to the major part of the midgut in all spiders. While some of these cells enter the yolk mass by scattered invasion from a wide area of the blastoderm and others are, perhaps, persistent cleavage energids (Kautzsch, 1910; Holm, 1952; Rempel, 1957), many of the vitellophages originate as early cells passing inwards via the gastral groove. The later cells passing in via the gastral groove then resolve themselves into two types. The majority are mesoderm cells, which proliferate and spread beneath the developing germ band (Morin, 1887, 1888; Kishinouye, 1891b; Wallstabe, 1908; Kautzsch, 1909, 1910; L. and W. Schimkewitsch, 1911; Holm, 1941, 1953, 1954; Rempel, 1957; Yoshikura, 1958; Ehn, 1963), but at the posterior end of the gastral groove, a group of internal cells enlarges to form the posterior cumulus. The evidence of Holm (1952) indicates that the posterior cumulus cells later develop as posterior midgut cells (see below). Thus, in most spiders, the cells entering the interior via the gastral groove include, not only mesoderm and posterior midgut cells, but also vitellophage components whose association with the gastral groove appears to be a secondary specialization evolved within the spiders. It must be emphasized, however, that araneid gastrulation requires more detailed investigation before a proper understanding can be achieved. The same is true to an even greater extent of some other groups. No satisfactory information on gastrulation in the Solifugae and Amblypygi is available at the present time and almost none for the Uropygi. Possibly, in *Thelyphonus*, the yolk mass is reinvaded from the blastoderm by scattered vitellophages which subsequently give rise to the midgut epithelium (Schimkewitsch, 1906), but even this point is in need of confirmation. In the opilionids, Faussek (1891) found that the first cells to move into the interior at the site of a transient gastral groove accumulate as a genital rudiment. Schimkewitsch (1898) suggested that vitellophages and mesoderm enter the interior from the same source and confirmed Faussek's view that the vitellophages later give rise to the midgut epithelium. In view of the antiquity of these observations, however, no useful comments can be made on opilionid gastrulation without renewed detailed studies.

Acarine gastrulation is slightly better known and seems to bear some resemblance to that of spiders. Several workers (Wagner, 1894 a, b; Bonnet, 1907; Reuter, 1909; Hafiz, 1935; Klumpp, 1954; Aeschlimann, 1958; Langenscheidt, 1958; Anderson, 1972) have observed that the gastral groove of acarine embryos is the site of proliferation of cells into the interior, initially forming a compact group of large cells beneath the germ disc. From this group, some cells escape into the interior as vitellophages while the remainder stay grouped together as a posterior midgut rudiment. Other vitellophages are released into the yolk mass throughout the blastoderm, though the posterior midgut rudiment plays a major role and the vitellophages only a minor role in the later development of the midgut (see below). Finally, before the gastral groove is obliterated, numerous small mesoderm cells, probably derived from the walls of the groove, accumulate between the posterior midgut rudiment and the superficial cells of the germ disc. The cells show a distinct bilateral grouping on either side of the posterior midgut rudiment (Wagner, 1894 a, b; Bonnet, 1907; Klumpp, 1954; Aeschlimann, 1958; Langenscheidt, 1958; Anderson, 1972). As in other arachnids, the stomodaeum and proctodaeum arise as independent invaginations.

Gastrulation in the small embryos of pseudoscorpions is, as might be expected, markedly specialized (Weygoldt, 1964, 1965, 1968). As we have already seen, the vitellophage nuclei within the yolk mass gain this location during cleavage, though whether this is a primitive condition reminiscent of Xiphosura or a specialized condition evolved in association with a secondary reduction of yolk in pseudoscorpions cannot be decided at the present time. The yolk mass subsequently gives rise directly to the anterior part of the midgut. Further cells invade the interior from the gastral groove. Some of these cells spread forwards and give rise to prosomal mesoderm. Others accumulate beneath the site of the gastral groove and subsequently gives rise to both the opisthosomal mesoderm and the posterior part of the midgut, though the early separation of the two cell types cannot be discerned. The stomodaeal and proctodaeal invaginations display their usual independence of the gastral groove, the stomodaeal invagination being highly precocious, in association with the early onset of embryonic feeding.

Finally, only the most general comments can be made on gastrulation in the scorpions. Brauer (1894) and Abd-el Wahab (1954) have given descriptions of this phase of development for ovoviviparous buthids but the reliability of their interpretations is still in question. It is certain only that the stomodaeum and proctodaeum develop as independent invaginations. The origin ascribed to the midgut, mesoderm and primordial germ cells by these workers is less well supported. The germ disc is said first to release the vitellophages which spread around the surface of the yolk mass. A compact group of cells said to be proliferated inwards from the centre of the germ disc then forms the primordial germ cells. Finally, in succession, two layers of cells appear to be individuated within the inner part of the germ disc. The innermost layer is the midgut rudiment and the layer next to the columnar ectoderm is the mesoderm rudiment. Only renewed investigation can provide an adequate test of the accuracy of these observations. Gastrulation in viviparous scorpions is also little known and cannot be assessed in relation to that of ovoviviparous species. Mathew (1956) showed that, in *Heterometrus*, the small, spherical blastula enlarges and exhibits a unipolar ingression of cells which fills the central cavity. Since the inner mass subsequently segregated into mesoderm and midgut rudiments, the region of ingression is presumably the equivalent of the germ disc of ovoviviparous species. As in pseudoscorpions, the independent invagination of the stomodaeum is highly precocious in viviparous scorpions.

The lack of precise information on the origin and early development of the internal rudiments of arachnid embryos is a crucial difficulty in chelicerate embryology, since it makes the elucidation of presumptive areas in the chelicerate blastoderm highly problematical and weighs heavily against adequate comparisons both within the chelicerates and between these and other arthropods. Once gastrulation is complete, however, the further development of the midgut, mesoderm and ectoderm follows paths which have proved easier to trace and can be described with more confidence.

Further Development of the Gut

As mentioned in the previous section, the xiphosuran midgut develops wholly from the yolky cells which fill the interior of the blastoderm (Kishinouye, 1891a; Kingsley, 1893). As the embryo changes shape during later development, the mass of cells becomes, first ovoid, then hemispherical in the prosoma and cylindrical in the opisthosoma. When the mesoderm extends towards the dorsal midline of the prosoma, it cuts into the midgut mass, subdividing the lateral portions of the latter into six pairs of segmental lobes, the rudiments of the digestive diverticula. During the further elaboration of the musculature, the first, second, fifth and sixth pairs of these lobes become cut off from the central midgut and communicate only with the third and fourth pairs of lobes, which remain their connections with the midgut as two pairs of hepatic ducts. Throughout this series of shape changes, the midgut rudiment remains a mass of yolky cells. Yolk resorption, accompanied by differentiation of the cells as epithelial cells of the midgut and digestive diverticular, is completed only after hatching has taken place.

The stomodaeum flexed backwards as it grows into the interior and comes into contact with the anterior face of the midgut. The stomodaeal wall differentiates as the epithelium of the buccal cavity, eosophagus and proventriculus and secretes a cuticular lining. The proctodaeum forms only a short, posterior termination of the gut.

Among arachnids other than scorpions, it now seems clear that the midgut develops in two parts (Fig. 155). A typical instance was described in the liphistiomorph spider *Heptathela* by Yoshikura (1955). The major part of the midgut arises directly from the yolk mass, whose vitellophages migrate to the surface of the yolk and give rise to midgut epithelium. During dorsal closure of the opisthosoma, the mesoderm cuts into the sides of the midgut mass to form the five pairs of lateral lobes which become the digestive diverticula. Meanwhile, the posterior midgut rudiment lying within the growth zone proliferates a separate, posterior portion of the midgut, devoid of yolk, which differentiates as the rectal sac and produces the Malphigian tubules as paired lateral outgrowths. The short proctodaeum invaginates after the rectal sac has formed. The stomodaeum, which invaginates much earlier, develops into the pharynx, oesophagus and sucking stomach.

Earlier workers on araneomorph spiders had also found that the vitellophages within the yolk mass give rise to the major part of the midgut epithelium and that the reactal sac and Malphigian tubules arise from a distinct posterior rudiment (Balfour, 1880; Morin, 1887; Schimkewitsch, 1887; Kishinouye, 1891b; Hamburger, 1910; Kautzsch, 1910; Rempel, 1957). Kautzsch (1910), working on *Agelena*, was the first to distinguish clearly between the posterior midgut rudiment and the proctodaeum. This observation was subsequently confirmed by Rempel (1957) for *Latrodectus*, but neither worker established the origin of the posterior midgut rudiment. Holm (1952), on the basis of marking experiments on *Agelena*

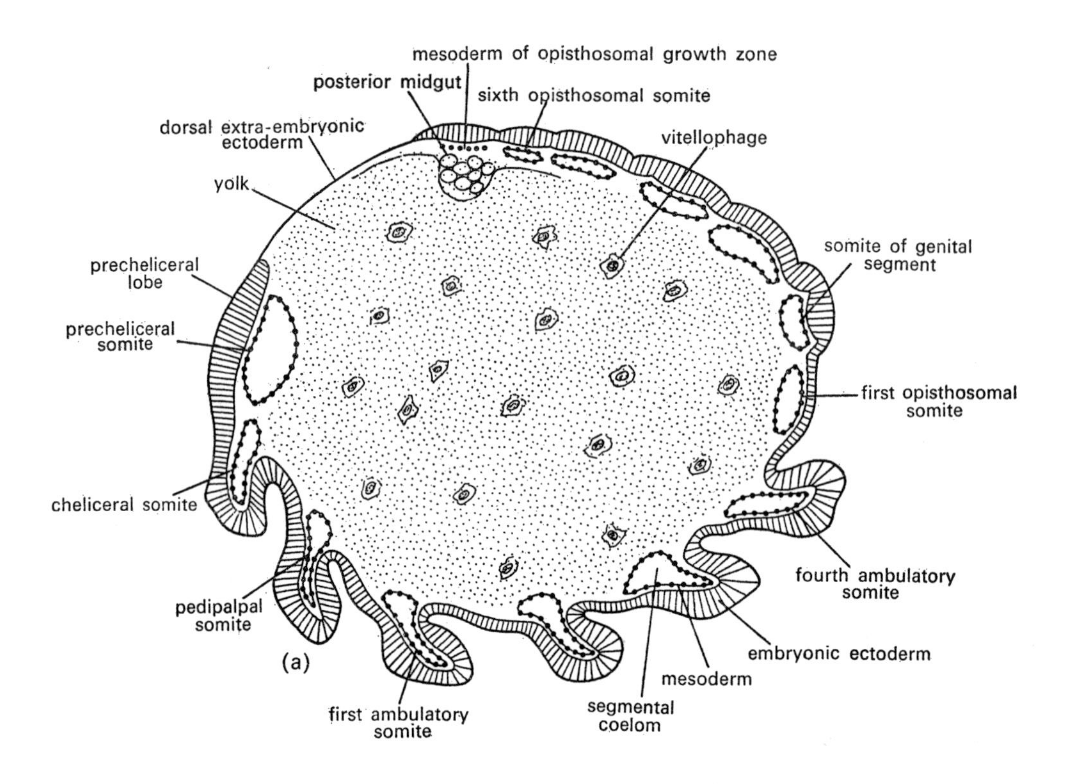

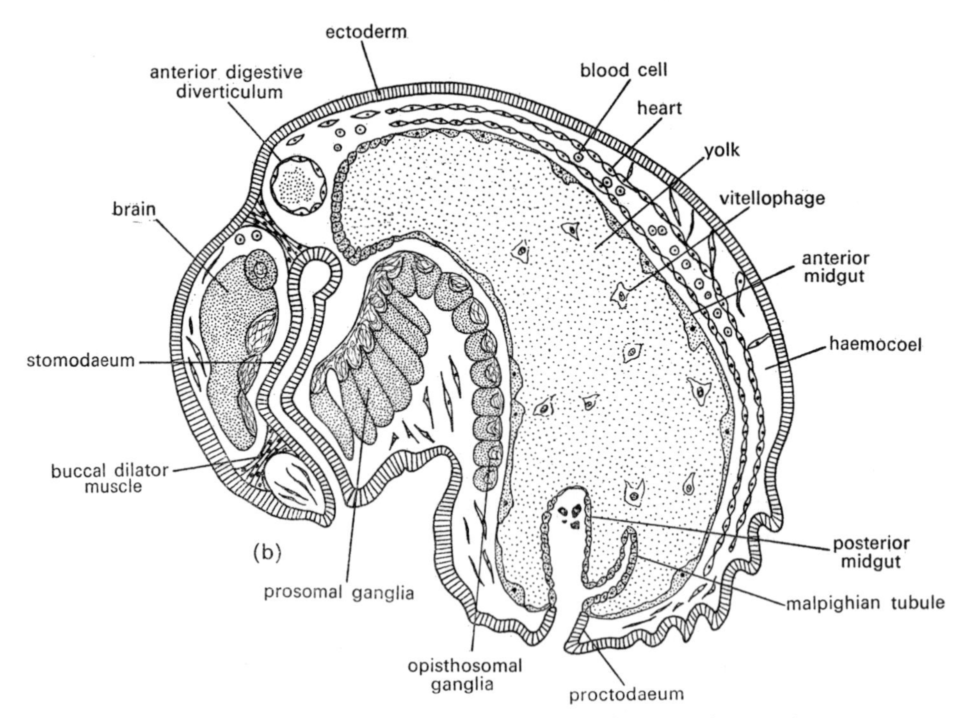

FIG. 155. Development of the araneid gut. (a) Diagrammatic parasagittal section through the embryo of *Agelena labyrinthica* during formation of the opisthosomal segments, modified after Wallstabe (1908); (b) sagittal section through late embryo of *Theridion*, after Dawydoff (1949).

embryos, suggested that the rudiment develops from the posterior cumulus cells, a result not incompatible with Yoshikura's later finding that the rectal sac of *Heptathela* is formed by posterior midgut cells lying within the opisthosomal growth zone. The posterior cumulus cells of entelegynids are associated with the opisthosomal growth zone when first differentiated, and terminate their migration at the location on the dorsal surface later attained by the posterior end of the fully segmented opisthosoma. Further detailed observations are needed, however, before the precise origin of the posterior midgut rudiment of spiders can be fully understood.

The situation is even less clear with regard to the Uropygi and Opiliones. In the former, Schimkewitsch (1906) observed that the major part of the midgut epithelium arises from vitellophages, as in spiders, and also identified a posterior midgut rudiment which develops into the rectal sac and Malphighian tubules, but did not trace the origin of the posterior rudiment. The same origin of the major part of the midgut has been found in opilionids (Faussek, 1891; Schimkewitsch, 1898; Holm, 1947; Moritz, 1957) but no description has yet been given of a posterior midgut rudiment. Together with a total lack of knowledge of these events in amblypygid and solifugid embryos, these fragmentary observations leave much to be done.

The development of the gut of acarine embryos is somewhat better known (Wagner, 1894 a, b; Bonnet, 1907; Klumpp, 1954; Langenscheidt, 1958; Aeschlimann, 1958; Anderson, 1972) and appears to be more specialized in that the vitellophages are not a major source of midgut epithelial cells (Fig. 156). The posterior midgut rudiment, formed during gastrulation as a distinct group of cells beneath the germ disc, retains its location beneath the germ disc during growth of the germ band. When elongation and segmentation of the germ band are complete, the posterior midgut rudiment lies beneath the telson. It maintains this position during contraction of the germ band until, in the contracted germ band, the rudiment lies beneath the posterior part of the opisthosoma. In ticks, and also in *Acarapis*, the rudiment now proliferates a pair of lateral arms which detach and hollow out as Malpighian tubules. The proximal ends of the tubules enlarge, unite in the midline and are joined by further cells from the posterior midgut rudiment in the production of a median vesicle, the rectal sac. This part of the posterior midgut rudiment is obviously equivalent to the entire posterior rudiment of spiders. The rectal sac establishes the usual connection with the proctodaeum but remains separate from the functional midgut sac. Accounts of the development of the midgut sac vary, but the weight of evidence seems to point to the remaining cells of the posterior midgut rudiment as the major source of anterior midgut cells. Towards the end of embryonic development, these cells proliferate and spread around the yolk mass to enclose the yolk and vitellophages (Wagner, 1894 a, b; Bonnet, 1907; Klumpp, 1954; Aeschlimann, 1958; Langenscheidt, 1958). Most authors also state that the vitellophages migrate to the surface of the yolk mass and participate in the formation of the midgut epithelium, but the evidence for this is inconclusive. As usual in chelicerates, the developing mesoderm presses into the sides of the midgut sac, dividing it into several pairs of diverticula. The stomodaeum gives rise to the lining epithelium of the pharynx and oesophagus (Wagner, 1894 a, b; Bonnet, 1907; Hafiz, 1935; Hughes, 1950; Klumpp, 1954; Aeschlimann, 1958; Langenscheidt, 1958; Anderson, 1972). In spite of gaps in the evidence, it seems likely that a common basis underlies the development of the midgut in spiders and

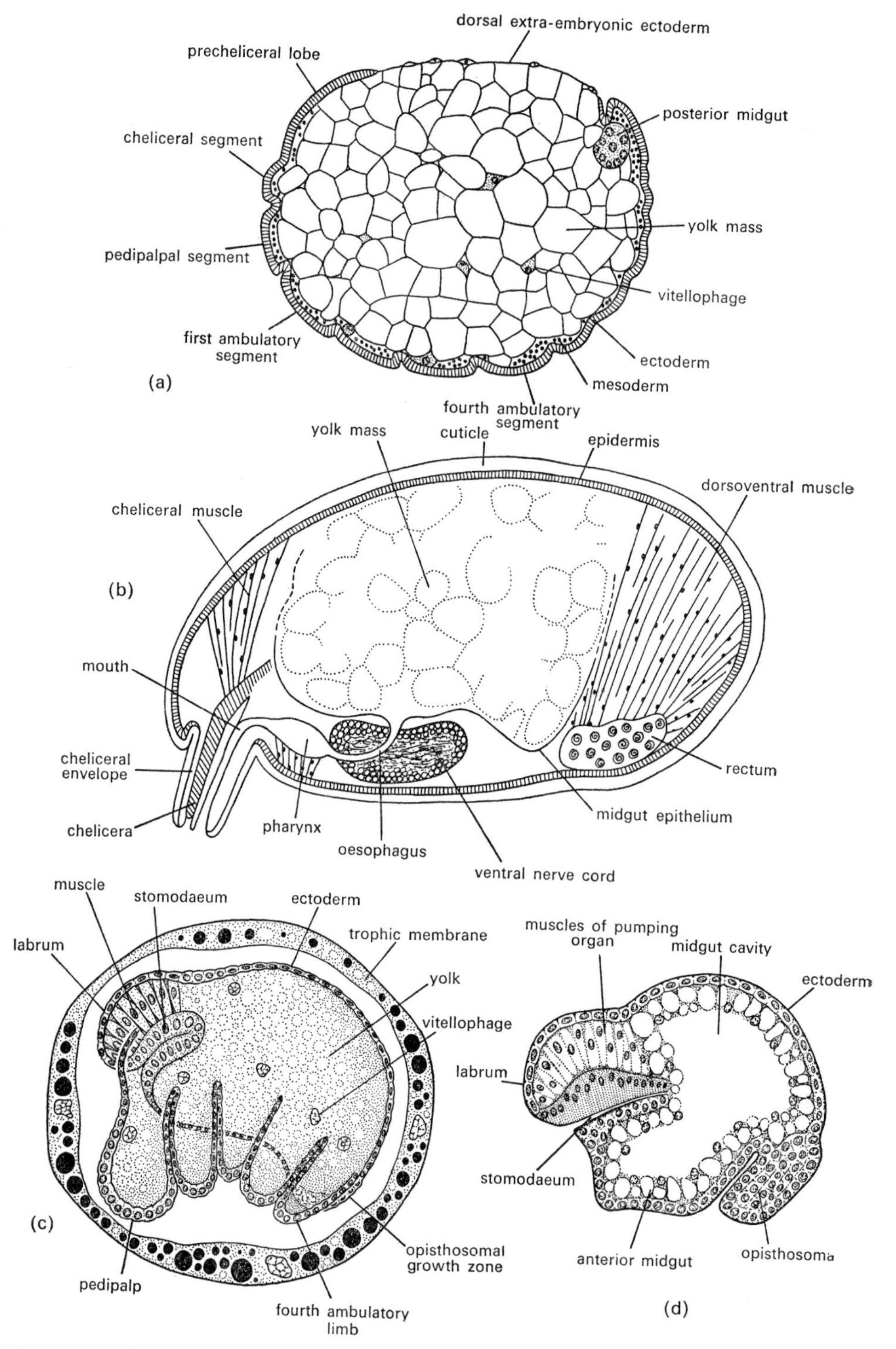

FIG. 156. (a) Sagittal section through segmented germ band stage of *Ornithodorus*, showing vitellophages and posterior midgut rudiment; (b) sagittal section through late embryo of *Ornithororus*; (c) lateral view of embryo of *Pselaphochernes* at onset of growth of opisthosomal rudiment; (d) sagittal section through the embryo of *Neobisium* as opisthosomal proliferation proceeds. [(a) and (b) after Aeschlimann, 1958; (c) and (d) after Weygoldt, 1964, 1965.]

in those other arachnid groups which a consideration of the external features of development associates with spiders. In principle, the yolk mass with its vitellophages is the rudiment of the major part of the midgut, while the rectal sac and Malphighian tubules develop from a separate, posterior midgut rudiment closely associated with the opisthosomal growth zone.

In spite of extreme specialization, the same mode of development is detectable in the pseudoscorpions (Fig. 156). As Weygoldt (1964, 1965) has recently shown, the small yolk mass develops precociously into a blind-ending midgut sac connected anteriorly with the precociously developed stomodaeum. Diverticula of the midgut sac protrude into the limb buds. The midgut sac receives the nutrient fluid ingested by the embryonic pumping organ. In *Neobisium*, the midgut sac develops directly into the definitive midgut and digestive diverticula, although *Pselaphochernes* is more specialized. Here, an unpaired anterior plate developed in the wall of the sac and a pair of posterolateral plates budded from the mesendoderm of the opisthosoma proliferate to form the definitive midgut wall in the late embryo, when the functional, embryonic midgut sac is resorbed. In both, however, a separate rudiment formed within the opisthosoma develops into a posterior midgut tube, including the rectal sac and Malphighian tubules.

Turning now to scorpions, the development of the gut in the yolky embryos of ovoviviparous species was described by Kowalevsky and Shulgin (1886), Laurie (1890) and Brauer (1894, 1895). In spite of the lack of more recent evidence, the agreement between the three accounts provides some indication of their reliability. The vitellophages derived from the germ disc spread around the surface of the yolk mass to some extent, but do not invade the yolk mass or take part in the formation of the definitive midgut epithelium. The latter is formed by proliferation and spread of the midgut rudiment around the yolk mass, accompanying the growth and segmentation of the germ band. As the forwardly flexed metasoma develops, a tubular extension of the midgut epithelium penetrates into the rudiment. The short proctodaeum arises late in development, as in invagination on the telson. The stomodaeum forms earlier, in the usual way. During dorsal closure, the upgrowing mesoderm cuts into the midgut sac, defining the digestive diverticula. Some yolk is still present in the midgut at birth.

The mode of development of the midgut in scorpions is thus unique among arachnids with yolky eggs. Furthermore, the specialized development of the gut in the yolkless embryos of viviparous scorpions is a modification of the same unique pattern (Figs. 153, 154 and 156). Preliminary observations on the precocious development of the stomodaeum and midgut sac in viviparous species were made by Poliansky (1903), Pflugfelder (1930) and Vachon (1950) but the fullest account is that given by Mathew (1956) for *Heterometrus*. When the small gastrula begins to increase in length, the internal mass of cells withdraws from the anterior end, leaving a space. The anterior part of the external wall then invaginates into this space, forming a large stomodaeum with the mouth directed towards the axial canal of the appendix. With further enlargement of the embryo, some of the internal cells accumulate on the outer surface of the stomodaeum and differentiate as stomodaeal muscle. An opening forms at the free end of the stomodaeum, through which nutritive material from the appendix enters the embryo. The internal cells immediately around the space receiving this fluid unite to form a midgut sac attached to the end of the stomodaeum. The remainder of the internal cells form mesoderm (see below). As the embryo grows longer, the midgut sac extends with it and becomes tubular. The stomodaeum has a suctorial action, pumping nutrient fluid from the appendix into the midgut. In later development, the midgut develops six pairs of lateral diverticula. The diverticula become branched, the first pair forming an anterior pair of digestive glands opening into the stomach, the last five pairs merging to form a pair of posterior digestive glands with five pairs of ducts opening into the midgut. Feeding continues until embryonic development is completed and the young are born. The short proctodaeum invaginates late in development but does not become connected with the midgut until after birth. In all respect, the development of the gut in yolkless scorpion embryos proceeds in the same manner as in yolk-filled embryos, except that the midgut epithelium develops precociously around a central mass of nutrient fluid rather than around a yolk mass. Malpighian tubules in both types of embryo arise as evaginations of the midgut wall at the junction of the mesosoma and metasoma, behind the last pair of digestive diverticula (Laurie, 1890; Brauer, 1895; Abd-el Wahab, 1956; Mathew, 1956). The significance of the differences between development of the midgut in xiphosurans, scorpions and other arachnids will be discussed in the final section of the present chapter.

Further Development of the Mesoderm

As far as we know at the present time, mesoderm enters along the ventral midline of each successive segment of the xiphosuran germ band and spreads laterally beneath the ectoderm of the segment as a pair of transverse somite rudiments. These then split internally, forming paired hollow somites (Kishinouye, 1891a; Kingsley, 1892, 1893; Iwanoff, 1933). The prosomal pairs develop well in advance of the opisthosomal pairs, which are successively serially delayed. According to Iwanoff (1933), the embryo of *Tachypleus gigas* develops three pairs of small opisthosomal somites behind the pair in the last gill bearing (seventh opisthosomal) segment and has a terminal unsegmented mass of mesoderm at the posterior end. Anteriorly, the cheliceral pair of somites extends forwards beneath the pre-cheliceral lobes and becomes partially subdivided into cheliceral and pre-cheliceral portions, but there is no evidence at present that separate pre-cheliceral somites are developed.

The somites have thick somatic and thin splanchnic walls (Fig. 157). In the segments in which limb buds are developed, each somite develops an appendicular lobe which penetrates into the corresponding limb bud. The further development of the somites is not yet well understood but it is known that the somites extend upwards towards the dorsal midline during later development and that the dorsal margins of all somites behind the second ambulatory segment participate in the formation of the heart. The paired lateral ostia of the heart develop at the intersegmented junctions between successive somite pairs. Commensurate with the pattern of dorsal closure, the development of the heart begins at the posterior end of the opisthosoma and proceeds in an anterior direction. On either side of the heart, the somatic walls of the somites separate from the overlying ectoderm and merge with the splanchnic walls to form the pericardial floor with its residual pericardial coelomic cavities. The appendicular lobes and adjacent somatic mesoderm give rise to somatic musculature, while the major part of the splanchnic wall of each somite presumably develops as splanchnic musculature, but the details of these processes are not known. Only the

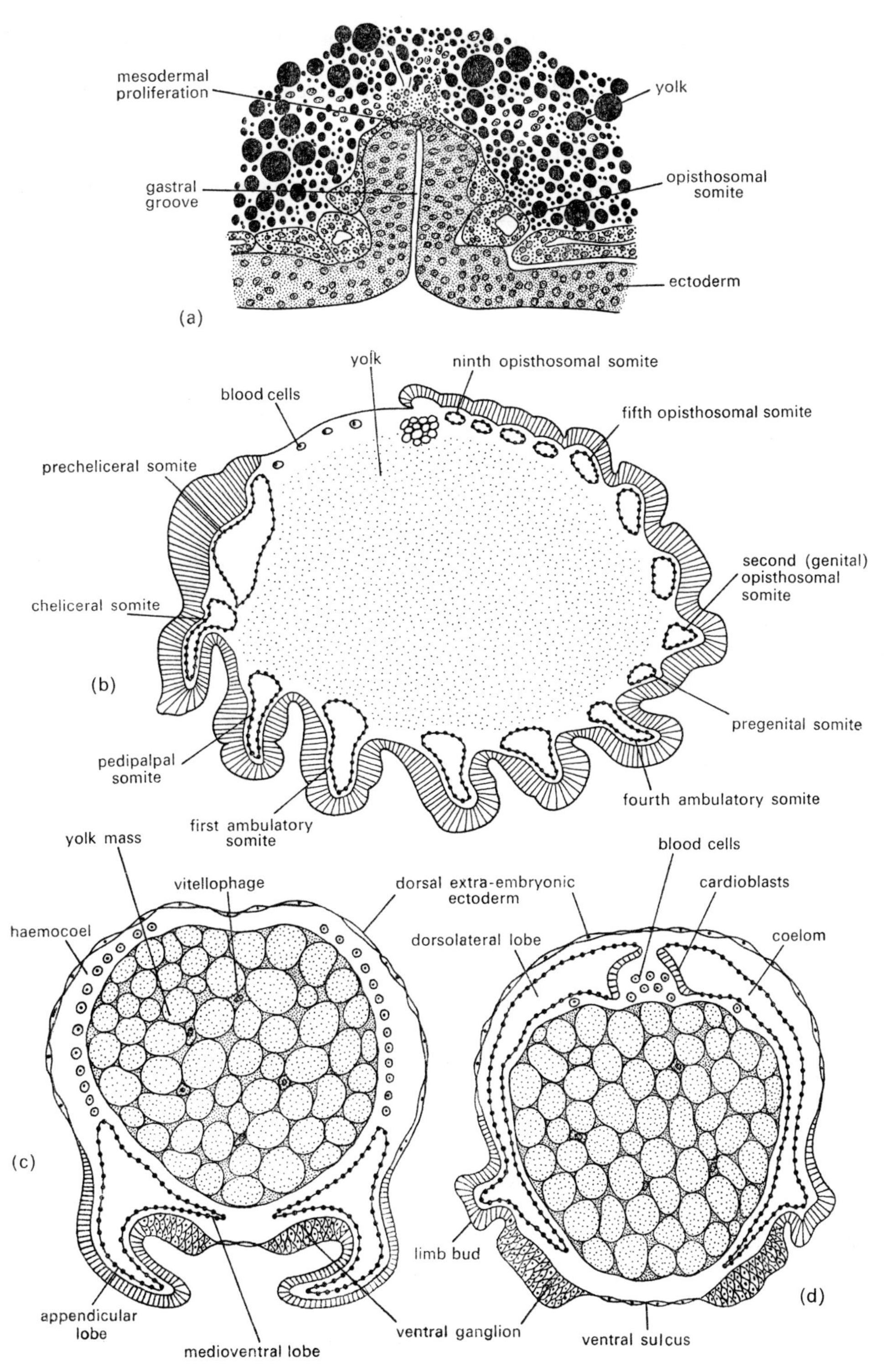

FIG. 157. (a) Transverse section through the gastral groove during the proliferation of opisthosomal somites in *Tachypleus*, after Iwanoff (1933); (b) parasagittal section through the segmented germ band of *Agelena*, after Wallstabe (1908); (c), (d) transverse sections showing the further development of the prosomal (c) and opisthosomal (d) somites of *Theridium* as inversion begins, after Dawydoff (1949).

development of the coxal glands, which will be discussed below, has been examined in any degree of detail.

The further development of the mesoderm is better known in the arachnids, though observations are scattered and some important aspects have been neglected. The mesoderm spreads as a single layer beneath the germ band, extending along the growing opisthosoma. Bilateral separation into paired mesodermal bands is then followed by transverse subdivision into paired somites. In spiders, the somites become hollow by inrolling of the lateral margins (Yoshikura, 1955; Rempel, 1957), as they do in some pterygotes (Chapter 7). Each segment gains a pair of coelomic sacs (Schimkewitsch, 1887; Morin, 1887, 1888; Wallstabe, 1908; Montgomery, 1909a; Kautzsch, 1909, 1910; L. and W. Schimkewitsch, 1911; Yoshikura, 1955; Rempel, 1957; Legendre, 1959), giving six prosomal pairs and twelve opisthosomal pairs, including a pre-genital pair (Fig. 157). In those segments which develop limb buds, the somites produce appendicular lobes. The development of the coelom in front of the cheliceral segment was a point of controversy among early workers. Wallstabe (1908) interpreted the single pair of pre-cheliceral coelomic pouches as forward extensions of the cheliceral somites. Montgomery (1909a), on the other hand, found that they were separate from the cheliceral somites. Yoshikura (1955) supported Wallstabe's view in his study of the primitive spider *Heptathela*, but recently, Pross (1966) has found two distinct pairs of precheliceral coelomic sacs in the lycosid *Pardosa* (Fig. 158).

The further development of the somites in spiders (Fig. 157) has been described principally by Morin (1888) and Rempel (1957). During dorsal closure, accompanied by inversion in most spider embryos, the dorsolateral portions of the somites extend towards the dorsal midline, where their dorsal margins give rise to the heart (Locy, 1886; Schimkewitsch, 1887; Morin, 1888; Kishinouye, 1891b; Montgomery, 1909a; Kautzsch, 1910; Yoshikura, 1955; Rempel, 1957). As in the Xiphosura, the ostia of the heart are persistent intersegmented spaces, five pairs being retained in *Heptathela*, four in mygalomorphs and three (sometimes two) in araneomorphs. Blood cells are present in the heart cavity, but the origin of these cells is not resolved. Morin (1887) and L. and W. Schimkewitsch (1911) thought that the blood cells were released from the walls of the somites. Balfour (1880), Locy (1886), Schimkewitsch (1887), Kishinouye (1891b) interpreted them as derivatives of vitellophages. Montgomery (1909a), on the other hand, trace the origin of the blood cells of *Theridion* to a proliferative source in the dorsal extra-embryonic ectoderm and this view was supported by Rempel (1957) for another theridiid, *Latrodectus*. None of these interpretations is based on satisfactory evidence.

Mesoderm on either side of the heart forms the pericardial floor. The somatic musculature is mainly a product of the appendicular lobes of the somites, while the medio-ventral portions of the somites of certain segments contribute to the coxal glands and gonads (see below). In general, these medio-ventral portions extend towards the ventral midline above the developing ganglia, where their somatic walls form the neurilemma. The splanchnic walls of the somites are mainly applied to the surface of the midgut and develop as splanchnic muscle. It is this layer which cuts into the midgut in such a way as to individuate the paired rudiments of the digestive diverticula.

Paired coelomic sacs have been observed in the embryonic segments of Solifugae (Heymons, 1904), Uropygi (Schimkewitsch, 1906) and Opiliones (Winkler, 1957) and also in the

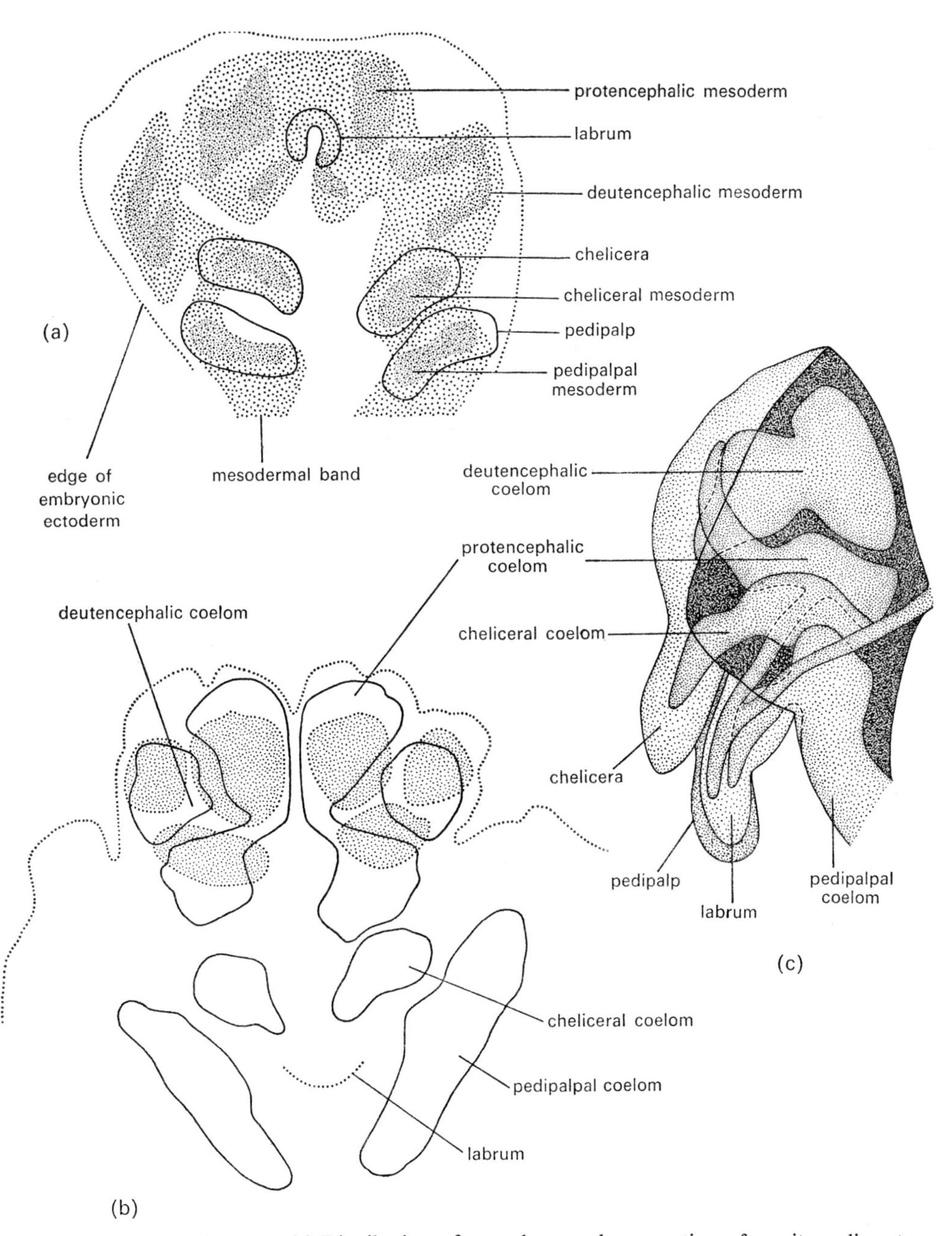

FIG. 158. *Pardosa hortensis.* (a) Distribution of mesoderm and aggregation of somite rudiments beneath the anterior end of the segmenting germ band; (b) distribution of anterior somites during inversion; (c) diagrammatic reconstruction of anterior end of embryo after inversion. (After Pross, 1966.)

opisthosomal segments of pseudoscorpions (Weygoldt, 1964, 1965), though they do not develop in the prosomal segments of the specialized pseudoscorpionid embryo. No useful comments can be made at the present time on the later development of the somites in these groups, other than that they contribute in *Neobisium* to the heart (Weygoldt, 1965).

In acarine embryos, the mesoderm forms paired bands which break up into somite rudiments, but the formation of coelomic spaces within the somites is relatively vestigial (Wagner, 1894 a, b; Bonnet, 1907; Klumpp, 1954; Aeschlimann, 1958; Langenscheidt, 1958). There are six prosomal and five opisthosomal pairs, the pre-cheliceral mesoderm being diffuse. Each prosomal somite sends the usual appendicular lobe into the corresponding limb bud. After contraction of the germ band, the somites break down and develop as limb musculature (from the appendicular lobes) and dorsoventral somatic muscles (Aeschlimann, 1958), but the details of this process are still unknown.

The most careful and detailed studies of mesoderm development in arachnids are those of Brauer (1895) and Mathew (1956) on scorpions. In the buthid embryo, the mesoderm in each segment rudiment of the germ band separates in the usual way, first longitudinally, then transversely, into paired somite rudiments. The six prosomal and seven mesosomal pairs become hollow and trilobed, but the six metasomal pairs lack appendicular lobes. Brauer interpreted the pre-cheliceral mesoderm as a pair of hollow, forward extensions of the cheliceral somites, but Abd-el Wahab (1952) described them as a separate pair of pre-cheliceral somites. During dorsal upgrowth, the dorsolateral lobes of the somites give off blood cells. The dorsal edges of the somites then combine to form the heart, as in all chelicerates, and the adjacent somatic mesoderm gives off cells which differentiate as dorsal longitudinal muscle, before separating from the ectoderm and merging with the underlying splanchnic mesoderm to form the pericardial floor. Narrow pericardial coelomic tubes persist in the pericardial floor. The principal product of the splanchnic mesoderm is splanchnic muscle, and of the appendicular lobes, somatic muscle, but the details of these aspects of development are not clear.

As might be expected, mesodermal development in yolkless scorpionid embryos shows a number of specializations (Mathew, 1956). Paired mesodermal bands develop ventrolaterally on either side of the gut, but do not subdivide directly into paired somites. The cheliceral pair of somites individuates precociously at the anterior ends of the mesodermal bands and develop as cheliceral and labral musculature, but the next emphasis is on development of the mesoderm of the enlarged mesosoma. Here, the mesodermal bands give off numerous blood cells and then, instead of forming paired somites, disperse to form a series of trabeculae connecting the body wall to the midgut wall. The functional significance of the trabeculae is not known. In the later development of the mesosoma, the trabeculae differentiate as somatic and splanchnic musculature and, dorsally, as the walls of the heart.

When the limb buds of the prosoma develop, the underlying mesodermal bands segment to form the corresponding hollow somites with limb bud diverticula. Similarly, at the anterior end of the mesosoma, the genital and pectinal segments also develop paired, ventrolateral somites before they develop limb buds. A small, transitory pair of pregenital somites accompanies them as a vestige of an almost suppressed pregenital segment.

Coxal glands

Almost all chelicerates have a pair of prosomal coxal glands, developed from certain prosomal somites (Goodrich, 1946). Those of *Limulus* originate from paired, hollow rudiments derived from the ventral portions of the somites of the six prosomal segments (Kingsley, 1893; Patten and Hazen, 1900). The cheliceral and fourth ambulatory rudiments are transient, but the remaining pairs become linked together longitudinally to form the lobes of the two coxal glands (Table 3). The pair of the third ambulatory segment also develo p looped, tubular extensions which form the ducts of the glands, opening to the exterior through short ectodermal ducts at the bases of the third ambulatory limbs.

Among arachnids, the number of pairs of coxal gland rudiments is usually reduced, though the glands always develop from coelomic end sacs, mesodermal ducts and ectodermal exit ducts. In the primitive spider *Heptathela*, Yoshikura (1955) found that paired end sacs and duct rudiments develop in the pedipalpal and first three ambulatory segments (Table 3). The pedipalpal pair remains separate. They become slightly coiled and gain openings to the exterior on the pedipalpal coxae just before hatching takes place. Since the glands degenerate after hatching, Yoshikura suggests that they may secrete a hatching fluid. The rudiments in the first three ambulatory segments link up to form the definitive coxal glands and become connected with short exit ducts developed as ectodermal invaginations on the posterior surfaces of the coxae of the three corresponding pairs of limbs. Later, the exit ducts of the second ambulatory segment close, leaving two pairs of definitive exit ducts, on the first and third ambulatory segments respectively. This arrangement is also retained in mygalomorph spiders, but in araneomorphs only the glands and ducts of the first ambulatory segment persist (Buxton, 1917).

The opilionids display a slightly different arrangement, recently investigated by Moritz (1959). Rudiments of end sacs and ducts develop in the ventral parts of the somites of the four ambulatory segments, but only those of the first segment persist, forming a pair of end sacs and ducts (Fig. 159). The ducts open temporarily into ectodermal exit ducts at the bases of the first pair of ambulatory limbs, but during later embryonic development, these openings are lost and a definitive pair of exit ducts invaginates at the bases of the third pair of ambulatory buds. Thus the completed glands originate partly from the first ambulatory segment (end sacs and mesodermal ducts) and partly from the third (ectodermal exit ducts).

Information on other groups is more fragmentary (Table 3), except for the scorpions, to which we will return below. In the acarines, the glands open at the bases of the second pair of ambulatory limbs, but their development has not been studied. In the Uropygi, the glands have two end sacs and open at the bases of the first pair of ambulatory limbs (Schimkewitsch, 1906). Amblypygids retain two separate pairs of end sacs and ducts, opening at the bases of the first and third pair of ambulatory limbs, as they do in primitive spiders (Pereyaslawzewa, 1901; Gough, 1902). Further studies are needed on these groups and on the Solifugae. The pseudoscorpions retain only a single pair of end sacs and mesodermal ducts (Fig. 160), developed from the somites of the third ambulatory segment (Weygoldt, 1964, 1965).

Scorpionid coxal glands are better described, the classical account being that of Brauer (1895) for *Buthus carpathicus*. Duct rudiments develop as the usual funnel-like outgrowths of the posteroventral walls of the four pairs of coelomic sacs in the ambulatory segments.

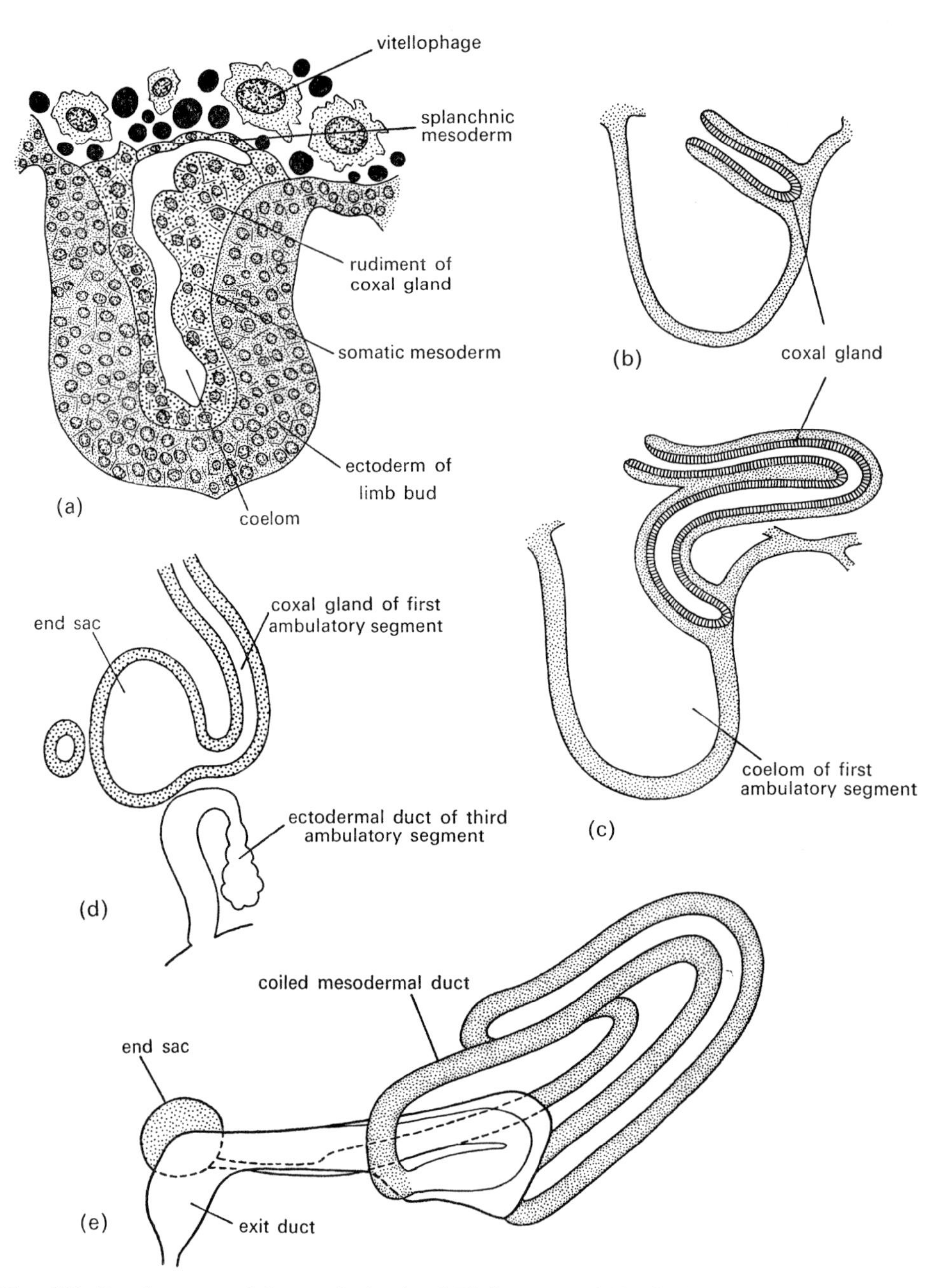

FIG. 159. Development of the coxal glands of *Phalangium*. (a) Vertical section through the first ambulatory limb bud and associated somite; (b), (c) diagrams to show the development of the mesodermal duct of the first ambulatory coxal gland; (d) development of the secondary ectodermal duct of the coxal gland on the third ambulatory segment; (e) the fully developed coxal gland. (After Moritz, 1959.)

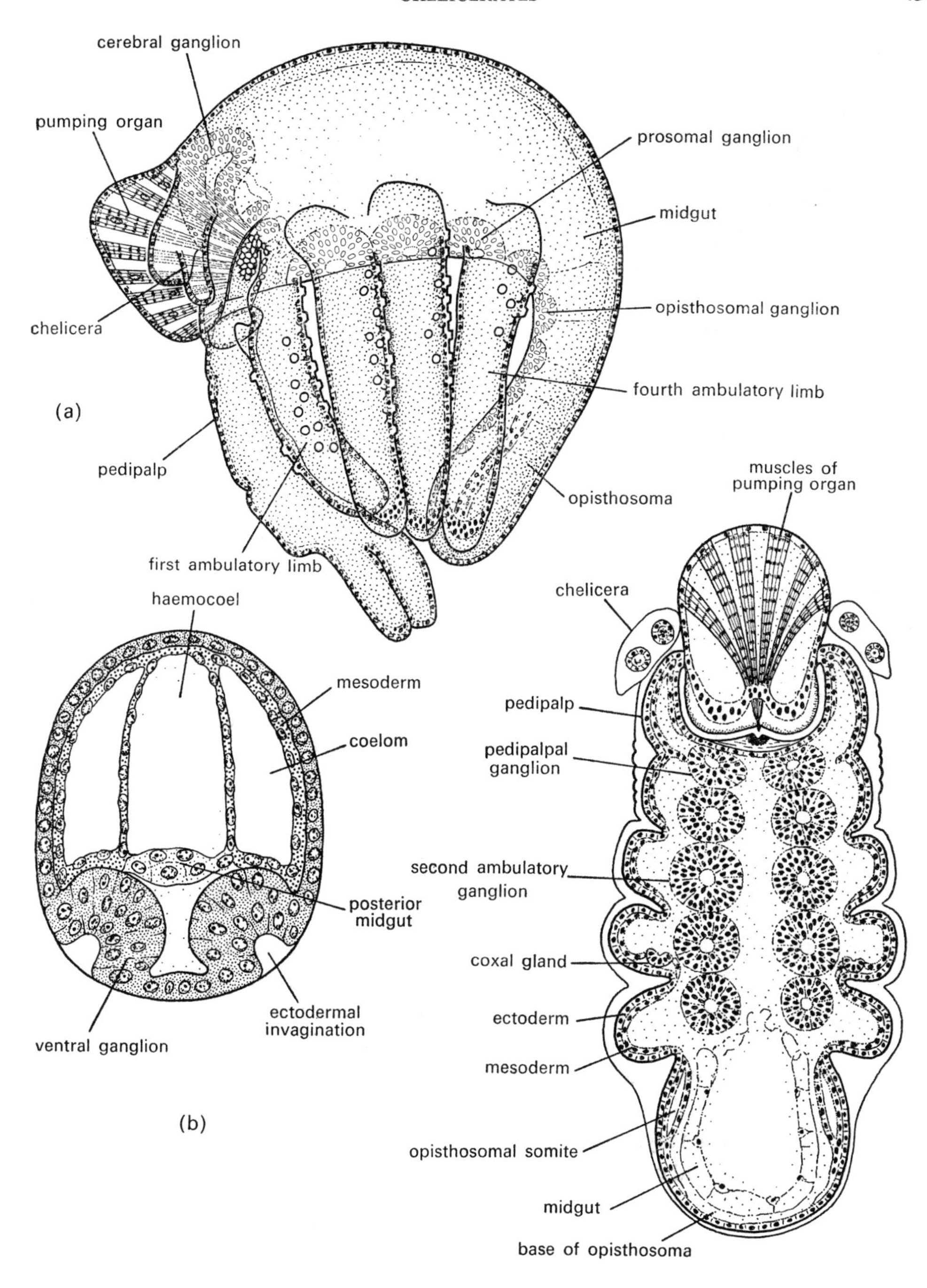

FIG. 160. Development of prosomal ventral ganglia in *Neobisium*. (a) Lateral view of embryo with well-developed opisthosoma and developing ventral ganglia; (b) transverse section through middle region of opisthosoma of the same embryo; (c) frontal section through embryo at the level of the prosomal ganglia. (After Weygoldt, 1965.)

Those of the first, second and fourth segment are transient vestiges, but those of the third ambulatory segment develop as coiled ducts with coelomic end sacs. Typical, short ectodermal exit ducts arise at the bases of the third pair of ambulatory limbs. The paired coelomic sacs of the third ambulatory segment of *Heterometrus* give rise to coxal glands in a similar way (Mathew, 1956).

It seems likely that primitive chelicerates had a pair of segmental organs opening on the coxae in all six prosomal segments, and that these have become reduced and specialized in different ways in the Xiphosura and arachnids (Table 3). The pattern of specialization in

TABLE 3. DEVELOPMENT OF COXAL GLANDS IN CHELICERATA

Group	Prosomal segment					
	Cheliceral	Pedipalpal	1st Ambulatory	2nd Ambulatory	3rd Ambulatory	4th Ambulatory
Ancestral	EDd	EDd	EDd	EDd	EDd	EDd
Xiphosura	V	E	E	E	EDd	V
Araneida, primitive	—	V	EDd	E	EDd	—
Araneida, specialized	—	—	EDd	V	V	—
Uropygi	—	—	EDd	E	—	—
Amblypygi	—	—	EDd	—	EDd	—
Solifugae	—	—	EDd	—	—	—
Opiliones	—	—	ED(d)	V	dV	V
Acarina	—	—	—	EDd	—	—
Pseudoscorpionida	—	—	—	—	EDd	—
Scorpionida	—	—	V	V	EDd	V

E, end sac; D, duct; V, transient vestige in embryo; d, ectodermal exit duct.

Xiphosura, including transience of the first and last pair of glands and of all ducts except those of the fifth pair, could provide the basis for the further specialization seen in scorpions, but not for those of other arachnids, which usually retain the ducts of the third pair of rudiments as functional ducts. The primitive spider *Heptathela*, however, retains a generalized pattern of coxal gland development from which those of all arachnids, including the scorpions, can be derived. The development of the coxal gland thus tells us almost nothing about chelicerate phylogeny, since the amount of information is insufficient to exclude the possibility of parallel and convergent evolution.

Gonads and genital ducts

The embryonic development of the reproductive system is not well known for the Chelicerata. In particular, the system awaits investigation in the Xiphosura, though the uniformity of the results available for Arachnida and the fact that all chelicerates share the second opisthosomal segment as the genital segment suggest that the development of the xiphosuran reproductive system is not likely to be much different from that of scorpions.

The work of Laurie (1890), Brauer (1895) and Mathew (1956) on scorpions and Purcell (1895) and Kautzsch (1910) on spiders shows that the gonads are formed mainly from the coelomic sacs of the genital segment, which unite and receive contributions from adjacent somites in producing the genital coelom. The gonoducts are coelomoducts which grow out from the somatic walls of the genital coelomic sacs, curve down on either side of the gut and meet a midventral ectodermal invagination at the genital pore. The serial homology of the coxal glands and gonads of chelicerates is quite clear.

The origin and development of acarine gonads is more specialized. According to Bonnet (1907), the gonads of *Hyalomma aegyptium* are first observed in the late embryo as a pair of small groups of cells of unknown origin (probably the mesoderm of the second opisthosomal segment), flanking the rectal sac. The paired rudiments increase in length, curving upwards so that the anterior ends meet above the rectal sac. The posterior end of each gonad then proliferate the gonoduct of its side, connecting in the usual way with a midventral, ectodermal exit duct.

Primordial germ cells in the chelicerates need further study. They become recognizable in spiders only at a late stage of development, as large cells in the walls of the third to sixth pairs of opisthosomal coelomic sacs (Schimkewitsch, 1887, 1898; Kishinouye, 1891b; Montgomery, 1909a; Kautzsch, 1910). As the genital coelom develops, the germ cells congregate in its walls. The scorpions and opilionids, however, show a precocious segregation of primordial germ cells as a compact group of large cells beneath the germ disc (Brauer, 1895; Moritz, 1957; Juberthie, 1964). In *Buthus carpathicus*, the germ cells remain grouped in the mesoderm of the growth zone during proliferation of the prosomal and mesosomal segments, until the growth zone turns forwards and forms the caudal papilla. After a temporary sojourn in the last mesosomal segment, the germ cells migrate forwards and enter the genital coelom. The opilionid germ cells similarly remain beneath the growth zone during formation of the first three opisthosomal segments, but are then left in the fourth opisthosomal segment as the activity of the growth zone continues. From here, they subsequently migrate forwards to the genital coelom. As in other arthropod groups, an early or late segregation of primordial germ cells in chelicerates has no phylogenetic significance.

Further Development of the Ectoderm

The general course of development of the ectoderm as epidermis has been given little attention in papers on chelicerate embryology, although the development of the important ectodermal derivatives (the nervous system, eyes, respiratory organs, glands and ducts) is quite well known. A few comments on the general development of the ectoderm are necessary, however, before we proceed to a consideration of organ systems.

In the Xiphosura, it is evident from the work of Kingsley (1892, 1893) and Iwanoff (1933) that blastoderm cells outside the area of the embryonic primordium participate in the initial development of embryonic ectoderm and that only a proportion of them pass through a phase of temporary attenuation as dorsal yolk-sac ectoderm. We have already seen that the ventral embryonic primordium elongates due to proliferation by a posterior growth zone, and that the anterior part of the primordium, in front of the growth zone, constitutes the rudiment of the cephalic lobe (pre-cheliceral lobe and cheliceral segment) only. Two facts are important about the embryonic primordium at this stage. Firstly, it is narrow as well as short. Secondly, it develops in the midline a gastral groove, whose anterior end is formed by the invagination of presumptive stomodaeum, while the remainder of the groove results from the invagination of presumptive mesoderm. The embryonic ectoderm of the embryonic primordium, i.e. of the pre-cheliceral lobe, cheliceral segment and growth zone, is bilateral on either side of the invaginating mesoderm. Now, as the growth zone becomes active in the forward proliferation of the rudiments of the pedipalpal and succeeding segments, mesoderm continues to be produced mesoderm midventrally, with embryonic ectoderm on either side. The mesoderm invaginates, continuing the gastral groove as far back as the twelfth opisthosomal segment in the manner already discussed in a previous section. At the same time, the embryonic ectoderm of the cephalic lobe, pedipalpal and succeeding segments and of the growth zone itself broadens by the incorporation of blastoderm cells on either side of it, with compensatory attenuation of the dorsal to lateral blastoderm as dorsal yolk sac ectoderm. By the time that the last prosomal segment has been formed, the ectoderm of each segment of the prosomal region of the germ band is broader than long and the growth zone is as wide as the segments in front of it. The lateral margins of the prosomal ectoderm lie some way up the sides of the embryo, where they meet the edge of the extra-embryonic ectoderm. The prosomal ectoderm covers much of the ventral surface of the embryo at this stage, leaving a disproportionately small area available at the posterior end for the development of the opisthosoma. The proliferation of opisthosomal ectoderm by the growth zone now follows a diminishing sequence. Obviously this is the reverse of the growth pattern of the prosoma. Instead of the incorporation of additional blastoderm cells into the lateral parts of the segment rudiments, the growth zone ectoderm itself is reduced in width as it gives rise to the ectoderm of the successively smaller segments. With closure of the gastral groove, the two ectodermal halves of each segment come together in the ventral midline. At the same time, limb buds and ganglia begin to develop and the cheliceral ectoderm behind the mouth becomes delineated from the pre-cheliceral ectoderm in front of and around the mouth.

Before going into the organogenetic elaboration of the embryonic ectoderm, it is advantageous to give brief consideration to the mode of formation of the segmental ectoderm of arachnid embryos. Once the events in Xiphosura are appreciated, this becomes a simple matter. The scorpions exhibit one modification, the remaining arachnids variations on another.

The development of the embryonic ectoderm of scorpions is seen at its most generalized in buthids (Brauer, 1895). The most important difference as compared with Xiphosura is that the buthid embryonic primordium separates from the remainder of the blastoderm, which overgrows the primordium to form the serosa. Thus the incorporation of additional blastoderm cells to form the lateral ectoderm of the prosomal segments cannot occur. All of the embryonic ectoderm arises from the embryonic primordium and the germ band remains narrow throughout its length. Given this difference, however, other modifications are minor. It seems likely that there is no gastral groove in scorpions. Thus the bilateral halves of the embryonic ectoderm are conjoined midventrally from the moment of their formation. The embryonic ectoderm of the embryonic primordium includes the pedipalpal ectoderm rudiments as well as the cephalic lobe ectoderm anteriorly and the growth zone ectoderm posteriorly. Thus the growth zone does not proliferate the pedipalpal ectoderm, though it buds off the ectoderm of the remaining segments as a succession of short blocks along the ventral surface of the yolk mass. The last six rudiments, making up the metasoma, are not concentrically compressed onto the posterior face of the yolk mass, but are carried beyond the yolk as a tubular, forwardly flexed, caudal papilla. Finally, the peripheral ectoderm of the embryonic primordium has a specialized development as an amnion. The embryonic ectoderm of yolkless scorpion embryos follows a similar sequential development (Mathew, 1956) except that the amnion and serosa are not formed and the ectoderm, in the absence of a large yolk mass, is budded off as a tubular wall throughout the length of the embryo.

In other arachnids, in contrast, the basic modification of development of the embryonic ectoderm is formation of the entire bilateral, prosomal ectoderm by ventral aggregation and division of blastoderm cells, with the growth zone ectoderm retaining only the capacity to proliferate opisthosomal ectoderm. During this process, which emphasizes what, in Xiphosura, is only an additive process, the dorsal blastoderm becomes attenuated as dorsal extra-embryonic ectoderm. The early development of the prosomal ectoderm in this way has not been described in detail for any species, but is indicated in all those papers which give accounts of the development of the arachnid germ band. The growth zone ectoderm then proliferates a succession of bilateral rudiments of opisthosomal, segmental ectoderm either on the posterior face of the yolk mass or, in primitive spiders, partially as a forwardly flexed, tubular caudal papilla. There is no reason to suppose that a caudal papilla ever existed in opilionids or acarines, but it is a basic feature of spiders. A caudal papilla may also have occurred ancestrally in other arachnids whose embryos show inversion (uropygids, amblypygids, perhaps solifugids) and is retained in the specialized embryos of pseudoscorpions.

As we have seen, many arachnid embryos show bilateral separation of the germ band and the development of a temporary ventral sulcus during later development. This occurs as a secondary feature in mygalomorph and especially in araneomorph spiders, in uropygids and amblypygids and in acarine embryos. In terms of the development of ectoderm, the formation of a ventral sulcus means the temporary attenuation of midventral ectoderm as ventral extra-embryonic ectoderm. Primitively, as in liphistiomorph spiders and opilionids, this does not occur. The specialized embryos of pseudoscorpions also lack a ventral sulcus.

Among other significant specializations of the early development of segmental ectoderm in arachnid embryos we can mention those of entelegynid and ixodid ticks. In the former, studied with particular care by Holm (1952), the aggregation of blastoderm cells in front of the growth zone takes in most of the blastoderm. Marking experiments by Holm showed, for example, that the ectoderm of the cephalic lobe derives from the anterodorsal part of the blastoderm. At the same time, the growth zone, unlike that of primitive spiders, proliferates the ectoderm of the last three prosomal segments as well as the opisthosomal ectoderm. The specialization displayed in ixodid ticks, e.g. *Hyalomma*, concerns the formation of the ventral sulcus (Anderson, 1972). Instead of aggregating midventrally and then separating, as in the argasid *Ornithodorus* (Aeschlimann, 1958), the two halves of the anterior part of the germ band, as far back as the second ambulatory segment, aggregate in *Hyalomma* directly in their bilateral positions, with the ventral sulcus between them.

In all those arachnids whose prosomal ectoderm is formed by direct blastodermal aggregation, the gastral groove is retained as a site of mesodermal entry into the interior. We have also seen, in discussing gastrulation, that a proportion of the presumptive midgut often finds its way into the interior by the same route. In relation to the embryonic ectoderm, the gastral groove is much more restricted than it is in the Xiphosura. The groove forms, in primitive spiders, between the two ectodermal halves of the last three prosomal segments (Yoshikura, 1955). In other spiders, it occupies a simular position but in opilionids and acarines, the gastral grooves is confined to the ventral midline of the germ disc, that is, the opisthosomal growth zone. What this means in terms of presumptive areas of the chelicerate blastoderm will be discussed in a subsequent section. The main point to be made in the present context is that the gastral groove is of little significance in arachnids in creating a temporary separation between the two ectodermal halves of the germ band.

Nervous system

The basic development of the nervous system is alike throughout the chelicerates. The ventrolateral embryonic ectoderm, median to the limb buds in the segments which develop limb buds, and in the corresponding position in limbless segments, thickens bilaterally to form a pair of segmental ganglion rudiments (Fig. 160). Simultaneously, a pair of more complex ganglia develops from the ectoderm of the precheliceral lobes (Fig. 161). In later development, the ganglia of the cheliceral segment move forwards on either side of the mouth and unite with the precheliceral ganglia to complete the supra-oesophageal ganglion, but retains a post-oral commissure. Most or all of the remaining ventral ganglia also concentrate into the prosoma to form a large suboesophageal ganglion. In the Xiphosura, the suboesophageal ganglion includes the five prosomal pairs and the pregenital and genital pairs. The ganglia of the first three gill-bearing segments remain separate and retain their segmental location, but those of the sixth and succeeding opisthosomal segments merge together to form a terminal opisthosomal ganglion. Scorpions develop a slightly modified arrangement. The suboesophageal ganglion is composed of the five prosomal and first five mesosomal segments (pregenital, genital, first three respiratory), but the last two mesosomal and six metasomal ganglia remain within their segments, except for fusion of the fifth and sixth metasomal pairs. Condensation of ganglia is more pronounced in the Uropygi, Solifugae, Opiliones, and liphistiomorph spiders, in which most of the ventral ganglia are merged into the suboesophageal ganglion, and reaches an extreme in the Amblypygi, mygalomorph and araneomorph spiders, Pseudoscorpiones and Acarina, in which none of the ventral ganglia remain separate from the suboesophageal ganglion.

The formation of ventral ganglia in xiphosurans was described by Kishinouye (1891a) and Iwanoff (1933) as a process of generalized proliferation of ventral ectoderm cells to produce paired segmental thickenings. In later development, the ganglionic thickenings separate into the interior, leaving epithelial ectoderm at the surface, before undergoing forward migration and fusion. The ventral ganglia of scorpions develop in a similar way (Kowalevsky and Shulgin, 1886; Laurie, 1890; Brauer, 1895; Abd-el Wahab, 1954). McClendon (1905) showed that in buthid scorpions, proliferation of ganglion cells is associated with the formation of numerous small pits, which indent the surface of the ganglion rudiments. The pits are obliterated as the rudiments become internal. Mathew (1956) found that ganglion formation in the yolkless embryo of *Heterometrus* is preceded by a segregation of large

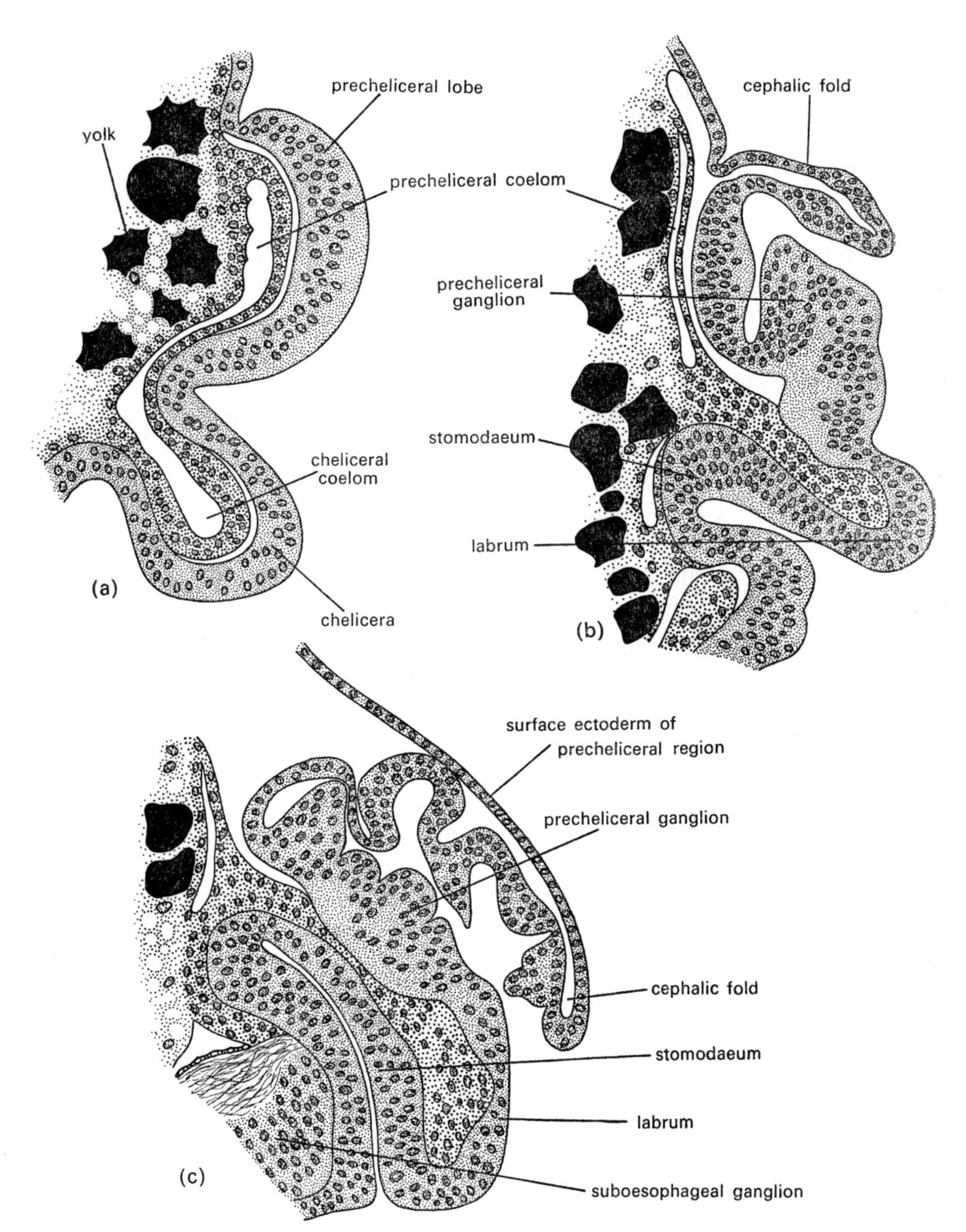

FIG. 161. Development of the precheliceral ganglia of *Heptathela*. (a) Parasagittal section through precheliceral lobe and chelicera before the onset of formation of semilunar grooves; (b) sagittal section through anterior end after onset of overgrowth of semilunar grooves; (c) sagittal section through anterior end during closure of semilunar grooves. (After Yoshikura, 1955.)

neuroblasts from an external layer of epithelial cells in the ventral ectoderm of the embryo. The neuroblasts then proliferate the ganglion cells. Neuroblasts are also reported to be developed in the ventral segmental ectoderm of the liphistiomorph spider *Heptathela* as a preliminary to ganglion formation (Yoshikura, 1955). Other accounts of the development of the ventral ganglia of spiders (Balfour, 1880; Morin, 1887, 1888; Schimkewitsch, 1887, 1898; Kishinouye, 1891b; Pappenheim, 1903; Montgomery, 1909a; Kautzsch, 1910; Rempel, 1957), all concerned with araneomorph species, make no mention of neuroblasts, though their presence in the embryo of *Dolomedes* is suggested by Legendre (1959). Associated with the process of inversion in arachnid embryos, the ganglionic thickenings pass up the sides of the yolk mass at the ventral edges of the segments before returning to the ventral midline as ventral closure proceeds, first in the prosoma, then in the opisthosoma (Rempel, 1957). The ganglia begin to become internally placed only during the ventral return movement, but eventually come together in the ventral midline in the usual way.

The same separation and subsequent reunion of ganglion rudiments occurs during inversion in uropygids (Kästner, 1950). Neither Kästner nor workers on other groups of arachnids (Opiliones, Schimkewitsch, 1898; Moritz, 1957; Acarina, Wagner, 1894a; Bonnet, 1907; Klumpp, 1954; Langenscheidt, 1958; Aeschlimann, 1958) mention neuroblasts in their accounts of the development of ventral ganglia. The only unusual mode of ganglion formation among arachnids is that recently described by Weygoldt (1964, 1965) in pseudoscorpions, in which the paired ventral ganglia of the prosomal segments develop as bilateral invaginations of the ventral segmental ectoderm (Fig. 160). In *Neobisium*, a distinct pair of invaginations in front of the pedipalpal pair gives rise to the cheliceral ganglia, but in *Pselaphochernes*, the cheliceral invaginations are merged with those of the brain.

The development of the precheliceral ganglia of chelicerates and their union with the cheliceral ganglia to form the supra-oesophageal ganglion is a more complex process than the development of the ventral ganglia. The pre-cheliceral ectoderm also gives rise to the eyes, which adds further complications to the interpretation of this part of the body. Fortunately, the arachnids have remained conservative in their mode of brain development, and recent studies have confirmed a remarkable similarity in the details of this process in groups as divergent as spiders, uropygids and opilionids. Probably, when more is known of the development of the xiphosuran brain, the same mode of development will emerge there. Kishinouye (1891a) showed that the precheliceral ganglia of *Tachypleus* develop through a pair of shallow, transient invaginations of pre-cheliceral ectoderm, the walls of which then subdivide into optic and other ganglia, but the details of the process need further study.

The spiders form a useful starting point for a consideration of arachnid brain development, since the several useful classical studies on this group (Salensky, 1871; Balfour, 1880; Kishinouye, 1891b; Pappensheim, 1903; Lambert, 1909; Montgomery, 1909a) have recently been supplemented by detailed new studies by Yoshikura (1955), Legendre (1959) and Pross (1966). The walls of the semilunar invaginations and lateral vesicles are the sources of the pre-cheliceral ganglia (Fig. 161). After the anterior margins of the semilunar grooves have spread back as folds over the posterior margins, the inner walls of the folds thicken, as does the tissue covered by the folds. The two thickened layers merge and give rise to the optic and cerebral ganglia. The walls of the lateral vesicles also thicken and contribute to these ganglia. According to Legendre (1959) and Pross (1966), a median, unpaired ectodermal thickening is also formed in front of the labral rudiment and sinks in to lie between the optic ganglia as an archicerebrum.

Formation and overgrowth of semilunar grooves, followed by thickening of the ectodermal layers which become internal during this process, is also the basic mode of development of the pre-cheliceral ganglia in the scorpions (Brauer, 1895; McClendon, 1905; Abd-el Wahab, 1952), uropygids (Kästner, 1949) and opilionids (Moritz, 1957). The pre-cheliceral ganglia of pseudoscorpions develop as a pair of simple invaginations of pre-cheliceral ectoderm (Weygoldt, 1964, 1965), while those of acarines are formed as ectodermal thickenings, without the intervention of invaginations (Aeschlimann, 1958; Langenscheidt, 1958; Anderson, 1972).

The median eyes of scorpions, spiders and uropygids arise as ectodermal invaginations associated with but independent of the semilunar invaginations (Brauer, 1895; McClendon, 1905; Abd-el Wahab, 1952; Homann, 1953; Yoshikura, 1955; Rempel, 1957). The optic ganglia usually develop from the deepest parts of the eye invaginations. The lateral eyes of arachnids, in contrast, develop as superficial ectodermal thickenings, without invagination.

As we shall see below, the interpretation of the composition of the pre-cheliceral part of the chelicerate head is still not satisfactorily resolved.

Respiratory organs

The development of the book-gills of Xiphosura from the limb buds of the third to seventh opisthosomal segments was described by Kishinouye (1891a) and Kingsley (1893). Transverse lamellae arise as outpouchings of the ectoderm of the posterior face of each limb bud, the first lamella being the most distal, the remainder progressively more proximal.

Among the arachnids, the development of the book-lungs of scorpions and spiders has been studied quite extensively, though the development of the book-lungs of uropygids and amblypygids and of the tracheae which many arachnids possess in addition to or in place of book-lungs is less well known. The book-lungs of scorpions and spiders all develop in the same way (see Metschnikoff, 1871; Kowalevsky and Shulgin, 1886; Laurie, 1890; Brauer, 1895; Pavlovsky, 1926, for scorpions; Locy, 1886; Morin, 1887; Bruce, 1887; Kishinouye, 1891b; Montgomery, 1909 a, b; Purcell, 1910; Ivanic, 1912, for spiders). Each book-lung originates as an ectodermal invagination behind an opisthosomal limb bud (Fig. 162). The invagination increases in size, forming a pulmonary sac, while the limb bud merges back into the general ectoderm of the segment. The anterior wall of the pulmonary sac, which is a direct continuation of the posterior wall of the limb bud, develops projecting lamellae the same distal-proximal sequence that is shown by the developing lamellae of xiphosuran book-gills. Thus there seems no reason to doubt that arachnid book-lungs are the invaginated homologues of xiphosuran book-gills. The most important consideration, therefore, is the segmental distribution of the book-lungs in different groups of arachnids and the relationship between book-lungs and arachnid tracheae.

We can see again, in this aspect of development, a fundamental distinction between scorpions and other arachnids. Scorpions develop four pairs of book-lungs, on the fourth to seventh opisthosomal segments. The limbs of the third opisthosomal segment develop as pectines, with no indication of transient book-lung formation. In spiders, in contrast, the limb buds of the third opisthosomal segment give way to a typical pair of book-lungs, a modification alternative to the pectines of scorpions. Liphistiomorph and mygalomorph spiders develop a second similar pair of book-lungs on the next opisthosomal segment, the fourth, but show no sign of vestiges of respiratory organs on the following segments. The limb buds of the fifth and sixth opisthosomal segments, as we have seen, persist and develop in spiders as spinnerets, while the seventh opisthosomal segment retains no limb buds. The same two pairs of book-lungs on the same two segments are developed in uropygids and

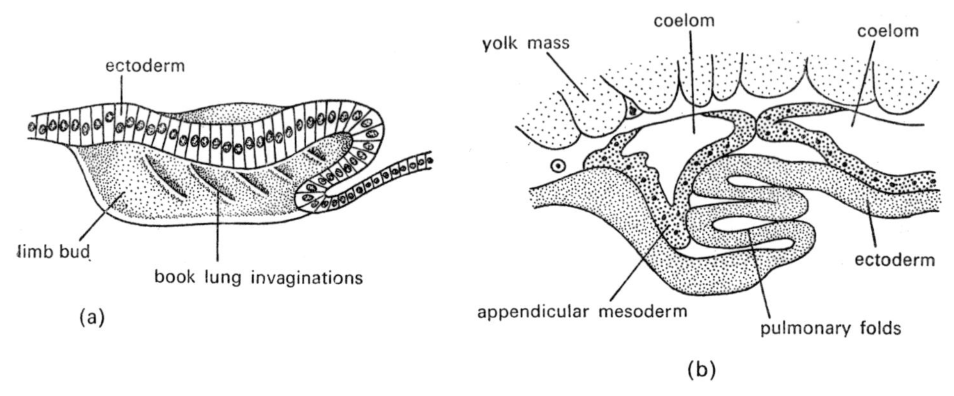

FIG. 162. Development of book-lungs in the spider *Attus*. (a) First abdominal appendage at the onset of formation of the folds of the book-lung, posterior view; (b) sagittal section of the same appendage at a later stage in infolding of the book-lung. (After Dawydoff, 1949.)

amblypygids. Furthermore, Weygoldt (1964) has shown that the two pairs of tracheal invaginations developed in pseudoscorpions arise on the same segments and have slight transient protrusions in front of their openings which are perhaps the vestiges of limb buds.

The second pair of respiratory organs in araneomorph spiders is also tracheal in structure. In some species, they are recognizably modified book-lungs, but in most spiders they diverge from a median ectodermal invagination on the ventral surface of the fourth opisthosomal segment, and may represent phylogenetically new structures. The interpretation of the tracheal invaginations in those species which have lost all traces of book-lungs and opisthosomal limb buds, namely the opilionids and acarines and the Solifugae, still remains in doubt and requires further study.

Extra-embryonic ectoderm

Except for secondarily yolkless embryos, all chelicerate embryos develop an attenuated, dorsal, extra-embryonic ectoderm from their dorsal blastoderm cells. Furthermore, associated with inversion of curvature of the germ band in many araneids, uropygids and amblypygids and with bilateral separation of the segmented germ band in acarines, an attenuated, ventral, extra-embryonic ectoderm is also developed. Dorsal closure of the body wall of all embryos, and ventral closure in many, is thus an essential step in development. How it takes place is not clear for any species, but the implication of all available descriptions is that the attenuated cells, both dorsal and ventral, are transformed into definitive cells of the surface epithelium and incorporated into the margins of the spreading embryonic ectoderm. There is little indication of dorsal organs in chelicerate embryos.

Hatching

A few observations are available on the hatching of chelicerate embryos. Sekiguchi (1960) described how the embryo of *Tachypleus* loses its chorion and is enclosed by a blastodermal cuticle throughout most of its development. Among arachnids, the spiders (Holm, 1954; Yoshikura, 1955, 1958) and uropygids (Yoshikura, 1961) develop egg teeth at the bases of the pedipalps of the advanced embryo. The egg teeth are used to break the egg membranes, facilitating escape of the embryo, and are shed after hatching. Yoshikura (1955) also suggested that the embryo of *Heptathela* secretes a hatching fluid from its transient, pedipalpal coxal glands.

Development of the Pre-oral Region of the Chelicerate Prosoma

The evidence bearing on the composition of the pre-oral region of the chelicerate prosoma has been set out in detail in the preceding sections of this chapter. Applying the usual embryological criteria of paired ganglia, mesodermal somites and limb buds as components of a segment, there is no doubt that the cheliceral segment can be identified as a segment by these criteria. Furthermore, the cheliceral segment always forms in a post-oral position and moves forward on either side of the mouth during later development to attain a pre-oral position. The chelicerae come to lie in front of the mouth and the cheliceral ganglia form components of the supraoesophageal ganglionic complex, but retain a post-oral commissure. The cheliceral somites also become pre-oral, though their contribution to the musculature and other components of the pre-oral region of the prosoma requires further study. Only the Xiphosura retain a transient vestige of segmental organs in the development of the cheliceral somites, and no species develops functional segmental organs from these somites.

The development and pre-oral migration of the cheliceral segment has been described in varying degrees of detail by many workers (Xiphosura, Kishinouye, 1891a; Kingsley, 1892, 1893; Iwanoff, 1933; Araneae, Kishinouye, 1891b, 1894; Wallstabe, 1908; Kautzsch, 1909; Montgomery, 1909a; Yoshikura, 1955, 1958; Rempel, 1957; Pross, 1966; Uropygi, Schimkewitsch, 1906; Kästner, 1948, 1949; Yoshikura, 1961; Solifugae, Heymons, 1904; Opiliones, Moritz, 1957; Acarina, Wagner, 1894a; Reuter, 1909; Aeschlimann, 1958; Anderson, 1972; Pseudoscorpiones, Weygoldt, 1964, 1965; Scorpiones, Brauer, 1894, 1895; Abd-el Wahab, 1952). The development of the pre-cheliceral region of the chelicerate prosoma, in contrast, is less clearly resolved and remains controversial. The penetration of mesoderm beneath the pre-cheliceral ectoderm and its development as pre-cheliceral somites needs further study. Some workers have taken the view that a pair of forward extensions of the cheliceral somites constitute the pre-cheliceral mesoderm (Xiphosura, Kishinouye, 1891a; Kingsley, 1892, 1893; Araneae, Wallstabe, 1908; Yoshikura, 1955; Uropygi, Schimkewitsch, 1906; Scorpiones, Brauer, 1895). Others, however, have reported that the pre-cheliceral mesoderm forms a separate pair of pre-cheliceral somites (Araneae, Kishinouye, 1891b, 1894; Kautzsch, 1909; Montgomery, 1909a; Scorpiones, Abd-el Wahab, 1952). In a recent study directed specifically towards this problem, Pross (1966) has now found that the lycosid spider *Pardosa* has two pairs of somites in front of the cheliceral pair and suggests that they constitute the somites of two precheliceral segments, prosocephalic (or labral) and deutocephalic, the cheliceral segment then being the tritocephalic segment. An essential corollary of this interpretation is that the paired ganglia of the two segments can also be identified. Pross equates the archicerebrum with prosocephalic ganglia and one of the smaller pairs of cerebral ganglia with deutocephalic ganglia, but further comparative investigation is required to test this interpretation. At the moment, all we can be certain of is that the components of the supraoesophageal ganglion other than the cheliceral ganglia arise from the pre-cheliceral ectoderm (Xiphosura, Kingsley, 1893; Araneae, Kishinouye, 1891b; Pappenheim, 1903; Montgomery, 1909a; Yoshikura, 1955; Legendre, 1959; Pross, 1966; Uropygi, Kästner, 1949; Opiliones, Moritz, 1957; Acarina, Aeschlimann, 1958; Pseudoscorpiones, Weygoldt, 1964, 1965; Scorpiones, Brauer, 1895; McClendon, 1905) and that somite mesoderm penetrates beneath this ectoderm. As I have suggested in previous chapters, the earlier emphasis on the comparative segmental composition of the arthropod head in assessment of the phylogenetic relationships between major arthropod groups can now be seen to have been misplaced. It is becoming increasingly evident that the cephalization of the anterior end has occurred independently in each of the three major groups of extant arthropods, from simple beginnings which are unknown and probably unknowable for each. In this case, the development of the pre-oral region of the chelicerate prosoma has a bearing on the interrelationships between chelicerate orders but not on the relationship between chelicerates and other arthropods.

Presumptive Areas of the Chelicerate Blastoderm

The detailed comparisons of chelicerate development given in the preceding sections of thi chapter make it clear that the patterns of development of the major groups of extant chelicerates are variations on a common theme. The most generalized expression of this theme occurs in the Xiphosura, which fail to satisfy the requirements for an ancestral developmental pattern for arachnids in only one respect, the absence of functional exit ducts of the coxal glands on the first ambulatory segment. Within the arachnids, the variations expressed in cleavage, gastrulation, the mode of development of the segmented germ band and the later development of such features as the opisthosomal limb buds and coxal glands, point to a fundamental divergence between the scorpions and the other arachnids. The same evidence also suggests that the other arachnids whose embryology is known fall into two groups, the Opiliones and Acarina being of one ancestry, the Araneae, Uropygi, Amblypygi, Solifugae and Pseudoscorpiones of another. These are tentative interpretations, which require further testing by new investigations, seeking new evidence. If correct, they must be commensurate with the summarized evidence provided by the construction of fate maps of the blastoderm. Unfortunately, it is at this level that the present knowledge of chelicerate embryology is least adequate. As we have seen, the origin and early development of the basic rudiments of the chelicerate embryo, especially during gastrulation, requires further study in all groups. The present confusion of interpretation is reminiscent of that which prevailed in crustacean embryology in the early part of this century. Modern studies on crustacean embryos have provided a much more precise understanding of the formation and

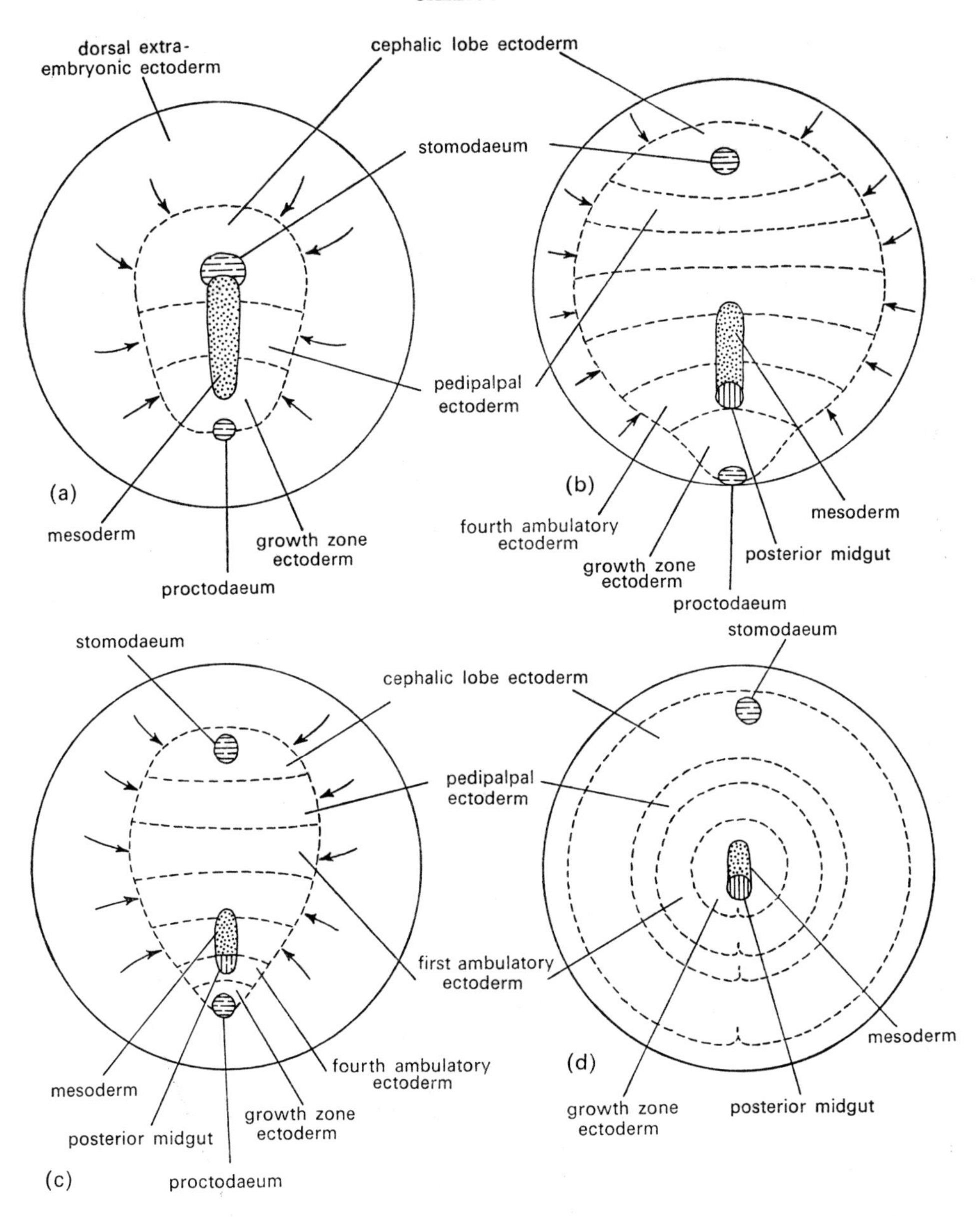

FIG. 163. Fate maps of the chelicerate blastoderm. (a) Xiphosura; (b) liphistiomorph spiders; (c) some mygalomorph and haplogyne spiders; (d) other spiders, including all entelegynes. (Based on the data summarized in the text.)

segregation of the basic rudiments and it is to be expected that work of an equivalent standard on the Chelicerata will do the same. At the present time we can do no more than indicate broad outlines and point the way to future expectations (Figs. 163 and 164). On the basis of present evidence, the following presumptive rudiments can be identified in chelicerate embryos:

1. Presumptive midgut, subdivided between presumptive anterior midgut and presumptive posterior midgut in most arachnids.
2. Presumptive stomodaeum and presumptive proctodaeum.
3. Presumptive mesoderm.
4. Presumptive embryonic ectoderm and extra-embryonic ectoderm.

The question will now be examined, whether these components conform to the idea of a basic expression in the Xiphosura and a number of derived expressions among the arachnids.

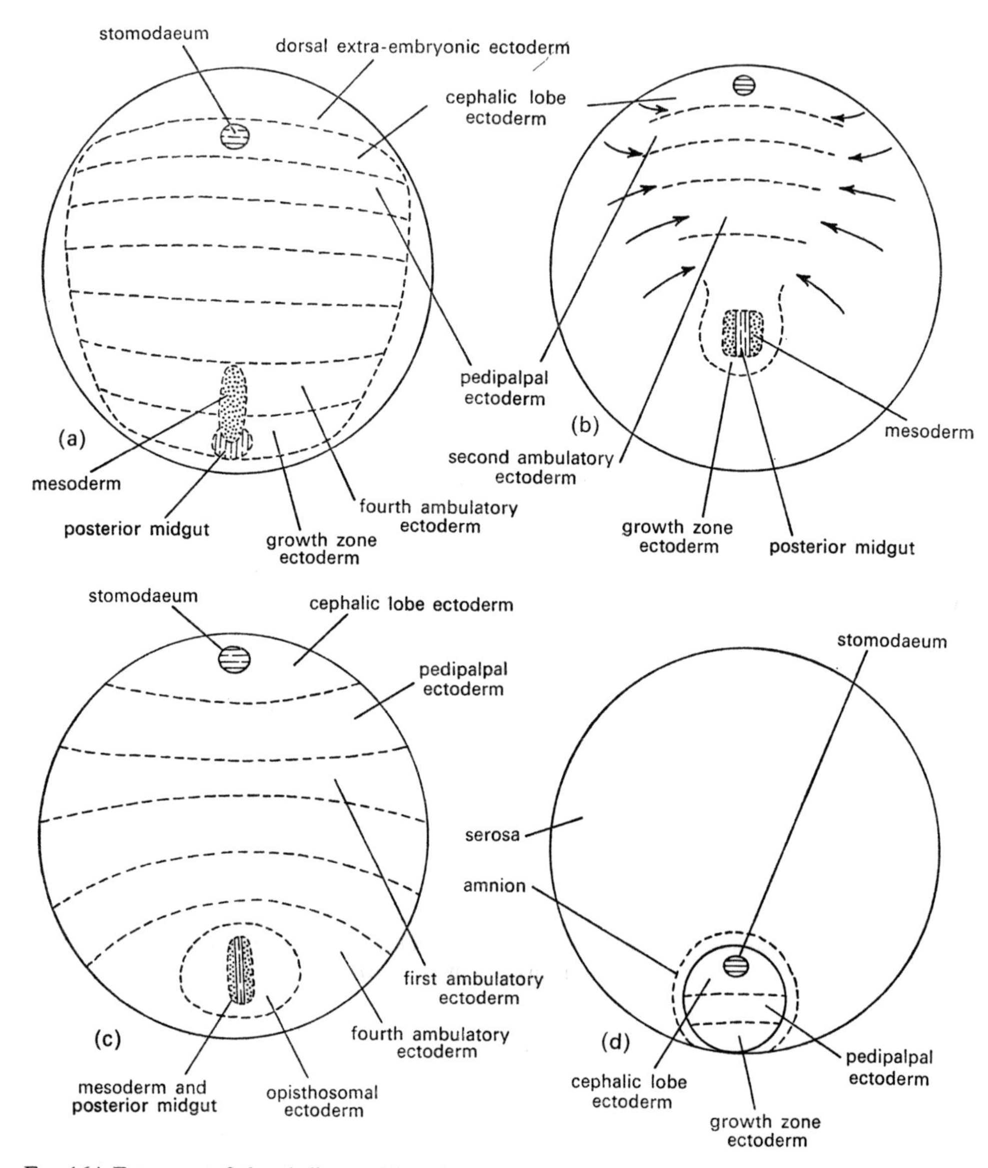

FIG. 164. Fate maps of the chelicerate blastoderm (continued). (a) Opilionids; (b) acarines; (c) pseudoscorpions; (d) scorpions. (Based on the data summarized in the text.)

Presumptive midgut

The xiphosuran presumptive midgut, which comprises the mass of large yolky cells enclosed by the blastoderm during total cleavage, has no place in the blastoderm and no gastrulation movement. On the reasonable presumption that total cleavage in the Xiphosura is primitive, this is the neatest solution to the yolk *versus* midgut problem seen in any extant arthropod (see Chapter 10 for further discussion). Among the arachnids, the scorpions retain a unitary yolk mass during discoidal cleavage and later surround it with vitellophage and midgut epithelia derived from the embryonic primordium. Other arachnids have a unitary yolk mass within the blastoderm, a subsequent invasion of vitellophages into the yolk mass and the additional development of a posterior midgut component behind the yolk mass. We are thus faced with a variety of specializations whose connection with the xiphosuran condition is far from clear.

Present information on the scorpions is insufficient to reveal the location of the presumptive midgut cells. All we know is that they form a part of the embryonic primordium and that the yolk mass remains acellular while the midgut encloses and digests it. This specialized mode of development could be functionally derived from that of the Xiphosura, but the actual path of derivation is obscure. For the other arachnids, the information is better but the interpretation is still problematical. We can be certain only of two things, a greater level of specialization than in the Xiphosura and a different mode from that of scorpions. The liphistiomorph spiders provide a useful starting point. The presumptive midgut is in two parts. The major part is presumptive anterior midgut, which consists of cells scattered throughout the embryonic primordium, with some concentration towards the posterior end. These cells migrate into the yolk mass, acting as vitellophages, and subsequently develop as the anterior midgut epithelium. The development of this part of the liphistiomorph midgut is thus only a minor functional variation on the xiphosuran mode. In addition, however, a group of presumptive posterior midgut cells is situated in the ventral midline of the embryonic primordium, just in front of the presumptive opisthosomal ectoderm (see below). These cells become internal and develop directly as the tubular posterior part of the midgut, without intervening vitellophage specialization. The posterior midgut component has no equivalent in the Xiphosura, but is a typical adaptation to the problem of combining dense yolk with midgut formation. We have seen analogous modifications in the annelids, onychophorans and crustaceans in earlier chapters.

Other araneid embryos exhibit only minor modifications of the liphistiomorph presumptive midgut. The posterior midgut has the same origin, but undergoes a preliminary posterior migration as the posterior cumulus, immediately after gastrulation. The origin of the vitellophages of the anterior midgut is partly from cells scattered throughout the embryonic primordium, but partly also from cells intermingled with the presumptive mesoderm (see below) in the ventral midline in front of the presumptive posterior midgut.

No useful comment can be made at the present time on the presumptive midgut of uropygids, amblypygids or solifuges, but something is known of other groups. The presumptive anterior midgut of the opilionids is represented partly by cleavage energids left within the yolk and partly by midventral cells associated with the presumptive mesoderm. Both components act temporarily as vitellophages before developing as anterior midgut epithelium. The Acarina have a similar dual source of vitellophages, but in this case the vitellophage specialization is performed to the exclusion of subsequent midgut formation. The acarine midgut develops from the presumptive posterior midgut alone (see below). Alternatively, in the pseudoscorpions, the vitellophages which subsequently develop as the anterior midgut are exclusively persistent cleavage energids and the presumptive anterior midgut is not represented in the blastoderm. In all three groups, the anterior midgut develops from the yolk mass as it does in spiders. The presumptive posterior midgut also appears to occupy much the same location in the opilionid, acarine and pseudoscorpion blastoderm as in the spiders, midventrally behind the presumptive mesoderm. The forward spread of the posterior midgut and resorption of the vitellophages in the later development of the acarine gut is clearly a specialization.

We can therefore arrive at the following conclusions. In respect of presumptive midgut, the arachnids are more specialized than the Xiphosura, although they could be functionally modified from the latter. They fall into two groups, scorpions and others. Beyond this point, the subject requires further investigation.

Presumptive stomodaeum and proctodaeum

In contrast, the origins of the stomodaeum and proctodaeum vary little throughout the Chelicerata. The presumptive stomodaeum lies in the ventral midline of the presumptive head lobe ectoderm in all species. In the Xiphosura, the presumptive stomodaeum is immediately in front of the anterior end of the presumptive mesoderm and invaginates as the anterior end of the wall of the gastral groove. In most arachnids, the presumptive mesoderm and gastral groove occupy a more posterior location along the ventral midline (see below). In scorpions, the gastral groove appears to have been eliminated. The presumptive stomodaeum of arachnids therefore invaginates into the interior independently of the gastral groove.

The presumptive proctodaeum comprises a posterior area of cells in the ventral midline behind the ectodermal growth zone and makes a late, independent invagination into the interior in all chelicerates.

Presumptive mesoderm

The chelicerate presumptive mesoderm is of interest in two ways. The xiphosuran condition is unique among arthropods and the arachnid presumptive mesoderm, in so far as it is understood, is a functional derivative of the xiphosuran condition. It must be emphasized, however, that we cannot include the scorpions in this generalization, since the origin of their mesoderm is not adequately known.

In the Xiphosura, the presumptive mesoderm occupies the ventral midline of the embryonic primordium, extending from the presumptive stomodaeum to the presumptive growth zone ectoderm (see below) as a narrow band of cells. At gastrulation, these cells invaginate, forming a gastral groove with the invaginating stomodaeum at its anterior end. The unique feature of the xiphosuran presumptive mesoderm is that it remains superficial in the growth zone and continues the process of gastral groove formation progressively along the growing germ band.

The liphistiomorph spiders differ from this condition in two ways. The presumptive mesoderm is midventral, but is more posteriorly placed relative to the stomodaeum and ectoderm and lies between the ectodermal halves of the presumptive third and fourth ambulatory segments, just in front of the presumptive posterior midgut. Furthermore, the mesoderm invaginates rapidly, with ensuing closure of the gastral groove and continues its posterior proliferation after it has become internal. In other spiders, opilionids and pseudoscorpions the presumptive mesoderm is similarly midventral, posterior in location and rapidly invaginated. Here, however, the presumptive mesoderm cells are intermingled with presumptive anterior midgut cells. Finally, in the Acarina, the short area of presumptive mesoderm is bilaterally separated on either side of a midventral area of presumptive midgut, a distinctly specialized arrangement. There is no doubt that the presumptive mesoderm of the arachnid groups mentioned above can be interpreted as a specialization of the presumptive mesoderm of the Xiphosura, but the present level of knowledge provides no useful indications of interrelations within the Arachnida.

Presumptive ectoderm

Finally, we come to the presumptive ectoderm, somewhat indistinctly divisible between the presumptive embryonic ectoderm and extra-embryonic ectoderm. These components together make up the great majority of the blastoderm of all chelicerates. The presumptive ectodermal areas of the Xiphosura first occupy only a relatively small part of the ventral surface of the blastoderm, but are gradually augmented by the incorporation of many of the lateral and anterior blastoderm cells, with simultaneous attenuation and spread of the dorsal part of the blastoderm as extra-embryonic ectoderm. Present knowledge, although unsatisfactory in some respects, indicates that the xiphosuran presumptive ectoderm comprises a presumptive head lobe area around the presumptive stomodaeum and anterior end of the mesoderm, a bilateral area of presumptive pedipalpal ectoderm on either side of the mesoderm and an area of presumptive growth zone ectoderm around the posterior end of the mesoderm. Among the arachnids, two modifications of this pattern are observed. The scorpions retain the same three components of presumptive ectoderm as the Xiphosura, but with sharp boundaries at the margin of the embryonic primordium. The remainder of the blastoderm is extra-embryonic and develops as the amnion and serosa. The other arachnids, in contrast, develop a longer series of presumptive embryonic ectodermal components directly in the blastoderm, as presumptive head lobe, pedipalpal and ambulatory ectoderm of the prosoma, and a growth zone which is presumptive opisthosomal ectoderm. As indicated above, the presumptive stomodaeum remains in the midline of the presumptive head lobe ectoderm, but the presumptive mesoderm is confined to the midline of the posterior ambulatory segments and growth zone. This arrangement of presumptive ectoderm is clear in the liphistiomorph spiders, with some secondary specialization in other spiders, and is also observed in the uropygids, opilionids, acarines and pseudoscorpions. Further incorporation of the lateral and anterior blastoderm cells into the presumptive ectoderm, accompanied by attenuation and spread of dorsal blastoderm as extra-embryonic ectoderm, also characterizes these species.

In general, then, a comparison of chelicerate fate maps favours the idea that xiphosuran embryonic development is more generalized than that of any arachnid and supports the concept of a fundamental divergence between the scorpions and the other arachnid orders. More thorough investigation may reveal unexpected specializations in the Xiphosura and force a reappraisal of the phylogenetic placement of the arachnids, but we can see already that the chelicerates have a distinctive theme of embryonic development, lacking anything in common with either the Onychophora or the Crustacea and being devoid of any reminiscence of spiral cleavage development. In the following chapter, I shall appraise the significance of these conclusions for arthropod phylogeny.

References

ABD-EL WAHAB, A. (1952) Some notes on the segmentation of *Buthus*. *Proc. Egypt. Acad. Sci.* **7**, 75–91.

ABD-EL WAHAB, A. (1954) The formation of the germ layers and their embryonic coverings in the scorpion *Buthus quinquestriatus* (H.E.). *Proc. Egypt. Acad. Sci.* **10**, 110–18.

ABD-EL WAHAB, A. (1956) Some observations on the embryology of scorpions. *Proc. R. phys. Soc. Edinb.* **25**, 7–9.

AESCHLIMANN, A. (1958) Développement embryonnaire d'*Ornithodorus moubata* (Murray) et transmission ovarienne de *Borrelia duttoni*. *Acta trop.* **15**, 15–64.

AESCHLIMANN, A. (1961) Complément a l'étude de l'embryologie d'*Ornithodorus moubata* (Murray). *Acta trop.* **18**, 58–60.

ANDERSON, D. T. (1972) The embryology of *Hyalomma dromedarii* (Acarina) (in preparation).

BALBIANI, M. (1873) Mémoire sur le développement des aranéides. *Ann. Sc. nat., Ser.* **5**, 18.

BALFOUR, F. M. (1880) Notes on the development of the Araneina. *Q. Jl. microsc. Sci.* **20**, 167–89.

BARROIS, J. (1877) Recherches sur le développment des Araignées. *J. Anat. Phys., Paris* **14**, 527–43.

BONNET, A. (1907) Recherches sur l'anatomie comparée et le développement des Ixodidés. *Ann. Univ. Lyon*, N.S. **20**, 1–185.

BRAUER, A. (1894) Beiträge zur Kenntnis der Entwicklungsgeschichte des Skorpions. I. *Z. wiss. Zool.* **57**, 402–32.

BRAUER, A. (1895) Beiträge zur Kenntnis der Entwicklungsgeschichte des Skorpions. II. *Z. wiss. Zool.* **59**, 351–433.

BRUCE, A. T. (1887) *Observations on the Embryology of Insects and Arachnids*, Johns Hopkins University, Baltimore.

BRUCKER, E. A. (1900) Monographie de *Pediculoides ventricosus* Newport et theorie des pièces buccales des Acariens. *Bull. Sci. Fr. Belg.* **35**, 335–442.

BUXTON, B. H. (1917) Notes on the anatomy of arachnids. The coxal glands of the arachnids. The ganglia of the arachnids. *J. Morph.* **29**, 1–25.

CAMIN, J. H. (1953) Observations on the life history and sensory behaviour of the snake mite *Ophionyssus natricis* (Gervais) (Acarina, Macronyssidae). *Spec. Publ. Chicago Acad. Sci.* **10**, 1–75.

CLAPARÈDE, E. (1862) Recherches sur l'évolution des Araignées. *Natuurk Verh. l. Utrecht Genoot. Kunst. Wiss.*

CLAPARÈDE, E. (1868) Studien an Acaridien. *Z. wiss. Zool.* **18**, 445–546.

COOPER, K. W. (1940) The nuclear cytology of the grass mite, *Pediculopsis graminum* (Reut.), with special reference to karyokinesis. *Chromosoma* **1**, 51–103.

CROME, W. (1963) Embryonalentwicklung ohne "Umrollung" (= Reversion) bei Vogelspinnen (Araneae, Orthognatha). *Deut. Entomol. Z.* **10**, 83–95.

CROME, W. (1964) Eikokon, Embryonalstadien und frühe Jugendformen von *Conothele arboricola* Pocock (Araneae: Ctenizidae). *Zool. Jb. Syst.* **91**, 411–50.

DAWYDOFF, C. (1949) Développement embryonnaire des arachnides. In GRASSÉ, P. P. (ed.), *Traité de Zoologie*, **6**, Masson, Paris.

EDWARDS, A. R. (1958) Cleavage in *Cheyletus eruditis* (Acari). *Nature, Lond.* **181**, 1409–10.

EHN, A. (1962) The development of spider embryos after centrifugation. *Zool. Bidr. Uppsala*, **35**, 339–68.

EHN, A. (1963) The embryonic development of the spider *Torania variata* Poc. (Sparassidae). *Zool. Bidr. Uppsala* **36**, 37–48.

FAUSSEK, V. (1891) Zur Embryologie von *Phalangium*. *Zool. Anz.* **14**, 3–5.

FULINSKI, B. (1912) Ein Beitrag zur Keimblätterbildung der Araneinen. *Bull. int. Acad. Cracovie* **7B**, 769–90.

GOODRICH, E. S. (1946) The study of nephridia and genital ducts since 1895 (continued). *Q. Jl. microsc. Sci.* **86**, 303–92.

GOUGH, L. H. (1902) The development of *Admetus pumilio* Koch; a contribution to the embryology of the pedipalps. *Q. Jl. microsc. Sci.* **45,** 595–630.

HAFIZ, H. A. (1935) The embryological development of *Cheyletus eruditus* (a mite). *Proc. R. Soc.* B, **117,** 174–201.

HAMBURGER, C. (1910) Zur Anatomie und Entwicklungsgeschichte der *Argyroneta aquatica* Cl. *Z. wiss. Zool.* **96,** 1–31.

HEINEMANN, R. L. and HUGHES, R. D. (1970) Reproduction, reproductive organs and meiosis in the bisexual, non-parthenogenetic mite *Caloglyphus mycophagus*, with reference to oocyte degeneration in virgins (Sarcoptiformes; Acaridae). *J. Morph.* **130,** 93–102.

HENKING, H. (1882) Beiträge zur Anatomie, Entwicklungsgeschichte und Biologie von *Trombidium fuliginosum* Herin. *Z. wiss. Zool.* **37,** 551–663.

HEYMONS, R. (1904) Über die Entwicklung und Morphologie der Solifugen. *C.R. Congr. Int. Zool. Bern* **1904,** 429–36.

HOLM, A. (1941) Studien über die Entwicklung und Entwicklungsbiologie der Spinnen. *Zool. Bidr. Uppsala* **19,** 1–214.

HOLM, A. (1947) On the development of *Opilio parietinus* Deg. *Zool. Bidr. Uppsala* **25,** 409–22.

HOLM, A. (1952) Experimentelle Untersuchungen über die Entwicklung und Entwicklungsphysiologie des Spinnenembryos. *Zool. Bidr. Uppsala* **59,** 293–424.

HOLM, A. (1954) Notes on the development of an orthognath spider, *Ischnothele karschi* Bos. and Lenz. *Zool. Bidr. Uppsala* **30,** 199–222.

HOMANN, H. (1953) Die Entwicklung der Nebenaugen bei den Araneen. I. *Biol. Zbl.* **72,** 373–85.

HUGHES, T. E. (1950) The embryonic development of the mite *Tyroglyphus farinae* Linnaeus 1758. *Proc. zool. Soc. Lond.* **119,** 873–86.

IVANIC, M. (1912) Die Lungentwicklung bei der dipneumonen Araneinen. *Zool. Anz.* **40,** 283–9.

IWANOFF, P. P. (1933) Die embryonale Entwicklung von *Limulus molluccanus*. *Zool. Jb. Anat. Ont.* **56,** 163–348.

JANECK, R. (1909) Die Entwicklung der Blättertracheen und der Tracheen bei den Spinnen. *Jena. Z. Naturwiss.* **44,** 587–646.

JUBERTHIE, C. (1961) Les phases du développement embryonnaire et leurs relations avec la temperature et l'humidité chez un opilion Palpatores. *C.R. Acad. Sci., Paris* **252,** 2142–4.

JUBERTHIE, C. (1964) Recherches sur la biologie des Opilions. *Ann. Speleol.* **19,** 1–237.

KÄSTNER, A. (1948) Zur Entwicklungsgeschichte von *Thelyphonus caudatus* (Pedipalpi). 1. Teil. Die Ausbildung der Körperform. *Zool. Jb. Anat. Ont.* **69,** 493–506.

KÄSTNER, A. (1949) Zur Entwicklungsgeschichte von *Thelyphonus caudatus* L. (Pedipalpi). 2. Teil. Die Entwicklung der Mundwerkzeuge, Beinhuften und Sterna. *Zool. Jb. Anat. Ont.* **70,** 169–97.

KÄSTNER, A. (1950) Zur Entwicklungsgeschichte von *Thelyphonus caudatus* L. (Pedipalpi). 3. Teil. Die Entwicklung des Zentralnervensystems. *Zool. Jb. Anat. Ont.* **71,** 1–55.

KAUTZSCH, G. (1909) Ueber die Entwicklung von *Agelena labyrinthica*. *Zool. Jb. Anat. Ont.* **28,** 477–538.

KAUTZSCH, G. (1910) Über die Entwicklung von *Agelena labyrinthica* Clerck. Tl. 2. *Zool. Jb. Anat. Ont.* **30,** 535–602.

KINGSLEY, J. S. (1892) The embryology of *Limulus*. *J. Morph.* **7,** 36–66.

KINGSLEY, J. S. (1893) The embryology of *Limulus*. *J. Morph.* **8,** 195–268.

KISHINOUYE, K. (1891a) On the development of *Limulus longispina*. *J. Coll. Sci. Imp. Univ. Japan* **5,** 53–100.

KISHINOUYE, K. (1891b) On the development of the Araneina. *J. Coll. Sci. Imp. Univ. Japan* **4,** 55–88.

KISHINOUYE, K. (1894) Note on the coelomic cavity of the spider. *J. Coll. Sci. Imp. Univ. Japan* **6,** 287–96.

KLUMPP, W. (1954) Embryologie und histologie der Bienenmilbe *Acarapis woodi* Rennie 1921. *Z. Parasitenk.* **16,** 407–42.

KOWALEVSKY, A. and SCHULGIN, M. (1886) Zur Entwicklungsgeschichte des Skorpions *(Androctonus ornatus)*. *Biol. Centralb.* **6,** 525–32.

LAMBERT, A. E. (1909) History of the procephalic lobes of *Epeira cinerea*. *J. Morph.* **20,** 413–59.

LANGENSCHEIDT, M. (1958) Embryologische, morphologische und histologische Untersuchungen an *Knemicoptes mutans* (Robin et Lanquetin). *Z. Parasitenk.* **18,** 349–85.

LAURIE, M. (1890) The embryology of the scorpion *(Euscorpius italicus)*. *Q. Jl. microsc. Sci.* **31,** 105–41.

LAURIE, M. (1891) Some points in the development of *Scorpio fulvipes*. *Q. Jl. microsc. Sci.* **32,** 105–41.

LEGENDRE, R. (1958) Contribution a l'étude du développement embryonnaire des Araignées. *Bull. Soc. Zool. Fr.* **83,** 60–75.

LEGENDRE, R. (1959) Contribution a l'étude du systems nerveux des Araneides. *Ann. Sci. Nat. Zool.* **1,** 339–473.

Locy, W. A. (1886) Observations on the development of *Agelena naevia*. *Bull. Mus. comp. Zool. Harvard* **12,** 63–103.

McClendon, J. F. (1905) On the anatomy and embryology of the nervous system of the scorpion. *Biol. Bull.* **8,** 38–55.

Mathew, A. P. (1956) Embryology of *Heterometrus scaber* (Thorell) Arachnida. *Zool. Mem. Univ. Travancore* **1,** 1–96.

Metschnikoff, E. (1871) Entwicklungsgeschichte des *Chelifer*. *Z. wiss. Zool.* **21,** 513–27.

Montgomery, Th. H. (1909a) The development of *Theridium*, an araneid, up to the stage of reversion. *J. Morph.* **20,** 297–352.

Montgomery, Th. H. (1909b) On the spinnarets, cribellum, colulus, tracheae and lung-books of araneids. *Proc. Acad. Nat. Sci. Phil.* **61,** 299–320.

Morin, J. (1887) Zur Entwicklungsgeschichte der Spinnen. *Biol. Centralbl.* **6,** 658–63.

Morin, J. (1888) Studien über die Entwicklungsgeschichte der Spinnen. *Abh. Russ. Ges. Naturf. Odessa* **13,** 93–204.

Moritz, M. (1957) Zur Embryonalentwicklung der Phalangiiden (Opiliones, Palpatores) unter besonderer Berücksichtigung der ausseren Morphologie, der Bildung der Mitteldarmes und der Genitalanlage. *Zool. Jb. Anat. Ont.* **76,** 331–70.

Moritz, M. (1959) Zur Embryonalentwicklung der Phalangiiden (Opiliones, Palpatores) II. Die Anlage und Entwicklung der Coxaldrüse bei *Phalangium opilio* L. *Zool. Jb. Anat. Ont.* **77,** 229–40.

Münchberg, P. (1935) Untersuchungen über den Laich und dessen Entwicklungsdauer bei der Hydracarinenunterfamilie der *Arrhenurinae* Wolcott. *SitzBer. Ges. naturf. Fr. Berlin* **1936,** pp. 213–32.

Pappenheim, P. (1903) Beitrag zur Kenntnis des Entwicklungsgeschichte von *Dolomedes fimbriatus* Clerk, mit besonderer Berücksichtigung der Bildung des Gehirn und den Augen. *Z. wiss. Zool.* **74,** 109–54.

Patten, W. and Hazen, P. (1900) The development of the coxal gland, branchial cartilages and genital ducts of *Limulus polyphemus*. *J. Morph.* **16,** 459–502.

Pavlovsky, E. N. (1926) Studies on the organization and development of scorpions. 5. The lungs. *Q. Jl. microsc. Sci.* **70,** 221–61.

Pereyaslawzewa, S. (1901) Développement embryonnaire des Phrynes. *Ann. Sc. Nat* (8) **13,** 117–304.

Perron, R. (1954) Untersuchungen über Bau, Entwicklung und Physiologie der Milbe *Histiostoma laboratorium* Hughes. *Acta. Zool., Stockh.* **35,** 71–176.

Pflugfelder, O. (1930) Zur Embryologie des Skorpions *Homurus australasiae* (F.). *Z. wiss. Zool.* **137,** 1–29.

Pokrowsky, S. (1899) Beobachtungen über das Eierablegen bei *Pholcus*. *Zool. Anz.* **22,** 270.

Poliansky, I. (1903) Zur Embryologie des Skorpions. *Zool. Anz.* **27,** 49–58.

Police, G. (1933) Sullo sviluppo pari des rostio nell embrione degli Araneidi. *Boll. Soc. Nat., Napoli* **44,** 3–49.

Prasse, J. (1968) Untersuchungen über Oogenese, Befruchtung, Eifurchung und Spermatogenese bei *Caloglyphus berlesei* Michael 1903 und *C. michaeli* Oudemans 1924 (Acari, Acaridae). *Biol. Zbl.* **87,** 757–75.

Pross, A. (1966) Untersuchungen zur Entwicklungsgeschichte der Araneae (*Pardosa hortensis* (Thorell)) und besonderer Berücksichtigung des vorderen Prosoma-abschnittes. *Z. Morph. Ökol. Tiere* **58,** 38–108.

Purcell, W. F. (1895) Note on the development of the lung, entapophyses, tracheae and genital duct in spiders. *Zool. Anz.* **18.**

Purcell, W. F. (1910) Development and origin of the respiratory organs in Araneae. *Q. Jl. microsc. Sci.* **54,** 1–110.

Rempel, J. G. (1957) On the embryology of the black widow spider, *Latrodectus mactans*. *Can. J. Zool.* **35,** 35–74.

Reuter, E. (1909) Zur Morphologie und Ontogenie der Acariden mit besonderer Berücksichtigung von *Pediculopsis graminum* (Reut.). *Acta Soc. Sci. Fennica* **36** (4), 1–285.

Sabatier, A. (1881) Formation du blastoderme chez les Aranéides. *C.R. Acad. Sci., Paris* **92,** 200–2.

Salensky, W. (1871) Die Entwicklungsgeschichte der Araneinen. *Mem. Kieff Soc. Nat.* **2.**

Schimkewitsch, W. (1887) Étude sur le développement des Araignées. *Arch. Biol.* **6,** 515–84.

Schimkewitsch, W. (1898) Entwicklung des Darmcanals bei Arachniden. *Trav. Soc. Nat. St. Peterbourg* **29,** 16–18.

Schimkewitsch, W. (1906) Über die Entwicklung von *Thelyphonus caudatus* (L), vergleichen mit derjenigen einiger anderer Arachniden. *Z. wiss. Zool.* **81,** 1–95.

Schimkewitsch, L. and W. (1911) Ein Beitrag zur Entwicklungsgeschichte der Tetrapneumones. I, II, III. *Bull. Acad. Imp. Sci. St. Petersb.*, ser. 6, 1911, pp. 637–54, 685–705, 775–90.

SEITZ, K.-A. (1966) Normale Entwicklung des Arachniden-Embryos *Cupiennius salei* Keyserling und seine Regulations befähigung nach Röntgenbestrahlungen. *Zool. Jb. Anat. Ont.* **83**, 327–447.

SEKIGUCHI, K. (1960) Embryonic development of the horse-shoe crab studied by vital staining. *Bull. Mar. Sta. Asamushi, Tohoku Univ.* **10** (2), 161–4.

SIMMONS, O. L. (1894) Development of the lungs of spiders. *Ann. Mag. nat. Hist.* (6), **14**.

SOKOLOV, I. I. (1952) Observations on the embryonic development of the granary mites. I. Construction of the egg and segmentation. *Trudy Leningrad Soc. Nat.* **71**, 245–60.

THOR, S. (1925) Ueber die Phylogenie und Systematik der Acarina, mit Beiträgen zur ersten Entwicklungsgeschichte einzehner Gruppen. *Nytt. Mag. Oslo* **62**, 123–6.

VACHON, M. (1950) Remarques préliminaire sur l'alimentation, les organs cheliceriens, le biberon et la tetine de l'embryon du scorpion: *Ischnurus ochropus* C. L. Koch (Scorpionidae). *Arch. Zool. éxp. gen.* **86**, 137–56.

VACHON, M. (1953) The biology of scorpions. *Endeavour* **12**, 80–89.

WAGNER, J. (1894a) Die Embryonalentwicklung von *Ixodes calcaratus* Bir. *Trav. Soc. Nat. St. Petersb.* **24** (2), **5**, 214–46.

WAGNER, J. (1894b) On the embryology of mites: segmentation of the ovum, origin of the germinal layers and development of appendages in *Ixodes*. *Ann. Mag. Nat. Hist.*, Ser. 6, **11**, 220–4.

WALLSTABE, P. (1908) Beitrag zur Kenntnis der Entwicklungsgeschichte der Araneinen. Die Entwicklung der äusseren Form und Segmentierung. *Zool. Jb. Anat. Ont.* **26**, 683–712.

WEYGOLDT, P. (1964) Vergleichend-embryologische Untersuchungen an Pseudoscorpionen (Chelonethi). *Z. Morph. Ökol. Tiere* **54**, 1–106.

WEYGOLDT, P. (1965) Vergleichend-embryologische Untersuchungen an Pseudoscorpionen. III. Die Entwicklung von *Neobisium muscorum* Leach (Neobisiinea, Neobisiidae). Mit dem Versuch einer Deutung der Evolution des embryonalen Pumporgans. *Z. Morph. Ökol. Tiere* **55**, 321–82.

WEYGOLDT, P. (1968) Vergleichend-embryologische Untersuchungen an Pseudoscorpionen. IV. Die Entwicklung von *Chthonius tetrachelatus* Preyssl., *Chthonius ischnocheles* Hermann (Chthoniinea, Chthoniidae) und *Verrucaditha spinosa* Banks (Chthoniinea, Tridenchthoniidae). *Z. Morph. Ökol. Tiere* **63**, 111–54.

WEYGOLDT, P. (1971) Vergleichend-embryologische Untersuchungen an Pseudoscorpionen: V. Das Embryonal-stadium mit seinem Pumporgan bei verschiedenen Arten und sein Wert als taxonomisches Merkmal. *Zool. Syst. Evolutionsforsch.* **9**, 3–29.

WINKLER, D. (1957) Die Entwicklung der äusseren Körpergestalt bei den Phalangiidae (Opiliones). *Mitt. Zool. Mus. Berlin* **33**, 355–89.

YOSHIKURA, M. (1954) Embryological studies on the liphistiid spider *Heptathela*. *Kumamoto J. Sci.* **3B**, 41–50.

YOSHIKURA, M. (1955) Embryological studies on the liphistiid spider *Heptathela kumurai*. II. *Kumamoto J. Sci.* **B2**, 1–86.

YOSHIKURA, M. (1958) On the development of the purse-web spider, *Atypus karschi* Donitz. *Kumamoto J. Sci.* B (2), *Biol.*, **3**, 73–85.

YOSHIKURA, M. (1961) The development of the whip-scorpion *Typopeltis stimpsonii* Wood. *Acta arachnol.* **17**, 19–24.

ZUKOWSKI, K. Z. (1964) Investigations into the embryonic development of *Pergamasus brevicornis* Berl. *Zool. Pol.* **14**, 247–68.

CHAPTER 10

A NEW SYNTHESIS?

ALTHOUGH some of the arguments presented in the preceding chapters are, perhaps, tenuous and some of the "facts" will in due course prove to be wrong, I venture to suggest that a sufficient basis is now available to provide an embryological test of the proposition set out at the end of Chapter 1. The conceptual approach employed in this book has made it possible to elucidate the basic mode of embryonic development of each major group of annelids and arthropods. The polychaetes have their own distinctive theme of development (Chapter 2). The same theme appears again in a modified form in the clitellate annelids (Chapter 3). A different theme can be discerned in the development of the Onychophora (Chapter 4). The modern myriapods exhibit two modes of development (Chapter 5), indicative of an early divergence between the ancestors of the Chilopoda and the ancestors of the Symphyla, Pauropoda and Diplopoda. Two patterns of development also occur in the Hexapoda (Chapters 6 and 7), one being shared by the Diplura and Collembola, the other by the Thysanura and Pterygota. The specializations of the latter, however, are obviously modifications of the more generalized dipluran mode, which thus constitutes a general theme for all hexapods. A quite different theme of development can be identified beneath the multitudinous developmental diversity of the Crustacea (Chapter 8) and another, in so far as present evidence permits, in the Chelicerata (Chapter 9). It therefore becomes possible to make comparisons between the various modes of development of annelid and arthropod embryos, giving consideration to the functional feasibility of a phylogenetic common ancestry between any or all of them. The important questions that must be examined are:

(a) whether a common basis underlines the embryonic development of the Onychophora, Myriapoda and Hexapoda, as their grouping into a phylogenetic assemblage would imply;
(b) how this common basis, if it exists, compares with the themes of development detectable in the Crustacea and the Chelicerata;
(c) how each of these modes of arthropodan development compares with the embryonic development of annelids.

In anticipation of the discussion which follows, it must be pointed out that certain parts of the present chapter reiterate arguments already put forward on previous occasions. In relation to (a) above, I have already argued briefly (Anderson, 1972) in favour of a common ground for onychophoran–myriapod–hexapod development, based on a primitive mode now retained only in the yolky Onychophora. The discussion will be expanded here to embrace the additional evidence gathered together in Chapters 4–7. Similarly, in relation to (b) and (c) above, I have suggested (Anderson, 1969) that the mode of development of crustacean embryos is fundamentally different from that of the Onychophora and that of annelids. Thirdly, again in relation to (c) above, I discussed fully in an earlier paper (Anderson, 1966) the case for interpreting onychophoran embryos as a modification of clitellate-like embryos, with the corollary that the Onychophora probably have a direct phylogenetic link with the segmented ancestors of the annelids. The opportunity is taken here to bring these lines of argument together, to comment on the Chelicerata and to adduce the level of embryological support that can be offered for the concept of annelid and arthropod phylogeny propounded by Tiegs and Manton.

The Onychophoran–Myriapod–Hexapod Assemblage or Uniramia

Manton (1972) has set out the functional morphological reasons which necessitate recognition of the Onychophora, Myriapoda and Hexapoda as a unitary phylogenetic assemblage, the Uniramia. In order to test the feasibility of this idea embryologically, the search for a common basis underlying the diversity of embryonic development in the Onychophora, Myriapoda and Hexapoda must reveal the following links:

(a) a link between the development of the Onychophora and that of the Myriapoda;
(b) a link between the two developmental alternatives within the Myriapoda;
(c) a link between the development of the Onychophora and that of the Hexapoda, not necessarily via the Myriapoda.

We can now apply to this search the concept of a comparison of fate maps of the groups involved.

The Onychophora and the Chilopoda

Of the two themes of development discernible in the Myriapoda, it is obvious that the chilopod theme stands close to the Onychophora. Every feature of chilopod development can be interpreted as a functional modification of basic onychophoran development. Both show intralecithal cleavage of a yolky, centrolecithal egg and formation of a uniform blastoderm of low, cuboidal cells. The fate map of the blastoderm exhibits the same basic configuration in both (compare Figs. 45 and 61) and the further development of each presumptive area can be compared as follows:

The *presumptive midgut* of the Onychophora comprises a long, midventral, anterior component, together with a compact posterior component lying behind the anterior component in the posteroventral midline. At gastrulation, the sheet of anterior midgut cells separates into two halves along the ventral midline, between the presumptive stomodaeum and proctodaeum, producing a long slit through which the internal yolk mass is temporarily exposed to the exterior. The bilateral halves of the presumptive anterior midgut turn slightly inwards on either side of the midventral slit and proliferate cells which migrate around the periphery of the yolk mass. As proliferation of the anterior midgut cells takes place, the adjacent ventrolateral cells of the blastoderm spread towards the ventral midline as an attenuated sheet of ventral, extra-embryonic ectoderm. The exposed surface of the yolk mass is covered once more as a result of this process and the two halves of the anterior midgut are also brought together in the ventral midline.

The migrating anterior midgut cells take up yolk, enlarging to form a deep, yolky epithelium which encloses the yolk mass. In this phase of development, therefore, the anterior midgut cells have a temporary vitellophage function. As the yolk is progressively utilized, the epithelium becomes the anterior part of the definitive midgut epithelium. Behind the yolk mass, and thus behind the anterior midgut, the posterior midgut rudiment proliferates cells which form a yolk-free, posterior midgut tube in the forwardly flexed, growing trunk. This tube gives rise directly to the remainder of the midgut epithelium. As the proliferation of posterior midgut cells is completed, the area from which they arise becomes indistinguishable at the surface of the embryo.

The presumptive midgut of the Chilopoda is a long, broad area of ventral blastoderm cells, occupying the same relative position as the presumptive anterior midgut of the Onychophora. In chilopods, however, vitellophage function and midgut formation are partially separated. Through a specialization of cleavage, numerous energids remain within the yolk mass during blastoderm formation and subsequently act as temporary vitellophages, which are eventually digested. Concomitantly, when the cells of the presumptive midgut become internal and spread as an epithelium around the yolk mass, the epithelium remains cuboidal and takes only a small amount of peripheral yolk into its cells. Further development to form the definitive midgut epithelium then proceeds more or less directly as the yolk is resorbed. Furthermore, associated with the fact that the entire complement of chilopod trunk segments is formed precociously along the sides of the yolk mass rather than as a sequence completed by proliferation of a tubular extension behind the yolk mass, the posterior midgut component present in the Onychophora is absent in chilopods. A new relationship also exists between the presumptive midgut and the ventral ectoderm. The presumptive midgut cells are intermingled with presumptive ectoderm cells throughout the entire midgut area and migrate diffusely into the interior before spreading around the yolk mass. As a result, when the midgut cells move inwards, presumptive ectoderm is left as a continuous sheet of cells at the surface, eliminating the complex onychophoran sequence of ventral exposure of the yolk, followed by ventral spread of ectoderm to restore a continuous surface epithelium.

Associated functional specializations attend the development of the *presumptive stomodaeum* and *proctodaeum* of the Chilopoda as compared with the Onychophora. In the latter, the stomodaeal and proctodaeal rudiments invaginate with the anterior midgut invagination and subsequently become tubular when the aperture of the invagination closes. The presumptive stomodaeum and proctodaeum of chilopods occupy the same relative locations in the blastoderm as they do in the Onychophora, but in the absence of a midventral midgut invagination, they grow directly into the interior as epithelial tubes.

The *presumptive mesoderm* of the Chilopoda retains the form and mode of development of the presumptive mesoderm of the Onychophora almost in its entirety. In both, the presumptive mesoderm is a small, posteroventral area of blastoderm cells behind the presumptive midgut and proctodaeum. Cells proliferated from this area spread forwards as bilateral bands between the midgut epithelium at the surface of the yolk mass and the ectodermal epithelium at the surface of the embryo. The anterior ends of the mesodermal bands push forwards on either side of the stomodaeum and meet in front of the stomodaeum. Posteriorly, the mesodermal bands of the Onychophora continue on either side of the posterior midgut tube formed behind the yolk-filled anterior midgut. In the Chilopoda, due to the rapid completion of all segment rudiments at an early stage of development, the mesodermal bands flank the yolk-filled midgut throughout their entire length.

Chilopods exhibit one further significant difference from the Onychophora in their mesoderm formation. Along the length of each mesodermal band, cells are added to the mesoderm by immigration from the overlying blastoderm, which subsequently develops as embryonic ectoderm. To some extent, therefore, the chilopod presumptive mesoderm is spread diffusely along the bilateral bands of presumptive embryonic ectoderm and becomes internal directly in its organ-forming position. The forward migration of mesoderm from a posterior proliferative area is thus partially replaced by a functionally more direct mode of development.

In the *further development of the mesoderm*, the chilopods exhibit an obvious specialization of the onychophoran condition. In both, the mesodermal bands break up into paired, hollow somites, each of which becomes trilobed, with appendicular, dorsolateral and medioventral lobes. Initially, one pair of somites, the antennal of Onychophora and the preantennal of chilopods, lies in front of the stomodaeum. The dorsolateral and medioventral lobes gradually extend towards the dorsal and ventral midlines, and the walls of the somites become further elaborated in the same general manner in both groups, with the following exceptions:

(a) Chilopods lack the development of segmental organs, except for paired posterior coelomoducts developing as genital ducts and for a pre-mandibular vestige developing as lymphoid tissue.
(b) Some of the cells at the base of the appendicular lobe develop in chilopods as fat-body, an organ system which is lacking in the Onychophora.
(c) The somatic walls of the dorsolateral and medioventral lobes contribute to the somatic musculature in chilopods. The appendicular lobe is the exclusive source of this component in the Onychophora.
(d) Development of the splanchnic musculature and gonads is more specialized in the Chilopoda than in the Onychophora. In the latter, the splanchnic walls of the somites contribute extensively to the pericardial floor and gonads, which are also formed in part by the somatic walls of the dorsolateral lobes. In chilopods, the splanchnic walls of the somites develop entirely as splanchnic musculature and the pericardial floor and gonads arise exclusively from the somatic walls of dorsolateral lobes.

All of these differences are functional modifications in the Chilopoda as compared with the Onychophora. Like the development of the midgut, the development of the mesoderm has a common basis in both groups and displays a fundamental similarity. Not surprisingly, the same can be said for the *presumptive ectoderm*. The dorsal, extra-embryonic ectoderm is the same in both, while the bilateral bands of presumptive embryonic ectoderm in the Chilopoda differ from those of the Onychophora in only two ways. They lie close together on either side of the ventral midline, with a narrow band of ventral extra-embryonic ectoderm between them, and they include the presumptive rudiments of all except the last few body segments. The first difference is a functional corollary of the way in which the presumptive midgut develops in the Chilopoda as compared with the Onychophora. Invagination is replaced by diffuse immigration and the presumptive embryonic ectoderm overlaps the presumptive midgut almost to the midline, restricting the ventral extra-embryonic ectoderm to a midventral position. Then, in the wide bands of presumptive embryonic ectoderm, numerous short, wide segment rudiments are present. Associated with this specialization, the posterior, ectodermal growth zone has only a brief activity and the trunk is not prolonged as a tube beyond the end of the yolk mass. As the segments grow longer, the two bands of embryonic tissue (ectoderm + somite mesoderm) flex dorsally up the sides of the yolk mass and the ventral extra-embryonic ectoderm now attenuates and spreads. In both the Onychophora and the Chilopoda, the extra-embryonic ectoderm is incorporated into the definitive ectoderm when the embryonic tissue spreads towards the dorsal and ventral midlines.

Only two points concerning the further development of the embryonic ectoderm need special mention. Firstly, ventral organs are prominent in the Onychophora, but not in chilopods. Since these organs are transient, however, and do not give rise to any definitive structures in the Onychophora, their absence in chilopods cannot be regarded as a matter of great significance. Secondly, a more difficult question is raised by differences in the development of the brain. The Chilopoda have prominent protocerebral ganglia, small pre-antennal ganglia and the usual deutocerebrum and tritocerebrum. On present evidence, admittedly based on nineteenth-century studies, the Onychophora lack protocerebral ganglia, the brain being formed by the ganglia of the first or antennal segment. The question is, how do we equate this difference with a common ancestry for the two groups? Only renewed investigation can provide an answer.

With the exception of the unresolved problem of development of the brain, a comparison of the total course of integrated embryonic development in the Onychophora and Chilopoda reveals a basic similarity, epitomized in similar fate maps of the blastoderm. A reasonable presumption can therefore be made that the ancestors of the Onychophora and the Myriapoda were a common stock. It should not be inferred from this, however, that the modern Onychophora can be envisaged as surviving ancestral chilopods. In addition to the adult morphological specializations which preclude this possibility, the divergent specialization of the Onychophora is also obvious in the development of the head. Setting aside the gnathal segments of the Chilopoda, whose homologues in the Onychophora are uncephalized trunk segments, we can see that the antennal segment of the Onychophora might have become vestigial as the pre-antennal segment of the Chilopoda, but that the jaw segment of Onychophora and the antennal segment of Chilopoda are alternative, not sequential, specializations. Furthermore, the chilopod expression of this segment is more generalized than the onychophoran expression. Similarly, the slime-papilla segment of the onychophoran head is developmentally and functionally specialized in ways related to the whole head structure and mode of life of the animal. There is no reason to assume that the chilopod pre-mandibular segment has been reduced and further cephalized from an earlier condition showing slime papilla specialization. We shall return to a further consideration of the myriapod head later in the present chapter, bearing in mind that comparative embryology provides strong support for the concept of a divergence of Onychophora and Myriapoda from a common ancestral stock of terrestrial, segmented lobopods with uniramous appendages.

The Symphyla, Pauropoda and Diplopoda

On morphological grounds, the Myriapoda are a monophyletic assemblage (Manton, 1964), but there are still deep-seated differences to be accounted for between the embryonic development of the Chilopoda and that of the remaining myriapods. The precocious hatching, progoneate condition and anamorphic post-embryonic growth characteristic of the Symphyla, Pauropoda and Diplopoda suggest, of course, a secondary reduction in the size and yolk content of the egg. This, in turn, could have led to the evolution of secondary total cleavage prior to blastoderm formation, with associated modifications of the mode of vitellophage formation.

At the same time, not all of the differences between the development of chilopods and other myriapods can be explained as a consequence of a reduction in yolk. A much more fundamental difference is observed in the development of fat-body from vitellophages rather than from somite mesoderm. Trivial though it may appear, this difference involves an associated placement of the midgut rudiment in the centre of the yolk mass, rather than around the yolk mass, and a new relationship between the yolk mass, midgut and somites, three basic components of the developing animal. Then, too, the symphylan–pauropod–diplopod group of myriapods retains a greater expression of segmental organs than the Chilopoda and has a different pattern of elaboration of its somite derivatives. The present requirement, therefore, is to explain the basic features of embryonic development of the Symphyla, Pauropoda and Diplopoda in terms functionally compatible with the pattern of development underlying the Chilopoda and Onychophora.

If we assume that a separation of vitellophage function from midgut-forming function was a feature of protomyriapods, the vitellophages being left within the yolk mass during cleavage, it can be inferred that the non-chilopod line of myriapod evolution became specialized in its embryonic development in the following important ways, associated with but functionally independent of the evolution of anamorphosis:

The *presumptive midgut* became reduced to a small proliferative area of blastoderm cells in front of the presumptive proctodaeum, with cells budding off from the area to migrate through the centre of the yolk mass and give rise directly to the midgut tube. Associated with this change to direct midgut formation, the vitellophages evolved as fat-body cells. This sequence of evolution is the exact opposite of the specialization observed in chilopods, but is functionally compatible with midgut development in the Onychophora, the anterior midgut being modified in development and the posterior midgut being lost. The stomodaeum and proctodaeum, concomitantly, develop directly from their anterior and posterior presumptive areas, as they do in chilopods.

The *presumptive mesoderm* was shifted entirely from its primitive posterior position to a bilateral distribution throughout the presumptive embryonic ectoderm, eliminating forward spread from the posterior end in favour of direct immigration in the organ-forming position. This trend is partially expressed in chilopods and might be expected to become more obvious in short-bodied embryos. In the further development of the mesoderm, the somites retained typical trilobation, but evolved a variety of specializations different from those of chilopods. Associated with the development of the midgut as a narrow, central tube, the dorsolateral lobes of the somites underwent a median shift, maintaining contact between their splanchnic walls and the external surface of the midgut tube. This shift is still retained in the development of the Symphyla, but is replaced in the pauropods and diplopods by a number of further specializations. With the specialization of development of the dorsolateral lobes, gonad formation devolved upon the medioventral lobes. Only the formation of the heart, pericardial septum, dorsal and ventral longitudinal muscle, appendicular muscle and splanchnic mesoderm remain the same in symphylans as in chilopods.

In one respect, however, the symphylan–pauropod–diplopod stock has retained a primitive feature of somite development, conspicuous in the Onychophora but lost in chilopods. There is considerable evidence, especially in the Symphyla, for the development of segmental organs from the medioventral lobes of the somites, transiently in the trunk segments but functionally in the premandibular and maxillary segments of the head (Table 4).

The *presumptive ectoderm* of the Symphyla, Pauropoda and Diplopoda, if one allows for the fact that all but the first few trunk segments are represented only by a growth zone, differs from that of chilopods in its formation, distribution and subsequent development merely in the vestigiality of the ventral, extra-embryonic ectoderm. Since the Symphyla, Pauropoda and Diplopoda lack a large, ventral area of presumptive midgut, develop only a few trunk segments in the embryo and have a relatively small yolk mass, this difference is not surprising. Indeed, a parallel vestigiality of the ventral extra-embryonic ectoderm obtains in the scutigeromorph centipedes. Just as they retain a number of segmental organs, however, the Symphyla, Pauropoda and Diplopoda retain one feature of ectodermal development that has been lost in the Chilopoda, namely, the development of ventral organs. Myriapod ventral organs differ from those of the Onychophora in contributing cells to the ganglia and persisting in part, in the Symphyla and perhaps the Diplopoda, as exsertile vesicles. It seems likely that the protomyriapods had ventral organs, but whether they gave rise to exsertile vesicles cannot be decided.

Since comparative development indicates that two alternative modes of embryonic development observed in the modern myriapods could both have a basis in an ancestral mode of development similar to that of Onychophora, it should be possible to discern, from the same evidence, something of the composition and structure of the protomyriapod head. The case for a basic dignathous head for all myriapods has been presented in Chapter 5. Taking the question of segmental organs into account, we can now envisage a protomyriapod head with two pairs of mainly sensory limbs at the front, one anterior to the mouth (pre-antennal), one flanking the mouth (antennal); a third pair of limbs (premandibular) at the sides of the mouth, reduced to papillae bearing the exit ducts of segmental organs modified as salivary glands; a fourth pair of limbs (mandibular) immediately behind the mouth, somewhat specialized as food-manipulating limbs and lacking associated segmental organs; and a fifth pair (maxillary), also playing a part in the manipulation of food, with associated segmental organs secreting a supplementary lubricating fluid. All extant myriapod heads can be derived from this basic, hypothetical head. Carrying the argument a step further, a simpler anterior end with generalized limbs, comprising a mainly sensory pair in front of the mouth, a sensory and food-manipulating pair flanking the mouth, a more

TABLE 4. SEGMENTAL ORGANS IN THE UNIRAMIA

Segment	Onychophora	Chilopoda	Symphyla	Pauropoda	Diplopoda	Hexapoda
1	ANTENNAL			PRE-ANTENNAL		
	Transient coelomoduct			Absent		
2	JAW			ANTENNAL		
	Transient coelomoduct			Absent		
3	SLIME PAPILLA			PRE-MANDIBULAR		
	Salivary gland	Lymphoid tissue	Embryonic segmental organ	Pre-mandibular salivary gland	Absent	Suboesophageal body
4	FIRST TRUNK			MANDIBULAR		
	Segmental organ			Absent		
5	SECOND TRUNK			MAXILLARY		
	Segmental organ	Absent	Maxillary salivary gland	Maxillary salivary gland	Maxillary salivary gland	Absent
6	THIRD TRUNK			POST-MAXILLARY		
		SECOND MAXILLARY	LABIAL	COLLUM	COLLUM	LABIAL
	Segmental organ			Absent		
7	FOURTH TRUNK			FIRST (LIMBED) TRUNK		
	Segmental organ			Absent		
8	Segmental organ			Absent		
9	Segmental organ	Absent	Transient coelomoduct	Absent	Absent	Absent
10	Segmental organ	Absent	Transient coelomoduct	Absent	Absent	Absent
11	Segmental organ	Absent	Transient coelomoduct	Absent	Absent	Absent
12	Segmental organ	Absent	Transient coelomoduct	Absent	Absent	Absent
13	Segmental organ	Absent	Transient coelomoduct	Absent	Absent	Absent
14	Segmental organ	Absent	Transient coelomoduct	Absent	Absent	Absent
15	Segmental organ	Absent	Absent		Absent	Gonoduct
16	Segmental organ	Absent	Absent		Absent	Gonoduct
17	Segmental organ	Absent	Absent		Absent	Gonoduct
18	Segmental organ	Absent	Absent		Absent	Absent
19	Segmental organ	Absent	Absent		Absent	Absent
20			Absent		Absent	Absent
Preanal	Gonoduct	Gonoduct			Absent	

vestigial pair behind them carrying the exit ducts of salivary glands, and two succeeding pairs of generalized trunk limbs with locomotory and food-manipulating functions (many arthropods use their anterior trunk limbs in both ways), would suffice as an ancestral anterior end for both the Onychophora and the Myriapoda. The case for a common ancestry of the Onychophora and the Myriapoda, based on embryological evidence, is thus encouraging. There seems little doubt that the protomyriapods which gave rise divergently to the modern myriapod classes had an onychophoran basis to their embryonic development.

The Hexapoda

With this conclusion in mind, it is now possible to examine the question, whether the basic pattern of embryonic development of the Hexapoda has its roots in onychophoran or myriapod development, or is distinct from either? Until quite recently, the Symphyla were much favoured as progenitors of the hexapods. The common possession of fourteen trunk segments and a labiate trignathous head, together with the apparent similarities in the embryonic development of the Symphyla, Diplura and Collembola, were among the factors which prompted this view (Tiegs, 1940, 1947; Tiegs and Manton, 1958). Detailed studies of functional morphology carried out by Manton (e.g. 1964, 1970, 1972), however, have now shown that a symphylan ancestry for hexapods is not acceptable. The embryological evidence gathered in the present account supports this view. When the comparison is based on the formation and fates of presumptive areas (compare Figs. 61, 76 and 88), the resemblances between the embryos of myriapods and hexapods can be seen to be superficial, convergent and misleading. In exploring this line of thought, we can also examine the minor question, raised in Chapter 7, of the relationship between the Pterygota and the Thysanura.

The *presumptive midgut* of the hexapods provides the initial clue to the peculiarities of this group. There is no doubt that the basic presumptive midgut of hexapods comprises the vitellophages in the yolk mass enclosed by the blastoderm. In general terms, this implies a diffuse spread of presumptive midgut cells throughout the blastoderm, their generalized immigration into the yolk mass and temporary function as vitellophages and their later entry into direct differentiation as midgut cells, as seen in the Diplura. Assuming the abandonment of a presumptive posterior midgut in hexapods, as in myriapods, in association with a changed relationship between the segmental growth zone and the yolk mass, the basic pattern of midgut development in the Hexapoda can be envisaged as a functional modification of anterior midgut development in the Onychophora. Significantly, the hexapod modification is different from either of the presumptive midgut modifications observed in myriapods. The simple, direct development of the presumptive stomodaeum and proctodaeum which the Hexapoda share in common with the Myriapoda does not argue against this, since such simplicity is a functional corollary of the loss of a ventrally invaginating, presumptive midgut.

Within the hexapods themselves, the Collembola and Thysanura show a precocious segregation of presumptive midgut cells as vitellophages in the yolk mass during cleavage (Chapter 6). The Pterygota also segregate their vitellophages during cleavage, in the same manner as the Thysanura, but then develop their midgut from bipolar presumptive rudiments in the blastoderm, later associated with the internal ends of the stomodaeum and proctodaeum. Various possibilities exist for the origin of this specialization. It could be a direct specialization of a prior onychophoran condition, independent of the apterygotes, in which the vitellophage function has become precocious and separate and the long anterior midgut area has become restricted to anterior and posterior areas. The latter separation is functionally correlated with the concentration of presumptive mesoderm in the ventral midline of the pterygote blastoderm. However, the pterygote midgut pattern is more likely to be a specialization of a prior thysanuran condition. When the thysanuran midgut develops from the vitellophages, the midgut epithelium is first formed at the ends of the yolk mass, in contact with the inner ends of the stomodaeum and proctodaeum and gradually spreads towards the middle of the yolk mass. The bipolar presumptive midgut rudiments of pterygotes could be a precocious manifestation of this phenomenon. Some hints have indeed been given of a contribution by the vitellophages to the middle region of the midgut in certain pterygotes (Odonata, Tschuproff, 1903; Heteroptera, Mori, 1969, 1971), but the evidence is not strong enough to be compelling. A decision on which of the two possibilities suggested above is the more likely is facilitated, as will be seen below, by other aspects of pterygote development.

The *presumptive mesoderm* of hexapods in its most generalized form (Diplura and Collembola) has a diffuse bilateral distribution throughout the presumptive embryonic ectoderm. This pattern of distribution is also observed in the symphylan–pauropod–diplopod stock of myriapods, as a specialization facilitating direct mesodermal entry in the organ-forming position. The midventral concentration of presumptive mesoderm in the Thysanura is clearly a secondary adaptation to the sharp separation and further development of the ectoderm and mesodermal rudiments. The same specialization characterizes the Pterygota. In the further development of the mesoderm, however, we find that the similarity between the Diplura and Symphyla in their early mesodermal segregation is misleading. Later mesodermal development in the hexapods as a whole has its own peculiarities, different from those of any myriapod, but closer to the Chilopoda than the Symphyla. After the common trilobation of the somites, the appendicular lobe has the usual development as limb musculature, the dorsolateral lobe produces the dorsal longitudinal muscle, pericardial septum and heart and the medioventral lobe develops in part as the ventral longitudinal muscle. These features are the same as in all myriapods. Yet the walls of the somites also give rise to the fat-body cells, as they do in the Chilopoda, while the splanchnic walls of the dorsolateral lobes of certain abdominal segments form the gonads as well as splanchnic musculature, as they do in the Onychophora. Furthermore, the genital ducts develop from posterior coelomoducts and lead to an opisthogoneate gonopore, as they do in the Onychophora and Chilopoda (Table 4).

In all features of somite development, therefore, other than those shared in common with all Myriapoda, the hexapods differ from the Symphyla, Pauropoda and Diplopoda. Furthermore, the single feature of mesodermal development which the Hexapoda share exclusively with the Chilopoda, namely, the proliferation of fat-body cells from the walls of the somites, could have evolved independently in the two groups. Mesodermal development in the Myriapoda and Hexapoda shares sufficient common ground to indicate that the hexapods evolved from ancestral lobopods related to the ancestors of the Myriapoda, but divergence must have occurred at an early stage of evolution, near the ancestry of the Onychophora.

This conclusion bears out the evidence of the midgut cited above and finds further support in the development of the hexapod head.

Although the Hexapoda share with the Myriapoda the same basic segmental composition of the dignathous head and exhibit cephalization and functional specialization of the post-maxillary segment as a labial segment, there are good functional reasons for regarding this similarity with myriapods as deceptive (Manton, 1964, 1972). There is no evidence that dignathy was antecedent to trignathy in the evolution of the Hexapoda, as it was in the trignathous myriapods. The labiate specialization of the post-maxillary segment in the Hexapoda, with its corollary of posterior tentorial arms, is fundamentally different from the convergent labiate condition in the Symphyla. Functional considerations dictate that the labiate hexapod head evolved as an integrated trignathous unit, since the hexapod tentorium and its associated musculature could not have evolved through a sequence which included the secondary incorporation of posterior tentorial arms. The embryological evidence supports this view. The head develops as an integrated trignathous unit in all hexapods and its total pattern of development is distinctly different from that of any myriapod. The hexapod head, like that of chilopods (Table 4), lacks cephalic segmental organs, develops glandular tissue from its premandibular somites (the suboesophageal body) and has ectodermal post-maxillary (labial) salivary glands, but these are secondary features in the Chilopoda as compared with other Myriapoda and can similarly be interpreted as secondary features in the Hexapoda. The protomyriapod head postulated on a previous page has no vestigial or recapitulatory representation in the development of the hexapod head.

Finally, we can consider the hexapod *presumptive ectoderm*. The dorsal extra-embryonic ectoderm is retained. Since the Symphyla are clearly unrelated to the Hexapoda, it is obvious that the remarkable dorsal organ of the Collembola, also present more vestigial in the Diplura, has evolved convergently with that of the Symphyla, in association with the attainment of dorsoventral flexure in a secondarily small, spherical egg. The Thysanura seem to have gone about dorsoventral flexure in another way, eschewing a dorsal organ but evolving an alternative specialization of their dorsal extra-embryonic ectoderm as an amnion and serosa, with associated katatrepsis. This feature, like many others, they share with the Pterygota. Indeed, there is reason to believe that the long embryonic primordium of the Dictyoptera (Fig. 94) is less specialized than the short embryonic primordium of the Thysanura (Figs. 81 and 82), which has evolved convergently with that of several pterygote orders.

In the absence of an extensive presumptive midgut in the blastoderm, the bilateral presumptive embryonic ectoderm of the hexapods, like that of the Symphyla, Pauropoda and Diplopoda, extends to the ventral midline. A ventral component of extra-embryonic ectoderm is absent. Basically, however, the presumptive embryonic ectoderm of the hexapods remains relatively long, as in the Chilopoda, and includes the rudiments of the ectoderm of all except the last few abdominal segments.

In the further development of the hexapod ectoderm, ventral organs are absent, as in the Chilopoda, and the development of the ganglia proceeds through the activity of specialized neuroblasts, a unique hexapod feature. Thus in the ectoderm, as in other components of the embryo, the hexapod affinities seem to lie more with the Chilopoda than with other myriapods, but the differences between chilopod and hexapod development are such as to necessitate postulation of a remote and generalized common ancestry in a lobopod stock close to the ancestors of the Onychophora. Within the ancestors of the hexapods, an early divergence must also have occurred between the ancestors of the Diplura, Collembola and Thysanura, probably before any of them had attained the hexapod state (Manton, 1972). There seems little doubt, however, that the ancestors of the Pterygota were hexapodous relatives of the Thysanura.

A consideration of comparative embryology thus supports Manton's conclusion, based on comparative functional morphology, that the Onychophora, Myriapoda and Hexapoda constitute a unitary phylogenetic assemblage, the Uniramia, but that the Hexapoda are not descended from the Myriapoda, nor the Myriapoda or Hexapoda from the modern Onychophora. From the point of view of comparative embryology, the most significant result of the present appraisal is the demonstration that the theme of embryonic development in the Onychophora can be regarded as a basic theme for all Uniramia. It thus becomes a simple matter to make meaningful comparisons between the embryonic development of the Uniramia and those of other arthropods (Crustacea and Chelicerata) on the one hand and the annelids on the other. There are advantages in proceeding first with the latter comparison.

The Annelida and the Onychophora

As the functional morphology of annelids and arthropods has become better known, the likelihood of a phylogenetic relationship between annelids and arthropods has gradually receded (e.g. Manton, 1967). Parapodium and lobopodium are mutually functionally exclusive and the similarities in adult morphology between the annelids and the Onychophora rest with little beyond metameric segmentation, a dorsal blood vessel and a ventral nerve cord, a combination which could have evolved more than once. Furthermore, until recently, there has been little evidence for spiral cleavage among the arthropods and still is none outside the Crustacea (see below). The Onychophora, and consequently the Myriapoda and Hexapoda, lack spiral cleavage entirely, having as their basis of development an intralecithal cleavage leading to the formation of a uniform blastoderm around a yolk mass. Since spiral cleavage is a fundamental feature of annelids, the classical approach is quite inadequate for a meaningful comparison between annelid and onychophoran development. As I have discussed in an early paper, however (Anderson, 1966), comparisons based on the formation and fates of presumptive areas (Figs. 29 and 45) reveal a remarkable similarity between onychophoran development (Chapter 4) and the particular modifications of annelid development expressed in the primitive development of clitellate annelids (Chapter 3). The same pattern of presumptive areas is exhibited in both and the detailed differences between corresponding areas are, in every case, further adaptations to yolk in the onychophoran embryo as compared with the clitellate embryo.

The *presumptive midgut* of clitellates, derived from 3A–3C and 4D, consists of a mass of large yolky cells occupying the ventral half of the blastoderm. The presumptive stomodaeum, ectoderm and mesoderm cells sit on top of this mass and there is no blastocoel. In further development, the midgut mass becomes enclosed by ectodermal overgrowth and develops directly into the epithelium of the midgut after a period during which the cells act in yolk digestion (i.e. as vitellophages). The onychophoran presumptive midgut is, as we have seen above, a ventral sheet of small cells at the surface of the yolk mass, subdivided into a long anterior midgut component and a small, compact, posterior midgut component, which proliferates the part of the midgut behind the yolk mass. There is no evidence for this happening in clitellates, but a small posterior midgut component cut off from the presumptive mesoderm has been reported in certain yolky polychaete embryos (see Chapter 2). The anterior midgut component of Onychophora splits open along the ventral midline, exposing the yolk. The two halves then turn inwards and proliferate cells which migrate around the yolk, acting as vitellophages before differentiating into definitive anterior midgut epithelium. Later, the two proliferative zones merge in the ventral midline as the ventral opening closes once more. This sequence of events is interpretable as a functional corollary of the transition from a massive, cellular presumptive midgut to a small-celled presumptive midgut at the surface of a yolk mass.

The *presumptive mesoderm* of clitellates is a pair of large M-cells formed by equal bilateral division of 4d. These cells lie in a posterior position, behind the presumptive midgut and become internal mainly by ectodermal overgrowth. They then bud off primordial germ cells before proliferating a pair of mesodermal bands forwards, on either side of the large midgut mass and beneath the spreading ectoderm. The mesodermal bands in turn give rise to the paired mesodermal somites. In the Onychophora the presumptive mesoderm is, like the presumptive midgut, a group of small cells at the surface of the yolk mass, but in the same relative position as 4d in clitellates, posteroventrally behind the presumptive midgut. Paired mesodermal bands are proliferated forwards beneath the blastoderm, external to the yolk mass and become subdivided into paired somites. Before the proliferation of mesodermal bands begins, the mesodermal presumptive area releases a group of primordial germ cells into the interior.

The *presumptive ectoderm* of clitellates consists of a large 2d cell surrounded anteriorly and laterally by an arc of micromeres. The 2d cell divides to give a transverse row of four ectoteloblasts on each side of the embryo and these then proliferate ectoblast bands (embryonic ectoderm) forwards external to the mesodermal bands. The two ectoblast bands meet in the anterior midline after growing forwards around the large midgut mass. During the divisions through which the 2d cell gives rise to the ectoteloblasts, the arc of micromeres proliferates and spreads back, both dorsally between the bilateral groups of 2d cells and laterally below the same cells. The ends of the lateral arms of the micromere tissue turn medially behind the 2d cells to meet each other and the dorsal backgrowth of micromeres. In the centre of this region of junction, the proctodaeum subsequently forms. Proliferation and spread of the micromere tissue continues while the ectoteloblasts are producing the forwardly growing ectoblast bands. The main area of spread is dorsally, between the two bands, producting a dorsal, temporary yolk-sac (or extra-embryonic) ectoderm which pushes the ectoblast bands and their underlying mesodermal bands downwards over the sides of the midgut mass towards the ventral midline, and at the same time effects gastrulation enclosure of the midgut. Narrow bands of micromere tissue also persist along the lower margins of the ectoblast bands and come together as ventral, temporary yolk-sac (or extra-embryonic) ectoderm as the two ectodermal edges meet in the ventral midline. The ventral, temporary yolk-sac ectoderm is soon incorporated into the definitive ectoderm and the dorsal, temporary yolk-sac ectoderm more gradually suffers the same fate.

The presumptive ectoderm of the onychophoran blastoderm is established directly in the pattern attained in clitellates as a result of the partial completion of overgrowth and spread. The dorsal, extra-embryonic ectoderm is formed presumptively as a broad area of blastoderm cells covering the dorsal half of the yolk mass. The ectoblast bands (embryonic ectoderm) are formed laterally as paired bands of blastoderm cells, meeting in the midline anteriorly and converging on the presumptive mesoderm posteriorly, where each band has a presumptive growth zone. Beneath the lower edges of the ectoblast bands, paired narrow strips of ventral, extra-embryonic ectoderm separate these bands from the midventral presumptive midgut. Gastrulation spread of the ectoderm is confined to the ventral, extra-embryonic components, which spread ventrally to meet in the ventral midline as the anterior midgut undergoes its invagination. In due course, the dorsal and ventral extra-embryonic ectoderm are incorporated in the definitive ectoderm.

The *presumptive stomodaeum* of clitellates is a group of small cells in the midline at the anterior margin of the arc of micromeres, mainly derived from 2b. This component is carried to an anteroventral position as a result of ectodermal spread, and lies between the anterior ends of the two ectoblast bands. The stomodaeum forms by direct ingrowth as a solid mass of cells, which becomes hollow and forms the mouth secondarily. The onychophoran presumptive stomodaeum is formed as a small group of blastoderm cells in the corresponding anteroventral position, between the anterior ends of the two presumptive ectoblast bands. The cells form the anterior wall of the invagination opening resulting from midgut invagination, and subsequently round off to form a tube with a lumen and mouth when the invagination opening is closed by spreading and juncture of the ventral, extra-embryonic ectoderm. The *presumptive proctodaeum* and its further development in the Onychophora have similarly become associated with the posterior end of the midgut invagination opening. In the clitellate embryo, the presumptive proctodaeal cells lie at the ends of the arc of micromeres surrounding the anterior and lateral faces of the ecto-teloblast mother cell. From here, as we have seen, they migrate to a posterior location before giving rise directly to the proctodaeum and anus. The onychophoran presumptive proctodaeum is similarly bilateral on either side of the posterior end of the presumptive anterior midgut. Here the two halves of the presumptive proctodaeum come together in a new way as the anterior midgut component invaginates, so that they form the posterior wall of the invagination opening. Tubulation, with formation of the anus, then takes place as the invagination opening closes. It can be seen that this modification places the proctodaeum in front of the site of mesodermal and ectodermal proliferation, so that segments arising from the growth zone in the Onychophora have to move forwards bilaterally past the anus. The same arrangement persists in the Chilopoda.

Since the mode of formation and subsequent development of the presumptive areas of the onychophoran blastoderm is so consistently a functional modification of the clitellate mode, further adapted to the development problems engendered by yolk, it is reasonable to postulate that the ancestors of the Onychophora, and therefore of the Uniramia, had a clitellate-like mode of spiral cleavage development and were related to the annelids (Anderson, 1966). Furthermore, a clitellate-like mode can be reasonably presumed to have had its origins in the mode of embryonic development retained today as an underlying theme in polychaete development, as discussed in Chapter 3. If we accept the proposition of Clark (1964) that the ancestral annelids had a general adult morphology more like that of oligochaetes than polychaetes, lacking segmental parapodia, the possibility of the Onychophora being related to the annelids in the way that the embryology suggests seems less implausible than if a polychaete ancestor had to be envisaged.

The Crustacea

Unlike the Uniramia, the Crustacea retain in many species a recognizable spiral pattern of total cleavage. Surprisingly, however, this does not lead to the conclusion that one might expect, namely, that the Crustacea are related to the annelids or to the Onychophora. Indeed, the reverse is the case (Anderson, 1969). Spiral cleavage is widespread among many phyla and can only be assessed phylogenetically in conjunction with the formation and subsequent development of the presumptive areas of the resulting blastula. When viewed in this way, the basic theme of crustacean embryonic development is seen to differ fundamentally, not only from the clitellate-like mode of development underlying the development of onychophoran embryos, but also from the basic annelid mode of development itself. The interpolation of a nauplius larva in the developmental sequence of Crustacea is not the reason for this difference. The sequence itself is fundamentally different and could not possibly be a modification of any annelid mode of development, as the following comparison shows (Figs. 6 and 111).

The polychaete *presumptive midgut* comprises the four ventral to posteroventral stem cells of spiral cleavage, 3A–3C and 4D. The crustacean presumptive midgut comprises the single posteroventral cell, 4D. This in itself is not a significant difference, but becomes so when taken together with the segregation and arrangement of the *presumptive mesoderm*. The polychaete presumptive mesoderm is the single posterior cell, 4d, lying above and behind the presumptive midgut. The crustacean presumptive mesoderm, in contrast, comprises the three stem cells 3A, 3B and 3C, which lie as a ventral arc anterior to the presumptive midgut. It is not the fact that the mesoderm is segregated differently in the Crustacea that makes this difference important, but the fact that the polychaete specialization and the crustacean specialization are spatially so dissimilar. Presumptive mesoderm in the form of a single 4d cell is a fundamental feature of several phyla of spiral cleavage invertebrates (e.g. molluscs, nemerteans, sipunculids and, probably, polyclad Platyhelminthes). In all cases the presumptive mesoderm lies behind the presumptive mesoderm. This relative juxtaposition persists in the modified spiral cleavage of clitellate annelids, even though the presumptive mesoderm cell may be segregated as 3d or in other specialized ways, and is retained in the blastoderm of the Onychophora, in which all trace of spiral cleavage is lost. The crustacean location of presumptive mesoderm anterior to presumptive midgut and its segregation via 3A, 3B and 3C cannot be regarded as a functional modification of the annelid pattern. 4d in Crustacea is merely a posterior ectoderm cell. Since spiral cleavage occurs in the Crustacea, we must contemplate the possibility of an ancestral condition of spiral cleavage in which the presumptive mesoderm was segregated as a ring around the presumptive midgut (i.e. radially symmetrically) at the posterior end of the blastula, and think of 4d mesoderm as one specialization, and crustacean mesoderm as another specialization, of this ancestral condition.

The fact that the crustacean presumptive stomodaeum originates mainly from 2b cells, as it does in annelids, does not affect this conclusion. The same origin of presumptive stomodaeum is widespread in the spiral cleavage phyla and can be regarded as a basic spiral cleavage feature retained in both annelids and crustaceans. Furthermore, the Crustacea are not protostomatous. Stomodaeal invagination occurs entirely independently of the gastrulation movements which enclose the mesoderm and midgut. Finally, the total dissimilarity in origin and development of the *presumptive ectoderm* in the polychaetes and Crustacea is further evidence that the two are not related. The crustacean presumptive ectoderm is highly specialized. Even though basically the egg is small and cleavage is total and spiral, there is no trace of presumptive prototroch or the subdivision between presumptive anterior and posterior ectoderm seen in the annelids (and molluscs, etc.). The ectoderm of the post-naupliar growth zone in the Crustacea does not have the 2d origin and association with 4d mesoderm found in the annelids. The probable origin is from 3d and 4d cells. The crustacean presumptive ectoderm is zoned in the characteristic pattern of the nauplius and thus specialized in a unique way, with presumptive pre-antennulary, antennulary, antennal, mandibular and post-naupliar zones laterally and a small area of dorsal extra-embryonic ectoderm. Nothing in this fate-pattern suggests any derivation from a prior annelid condition. It is also clear that the spirally based cleavage and presumptive area pattern of the Crustacea does not provide a functional basis for the blastodermal pattern of the Onychophora in the way that the annelid mode of development does. We have seen in Chapter 8 that the evolution of a blastoderm in the Crustacea has occurred several times without disruption or rejuxtaposition of the presumptive area pattern or a transition to anything resembling the onychophoran fate map. The utility of the Crustacea in considerations of the evolution of a blastoderm is, in fact, that they emphasize the stability of fate maps during such evolution. This, in turn, lends weight to the idea that the similarity between the fate maps of the Onychophora and the clitellate annelids is not just fortuitous. Comparative embryology, therefore, strongly supports Manton's view that the Crustacea are not linked phylogenetically with the Uniramia or Annelida, and further suggests that they have a remote spiral cleavage origin independent of the segmented ancestors of the Annelida and Uniramia.

The Chelicerata

Having given support to the existence of at least two independent groups of animals that have evolved as arthropods, it now remains to attempt an assessment of the relative position of the Chelicerata. Are they related to the Uniramia or the Crustacea or neither? Are they related to the Annelida or to the 4d branch of the spiral cleavage assemblage in any way? Are they, indeed, descendants of the spiral cleavage assemblage or not?

It is probably simplest to give the answer to these questions and then to discuss briefly why this answer holds. On present evidence, there is nothing in the basic embryonic developmental pattern of the Chelicerata that indicates relationship to any other group of segmented invertebrates. The Chelicerata retain no evidence of spiral cleavage, even in a modified or vestigial form. Taking the xiphosuran developmental sequence as basic, cleavage remains total but leads to a blastoderm around a central mass of yolky, presumptive midgut cells. The fact of total cleavage retained in association with increased yolk immediately excludes any direct relationship with the Uniramia. These, as we have seen, show a basic condition of intralecithal cleavage. Possible comparisons then remain with either the clitellate-like modification of spiral cleavage development which probably occurred in the ancestors of the Uniramia, with the spiral cleavage development of the polychaetes or, finally, with the spiral cleavage development of the Crustacea. In the absence of spiral cleavage itself within the Chelicerata, we must rely once again on a comparison of fate maps (Figs. 6, 29, 111 and 163).

The basic *presumptive midgut* of the Chelicerata, segregated internally as a result of cleavage, and retaining no representation in the blastoderm, could be a functional modification of the presumptive midgut of any of the groups mentioned above. Some polychaetes show a tendency in this direction in association with increased yolk, e.g. *Neanthes*, as do the cirripedes among the Crustacea. The transition from spiral cleavage and a blastula to intralecithal cleavage and a blastoderm in the early evolution of the Uniramia could have involved an intermediate stage of functional modification of the presumptive midgut similar to that of the Chelicerata. The possibilities are thus too numerous. A firm selection cannot be made between them and the likelihood of other unknown ancestries cannot be excluded.

The absence of any blastodermal representation of the presumptive midgut in the Chelicerata also poses another problem. The basic *presumptive mesoderm* of the Chelicerata is located in the blastoderm as a small, elongate area in the ventral midline. An essential component of the argument for the relationship of the Uniramia to the annelids and the independence of the Crustacea from the annelids or Uniramia lies in the relative placement and modes of segregation of the presumptive mesoderm and presumptive midgut in the blastula or blastoderm wall. Due to the precocious internal segregation of the presumptive midgut in the Chelicerata, this relationship is lost. We are unable to say whether the presumptive mesoderm in the ancestors of the Chelicerata lay anterior to or posterior to the presumptive midgut, or whether it had some other, unknown spatial relationship. We are also unable to nominate it either as a functional modification of 4d mesoderm or a modification of anything resembling the crustacean 3A–3C mesoderm. The subsequent development of the basic presumptive mesoderm of the Chelicerata during gastrulation and somite formation is unique among arthropods. The presumptive mesoderm sinks inwards slightly, forming a shallow gastral groove, and proliferates mesoderm cells which spread laterally in both directions beneath the presumptive ectoderm as incipient somite cells. Analogous modes of mesoderm proliferation are seen in the Onychophora and some of the blastodermal Crustacea. The unique feature of the chelicerate presumptive mesoderm, however, is the mode of action of its posterior growth zone. The growth zone, as far as can be ascertained, remains superficial. It proliferates the mesoderm of all segments behind the pedipalpal segment as a midventral continuation of the initial presumptive mesoderm. The newly pro-liferated mesoderm then enters into the same activity as that observed in the initial (pre-cheliceral, cheliceral and pedipalpal) mesoderm, sinking in midventrally to create a temporary continuation of the gastral groove and proliferating somite cells to either side beneath the associated presumptive ectoderm. Analagous prolongation of a midventral gastral groove in association with the posterior proliferation of presumptive mesoderm is seen in the Coleoptera and other generalized holometabolan pterygotes (Chapter 7) as a secondary condition evolved within the Pterygota. The coleopteran expression of this mode of mesodermal gastrulation and proliferation is based, like that of the Chelicerata, on the localization of presumptive mesoderm in the ventral midline of the blastoderm. In other words, the same functional modification of presumptive mesoderm in relation to yolk has evolved convergently in the Pterygota and Chelicerata. As in the case of the presumptive midgut, its ancestry in the Chelicerata could lie in an annelid-like precursor, a crustacean precursor or neither.

Since the presumptive mesoderm and midgut of the Chelicerata are so uninformative about phylogenetic relationships, other than emphasizing the lack of affinity between the Chelicerata and the Uniramia, there is little hope of adequate information from the *presumptive ectoderm*. In the same manner as the crustacean presumptive ectoderm, the presumptive ectoderm of the Chelicerata has a zonation which is chelicerate only, with no vestige of resemblance to any other presumptive ectoderm pattern. The large area of dorsal extra-embryonic ectoderm is a functionally corollary of the large mass of internal presumptive midgut cells. The presumptive embryonic ectoderm comprises cephalic (precheliceral and cheliceral), pedipalpal and growth zone units which could conceivably bear a relationship to the naupliar presumptive ectoderm of the Crustacea, but seem to be entirely remote from the presumptive ectoderm and 2d ectoteloblasts of annelids.

We are left, then, with only the functional association of the *presumptive stomodaeum* in chelicerates with the anterior end of the gastral groove as the mouth. Whether this association is primary (i.e. protostomatous in the polychaete sense) or secondary cannot be decided. In the normal condition of protostomy, the presumptive stomodaeum gastrulates in conjunction with the presumptive midgut. In chelicerates, it is associated with the specialized presumptive mesoderm.

In sum, the basic pattern of chelicerate embryonic development as epitomized in the fate map of the Xiphosura is enigmatic. Neither the mode of segregation, the relative juxtaposition, nor the subsequent development of the presumptive areas yield enough information to sustain any useful comparison with annelids or other arthropods. It may be, however, that the lack of similarity is significant. The embryological evidence makes it certain that the Chelicerata have no direct link with the Uniramia. Possibilities of a relationship with either the Annelida or the Crustacea cannot be disproved, but neither can they be proved. Perhaps the most useful conclusion to be reached at the present time is a negative one. The embryological evidence does not argue against the evolution of the Chelicerata from an ancestry independent of those of the annelids or other arthropods.

The Arthropods are Polyphyletic

Interpretation of the comparative embryology of the annelids and arthropods by means of fate maps thus yields the following general conclusions. The polychaetes exhibit a generalized development in which spiral cleavage produces a configuration of presumptive areas whose subsequent development is as a trochophore larva, then a metamerically segmented annelid. The clitellate annelids display a particular modification of the polychaete pattern of development, in which spiral cleavage is retained but the trochophore is eliminated in favour of the direct, lecithotrophic development of the metamerically segmented body.

The clitellate modification of annelid development provides a functional intermediate to the intralecithal cleavage and blastodermal development of the Onychophora, indicating at the very least that the Onychophora are related to the segmental ancestors of the annelids. The Onychophoran pattern of development underlies all of the developmental diversity and specialization exhibited within the Myriapoda and Hexapoda, supporting the view that the onychophoran–myriapod–hexapod assemblage is a unitary phylogenetic group, the Uniramia. Within the Uniramia, however, the Onychophora are not ancestral to either the Myriapoda or Hexapoda and the Hexapoda are not descended from any of the subgroups of the modern Myriapoda. The embryological evidence gives clear indications that the Onychophora, Myriapoda and Hexapoda have each diverged independently from a common lobopod ancestry.

Among the Myriapoda, the Chilopoda must have diverged early, but the Symphyla, Pauropoda and Diplopoda seem likely to have had a common ancestry, with later divergence of the symphylan line of evolution from a line which subsequently gave rise to the Pauropoda and Diplopoda. Among the hexapods, the Diplura and Collembola share more embryological similarities than either shares with the Thysanura. There is also evidence that Pterygota are more closely related to the Thysanura than to the Diplura and Collembola.

The Crustacea are a second unitary group of arthropods, with a much more distinct unity in their embryonic development than the Uniramia. The mode of development in Crustacea is based on spiral cleavage and a configuration of presumptive areas whose subsequent development is as a nauplius. The details of this sequence leave no doubt that the Crustacea are unrelated to the Annelida or to the Uniramia, except in so far as all three groups are members of a larger, spiral cleavage assemblage of invertebrates. The metamerically segmented coelomates which gave rise to the Crustacea cannot be identified.

The Chelicerata comprise a third unitary group of arthropods, again with a distinctive theme of embryonic development. Their specialized total cleavage and configuration of presumptive areas obscure any evidence of past relationships, except that a link with the Uniramia can be ruled out. Possible relationships with either the Annelida or the Crustacea cannot be confirmed or denied embryologically. A phylogenetic origin of Chelicerata independent of those of the annelids and other arthropods is also possible, since there is no certainty that their ancestors even had spiral cleavage.

In every way, therefore, the evidence of comparative embryology favours the view of Manton (1972) that the arthropods are a polyphyletic assemblage and that arthropodization has occurred at least twice, probably three times and possibly more than three times. The embryological differences between the three major groups of arthropods are such as to fully justify Manton's conclusion that each should be raised to the status of a phylum. Each has the underlying unity of embryonic development that a phylum should display. Each has a uniquely different pattern of development. The evidence of embryology therefore bears out the evidence of comparative morphology that a phylum Uniramia, a phylum Crustacea and a phylum Chelicerata are more meaningful entities than a "phylum Arthropoda". It also supports the view that the Onychophora, Myriapoda and Hexapoda are sub-phyla of the phylum Uniramia. The status of the subordinate ranks of the Phylum Crustacea and the Phylum Chelicerata remains an open question.

References

ANDERSON, D. T. (1966) The comparative early embryology of the Oligochaeta, Hirudinea and Onychophora. *Proc. Linn. Soc. N.S.W.* **91**, 10–43.

ANDERSON, D. T. (1969) On the embryology of the cirripide crustaceans *Tetraclita rosea* (Krauss), *Tetraclita purpurascens* (Wood), *Chthamalus antennatus* Darwin and *Chamaesipho columna* (Spengler) and some considerations of crustacean phylogenetic relationships. *Phil. Trans. R. Soc.* B, **256**, 183–235.

ANDERSON, D. T. (1972) The development of hemimetabolous insects. In COUNCE, S. A. (ed.), *Developmental Systems—Insects*, Academic Press, New York.

CLARK, R. B. (1964) *Dynamics in Metazoan Evolution*, Clarendon Press, Oxford.

MANTON, S. M. (1964) Mandibular mechanisms and the evolution of the arthropods. *Phil. Trans. R. Soc.* B, **247**, 1–183.

MANTON, S. M. (1967) The polychaete *Spinther* and the origin of the Arthropoda. *J. nat. Hist.* **1**, 1–22.

MANTON, S. M. (1970) Arthropods: Introduction. *Chemical Zoology* **5**, 1–34.

MANTON, S. M. (1972) The evolution of arthropod locomotory mechanisms, Part 10. *J. Linn. Soc. Zool.* (**51**, 203–400.

MORI, H. (1969) Normal embryogenesis of the water strider *Gerris paludaris insularis* Motchulsky, with special reference to midgut formation. *Japanese J. Zool.* **16**, 53–67.

MORI, H. (1971) Experimental investigations on midgut formation in *Gerris* (Heteroptera) (in press).

TIEGS, O. W. (1940) The embryology and affinities of the Symphyla, based on a study of *Hanseniella agilis*. *Q. Jl. microsc. Sci.* **82**, 1–225.

TIEGS, O. W. (1947) The development and affinities of the Pauropoda, based on a study of *Pauropus sylvaticus*. *Q. Jl. microsc. Sci.* **88**, 165–257 and 275–336.

TIEGS, O. W. and MANTON, S. M. (1958) The evolution of the Arthropoda. *Biol. Rev.* **33**, 255–337.

TSCHRUPROFF, H. (1903) Ueber die Entwicklung der Keimblätter bei den Libellen. *Zool. Anz.* **27**, 29–34.

INDEX OF GENERIC NAMES

Page references in *italic type* indicate text figures

AUTHOR INDEX

SUBJECT INDEX

Page references in *italic type* indicate text figures